The Grammar of Inalienability

Empirical Approaches to Language Typology

14

Editors

Georg Bossong

Bernard Comrie

Mouton de Gruyter

Berlin · New York

The Grammar of Inalienability

A Typological Perspective on Body Part Terms and the Part-Whole Relation

edited by

Hilary Chappell
William McGregor

Mouton de Gruyter
Berlin · New York 1996

Mouton de Gruyter (formerly Mouton, The Hague)
is a Division of Walter de Gruyter & Co., Berlin.

♾ Printed on acid-free paper which falls within the guidelines of the ANSI to ensure permanence and durability.

Library of Congress Cataloging-in-Publication-Data

The Grammar of Inalienability : a typological perspective on body part terms and the part-whole relation / edited by Hilary Chappell, William McGregor.
p. cm. – (Empirical approaches to language typology ; 14)
Includes bibliographical references and index.
ISBN 3-11-012804-7 (alk. paper)
1. Grammar, Comparative and general – Possessives. 2. Human anatomy – Nomenclature (Popular) 3. Semantics, Comparative. 4. Language and culture. I. Chappell, Hilary, 1955– . II. McGregor, William. III. Series.
P299.P67G73 1995
415–dc20 95-37989
CIP

Die Deutsche Bibliothek – Cataloging-in-Publication-Data

The **grammar of inalienability** : a typological perspective on body part terms and the part-whole relation / ed. by Hilary Chappell and William McGregor. – Berlin ; New York : Mouton de Gruyter, 1996
(Empirical approaches to language typology ; 14)
ISBN 3-11-012804-7
NE: Chappell, Hilary [Hrsg.]; GT

Disk conversion with T_EX: Lewis & Leins GmbH, Berlin
Printing: Gerike GmbH, Berlin
Binding: Lüderitz & Bauer-GmbH, Berlin
Printed in Germany.

Dedication

We dedicate this book to the late Steve Johnson. His energetic support, wholehearted backing and assistance in setting up the 'Body Parts in Grammar' workshop and general enthusiasm for the proposal contributed greatly to its success at the annual conference of the Australian Linguistic Society held in his department at the University of New England, Armidale, New South Wales in August, 1988.

Preface

The present volume grew out of a Workshop on the theme 'Body parts in grammar', which was held during the Australian Linguistic Society Annual Conference at the University of New England in August 1988. Due to the success of the workshop, we decided to ask the presenters to write up their papers for publication. To broaden the scope of languages represented, we also invited several other linguists to contribute to the project. The papers cover a number of topics under the headings of inalienability and the personal domain, and draw on data from widely distributed and, in some instances, previously undescribed languages. There is a particular focus on the Pacific region, with contributions on Oceanic, Australian, Asian, and American languages.

We are grateful to Verlag Sauerländer (Aarau, Switzerland) for permission to publish Christine Béal and Hilary Chappell's translation of Charles Bally's 1926 article.

Melbourne
September 1994

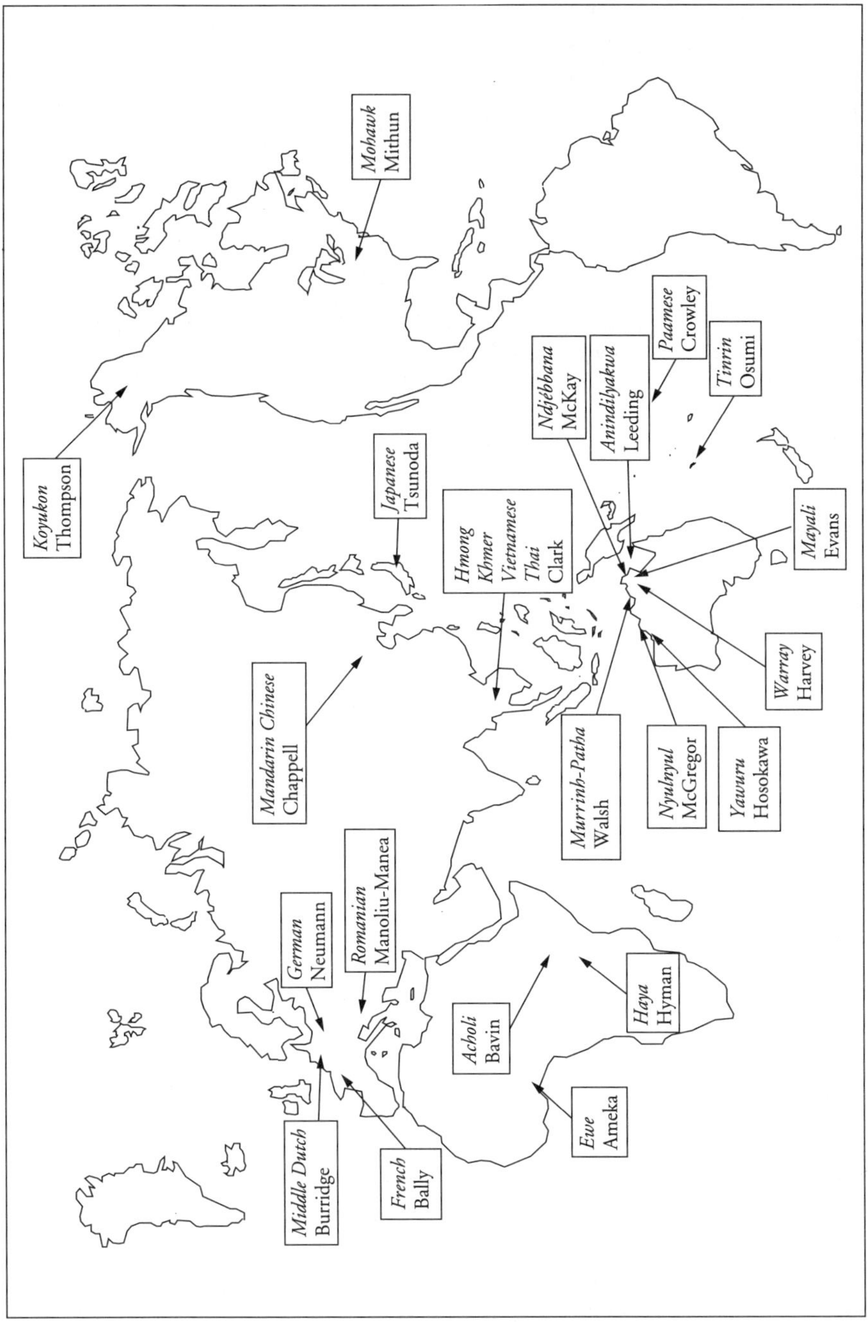

Main languages discussed in this volume

Contents

Part VIII: Bibliography

Part I

Introduction

Prolegomena to a theory of inalienability

Hilary Chappell and William McGregor

For most linguists, the term "inalienability" evokes the complementary term "alienability" and brings to mind the existence of different ways of expressing possession in many "exotic" languages of Australia, the Pacific, Africa and America. The contrasting semantics of these two main coding possibilities for possession was remarked upon early this century by Lévy-Bruhl who noted (1914: 97-98) that in Melanesian languages there were typically two classes of nouns, distinguished by the method used to mark possession. One class comprised suffix-taking nouns designating parts of the body, kin, spatial relations, objects closely associated with a person such as weapons and fishing nets and also inanimate parts, with the suffix indicating the person and number of the possessor. The second class comprised all other nouns; for these nouns, possession was represented by a free possessive morpheme to which the same set of pronominal suffixes was attached. Remarkably, this dichotomy represents a basic semantic pattern that recurs across many languages, regardless of genetic affiliation or grammatical type.

The two classes of nouns so defined are not, however, necessarily disjoint. In a number of languages, both possession constructions may be possible for certain nouns, with a concomitant change of meaning. This is shown by example (1) – from the Melanesian language Patpatar (Pala dialect), spoken in New Ireland (Peekel 1909: 18, cited in Lévy-Bruhl 1914: 99)[1] – where the same noun stem *kat-* 'liver' may refer either to the possessor's own body part or to a separated body part, as of an animal, viewed as an item of food rather than as part of a living being.[2]

(1) *a **kati**gu* 'my liver' (inalienable possession)

versus

***agu** kat* 'my liver that I am going to eat' (alienable possession)

From Lévy-Bruhl's description of this basic division for Melanesian nouns, it is already apparent that the first type of possession – inalienable possession – groups together items which are closely connected with the person either because the relationship is inherent, as with spatial relations such as 'front', 'top' or 'side'; or because it is integral to the person, as with body parts (the same applies for parts of inanimate wholes); or because there is a close biological or social bond between two people, as in the case of kin. The fourth type Lévy-Bruhl mentions, inalienably possessed material objects, is restricted to just those items which are essential for one's livelihood – again, closely connected to a person's survival. All four types of inalienable possessions comprise then either inextricable, essential or unchangeable relations between "possessor" and "possessed" – that is, relations over which possessors exercise little choice or control: every person is born into a kin network, their very existence implying a biological mother and father; and every person has a body made up of parts that in the normal course of events remain indivisible from the whole and which can be viewed in terms of unchanging (non-deictic) spatial dimensions regardless of a person's position or speaker's reference point. Standing on our heads or lying down, *back*, *sides* and *front* all refer to the same places on the body. (They do not shift as do the spatial deictics *in front of X* and *behind X* which depend on the speaker's position relative to an object, if not to some other chosen spatial point of reference.)

Whereas inalienability denotes an indissoluble connection between two entities – a permanent and inherent association between the possessor and the possessed – the complementary notion of alienability refers to a variety of rather freely made associations between two referents, that is, relationships of a less permanent and inherent type (cf. Chappell – McGregor 1989: 25), including transient possession and right to use or control an object. In Melanesian languages – in fact, in Oceanic languages generally – alienable possession is further subdivided into several types depending on the purpose of the possession: for example, whether it is for eating, drinking, planting, a means of livelihood, or for use as a weapon. The alienable category can thus be viewed as the general category of possession, even though it is typically the one which receives overt morphological marking (cf. Chappell – McGregor 1989: 25).

As Haiman (1985: 130) points out, the conceptual distance between an inalienable possession and its possessor is less than that between an alienable possession and its possessor, and this is iconically reflected in many languages (see also Croft 1991: 174-176). Thus, inalienable possession is realised by juxtaposition of the nominals referring to the possessor and the possessed, in that order, as in Djaru (Tsunoda 1981: 179), Yidiny (Dixon 1977: 360), Mandarin Chinese (Chappell – Thompson 1992) and Ewe (Ameka 1995). Another commonly found inalienable construction involves affixation of a pronominal cross-referencing the possessor on the possessed nominal; this is found in Manam (Lichtenberk 1983), Paamese (Crowley 1995), Tinrin (Osumi 1995), Nyulnyul (McGregor 1995), and Ndjébbana (McKay 1995). In most languages, alienable possession is morphologically marked by genitive markers (see further Chappell – McGregor 1989, Haiman 1985: 131); linker morphemes, which may be either separate words, or bound to either or both of the phrase constituents, as in Ewe (Ameka 1995), Mandarin (Chappell 1995; Chappell – Thompson 1992) and Acholi (Bavin 1995); possessive classifiers, as in Paamese (Crowley 1995) and Tinrin (Osumi 1995); and possessive pronominal linkers, as in Nyulnyul (McGregor in preparation).

Lévy-Bruhl's perceptive interpretation of data collated from several grammars of Melanesian languages available at the time provided the inspiration for Bally's famous 1926 article on the expression of the personal domain in Indo-European languages. Lévy-Bruhl doubted that the semantic distinction holding for possession in Melanesian languages could be conveyed in European languages, or at best, could only be conveyed in an implicit manner. Bally (1926 [1995]) challenged Lévy-Bruhl on this point, and showed that an almost identical distinction, which he termed "sphère personnelle" or personal domain, was coded by many Indo-European languages, not at the noun phrase level, but rather at clause level, typically by *dative of involvement* constructions ("le datif de participation"). Thus, argued Bally, dative constructions in many European languages code the indivisibility of a person and an associated body part or possession by indicating the affectedness of the owner as the outcome of an event involving the part. Such dative constructions are exhibited by the three Indo-European languages discussed in Part VI of this book: Romanian, German and Middle Dutch.

It follows from Bally's arguments that inalienability cannot be narrowly circumscribed to just word or phrase level possessive constructions but should be broadened to the clause level to include, first of all, the dative of involvement constructions (see also Fillmore 1968: 61ff on deep structure assignment of the dative for possession). The collection of language-specific studies in this book shows clearly, however, that languages employ various other clause-level constructions for the expression of inalienability or the personal domain in addition to the dative of involvement. These include the *double subject constructions* of Chinese, Japanese and Yawuru which give identical semantic or grammatical case roles to nouns coding the person and their "part", and the *body part locative constructions* of languages such as Romanian and German in which the "part" noun is represented by a locative prepositional phrase while the possessor is retained in a core grammatical role. A fourth phenomenon is the verb phrase level construction of *noun incorporation* (Mithun 1984, Baker 1987), in which a nominal is incorporated into a verbal complex. In a number of languages examined in this book – Koyukon Athabaskan, Mayali, Mohawk, Murrinh-Patha, Warray and several Southeast Asian languages such as Hmong, Thai, Khmer and Vietnamese – an inalienable part or associated aspect of an argument NP, typically a transitive object or an intransitive subject, may be incorporated into the verb. Several of the papers, including those on Japanese, Koyukon Athabaskan and German (also for Indo-European languages in Bally (1926 [1995])) discuss, in addition, *verbs of possession* while others, discuss *proprietive markers* ("having" affixes) in Anindilyakwa, Warrungu and Djaru (Tsunoda 1995). These languages show differences in choice of verb of possession and/or use of proprietive marker according to whether the possession is alienable or inalienable. Further cross-linguistic study of a larger sample of languages of the world would be likely to reveal many more clausal constructions for the expression of inalienability than mentioned in the brief discussion above. (See, for example, Hale 1981 and McGregor 1985 for analysis of what they refer to as the "favourite construction", which embraces a set of double subject, double object, double locative, double dative, etc. constructions in the Australian languages Warlpiri and Gooniyandi respectively.)

It is widely believed that constructions representing inalienability, such as the four main types described above, derive by a syntactic process of *possessor ascension* from underlying structures (initial strata in relational grammar parlance) in which the part occurs in a possessive phrase and fulfils an argument role in the clause. Thus, the English body part locative construction of (2) is often regarded as deriving from the underlying structure represented by (3) (e.g. Frantz 1981: 30; Fox 1981: 323). The possessor NP appears to have "ascended" or to have been "raised" from its original position as a constituent of the object NP to take on the object role itself:

(2) *The dog bit Cliff on the ankle.*
(3) *The dog bit Cliff's ankle.*

The major problem with the possessor ascension analysis is, as Blake (1990: 102) points out, that it is based on an assumption that the two constructions have the same meaning. This is clearly false. The "possessor ascension" construction in (2) represents the bite as more intimately affecting Cliff than does (3), which represents the part, Cliff's ankle, as though it were disembodied from the person, that is, as though it were a separate entity (see also Wierzbicka 1979). In the first, the action is viewed as being directed at the person, who is clearly the patient, but taking effect through a body part, whereas, in the second, the action is viewed as being directed at the part to the exclusion of the person. In other words, (2) represents a type of inalienability in contrast to the conceptual separateness expressed by (3): the part is treated as a part of Cliff's personal domain in (2), but not in (3).

Given their different semantics, the two construction types must be regarded as equally "basic", a point made again and again by many of the contributors to this volume. A number of them, however, retain the term "possessor ascension" – or a synonymous term such as "possessor promotion" or "possessor raising" – as a convenient descriptive label, without necessarily implying that one of the contrasting structures is more basic.

Possessor ascension conveniently groups together a number of formally quite different construction types which share some semantic similarities: they often represent inalienability in contrast to the alienability of the non-ascension construction, in which the possessed noun occurs in a genitive NP with the possessor. The interpretive association is, however, sometimes reversed, as, for example, in Ewe (Ameka 1995) where the "possessor ascension" construction increases in acceptability the more alienable the possessed noun becomes. The label can thus be misleading to the uninitiated. Furthermore, it is important to note that there are two quite different types of possessor ascension (see e.g. Blake 1984: 438). In one, the ascending possessor assumes the former role of the possessed noun, which then becomes a "chômeur" – as in Haya and English. In the other, the possessor takes on status as an indirect object or oblique, the possessed noun apparently remaining in the same grammatical role, as in the dative constructions of many Indo-European languages (see also Fox 1981; Seiler 1983: 39-45), and a similar construction in Ewe (Ameka 1995).[3]

From the semantic perspective, one can take Bally's analysis (1926 [1995]) as a starting point since it transcends the problems discussed below inherent in attempting to set up a universal hierarchy of inalienable categories. Bally proposed the concept of the personal domain as an analytical tool and defined it to include anything associated in "an habitual, intimate or organic way" with the person, viewing it primarily as a socio-cultural construct. Crucially, the extent of this domain was not to be understood as a pre-given fact of the world such as one that could be limited to description in terms of a person-body part relationship. Bally observed that it varies not only from language to language, but even within a single language, according to the way in which a particular real-world phenomenon is construed from among the possible alternatives.

In the more recent literature on inalienability, however, a different approach can be discerned. Many empirically-based studies attempt to ascertain which semantic category (or categories) represents prototypical inalienable possession, and following from this, attempt to construct an alienability scale or hierarchy. It has been proposed, for example, that intuitively speaking the prototypical inalienable category should comprise body parts (Haiman 1985: 130). On the basis of cross-linguistic evidence, however, Haiman (1985: 136) modifies this assumption to disjunctively place body parts and kin together as prototypical inalienables, as do Nichols (1988: 572 and 1992: 160) and Chappell – McGregor (1989: 26). By contrast, Seiler (1983: 13) suggests that the ranking might be of the order of kinship followed by body parts, although he remains in basic agreement with Bally's concept of the *sphère personnelle*. Nichols specifically proposes the following hierarchy, on the basis of a large and comprehensive sample of nominal inalienable constructions in North American languages combined with a smaller number of languages from other regions and language families:

1. kin terms and/or body parts
2. part-whole and/or spatial relations
3. culturally basic possessed items (e.g. arrows, domestic animals)

Unfortunately, in some languages spatial orientation terms appear alone at the top of the hierarchy as the most inalienable category, as in Ewe (Ameka 1995) and Mandarin (Chappell – Thompson 1992). At the very least terms for spatial relations would need to be placed in disjunctive inclusion with both body parts and kin in an implicational hierarchy. Hence, it appears that differences between languages as to which categories they treat as inalienable may not be reconciled in terms of a universal hierarchy.

We are forced to this conclusion also by virtue of the fact that in some languages all three categories of kin, body part and spatial relation terms may be formally treated as inalienables (as in Paamese (Crowley 1995) and Tinrin (Osumi 1995)); in others, body parts but not kin are inalienable (as in many Australian languages (Dixon 1980: 293)); in other languages, kin and spatial terms but not body parts are treated as inalienable (e.g. Ewe (Ameka 1995)); and in yet other languages just kin and body parts are treated as inalienable, (as in most Athabaskan languages – Thompson (1995)). This remains problematical even if we permit certain steps in the hierarchy to be "skipped", following Nichols (1988: 573), and the hierarchy to be "further elaborated" without certain of these core or prototypical categories such as kin and body part terms, depending on the particular language. Nichols admittedly views inalienability as a lexical property of nouns (1988: 574) for which, at best, only generalisations in the form of the implicational hierarchy she devises can be made.

Therefore it does not seem useful to set up a universal hierarchy to account for these cross-linguistic differences in the classes of inalienable nouns, since if body parts, kin terms and spatial terms are placed in the same relative position on the hierarchy, this results, first of all, in loss of predictive power (see also Haiman 1985: 135-136). Secondly, it glosses over the further complication that it is frequently only subsets of kin terms, body part and spatial terms that are selectively represented as inalienable in a given language. Thirdly, the treatment of categories such as personal representations, bodily fluids, exuviae and personal attributes varies from language to language, in ways which appear to be quite independent of the treatment of other semantic categories.

In contradistinction to Nichols, however, who states:

> No account of the semantics of possession types will accurately predict the membership of the "inalienable" set of nouns, either within one language or cross-linguistically. (1988: 568)

we believe that predictions for each language can be made on the basis of cultural and pragmatic knowledge, as Bally (1926 [1995]) suggests, and that this works in with the semantic component of a grammar to precisely characterise the personal domain, or inalienability, as the language-specific descriptions of inalienability in this volume amply demonstrate.

The contributions to this volume provide detailed descriptions of inalienable constructions and related phenomena in some twenty four genetically and typologically diverse languages of Africa, Asia, Australia, Europe, North America and the Pacific (see map on page ix), with particular focus on the Australian, Asian and Pacific region. They attempt not just to identify and characterise the range of construction types found in the languages, but also provide careful investigations of their semantics, highlighting – and attempting to explain – the ranges of entities which may be inalienably possessed in the various constructions.

Before summarising the contributions, we comment on two important methodological attributes shared by the papers in this volume. First, each author develops language internal arguments for the analyses they propose, rather than assumes a universalist position. Second, each paper is concerned with adducing meaning differences associated with observed formal grammatical differences, and uncovering (if possible), motivations for these associations; they do not provide mere structural descriptions of grammatical constructions in particular languages. Unsurprisingly, then, a recurrent methodological tool is agnation: the elucidation of minimal or near minimal grammatical pairs, coupled with attempts to specify the meaning differences thereby encoded.

It should not be supposed, however, that because the contributions are staunchly empirical in orientation, the authors take atheoretical stances, or are not interested in the implications of their findings to linguistic theories. Quite the contrary, in fact. Various differing theoretical orientations are exhibited by the authors, including natural semantic metalanguage advocated by Wierzbicka (1972, 1981); Diverian (1964) and cognitive linguistics approaches (Langacker 1987); systemic functional grammar (Halliday 1985); and cognitive discourse analysis (Chafe 1987). In most cases the authors employ various blends of these theoretical approaches, as required by descriptive needs.

The bulk of this book is divided into six parts, according to the geographical provenance of the language. Part II contains seven papers on Australian Aboriginal languages. Through no particular design of the editors, the languages represented all belong to the rather poorly known northern prefixing or non-Pama-Nyungan families (Capell 1940, Wurm 1972, Dixon 1980: 21); just two members of the better-known southern suffixing Pama-Nyungan family are briefly discussed in Tsunoda's contribution (Part IV). The seven non-Pama-Nyungan languages represented belong to five distinct families, according to the classification of O'Grady – Voegelin – Voegelin 1966 (also adopted in Wurm 1972 and Yallop 1982: 45-47): Mayali and Warray belong to the Gunwinjguan family of Arnhem Land (Northern Territory); Nyulnyul and Yawuru to the Nyulnyulan family of Dampier Land (Kimberley, Western Australia); the other three Northern Territory languages, Anindilyakwa,

Ndjébbana and Murrinh-Patha, constitute the single members of their families.

Evans discusses in detail the syntax and semantics of body part noun incorporation into *Mayali* verbs. He argues that incorporation of body part nouns, contra Baker 1987: 4, is as regular as other types of syntactic incorporation, and is governed by the same accessibility hierarchy for thematic role: basically, incorporation in Mayali primarily involves parts whose whole functions as intransitive subject or transitive object (see Mithun 1984, 1986). For other roles, a different construction is required involving the body part represented in an external, case-marked nominal. A number of nouns other than body part nouns incorporate in the same construction, including terms for personal representation such as 'spirit', 'speech' and 'name', products and inanimate parts of wholes but not kin terms or 'country'. According to Evans, the entire class may be characterised as those entities which imply the existence of some other entity, the "whole" to which they belong, or with which they are associated. None of these nouns is, however, necessarily incorporated, and Evans argues that discourse factors condition their incorporation: they normally incorporate unless there is special focus on the part, indicating its status as an independent discourse participant. Evans points to various grammatical parallels between body part incorporation and another type of syntactic incorporation in Mayali, namely generic incorporation, which he relates to morpho-syntactic similarities in the encoding of part-whole, generic-specific and secondary predicate constructions in Australian languages: all involve apposition of the part and the whole nominal. Evans further suggests that this formal similarity is semantically motivated: different aspects of the same entity are juxtaposed. No change occurs in argument structure of the clause, and thus Evans argues against a "possessor raising" analysis for body part incorporation.

Harvey examines inalienability in relation to three morpho-syntactic phenomena in *Warray*: noun classes, nominal compounding and noun incorporation. He argues that noun class marking distinguishes alienably from inalienably possessed body parts and products, the alienable class including terms for exuviae and regenerative (or replaceable) "parts" including hair and parts of inanimate wholes such as leaves. Several apparent anomalies in the class marking of body part and related terms can, he argues, be accounted for if a notion of "person" rather than "body" is taken to be the prototypical "whole" in relation to human beings. Harvey next describes the process of synecdoche through which names for plant and animal species, as well as traditional nicknames, are formed through a type of nominal compounding, productive only for inalienably possessed parts, in which a condition of the part is represented as an enduring quality of the whole. The main part of Harvey's contribution analyses noun incorporation, which is restricted to the personal domain for animates, that is, body part nouns and personal attributes. Absolutive nouns of these types may be incorporated, in which case they indicate the extent or locus of an event. Only the whole noun, and not the part, is cross-referenced by means of a pronominal prefix on the verb; the incorporated part noun thus has no argument status, and functions as a "range" (Halliday 1985). Discourse factors such as lack of individuation or low discourse status of the "part" noun motivate use of an incorporated rather than external nominal. Harvey also points out the difficulties of analysing this Warray construction in terms of "possessor ascension", it not being possible in all cases to derive incorporated from unincorporated structures. Moreover, Harvey views the clause with the incorporated "part" term as being the unmarked one both in terms of construction type and discourse status.

Hosokawa sets up a typology of body part syntax in *Yawuru*, contrasting an array of four non-basic syntactic constructions: the double subject transitive, the double subject intransitive, the double object and the quasi-passive. All four constructions code a whole-part relation between a person and part of their body; cross-referencing in the verb is restricted to the "whole" noun. By extension, other entities which are regarded in Yawuru culture as inalienably possessed – primarily personal representation such as names, shadows, footprints, personal dreamings, etc. – are also coded in the same way. Accordingly, Hosokawa characterises these four constructions as "identity-sensitive". Nouns from these semantic domains are, however, not necessarily incorporated into clauses by means of the "identity-sensitive" constructions: they may also be represented by one of the three basic clause types. Hosokawa goes on to argue that the two possibilities contrast semantically: in the identity sensitive constructions the referents are treated as inalienable parts of the person's whole existence; otherwise, they are represented as alienable, and thus as not essential to a person's identity. Further, Hosokawa shows that each of the "identity-sensitive" clause types contrasts semantically as well: both double-subject constructions foreground the part and defocus the whole; the double-object construction backgrounds the part; and the quasi-passive indicates non-volitionality and inadvertency of the event with the focus being on a typically inanimate agent.

Three phenomena relating to the morphosyntax of body part expressions in *Anindilyakwa* (Groote Eylandt) are discussed in depth in Leeding's contribution: noun classes, possession types and noun incorporation. Leeding argues that noun classes are semantically based, and may be characterised by the features [± personified], [± singular], [± feminine human], [± masculine human] (the binary opposition for the two genders being humans versus nonhuman animates and inanimates with supernatural powers), [± visible] and [± lustrous]. Most body parts fall into the non-personified classes according to their appearance as lustrous or not, while some belong to the non-human masculine class when associated with the spirit world (e.g. through association with ceremony or sickness and death). Leeding next distinguishes four morphologically distinct types of nominal possession and shows how they may be ranged along a continuum from inalienable to semi-alienable to alienable; with kin relations forming a fourth and special category of their own. In the final part of the analysis, three types of noun incorporation are identified and investigated. Syntactic incorporation is restricted to body parts that are either grammatical objects of transitive action verbs or subjects of reflexive ones. Interestingly, incorporation of body part nouns into transitive verbs is in complementary distribution with pronominal cross-referencing of the whole, suggesting that the two are functionally akin. Leeding thus suggests that incorporation represents argument status of the body part noun, and hence that noun incorporation, unlike other noun incorporating languages discussed in this book, represents alienability, rather than inalienability. This type is productive, and usually has a non-metaphorical interpretation. In lexical compounding, by contrast, a body part noun may be incorporated into either an intransitive verb to which it holds a subject role, or an adjective or noun root. Although nonproductive, this type frequently engenders metaphorical meaning extensions. The third type involves incorporation of body part nouns into shape adjective roots to form species names through the coding of salient physical features. This is similar in function to nominal compounding in Warray (see above).

McGregor's contribution describes inalienable possession in *Nyulnyul*. This language employs a system of obligatory pronominal prefixes coding person and number of the possessor, which are attached to a small set of around forty nominal roots and stems referring to body parts, personal representation (such as names, images and footprints), and protective coverings. Prefixing is shown to be restricted to just those body part and attribute nouns which are regarded as essential to the normal functioning of a human being as reflected in Nyulnyul culture, where the term "human being" is interpreted in this particular linguistic context as a generalised, non-sex-specific person without any of the salient characterising features of the individual. A semantic principle is thus shown to account for the pattern of prefixing in a regular and inclusive manner; prefixing is not arbitrarily restricted to vowel initial nouns, as suggested by Capell (1972). Comparison is made of prefixing nouns in neighbouring Nyulnyulan and Worroran languages, showing that these largely coincide. McGregor also suggests a potential diachronic source for nominal prefixing in Nyulnyul, proposing that it is the outcome of morphologisation of the double object construction in which possessor and part nouns are identically case marked as absolutive. Support for this proposal is found in evidence that prefixing nouns in Nyulnyul correspond by and large to nouns which are frequently found in the double object construction in the genetically unrelated language Gooniyandi (McGregor 1985).

The Arnhem Land language *Njébbana* shows four basically disjoint possession classes defined according to the means of marking the possessor. These are structurally and semantically characterised as follows: (A) a free cardinal pronoun in juxtaposition with the possessed noun, the latter being drawn from an open class of nouns referring to objects viewed as independent items, including bodily products, internal organs, bones, kin, as well as artefacts, locations and loan words; (B) a pronominal subject prefix attached to a form of the verb *réndjeyi* 'stand, be' preceded by the possessed noun which codes mainly external body parts; (C) a possessive pronoun suffixed to the possessed noun coding a closed set of body parts; and (D) a pronominal prefix to the possessed noun which belongs to a closed class of nouns indicating body parts, types of people and attributes (qualities). McKay argues for a semantic motivation in determining the choice of possession marking, based on the degree of inalienability from most alienable (separate pronoun) to the least alienable (pronominal affixes to the noun). The degree of morphological "closeness" is an iconic reflection of inalienability (cf. Haiman 1985; Croft 1991: 174-176): at the inalienable end of this continuum, for classes C and D, affixation codes the conceptual identity of possessor and possessed; at the alienable end, the separate word status reflects the cognitive status of the referents as separate entities. McKay suggests that the more alienable possession class B with the positional verb *réndjeyi* 'stand, be' is a somewhat anomalous intermediate type, coding mainly external and visible parts of the body. The use of such an existential verb of posture supports, however, Lyons (1967) proposed universal link between possessive, locational and existential constructions.

Walsh describes the metaphorical extension of body part terms in noun incorporation in *Murrinh-Patha* (Wadeye (Port Keats)), which, argues Walsh, plays a central role in Murrinh-Patha grammar. Approximately thirty five body part roots may be incorporated into verbs, nouns and adjectives, coding a continuum of senses from the literal to the metaphorical. Walsh first describes adjectival roots with incorporated body part nouns which may be suffixed by bound pronouns cross-referencing the possessor in a similar manner to Njébbana possession class C (see previous paragraph), thereby identifying the person with a certain quality. Next, in the main part of the analysis, he outlines a large number of metaphorical themes associated with noun incorporation, thus displaying the richness of this linguistic device in Murrinh-Patha. These metaphors include spatial orientation, emotions, shape, attitudes and knowledge; for example, *-rdarri-* 'back' metaphorically extends to 'behindness', 'anteriority', 'plenty' and 'solidarity'. Thirdly, he further suggests that analysis of the processes of metaphor may prove relevant for other parts of the lexico-grammar: they may suggest analysis as complex nominal stems forms which might otherwise appear to be simple noun roots.

Part III contains two contributions which deal with a pair of distantly related Oceanic languages of the Austronesian family, a language family which is scattered across the Pacific region. One is Paamese, a southern Melanesian language spoken on the island of Paama in the Republic of Vanuatu (formerly the condominium of the New Hebrides). The other is Tinrin, a language spoken in New Caledonia.

In the first paper by Crowley, a comprehensive treatment is provided of the two main strategies used for marking nominal possession in *Paamese*. Inalienable possession is coded by pronominal possessive suffixes on the possessed noun, whereas alienable possession is marked by a set of free form possessive constituents to which the same set of pronominal suffixes cross-referencing the possessor is attached. Crowley undertakes a detailed investigation of a number of apparent exceptions to this semantically-based alternation, arguing that it is ultimately possible to account for the possession type associated with almost all nouns, although this requires a reformulation of the notions of alienability and inalienability in Paamese on the basis of culture-specific knowledge. The categories of consanguineal kin, inseparable body parts including internal organs essential to life, personal representation, body products which are exuded through normal bodily functions or are permanently associated with a person, as well as some nouns for part-whole relations, imprints, spatial orientation and "best food" are expressed by means of the inalienable strategy. For many of these, suggests Crowley, the referent of the possessed noun does not exist independently of the possessor. By contrast, internal body parts which are not seen as central to emotions or life, and temporary manifestations of the body such as swellings and infections associated with abnormal activity are normally alienably marked, these existing independently of the possessor – internal organs, for example, are normally encountered in the context of butchery. In addition, vocatives for kin and many affinal kin terms, also loan words from the English-lexifier pidgin, Bislama, in which most Paamese speakers are bilingual, are treated as alienables. Crowley concludes his analysis by showing that it is possible to treat inalienably marked nouns as unpossessed, that is, as unrelated to any possessor, when, for example, a body part noun is dissociated from its possessor; used generically; viewed as a source of food or forms part of a metaphor. For this purpose, some suffixing nouns show a derived unsuffixed free form; but most use a "dummy" third person singular suffix.

In the second paper, Osumi discusses the semantics of morphosyntactic strategies for the expression of possession, similar to those in Paamese, and distinguishes several different types of inalienable and alienable possession in *Tinrin* showing how these can be ranged along a continuum. Inalienable possession is expressed by means of possessed-possessor word order: the possessed noun belongs to a restricted set of bound nominals while the possessor noun may be coded either by a pronominal suffix or a full noun. Bound nouns represent inherent relations such as kin, body parts and products, personal representations, and also certain inanimate parts. This same nominal construction may also be used with bound location nouns referring to a spatial attribute of an entity (e.g. beside, inside, etc.), or to one of seven possessive classifiers (whose primary function is, however, to form an alienable construction). Alienable possession falls into two types: (a) a construction employing one of seven possessive classifiers procliticised to the possessor noun or suffixed by a pronominal and indicating the type of possession – as fruit, meat, starch food, objects for chewing, drinks, plants or belongings in conjunction with a specific possessed NP that may precede or follow this unit;[4] (b) a construction in which one of three possessive prepositions – associated with temporary possession, essential means of livelihood, or fire – is placed invariably after the possessed noun and precedes the possessor. Osumi also identifies a third type of possessive construction coded by a link morpheme *-nrâ-*. This, she argues, is intermediate between inalienable and alienable possession, and typically includes possessed nouns referring to affinal and more distant kin, transient personal attributes, internal organs and bodily exuviae.

Part IV consists of three contributions dealing with a number of languages of Asia from several unrelated families. The first, Mandarin Chinese belongs to the Sinitic subgrouping within Sino-Tibetan, while the genetic affiliation for the second, Japanese, remains a subject of debate, some linguists arguing for its inclusion in Altaic, others for a remote connection to Austronesian. Tsunoda's analysis of Japanese makes comparisons of different construction types in that language with their correlates in two Australian languages – Djaru and Warrungu – and also with English. In the third paper by Clark, a large range of unrelated languages spoken in the mainland Southeast Asian countries of Thailand, Laos, Cambodia and Vietnam is represented, namely, Austroasiatic languages such as Vietnamese (Viet-Muong) and Khmer, Chrau and Rengao (all part of the Mon-Khmer branch); White Hmong of Laos (Hmong-Mien); also Thai and Nung (Tai). In languages of the Asian region, alienable-inalienable distinctions are not generally morphologically marked on nouns. Instead, reflexes of this semantic phenomenon are typically found at the clause-level, and it is these which are examined in each of the three papers in part IV.

Chappell presents a discourse-based semantic study of double subject constructions with intransitive predicates in *Mandarin Chinese*, using data from both spoken and literary narratives. After briefly reviewing the nominal syntax of genitive and appositional noun phrases as a potential strategy for the expression of inalienability, she distinguishes three types of topic-comment constructions showing that one in particular – the double subject construction – expresses the relation of inalienability and the personal domain in Mandarin. In this construction there are two utterance-initial NPs, the first representing the whole and the second, the part or relational noun; the NPs are simply juxtaposed, without any morphological marking. The semantic function of this construction is to characterise a person (the whole) in terms of a predicate that can refer to either physical and psychological states or conditions of a part, (but not intense transient emotions); to a kin relation; or to the social self as embodied in collectives such as nation, workplace or institute of study (viewed as a community of people the individual may identify with). Analysis of the intonational properties of the double subject construction shows that if the possessor is referred to by a lexical NP, it is usually set off in its own intonation unit from the remainder of the utterance. If it is coded by a pronominal, however, there is usually no intonation break between possessor and possessed. The possessor NP may of course be ellipsed, if given. Chappell refers to this construction as a "reduced form of the double subject construction". Thus, there is an overall tendency not to mention the possessor in the same intonation unit unless in pronominal form which tallies well with Chafe's "one new concept at a time" constraint (1987: 32). A second topic concerns the function of complex genitive noun phrases in Mandarin which are shown to individuate and focus upon – as opposed to relate – aspects of the personal domain to their possessors.

Clark examines the semantics and syntax of a construction involving stative verbs and predicative body part expressions in a number of Southeast Asian languages, specifically, *Vietnamese*, *Khmer*, *Chrau*, *Rengao*, *White Hmong*, *Thai* and *Nung*. In this construction, the term referring to the person is subject and precedes the verb, while the body part term is predicative, generally following the verb. This structure codes a physical state undergone by the subject possessor but which is located in a body part, and can metaphorically extend to emotions which are considered to be located in various organs, depending on the language. This contrasts semantically with a construction in which the part and whole together form a NP, the subject of the clause and in combination with the stative predicate, simply indicates a state of the body part. Clark suggests that the predicative construction may involve some form of noun incorporation, consequent upon possessor raising of the "whole" NP to subject role. However, if it is in fact noun incorporation, then the noun is not as tightly bound as in languages such as Mohawk and Mayali. This notwithstanding, it is true that the body part noun and the stative verb function as a single grammatical unit, the body part noun having foregone its argument status to be coded as a single unmodified noun within the predicate – although classifier modification is permitted in some languages. Clark emphasises that it is this predicative construction with an incorporated body part noun which is the preferred strategy as it permits focus upon the subject possessor undergoing the state.

Tsunoda argues that inalienability must be analysed in the form of a possession cline for *Japanese* representing the degree of closeness of possessor and possessed nouns. Two common strategies for coding nominal and verbal possession in Japanese are first identified and contrasted with English and two Pama-Nyungan languages of Australia, Djaru and Warrungu. The main part of Tsunoda's analysis concerns, however, honorifics in Japanese, for which it is well-known that politeness or respect may be expressed vis-à-vis either the addressee or some other third person referent typically coded as a subject or object argument of the clause. In addition to these two categories of performative and propositional honorifics, Tsunoda shows that the possessor of either a subject or object argument may be accorded a respect honorific. This phenomenon he designates "possessor respect", arguing that the acceptability of possessor respect correlates with a cline relating to the semantic domain of the possessed item in the following way: body part > inherent attribute > clothing > (kin) > pets > products > other possessions.[5] The higher on the cline the possessed item is, the correspondingly more acceptable is possessor respect. Moreover, this interacts with the syntactic role of the possessed item in Japanese: possessor respect can only be coded when the possessed noun is direct object or indirect object of a transitive verb, or subject of an intransitive verb, but not, when it is subject of a transitive verb. Tsunoda shows that this cline is also applicable to other phenomena in Japanese, such as the double subject construction, the five different possessive verbs taking different case frames, and the genitive construction, at the same time, observing correlations with the acceptability of possessor ascension constructions in other languages. He argues further that it accounts for the choice between inalienable and alienable possession constructions in Djaru and Warrungu, where inalienable possession, expressed by juxtaposition, is limited to body parts and attributes; lower down the cline, possession is indicated by the genitive. In a final section, he discusses the cline with respect to attributive constructions in Warrungu and Djaru with the 'having' suffix and compares these with the counterpart English constructions.

Part V contains contributions describing inalienability in two unrelated polysynthetic Amerindian languages of North America, both of which show alienable versus inalienable contrasts in nominal possession constructions, and noun incorporation for inalienables. These are Mohawk, an Iroquoian language spoken in the northeastern part of North America and Koyukon, a NaDene language belonging to the Athabaskan family whose speakers live in the interior regions of Alaska.

In the first paper, Mithun distinguishes two classes of nouns in *Mohawk* on formal and semantic grounds. The first class, which take agentive pronominal prefixes indicating the possessor, together with a locative suffix on the possessed noun, contains mainly nouns for attached, controllable and visible body parts of humans and animates. Most other nouns including material possessions, internal organs, separated body parts and some external parts and coverings of the body are marked by patient pronominal prefixes, which signal alienability; these nouns occur without the locative suffix. (Kin relations are typically coded by verbal, not nominal, means.) Mithun interprets the contrast between inalienable and alienable nominal possession in Mohawk in terms of, respectively, inseparability and conceptual identity of possessor with the part as opposed to status as a discrete possession.

A second construction relating to inalienability in Mohawk is noun incorporation. Body part nouns may be incorporated into transitive and intransitive verbs, the possessor being the grammatical patient for both verb types, with cross-referencing by the patient pronominal prefix in the verb. The set of nouns that may be incorporated into Mohawk verbs includes, however, alienable possessions as well. But for alienable possessions – as distinct from inalienable possessions – the possessor of the incorporated noun cannot usually be cross-referenced as a core argument of the verb by the patient prefix. Interestingly, the class of nominals that requires agent possessors is not identical to the class whose possessors are cross-referenced in the verb. For example, the internal organs and outer coverings that are alienably possessed in the nominal construction, are often incorporated like inalienables into the verb. Mithun accounts for the discrepancy between these two classes of "inalienable" nouns in terms of grammatical differences between the two constructions that realise them: incorporation into the verb signals the possessor's involvement in the event and their affectedness as opposed to the inalienable nominal construction which marks identity between the possessor and the possessed.

Thompson's contribution deals primarily with nominal inalienability in *Koyukon Athabaskan*. He distinguishes three different possession classes for Koyukon nouns: inalienably possessed nouns, which require an overt possessor; alienably possessed nouns, which may or may not have an overt possessor; and unpossessible nouns, which cannot have an overt possessor without some morphological modification. In inalienable possession, the possessor is represented by an accusative pronominal prefix, and potentially by a free noun as well, in apposition to the possessed nominal, which follows. Terms for kin, body parts and personal representation are inalienably possessed; however, body part terms differ from kin and other inalienable terms in that they may occur unbound, that is, without the prefix, in the presence of a full lexical noun coding the possessor. A body part of one entity may be possessed by another entity, in which case the term for the detached body part appears in a double possessive construction with the indefinite third person "alienating" prefix *-k'e-* as well as a prefix referring to the new owner of the part. The function and status of a possessive suffix is also discussed with reference to the different possession classes in Koyukon.

The contrast between alienable and inalienable possession also finds realisation at clause-level; only alienable possession corresponds to clauses with the verb for 'have' whereas a form of the verb 'be, exist' is used for kinship and occasionally for body parts. Furthermore, Koyukon permits body part noun incorporation, although it is not as productive as in some languages. The part may belong to the subject of a transitive clause indicating the normal movement or stasis of the part, or of the object of a verb of violence; rarely, body part instruments are incorporated. Thompson does not discuss the semantic contrast between clauses with incorporated versus unincorporated body part terms; he does suggest, however, that certain concomitant verbal classifier changes may be conditioned by the relationship of inalienable possession, specifically for body part incorporation where the possessor is subject and a change of state is understood to occur in the body part. He describes three factors as being relevant to inalienability in Koyukon: degree of connectedness of entity to possessor; degree of association with possessor; and degree of salience apart from the possessor.

The papers of Part VI deal with three Indo-European languages, two Germanic languages – German and Middle Dutch – and one Romance language – Romanian. Each of these languages exhibits the dative of involvement construction type discussed by Bally (1926 [1995]), as well as a number of other clausal constructions which may be employed in the representation of part-whole relations or inalienability. The three contributions discuss in considerable depth the semantic contrasts between the various modes of expression.

Burridge's paper investigates various means of incorporating body part nouns into clauses in 14th and 15th century *Middle Dutch* medical treatises (Brabantish and Hollandish dialects). Although some medical writers of the time treat body parts as separate entities distinct from their wholes, most use the dative construction of inalienability in which the person is coded as a separate dative argument, and the body part appears in a phrase with the definite article. The range of items that may be treated as inalienable by virtue of occurrence in this construction type embraces anything that is in close association with a person's body, including not only parts of the body, but also speech, mind, bodily coverings, exuviae, and temporary manifestations such as sickness and sores. A variety of other body part constructions were available in Middle Dutch, including one in which the part and possessor are accorded identical case marking (either dative or nominative) and another in which the part occurs in the dative and the possessor in the nominative. Burridge argues that this range of constructions was used to code subtle semantic distinctions, contrasting the prominence, topicworthiness, lack of volitionality and relative degree of involvement of the part and whole in the event and should not be viewed merely as illustration of the disintegration of the case-marking system (as is frequently supposed), nor purely in terms of inalienability. She concludes by briefly considering the replacement of genitive case-marking by the dative in two "transplanted" Germanic languages, as well as in non-standard colloquial German, as an example of reanalysis of dative clausal constituents as phrasal ones.

Manoliu-Manea's contribution focuses on four construction types associated with expressing the part-whole relationship in *Romanian* clauses. Two are dative constructions, one containing transitive verbs and the other, intransitive verbs. In both types, the whole appears in the dative case while the part appears respectively in either accusative or nominative case. The other two contain prepositional constructions in which the whole functions in its turn as subject or object, for intransitive and transitive predicates respectively, while the part appears in a locative prepositional phrase. Manoliu-Manea proposes that these construction types contrast primarily with respect to the discourse feature of topicality and the cognitive feature of centrality. In the prepositional construction, the whole noun, irrespective of its case marking, is topical and central whereas the part, appearing in a prepositional phrase represents a restrictive specification of the affected whole and may not be modified. Here the part noun has the lowest degree of autonomy. Conversely, in dative constructions, the whole is assigned a non-argument experiencer or benefactive role as opposed to the part which is accorded participant status and is viewed as central and topical. The higher degree of autonomy of the part in the dative constructions correlates with the possibility of its modification. Manoliu-Manea also discusses the semantic range of these constructions showing that the prepositional construction is not restricted to body parts but may also be used with clothing in contact with the body, domestic objects and vehicles, and inanimate part-whole relations, while the dative construction can be used with kinship and other possessions (see also Malczewska – McGregor 1990 on Polish datives). Interestingly, metaphorical extension is largely restricted to the prepositional construction, while nonvolitional change is represented by the dative.

Neumann adopts an approach based on Natural Semantic Metalanguage (Wierzbicka 1972, 1981) with adaptations from Diver's theory of inference (1964) in her analysis of a range of constructions representing the part-whole relation in standard *German*, including the dative. One of her main goals is to characterise the meaning of the dative in contrast to other construction types available in the language. She begins this task by showing that the genitive construction encodes a view of the part as isolated or separate from the person, or whole, in the sense that the action involving it does not affect the whole. By contrast, the seven main dative constructions indicate not just that the person is possessor of the part, but also that they are related through the action or event: what happens to the part also happens to the whole – or, as Neumann puts it, what can be said of the part can also be said of the whole. Neumann argues her thesis through an examination of the contrasting construction types – involving genitive, accusative and nominative case marking of the possessor – in the context of various types of event (including movement, impact, state, change of state, posture, cognition, perception, and so on), and various conditions of the animate being in relation to these event types. The dative construction is not, however, restricted to action on parts of a person's body, but is also used of other objects, where a change in them affects the person by virtue of its fortunate or unfortunate nature. Neumann argues that the semantic component of experiencer in the German dative constructions is an inferential possibility rather than a core and invariant part of their meaning. The same applies for the case of the positive and negative connotations of the *dativus commodi* and the *dativus incommodi*.

The three papers in part VII discuss nominal phrase level and clause-level reflections of alienable and inalienable possession in three African languages belonging to different genetic groups: Ewe belongs to the Kwa branch of the Niger-Congo family and is spoken in a region of West Africa including southern parts of Ghana and Togo and a small area of Benin, Acholi belongs the Luo branch of Western Nilotic and is spoken in Uganda, while Haya is a Bantu language of Tanzania.

Ameka examines the semantics of possessive constructions in *Ewe* nominal and clausal syntax, employing the reductive paraphrase method of Natural Semantic Metalanguage in his semantic representations (Wierzbicka 1972, 1981). Alienable possession is coded at noun phrase level by NP *Φe* NP in contrast to inalienable possession which is coded by NPs in apposition. Contrary to cross-linguistic semantic generalisations for inalienable possession, in Ewe body parts and part-whole relations are normally expressed by the alienable construction,[6] as are, predictably, material possessions. Ameka explains this in terms of the feature of control – one can do things with parts of the body in the same manner as one uses material possessions. In addition, Ameka provides a functional-historical explanation as to the alienable treatment of body part terms by virtue of the need to grammatically distinguish these from cognate spatial orientation terms. Coded by the inalienable construction are terms for spatial orientation, kin and certain social relationships which are inherently relational. In contrast to alienables, these represent relations over which possessors exercise little control. The alienable and inalienable possessive constructions in Ewe are also contrasted with classificatory nominal compounds taking a high tone suffix, viewed by Ameka as expressing a permanent relationship. These compounds code habitual association between two persons or a generic-specific relationship.

The second part of Ameka's study concerns the clausal syntax of body parts and other possessions in two types of transitive and intransitive constructions. Part nouns may be found in either alienable genitive noun phrases taking subject or object role with a dependent possessor or in what Ameka terms "possessor ascension" constructions. In Ewe "possessor ascension" constructions, although the body part term retains its core grammatical role, the possessor is "promoted" from genitive to dative-oblique in an adjunct phrase, taking on a recipient role. This results in the part being viewed as an individuated and central participant which acts as the locus of the event whereas the whole is viewed as only indirectly affected through its connection to the part, this being reflected in its peripheral grammatical status. Ameka shows that acceptability in the possessor ascension constructions increases with degree of alienability of the possession. The concluding sections show that various syntactic constructions in Ewe treat body part terms as prototypical nouns, thus mirroring their conceptualisation in possessive constructions as alienable individuated entities rather than marginal and undifferentiated in status, the latter being the usual case in many other languages (cf. Hopper – Thompson 1984).

Bavin treats nominal possession in *Acholi*, which shows the general syntactic order of possessed noun preceding possessor noun for all types. Inalienable possession with a nominal possessor is marked by simple juxtaposition of the two nouns, whereas alienable possession employs the possessive marker *pa* as a linker. In the case of pronominal possessors, both types of possession are marked by a bound suffix identical in many instances to the object suffix on verbs. In certain restricted environments – namely, for a singular possessor and consonant final possessed noun roots – the alienable versus inalienable distinction is still made. Elsewhere the distinction is neutralised, for example, for all plural pronominal possessors. The semantic range of the inalienable construction in Acholi includes body part terms – internal and external, body fluids, name, character and parts of inanimate objects, none of which is subject to a constraint on physical contiguity with the whole. The alienable construction is used for kin, shadow, speech, as well as property and places owned by an individual. Bavin also discusses nominal classification in Acholi, which is formally identical with the inalienable construction but semantically codes a single entity in terms of a generic-specific relation. Metaphorical uses of body part terms in various construction types are shown to include extension in meaning to parts of inanimates, spatial relationships (e.g. 'forehead' for 'in front of'), emotions and other human qualities achieved through the attribution of physical states to the body part.

In the final paper, Hyman investigates the syntax and semantics of two grammatical phenomena in *Haya* which he dubs "possessor deletion" and "possessor promotion". The first refers to deletion under identity of possessive pronouns, specifically for Haya transitive action clauses expressing subject possessors who act upon their body parts (coded as grammatical objects). As in similar body part constructions in Igbo (a Kwa language of Eastern Nigeria) and French (Bally 1926 [1995]), the body part nominal may not be modified by a possessive pronoun. The second refers to a construction in which the possessor is "promoted" out of a genitive NP and is coded as either direct object of a transitive verb or subject of an intransitive one while the possessed body part noun loses its argument status. In fact, the part noun may only bear a participant role in Haya as head noun of an object or subject NP when it is interpreted as detached or dissociated from the whole.

Hyman further argues that the possessor promotion construction is employed only when the possessor is an experiencer, that is, has been physically affected by the event. This results, first of all, in a restriction on possessed nouns to those which are parts of the human body or else, closely associated objects such as clothing worn on the body at the time of the event. Secondly, the possessor promotion construction is more likely to be associated with processes involving physical effect (e.g. actions) than with those that do not (e.g. sensory processes) since here action on the part tends to be restricted to that part, without affecting the whole. Thirdly, as Hyman observes, a possessor hierarchy appears to be in operation: the degree of acceptability of possessor promotion is higher for first person possessors than for second and third person, and for human possessors as opposed to animate possessors. Hyman argues that persons rather than their parts show more prototypical discourse and semantic properties of direct objects, thus providing an explanation for the motivation behind possessor promotion.

One thing which emerges clearly from the various contributions to this volume is the cross-linguistic significance and generalisability of Bally's (1926) original insights. However, by no means all of the rich set of ideas contained in this article have been subsequently explored. Bally regarded the personal domain as a socio-cultural construct whose extent could be determined by cultural knowledge as much as by linguistic (1926: 77) and it is this view which is given validity by most of the analyses in this collection. He further suggested that there may be some relationship between its extent and the social networks of communication within the linguistic community: the expression of the personal domain might be more "established" and frequent in informal colloquial speech than in formal speech and writing. This is an interesting observation, and would seem to contain a certain amount of truth, though perhaps for reasons rather different to those advanced by Bally. We suspect that it relates to differences in world views associated with the different socio-cultural situations, rather than the nature of communication networks. In face to face informal interaction persons generally emerge as of primary importance. In the more formal contexts, for example, especially medical, persons emerge as less important: they become objectified entities, on whose bodies – as distinct from persons – various acts are performed. This represents an area of study which would repay careful investigation.

The above summaries simplify the details of the contributions considerably, glossing over many complexities in the formal means of expressing inalienability, in the different nuances of its expression and in the semantic domains encompassed by the various languages under investigation. We hope that the reader's curiosity will have been sufficiently roused to now wish to grapple with this imbroglio by engaging with the individual contributions themselves.

Notes

1. See also Beaumont (1972) on the languages of New Ireland, a colony of Germany from 1885 to 1914 named "Neu-Mecklenburg" – whence the title of Peekel's 1909 grammar.
2. The original French text summarised above goes as follows: "La première classe (les noms qui prennent le suffixe) se compose des noms qui désignent les membres du corps, les parties d'une chose, les objets en relation étroite avec un homme (ses armes, son filet de pêche, etc.), les relations de parenté, et de quelques prépositions exprimant des rapports dans l'espace: à côté de, au dessus de, près de, loin de, etc. La second classe comprend tous les autres noms."
3. It is interesting to speculate whether the fact that Haya and Ewe respectively employ these two constructions may correlate with the fact that in Haya but not in Ewe, the possessor ascension construction represents inalienability. See also Manoliu-Manea 1995 on the contrast between dative and locative preposition constructions in Romanian, which adds further support to this possibility.
4. Osumi regards the alienable possessive classifiers themselves as being inalienably possessed, since the activities associated with the possessive types they represent are essential to existence.
5. The category of kin is bracketed because Tsunoda did not include it in his investigation, and placed it here according to his intuitions, based on Japanese, Djaru and Warrungu – see page 8 above.
6. This generalisation holds for Ewe except in the case of first and second person pronominal possessors for which both alienable and inalienable forms are possible.

References

Ameka, Felix

1995 "Body parts in Ewe grammar", [this volume].

Bach, Emmon – Robert T. Harms (eds.)

1968 *Universals in linguistic theory.* New York: Holt, Rinehart & Winston.

Baker, Mark

1987 "Noun incorporation and the nature of linguistic representation." Paper presented at Wenner-Gren conference on "The role of theory in language description", Ocho Rios, Jamaica.

Bally, Charles

1926 "L'expression des idées de sphère personnelle et de solidarité dans les langues indo-européennes", in: F. Fankhauser – Jakob Jud (eds.), 68-78.

[1995] [Translated and reprinted in this volume.]

Bavin, Edith

1995 "Body parts in Acholi: alienable and inalienable distinctions, and extended uses", [this volume].

Beaumont, Clive H.

1972 "New Ireland languages", in: Clive H. Beaumont – Darrell T. Tryon – Stephen A. Wurm (eds.), 1-41.

Beaumont, Clive H. – Darrell T. Tryon – Stephen A. Wurm (eds.)
1972 *Papers in Linguistics of Melanesia 3.* (Pacific Linguistics Series A Number 35.) Canberra: Pacific Linguistics.
Blake, Barry J.
1984 "Problems of possessor ascension: some Australian examples", *Linguistics* 22: 437-453.
1990 *Relational grammar.* London: Routledge.
Capell, Arthur
1940 "The classification of languages in north and north-west Australia", *Oceania* 10: 241-272, 404-433.
1972 "The languages of the northern Kimberley, W.A.: some structural principles", *Oceania* 43: 54-65.
Chafe, Wallace
1987 "Cognitive constraints on information flow", in: Russell S. Tomlin (ed.), 21-51.
Chappell, Hilary
1995 "Inalienability and the personal domain in Mandarin Chinese discourse", [this volume.]
Chappell, Hilary – William McGregor
1989 "Alienability, inalienability and nominal classification", *Proceedings of the Berkeley Linguistic Society* 15: 24-36.
Chappell, Hilary – Sandra A. Thompson
1992 Semantics and pragmatics of associative *de* in Mandarin discourse. *Cahiers de Linguistique Asie Orientale* 21.2: 199-229
Croft, William
1991 *Typology and universals.* Cambridge: Cambridge University Press.
Crowley, Terry
1995 "Inalienable possession in Paamese", [this volume.]
Diver, William
1964 "The system of agency of the Latin noun", *Word* 20: 178-196.
Dixon, Robert M.W.
1977 *A grammar of Yidiny.* Cambridge: Cambridge University Press.
1980 *The languages of Australia.* Cambridge: Cambridge University Press.
Dixon, Robert M.W. (ed.)
1976 *Grammatical categories in Australian languages.* Canberra: Australian Institute of Aboriginal Studies.
Fankhauser, Franz – Jakob Jud (eds.)
1926 *Festschrift Louis Gauchat.* Aarau: Verlag Sauerländer.
Fillmore, Charles
1968 "The case for case", in: Emmon Bach – Robert T. Harms (eds.), 1-90.
Fox, Barbara
1981 "Body part syntax: towards a universal characterization," *Studies in Language* 5: 323-342.
Frantz, Donald
1981 *Grammatical relations in universal grammar.* Bloomington: Indiana University Linguistics Club.
Haiman, John
1985 *Natural syntax.* Cambridge: Cambridge University Press.

Haiman, John (ed.)
1985 *Iconicity in syntax*. Amsterdam: John Benjamins.

Hale, Ken
1981 "Preliminary remarks on the grammar of part-whole relations in Warlpiri", in: John Hollyman – Andrew Pawley (eds.), 333-344.

Halliday, Michael A.K.
1985 *An introduction to functional grammar*. London: Edward Arnold.

Hollyman, John – Andrew Pawley (eds.)
1981 *Studies in Pacific linguistics in honor of Bruce Biggs*. Auckland: Linguistic Society of New Zealand.

Hopper, Paul – Sandra A. Thompson
1984 "The discourse basis for lexical categories in universal grammar", *Language* 60: 703-752.
1985 "The iconicity of the universal categories of "noun" and "verb"—", in: John Haiman (ed.), 151-183.

Langacker, Ronald W.
1987 *Foundations of cognitive grammar, Volume 1. Theoretical prerequisites*. Stanford: Stanford University Press.

Lévy-Bruhl, Lucien
1914 "L'expression de la possession dans les langues mélanésiennes," *Mémoires de la Société de Linguistique de Paris* 19(2): 96-104.

Lichtenberk, Frans
1983 *A grammar of Manam*. (Oceanic Linguistics Special Publication Number 18.) Hawaii: University of Hawaii Press.

Lyons, John
1967 "A note on possessive, existential and locative sentences", *Foundations of Language* 3: 390-396.

McGregor, William
1985 "Body parts in Kuniyanti clause grammar", *Australian Journal of Linguistics* 5: 209-232.
1995 "Nominal prefixing in Nyulnyul", [this volume.]
in prep. *A grammar of Nyulnyul*.

McKay, Graham
1995 "Body parts, possession marking and nominal classes in Ndjébbana", [this volume.]

Malczewska, Beata – William McGregor
1990 "Body parts in Polish grammar", a paper presented at the Annual Conference of the Australian Linguistics Society held at Macquarie University, Sydney in September, 1990. [Unpublished manuscript.]

Manoliu-Manea, Maria
1995 "Inalienability and topicality in Romanian: pragma-semantics of syntax", [this volume.]

Mithun, Marianne
1984 "The evolution of noun incorporation", *Language* 60: 847-894.
1986 "On the nature of noun incorporation", *Language* 62: 32-37.

Nichols, Johanna
1988 "On alienable and inalienable possession", in: William Shipley (ed.), 557-609.
1992 *Linguistic diversity in space and time.* Chicago: University of Chicago Press.
O'Grady, Geoffrey N. – Voegelin, Charles F. – Voegelin, Florence M.
1966 *Languages of the world: Indo-Pacific fascicle* 6 (= *Anthropological Linguistics* 8, ii).
Osumi, Midori
1995 "Body parts in Tinrin", [this volume.]
Peekel, Gerhard
1909 *Grammatik der Neu-mecklenburgischen Sprache, speziell der Pala-Sprache.* (Archiv für das Studium deutscher Kolonialsprachen, Band 9.) Berlin: Dietrich Riemer.
Seiler, Hansjakob
1983 *Possession as an operational dimension of language.* (Language Universals Series, Volume 2.) Tübingen: Gunter Narr Verlag.
Shipley, William (ed.)
1988 *In honor of Mary Haas: from the Haas festival conference on native American linguistics.* Berlin: Mouton de Gruyter.
Tomlin, Russell S. (ed.)
1987 *Coherence and grounding in discourse.* Amsterdam: John Benjamins.
Tsunoda, Tasaku
1981 *The Djaru language of Kimberley, Western Australia.* Canberra: Pacific Linguistics.
1995 "The possession cline in Japanese and other languages", [this volume].
Wierzbicka, Anna
1972 *Semantic primitives.* Frankfurt: Athenäum.
1979 "Ethnosyntax and the philosophy of grammar", *Studies in Language* 3(3): 313-383.
1981 *Lingua mentalis.* Sydney: Academic Press.
Wurm, Stephen
1972 *Languages of Australia and Tasmania.* The Hague: Mouton.
Yallop, Colin
1982 *Australian Aboriginal languages.* London: André Deutsch.

The expression of concepts of the personal domain and indivisibility in Indo-European languages

Charles Bally

Translated by Christine Béal and Hilary Chappell, "*L'expression des idées de sphère personnelle et de solidarité dans les langues indo-européennes*"[1]

Foreword

This article by the Geneva School linguist, Charles Bally (1865-1947), represents a seminal work on the theme of this book – the grammatical strategies used in the expression of inalienability and the scope of the personal domain, here, with the focus on Indo-European languages.[2]

Bally was a student of the Swiss linguist, Ferdinand de Saussure (1857-1913) whose approach to language description and analysis laid the foundation of the Geneva School of Linguistics and European structuralism. This includes the celebrated *langue – parole* distinction; emphasis on the separation of the synchronic state of language from diachronic phases of development; the view of language as a system including the study of syntagmatic versus paradigmatic relations and the arbitrary nature of the form-meaning nexus (*signifiant -signifié*). Two other well-known linguists from this School, apart from de Saussure and Bally, were Albert Séchehaye (1870-1946) and Henri Frei. It was Bally and Séchehaye, in fact, who were responsible for editing the book *Cours de linguistique générale* (Geneva, 1916) on the basis of course notes taken by students of de Saussure's lectures. This was posthumously published in de Saussure's name.

In this article, Bally takes issue with Lévy-Bruhl's claim that French does not code inalienable possession in contrast to Melanesian languages by examining the semantic function of dative case marking in French and contrasting its use with the accusative and genitive cases. One of his main descriptive goals is to consider restrictions on dative case marking in Romance languages from both diachronic and synchronic perspectives, comparing this at various points with German, Russian, Ancient and

Homeric Greek. The historical development of the syntactic means used to code inalienable possession in Indo-European languages such as Greek and Latin is also traced. This informative and revealing aspect of the study clearly diverges from traditional structuralist practices regarding the use of historical criteria in synchronic description.

Further tenets of the Geneva School particularly associated with Bally's work, were the importance given to the affective or emotive nature of language which was to become subsumed under the study of stylistics as well as the social function of language. The relevance of both to the expression of inalienability is, in fact, raised by Bally in the introduction to his paper as well as in the main body. Significantly, he attempts to account for restrictions on the use of the dative, that is, why it cannot be used to mark nominal paradigms in many languages, in terms of the communicative needs of restricted social networks.

Bally devoted much research to the stylistic possibilities and expressive resources of language, viewed in terms of the range of different modes of expression afforded speakers by each language and analysed in terms of their stylistic (or semantic) values. This method is clearly apparent in the following article, where Bally explicitly states (p.34) that "every subjective idea ... crucial to the vital functions of the individual tends to be thought of in terms of affectedness and consequently to be rendered in an expressive manner". He proceeds to compare syntactic and morphological means, particularly in French and Ancient Greek, for expressing different ways of viewing the relation between a person and the parts of the body; the psyche; aspects of the personality; kin relations, and simple possessions; all or some of which may be used to define the scope of the personal domain, depending on the language.

Bally's article has given direct and indirect inspiration to work on this topic by Frei (1939), Hatcher (1944), Diffloth (1974), Wierzbicka (1979), Seiler (1983), Kliffer (1984), McGregor (1985) and Chappell (1986), among others. The fact that the Festschrift in which it is published is not readily available has provided the motivation to render it into English, and thus enable Bally's work to reach a wider audience.

The translation

The concept of indivisibility is closely related to the personal domain.

The personal domain includes or can include objects and beings associated with a person in an habitual, intimate or organic way (e.g. the body and its parts, clothes, the family, etc.). Each constitutive element of the domain is regarded, not as a simple property, but as an integral part of the person. A French person, for example, does not think of *ma tête* 'my head' in the same way as *mon jardin* 'my garden'. As Lévy-Bruhl points out with some justification in his 1914 article which describes this concept neatly – 'When I talk about my head, I do not intend to say that it belongs *to me*, but that it is *me*.' (1914: 98).

The idea of indivisibility or of being part of a whole follows on directly from the above: each phenomenon, action, state or quality which affects any part whatsoever of the personal domain, automatically affects the whole person. The part of the body directly affected is only the medium for a condition which spreads to the whole system. For the French, *se laver les mains* 'to wash one's hands' (literally 'to wash to oneself the hands')[3] is not the same thing as *laver ses mains* 'to wash one's hands' and they would be unlikely to use the counterpart of the English 'I've broken *my leg*': *J'ai cassé ma jambe* rather than *Je me suis cassé la jambe* (literally 'I to myself have broken the leg') unless they were talking about a wooden leg.

Needless to say a person involved in an action may be viewed as a whole regardless of the constituent parts (cf. *se laver les mains* literally 'wash to oneself the hands' but also *se laver* 'wash oneself').

In the Indo-European languages, the concept of personal domain is related to organic life since only living beings or objects personified in the imagination can be represented in this way. To say of a table that *on lui casse une jambe* (literally: 'one to it breaks a leg') 'One breaks its leg' is to personify it. This example clearly shows that the idea is closely related to the animate category which continues to play such an important role in Indo-European languages.

The concept of personal domain is an entirely subjective one. Nothing prevents the collective imagination from attributing to the self objects that normally have their own independent existence or, conversely, of detaching those things which in reality cannot be. The extent of the domain is determined by the cultural outlook of each linguistic group. Its limits may vary from language to language and vary within the same language during the course of its evolution. It generally includes the body, its parts and sometimes its dimensions , the soul of the individual, and in some cases, the voice and the name. It may also include to a varying degree everything which holds an habitual relationship to it: clothing, familiar objects, utensils; people in one's social circle, family, servants and friends. Moreover, in each of these subcategories, there is further opportunity for subtle distinctions. Finally, and most importantly, these beings and objects may be viewed either as being a part of the self or as being detached from the person (see below).

The above shows that the concepts of personal domain and indivisibility are closely linked to the social function of language, and that their evolution may depend on that of society itself. A priori, it seems likely that the denser communication is and the more numerous and complex social relations are, the more the domain of the individual shrinks which results in thinking from the other's point of view rather than from the perspective of the self. Needless to say, such links between language and society should be carefully qualified, language being, of all institutions, the most conservative yet simultaneously, the most easily influenced from outside. A linguistic form does not necessarily correspond to the actual state of society but may either reflect a tradition long past or be the result of foreign influences.

Finally, let us note that every subjective idea, every notion that is crucial to the vital functions of the individual tends to be thought of in terms of affectedness, and consequently to be rendered in an expressive manner. The concepts defined above are then those which language rarely presents in a purely intellectual, impersonal way. In other words language constantly seeks to renew linguistic processes which have lost their expressive sharpness through overuse.

The present study does not claim to give an exhaustive description of all the linguistic means used by Proto-Indo-European and the Indo-European languages to express these ideas. We are merely trying to set up some guidelines and to show in which direction evolution seems to have occurred, emphasizing innovations belonging to the modern languages.

The oldest ways of expressing the person involved in the action, are, as we know, the middle voice of the verb and the dative of personal pronouns, rather than the active voice or its deponents with the genitive or possessive pronouns.

The middle voice, later replaced by the reflexive verb, presents the action as taking place in the domain of the subject of the verb (Greek: νίζομαι *nizomai* [Pres:wash:middle:1sg] 'I wash myself' (all glossing abbreviations are given at the end of note 1) in opposition to νίζω τράπεζαν *nizō trapezan* [Pres:wash:active:1sg table] 'I wash a table') while for the expression of indivisibility between the self and parts of the body, Greek uses: νίζομαι τὰς χεῖρας *nizomai (tas) cheiras* (Pres:wash:middle:1sg the:ACC hands:ACC; literally 'I wash myself the hands') 'I wash my hands'.

The dative indicates the person involved when it is not the subject of the sentence:

Greek:

(1)	βοῦς	ἱερεύειν	θεῷ
	bous	*hiereuein*	*theōi*
	bull:ACC:pl	sacrifice:active:INF	god:DAT:sg

'to sacrifice cattle in honour of a god'

Homeric Greek:

(2)	οὐκ	ἄῤ	ἔμελλες	οὐδὲ
	ouk	*ar'*	*emelles*	*oude*
	Neg	then	Past:intend:2sg:Impf	not:even

θανὼν	λήσεσθαι	ἐμοὶ
thanōn	*lēsesthai*	*emoi*
Aor:part:NOM:sg:die	forget:middle:INF	1sg:DAT

χόλου
cholou
anger:GEN:sg

'Not even in death then, were you going to forget the memory of my anger.'

(Odyssey XI: 553: *mihi oblivisci* 1sg:DAT forget:Past:1sg 'I forgot myself.')

For the expression of indivisibility, examples such as the following can be found:

(3)	νίζω	σοι	(τὰς)	χεῖρας
	nizō	*soi*	*(tas)*	*cheiras*
	wash:1sg	2sg:DAT	(the:ACC)	hands:ACC:pl:(active)

'I wash your hands.' (literally: 'I wash to you the hands.')

Homeric:

(4)	θυμός	μοι	ἔσσυται
	thumos	*moi*	*essutai*
	desire:NOM:sg	1sg:DAT	Pres:rush:3sg:(middle)

'I am overcome with desire.' (literally: 'spirit to me soars')

Examples (3) and (4) can be contrasted with (5), (6) and (7) which do not use the dative:

(5)	νίζω	πόδας	τραπέζης
	nizō	*podas*	*trapezēs*
	Pres :wash:1sg (active)	foot:ACC:pl	table:GEN:sg

'I wash the legs of a table.'

(6)	νίζω	τὰς	σὰς	χεῖρας
	nizō	*tas*	*sas*	*cheiras*
	Pres:wash:1sg (active)	the:ACC:Pl	your:ACC	hands:ACC:pl (2pl:poss)

'I wash your hands.'

(7)	(νίζω)	τὰς	χεῖρας	σου
	(nizō)	*tas*	*cheiras*	*sou*
	(Pres:wash:1sg) (active)	the:ACC:Pl	hands:ACC:Pl	of:you (2sg:GEN)

'I wash your hands.' (literally 'I wash the hands of you.')

Hence the middle voice is merely an alternative to the active with the dative, in the same way as the passive is an alternative to the active with the accusative.

The mechanism of the middle voice is relatively well-known. Returning to the dative of involvement, although there is the important study by Havers (1911), very little research has been done into the question of any affinity between the two processes. Furthermore, the means of expression which encodes this complex of meanings has not been examined as a whole.

For there are, in fact, others: Within classical languages, it is easy to see that from the time of Indo-European (cf. Brugmann, 1886-1892: 2, II, 2, 633), the following type of construction in the accusative always expresses the self and one of its integral parts, not simply possessed:

(8)

καθ'	ὅλον	καὶ	μέρος
kath'	holon	*kai*	meros
in:accordance:with	whole:ACC	and	part:ACC

'in accordance with a whole, also a part'

However, this is also found in other grammatical cases, cf. Iliad I,150:

(9)

πῶς	τίς	τοι	πρόφρων
pōs	*tis*	toi	*prophrōn*
how	INDEF:NOM	2sg:DAT	thinking:NOM:sg

ἔπεσιν	πείθηται
epesin	*peithētai*
word:DAT:pl	Pres:persuade:Pass:3sg:Subj

'How should a thinking person be persuaded by your words?'

Another example of the first type comes from Homeric:

(10)

τὸν	δ'	ἄορι	πλῆξ'	αὐχένα
ton	*d'*	*aori*	*plēx*	*auchena*
he:ACC:sg	and	sword:DAT:sg	Aor:strike:3sg	neck:ACC

'and he struck him with his/a sword upon the neck'

Compare the following three examples: The first and third contain double accusatives [(11) and (13)] while the middle example [(12)] shows a combination of *dative* + *accusative*:

(11)

νίζω	τινὰ	χεῖρας
nizō	*tina*	*cheiras*
Pres:wash:1sg	Indef:Pro:3sg:ACC	hands:ACC:Pl

'I wash someone's hands.'

(12)

νίζω	τινὶ	χεῖρας
nizō	*tini*	*cheiras*
Pres:wash:1sg	Indef:Pro:3sg:DAT	hands:ACC:Pl

'I wash someone's hands.' (literally: 'I wash to someone the hands.')

(13)

ἥ	σε	πόδας	νίψει
hē	*se*	*podas*	*nipsei*
Def:NOM:sg	2:ACC:sg	feet:ACC:Pl	wash:Fut:3sg

'She will wash your feet.' (Odyssey, XIX: 356)

The same idea of involvement can also be rendered by the *accusative* + *dative* or, in an even more personal and expressive manner, by the double accusative. The latter ceased to be productive at an early stage and its usage then became restricted to poetic language in Ancient Greek.

When changed into the passive, we know that this syntactic form takes as subject the noun referring to the person involved: cf. :

Active:

(14) ἐκκόπτω τινὰ τὸν ὀφθαλμόν
ekkoptō tina ton ophthalmon
Pres:knock:out:1sg Indef:3sg:ACC the:ACC:sg eye:ACC:sg
'I knock out (literally: cut out) someone's eye.'

Passive:

(15) ἐξεκόπην τὸν ὀφθαλμὸν λίθῳ
exekopēn ton opthalmon lithōi
out:Past:knock:1sg:Pass the:ACC:sg eye:ACC:sg stone:DAT:sg
'My eye was knocked out by a stone.' (Aristophanes, *The Clouds*, 24)

This explains why these intransitive verbs and adjectives have the same construction:

(16) ἀλγῶ τοὺς ὀφθαλμούς
algō tous opthalmous,
Pres:hurt:1sg the:ACC:pl eyes:ACC:Pl
'My eyes hurt.'

(17) πηρὸς τὴν χεῖρα
pēros tēn cheira
about the:ACC:sg hand:ACC:sg
'incapacitated with regard to his hand' (see Brugmann in *Indogermanische Forschungen* 27: 121 ff)

Plato admirably defined the value of this accusative in his *Republic* (462 c-d).

It is generally called in this case, the accusative of specialisation, as in the following example:

(18) νικᾶν μεγάλην μάχην
nikan megalēn machēn,
win:Pres:INF big:ACC:sg battle:ACC
σοφὸς πᾶσαν τέχνην
sophos pasan technēn
wise:NOM:sg all:ACC:sg skill:ACC:sg
'to win a great battle, wise in regard to every skill'

However, between the two uses there is a noticeable difference: In the first, the focus is on the direct object referring to a person indirectly by means of an integral part. In the second, it is the verb or adjective which is being focused on. It is surprising that two examples as diverse as the following can be found side-by-side in the grammar of Greek by Kühner – Gerth, *Ausführliche Grammatik der griechischen Sprache* (Leverkusen, 410, 6), two examples as divergent as:

(19) τὸ σῶμα οὐκ ἀφυής
to sōma ouk aphuēs
the:ACC:sg body:ACC:sg Neg Priv:natural:talent:NOM:sg
'not lacking physical ability' (literally: 'not lacking in natural talent as regards his body') (*Xenophon*, The education of Cyrus, 2, 3, 7)

(20) δεινὸς ταύτην τὴν τέχνην
deinos tautēn tēn technēn
clever:NOM:sg this:ACC:sg the:NOM:sg skill:ACC:sg
'clever at this skill' (ibid, 8, 4, 18).

In fact, in (21), the accusative is partially adnominal, being in substance equal to (22):

(21) Κῦρος οὐκ ἀφυής ἐστι
Kuros ouk aphuēs esti
Cyrus:NOM:sg Neg Priv:natural:talent:NOM:sg Pres:be:3sg
τὸ σῶμα
to sōma
the:ACC:sg body:ACC:sg
'Cyrus was no slouch physically.'

(22)

τὸ	σῶμα	τοῦ	Κύρου
to	*sōma*	*tou*	*Kurou*
the:NOM:sg	body:NOM:sg	the:GEN:sg	Cyrus:GEN:sg

οὐκ	ἔστιν	ἀφυής
ouk	*estin*	*aphuēs*
Neg	Pres:be:3sg	Priv:natural:talent:NOM:sg

'The body of Cyrus was not lacking in natural ability.'

Note, however, that through the accusative, Greek marks the involvement of the person in the action or the state which affects one of its parts, as in (23):

(23)

ἀλγῶ	τοὺς	ὀφθαλμούς
algō	*tous*	*opthalmous*
Pres:have:pain:1sg	the:ACC:Pl	eyes:ACC:Pl

'I have a pain in my eyes.'

This can be opposed to the use of the nominative in example (24) which emphasises the effect of the eye disease on the self as a whole:

(24)

οἱ	ὀφθαλμοί	μου	ἀλγοῦσι
hoi	*opthalmoi*	*mou*	*algousi*
the:NOM:Pl	eyes:NOM:Pl	1sg:GEN	Pres:hurt:3Pl

'My eyes hurt.'

To the contrary, (25) acts as a specification of (26) with the noun in the accusative acting as the adjective, exactly as in (27) where it is cognate to the verb:

(25)

δεινὸς	τέχνην	τινά
deinos	*technēn*	*tina*
clever:NOM:sg	skill:ACC:sg	Indef:3sg:ACC

'clever at some skill'

(26)

δεινὸς	δεινότητά	τινα
deinos	*deinotēta*	*tina*
clever:NOM:sg	cleverness:ACC:sg	Indef:3sg:ACC

'clever at some cleverness'

(27)

νιχᾶν	νίχην
nikan	*nikēn*
Pres:win:INF	victory:ACC:sg

'win a victory'

Apparent convergences show this difference clearly. In the double accusative structure in (28), τραύματα *traumata* 'wounds' is adverbial and has nothing to do with the idea of personal involvement whereas the passive in (29) refers to persons whose wounds have been dressed and encodes that from this moment on these wounds are viewed as part of the warriors themselves.

Double accusative:

(28)

τραύματα	πλήττειν	τινά
traumata	*plēttein*	*tina*
wounds:ACC:Pl	Pres:strike:INF	Indef:3sg:ACC

'to wound someone'

Passive:

(29)

τὰ	τραύματα	ἐπιδεδεμένοι
ta	*traumata*	*epidedemenoi*
the:ACC:Pl	wounds:ACC:Pl	bind:upon:Perf:Pass:Part:NOM:Pl

'their wounds being bandaged' (Xenophon, *The education of Cyrus*: 5, 2, 32)

The contrast appears even more clearly when the two types of specification occur in the same sentence:

Double accusative and passive:

(30)

πληγεὶς	τὴν	κεφαλὴν
plēgeis	*tēn*	*kephalēn*
strike:Aor: Pass:Part:NOM:sg	the:ACC:sg	head:ACC:sg

χαιρίαν	(πληγήν)
kairian	(implied: *plēgēn*)
critical:ACC:sg	strike:ACC:sg (noun)

'having been struck a mortal blow to the head'

This interpretation is a natural consequence of the very idea of indivisibility.

The association between the whole and the part is one of substance to substance – an internominal relation. At the same time, it is also a verbal relation which makes sense since the idea of involvement only manifests itself through the occurrence of phenomena or states that in affecting a part of the person, affect the person in their entirety (see p. 33 above). It is precisely this syntactic duality which gives the structure its synthetic characteristics (see p. 49 below). However, the accusative and the dative of involvement tend to function like adnominals. Sometimes only a slight syntactic change is enough for this adnominal role to resurface. This explains how the construction type in German, exemplified by (31) has come into use in colloquial language in the same way as in French (see p. 46 below), exemplified by (32):

(31)	*dem*	*Vater*	*sein*	*Hut*
	the:DAT	father	his:NOM	hat

'the father's hat'

(32)	*le*	*chapeau*	*à*	*papa*
	the	hat	to:DAT	daddy

'daddy's hat'

What kinds of changes have taken place in modern languages in this respect? Once again, putting aside details of isolated examples in specific languages, we will identify some major points which might be conducive to the orientation of a more complete study.

We can distinguish between syntactic reflexes, modifications of an earlier stage, and innovations. For example, the dative has been preserved, the middle has changed to the reflexive while new constructions have been introduced in Indo-European languages.

The use of the reflexive is a move towards more analytical processes. One of its consequences is a more generalised use of the dative: Compare German and Ancient Greek in (33) and (34) respectively:

(33)	*Ich*	*wasche*	*mir*	*die*	*Hände*
	I	wash	1sg:DAT	the:ACC:Pl	hands

'I wash my hands.'

(34) νίζομαι τὰς χεῖρας
nizomai *tas* *cheiras*
Pres:wash:1sg (middle) the:ACC:Pl hands:ACC:Pl
'I wash my hands.'

However, we will see that for other reasons, the dative has also been replaced in its turn by new processes.

The development of the definite article – which has already been foreshadowed in the examples cited above – has provided a new means for the expression of indivisibility, by indicating the part directly involved. The constructions which have evolved in the Romance and Germanic languages are analogous to those which can be observed in Attic Greek: for example, French forms such as:

(35) *la* *tête* *lui* *tourne*
the head 3sg:DAT turn
'S/he is dizzy.'

(36) *la* *main* *lui* *démange*
the hand 3sg:DAT itch
'His/her hand is itchy.'

In the following French examples, the absence of the dative reflexive in the first group encodes more instinctive actions than the construction with the dative reflexive pronoun in the second group:

I. Verb + Definite Article + Body Part Term

(37) *baisser* *les* *yeux*
lower the eyes
'cast one's eyes down'

(38) *froncer* *les* *sourcils*
crease the eyebrows
'knit one's brow, frown'

(39) *hausser* *les* *épaules*
lift the shoulders
'shrug'

II. *Dative reflexive* + Verb + Definite Article + Body Part Term

(40) *se* *casser* *la* *jambe*
DAT:Refl break the leg
'break one's leg'

(41) *se* *frotter* *les* *mains*
DAT:Refl rub the hands
'rub one's hands'

The semantic value of the definite article is striking in contrast to the possessive pronoun: Compare *baisser* les *yeux* [lower the eyes] 'cast one's eyes down' with *abaisser* ses *regards sur quelqu'un* [lower one's (3sg:Poss:PL) looks on someone] 'gaze down at somebody' and *se tacher* les *mains* [DAT:Refl stain the hands] 'dirty one's hands' with *tacher* ses *habits* [stain 3sg:Poss:PL clothes:PL] 'stain one's clothes'.

The combination of the pronominal dative with the possessive adjective allows the expression of a looser involvement:

(42) *Qui* *est-ce* *qui* *m'a* *pris* *ma* *plume*?
who is-it who DAT:1sg:has taken my pen
'Who has taken my pen from me?'

As for the dative of involvement, it is gradually going out of use in modern languages which have undergone the most marked evolution (e.g. French and English). This trend can be explained by complex historical reasons. While it is still common in Germanic and Slavonic, its usage seems to have been inexplicably restricted in the Romance languages, clearly showing its special status as a syntactic reflex. This restriction goes back to Proto-Indo-European. Havers (1911) has shown that from the beginning its use has been regular only with the personal pronouns of the three persons **moi* 'to:1sg' **toi* 'to:2sg' **soi* 'to:3sg' cf. Greek μοι *moi* '1sg:DAT', τοι *toi* '2sg:DAT', οἱ *hoi* '3sg:DAT'. The nouns (apart from proper nouns to a lesser extent) exclude it. The dative of involvement applied to nouns, is, strictly speaking, an innovation which developed in different ways in different languages. Its usage never became a regular feature of Latin or Romance.

It is interesting to observe how this tendency of Indo-European languages not to extend the dative beyond the personal pronouns is already exemplified in Homeric Greek by a curious irregularity. We find examples such as (43) from *Odyssey* X: 484 and (44) from *Odyssey* IX: 256:

(43) θυμὸς δέ μοι ἔσσυται ἤδη
thumos de moi essutai ēdē,
spirit:NOM:sg and 1sg:DAT Pres:rush:3sg already
ἠδ' ἄλλων ἑτάρων
ēd' allōn hetarōn
and other:GEN:Pl companion:GEN:Pl
'and already my spirit is eager (to go), and (so is that) of my other companions'

(44) ἡμῖν δ' αὖτε κατεκλάσθη
hēmin d' aute kateklasthē
1Pl:DAT and in:turn break:down:Aor:Pass:3sg
φίλον ἦτορ δεισάντων
philon ētor deisantōn
dear:NOM:sg heart:NOM:sg fear:Aor:Part:GEN:Pl
'and in turn, our dear heart was shattered, since we were afraid'

Havers (1911: 65 *et passim*) cites many similar cases where the syntactic parallelism is forgone rather than extend the domain of the dative.

It was precisely this restriction in usage which persisted in the Romance languages, for example, the dative of the affected person: In French, the rules of syntax usually prohibit the replacement of the dative pronouns *me* '1sg', *te* '2sg', *lui* '3sg' etc. by nominal groups formed with *à* 'to'. No-one would hesitate to say (45) and (46) but (47) and (48) are very awkward:

(45) *Je vous obtiendrai cette place*
I you:DAT obtain:Fut this job
'I will get this job for you.'

(46) *Je vous trouverai une chambre*
I you:DAT find:FUT a room
'I will find a room for you.'

(47) *?J'obtiendrai cette place à mon neveu*
I:obtain:Fut this job to:DAT my nephew
'I will get this job for my nephew.'

(48) *?Je trouverai une chambre à mon ami*
I find:Fut a room to:DAT my friend
'I will find a room for my friend.'

If the Dative case with *à* 'to' does not jar, then it is due to the verb undergoing some analogical influence whereby it loses the meaning of close involvement:

(49) *J'achète une montre à mon fils*
I:buy a watch to:DAT my son
'I buy a watch for my son.'

If (49) is possible, it is because *acheter* 'buy' has formed a construction similar to that of *donner* 'give' (< *donner quelque chose à quelqu'un* 'give something to someone').

With certain verbs of movement (*courir* 'run', *tomber* 'fall', etc.), the direction may be indicated by an adverb (for example, *dessus* 'above') and the person who is the goal of the movement may be placed in the dative, provided that it shows the extent of involvement in the action. As long as this is only a question of the personal pronouns, it is acceptable. It is common to find constructions such as the following in (50) and (51) and even (52), especially in colloquial language:

(50) *On lui court après*
one 3sg:DAT runs after
'S/he is being chased.'

(51) *On lui tombe dessus*
one 3sg:DAT falls on:top
'They are falling on top of him/her.'

(52) *Le couteau lui entre dedans*
the knife 3sg:DAT enters inside
'The knife goes into him/her.'

This is how French is able to make a subtle distinction between the animate and the inanimate, the latter being marked by the absence of a pronoun, for example, as in (53):

(53) *Le chat s'approche de la table et*
the cat DAT:come:close of the table and

(**il lui) monte dessus*
(*3sg 3sg:DAT) climbs on:top
'The cat approaches the table and climbs on top.'[4]

However, where the noun indicates a person, we have to use constructions such as (54), (55) and (56):

(54) *On court après un voleur (On lui court après)*
one runs after a thief (one 3sg:DAT runs after)
'A thief is run after.' ('S/he is run after.')

(55) *On tombe sur un ennemi (On lui tombe dessus)*
one falls upon a foe (one 3sg:DAT falls upon)
'A foe is fallen upon.' ('S/he is fallen upon.')

(56) *On passe devant un inférieur*
one pushes in:front:of a subordinate
'One pushes in front of a subordinate.'
(in the context of using superior status to get served first)
(On lui passe devant)
(one 3sg:DAT pushes in:front)
('One pushes in front of her/him.')

Similarly, when the whole is associated with its part, we would hesitate to use (58) in French rather than (57):

(57) *Je lui casse la jambe*
I 3sg:DAT break the leg
'I am breaking his leg.'

(58) *?Je casse la jambe à Paul*
I break the leg to:DAT Paul
'I am breaking Paul's leg. '

As we have seen, this awkwardness is no doubt due to interpreting *la jambe à:DAT Paul* to mean the same as *la jambe de:GEN Paul* 'Paul's leg'. The interpretation itself arises from the tendency of Romance languages to position elements in the sentence so that word order determines the meaning.[5] In this case, the tendency has been favoured by a reluctance to mark the nominal group by the dative of involvement. Note that German, which retained its inflectional morphology and is thus less dependent upon word order than French, has been able to extend the dative to nouns, which adds much to the stylistic use of this syntactic construction:

(59) *den* *Mädchen* *nachlaufen*
the:DAT:Pl girls run:after
'be after, chase girls'

(60) *dem* *Feinde* *das* *Herz* *durchbohren*
the:DAT enemy the:ACC heart pierce
'pierce the heart of the foe'

But what accounts for the difficulty of using nominals in the dative of indivisibility? I would venture to suggest that this means of expression has always been used in situations such as family life, where the needs of communication are specified by the context in each case, and where the speakers are part of a small and stable circle of subjects. The beings and the objects talked about occur naturally in conversation without the need for explicit reference. The persons are nearly always *moi* 'me', *toi* 'you' and *lui* 'him' or 'her', in other words, the interlocutors – individuals and their immediate circle of acquaintances who do not need any further specification. Beyond these cases, the reference to persons, without use of proper names requires varied and subtle grammatical marking belonging to a more extended and complex communicative context. However, when the network of relations spreads out, the older system becomes more awkward, so that if the language maintains the expression of indivisibility, it will seek to replace it by other constructions where the traditional form cannot be used (as for the nominal group in the Dative).

The forms of pronouns, unique in their kind, which can be used to express the dative of involvement confirm this hypothesis. It is very likely that this kind of syntax was first applied to uninflected words of proto Indo-European origin (**moi* '1sg', **toi* '2sg', **soi* '3sg'), perhaps also to the direct or indirect cognates in Sanskrit *nas* '1pl:ACC/DAT/GEN pronoun (accentless)', *vas* '2pl:ACC/DAT/GEN pronoun (accentless) and Greek σφι *sphi* '3Pl:DAT' etc). Because these forms became mere morphological archaisms at a very early stage (despite their heavy usage), it prevented an analogical extension of this type of dative to the nominals and the inflected pronouns (Sanskrit *ayam* 'sg:indefinite pronoun:NOM' etc.) and even to the analytical personal pronouns (Sanskrit *túbhyam* 2sg:DAT 'to thee' etc.) cf. Havers (1911:30, 44). It could be conceded that the rules of this restricted framework were partially violated in certain languages. Nearly everywhere the usage of the dative has been extended to at least all the personal pronouns. However, it has rarely gone beyond this, because of the considerable difference in form in the case of the nominals of Greek μοι *moi* '1sg:DAT',σοι *soi* '2sg:DAT', οἱ *hoi* '3sg:DAT',υἱῷ *huiōi* 'son:DAT', πατρί *patri* 'father:DAT', θεοῖς *theois* 'gods:DAT:Pl', ἀνδράσι *andrasi* 'men:DAT:Pl' etc.; French *me* '1sg:DAT' *lui* '3sg:DAT', *leur* '3Pl:DAT' and *à mon père* 'to my father' etc.

The modern languages have another reason for not generalizing this use of the dative. They tend towards analyticity, whereas dative constructions are a synthetic device. We have seen above that the dative case enters into a double relation with both the noun and the verb. Thus in the German example (61) below, *sich* 'to herself' relates to *Rock* 'skirt' as much as to *zerriss* 'tore':

(61)	*Sie*	*zerriss*	*sich*	*beim*	*Fallen*	*den*	*Rock*
	she	tore	3sg:DAT	at:DAT	fall	the:ACC	skirt

'She tore her skirt falling over.' (literally: 'She tore to herself at the fall the skirt')

Furthermore, the dative of unstressed pronouns of the kind **moi*, **toi*, **soi* etc. has from the time of Proto-Indo-European gradually moved as close as possible to the beginning of the sentence (Wackernagel's law, see Wackernagel 1953), which separates it from the word expressing the affected part, although the two nouns are semantically connected. The preverbal position of unstressed conjunctive pronouns[6] in Romance languages is a weakened form of this rule. Compare French examples (62) and (63):

(62)	*On*	*lui*	*tranche*	*la*	*tête*
	one	3sg:DAT	cut	the	head

'She had her/his head cut off.'
or: 'He had her/his head cut off.'

(63)	*On*	(**lui*)	*déchire*	*ses*	*habits*
	one	(*3sg:DAT)	tears	3sg:Poss	clothes

'Someone is tearing her/his clothes.'

Also compare French example (64) with its English translation:

(64) *On lui coupe la jambe*
one 3sg:DAT cuts the leg
'Someone cuts his/her leg off.'

For all of these reasons, some modern languages have diverged from the older process. It suffices to compare Latin with the Romance languages:

The general change consisted in altering certain kinds of sentences containing a dative of involvement *in a way in which the affected person became subject of the sentence*. Since this use did not come into competition with any other ones in existence, the nominals were allowed to fill this position. It is thus that the Latin *mihi sunt capilli nigri* (1sg:DAT are (be:3pl) hair:Pl:NOM black:Pl:NOM) is translated into French as (65):

(65) *J'ai les cheveux noirs*
I:have the:Pl hair:Pl black:Pl
'I have black hair.'

Thanks to the use of the definite article (cf. p. 43 above), the idea of being a part of a whole has remained distinct from that of just belonging. Compare (66) with (67):

(66) *J'ai des cheveux noirs*
I:have some:Pl hair:Pl black:Pl
'I have black hair.'

(67) *Mes cheveux sont noirs*
my:Pl hair:Pl are black:Pl
'My hair is black.'

We can guess that this innovation was greatly favoured by the generalisation in use of the verb *avoir* 'have', which, instead of indicating simple possession (*J'ai une maison* 'I have a house') has come to express a mere relation in many other cases other than those which interest us here: *J'ai une mission* 'I have a mission'; *J'ai de l'espoir* 'I have hope'; *J'ai soixante ans* 'I am sixty' etc. Meillet has brought this important fact to light in his article in *Festschrift J. Wackernagel* p9ff.

The use of *avoir* 'have' to render the notion of involvement and of indivisibility is full of nuances. We can say *avoir de la grâce* 'have grace' in the sense of *être gracieux* 'be gracious'. *Avoir* can express, apart from the quality associated with a person, a momentary state undergone by someone: *Paul a la fièvre* 'Paul has fever' or *Paul a les yeux rouges (pour avoir pleuré)* 'Paul has red eyes (from having cried)'; *Il a le sang à la tête* 'He is flushed in the face'. In addition to this, it can express the state in which a person is found that affects another. Such is the difference between (68) and (69):

(68) *Un de mes fils est malade*
'One of my sons is sick.'

(69) *J'ai un fils malade*
I:have a son sick
'I have a sick son.'

Furthermore, *avoir* 'have' may express the action itself insofar as it intimately affects the subject. This produces a connotation different from the rather impersonal meaning of the passive. Compare (70) with (71):

Passive:

(70) *La mâchoire du soldat fut fracassée*
the jaw of:the soldier was fractured
'The soldier's jaw was fractured.'

Avoir 'have':

(71) *Le soldat eut la mâchoire fracassée*
the soldier had the jaw fractured
'The soldier had his jaw fractured.'

Other sentences show an even more liberal use with the same basic value. Contrast the use of (72) with (73):

(72) *Avez-vous des malades chez vous?*
have-you:Pl some sick:people by you
'Is anyone sick at home?' (literally: 'Do you have any sick people at your place?')

(73) *Y a-t-il des malades ...?*
there is-it some sick:people
'Are there sick people?

(74) *Qu'avez- vous?*
what:have you
'What's the matter with you?'

J'ai que je suis dans une rage folle
I:have that I am in an anger mad
'I'm furious – that's the problem.'

Even sentences such as (75) are heard:

(75) *J'ai ma femme qui est très malade*
I:have my wife who is very sick
'The fact is that my wife is very sick.'

We can measure how relatively new this syntax is in Indo-European languages by trying to express it in the older languages, or in the modern languages which have changed less markedly as in the Russian example (76); cf. French: *L'enfant a mal à la tête* (the:child has pain at the head):

(76) *U rebënka bolít golová*
at child:GEN ache head:NOM
'The child has a headache.'

As for Latin, we know that literal translation with *avoir* leads to many solecisms. However, it should be noted that our knowledge of Latin is largely based on literary texts. It is difficult to say to what degree Romance forms were used in the colloquial language of Rome. Cicero risked *habere febrim* (have:INF fever:ACC) 'have fever' and *habere oculos acres* (have:INF eye:ACC:pl sharp:ACC:pl) 'have sharp eyes'. The studies in this area are too limited to draw any conclusions.

There are other verbs of less frequency than *avoir* 'have' which perform the same function but are more colourful or contain different aspectual nuances. For example *porter* 'carry, wear' in the phrase which follows in (77):

(77) *porter les cheveux longs*
wear the:Pl hair:Pl long:Pl
'have long hair' (referring to hairstyle)

Prendre 'take', *recevoir* 'receive' and colloquially *attraper* 'catch' all express the completive aspect corresponding to *avoir* 'have':

(78) *prendre un rhume*; *attraper la grippe*
take a cold catch the influenza
'catch a cold' 'catch the 'flu'

recevoir des coups
receive some blows

'receive blows'

cf. German *bekommen* 'receive', *kriegen* 'get', Alemannic *übercho* 'be overcome' etc.[7] Others convey in an expressive manner the experiential passive. Thus opposite the regular passive in French with *être* + *Past Participle* in (79) there is also *s'être vu* + *Infinitive* as in (80):

(79) *Ses biens ont été confisqués*
his:Pl goods have been confiscated

'His goods have been confiscated.'

(80) *Il s'est vu confisquer ses biens*
he DAT:Refl:was seen confiscate his:Pl property:and:wealth

'He had to face his property and wealth being confiscated.'

Similarly, alongside the regular passive in the following example (81), an alternative structure with *s'être laissé* or *s'être fait* can be used as in (82):

(81) *Le renard a été pris au piège*
the fox has been taken to:the trap

'The fox has been caught in the trap.'

(82) *Le renard s'est laissé* (or even: *s'est fait*) *prendre au piège*
the fox DAT:Refl:is let (DAT:Refl:is made) take to:the trap

'The fox got itself caught in a trap.'

Undoubtedly this is the same tendency which leads to the formation of (83) with *s'est tué* 'got oneself killed' rather than using *a péri* 'has died':

(83) *Cet alpiniste s'est tué au Cervin*
this mountaineer DAT:Refl:is killed at:the Cervin

'This mountaineer got himself/herself killed at Cervin.'

Note that modern languages can also use a similar process to turn the ethic dative into the nominative which indicates personal involvement without any notion of indivisibility between body and part. This is how I would explain constructions such as (84) and (85):

(84) *Vous me voyez aux regrets*
you:pl 1sg:ACC see to:the:pl regrets
'I'm terribly sorry.'

(85) *Vous voyez un homme bien ennuyé*
you:pl see a man very embarrassed
'I'm most embarrassed.'[8]

Examples (84) and (85) are equivalent in substance to (86) and (87) respectively:

(86) *Je regrette beaucoup*
1sg regret much
'I am very sorry.'

(87) *Je suis bien ennuyé*
I am very embarrassed
'I am very embarrassed.'

Compare also the use of *s'être laissé* in (88) with the impersonal passive in (89):

(88) *Je me suis laissé dire que ...*
I DAT:Refl am let say that
'I have it from a good source that ...'
(that is, in the context where I've been told confidential information)

(89) *On m'a dit que ...*
one me:DAT:has said that
'Someone told me that ...' (OR: 'I've been told that ...')

The same situation applies in German. Compare the following two examples in (90) and (91):

(90) *Da hast du mich*
there have you:NOM me:ACC
'There you have me.'

(91) *Da bin ich*
there am I:NOM
'There I am.'

We can see from what precedes that the processes through which a grammatical concept is expressed can be very 'internal' or affective in sense without ceasing to be real and linguistic in substance. We need to distinguish between combinatory indirect processes and mere implication aided by context and situation, in which case, language is not at issue. In the article, cited at the beginning of this study, Lévy-Bruhl has not been completely justified in believing that we do not distinguish the expressions *mon bras* 'my arm' and *mon billet de chemin de fer* 'my train ticket' or that our languages dispense with the distinctions made by Melanesian languages in expressing involvement. In fact, we express these nuances with the same precision but by means of slightly more indirect strategies. Lévy-Bruhl shows that, in Melanesian, *ma tête* 'my head' is marked in two ways according to whether it involves a part of my body or a piece of meat that I am about to eat. We have seen in the first case, French chooses the syntax exemplified in (92) and in the second, the construction given in (93):

(92) *On me rase la tête*
one me:DAT shave the head
'I am being shaven.'

(93) *Je mange ma tête*
I eat my head
'I am eating my head.'

Moreover, this does not represent the limits of the distinction in French. It is possible by the same means to indicate that a part of the body is either an integral part of the person or is viewed as detached from it. The first idea is expressed when we use sentences such as (94):

(94) *Quelqu'un me prend le bras*
someone me:DAT takes the arm
'Someone takes me by the arm.'

However, treating the limb purely as an object, comparable to a walking stick or crutch, is evident when we say to someone who is tired:

(95) *Prenez mon bras*
take:IMP my arm
'Take my arm!'

We may also contrast (96) with (97):

(96) *avoir le sang chaud*
have the blood hot
'be hot-blooded; be hot-headed'

(97) *donner son sang pour la patrie*
give one's blood for the motherland
'give one's blood for the motherland'

It is for a similar reason that the concept of the integral part disappears once the object is characterised in any way. Hence, French example (98) with the definite article determining the body part noun is changed to the possessive pronoun in (99) upon adjectival modification of the body part term:

(98) *serrer la main à quelqu'un*
squeeze the hand to someone
'shake someone's hand'

(99) *serrer sa main robuste*
squeeze one's hand strong
'shake his/her strong hand'

The fact is that in all these cases (with possessive modification), the object acquires a personality of its own, detached from the individual of which it forms a part. Hence, example (100) can mean 'the leg which always aches':

(100) *J'ai mal à ma jambe*
I:have pain at my leg
(Literally: I have an ache in my leg)
'My leg is aching.'

Once the diverse processes are known through which a language expresses the concept of the personal domain, it would be interesting for the sociologist as much as for the linguist to delimit the extent of this domain with the aid of grammar. We have already mentioned that variation depends in great part on the relative intensity of communication.

Could it also be assumed that in the same language, the expression of this tendency would be more established and frequent in informal colloquial speech than in the written language, more apparent in the rural than in the urban dialects? In the same way, the more generalised in use a language becomes, the more it will try to express impersonally what it has formerly conveyed subjectively.

Thus, French today restricts the domain of the subjective self to the body and mind of the subject (cf. *se fatiguer les jambes* 'tire one's legs out' (literally: 'tire to:oneself the legs'), *se torturer l'esprit* 'torment oneself' (literally: 'torment to:oneself the mind'). It excludes individuals in a person's social circle or any other. Furthermore, we cannot use the French counterparts of certain Provençal examples, shown by (101) and (102) (Ronjat *Syntaxe des parlers provençaux* [Syntax of the dialects of Provence] p.135 f):

(101) **Ôte la veste* (cf. Modern French: *Ôte ta veste!*)
take:off the coat (take:off your[2sg] coat)
'Take off your coat.'

(102) **Il s'est embourbé le cheval*
he DAT:Refl:is stuck:in:mud the horse
'His horse got bogged in the mud on him.'
(cf. Modern French: *Il a embourbé son cheval* [he has stuck:in:mud his horse] 'He has got his horse bogged in the mud.')

Similarly, the Italian counterparts of the following French examples in (103), (104) and (105) are acceptable, as opposed to the French:

(103) **Lève (-toi) le chapeau*
raise you:DAT:Refl the hat
'Take off your hat!'

(104) **Cet homme ne m'est pas père*
this man NEG me:DAT:Refl:is NEG father
'This man is not a father to me.'

(105) **La mère m'est morte*
the mother me:DAT:Refl:is dead
'My mother is dead.'

Finally, the German counterpart of (106), as in (61) above, is also acceptable:[9]

(106) **Elle s'est déchiré la robe*
she DAT:Refl:is torn the dress
'She has torn her dress.'

Nonetheless, it was not always so in French, a fact that we can still indirectly ascertain. It is known that in every language, fixed idioms and restricted grammatical types are often reflexes of older generalised usages. Certain fixed expressions lead us to conclude that there was a much greater extension of the 'self' in Old French.[10] Consider the examples of idioms in (107) and (108):

(107) *tirer l'épée* (cf. *dégaîner son épée*
draw the:sword 'unsheathe his/her sword')
'draw swords'

(108) *avoir la tête près du bonnet* (cf. *mettre son bonnet*
have the head near to beanie 'put on one's beanie')
'be quick-tempered'

In the same way, complements of manner without any prepositions, for example, *marcher la tête haute* 'walk (with) head held high', enable avoidance of the possessive, where it would generally be required by the rules of grammar as also in (109) and (110);

(109) *s'avancer l'épée au clair*
Refl:go:forward the:sword to:the open
'go forward brandishing one's sword'

(110) *se promener la pipe allumée*
Refl walk the pipe lit
'go for a walk, pipe lit' etc.

This sketch is admittedly incomplete but does not claim to be an exhaustive study of the topic. It is merely trying to show different aspects of a subject, much of which remains to be explored.

Notes

1. In: Franz Fankhauser – Jakob Jud (eds.) *Festschrift Louis Gauchat* 1926: 68-78. Aarau, Switzerland: H.R. Sauerländer. We gratefully acknowledge permission from Verlag Sauerländer (Aarau, Switzerland) to publish this translation of the original article in French by Bally.
   We would like to thank Gregory Horsley of La Trobe University for his transliteration, glossing and translation of the more than numerous Ancient Greek examples in Bally's original article; also Gregory Bailey and Barry Blake, both of La Trobe University, for their respective translations of the Sanskrit and Latin examples, Klaus Ecker, University of Melbourne, for the German examples and Werner Drossard, University of Cologne for the Russian. Any errors of interpretation remain of course our responsibility. All references noted in the translation are to be found in the reference list at the end of this article, added in by the translators.
   Abbreviations used in the glossing:
   Abbreviations used in the glossing are as follows: ACC = accusative ; Aor = Aorist; DAT = dative; Fut = future; GEN = genitive; IMP = imperative; Impf = imperfect; Indef = indefinite; INF = infinitive; NEG = negative; NOM = nominative; Part = participle; Pass = passive; Perf = perfect; Pl = plural; Poss = possessive; Pres = present; Priv = privative; Pro = pronoun; Refl = reflexive; sg = singular.
2. Background material on the Geneva School of Linguistics can be found in Bally (1932), de Saussure (1916), Godel (1961), Ivíc (1965), Leroy (1963) and Lepschy (1970). We base the foreword on these readings.
3. The translators have taken the following liberties, among others, in order to enhance readability of this article:
   (i) Addition of free translations of all the language examples, including transliteration of the Ancient Greek examples into the Latin alphabet.
   (ii) Language examples within the text are also given literal translations where this deviates widely from the free translation into English.
   (iii) Indented language examples, set off from the text, have all been given an interlinear morpheme-by-morpheme glossing as is standard practice in linguistic analyses. Note that in the original manuscript, there were no indented examples, the majority being presented untranslated within the text.
   (iv) Numbering of the indented examples
   (v) Roman font for focused morphemes.
4. The inanimate noun *table* cannot be referred to pronominally as *lui* 3sg:DAT in example (53).
5. Bally names this *la construction progressive* 'the progressive construction', by which he means the analytical use of word order to determine meaning. See Bally (1932: 208 ff.)
6. The French term used here – *pronoms conjoints* – refers to unstressed pronouns which are generally placed immediately preceding the verb. These are *le, la, lui* , *les* and *leur* for the third person in Modern French. In contrast to this, the term *pronoms disjoints* refers to pronouns separated by other elements from the verb which are usually stressed (*tonique*) such as *lui*, *elle*, *eux* and *elles* for the third person.
7. Thanks to Peter Paul, Linguistics Department, Monash University, Melbourne for his enlightenment with respect to Alemannic.

8. This point of Bally's only appears to make sense if he is referring to the Latin counterparts of these two examples.
9. Wierzbicka (1979) also discusses the extent of the personal domain in French, German, Italian and Polish dative constructions.
10. Note, however, that the following combinations are possible in contemporary French, see also Diffloth (1974):
    Tombe la veste 'Take off your vest' (literally: 'Take off the vest')
    Ils m'ont cassé le parebrise 'They've broken my windscreen' (literally: 'They to me have broken the windscreen.')
    Les policiers m'ont fouillé les poches 'The police have searched my pockets' (literally: 'The police to me have searched the pockets')

References

Bally, Charles
1932 *Linguistique générale et linguistique française*. Berne: A. Franke Verlag.

Brugmann, Karl – Berthold Delbrück
1886-1892 *Grundriss der vergleichenden Grammatik der indogermanischen Sprachen. Kurzgefasste Darstellung der Geschichte des Altindischen, Altiranischen, (Avestischen und Altpersischen)*. 5 vols. Straßburg: K.J. Trubner.

Chappell, Hilary
1986 "The passive of bodily effect in standard Chinese", *Studies in Language* 10.2: 271-296.

de Saussure, Ferdinand
1916 *Cours de linguistique générale*. Publié par Ch. Bally et A. Séchehaye avec la collaboration de A. Riedlinger. Lausanne-Paris: Payot.

Diffloth, Gérard
1974 "Body moves in Semai and French", *Papers from the 10th Regional Meeting of the Chicago Linguistic Society* 10: 128-138.

Frei, Henri
1939 "Sylvie et jolie des yeux", in: Albert Séchehaye et al (eds.), 185-192.

Godel, Robert
1961 "L'école saussurienne de Genève", in: Christine Mohrmann – Alf Sommerfelt – Joshua Whatmough (eds.), 294-299.

Hatcher, Anna Granville
1944a "Il tend les mains vs Il tend ses mains", *Studies in Philology* 41: 457-481.
1944b "Il me prend le bras vs Il prends mon bras", *Romanic Review* 35: 156-164.

Havers, Wilhelm
1911 *Untersuchungen zur Kasussyntax der indogermanischen Sprachen*. Straßburg: K. J. Trubner.

Kliffer, Michael D.
1984 "Interpenetration of linguistic levels: French inalienable possession", *Lingua* 62: 187-208.

Ivíc, Milka
1965 *Trends in linguistics*. The Hague: Mouton & Co.

Kühner, Raphael – Bernhard Gerth
1898-1904 *Ausführliche Grammatik der griechischen Sprache*. Part II: *Satzlehre* [revised by Berthold Gerth]. 3 vols. Hannover & Leipzig: Hahn.
Lepschy, Giulio
1970 *A survey of structural linguistics*. London: Faber & Faber.
Leroy, Maurice
1963 *Les grands courants de la linguistique moderne*. Bruxelles: Presses Universitaires de Bruxelles.
Lévy-Bruhl, Lucien
1914 "L'expression de la possession dans les langues mélanésiennes", *Mémoire de la Société de Linguistique de Paris* 19.2: 96-104.
McGregor, William
1985 "Body parts in Kuniyanti grammar", *Australian Journal of Linguistics* 5: 209-232.
Meillet, Antoine et al (eds.)
1923 *Festschrift Jacob Wackernagel zur Vollendung des 70. Lebensjahres am 11. Dezember 1923, gewidmet von Schülern, Freunden und Kollegen*. Göttingen: Vandenhoeck & Ruprecht.
Mohrmann, Christine – Alf Sommerfelt – Joshua Whatmough (eds.)
1961 *Trends in European and American Linguistics: 1930-1960. Edited on the occasion of the 9th International Congress of Linguists, Cambridge, Mass., 27 August – 1 September, 1962*. Utrecht: Spectrum.
Ronjat, Jules
1913 *Essai de syntaxe des parlers provençaux modernes*. Maçon: Protat Frères, imprimeurs.
Séchehaye, Albert et al (eds.)
1939 *Mélanges de linguistique offerts à Charles Bally*. Geneva: Georg.
Seiler, Hansjakob
1983 *Possession as an operational dimension of language*. Tübingen: Gunter Narr Verlag.
Wackernagel, Jacob
1953 "Über ein Gesetz der Indogermanischen Wortstellung", in: *Kleine Schriften*. Göttingen: Latte, 1: 1-103.
Wierzbicka, Anna
1979 "Ethnosyntax and the philosophy of grammar", *Studies in Language* 3.3: 313-383.

Part II

Languages of Australia

The syntax and semantics of body part incorporation in Mayali[1]

Nicholas Evans

1. Introduction

In this paper I examine in some detail the syntax and semantics of body part noun incorporation in Mayali, a Gunwinyguan (non-Pama-Nyungan) language of North Western Arnhem Land, Australia. Noun incorporation is definable as the morphological combination of a nominal root with a predicate to make a single complex predicate word. Illustrative Mayali examples are (1) and (2), in which a generic inanimate noun is incorporated, and (3) and (4) in which a body part noun is incorporated; to each incorporated construction there corresponds an unincorporated construction in which the nominal is 'external', that is, outside the verbal word. As these examples illustrate, incorporated nominals drop their noun-class prefixes.

(1) *Ngarradj* *ba-djobke-ng* *gun-warde.*[2]
white:cockatoo 3P-split-PP IV-rock[3]
'White cockatoo split the rock.'

(2) *Ngarradj* *ba-warde-djobge-ng.*
white:cockatoo 3P-rock-split-PP
'White cockatoo split the rock.'

(3) *Bamurru* *a-bom* *gun-godj*
magpie:goose 1-shoot:PP IV-head
'I shot the magpie goose in the head.'

(4) *Bamurru* *a-godj-bom*
magpie:goose 1-head-shoot:PP
'I shot the magpie goose in the head.'

Incorporation of body part nouns is widespread among the world's languages, being attested in languages of the Americas (e.g. Tupinamba: Tupi-Guarani (Mithun 1984), Blackfoot (Frantz 1971), Nahuatl (Merlan

1976) and Mohawk (Mithun 1995)), of North-East Asia (e.g. Chukchi and Koryak), and of Australia (cf. the papers in this volume by Harvey and Leeding on Warray and Anindilyakwa respectively). Most available descriptions, however, fail to provide full and explicit information on either the syntax or the semantics of this construction; so much so that Baker (1993: 15) deliberately omits body-part data from his account of noun-incorporation on the grounds that it is either unsystematic, or inadequately described:

> I also exclude from consideration incorporated body part terms; these are very common, but their role in the sentence as a whole is often far from clear. This type of idealization is simply narrowing in on the natural class of things about which something meaningful can be said.

In this paper, then, I intend to rectify this gap in our knowledge; one conclusion will be that body part incorporation in Mayali is just as syntactically tractable as other types of incorporation. In section 2 I provide background grammatical information on Mayali, including the noun-class membership of body-parts; in section 3 I survey some basic facts about noun incorporation in Mayali that will be relevant to my argument. In section 4 I discuss the intricate syntax of body part incorporation, and argue that it is governed by the same accessibility hierarchy that controls other types of noun incorporation; basically, intransitive subjects or transitive objects incorporate, with the less topical argument incorporated in double-object constructions. The crucial rule for body-part nominals is that they can incorporate if, following possessor raising, their 'wholes' occupy grammatical relations that sanction generic incorporation. Body parts in other roles require a different construction involving case-marked external nominals. The term 'possessor raising' is often used in a rather loose way and in section 5 I assess its validity and range for Mayali; I conclude that body part incorporation does not change the argument structure of the clause, but should rather be treated as a type of syntactic apposition.

In section 6 I survey the semantic range of incorporated part nominals in order to characterise the meaning of the covert class defined by incorporability in the body part slot. In addition to body parts proper, this includes a range of other manifestations of an entity including speech, spirits, excreta, nests and tracks. The crucial generalisation for Mayali, I argue, is that incorporable 'body parts' are entities that imply the sometime existence of their 'wholes', not that they are 'inseparable from the body' or in a 'one-to-one exclusive relationship with the owner'. In section 7 I examine some interesting cases in which the separation of part from whole is clearly perceptual rather than physical; in such circumstances the part is able, exceptionally for incorporated body parts, to govern number marking. In section 8 I discuss the discourse conditions under which a body part fails to be incorporated. Finally, in section 9, I survey some interesting recurrent similarities between the syntax of body-part, generic-specific, and secondary-predication constructions, and suggest that this reflects shared semantic similarities of these three situations.

2. Typological background

Mayali is a polysynthetic language with a system of four noun classes but relatively undeveloped nominal case morphology. Consistent with its 'head-marking' typology (Nichols 1986), a large amount of information is carried on a multimorphemic, one-word 'verb complex' which may include pronominal prefixes for up to two arguments, various manner, directional and aspectual prefixes, applicatives, incorporated nominals, and suffixes for reflexive/reciprocal and tense/aspect/mood. Verbal affix positions are the only obligatory overt realisation of arguments and overt external NPs are only present for specialised discourse purposes such as introducing new participants. A typically elaborate verbal complex is (5):

(5) *Aban-yawoyh-warrgah-marne-ganj-ginje-ng.*
1/3a-again-wrong-BEN-meat-cook-PP
'I cooked the wrong meat for them again.'

Pronominal prefixes represent the person and number of up to two arguments; these may be underlying arguments of the lexical verb, or derived arguments introduced by applicative prefixes (as with 'them' in (5)). The first prefix slot represents the subject, and the second basically represents the object, though with some complications. With transitive (6) or derived semi-transitive verbs (7) the second slot represents the remaining argument, whether object or indirect object; with ditransitive verbs it represents the indirect object, whether underlying (8) or derived (9).

(6) *Ngan-na-ng.*
3/1-see-PP
'(S)he saw me.'

(7) *Ngan-marne-dowe-ng.*
3/1-BEN-die-PP
'(S)he died on me.'

(8) *Ngan-wo-ng gun-ganj.*
3/1-give-PP IV-meat
'(S)he gave the meat to me.'

(9) *Ngan-marne-ginje-ng.*
3/1-BEN-cook-PP
'(S)he cooked it for me.'

Underlyingly transitive verbs with a third argument added by the comitative applicative yield derived 'double-object' verbs. The principles of choice here are comparable to those found with double-object verbs in Bantu languages (e.g. Duranti 1979). The choice of which object is represented pronominally here is lexically determined, although there are underlying functional and semantic reasons discussed in Evans (1995b).[4] In (10) and (11) the original object is represented by a pronominal prefix, while in (12) and (13) it is the derived object (corresponding to the English 'with' phrase) that is so represented.

(10) *Abanmani-yi-na-ng.*
1/3ua-COM-see-PP
'I saw them two with something. (*I saw him/her with those two).'

(11) *Ngan-madj-yi-gadju-ng.*
3/1-swag-COM-follow-NP
'He follows me in the swag (that is, we take turns to sleep with the same woman).'

(12) *Ngan-yi-bawo-ng.*
3/1-COM-leave-PP
'He left it with me.'

(13) *Aban-bo-yi-warlkga-ng yerre.*
1/3a-liquid-COM-hide-PP behind
'I hid the drink with them out the back.'

Note that the question of which argument is represented by the pronominal prefix cannot be answered by invoking thematic roles – in (11), for example, a 'theme' outranks a 'location' for the pronominal slot, while in (13) a 'location' outranks a 'theme'.[5] We shall see in section 4 that the rules governing which argument undergoes noun-incorporation are basically complementary to those for the object prefix slot.

Most free nominals have one of the four noun-class prefixes *na-* (class I, basically masculine), *(ng)al-* (class II, basically feminine), *(ng)an-* (class III, basically plants) and *gun-* (class IV, basically things, abstracts and body parts). The sister dialect Kunwinjku has an identical system except that class III is *man-* rather than *(ng)an-*. Mayali examples are *na-marrgon* 'lightning', *(ng)al-gohbanj* 'old woman', *(ng)an-larrh* 'callitris pine' and *gun-wok* 'language'. (Velar nasals are commonly dropped phrase-initially, including in citation form, for all words except a few interjections.) There is noun-phrase concord in noun classes I-III, affecting demonstratives, numerals and adjectives among others:

(14) *na-mege na-gohbanj na-gimuk*
I-that I-old:person I-big
'that big old man'

(15) *al-gudji ngal-gohbanj*
II-one II-old:person
'one old woman'

(16) *an-dehne an-larrh ngan-mak*
III-that III-callitris:pine III-good
'that good callitris pine'

Gun- class nouns take *ngan-* class concord: *an-dehne gun-rurrk ngan-gimuk* 'that large dwelling'. Here Mayali differs from its better-known sister dialect Kunwinjku which has *kun-* class concord,[6] e.g. *kun-rurrk kun-kimuk* 'big dwelling'.

Some nominals govern an agreement class different from the 'inherent class' expressed by their prefix, e.g. *an-djewk na-gimuk* (III-rain I-big) 'heavy rain'. Because of these complications, nouns lacking class prefixes can be assigned an agreement class only (I, II or III), not an 'inherent class'; examples are *bininj na-gimuk* 'big man (I)', *al-ehne daluk* 'that woman (II)', and *barradjungga an-mak* 'good purple lily (III)'. Noun-classes in Mayali are restricted to nominals – the noun class of core participants is not cross-referenced in the verb complex.

2.1. Body parts and noun class

Over 90% of animate body-part nominals take the *gun-* class prefix, e.g. *gun-bard* 'knee', *gun-djen* 'tongue', *gun-bid* 'hand, finger', *gun-rakmo* 'backbone'. Parts of trees usually employ the same root as human and animal body parts, but with the *an-* 'vegetable class' prefix, e.g. *an-berl* 'branch' (cf. *gun-berl* 'arm, wing'), *an-dad* 'root' (cf. *gun-dad* 'thigh, leg'), *an-godj* 'tuber' (cf. *gun-godj* 'head'), *an-mim* 'seed, seed-pod' (cf. *gun-mim* 'eye').

There is also a small, semantically homogeneous subclass of human and animal body-parts which take the *(ng)an-* 'vegetable class' prefix. This class contains:

(a) eliminatory exuviae (*an-dili* 'piss', *an-gord* 'shit') and sexually-produced fluids (*an-gurrk* 'vaginal juices', *an-duk* 'semen'). Note that other excretions and body fluids take the *gun-* prefix or no prefix: *gun-gurrnginj* 'sweat', *gun-mimnalk* 'tear', *gun-nunj* 'spittle', *gun-nud* 'pus', *gun-gurlh* 'vomit', *gurlba* 'blood' (*kun-kurlba* in easterly dialects of Kunwinjku), *gurrula* 'fluid from afterbirth'.

(b) about half the nouns referring to genitalia, particularly internal:[7] *an-barle* 'vagina' (cf. Kunwinjku *man-barle* 'labia'), *an-berd* 'penis' (a doublet with *gun-berd* 'tail'), *an-gorn* 'crotch, vagina' (cf. Kunwinjku *kun-korn* 'testicle', and note that in both this and the *barle* pair a Kunwinjku word naming external genitalia corresponds to a Mayali *an-* word naming internal genitalia).

Genitalia in the *gun-* class include *gun-djin* 'female genitalia' (again significantly for the Class III 'internal' / Class IV 'external' correlation, Kunwinjku has *man-djin*, with the slightly different meaning 'inside of vagina, walls of vagina'), *gun-djin.gurruk* 'clitoris', *gun-djinwel* 'labia', *gun-gornben* 'testicles', *gun-goy* 'glans penis'.

A scattering of body parts take no prefix: *garderregangalh* 'top of head, fontanelle', *guk* 'body', and *gurrula* 'watery part of afterbirth' (also 'tide, current)' and *gurlba* 'blood' (*kun-kurlba* in Eastern Kunwinjku). I doubt there is any semantic basis to this group.

A clearer delineation of body-part nouns as a covert morphological category is given by a special sort of 'whole-part' compounding. In Gundjeyhmi one can distinguish three types of noun compound on both formal and semantic grounds:

(a) bahuvrihi compounds have the form N-Adj, with no class prefix, and mean 'having N which is Adj'. Examples are:

(17) *rakmo-warre*
hip-bad
'lame' (cf. *gun-rakmo* 'hip')

(18) *njam-gimuk*
guts-big
'fat-bellied' (cf. *gun-njam* 'guts')

(b) predicative compounds have the form *an*-N-Adj, and mean 'an X which is Y':

(19) *an-dulk-rayek*
III-wood-hard
'hard wood, wood which is hard' (cf. *gun-dulk* 'tree, wood')

(20) *an-boy-gilelh*
III-cooking:stone-soft
'antbed used in ground oven' (Literally: 'cooking-stone which is soft' – cf. *gun-boy* 'cooking stone')

(c) whole-part compounds have the form CL-N_{whole}-N_{part}. They take the class prefix appropriate to their head noun, the first element. The semantic range of this construction parallels that of the body-part incorporation, discussed in section 6.

(21) *gun-denge-gurlah*
IV-foot-skin
'sole of foot' (cf. *gun-denge* 'foot', *gun-gurlah* 'skin')

(22) *gun-bid-ngalanj*
IV-hand/finger-nail
'finger nail' (cf. *gun-bid* 'hand, finger', *gun-ngalanj* 'claw, nail')

(23) *gun-dulk-berl*
IV-tree-branch
'tree branch' (cf. *gun-dulk* 'tree', *an-berl* 'branch')

(24) *an-gole-bard*
III-bamboo-knee
'node of bamboo' (cf. *an-gole* 'bamboo', *gun-bard* 'knee')

3. Morphology of noun incorporation

When incorporated, nominals drop the noun class prefix: cf. *gun-ganj* (IV-meat), *ba-ganj-nguneng* (3P-meat-ate) 'he/she ate the meat'; *an-yiwk* (III-honey), *a-yiwk-mang* (1-honey-getNP) 'I will get the honey'. There are also some minor morphophonemic changes, such as initial *d* becoming *rr* when the preceding prefix is an open monosyllable: *gun-dulk* 'tree' but *ba-rrulk-doy* (3P-tree-struck) 'it (lightning) struck the tree'.

There is one suppletive incorporated root: *gukku* 'water' but *ba-bo-barlmeng* (3P-liquid-overflowed) 'the water overflowed'. The form *bo-* is also used in nominal compounds, e.g. *an-bo-gimuk* 'lots of water'.[8]

Both these features of noun incorporation – the dropping of affixes from the incorporated nominal, and the existence of suppletive incorporating forms – are common cross-linguistically (Mithun 1984).

3.1. Three types of noun incorporation

Three types of noun incorporation can be distinguished in Mayali: lexical compounding (LI), body part incorporation (BPI), and generic incorporation (GI). Each involves incorporation of a noun into a different morphological position, as follows:

(25) pronominal prefixes (aspectuals/quantifiers) (implicating applicative) (GI) (BPI) (comitative applicative) [LI [ROOT]] TAM inflection

Lexical incorporation involves the compounding of a root (usually nominal or adjectival but sometimes verbal or adverbal) with a verbal root to form a new verb lexeme, e.g. *danj+bu-* (spear+hit) 'spear', *marri+dowe-* (be hungry+die) 'be starving'. The vast majority of verb lexemes in Mayali involve such lexical compounds; I shall indicate such lexical incorporation by a plus symbol (+) between compounding morphemes, as opposed to a hyphen (-) between syntactically incorporated nominals and their verbs. Since lexical incorporation derives new predicates (whose meaning may be compositional, as with *bo+ngu-* (water+eat) 'drink' or non-compositional, e.g. *ngey+bu-* (flower+hit) 'flower, break into flower'), there is typically no other way of expressing the resultant meaning, and no option of paraphrase by omitting the lexically incorporated nominal or having it appear as an external nominal.

In contrast to lexical incorporation, the two types of syntactic incorporation – *generic* and *body part* incorporation – are 'optional' in the sense that paraphrases with external nominals exist, as in (1) and (3). With some constructions (e.g. body part incorporation) the incorporated form is more usual, with others (e.g. generic incorporation of objects) it is the unincorporated construction which is more usual. In both cases the choice between incorporated and unincorporated forms depends on discourse factors. As an initial characterisation, new, stressed, contrasted or conjoined nominals are external, while given or unstressed nominals are incorporated. I shall substantiate this claim with respect to body part nominals in section 8.

There is a major difference between lexical incorporation on the one hand, and generic and body part incorporation on the other. Lexical incorporation is a non-productive, non-compositional and obligatory closed system. Stems with lexically incorporated nominals lack unincorporated paraphrases, and thus occupy quite a different position in Mayali grammar to the optional syntactic incorporation of generic or body part nominals. Below I outline specific morphosyntactic tests distinguishing the three types of incorporation.

3.1.1. Tests distinguishing lexical from other types of incorporation

As an example illustrating the difference between lexical and syntactic incorporation, consider the verb complex *arrimimbowoni* 'we used to put (OBJ's) fruit in the water', from the following sentence:

(26) *An-barnadja* *ngarri-mim-bo+wo-ni*
III-owenia:vernicosa 1a-fruit-water+put-PI
'We used to put the fruit of owenia vernicosa in the water (to poison the fish).'

This contains the incorporated nominals *-mim* 'fruit, seed' and *bo-* 'liquid'. The first (and outermost) is syntactically incorporated, the second is lexically incorporated. The omissibility test gives quite different results for the two incorporated nominals. *-Mim* can readily be omitted from the complex, with optional replacement by an external nominal:

(27) *An-barnadjdja* *(an-mim)* *ngarri-bo+wo-ni*
III-owenia:vernicosa III-fruit 1a-water+put-PI
'We used to put owenia vernicosa fruit in the water.'

However, omission of *bo-* is quite unacceptable, even when there is an external nominal like *gukku* 'water':

(28) **An-barnadjdja gu-wukku ngarri-mim-wo-ni.*
III-owenia:vernicosa LOC-water 1a-fruit-put-PI

In addition to optionality, several other features distinguish syntactic incorporation from lexical incorporation:

Full productivity: syntactically incorporated nominals can appear with any semantically compatible verb lexeme; lexical incorporated nominals cannot.
Word position: as mentioned above, incorporated nominals are ordered as generic incorporated nominal before body part incorporated nominal before lexically incorporated nominal. This is illustrated by the following example in which the 'generic' *yaw-* 'child' precedes the 'body part' *guk-* 'body' which precedes the lexically incorporated nominal *girri-* 'ground oven' which is part of the complex verb stem *girri+bu-* (ground:oven+hit) 'cook in a ground oven'.

(29) *Na-marnde ba-yaw-guk-girri+bo-m.*
I-devil 3P-child-body-ground:oven+hit-PP
'The devil cooked the child's body in a ground oven.'

Potential pause: pauses may be taken immediately before the verb stem on long verb complexes, that is, after syntactically incorporated nominals but before lexically incorporated nominals. For example, *gaban-marne-madj...ga-n* (3/3a-BEN-swag...take-NP) 'he will carry their swags' but *gabanmarne ... girribun* (3/3a-BEN ... ground.oven+hit-NP) 'he will roast (it) for them in a ground oven'.
Accessibility to external modification: syntactically incorporated nominals may be modified by external demonstratives, adjectives, numerals and relative clauses. Examples are:

(30) *Al-daluk gabi-yaw-garrm-e.*
II-female 3/3anim-child-have-NP
'She has a female child / a baby daughter.'

(31) *Gu-gun nga-mim-baba-ng.*
LOC-right 1-eye-hurt-NP
'My right eye hurts.'

Lexically incorporated nominals cannot be, as shown by the unacceptability of trying to modify 'pouch' in (32):

(32) **An-gimuk ga-yaw-djol+ga-n.*
III-big 3NP-child-pouch+take-NP
'It is carrying a baby in its big pouch.'

Grammatical relations between incorporated nominal and verb: syntactically incorporated nominals, as I shall show in section 4, are entirely predictable in the grammatical relations they contract with their verb. Lexically incorporated nominals, on the other hand, may bear a variety of grammatical relations to the incorporating verb. With some, the grammatical relation is one of those permitted in syntactic incorporation, e.g. intransitive subject (*+yo* 'track:lie') or transitive object (*+ngu* 'liquid:eat' i.e. 'drink'). With others, the relation is disallowed in syntactic incorporation but occurs in a limited number of lexical incorporations. Examples of the latter are instrument, location, destination and source complements, and manner nominals, as in the following examples.

- Body part of intransitive subject: *bok+yo* (track+lie) 'track be present'; *bengh+ngukme* (ear-become:shit) 'forget'.
- Body part of object: *beng+ga-* (ear-take-) 'know', *bid+marre-* (hand-take-RR) 'shake hands', *bok+ga-* (track-take) 'follow track', *dile+bu-* (piss-hit-) 'urinate', *goy+bu-* (glans:penis-hit-) 'urinate (used by a man to his classificatory wives)', *duk+ma-* (semen-take) 'receive OBJ's semen', *djikka+wo-* (breast-give-) 'suckle, breast-feed'.
- Object: *bo+ngun* (liquid+eat) 'drink' , *werrk+we* (outside (part)+ throw) 'peel off (e.g. sheath from bamboo)'.
- Body part of transitive subject: *djol+ga* (pouch+carry) (SUBJ) 'carry (OBJ) in (SUBJ's) pouch'.
- Instrument : *danj+bu* (fish:spear+hit) 'spear with a fish spear' ; *barrk+bun* (covering+hit) 'wrap'.
- Location : *girri+bu* (ground:oven+hit) 'cook in a ground oven'.
- Destination : *da+gurrme* (sun+put) 'put out to dry'; *da+yo* (sun+lie) 'lie in the sun'; *bo+wo* (liquid+give/put) 'put into water'.
- Source : *wilk+mang* (ashes+get) 'take out from ashes'.
- Manner: *yirri+yo* (spread+lie) 'spread, be spread'; *bulurru+yo* (slide+ lie) 'slither'.

The wide range of grammatical relations between incorporated nominal and verb here parallels the situation in languages whose only form of incorporation is lexical, such as Kayardild (Evans 1995a) or Yidiny

(Dixon 1977). The variable and implicit relations between incorporated noun and verb in such compounds conforms to Sapir's (1911: 257) famous analogy between noun incorporation into verbs and noun-noun compounds:

> The grammatical expression of a logical relation, in other words a syntactic process, is sacrificed to a compositional process in which the logical relation is only implied. The sacrifice of syntax to morphology or word-building is indeed a general tendency in more than one American language.

It is clear that if lexical incorporation is not systematically distinguished from syntactic incorporation, the range of grammatical relations would be wide enough to support the view that there are no syntactic constraints on generic or body part incorporation. I shall argue against this anarchic position in section 4.

Since one aim of this paper is to prove that the syntactic incorporation of body part nominals involves predictable grammatical relations between incorporated nominal and predicate, it would be circular to use grammatical relations to decide whether a given incorporated nominal is lexical or syntactic. All potential counterexamples to the hypothesis of predictable grammatical relations – e.g. *djol+ga-* 'carry in pouch', *duk+ma-* 'receive semen' and *danj+bu-* 'spear with a fish spear' must be shown to be lexical by other tests, such as the others given above.

3.2. Distinguishing body-part from generic incorporation

Usually the differences in reference of incorporated body part nominals and incorporated generics are sufficient to distinguish the two types.

Incorporated generics are drawn from a closed class of around 40 items, including *gun-bolk* 'place, ground', *gun-dulk* 'tree', *gun-dalk* 'grass', *gun-dorrh* 'vine', *gun-boy* 'cooking stone', *gun-warde* 'rock', *-barrarn* 'escarpment', *an-rud* 'road, track', *gun-bolh* 'track', *gun-rurrk* 'house', *an-yiwk* 'honey', *bo-* 'liquid' (incorporated only), *gun-bili* 'fire', *gun-djurrk* 'running water', *gun-yerrng* 'firewood', *gun-madj* 'swag, possessions', *gun-gorrk* 'material, clothes', *an-me* 'food'. Although most are semantically generic, some, such as *gun-ngale* and *gun-wabban*, both meaning 'axe-handle', are more specific. As the reader will have noticed, these are all inanimate; the only freely incorporable animate is *yaw* 'child, baby', although the nouns *daluk* 'woman' and *bininj* 'man' can be incorporated into certain verbs. Incorporated generics, as the name implies, serve to classify items that may be further specified by external nominals, as in (33), (35) and (36). Specific nominals, on the other hand, cannot be incorporated.

(33)	*Ga-rrulk-di*	*an-dubang*	/	*an-bernbern*
	3NP-tree-stand(NP)	III-ironwood:tree		III-ghost.gum
	'An ironwood/ghostgum tree is there.'			

(34) **Ga-rrubang-di* / **ga-bernbern-di.*

(35)	*Ga-yaw-garrm-e*	*al-daluk*
	3/3-baby-have-NP	II-female
	'She has a baby girl.'	

(36)	*Ba-bo-rrowe-ng*	*gukku*	/	*gun-gih.*
	3P-liquid-die-PP	water		IV-mud
	'The water/mud has dried up.'			

(37) **Ba-gih-dowe-ng.*

Incorporated body parts, by contrast, form an open class: any named body part of a human, animal or plant can be incorporated, no matter how specific. For example, detailed anatomical terms are regularly formed via whole-part compounding, e.g. *gun-berl-gal-murrng-yahwurd* (IV-arm-marrow-bone-small) 'radius', *gun-godj-mud* (IV-head-hair) 'head hair' or *gun-garre-mok* (IV-calf-sore) 'sore on calf', and these compounded body parts may be incorporated:

(38) *Ngan-garre-mok-bukka-ng.*
3/1m-calf+sore-show-PP
'He showed me his calf-sore.'

(39) *Nga-godj-mud-djobge-rre-n.*
1-head-hair-cut-RR-NP
'I'm going to cut my hair.'

Despite these differences in their semantic foci, the classes of incorporated generics and body parts show some overlap, resulting largely from polysemy of the relevant body part noun. Some examples are given in Table 1.

Others have the same reference, but may occur with either a generic function (when the possessor is irrelevant) or a body part function (when

Table 1. Some semantic contrasts in incorporated body part terms vs. incorporated generics

Term	Meaning as incorporated body part	Meaning as incorporated generic
gun-gurlah	skin	pelt, hide
an-mim	fruit, seed pod	fruit, seed pod
gun-ganj	flesh, muscle	meat

the possessor is involved, and named). Examples are *dabu* 'egg', *an-djud* 'butt (of yam'), *gun-yed* 'nest', *gun-marlaworr* 'leaf' and *gun-bok* 'track'.

For nominals permitting this semantic and functional overlap, it is not always obvious whether a generic or body-part construction is involved. Apart from the clues supplied by the presence of an external nominal denoting the 'whole' if a body-part construction is involved, or a 'specific' if a generic construction is involved, the most reliable test is to try double incorporation, since only a single incorporated nominal per slot type is allowed. In (29) above, for example, and in (40), the presence of *yaw* 'child' in the generic nominal slot means *guk-* and *-dang*, in the following slot, can only be interpreted as body part nominals.

(40) *Ba-yaw-dang-barrme-ng, ngal-badjan, gabi-wo-n.*
3P-baby-mouth-open-PP II-mother 3/3anNP-give-NP
'The baby opened its mouth, and (its) mother is feeding it.'

Another test is to see whether the affected possessor must be introduced with the implicating applicative (as happens with generics) or not (as with body parts). This will be described in section 5.

While on the subject of generic vs. body part nouns, it is worth pointing out that incorporated *guk-* 'body', though basically a body-part as in (41), is sometimes used as a sort of dummy generic to make up for the lack of a generic for adult humans, or animates. With trivalent verbs in which a human third argument is out-ranked for the second pronominal slot (e.g. (42)), it is almost invariably cross-referenced on the verb by incorporated *-guk-* which is here leached of its body-part meaning:[9]

(41) *Ba-guk-yi-golu-y gunj, ba-guk-yi-lobm-i.*
3P-body-COM-descend-PP kangaroo 3P-body-COM-run-PP
'He ran down carrying the kangaroo's body.

(42) *Abanmani-guk-wo-ng al-beywurd*
1/3ua-body-give-PP II-daughter
'He gave his daughter to them two.'

4. The syntax of body part incorporation

If one is careful to distinguish lexical from syntactic incorporation, the range of grammatical relations between generic incorporated nominal and predicate is remarkably homogeneous across a wide variety of languages of North East Asia (e.g. Chukchi, Koryak), Australia, Oceania and America (Mithun 1984; Baker 1988, 1993). Transitive objects and intransitive subjects are the usual candidates; additionally, instrument and locative complements may incorporate (cf. Mithun 1984). Transitive subjects and benefactives or indirect objects never incorporate. Explanations for these recurrent regularities have been advanced on the basis of thematic hierarchies (implicit in Mithun 1984) or phrase-structural principles limiting incorporation to phrase structure sisters of the verb (as argued in Baker 1988, 1993). I do not wish to evaluate these claims here, but will assume an account for Mayali (defended in more detail in Evans 1995b) in which the incorporated nominal is accounted for by a combination of thematic-role and prototypical animacy factors.

In contrast to the clear regularities with incorporated generics, various authors have concluded that body part incorporation is less regular in terms of grammatical or thematic relations to the verb. In section 1 we have already cited the views expressed by Baker (1993) alleging the irregularity of body-part incorporation. And in general it seems that noun-incorporating languages treat body part nominals in a variety of ways. In some languages, such as Ngan.gikurunggurr (Reid 1982), the only incorporable nominals are body parts. In other languages, such as Warray (Harvey 1995) incorporated body parts are treated as locations, a coding option also available to unincorporated body part nominals in such languages as Mohawk (Mithun 1995).

Mithun (1984) treats many examples of body part incorporation under 'Type II' incorporation, which serves to manipulate case relations within clauses: "Noun incorporation of body parts allows affected persons to assume a primary case role, such as subject or object, rather than merely oblique possessor". Implicit in this analysis is a view of Type II noun-incorporation as a formal marker of possessor raising (cf. section 5 below).

In this section I want to argue for a different analysis again: body-parts incorporate if their wholes occupy grammatical relations that sanction generic incorporation. Moreover part-whole NPs in Mayali, I shall argue, have a double-headed syntactic status and their two components should be treated as syntactically apposed, rather than treating the whole as a possessive adnominal, or the part as a locative. In general, I will argue, notional core NPs in Mayali potentially unify information from (a) pronominal prefixes (b) incorporated nominals (c) external nouns, demonstratives and pronominals. An example in which a notional NP in Mayali unifies information from all three is:

(43) *Gabani-dulk-di* *an-dubang.*
3uaNP-tree-standNP III-ironwood
'Two ironwood trees are there.'

Shifting part of the NP from external to incorporated nominal has no effect on the argument structure of the clause, merely on the locus of coding of an NP. (Of course, it may have significant effects on discourse emphasis). And both generic/specific constructions and part-whole constructions are susceptible to having part of the NP coded on the verb.

Let us begin by considering the rules for incorporating 'generic' nominals. In Mayali the grammatical relation of incorporated nominals to the verb is predictable from a hierarchy in which grammatical roles and a topicality hierarchy interact. If the clause is intransitive (44) or semi-transitive (45), its subject can incorporate. Note that, *contra* Baker (1988, 1993) this is true even if the verb is 'unergative' (46).

(44) *Ba-rrulk-man.ga-ng.*
3P-tree-fall-PP
'The tree fell.'

(45) *Ngan-marne-ganj-warrem-inj.*
3/1-BEN-meat-go:bad-PP
'The meat went bad on me.'

(46) *Na-mege* *wurdyaw* *ga-yaw-wage.*
I-that baby 3NP-baby-crawlNP
'That baby is crawling.'

If the clause is transitive – either underlyingly, or through addition of a new object to an underlying intransitive by the COMitative applicative – its object can incorporate, but never its subject:

(47) *A-ganj-ngune-ng.*
1/3-meat-eat-PP
'I ate the meat.'

(48) *A-madj-yi-rrurnde-ng.*
1/3-swag-COM-return-PP
'I took the swag back.'

If the clause is ditransitive, whether underlyingly (as in (49)) or through addition of a new indirect object by the benefactive applicative (as in (50)), its object can incorporate, but never its indirect object.

(49) *Abanmani-ganj-wo-ng.*
1/3ua-meat-give-PP
'I gave meat to the two of them.'

(50) *Abanmani-marne-ganj-ginje-ng.*
1/3ua-BEN-meat-cook-PP
'I cooked meat for the two of them.'

If the clause has two objects – one underlying, and one added by the COMitative applicative – the choice of which argument incorporates is lexically determined. Some verbs, such as *bawo-n* 'leave' (51) and *warlkga-n* 'hide' (52), incorporate their original object, while others, such as *na-n* 'see' (53) and *gadju-ng* 'hide' (54), incorporate the object added by the comitative applicative. For a given verb lexeme, it is always the argument that misses out on pronominal representation (see section 2) which is eligible for noun incorporation.

(51) *Ngan-madj-yi-bawo-ø*
3/1-swag-COM-leave-IMP
'Leave the swag with me!' (Again, note the oddness of 'leave me with the swag'.)

(52) *Aban-bo-yi-warlkga-ng yerre.*
1/3a-liquid-COM-hide-PP behind
'I hid the drink with the people behind.'

(53) *Ngan-bo-yi-na-ng.*
3/1-liquid-COM-see-PP
'He saw me with the drink.'

(54) *Ngan-madj-yi-gadju-ng.*
3/1-swag-COM-follow-NP
'He follows me in the swag (i.e. we share the same lover).'

The effect of the above rules is that whenever there are three argument slots, the incorporated nominal slot can absorb whichever argument has failed to be represented in the pronominal slot.

When there are fewer than three arguments, one argument is potentially represented twice: either the intransitive subject as in (44)–(46), or the object as in (47)–(48). Nonetheless, there is less formal redundancy than this last statement implies, since most incorporated nominals are inanimate, and inanimates are only represented pronominally in a reduced way. In fact, on one analysis, it could be argued that inanimates are not overtly represented at all by pronominals. I shall go into the details of this because they will be relevant to our discussion of overt cross-referencing of body parts in section 7. Firstly, their number is only represented pronominally if they are being personalised, as when being treated as personifications of mythological beings (this was the context for (43) above); in other cases the 'minimal' number is used:

(55) *Ga-mirnde-rri* *an-bernbern.*
3NP-many-standNP III-ghost.gum
'There are many ghost gum trees there.'

Even this prefix can plausibly analysed not as a person marker *per se*, but rather as a marker of tense that is suppressed with non-third person subjects (see Evans 1995b). Finally, inanimate objects are never formally marked, in the sense that the transitive prefix combinations coding action on a third person inanimate are never distinct from the corresponding intransitive prefixes ((56), (58)), while this is only sometimes the case with animate objects (no overt coding in (56), but overt coding in (57)):

(56) *Nga-rrolga-ng* *nga-na-ng.*
1-stand:up-PP 1/3-see-PP
'I stood up and saw him/her/ it.'

(57) *Ba-rrolga-ng* *bi-na-ng.*
3P-stand:up-PP 3/3anim-see-PP
'(S)he stood up and saw him/her.'

(58) *Ba-rrolga-ng* *ba-na-ng.*
3P-stand:up-PP 3/3inanim-see-PP
'(S)he stood up and saw it.'

If we accept this line of argument, which treats inanimates as absent from pronominal representation as opposed to present but with zero form, incorporated nominals directly complement pronominal prefixes in the representation of arguments: either they represent arguments of ditransitives that would miss out on the two available pronominal slots, or they represent arguments of intransitives or transitives which would not have received overt morphological representation.

Having examined the rules governing incorporation of generics, we now turn to the rules for body-part incorporation. For the moment I shall assume without justification an analysis in which the 'possessors' of body parts are 'raised' to occupy the same grammatical relations as the parts themselves, to which they are syntactically opposed. I justify this analysis in section 5. Assuming this analysis, the rule can be stated very simply:

(59) Body-part nominals can incorporate if, after possessor raising, their wholes occupy grammatical relations that sanction generic incorporation.

Thus for intransitive ((60), (61)) or semi-transitive ((62), (63)) verbs one can incorporate the body part of the subject. As with generics, incorporation can occur into an unergative, volitional verb (64). Note that, to understand this analysis, it helps to recast the English translation in a way that reflects the Mayali syntactic organisation, e.g. 'I cracked in the ribs' or 'He became big in the eyes for her' instead of 'I cracked my ribs' or 'He made his eyes big for her.'

(60) *A-bikbik-bakme-ng.*
1-rib-crack(intr.)-PP
'I cracked my ribs.'

(61) *A-mim-warremi-nj.*
1-eye-go:bad-PP
'My eyesight has gone.' (Literally: 'I went bad in the eyes.')

(62) *Bi-marne-mim-gimukm-inj.*
3/3anim-BEN-eye-become:big-PP
'He made eyes at her.' (Literally: 'He became big in the eyes for her.')

(63) *Ngan-marne-djen-bebme-ng.*
1/3-BEN-tongue-appear-PP
'She stuck out her tongue at me.' (Literally: 'She emerged in the tongue for me.')

(64) *Ga-bid-wayda-n.*
3NP-hand-wave(intr.)-NP
'He is waving (with) his hand.'

With transitive verbs, body parts of the object of the verb are incorporated:

(65) *Gun-dulk an-bid-djudme-n.*
IV-splinter 3/1-finger-stick:in-NP
'A splinter is sticking into my finger.' (Literally: 'A splinter is sticking into me, in the finger.')

(66) *Ba-rrang-barrme-ng yaw.*
3P-mouth-feed-PP baby
'(The bird) is putting food in its baby's mouth.'

(67) *Ba-milh-dulubu-ni na-bang burl*
3P-forehead-shoot-PI I-dangerous bull
'He would shoot dangerous bulls in the forehead.'

The incorporated body part is never construed with the transitive subject. In (68) the 'mouth' must belong to the fish, not the pelican, even though *-dang* may mean 'mouth' or 'beak' and either meaning is pragmatically plausible. In (69) the incorporated form 'hand' must be construed as the object's; to specify that the subject used his hand a free nominal plus the 'from' suffix must be used as in (70). If the subject's body part is seen as a 'location' rather than an 'instrument', it is represented by a free nominal with the locative prefix, as in (71).

(68) *Makkakurr ba-rrang-danjbo-m djenj.*
pelican 3/3-mouth-spear-PI fish
'The pelican 'speared' the fish in the mouth (*with its beak)'.

(69) *A-bid-garrme-ng daluk.*
1/3-hand-grasp-PP woman
'I touched the woman on the hand.'

(70) *Gun-bid-be nga-garrme-ng daluk.*
IV-hand-from 1/3-grasp-PP woman
'I touched the woman with my hand.'

(71) *Gun-ganj a-ngorrka-ni gu-garlang.*
IV-meat 1min-carry on shoulder-PI LOC-shoulder
'I carried the meat on my shoulder.'

With ditransitive verbs, whether underlying (72) or derived by adding an indirect object to an underlying transitive via the benefactive applicative (73), it is the body part of the object that is incorporated:

(72) *Ngaye wurdyaw aban-ganem-bukka-ng doctor.*
1sg child 1/3a-ear-show-PP doctor
'I showed my child's ear to the doctors.'

(73) *Ngabornang ba-wam barri-ngabed-marnb-om,*
my:daughter 3P-goPP 3a/3P-hair-make-PP
ngandi-marne-ngabed-marnb-om.
3a/1-BEN-hair-make-PP
'My daughter went (to the salon) so they would do up her hair, and they did up her hair for me.'

To show complete parallelism between body-part incorporation and generic incorporation we need to show that with double-object verbs, body-part incorporation follows the same pattern of lexically-determined choice as generic objects. Evidence for this is provided by the following two sentences: in (74), with the verb *na-n* 'see', the body part of the derived object incorporates (cf. (53)), while in (75), with the verb *bawo-n* 'leave', the body part of the underlying object incorporates (cf. (51)).

(74) *Aban-berd-yi-na-ng anabbarru, barri-berd-garrm-i.*
1/3a-tail-COM-see-PP buffalo 3a/3inanPST-tail-hold-PI
'I saw them with the buffalo's tail, they were holding it by the tail.'

(75) *Ngan-ganem-yi-bawo-ng anabbarru.*
3/1-horn-COM-leave-PP buffalo
'He left the buffalo's horns with me, he left me the buffalo's horns.'

Before concluding this section, I must mention one exception to the general isomorphism between generic incorporation and part-whole incorporation: for one transitive verb, a body part can be incorporated as subject in an 'impersonal verb construction'.[10] The verb *ga-n* 'take, carry' has an idiomatic meaning and argument frame '(body part:SUBJ) hurt (experiencer:OBJ)'; when used with this meaning it may incorporate the hurting body part:

(76) *Gun-yidme an-ga-n.*
IV-tooth 3/1-carry-NP
'My tooth is hurting me, is aching.'

(77) *An-yidme-ga-n.*
3/1-tooth-carry-NP
'My tooth is hurting me, is aching.'

Several other Gunwinyguan languages have similarly exceptional patterns allowing the incorporation of inanimate transitive subjects. In Kunwinjku there is an exactly parallel exception involving the verb *kadjung* 'chase', which has an idiomatic meaning and argument frame identical to Mayali *gan* 'carry'. Carroll (1976: 75) gives the following example:

(78) *Ngan-yidme-kadju-ng.*
3/1-tooth-chase-NP
'My tooth is hurting me.'

Other languages with parallel exceptions are Kunparlang (e.g. *djangal-ngaybu ga-ngany-djangal-gulangwany* (literally, 'foot-my it-me-foot-inflicts:pain') – Coleman 1982), and Nunggubuyu (Heath 1984: 473), which has a number of verbs allowing transitive subject incorporation, all of which "involve an inanimate agent, including pain or disease, acting on a human or other object".

We have shown in this section that grammatical relations between incorporated nominal and verb are essentially identical for generic incorporation and body part incorporation. The sole exception involves the idiomatic pain verb *ga-* 'carry', which can incorporate the locus of pain as a transitive subject.

5. Body part incorporation and the status of possessor raising

Accounts of body part incorporation often make use of the term 'possessor raising', defined by Mithun (1995) as a construction in which 'the possessor of an entity assumes the grammatical role of that entity.' In this section I assess the suitability of this account for Mayali body part incorporation.

If the term 'possessor raising' is taken to mean simply that in Mayali the 'whole' (or 'possessor') is coded as an argument equivalent to that representing the 'part' (or 'possessed') term in English, the term is quite appropriate. In each of (79)–(81) the 'possessor', represented as an adnominal in English, is represented as an argument in Mayali.

(79) *A-mim-warremi-nj.*
1-eye-go:bad-PP
'My eyesight has gone.' (Literally: 'I went bad in the eyes.')

(80) *A-bid-garrme-ng daluk.*
1/3-hand-grasp-PP woman
'I touched the woman's hand, I touched the woman on the hand.'

(81) *Aban-berd-yi-na-ng anabbarru, barri-berd-garrm-i.*
1/3a-tail-COM-see-PP buffalo 3a/3inanP-tail-hold-PI
'I saw them with the buffalo's tail (Literally: 'I saw them with the buffalo at its tail'), they were holding it by the tail.'

However, the term 'possessor raising' has analytical implications which are not appropriate to Mayali. It is sometimes argued for other languages (e.g. Mithun 1995 on Mohawk) that body part incorporation is a formal index of possessor raising, which is conceived of as a process of syntactic derivation rather than simply a descriptive label about semantic interpretation. I shall refer to this as the 'multistratal' conception of possessor raising, and shall argue that this is inappropriate for Mayali – that the same syntactic argument structure should be postulated for a clause whether or not the body part is incorporated. This raises the question of what that structure is. I shall suggest that the most suitable analysis is one in which part and whole are syntactically in apposition, with 'head' like properties shared between the part and the whole, and that this analysis is appropriate whether or not the part is incorporated.

The first question, of whether body part incorporation constitutes a formal index of possessor raising, is best approached by beginning with another Mayali construction in which there is a clear formal index of a possessor being raised to argument status: the use of the benefactive applicative to promote possessors to argument status, exemplified by (82) and (83). In (82) the possessor is shown by the possessive pronoun *ngarduk* 'my'; the verb *melme-* 'touch with foot, kick' is transitive, and its object is 'child'. In (83) the benefactive applicative prefix *marne-* promotes the possessor to indirect object status, as shown by its representation in the pronominal prefix slot. The resulting derived verb is ditransitive.

(82) *Bi-yaw-melme-ng* *ngarduk* *na-beywurd.*
3/3an-child-touch:with:foot-PP my I-child
'He kicked my child.'

(83) *Ngan-marne-yaw-melme-ng*
3/1-BEN-child-touch:with:foot-PP
*na-beywurd (*ngan-yaw-melme-ng)*
I-child
'He kicked my child, he kicked the child on me.'

In the case of (83), then, there is evidence that the argument structure of the clause has been changed, since different arguments get represented in the pronominal prefix slots. And there is a formal exponent of the change, in the form of the applicative *marne.* Finally, there are two alternative constructions available, a 'possessor-raised' construction (83), and an unraised construction in which the possessor is represented by an adnominal, not an argument. These three considerations justify treating 'possessor raising' as a productive rule of Mayali, with clear syntactic ramifications.

If we now turn to the relevant body part constructions, we find that none of the above three observations hold. Firstly, the argument structure of the clause, as shown by the choice of pronominal prefixes, is the same regardless of whether or not the body part is incorporated ((84) and (85)), so body-part incorporation cannot be claimed to be a formal index of possessor raising. This contrasts with the situation in Mohawk (Mithun 1995), where nominal incorporation can signal possessor raising both of body-parts and of other possessed items, subject to the semantic condition that the possessor must be the primary affected participant of the clause (as with 'he car-stole me' for 'he stole my car').

(84) *Ngan-melme-ng* *an-gorn.*
3/1-crotch-touch:with:foot-PP III-crotch
'He felt my crotch with his foot.'

(85) *Ngan-gorn-melme-ng (/ *ngan-marne-gorn-melmeng).*[11]
3/1-crotch-touch:with:foot-PP
'He felt my crotch with his foot.'

Secondly, the formal index *marne-*, found with other types of possessor raising, is not found in the case of body-part incorporation. Thirdly, there is no corresponding construction in which the whole is encoded as an adnominal possessor NP.[12] There is thus no motivation for postulating possessor-raising as an optional, formally registered rule with effects on argument structure in Mayali.[13]

An alternative account of body-part syntax in Mayali is to see part nouns (whether incorporated or not) as syntactically in apposition to their wholes. This analysis has been proposed for a number of dependent-marking Australian languages in which part nominals agree in case with, but are separate constituents, from their wholes (e.g. Warlpiri (Hale 1981) and Yankunytjatjara (Goddard 1982)). It has also been proposed for the Gunwinyguan language Ngalakan, fairly closely related to Mayali (Merlan 1983).

On this analysis, then, part nominals and whole nominals are syntactically in apposition, regardless of whether or not one is incorporated, and both are linked to the same argument position of the verb. It is possible to incorporate one of the apposed pair, for discourse purposes, but this does not alter the basic appositional argument structure. One advantage of this analysis is that it accounts for the syntactic constraints on body-part incorporation automatically: we already need a set of interpretation rules specifying which argument incorporates for generics (see section 3), and by applying these to all apposed nominals linked to the same argument position, we account for constraints on body-part incorporation as well. Another advantage is that it accounts for certain parallelisms, discussed in section 8, between the incorporation of body-parts, of generics, and of secondary predicates. We shall see there that, in dependent-marking Australian languages, there is a widespread tendency to syntactically appose all three types by the mechanism of case-agreement. The fact that all three types allow incorporation in Mayali is a significant parallel, that could be captured by saying that noun incorporation has the general characteristic of picking out one of two apposed nominals.

The apposition analysis, with its implication that information is spread over the two apposed elements, is also well-placed to account for certain splits in head-like semantic properties between the part and the whole. Apposition-based analysis of part-whole constructions in dependent-marking languages that have attempted to identify one or other element as head have generally concluded that some head-like properties are associated with the part, some with the whole (see, e.g., Hale (1981) on Warlpiri, Evans (1995a) on Kayardild). For example, number marking in argument positions indexes the number of the whole, not the part, as shown by the following examples from Warlpiri (86) and Mayali (87). On the other hand, semantic entailments about which entity the predicate holds apply sometimes only to the part, sometimes only to the whole: (88), from Warlpiri, entails only that the head, not the whole person, entered the cave. Similarly (89), from Mayali, entails only that the eyes, not the whole person, became big, and (90) only that the tongue, not the whole person, emerged.

(86) *Wati=ø=ø=rla* *karnta-ku* *yarnka-ja*
man=PST=3sg=3sgIOBJ woman-DAT grab-PST
jurru-ku *rdaka-jarra.*
head-DAT hand-two:ABS
'The man grabbed hold of the woman (by the) head (with his) two hands.'

(87) *Abanmani-bid-garrme-ng* *daluk.*
1/3ua-hand-grasp-PP woman
'I grabbed the two women by their hands.' (*'I grabbed the woman by her two hands.')

(88) *Jurru=ø=ø* *yuka-ja* *pirnki-kirra* *wati.*
head=PST=3 enter-PST cave-ALL manABS
'The man poked his head inside the cave.' (Laughren 1989)

(89) *Bi-marne-mim-gimukm-inj.*
3/3anim-BEN-eye-become:big-PP
'He made eyes at her.' (Literally: 'He became big in the eyes for her').

(90) *Ngan-marne-djen-bebme-ng.*
3/1-BEN-tongue-emerge-PP
'She stuck out her tongue at me.' (Literally: 'She, with her tongue, emerged at me.')

In all the examples so far the predicate holds only for the part, not for the whole, and this is the usual situation, but in (91) 'being afraid' is predicated of the whole, not the part (here, the hand), and in (92) the predicate 'ascending' holds for both (although only the back may actually be sticking out of the water).

(91) *Nga-melme-ng* *bigibigi* *gun-denge-be,*
1/3-touch:with:foot-PP pig IV-foot-from
nga-bid-gelem-inj.
1-hand-fear-PP
'I touched the pig with my foot, I was afraid for my hand.' ("I bin fright my hand").

(92) *Ginga* *ba-bodme-wayhme-ng.*
crocodile 3P-back-ascend-PP
'The crocodile's back has floated up.'

Given these complexities, it seems unlikely that the scope of the predicate could be decided at the level of syntax. Rather, detailed knowledge of the global semantics, in combination with real world knowledge (e.g. that tongues can stick out without their possessor moving, but that crocodile's backs cannot float up without the whole crocodile floating up as well) is probably used to reconstruct the scope of the predicate, starting from the general premise of body part involvement. As McGregor (1985: 210-11) has put it:

> the body part specifies the EXTENT or LOCUS of the participant's involvement in the action. That is, it specifies that part of the individual which is most directly and intimately involved in the action. And secondly, it expresses the fact that it is usually whole individuals rather than their body parts that are involved in the 'direction' of actions or processes.

By simply specifying which is the whole and which the part, and linking them to the same entity, while leaving open the question of whether the predicate had scope over whole only, part only, or both, the apposition analysis builds in about the right amount of syntactically-specified semantics.

To summarise this section, 'possessor raising' is a suitable term for Mayali body-part constructions only as a descriptive term that captures certain semantic facts about these constructions – namely, that the possessor of the body part is treated as if it were the main argument. There is no evidence in Mayali for the existence of a syntactic rule mapping between 'possessor raised' and 'whole-possessor' constructions, nor is there evidence that body-part incorporation indexes a change in argument structure as it does in a language like Mohawk. Instead, a more suitable analysis sees parts as syntactically in apposition to their wholes, and body-part incorporation as picking out one of a pair of apposed nominals.

6. The semantic range of incorporated body part nominals

Although I have been using the general term 'body part incorporation' in deference to descriptive tradition, body parts are only a subset, though perhaps the prototypical subset, of nominals incorporable in this construction.

Some clearcut examples of body parts proper being incorporated have already been given (e.g. (60)–(69), (79)–(81)), further examples are (93) and (94). The construction is not limited to parts of animates, but extends to parts of trees and plants (95):

(93) *A-yidme-baba-ng.*
1-tooth-hurt-NPST
'My tooth aches.'

(94) *A-berd-bakke-ng duruk*
1min-tail-break-PP dog
'I broke the dog's tail.'

(95) *Gaban-gorlebard-djudme-rre-n.*
3/3a-node:of:bamboo-join-RR-NP
'The bamboo's nodes join it together.' (cf. *an-gorle-bard* 'III-bamboo-knee')

If the verb is intransitive, the incorporated part will be interpreted as executing the action (64), undergoing it (60), or being the focus of it (91). If the verb is transitive or ditransitive, the incorporated part will be interpreted as the part of the object upon which the action is focussed. This semantic range parallels that found in body-part constructions in non-incorporating Australian languages like Warlpiri (Hale 1981) and Gooniyandi (McGregor 1985). All of the specific meanings just mentioned are essentially sub-senses of a more general meaning of 'body part involvement', as discussed above.

However, the semantic range of this construction is wider than what one would normally understand by 'body parts', or even by 'part-whole' constructions. To begin with, it includes 'replaceable' body parts like hair (96), and spirits (97), contents (98), and corpses (99); in all these cases action involving the part simultaneously involves action on the possessor or whole as with the canonical body part incorporations already discussed.

(96) *Ngabornang ba-wam barri-ngabed-marnb-om,*
my:daughter 3P-go:PP 3a/3P-hair-make-PP
ngandi-marne-ngabed-marnb-om.
3a/1-BEN-hair-make-PP
'My daughter went (to the salon) so they would do up her hair, and they did up her hair for me.'

(97) *Na-morrorddo gabi-waral-ma-ng*
I-shooting:star 3/3NP-spirit-take-NP
gabi-waral-yi-rrolga-n.
3/3NP-spirit-COM-go:up-NP
'The shooting star (believed to be an agent of death) takes his spirit, and goes up into the sky with his spirit.'

(98) *Djabbilarna ba-bo-warrkme-ng.*
billycan 3P-liquid-drop-PP
'The billycan of water dropped.'

(99) *Gabarri-guk-gurrm-e.*
3a/3NP-body-put-NP
'They are putting his body (onto the burial platform).'

But there is a second set of incorporable part nominals for which appearance in this type of construction does not entail any effects upon the whole. This set includes nouns of 'personal representation' (Chappell – McGregor 1995) such as speech (100), names (101), and tracks (102); 'products' such as nests (103) and eggs (104), and residues such as bones (105).

(100) *Aban-wok-bekka-n.*
1/3a-speech-hear-NP
'I hear their speech (on a tape recorder).'

(101) *David ga-ngey-burrbu-n.*
3/3-name-know-NP
'David will know its name.'

(102) *Gorlobbarr ga-berd-bok+yo-ø*
antilopine:wallaroo 3NP-tail-track+lie-NP
'A wallaroo's tail-track is there.'

(103) *Na-wurrkbil ga-yed-yo-ø.*
I-eagle 3NP-nest-lie-NP
'There is the eagle's nest .'

(104) *Gumugen ga-rrabu-yo-ø.*
freshwater:crocodile 3NP-egg-lie-NP
'The freshwater crocodile's eggs are there.'

(105) *Gubehne gulubarn ba-murrung-do-y.*
here flying:fox 3/3P-bone-crush-PP
'Someone crushed up the bones of a flying fox here.'

Is it possible to give a unified semantic characterisation to this set? Clearly the term "body part" is too narrow, as it would fail to include, e.g. nests, tracks, or names. Nor is inseparability a requirement – the bones in (105), for example, have long been separated from their one-time "possessor". (We shall see in section 7 that there is a way of encoding separation, but that this is not incompatible with incorporation.) There is no requirement of "unity of action" – in (103) the eagle is not lying "with" the nest, and in (102) the crocodile that has laid the eggs is not present. On the other hand, the term "inalienable" is too strong: such "inalienables" as core kin, and country, cannot be incorporated in this construction type.

What is important, it seems to me, is that one item is seen as a "clue" to the existence of the other. The incorporated noun is an "index" of the whole in the Peircian sense, or an "indexical", in the sense of Lyons (1977: 106) who takes as "criterial for the application of the term indexical that there shall be some known or assumed connexion between a sign A, and its significatum C such that the occurrence of A can be held to imply the presence or existence of C". In other words, the incorporated part noun, although it may in fact be separated physically from its 'possessor' at the time of the reported action, would nevertheless not exist unless its possessor exists or existed.

7. Encoding cognitive separation of body parts

As mentioned above, physical separation of the body part does not prevent it from being incorporated. However, there is a construction which allows cognitive separation to be shown by representing them as separate arguments of the verb, and encoding their number in the relevant pronominal prefix position.

In the overwhelming majority of body-part constructions, including all examples given so far, the pronominal prefix positions on the verb represent the person and number of the whole, not the part. Thus in (106) the subject prefix is singular since the man is singular; the fact that two eyes are affected is not overtly represented, and the unit-augmented (here equivalent to dual) prefix *bani-* is unacceptable. In (107) the prefix *ga-* represents the singularity of the tree, not the plurality of its pods.

(106) *Ba-(/*bani-)mim-barnibarnimi-nj.*
3P(/*3uaP)-eye-droop-PP
'His (the drunk man's) eyes drooped.'

(107) *Ga(/*gabarri-)-mim-barndi-ø*
3NP(/*3a)-seed:pod-hang-NP
'The tree's seed pods are hanging.'

However, I have a small set of examples in which body parts are cross-referenced for number and, in those cases where it would differ, for person (see (109d below)). An example is:

(108) *Barri-bok-barrkme-ng.*
3a-track/footprint-break-PP
'The tracks are broken/cracked.' ("Footprint they bin bust".)

The key semantic factor here is separation, a fact to which I was first alerted by Toby Gangele when he pointed out that the above sentence is only acceptable in the context of tracks set in concrete. These are sufficiently separated and 'distinct' to govern their own number agreement. Incorporated *bok-* here is best analysed as the incorporated intransitive subject itself, rather than 'part' of a subject.

Although in the case of (108) the separation is physical, this need not always be the case. The complexities of treating separability are well illustrated in the following fictitious dialogue between dentist (D) and patient (P), written by Judith Alderson:

(109) a. D: *Yi-m-ra-y* *yi-yerrga,* *ø-yidme-na-n.*
2-hither-go-IMP 2-sit(IMP) 1/2-tooth-look-NP
'Come and sit down. I'll look at your teeth.'

b. D: *Yi-rrang-barrme-n, gan-bukka ayega yi-baba-ng*
2-mouth-open-IMP 2/1-show(IMP) where 2-hurt-NP
'Open your mouth, and show me where it hurts.'

c. P: *nahni bogen yerre*
this two behind
'These two behind.'

d. D: *Bani-yidme-nudmi-nj, ø-yidme-durrkma-ng.*
3ua-tooth-go:rotten-PP 1/2-tooth-pull-NP
'The two teeth are rotten, I'll pull your teeth out.'

e. P: *Waw! Warddaw!*
'Ouch! Yow!'

f. D: *Okko ø-yidme-durrkme-y. Ma, yi-yidme-na-rre-men!*
already 1/2-tooth-pull-PP well 2-tooth-look-RR-IMP
'I've already pulled your teeth out. Well, take a look at your own teeth!'

g. P: *Wow, bani-nudmi-nj.*
yes 3uaP-go:rotten-PP
'Yeah, they've gone rotten all right.'

h. D: *Bani-yidme-guyeng-gen yiman dalkgen.*
3ua-tooth-long-GEN like dingo
'They're long teeth, like a dingo's.'

i. P: *Ma, abanmani-yidme-ga-n, aban-yidme-bukka-n.*
well 1/3ua-tooth-take-NP 1/3a-tooth-show-NP
'Well, I'll take the two teeth and show them to everyone.'

j. D: *Nja, bobo.*
here:you:are see:you:later
'Here you are, and see you later.'

As this dialogue shows, it is too simplistic to equate number-cross-referencing with physical separation. It is true that the 'intact' teeth in (a) and the second verb in (d) are not cross-referenced for number, and that the extracted teeth in (g), (h) and (i) are cross-referenced with the unit-augmented (two) prefix. However, in the first verb of (d) the still-attached teeth trigger number-marking, and in (f) the extracted teeth do not trigger number-marking; furthermore, 'tooth' and 'you' must be viewed as part of the same entity in (f) to account for the use of the reflexive. The crucial difference, it would appear, is not physical separation but cognitive separation: are the teeth being focussed on as separate entities, or as part of something else?

8. To incorporate or not to incorporate?

As Mithun (1986) has remarked, in incorporating languages there is considerable variation across construction types in whether incorporated or external nominals are the unmarked choice. With body part nominals in Mayali, incorporation is clearly the unmarked choice whenever this is grammatically sanctioned.[14] In this section I examine the relatively rare cases in which body parts are not incorporated, and argue that non-incorporation is controlled exclusively by discourse factors.

Before examining the discourse factors controlling body part incorporation it is worthwhile examining the factors controlling generic incorporation, which are somewhat different. In Mayali, incorporated generics typically carry given information, and thus generics are frequently encountered both incorporated and externally. Incorporated generic nominals typically progress from external to incorporated status through the discourse, as in the following short text about getting dingo pelts for bounty.

(110) *Ngaye gorrogo an-bang nga-gurrm-i, gun-gurlah*
I before III-dangerous 1/3-put down-PI IV-pelt

a-ma-ngi. gun-gurlah a-ga-ni djamun-djahdjam.
1/3-get-PI IV-pelt 1/3-take-PI dangerous-place

gun-gurlah-wo-ni, gun-warde an-wo-ni.
1-pelt-give-PI IV-money 3/1-give-PI

'In the old days I used to put down (dingo) baits to get their pelts. I would take the hides to the police station. I would give them to him and he would give me money.'

Note how *-gurlah* is first introduced as a free nominal, recurs once as a free NP, then is incorporated into the verb *-woni*. Note also how the *new* entity *warde-* 'rock, money' is not incorporated into the verb in the following clause, even though the argument frame is the same, and even though the verb complex *anwardewoni* 'he used to give me money' is quite grammatical.

With body part nominals, incorporation is by far the commonest choice whenever the correct thematic relations obtain between the possessor NP and the verb. This may partly reflect the unnaturalness of focussing on the effects of the action on the parts rather than on their possessors. As Mithun (1984: 858) points out in connection with the following Blackfoot examples from Frantz (1971), speakers find the first, unincorporated variant 'marginally grammatical but odd, since it implies that the effect of my action on the back is more important than its effect on the man'.

(111) a. ??*nit-ssiksii-hpa oma ninaawa okakini*
I-break-it that man his:back

b. *nit-ssik-o'kakin-aw oma ninaawa*
I-break-back-him that man
'I broke the man's back.'

However, since body parts in Mayali incorporate even when the part is cognitively separate, as in the examples in section 7, this cannot be the whole explanation. More important is the fact that incorporated nominals tend in general to be backgrounded with respect to other material (Mithun 1984). In all Mayali examples where body parts do not incorporate, there is clear evidence that, for one reason or another, they are being foregrounded. This may result from conjunction, contrast, and when emerging as a distinct discourse participant in their own right.

An example of non-incorporation under conjunction, which has the effect of focussing on successive attributes of individual body parts, is the following:

(112) *Barri-marnbom rowk,*
3aP-prepare:PP all
'They prepared it (the kangaroo),
barri-bebbe-gana-ga-ng, gun-berd, gun-dad,
3aP-each-ITER-take-PP IV-tail IV-thigh
'they each took their share, some part of the tail, some a thigh,
njanjuk namegebu barri-bebbe-gana-ga-ng.
anything all:that 3aP-each-ITER-take-PP
'they each took something like that.'

In the case of (112) it could be argued that the non-incorporation of conjuncts represents a grammatical rather than a discourse fact, since no language to my knowledge has been reported as allowing the incorporation of conjoined nominals. But in the next example, which directly follows (112) in the original text, non-incorporation is not grammatically forced, since the word *barrumeyiwarlkgarrinj* 'he hid himself away with the backbone' is perfectly grammatical. Here non-incorporation has a clear discourse motivation – contrastive pairing. The discourse salience of 'backbone' in an English translation would be indicated by double intonational emphasis – '[Some took part of the tail, some a thigh, but] *quail* hid himself away with the *backbone*'.

(113) *Na-gudji djirndih gun-dume ba-yi-warlkga-rr-inj.*
I-one quail IV-backbone 3P-COM-hide-RR-PP
'One bird, quail, hid himself away with the *backbone*.'

Another example of non-incorporation of a body part due to discourse contrast comes from the following textlet, where the speaker is describing his prowess at catching magpie geese with a throwing-stick, and enumerates a number of different 'shots', each hitting a different goose on a different body part:

(114) *Bamurru ba-m-re-y, an-barnba nga-rrelmi*
goose 3P-hither-go-PI III-goose:stick 1-throw:PI
gun-berl a-bakke-yi, yiga gun-gom a-bu-ni,
IV-wing 1/3-break-PI sometimes IV-neck 1/3-hit-PI
yiga gun-geb a-geb-badjdji-ni
sometimes IV-beak 1/3-beak-smash-PI
'A magpie goose would come up, and I'd throw a goose stick. Sometimes I might break a wing, sometimes I'd hit one in the neck, or I'd smash a beak.'

Non-incorporation may also signal the emergence of a body part as a discourse participant in its own right, as illustrated by the following excerpt from a mythical text about the origin of the birds. This episode involves *wirriwirriyak* the black-faced cuckoo shrike, who is unable to hunt due to his sore foot, and the other birds and in particular *ragul* the red-eyed pigeon, who undertakes to cure him by lancing the pus out of his foot.

(115) a. *Gun-dulk barri-me, barri-nud-gorrhge-ng.*
IV-stick 3aP-get:PP 3aP-pus-burst-PP
'They picked up a stick, and they burst his pus out.'

b. *Njamed na-wu , ragul,*
whatchamacallit I-that red-eyed.pigeon
'That whatchamacallit, the red-eyed pigeon,'

c. *nungga gun-dulk ba-me-y.*
he IV-stick 3P-get-PP
'he picked up a stick.'

d. *"Aye, a-nud-gorrhge" ba-yime-ng.*
me 1-pus-burstNP 3P-say-PP
' "Me, I'll burst the pus out," he said.'

e. *Bi-mok-garu-y, bi-nud-gorrhge-ng.*
3/3P-sore-dig-PP 3/3P-pus-burst-PP
'He dug in his sore and burst his pus out.'

f. *Gun-nud ba-rrolga-ng an-ege.*
IV-pus 3PST-arise-PP III-that
'All the pus rushed out.'

g. *Gurlba gun-nud bi-rrelkge-ng rowk, ragul.*
blood IV-pus 3/3P-spatter-PP all red:eyed:pigeon
'Blood and pus spattered him all over, the red-eyed pigeon.'

In the first few lines, disposing of the pus is being viewed as an action carried out for the benefit of *wirriwirriyak*, and it remains in the unmarked incorporated position. But by line (f) the pus emerges (literally) as an independent participant, spattering over *ragul*'s face and hence accounting mythologically for his 'red eyes', and at this point it becomes an external argument. Although in this case non-incorporation coincides with physical separation, we have seen already that physically and conceptually separate parts may still incorporate (e.g. (108), (109)) and conversely that non-incorporation may take place even when the part is not separate (e.g. (114)). It seems unlikely, then, that it is simply the physical separation of the pus which is responsible for non-incorporation here. An additional factor favouring non-incorporation, in line (g), is its conjunction with another body part nominal, *gurlba* 'blood'.

In summary, then, non-incorporation of eligible body-part nominals signals discourse salience due to various factors: conjunction, contrast, and independent interest as a discourse participant. Note that although physical or cognitive separation may on occasion go together with discourse salience, as in (115f, g) the two factors are in principle independent, and have different formal realisations: discourse salience is shown by non-incorporation; cognitive separation by encoding as a distinct argument and control of person and number marking by the body part.[15]

9. Relations of body-part incorporation to other types of incorporation

There are striking similarities between the semantic range of incorporated constructions in Mayali (and other Gunwinyguan languages) and of case-agreement constructions in the structurally very different dependent-marking languages of Australia, such as Warlpiri or Kayardild. Thus in Kayardild constructions in which two distinct entities receive identical case-marking are used (a) for generic-specific constructions (116), (b) for body-part constructions (117), and (c) for secondary predicates (118):

(116)	*mutha-a*	*yarbud-a*	*wuran-da*	*kaarrku,*
	many-NOM	game-NOM	food-NOM	seagull(NOM)
	wanikarr,	*jirrkuriida*	*kada*	*thaath,* *warrmara-y*
	pelican(NOM)	northward	again	return wind-LOC

‘Lots of game, food, seagulls, pelicans are coming back northward again in the wind.’

(117)	*ngada*	*kurrija*	*banga-ya*	*bartha-y.*
	I	saw	turtle-OBJ	track-OBJ

‘I saw a turtle track.’

(118)	*ngada*	*kurrija*	*dangka-ya*	*malangarrba-y.*
	I	saw	man-OBJ	drunk-OBJ

‘I saw the man drunk.’

The first two of these semantic relations will typically be expressed in Mayali using noun-incorporated constructions (provided the correct grammatical relations obtain), as we have seen. The third type, secondary predication, may also be expressed by incorporation in Mayali, although the number of incorporable secondary predicates is small and other secondary predicates can only be external.[16] Incorporable secondary predicates in Mayali are, furthermore, linked to the same grammatical relations as incorporable generics, namely intransitive subjects (119) and objects (120).

(119) *Ba-rrarrgid-wa-m.*
3P-alive-go-PP
'He got away alive.'

(120) *Barri-re-y gare ginga barri-barlah-na-ni. Barri-djuhm-i*
3aP-go-PI maybe crocodile 3aP/3-track-see-PI 3a-swim-PI
barri-yawa-ni barri-darrgid-ma-ngi.
3a/3P-look:for-PI 3a/3-alive-pick:up-PI
'They (novices in the Morak ceremony) might see some crocodile tracks. They'd have to go into the water, look around for the crocodile and pick it up alive.'

For noun incorporation to span generics, body parts and secondary predicates is not limited to Australian languages. Sapir (1911) includes examples of all three types occurring in Yana. It is therefore worth asking what these three constructions have in common semantically, for the use of different grammatical means to express the same range of types in widely differing languages can hardly be a coincidence.

The crucial similarity, it seems to me, is that all three constructions involve thinking of one thing as another thing. In the generic/specific construction, one thinks of a specific entity X as a (kind of) generic entity Y. In part-whole constructions, when one thinks of (part) X, one is bound to consider the present or former existence of (whole) Y; X is an index of Y. In secondary-predicate constructions, thinking of X *at a given time* (or 'stage' – Kratzer 1989) one thinks of X as Y (e.g. 'alive') in (120).

Of course, there are important semantic differences, as even these very provisional characterisations show – the differences between thinking of as a kind of, thinking of as an index, and thinking of as a stage. There are also differences in symmetry: what one might call the 'primary entity', i.e. the thing being thought of as something else, is the incorporated element with part-whole constructions, but the unincorporated element with generic/specific and secondary-predicate constructions. The existence of such differences helps explain why incorporating languages do not always use the same construction for all three types. Ngan'gikurunggurr (Reid 1982), for example, allows the incorporation of body parts but not of generics or secondary predicates.

Nonetheless, this analysis does show there are sufficient similarities to explain the recurrent similarity in morphosyntactic encoding of the three semantic types (cf. Chappell – McGregor 1989: 30-32). If this semantic rationale is well-founded, we may expect to find a number of other instances of morphosyntactic similarity between part-whole, generic-specific and secondary predicate constructions.[17]

10. Conclusions

Cross-linguistically, noun incorporation is a widely-used morphosyntactic resource for representing body-part involvement or affectedness. However, the syntactic treatment given to incorporated body part nominals varies considerably. In Warray, they are treated in the same way as other locations (Harvey 1995). In Mohawk, they parallel other possessed items, the effect on which is seen as less important than the effect on their possessor, which by incorporating change the argument frame of the verb (Mithun 1995).

In Mayali, body-part incorporation is grammatically parallel to the incorporation of generics, and has no effect on the argument structure of the clause. It is treated more naturally as incorporation of one of a pair of apposed nominals than as a manipulation of argument frames. Body-part nominals can incorporate if their 'wholes', following possessor raising, occupy grammatical relations that sanction generic incorporation. These relations are: subject of intransitive or semi-transitive verbs, and otherwise object, never transitive subjects (except for one idiomatic impersonal verb) or indirect object. Where the comitative applicative yields a double-object verb, eligibility for incorporation is restricted to one of the objects, with this choice determined lexically.

The semantic class of "body parts" defined by incorporability is somewhat wider than the prototypical "body part" class discussed in the introduction to this volume, and includes separated body parts, nests, corpses, bodily excretions, products and residues. The key semantic factor seems to be that the 'part' is an 'index' of the whole: it implies the presence or existence, current or former, of the whole.

Typically the body part is incorporated, and it is not treated as a distinct argument, as shown by its failure to govern number and person in the pronominal prefixes to the verb. These two formal features – incorporation, and non-argument status – are found whenever the part is less salient than its whole in discourse terms, and is not conceptually separated from the whole. Two types of departure from this norm can occur. When the part is conceptually separated from the whole, it is treated as an independent argument, and the person and number of the part, rather than the whole, are represented on the pronominal prefix. When the part is treated as more salient than the whole in the discourse, such as when it is contrasted, conjoined, or becomes a discourse participant in its own right, it fails to incorporate.

Australian languages exhibit recurrent morphosyntactic similarities in the grammatical encoding of part-whole, generic-specific and secondary-predicate constructions, all three tending to be represented by syntactic apposition in some typologically appropriate form. In dependent-marking Australian languages apposition takes the form of case agreement between distinct constituents; in those head-marking Australian languages with highly productive incorporation (the majority of Gunwinyguan languages, and Tiwi) it takes the form of syntactic apposition of an incorporated and an external element, both linked to the same grammatical relation of the verb. Inasmuch as syntactic apposition involves juxtaposing different formal representations of the same argument, while the semantics of these three constructions involves juxtaposing different countenances of the same entity, these morphosyntactic parallels are semantically based.

Notes

1. My fieldwork on Mayali (with particular emphasis on the Gun-djeyhmi dialect) was financially supported by Australian National Parks and Wildlife and the Gagudju Association, which bodies I thank for their manifold assistance. For their incisive, inventive and good-humoured instruction in Mayali I wish to thank the late Toby Gangele, Minnie Alderson, the late Nipper Kapirrigi, David Kanari, Eddy Hardy, Violet Lawson, Judith Alderson and Mandy Muir. I have also benefitted from discussions of noun incorporation with Marks Baker, Harvey and Durie, Marianne Mithun, and David Wilkins, and the editorial suggestions of Hilary Chappell and Bill McGregor; though I alone am responsible for any errors of fact or interpretation in this paper.
2. I employ the following practical orthography: stops (phonetically voiced syllable-initially, voiceless syllable-finally) are bilabial *b*, alveolar *d*, retroflex *rd*, palatal *dj*, velar *g* (syllable-initial) and *k* (syllable-final) and glottal *h*; nasals are bilabial *m*, alveolar *n*,

retroflex *rn*, palatal *nj*, and velar *ng*; laterals are alveolar *l* and retroflex *rl*, and rhotics are retroflex continuant *r* and alveolar tap/trill *rr*. Fortis stops (phonemically long, tense and voiceless) are written with doubled letters.

3. Abbreviations in glosses are a(ugmented number), BEN(efactive applicative), COM (itative applicative), GEN(itive), IMP(erative), LOC(ative), N(on)P(ast), P(ast) I(mperfective), P(ast)P(erfective), P(ast), u(nit)a(ugmented number). I, II, III and IV are noun-class prefixes. Minimal number is not glossed as such, so 3P means '3rd minimal past'. Prefix sets representing two or more pronominal arguments are written with the subject argument first, then the object argument; thus 3/1 means '3rd person minimal subject acting on first person minimal object'.
4. Basically, the distribution of lexical choices reflects functional factors in that the pronominally represented argument is the one prototypically higher in animacy of the two objects; conversely, the argument subject to noun incorporation (see section 4) is the one prototypically lower in animacy.
5. I employ the thematic roles used in Foley – Van Valin (1984); 'theme' is defined as 'that whose location changes or is at issue'. Other sets of thematic roles – e.g. those proposed by Ostler and subsequently used by Kiparsky, do not work any better. For example, on the Ostler-Kiparsky set, in (13) a 'locative' (with them) outranks a 'theme' (the drink), while in (11) a 'theme' (me) outranks a 'locative' (in the swag) .
6. The Kunwinjku orthography writes *k* instead of *g* for syllable initial velar stops.
7. There are several other Australian languages in which genitalia have anomalous noun-class membership, or are associated with edibility (perhaps mediated by the common use of 'eat' as a euphemism for 'fuck'). In Yanyuwa *-ragugu* 'penis', although taking a male possessive prefix, governs 'food-class' modifier agreement (Kirton 1971: 58). In Maung one obscene term for 'penis', *ma-jiradad*, involves the vegetable class prefix on the root *jiradad* 'meat food' (Capell – Hinch 1970: 52). In Tiwi 'parts of the body are of the same gender as their possessor, except for the genital organs, which are invariably of the opposite gender to the possessor' (Osborne 1974: 51).
8. Several related languages have the same suppletion, e.g. Ngandi *bun-* 'water' (compounded form), *(ku)djark* 'water' (free form); Ngalakan *binyi* 'water' (compounded form), *we?* 'water' (free form). In Ngalakan, as in Mayali, this is the only suppletive incorporated nominal in the language.
9. For arguments that the incorporated nominal slot is functionally complementary with the two pronominal prefix slots and as such can provide a third verbal actant slot, see Evans 1995b.
10. Walsh 1987 surveys impersonal constructions in Australian languages, but gives no examples of noun incorporation.
11. *Nganmarnegornmelmeng* is not ungrammatical, however – it could mean 'he touched (her) crotch with his foot' with the implication that 'her' refers to a close female relation (e.g. wife, sister) whose interests coincide with the speaker's.
12. It is possible to treat the part as an independent argument when stressing perceptual separation – see section 7 – but in this case the possessor is not overtly represented.
13. Of course one could always postulate an obligatory relation-changing rule, without morphological consequences. Blake (1984) essentially adopts this analysis with respect to Kalkatungu, a dependent-marking Australian language. He sets up two types of possessor-raising – one, applying to alienables, that is optional and registered by a 'goal' applicative suffix on the verb; one, applying to body parts, is obligatory, and not

registered on the verb. I prefer to avoid postulating syntactic rules with no language-specific justification, and do not adopt such a solution for Mayali. In any case, even if we were to adopt obligatory possessor-raising as a rule of Mayali syntax, it would not be connected to body-part incorporation, which as we have seen does not alter the argument structure of the clause.

14. Obviously there are also cases, such as body parts of transitive subjects, which do not allow incorporation for grammatical reasons (section 4); since non-incorporation in such cases is obligatory this is irrelevant to the present discussion.
15. In theory it should be possible for both things to happen at once – one should find clauses in which the part fails to incorporate, and also controls person and number marking. So far, however, I have no examples. Perhaps it will be necessary to translate Gogol's 'The Nose' and Philip Roth's 'The Breast' into Mayali.
16. The related language Rembarnga incorporates secondary predicates far more extensively – see McKay (1975: 292) for examples.
17. For example, it is interesting that the Russian instrumental case can have both a "body-part" meaning and a temporary predicate meaning. According to the explications contained in Wierzbicka (1980), a sentence like *Ivan kivnul golovoj* [Ivan nodded head-INSTR] contains the component "X did something that can be thought of a something that happened to part Y of his body". A temporary predicate sentence like *On (togda) rabotal saxterom* [he (then) worked miner-INSTR] contains in its explication the component 'thinking of X-at-that-time one could say "this Y"'. Both explications thus share the general characterisation of 'thinking of X as Y'.

References

Alsina, Alex – Joan Bresnan – Peter Sells (eds.)
1995 *Complex predicates.* Stanford: CSLI

Bach, Emmon – Angelika Kratzer – Barbara Hall Partee (eds.)
1989 Papers on quantification. University of Massachusetts and Amherst: National Science Foundation Report. [Unpublished MS.]

Baker, Mark
1988 *Incorporation: a theory of grammatical function changing*. Chicago: University of Chicago Press.
1993 "Noun incorporation and the nature of linguistic representation", in: William Foley (ed.), 13-44.

Blake, Barry J.
1984 "Problems for possessor ascension: some Australian examples", *Linguistics* 22: 437-453.

Capell, Arthur – Heather E. Hinch.
1970 *Maung grammar, texts and vocabulary*. The Hague: Mouton.

Carroll, Peter
1976. Kunwinjku: a language of Western Arnhem land. [Unpublished M.A. thesis, Australian National University.]

Chappell, Hilary – William B. McGregor
1989 "Alienability, inalienability and nominal classification", *Berkeley Linguistic Society* 15: 24-36.
1995 "Prolegomena to a theory of inalienability", [this volume.]

Coleman, Caroline
1982 Kunparlang morphology. [Unpublished B.A. (Honours) thesis, Australian National University.]

Dixon, Robert M.W.
1977 *A grammar of Yidiny*. Cambridge: Cambridge University Press.

Dixon, Robert M.W. (ed.)
1976 *Grammatical categories in Australian languages*. Canberra: Australian Institute of Aboriginal Studies.

Duranti, Alessandro
1979 "Object, clitic pronouns in Bantu, and the topicality hierarchy", *Studies in African Linguistics* 10: 31-46.

Evans, Nicholas
1991 A grammar of Mayali. [Unpublished MS.] de Gruyter.
1995a *A grammar of Kayardild.* Berlin: Mouton de Gruyter.
1995b "Role or cast? Noun incorporation and complex predicates in Mayali", in: Alsina, Alex – Joan Bresnan – Peter Sells. (eds.)

Foley, William (ed.)
1993 *The role of theory in language description.* Berlin: Mouton de Gruyter

Foley, William A. – Robert Van Valin
1984 *Functional syntax and universal grammar.* Cambridge: Cambridge University Press.

Frantz, Donald G.
1971 *Towards a generative grammar of Blackfoot.* Norman, Oklahoma: Summer Institute of Linguistics.

Goddard, Cliff
1982 "Case systems and case marking in Australian languages: a new interpretation", *Australian Journal of Linguistics* 2: 167-196.
1983 A semantically oriented grammar of Yankunytjatjara. [Unpublished Ph. D. thesis, Australian National University.]

Hale, Kenneth
1981 "Preliminary remarks on the grammar of part-whole relations in Warlpiri", in: John Hollyman – Andrew Pawley (eds.), 333-344.

Harvey, Mark
1995 "Body parts in Warray", [this volume.]

Heath, Jeffrey
1978 *Ngandi grammar, texts and dictionary.* Canberra: Australian Institute of Aboriginal Studies.
1984 *Functional grammar of Nunggubuyu.* Canberra: Australian Institute of Aboriginal Studies.

Hollyman, John – Andrew Pawley (eds.)
1981 *Studies in Pacific Linguistics in honor of Bruce Biggs.* Auckland: Linguistic Society of New Zealand.

Kiparsky, Paul
1989 Double objects. [Paper presented to a conference on non-standard case, Tucson, Arizona, July 1989.]
Kirton, Jean
1971 "Complexities of Yanyuwa nouns", in: Jean Kirton (ed.), 15-70.
Kirton, Jean (ed.)
1971 *Papers in Australian Linguistics Number 5.* Canberra: Pacific Linguistics.
Kratzer, Angelika
1989 Stage-level and individual-level predicates. [Unpublished MS, in: Emmon Bach – Angelika Kratzer – Barbara Hall Partee (eds.).]
Laughren, Mary
1989 Secondary predication as a diagnostic of underlying syntactic structure; evidence from some Pama-Nyungan languages. [Paper presented at Australian Linguistic Society conference, Monash University, September, 1989.]
Leeding, Velma J.
1995 "Body parts and possession in Anindilyakwa", [this volume.]
Lyons, John
1977 *Semantics, Volume 1.* Cambridge: Cambridge University Press.
McGregor, William B.
1985 "Body parts in Kuniyanti clause grammar", *Australian Journal of Linguistics* 5: 209-232.
McKay, Graham R.
1975 Rembarnga: a language of central Arnhem Land. [Unpublished Ph. D. thesis, Australian National University.]
Merlan, Francesca
1976 "Noun incorporation and discourse reference in modern Nahuatl", *International Journal of American Linguistics* 42: 177-191.
1983 *Ngalakan grammar, texts and vocabulary.* Canberra: Pacific Linguistics.
Mithun, Marianne
1984 "The evolution of noun incorporation", *Language* 60: 847-894.
1986 "On the nature of noun incorporation", *Language* 62: 32-37.
1995 "Multiple reflections of inalienability in Mohawk", [this volume.]
Nichols, Johanna
1986 "Head-marking grammar and dependent-marking grammar." *Language* 62: 56-119.
Osborne, Charles R.
1974 *The Tiwi language.* Canberra: Australian Institute of Aboriginal Studies.
Ostler, Nicholas
1979 *Case-linking: A Theory of Case and Verb Diathesis Applied to Classical Sanskrit.* [Doctoral dissertation, MIT, distributed by Indiana University Linguistics Club.]
Reid, Nicholas
1982 The basic morphology of Ngan'gikurunggurr. [Unpublished B.A. (Hons.) thesis, Australian National University.]
Sapir, Edward
1911 "The problem of noun incorporation in American languages", *American Anthropologist* 13: 250-82.

Silverstein, Michael
1976 "Hierarchy of features and ergativity", in: Robert M.W. Dixon (ed.), 112-171.

Steele, Ross – Terry Threadgold (eds.)
1987 *Language topics: essays in honour of Michael Halliday*. Amsterdam – Philadelphia: John Benjamins.

Walsh, Michael
1987 "The impersonal verb construction in Australian languages", in: Ross Steele – Terry Threadgold (eds.), 425-438.

Wierzbicka, Anna
1980 *The case for surface case*. Ann Arbor: Karoma.

Body parts in Warray

Mark Harvey

1. Introduction

This paper[1] examines body parts in the morphological and syntactic structure of Warray.[2] Warray is a member of the Gunwinjguan language family (a non-Pama-Nyungan family), and was originally spoken around the town of Adelaide River approximately 100 km south of Darwin. There are three main areas of interest.

(1) The pattern of noun-class marking. Class marking distinguishes alienable from inalienable body parts. This paper argues that apparent anomalies in the class marking of certain body parts proceed on a principled basis if a notion akin to that of "person" as opposed to "body" is taken to be the prototypical "whole" notion for the class.

(2) Nominal compounding of body parts. Body parts are by far the most commonly compounded nominals in Warray. There are two types of body part compounds. One type of body part compound is productively attested only with inalienable body parts. It presents the conditions of body parts as relatively enduring characteristics of the whole. This type of compound frequently functions as a synecdoche [a figure of speech where a part is used to mean the whole]. Synecdochic compounds have a wide range of use in Warray as natural species names. The other type of body part compounding creates new body part lexemes.

(3) Clauses involving body parts show two main areas of interest. One of these is the "possessor ascension" of cross-referencing (where the whole is cross-referenced rather than the part). This is the normal pattern of cross-referencing in Warray, but variations are possible. The significance of the normal pattern and the nature of the contrast with the variations are examined.

In possessor ascension clauses the part noun is also frequently incorporated into the verbal complex. Noun incorporation is examined in some detail in this paper. The nature of the distinction between noun incorporation and denominal verb formation is considered. The grammatical structure of noun incorporation is analysed and it is argued that noun incorporation is essentially adverbial/locative in nature in Warray.

Possessor ascension cross-referencing and noun incorporation permit us to "verbally" define an inalienable body parts class. This verbally defined inalienable class shows some differences to the nominally defined inalienable class. The nature of these differences are examined.

As Warray is a dying language with only a very few speakers, it would appear desirable to provide some basic information on the competence of my consultants. I worked on the language with two main consultants. They were both born in the mid-late 1920's, and appear to have used Warray as one of their main daily languages until their mid to late teens. However they have very little common life history. Though they knew about each other from their childhood, they did not actually meet until approximately 1950, and since that time they have not been in close or frequent contact. Moreover I did not normally work with them together. They speak slightly different dialects, but the differences are quite minor, and none are relevant to the matters covered in this paper. Both consultants have a good command of Warray, though their knowledge is somewhat rusty owing to lack of use, with English being their main language of daily use. Neither of my consultants gave lengthy texts and so it is frequently difficult or impossible to comment on discourse factors and their effects.

2. Typological background

Warray has three major parts of speech: nominals, particles and verbs. In Warray, unlike many Australian languages, it is possible to formally distinguish nouns and adjectives on three grounds. Firstly they take different inchoatives. Nouns take the inchoative verb *nayi-yn* 'to become', whereas adjectives take an inchoative auxiliary *-ji-yn*. Secondly adjectives take the Oblique suffix *-wu*, whereas nouns do not. Thirdly nouns and adjectives show significantly different patterns of class prefixing. Class marking of nouns distinguishes four reasonably semantically consistent classes.

I	*a-*	some human male nouns
II	*al-*	all human female nouns
III	*an-*	inalienable body parts, and some parts of the landscape.
IV	*ø-*	everything else, including alienable body parts.

Class marking of adjectives proceeds on a rather different basis. The general principles are as follows.

I	*a-*	is the unmarked form, being a possible modifier for referents of any class
II	*al-*	is only found with human female referents
III	*an-*	is the usual form for referents lower on the animacy hierarchy.
IV	*ø-*	is only very infrequently found. Usually there is also a preceding head nominal and such forms are probably better analysed as quasi-compounds.

There are however a number of lexically governed exceptions. *-murdek* 'big' does not have a Class III form, whereas *-wak* 'little' and *-jerriyn* 'one' occur only with Class III forms. Even if an adjective does occur with all three substantive prefixes, exceptions to the general patterns are found (e.g. *nal an-warru* 'a bad man'). Given the variable nature of class marking in Warray, the class markers are simply glossed by the Roman numeral identifying the particular marker.

Warray has a well developed case marking system, with the following case markers.

-ø	Absolutive
-yi	Ergative/Instrumental
-yiwu	Comitative
-(w)u	Dative
-lik	Allative/Locative
-ba	Perlative
-yang	Ablative/Causal

-yi is not obligatory in its ergative function. Transitive Subjects may alternatively take Absolutive case marking. Case marking is otherwise obligatory. The determination of NP constituents and occurrence patterns of case marking are complex (Harvey 1990). However NPs consisting solely of nouns and/or adjectives generally show a head nominal – immediately following modifier nominal constituent structure. Case marking attaches to the final constituent in such NPs.

Although case marking is an important system in Warray, the language is generally head-marking (Nichols 1986), like the other members of the Gunwinjguan family. The verbal complex (hereafter VC) may carry a large amount of information. It is the only obligatory constituent in verbal clauses, which are the unmarked clause type in Warray. The VC is by far the most morphologically complicated form, and has the following structure:

> Pronominal Prefix Complex + (Indirect Object Prefix) + (Adverb) + (Noun) + Root + (Auxiliary) + (Reciprocal/Reflexive) + Aspect/Mood/Tense Suffixes + (Directionals)

The pronominal prefix complex carries information as to the person and number of subject and object of the verb. Unlike other Gunwinjguan languages, the pronominal prefix complex in Warray is relatively agglutinative. In most cases subject and object prefixes may be factored out, though their ordering is somewhat complex (Harvey 1990). There is one set of object prefixes and three sets of subject prefixes; an unmarked set (found chiefly in the past realis), a non-past set (future intentional), and an irrealis set (past irrealis and future uncertain). The three sets of subject prefixes combine with conjugationally determined verbal suffixes to convey information about aspect, mood and tense.

The unmarked subject prefixes and object prefixes are as shown in the paradigm of Table 1 (the non-past and irrealis subject prefixes are based on the unmarked forms).

Table 1. The unmarked subject prefixes

		sg	dl	pl
1	inc		*ma-*	*i-*
exc	*at-*		*i-*	
2		*an-*		*a-*
3		*ø-*		*ba-*

Only the 1inc has a specific dual form, otherwise the plural is used. The 3rd plural is also used for specific indefinite meanings such as 'someone'.

Table 2. The object prefixes

	sg	pl
1	*ban-*	*in-*
2	*ana-*	*in-*
3	*ø-bin-/bun-*[a]	

a. These two forms are dialectal variants.

As in the other Gunwinjguan languages there are two major classes of verb stems, simple and compound. Simple stems consist of just a verb root (many of which are monosyllabic). Compound stems consist of a verbal or nominal root, followed by an auxiliary. Nearly all auxiliaries are monosyllabic, and many also occur independently as simple verbs (as it would appear to be historically the case for all auxiliaries). However the auxiliaries, even those which also occur as simple verbs, do not in most cases have any substantive predicate meaning in their function as auxiliaries. They simply act to verbalise the root they compound with.

In addition to these basic constituents the VC in Warray, as in other Gunwinjguan languages, may also contain reciprocal or reflexive suffixes, a valence increasing prefix (the indirect object prefix), incorporated adverbs and nouns, and directionals. As the following example shows, the VC may code a considerable amount of information.

(1) *gat-gubal-nabat-dayn?-mi-yi-n*
IRR:1sgS-nearly-hand-cut-Aux-refl-P[3]
'I nearly cut my hand.' [literally, I nearly hand-cut myself]

Cross-referencing of objects is not restricted to direct objects (objects not requiring a valence increasing affix) in Warray. Indirect objects may also be cross-referenced with the use of the valence increasing indirect object prefix *nat-*.

(2) *bat-mi* *muya* *nguny-u-wu*
1sgS:NP-get:A tucker 2sg-OBL-DAT
'I will get tucker for you.'

(3) *barriyn-nat-mi* *muya*
1sgS→2sgO:NP-IO-get:A tucker
'I will get you tucker.'

As a comparison of (2) and (3) shows the indirect object displaces the direct object as the entity cross-referenced by the object prefixes. The IO prefix *nat-* is the overt signal that the valence of the verb has been increased and that the object cross-referenced entity is an indirect, rather than a direct, object. A very wide range of entities may be coded as indirect objects: affected locatives, benefactives, ethical datives, goals, possessors, and purposives. Essentially any non-sub-categorised entity which is affected, or towards which the verbal predicate is directed may cross-referenced as an indirect object. When an entity is cross-referenced as an indirect object, it takes absolutive case.

(4) *mimi at-nat-ji-yi*
uncle 1sgS-IO-say-PP
'I told uncle.'

(5) **mimi-wu at-nat-ji-yi*
uncle-DAT 1sgS-IO-say-PP
'I told uncle.'

3. Noun class marking of body parts in Warray

In Warray body part nouns belong to either Class III (*an-*) or to Class IV (ø-). The great majority of body part nouns belong to Class III; only those listed in Table 3 belong to Class IV.

Table 3. Body parts belonging to Class IV

balmuk	'wing'	*bula*	'grey hair'
deklok	'semen'	*dum-mila*	'tears'
gijarniyn	'spit'	*gijewek*	'beard'
gine	'mucus'	*giyak*	'sweat'
gurratj	'blood'	*julmija*	'pubic hair'
mala	'leaf'	*mija*	'hair'
mel	'vagina'	*nguk*	'faeces, guts'
wurl	'urine'		

Additionally there is the noun *garle* 'cocoon, dillybag, pouch [of a marsupial], womb'. However this noun is only marginally, if at all, a member of the body parts category. For my consultants it appeared that its primary meaning was 'dillybag' with the other meanings being related to this primary meaning on the basis of perceived similarities of shape and function.

The set of Class IV body part nouns shows a reasonable degree of consistency. With the exceptions of *balmuk* 'wing', *garle* in the senses of 'pouch, womb' and *mel* 'vagina', all the nouns refer to bodily fluids, excretions, hair and leaves. Bodily fluids, excretions, hair and leaves share a common distinguishing characteristic as opposed to other body parts. Portions of them are visibly lost and replaced as the normal course of events (hereafter "replaceable/ability" will be used as a cover term for this concept). As such they may be viewed as being alienably attached to the body. This would suggest that the opposition between Class III and Class

IV marking for body parts reflects a semantically based inalienable vs. alienable opposition. The following two pairs provide some support for the hypothesis that replaceability is an important criterion determining class marking.

(6) *nguk* 'faeces, guts'
an-nguk 'guts'

(7) *mala* 'leaf on tree or ground'
an-mara 'leaf on tree'

The Class III forms can only refer to entities attached to the whole (that is, not lost to the whole), whereas the Class IV forms can refer to both attached and detached entities. However as the marking pattern in 6 and 7 is not productive (the difference between the free and bound root forms of 'leaf' should be noted), its evidentiary value is somewhat uncertain. Further there are some replaceable body parts which take Class III marking.

Table 4. Replaceable body parts belonging to Class III

an-bun	'feather'
an-dum-mija-wek	'eyebrow/lash'
an-gundu	'egg'
an-nabat-bidiyn	'fingernail'
an-ngar	'fur, body hair'
an-ngubat-bidiyn	'toenail'

Therefore replaceability clearly does not completely control class marking. The situation in Warray would appear to be that described by Dixon (1982: 182) in his discussion of noun classes "some [class memberships] are without explanation ... some may have had an explanation in terms of an earlier stage of the language, but the class assignment has been retained and the explanation lost as the language has altered." However there is another factor to be considered which shows that the occurrence of exceptions is less arbitrary than it initially appears.

This additional factor is the nature of the whole in the part/whole relationship. The term "body part" indicates that the body is the prototypical whole. However I would argue that the prototypical whole for the body part class is not the "body", but is rather a notion similar to that of "person". This viewpoint is argued explicitly by Bally ([1995]: 33 ff), also Wilkins (pers. comm.), and implicitly by McGregor (1985: 228). As such I would argue that the "body part" class is more accurately to be described as the "person part" class. I will continue to use the term "body part" generally, as it is the standard term. However, as we will see, there are situations where the term "person part" is more appropriate, and consequently I will use the term "person part" in these situations.

Clearly both "body" and "person" are complex notions and some consideration of the differences between the two is necessary. I do not propose to provide a complete semantic description of either notion here, but merely to set out the prototypical differences that appear to be relevant in determining the distribution of class marking. Firstly I think that it must be said that persons are conceived of as normally having bodies. However this does not mean that "body" and "person" are synonymous. Animals have bodies, but are not persons. Similarly creatures of the imagination such as doppelgangers and zombies have bodies, but they are not persons. Clearly then persons are human.

However this does not mean that "person" and "human" are synonymous. It is quite possible to be a human and an "unperson", as in Orwell's *1984*. I would suggest that the relevant difference is a social one. To be a person is to have some distinguishing individual characteristics in terms of capacity for action in, and membership of the social world. An "unperson" has no membership of the social world nor any capacity for action in the social world. The terms *impersonal*, *to depersonalise*, *to personify* and *the personal touch* show that it is not merely a capacity for action in the social world that is relevant in defining "person". These terms all involve concepts of recognition of, and capacity to act in terms of distinguishing individual characteristics. Therefore I would argue that the relevant differences between "person" and "body" are being human, and having and recognising distinguishing individual characteristics in terms of the capacity to act in the social world.

The preceding discussion has been conducted in terms of English semantics and sociocultural perspectives. Its transferrability to an Aboriginal sociocultural perspective is by no means unproblematic. However I lack the material with which to make the necessary semantic differentiations in Warray. Therefore I can only offer an English based set of distinctions and demonstrate that similar distinctions appear to be relevant in the class marking of body parts in Warray. I do not claim that the distinctions in Warray relate to a concept identical to that of "person" in English. They do however suggest that some notion similar to that of "person" involving the features that distinguish "person" from "body" is relevant in explaining the distribution of class marking in Warray. A number of Aboriginal languages do have lexemes translated as being equivalent to English "person" (Dixon 1980: 119). However Warray is not among them, and furthermore, the exact meanings of these lexemes have not been fully investigated.

If "person" is taken to be the relevant prototypical whole for the class then it is possible to see that the distribution of most of the anomalies in the class marking of body parts proceeds on a principled, though not completely predictable basis. This is the case not only for the anomalies we have already discussed, but also for some additional anomalies in class marking. No equivalent explanation is possible if "body" is taken to be the prototypical "whole" notion.

In discussing the distribution of apparent anomalies in class marking I will be following Lakoff's (1986: 17-18) analysis of the structure of noun classes in terms of prototype. I would argue that "person" is the central notion for the part/whole relationship, with less central members being linked to it by chains. The nature of the linking chains is variable, but the links are motivated. However while the links are motivated it is not possible to predict the exact constitution of a class, nor is it necessarily the case that all members of a class will share some common feature. In Lakoff's terms "person" is a typical prototype (1986: 33-34) for the whole in the part/whole relationship.

There are four sets of body part nouns whose class marking is anomalous, either in terms of the posited replaceability dichotomy (Class III vs. Class IV), or in terms of the "body" being the focal whole notion. Three sets of exceptions become less anomalous if the focal notion of the "body part" class is taken to be "person/individual". The first set consists of nouns referring to physically inalienable body parts which do not belong to Class III. In Warray this set consists of *mel* 'vagina' and the marginal body part term *garle* 'pouch, womb'. As Evans (1995) notes in this volume, anomalous class marking for genitalia is not uncommon in Australian languages. This anomalous class marking would appear to be a reflection of a rule set out by Dixon (1982: 179) in his discussion of the principles of noun class marking: "If a subset of nouns has some particular important property that the rest of the set do not have, then the members of the subset may be assigned to a different class from the rest of the set, to 'mark' this property".

Given the salience of sex and sexuality in human life, it is hardly surprising to find that body part nouns referring to genitalia may be marked distinctively from other inalienable body part nouns. If *garle* in its meanings 'pouch, womb' is to be treated as a member of the body part class, then its Class IV membership would appear to follow in the same way given the intimate connection of sex and reproduction. While the occurrence of anomalous marking for genitalia and reproductive organs may be explicable, its actual distribution is not, as predicted by Lakoff (1986: 18). There does not appear to be any immediately obvious reason why *mel* belongs to Class IV, but the other two nouns referring to genitalia *an-layn* 'penis, tail' and *an-mak* 'testicles' belong to Class III.

The second set of anomalously marked nouns are specifically animal body parts which are marked in a manner contrary to that which would be posited on the basis of the proposed replaceability criterion.

(8) *an-bun* 'feather'

(9) *an-gundu* 'egg'

(10) *an-ngar* 'fur, body hair'[4]

(11) *balmuk* 'wing'

The three Class III nouns all refer to replaceable alienable body parts, and the Class IV noun refers to a non-replaceable inalienable body part. If 'person' is taken to be the prototypical whole, then animals are less central wholes, because they are not persons. However they are linked to the notion of person by the fact that they have bodies. By virtue of this link, the class marking of specifically animal parts is motivated. The actual class marking of these parts is however irregular.

This may be simply because animal parts are peripheral to the part/whole class, and irregularities are to be expected in peripheral areas. Alternatively the irregular class marking may be explained by a principle of Lakoff's. He suggests (1986: 22) that when two categories are in minimal contrast, exceptions to one category will be assigned to the other category. In the marking of body parts in Warray, Classes III and IV are in a minimal contrast. The irregular marking of the animal body parts in (8)-(11) may function to indicate their peripheral/exceptional status.

Considerations of animacy may also be relevant as (8)-(11) all describe parts prototypically belonging to higher animates. Higher animates are the group most saliently opposed to humans in animacy terms. The assignment of "higher animate" parts to the class which is opposite to the one that would be predicted on the basis of "human" parts, may be motivated by the prototypical animacy opposition between humans and higher animates. The only other "higher animate" part, apart from those listed in (8)-(11), is *an-layn* 'tail'. However *an-layn* differs from the parts in (8)-(11) in that it also means 'penis', and as such has human reference.

I do not have any evidence as to whether either, and if so which, of the two meanings is the primary meaning of *an-layn*. It may be noted that the nouns referring to the parts of lower animates or inanimates are regularly class-marked; *an-ngak* 'shell', *an-durt* 'roots', *an-jatban* 'fork of a tree', *an-dan* 'point, tip'.

The third set consists of three nouns which do not refer to parts of the physical body as such, but which belong to Class III.

(12) *an-bong* 'print, track'

(13) *an-mewel* 'clothes'

(14) *an-nyi* 'name'

These nouns all refer entities which are to varying degrees individually distinctive of persons in terms of the human world. They fall within Bally's concept of the personal domain: "The personal domain includes or can include objects and beings associated with a person in an habitual, intimate or organic way (e.g. the body and its parts, clothes, the family, etc.) ... The concept of personal domain is an entirely subjective one ... The extent of the domain is determined by the cultural outlook of each linguistic group." (Bally [1995])

Names are particularly salient parts of a person's social identity for Aboriginal people. Stanner (1937: 301) states "the personal names by which a man is known are something more than names. Native statements suggest that names are thought to partake of the personality which they designate. The name seems to bear much the same relation to the personality as the shadow or image does to the sentient body ... Names are not symbols so much as verbal projections of an identity which is well known in the flesh." Similarly, prints are entities which are directly physically distinctive of the individual.

Clothes are however much less obviously individually distinctive of persons. Throughout the area in which Warray was spoken one of the major traditional funeral rites was the public burning, or burial with the corpse of a person's clothes and possessions (Berndt – Berndt 1985: 457). This destruction appears to be symbolic of the separation of the dead individual from the living. There was therefore an important traditional context in which clothes were individually distinctive. However we should note that the noun *mekmek* 'swag, possessions' belongs to Class IV. Further the precise pre-contact meanings of the terms *an-mewel* and *mekmek* (even the existence of the term *mekmek*) are unknown. Clearly while the inalienable class marking of *an-mewel* may be motivated, it is much less strongly motivated than that of 'name' or 'print'.

An-mewel is a rather marginal member of the person part class. As we shall see (section 5.6, examples (82) and (83)), while 'clothes' is marked as a member of the nominal inalienable person part class, it does not form part of the "verbal" inalienable person part class defined by possibility of incorporation within the verbal complex and cross-referencing of the whole rather than the part. It would seem likely that this difference in the treatment of 'clothes' from other Class III nouns is explicable in terms of the fact that clothes are not necessarily and directly associated with the individual.

The fourth set of exceptions is not semantically based. It involves three compound forms, and is examined in the following discussion of nominal compounding – see examples (29)-(31).

4. Nominal body part compounds

Nominal compounds have the following structure in Warray.

Class Marker + Modified Root + Modifier Root

This structure is parallel to that of NPs (see section 2). Compounds are right branching in Warray (Harvey 1990; Lieber 1981), and may formally be either nouns or adjectives, depending on the class of the rightmost root. Consequently compounds display rather variable class marking, as the class marking of nouns and adjectives is significantly different (section 2).

Body part compounds are the most common nominal compounds in Warray. There are two types of body part compounds in Warray. One type presents the condition of a particular body part as being a relatively enduring characteristic of the whole. This type is only productively attested with inalienable [non-replaceable] body parts. Presumably this is because the conditions of alienable [replaceable] body parts cannot in general be interpreted as relatively enduring characteristics of the whole.

(15) *a-/al-dum-jaminy-u*
I-/II-eye-bad-OBL
'a man/woman with bad eyes' [Literally, 'the bad-eyed one']

(16) *a-yel-gubam-u*
I-flesh-big-OBL
'a fat man' [Literally, 'the big-fleshed one']

This type of compound is frequently used as a synecdoche. Most traditional nicknames are synecdochic compounds.

(17) *a-murlu-nendu*
I-hip-horse
'He who was kicked in the hip by a horse'

(18) *al-garra-dren*
II-leg-train
'She whose leg was run over by a train'

The names of a number of animal and plant species are synecdochic compounds, with varying degrees of literalness. This appears to have been an areal feature as it is found to a much greater extent in the unrelated Limilngan language spoken to the north-east of Warray (though the forms in Limilngan are NP's rather than compounds).

(19) *a-bam-gubam(-u)*
I-head-big-OBL
'Brown snake'[5] [Literally: 'the big-headed one']

(20) *an-garra-bitbit*
III-leg-red
'plant species' (leea rubra)[6] [Literally: 'the red-legged one']

(21) *a-ngarndi-bit-u* I-throat-coloured-OBL
'Eucalyptus miniata'[7] [Literally: 'the colour-throated one']

(22) *an-nguk-buye-buye*
III-guts-stinking-stinking
'beetle species' [Literally: 'the stinking-gutted one']

(23) *(a-)ganim-bali-wu*
I-ear-big-OBL
'frill-necked lizard' [Literally: 'the big-eared one']

(24) *dum-dingding-u*
eye-sticky-OBL
'fish species' [Literally: 'the sticky-eyed one']

(25) *jen-bulang-u*
tongue-cheeky-OBL
'tree species' [Literally: 'the cheeky-tongued one']

The class marking of these compounds is somewhat variable. The compounds in (19) to (22) show the adjectival class marking (see section 2), that would be expected from their structure. However (24) and (25) show Class IV marking, and (23) shows Class IV marking as an alternant. Class IV is the correct noun class for the entities that these compounds designate. Over time it appears that adjectival synecdochic compounds describing natural species show a tendency to be re-analysed as nouns. As a result they also tend to show the class marking patterns of nouns. This is most commonly the case with compounds such as (24) and (25), where the synecdoche is highly obscure. This tendency is most clearly exemplified by the dialectal variants of the following lexeme.

(26) *an-bam-jili-wilek-wilek* and *bamjiwilewilek*
III-head-mouth-charcoal-charcoal
'brown snakes with a stripe on their heads.'

The first variant is still analysable as a compound (there are other examples of compounds involving more than two nominal roots in Warray), but the second is synchronically unanalysable.

The following synecdochic compound provides a particularly clear illustration of the essentially inalienable nature of this "enduring characteristic" type of body part compound.

(27) *mara-bali-wu*
leaf-big-OBL
'Leichhardt tree.'

This compound takes Class IV marking, arguing that it, like (24) and (25), shows a tendency to be reanalysed as a noun. However the important point is that the root for 'leaf' is the inalienable root *-mara*, and not the alienable root *-mala* (cf. (7)). Apart from the roots *-murlu* 'hip', *-bam* 'head', *-garra* 'leg', *-ngarndi* 'throat', *-nguk* 'guts', *-dum* 'eye', *-jen* 'tongue' and *-jili* 'mouth' already illustrated in the preceding examples, the roots *-jat* 'leg' and *-miyn* 'belly' also occur in synecdochic species name compounds. All of these are inalienable body parts. There is only one possible alienable "enduring characteristic" compound.

(28) *bun-nat-dingding*
feather-IO-sticky
'bird species'

There are a number of problems with this form. Firstly even though *an-bun* 'feather' is semantically alienable, it is formally marked as inalienable (cf. examples (8)-(11)). Secondly it is not entirely clear to what degree this compound is synchronically analysable in Warray. Thirdly this compound involves the otherwise verbal indirect object prefix *nat-* (only one other nominal compound involves this prefix). Therefore the compound in (28) is clearly a marginal form, and does not invalidate the inalienable character of this compound type.

The other type of body part compound in Warray are compounds which are themselves body part lexemes. Most of these compounds refer to inalienable body parts, and, as expected, belong to Class III. However there are three compounds which appear to refer to semantically alienable body parts, and these three compounds, contrary to expectations, also belong to Class III.

(29) *an-dum-mija-wek*
III-eye-hair-little
'eyebrow/lash'

(30) *an-nabat-bidiyn*
III-finger-nail
'fingernail'

(31) *an-ngubat-bidiyn*
III-toe-nail
'toenail'

These three compounds may be contrasted with another body part compound referring to a semantically alienable part, which belongs to Class IV.

(32) *dum-mila*
eye-tear
'tears'

I do not have any immediate explanation for the apparently anomalous class membership of the compounds in (29)-(31). One possibility which may warrant further investigation is the hypothesis that these compounds are not in fact anomalously class marked (i.e. that culturally their referents are not viewed as alienably possessed).

5. Syntax of clauses involving body parts

The major areas of interest in the syntax of body parts are the "possessor ascension" pattern of cross-referencing (where the whole is cross-referenced rather than the part) and noun incorporation. These two phenomena allow us to define a "verbal" inalienable body parts class which differs from the nominal inalienable class. However before examining the differences between these two inalienable classes, I will first examine the structuring of "possessor ascension" and noun incorporation in detail. I will also examine the other clause type involving body parts which is unusually coded, subjectless transitive clauses.

5.1. Body parts in subjectless transitive clauses

"Subjectless" transitive clauses are clauses based on transitive verbal predicates with passive or medio-passive interpretations. Subjectless clauses involving body parts do not show any overt marking of the subjectless status of the clause. Thus a clause such as (33) may have either of the two interpretations given, depending on context.

(33) *an-nebe ban-dayn?-mi*
III-hand 1sgO-cut-Aux:PP
'He cut my hand', or 'My hand is cut.'

Subjectless situations involving other types of entities require either overt indefinite subject marking ((34) – in Warray the 3pl category also marks indefinite entities), or the reflexive (35), or the use of an intransitive verb ((36) and (37)).

(34) *bun-ba-bun-miyn ginganawu ga-ba-yu-yu*
3plO-indefS-bury-Aux:PP over:there NP-3plS-PR-lie:NP
'They are lying buried over there.' [Literally: 'Someone buried them and they are lying over there.']

(35) *gujili-ʔmi-yi-yn wik*
fill-caus-refl-PP water
'It is filled with water.'

(36) *wik juluk-miyn*
water spill-Aux:PP
'He spilled the water.' [*juluk-marl* 'to spill (transitive)']

(37) *wik juluk-jang*
water spill-Aux:PP
'The water has spilled.' [*juluk-jang* 'to spill (intransitive)']

5.2. Cross-referencing of body parts

There are three possibilities for the cross-referencing of part/whole entities in intransitive subject (S), or transitive direct object (DO) function in Warray. The normal pattern is for the whole to be cross-referenced as S or DO. This pattern of cross-referencing is commonly known as "possessor ascension". However there are two alternative patterns; the part may be cross-referenced as S or DO, or the whole may be cross-referenced as an indirect object. The different constructions code differences in the discourse status of the part, and affectedness of the whole.

The term "possessor ascension", which refers to the usual pattern of cross-referencing, implies that at some "deep" level of structure the clause is generated with the body part cross-referenced, and that the possessor then displaces it at some later level in the derivation of the clause. While I use the term "possessor ascension" in this paper in its well known sense, I do not subscribe to the implications of the term in any way. There is no evidence in Warray for preferring the more complicated possessor ascension hypothesis over the simpler base-generation of cross-referencing of the individual. Indeed there is evidence which directly contradicts the possessor ascension hypothesis. In Warray there are possessor ascension constructions where the part occurs as a free nominal with peripheral case marking ((45)-(47)). This would appear to pose insuperable problems for any theory of possessor ascension, as the body part, being a peripheral NP, is presumably never cross-referenced at any level of derivation.

The possessor ascension construction cannot be viewed as a reflection of the real world physical contiguity of body parts and their possessors

(Fox 1981: 323), as it is possible to so code "parts" which are not in any sense physically contiguous with their possessors.

(38) *an-nyi* *amala* *gan-ban-mitj-na-n*
III-name Neg IRR-1sgO-know-Aux-P
'He does not know my name.' [Literally: 'He does not know me name.']

Nor is it possible to view it as a reflection solely of the fact that the individual is affected by things which affect the part. The alternative construction where the whole is coded as an IO also indicates that the whole is affected.

(39) *ban-nat-dayn?-mi* *an-garra* *an-mu* *an-geng-u*
1sgO-IO-cut-Aux:PP III-leg III-bone III-new-OBL
gut-miyn
put-Aux:PP
'(The doctor) cut my leg open and put a new bone in.' [Literally: 'The doctor cut open the leg, affecting me, and put a new bone in', not: *'The doctor cut me open, the leg, and put a new bone in.']

(39) is not an example of possessor ascension. The individual is coded as an IO (see the literal translation), and not as a DO (cf. also (33)) – the *nat-* prefix indicates that the object is indirect, rather than a Direct one (see section 2).

The most plausible explanation for the "possessor ascension" construction would appear to be that it is a reflection of the fact that parts are dependent, non-individuated entities in discourse, whereas their individual possessors are independent and individuated (Hopper – Thompson 1984: 724-726). Therefore the individual possessors are cross-referenced. However if a part is individuated and independent in discourse terms, then the alternative constructions with either the part itself cross-referenced, or the whole cross-referenced as an IO may be used. The choice between these two alternatives depends partly on whether there is another entity involved or not. If only the part/whole entity is involved then the part will take cross-referencing.

(40) *an-nebe* *at-dayn?-mi* *ngek-u-wu* *ngek*
III-hand 1sgS-cut-Aux:PP 1sg-OBL-DAT 1sg

amala amba-yang ngek-nanak an-nebe at-dayn?-mi
Neg where-ABL 1sg-only III-hand 1sgS-cut-Aux:PP

'I cut my hand. Me, not anybody else, only me, I cut (my) hand.'

In (40) the speaker wishes to emphasise her own sole and individual responsibility for the action affecting her hand. The hand is therefore individuated in discourse terms, and is therefore coded as an independent patient object (the verb has not been reflexivised – cf. (1).

(41) *bulbul ban-lagi-yn an-dum*
sick 1sgO-toss-PP III-eye

ga-gurditj-gurditj-marl ngek-u-wu an-doy
NP-R-go around-Aux:NP 1sg-OBL-DAT III-heart

mutjla an-dum an-doy a-wa-rru
too III-eye III-heart I-bad-OBL

'I feel sick. My eyes are going round, my heart too. My eyes and heart are bad.'

In (41) the speaker, after making a general statement that she is sick, focusses on the particular individuated parts which are feeling bad. This construction with the part taking cross-referencing is also possible when there is another entity involved.

(42) *an-nyi ngek-u-wu ban-ganimup-miyn amala*
III-name 1sg-OBL-DAT 1sgO-forget-Aux:PP Neg

ø-gan-mitj-na-n
3sgO-IRR-know-Aux-P

'He has forgotten my name. He does not know it.'

In (42) the second VC cross-references the part (3sgO prefix ø). (42) shows that the discourse saliency/individuation of a part is very much a matter of the speaker's perspective. The first VC in (42) cross-references the individual. Discourse structuring patterns almost certainly have an important part to play in determining the cross-reference patterns shown by "parts". It is worth noting that in both (41) and (42), the whole is introduced in the initial clause, and is cross-referenced in that clause. It seems likely that an immediately preceding reference to the whole is an important factor in many situations involving cross-reference of the part.

There certainly do not appear to be any strong real world reasons for the change in cross-referencing patterns between the two VCs. In (42) the whole does not appear to be greatly affected by the activity of not knowing affecting the part. However if the whole is strongly affected, and another entity is involved then there is a strong tendency for the whole to be cross-referenced as an IO, as in (39) and the following examples.

(43) *ngek an-letma ban-nat-wuli-wul-iyn*
1sg III-tooth 1sgO-IO-R-finish-PP
a-gudangyi yirrʔ-mayim wul-iyn ngek-u-wu
I-doctor pull:out-Aux:PI finish-PP 1sg-OBL-DAT
'My tooth was finished. The doctor pulled it out. My (tooth) was finished.'

(44) *an-letma ban-nat-yirrʔ-mi a-gudangyi*
III-tooth 1sgO-IO-pull:out-Aux: PP I-doctor
'The dentist pulled out my tooth.'

IO cross-referencing may be found even if the VC is semantically passive (e.g. *ban-nat-wuli-wul-iyn* in (43)). IO cross-referencing is not obligatory, as a comparison of (43) and (44) reveals. This shows that coding the affectedness of the whole is also a matter of the speaker's perspective. In summary the evidence suggests the following cross-referencing patterns in Warray.

1. Normal discourse status of part: dependent, non-individuated; individual cross-referenced as S or DO.
2. Marked discourse status of part: individuated; part cross-referenced as S or DO (unmarked construction); and individual cross-referenced as IO (marked construction – when another entity used is involved and individual is affected).

Thus it appears that the major parameter controlling cross-referencing patterns is discourse status of the part. The usual low discourse status of body parts is coded by the possessor ascension cross-referencing pattern. If a part has a marked higher discourse status, then two constructions exist to code this marked status. One of these constructions, cross-referencing the part, is the unmarked way of coding this marked status, as it may code any situation where the part has a higher discourse status. The other way, cross-referencing the whole as an IO, is a highly marked construction that is only used when there is another entity involved and when the whole is affected.

This analysis of cross-referencing patterns being controlled by discourse factors is to some degree hypothetical. The lack of a reasonable textual corpus in Warray prevents full testing of the hypothesis, and no doubt such a corpus would indicate the need of a more precise refinement of the discourse factors affecting the cross-referencing of body parts.

5.3. Case marking and functions of body part nominals in possessor ascension constructions

The following examples demonstrate the possibilities for case marking of body parts in possessor ascension constructions.

(45) *ngirri ban-na-wa-m wang an-nebe-yang*
dog 1sgO-take:off-Aux-PP meat III-hand-ABL
'The dog took the meat out of my hand.'[8] [Literally: 'The dog took the meat off me from the hand.']

(46) *an-ngube-yi ga-bun-mi-yi-n yok-lik*
III-foot-INS NP-bury-Aux-refl-NP sand-LOC
'He is drawing with his feet in the sand.' [Literally: 'He is burying himself with the feet in the sand.']

(47) *nginyang an-ga-dirriʔ-dirri-marl an-bart-yi*
what 2sgS-NP-R-crawl-Aux:NP III-knee-INS
'Why are you crawling about on your knees?' [Literally: 'Why are you crawling about with the knees?']

(48) *an-bam at-bu-m nal*
III-head 1sgS-hit-PP man
'I hit the man on the head.' [Literally: 'I hit the man head.']

The thematic roles of the substantively case marked body parts in (45)-(47) are made explicit by those case markers. The semantic function and thematic role of the Absolutive [-ø] case marked body part in (48) is less explicit. Here the construction type with the body part taking Absolutive case, is in terms of frequency the most usual construction for coding body parts in possessor ascension constructions. Therefore further investigation as to the semantic function and thematic role of such Absolutive case marked body parts is warranted. On the basis of (48), it might appear that Absolutive case marked body parts could be described as having a locative thematic role. However, this characterisation does not account for all Absolutive marked body parts. We have already seen that body parts such as 'name' which are not readily characterisable as having a locative role occur in this construction type – see (38) above.

A good description of the semantic function of Absolutive marked body parts in possessor ascension constructions is provided by Halliday's notion of Range. Halliday (1985: 134) states that the Range "expresses the domain over which the process takes place". While 'name' in (38) is not readily characterisable as having a locative role, it certainly expresses the domain over which the person's not knowing occurs. Similarly physical body parts taking Absolutive case, as in (48), are characterisable as providing the Range of the process. This view of the most common function of body parts is in essential agreement with McGregor (1985: 210-211) "the body part specifies the EXTENT or LOCUS of the participant's involvement in the action. That is, it specifies the part of the individual which is most directly and intimately involved in the action."

It is important to note however that it is not clearly the case that the converse is true, that body parts expressing the Range of a process will take Absolutive case marking. The Ablative and Instrumental case marked body parts in (45)-(47) are also interpretable as providing the domain of the process. As Ablative and Instrumental case marking are apparently obligatory in (45)-(47) it would appear that for physically locatable body parts the usual concepts of thematic roles are relevant in determining case marking. This hypothesis is supported by the following example:

(49) *at-li-yn* *an-nebe-yi* *at-bu-yi-yn* *an-bam*
1sgS-fall-PP III-hand-INS 1sgS-hit-refl-PP III-head
'I fell and hit myself on the head with my hand.' [Literally: 'I fell and hit myself head with the hand.']

Here there are two body parts, one in an instrumental role and one in a locative role. Determination of the Range is problematic. Quite possibly both body parts are Ranges. Whatever the Range is in (49), however, it is the body part with the locative role which receives Absolutive case marking.

In summary it appears that independent body part nouns in possessor ascension clauses function semantically as Ranges. The usual case marking for body part Ranges is Absolutive. This is found with entities which are not physically locatable (in terms of the analysis proposed in section 3, when the person part is not a body part), as 'name' in (38) (the exact thematic role of 'name' in (38) is unclear to me). It is also found in the great majority of cases with physically locatable parts (that is, when the person part is a body part). However case marking patterns argue that Absolutive case-marked physical body parts should be interpreted as having a locative role in Warray. If the part is to be interpreted as having another role it will receive overt case marking to mark this. These interpretations are to some extent language specific – witness the different thematic roles assigned in (47) to the body part by English and Warray.

If Absolutive case-marked physical body parts are to be interpreted as having a locative thematic role, this naturally raises the question of why they do not instead receive Locative case marking. An answer to this question requires a consideration of the syntactic function of Absolutive marked part nouns. Firstly, part nouns only take Absolutive case marking when the whole is in an S or DO function, where possessor ascension constructions are possible. When the whole is in a transitive subject function, possessor ascension constructions are not possible, and the part does take Locative case marking (see (68b)).

As such it is clear that the use of Absolutive case marking for part nouns in a locative thematic role does not result from some inherent lexical prohibition on the use of the Locative case marker with part nouns. Rather, following standard practice, it would appear most reasonable to analyse the occurrence of Absolutive case marking on part nouns as indicating that they form part of the S or DO NP. This in turn raises the question of the function of the part nouns in these NPs. This question is a very complex one, and involves issues concerning the nature of predication and modification – issues which are problematic and much debated. Consequently only a provisional answer may be given.

Hale (1981: 338) suggests that the body part is predicated as a secondary predicate of the whole "in the sense that what is true of the PART is seen to be true of the WHOLE". However, as McGregor (1985: 212) points out, what is true of the part is not necessarily true of the whole. In (38) the fact that the person does not know my name, does not necessarily mean that he does not know me. As McGregor further observes it is by no means clear on general grounds that the part/whole relationship involves predication. It certainly does not involve attribution or equation. Napoli (1988a: 7) analyses a predicate as an event which has role players. Within such an analysis of predication, part nouns would presumably have to be analysed as "events", with the wholes somehow being "role players". While such an analysis is not impossible, it is not immediately suggested.

The relationship between the part and the whole would appear to be more akin to modification than predication. It delimits the reference of the whole, in a manner similar to the way that prototypical modifiers such as adjectives are understood to delimit the reference of their head. Provisionally it is most economical to analyse Absolutive case-marked part nouns as modifiers, normally locative modifiers. Absolutive case marked part nouns are semantically and functionally similar to prototypical modifiers, and they show the same case marking pattern as prototypical modifiers. However it is necessary to recognise that the distinction between predication and modification is in itself problematic, most especially in relation to non-configurational languages, such as Warray. It is highly likely that further work on these topics within non-configurational languages will require review of the analysis posited for Absolutive case marked part nouns.

5.4. Noun incorporation and denominal verbs

Before proceeding with discussion of the particulars of noun incorporation in Warray it is necessary to consider the nature of noun incorporation, and the distinction between it and denominal verb formation.

Sapir (1911: 257), in the classic discussion of noun incorporation, defines it as "this process of compounding a noun stem with a verb that it is here proposed to call noun incorporation, no matter what the syntactic function of the noun logically is." Mithun (1984: 848) defines noun incorporation as "a particular type of compounding in which a Verb and Noun combine to form a new Verb. The Noun bears a specific relationship to its host Verb – as patient, location or instrument."

In effect Mithun defines "Noun incorporation" as the compounding of a noun with a verb to form a new verb subject to the condition that the noun must bear a specific argument relationship to the verb. Sapir is apparently in general agreement with this, stating (1911: 257),

> This type of verb, "to song-write" ... is best considered a particular class of the more general type of noun-verb compound verb. As a matter of fact, it is often just as difficult, at least in some American languages, to draw the line between the objective and non-objective use of an incorporated noun as it is to determine the precise syntactic value of the qualifying member of a compound noun. Thus "I hit his face" may often be interpreted locatively as "I hit him in the face", while even so transparent an example as "I eat meat" may often be understood instrumentally as "I feed on or with meat".

It appears that Sapir is operating on the implicit assumption that there will be an argument relationship between the noun and the verb. However he adopts a rather asystematic and lexicalist approach to the process of noun incorporation. Sapir and Mithun make the following statements in connection with denominal verb formation:

> Many American languages form denominative verbs from noun stems by means of various derivative affixes of verbal, generally transitive, meaning. Thus, from Paiute *qani-* "house" are formed *qanintcu-* "to build a house" and *qanix*^{y}*ai-* "to have a house", from Yana *hauyauba-* "deer fat" is formed *hauyauba*$^{\varepsilon}$*–inigui*$^{\varepsilon}$*a-* "to contain nothing but deer fat". In these derivative verbs the nouns "house" and "deer fat" can not be considered as incorporated, for the verbal elements *-ntcu-*, *-x*^{y}*ai-*, and *-i*$^{\varepsilon}$*nigui*$^{\varepsilon}$*a-* are not verb stems but verb-forming affixes morphologically comparable to English *-ize* in verbs of the type *materialize, pauperize.* (Sapir 1911: 254)

> [D]enominal verb formation is a different formal process. In noun incorporation, as commonly understood since Sapir 1911, a noun stem is compounded with a verb stem to yield a more specific, derived verb stem. The Greenlandic [Eskimo] construction is based on a single noun stem with a derivational suffix. It is not entirely clear why one would refer to this as noun incorporation, since it is not obvious what such nouns are incorporated into. In incorporating languages, a verb minus its incorporated noun is still a well-formed verb; but in Greenlandic, a denominal verb minus its noun stem would be no word at all. (Mithun 1986: 32)

It would appear that Mithun and Sapir are in substantial agreement, that if the verbal morpheme in a morphologically analysable verb stem, involving a noun root, cannot stand alone as a well formed verb, then the whole stem should be regarded as a denominal verb stem as opposed to a compound verb stem involving noun incorporation. This criterion, a negative criterion for noun incorporation, is intuitively appealing.

However Warray, and other Gunwinjguan languages, provide two major problems in relation to the use of this "independent occurrence" negative criterion for noun incorporation. These problems arise in relation to what is described as "lexical incorporation" (see Evans 1995, and Merlan 1983: 143-144). One problem involves verb stems which are morphologically analysable as consisting of noun + verb where both the noun and the verb can occur independently, but which one would wish to exclude from the scope of an analysis of noun incorporation by the positive definition of noun incorporation as being the compounding of noun + verb, with an argument-like relationship between the noun and the verb, that Mithun and Sapir are implicitly using.

As mentioned in section 2, Warray has two main types of verbs: simple verbs consisting of a verb root; and compound verbs consisting of a nominal or verbal root and an auxiliary. Most auxiliaries also independently occur as simple verbs. The essential function of the auxiliary system is to verbalise nominal or verbal roots (see further Harvey 1990). Denominal compound verbs are extensively and productively formed by the auxiliary system. Thus the independent verbs *bu-m* 'to hit', and *lam* 'to spear' function as auxiliaries to verbal roots:

(50) *nyi-bu-m* 'to cry' *gi-lam* 'to pour'

Bu-m functions productively as a factitive denominaliser meaning 'to make noun X', as in (51):

(51) *wek-bu-m* 'to make a fire' (from *wek* 'fire')

lam is also found in denominal compounds, though not productively.

(52) *giliyn-lam* 'to line up' (from *an-giliyn* 'line')

There is every reason in Warray to analyse these noun + verb compounds as classing with such prototypically denominal forms as the adjectival inchoative *-ji-yn*.

(53) *gamu-ji-yn* 'to become tough' (from *-gamu* 'tough')

The inchoative *-ji-yn* is an auxiliary, parallel in form and function to all other auxiliaries. Its only marked feature is that it verbalises nominal roots, whereas other auxiliaries chiefly verbalise verb roots. Nevertheless (51)-(53) are all Nominal Root + Verbalising Auxiliary constructions. The fact that *bu-m* and *lam* also occur as independent verbs is irrelevant. In all three cases the nominal provides the main predicate information in the compound, whereas the auxiliaries provide only very limited predicate information (it should be noted that there is no correlation between the auxiliary and independent meanings of *bu-m* and *lam*).

Following the "independent occurrence" criterion with respect to (51) and (52) would produce a most unsatisfactory result, as they would presumably have to be analysed as examples of noun incorporation. They are clearly denominal verbs formed by using the auxiliary system, whose major function in Warray is as a verbalising system. If the positive criterion requiring an argument relationship between the noun and the verb is followed then (51) and (52) will be excluded from the scope of noun incorporation.

The second major problem for "independent occurrence" involves verb stems which are morphologically analysable as consisting of noun + verb, where the verb is not attested independently, but which one would wish to analyse as being examples of noun incorporation because there is an argument-like relationship between the noun and the verb.

(54) *dum-bayʔ-marl*
eye-open-Aux
'to open one's eyes'

(55) *bam-berrngberrng-marl*
head-shake-Aux
'to shake one's head'

Bayʔ-marl and *berrngberrng-marl* are not found independently of the compound forms given ('to open' is otherwise expressed by the verb *del-bu-m*, and 'to shake' is otherwise *dolʔdol-marl*). Therefore these verbs must be analysed as denominal verbalisers by the independent occurrence criterion. In connection with this point Mithun in her discussion of noun incorporation (1984: 847-848) states "Interestingly, all languages which exhibit such morphological structures also have syntactic paraphrases. If we know that, in Koryak, one can say *tiqoyanmatekin* 'I-reindeer-slaughter' then we can correctly predict the existence of a sentence like *Tinmekin qoyawge* 'I-slaughter reindeer'."

It is not clear what status Mithun attributes to this statement. It may be a statement about a general tendency in the majority of cases. Alternatively it may be a qualification on the interpretation of noun incorporation in general (i.e. if a compound construction does not have a syntactic paraphrase then it is not to be treated as an example of noun incorporation). The following statements in Mithun (1986: 33) suggest that the first alternative is intended: "Speakers of an incorporating language may never hear or produce sentences about hand-washing where the noun for 'hand' appears outside of the verb 'wash', and may even find such sentences odd, because the situation calling for them would be odd. This does not mean that noun incorporation is arbitrary."

I see no principled reason why the existence of a syntactic paraphrase should be taken as a diagnostic for noun incorporation. The compounds in (54) and (55) arc both formally and functionally parallel to noun incorporation forms, and are quite distinct from the denominal compounds illustrated in (51)-(53). In the former pair both the noun and verb make significant contributions to the lexical meaning of the compound, and there is an argument relationship between the noun and the verb.

Overall for Warray it would appear that use of the formal "independent occurrence" criterion produces unsatisfactory results both from the point of view of positively defining denominal verb stems and from the point of view of negatively defining noun incorporation constructions. The positive criterion requiring an argument relationship between the noun and the verb on the other hand does provide satisfactory results.

Therefore any noun + verb compound, where the noun and the verb have an argument relationship will be treated as an example of noun incorporation. Noun + verb compounds which do not show an argument relationship will be treated as examples of denominal verb formation. This distinction correlates with a distinction in the lexical nature of the two compound types. In noun incorporation the lexical meaning of the compound is componential from the meaning of the noun and the verb. In denominal verb formation the verb is leached of its meaning as an independent form and makes only a minimal contribution to the lexical meaning of the compound. This explains why in Warray, as in other Gunwinjguan languages, denominal compounding shows a much greater degree of lexicalisation than noun incorporation compounding does.

Occasionally the boundary between noun incorporation and denominal verb formation may be somewhat difficult to draw. The verb *woy* occurs both as an independent verb meaning 'to give', and as a productive denominalising causative auxiliary.

(56) *dilmi-woy*
dry-caus
'to dry something'

When *woy* compounds with an adjective as in (56), its denominalising function is clear. However the analysis of the following compound on the noun *an-nyi* 'name' is more problematic.

(57) *an-nyi* *ba-nyi-wu-ji-yn-ʔlul*
III-name 3plS-name-caus/give-recip-PP-pair
'The pair of them gave each other their names.'

Transferral of names, under a system known as *ngirrwart*, was an important feature of Warray social organisation traditionally. The compound verb in this clause describes this *ngirrwart* system, and could be interpreted as a noun incorporation compound with the sole meaning 'to bestow/give a name'. Alternatively it could a causative denominal compound equivalent in meaning to the English denominal verb 'to name'. The English denominal verb means both 'to bestow/give a name' and 'to say/speak a name'. The following example shows the Warray compound verb *nyi-woy* has the same range of meaning as 'to name'.

(58) *amala gan-a-nyi-wu ju-m gaji-n*
Neg IRR-2plS-name-caus:NP die-PP that-PRM
'Don't name him! He has died that one!', but *'Don't give him a name! He has died that one!'

The second translation is not infelicitous merely in English; the Warray verb *woy* 'to give' requires transferral of the patient/theme entity, just as much as the English verb *to give* does. Clearly no transferral occurs in (58), and therefore the *woy* in the compound verb in this example is not acting as the verb 'to give', but must rather be acting as a causative auxiliary.

Warray does not normally productively incorporate patient objects (cf. (59)-(61)). Given that the compound *nyi-woy* conforms formally to a productive denominalising pattern (56), there is good evidence that in (58) the denominal causative interpretation is the correct interpretation of the compound verb.

5.5. The structure of noun incorporation

Noun incorporation in Warray is of general typological interest as it provides a clear counter-example to the generalisation of Mithun (1984: 875) "If a language incorporates nouns of only one semantic case, they will be patients of transitive verbs ... If a language incorporates only two types of arguments, they will be patients of transitive and intransitive verbs ... Many languages additionally incorporate instruments and/or locations".

Warray does have productive noun incorporation, but this is essentially adverbial/locative in nature, as we shall see in this section. Apart from one set of exceptions, Warray does not productively incorporate patient objects (cf. (57)), nor is there any evidence that it ever did so productively. The one set of exceptions presents a somewhat unusual aspect. All the exceptions involve the lexeme *warri* 'child', and they are all concerned with various aspects of reproduction, both physical and social.

(59) *amala gaji bipi ngek-u-wu ban-warri-gut-miyn*
Neg that father 1sg-OBL-DAT 1sgO-child-find-Aux:PP
'No that was my father who found me as a child (spirit).'

(60) *ba-warri-jalak-miyn*
3plS-child-show-Aux.PP
'They made him a father to the child.' [Literally: 'They showed him the child.']

(61) *girrangajerriyn ba-warri-gut-mi-yi-yn*
three 3plS-child-bear-Aux-detr-PP
'They had three children.'

These examples present problems in translation as they relate to Aboriginal conceptualisations of conception and parenthood, which lack analogues in European social organisation. Further my research in these areas is incomplete. Consequently it should be noted that the translations given for (59)-(61) are approximate. (59) relates to the very widespread set of conceptualisations concerning the creation of social individuality by the finding of spirit children (Berndt – Berndt 1985: 150-153). The primary meaning of the verb *gut-marl* in (59) is 'to put down'. However it has a wide range of secondary meanings concerned with reproduction: 'to give birth to', 'to lay', 'to be the progenitor of', 'to find the spirit of'. This verb is found in a detransitivised form in (61), where it appears to have a lexicalised meaning 'to have offspring'. (60) also involves lexicalisation, as a comparison of the literal and free translations reveals. ((60) describes a social process which broadens the range of close parenthood, somewhat akin to adoption.)

The productive incorporation of *warri* 'child' in an apparently patient/theme role is restricted to clauses expressing these reproductive concepts and practices in Warray. *Warri* cannot apparently be incorporated in a patient/theme role in clauses expressing other concepts and practices (e.g. it appears that the incorporated structure (60) could not be used in a clause such as "They showed the child to its grandparents."). Further investigation is required for a more complete understanding of the incorporation structures illustrated in (59)-(61).

The highly productive incorporation of body parts in Warray appears to be of a rather different nature to the incorporation of *warri*. Mithun (1984: 858) treats incorporation of body parts as an aspect of what she describes as noun incorporation Type 11: "Noun incorporation of body parts allows affected persons to assume a primary case role, such as subject or direct object, rather than merely oblique possessor." Essentially Mithun argues that incorporation of body parts occurs to permit possessor ascension. This is incorrect. While noun incorporation requires the possessor ascension pattern of cross-referencing, possessor ascension does not require noun incorporation. In Warray, and in the other northern prefixing languages of Australia, it is quite possible to cross-reference the whole without incorporating the part. Therefore incorporation cannot be said to license possessor ascension.

Incorporation of body parts in Warray is not restricted to animates:

(62) *mala-wuli-yi-yn* *yumbal*
leaf-finish-detr-PP tree
'The tree has lost its leaves.' [Literally: 'The tree has leaf-finished.']

(63) *an-jili* *at-jili-del-bu-m*
III-mouth 1sgS-mouth-open-Aux-PP
'I opened the tin lid.' [Literally: 'I lid-opened it, lid.']

Normally only one noun can be incorporated into the VC, but there are a few examples of incorporation of two noun roots.

(64) *bat-du-dum-bam-birri-miyn*
1sgS-PR-eye-head-feel:dizzy-Aux:A
'I feel dizzy.'

(65) *at-miyn-gubulung-yu-ng*
1sgS-belly-down-lie-PP
'I lay belly down.'

The degree of lexicalisation and internal structure of these forms are uncertain. The compound *miyn-gubulung* 'belly down' occurs independently (as does *gubulung*). This suggests that (65) involves the incorporation of a compound noun. The particular combination of noun roots in (64) is not found elsewhere, and it does not, in any obvious sense, display the usual modified-modifier structure of nominal compounds (see section 4). As such it would seem likely that it should be analysed as a lexicalised combination.

In Warray incorporation is possible whenever the whole noun is an S or DO:

(66) *at-nabat-gamu-ji-yn*
1sgS-hand-stiff-inch-PP
'My hand is stiff.' [Literally: 'I became hand-stiff.']

(67) *bun-lorrʔ-may*
feather-pluck-Aux:IMP
'Pluck the feathers out!' [Literally: 'Feather-pluck it!']

It is not possible to incorporate when the whole is a transitive Subject. Thus 'I carried the child on my shoulder' can only be expressed by (68b), not by (68a).

(68) a.	**at-murnak-wuk-mi*	*an-wak*
	1sgS-shoulder-carry-Aux:PP	III-little

(68) b.	*at-wuk-mi*	*an-wak*	*an-murnak-lik*
	1sgS-carry-Aux:PP	III-little	III-shoulder-LOC

(68a) could only have the rather bizarre meaning 'I carried the child's shoulder'. In clauses based on an intransitive verbal predicate with an IO, incorporated forms are taken to refer to the S.

(69)	*gakuy-u*	*an-ga-nat-gurrung-wayʔ-way-gayi-n*
	after-DAT	2sgS-NP-IO-arm-R-wave-Aux-NP

'Later you will wave to him.'

There are no definitive examples of incorporation on a transitive verb with an IO in the data available. The impossibility of incorporation on transitive Subjects appears to be a language universal restriction (Baker 1988: 81). Baker argues that the universal restriction on transitive Subjects is to be explained in terms of the Unaccusative Hypothesis. The Unaccusative Hypothesis (Perlmutter 1978: 162-163) divides intransitive verbs into two classes along essentially semantic lines. Verbs describing willed or volitional acts are unergative verbs with an underlying subject. Verbs describing other types of intransitive events are unaccusative verbs with an initial underlying object, which in Government Binding Theory moves to subject position at surface structure. The exact membership of the two classes varies language specifically.

Baker (1988: 88-89) argues that only unaccusative verbs permit incorporation. Therefore it is possible to explain the restriction on transitive Subjects as following on a universal restriction of incorporation to what in Government Binding theory are deep structure objects. While Unaccusativity may be relevant to the distribution of incorporation in some languages there is no evidence that it is of any relevance in Warray.

Firstly, as Napoli (1988b: 141-142) does, I do not view the unaccusative distinction as being a language universal property. I view it as a language specific property requiring evidence to substantiate it. There is no evidence that anything resembling the unaccusative distinction, as it is commonly understood in Government Binding theory (cf. Burzio 1986), operates in Warray. The prefixing patterns of intransitives do not correlate in any significant way with the unaccusative hypothesis. There are two verbs *ngarndi-dep-marl* 'to be thirsty' and *wulwul-mi* 'to ache' which take the object prefixes to cross-reference their inherent experiencer role. These two verbs are certainly unaccusative in the terms of Perlmutter's original formulation and the fact that they take object prefixing is clearly related in some way to this fact. However all other intransitive verbs take subject prefixing to describe their inherent roles, no matter what the thematic status of that role.

Neither does the pattern of incorporation among intransitives correlate in any way with the unaccusative hypothesis. Highly agentive, volitional and controlled intransitives may incorporate (cf. also (69)).

(70)	*at-nabat-nyim*	*an-bokbok-u-lik*	*yumbal-lik*
	1sgS-hand-enter	III-hollow-OBL-LOC	log-LOC

'I put my hand into the hollow log.' [Literally: 'I hand-entered the hollow log.']

Unless one is to take the vacuous position that all intransitives are unaccusative, it is clear that the unaccusative hypothesis is not of relevance in explaining the distribution of incorporation in Warray. Rather an explanation for the distribution of incorporation must be sought from the nature and functions of incorporation.

In Warray body parts form by far the most frequently incorporated class of nouns. This conforms with the observable generalisation for incorporating languages in Northern Australia that the class of incorporable nouns always includes body parts (Heath 1978: 118, 1980: 86, 1981: 280-201, 1984: 472; Merlan 1983: 144; McKay 1975: 299; Osborne 1974: 50). This is also observable for languages elsewhere in the world (Nahuatl – Merlan 1976: 188). In some languages only body parts are incorporable (Muskogean – Haas 1941, Ngan.gi Kurung.gurr – Reid, pers. comm.). These facts argue that body part incorporation is a primary rather than a secondary type of incorporation as Mithun treats it.

The possibilities for incorporation other than of body parts are much less extensive in Warray than in some other Gunwinjguan languages, such as Mayali (Evans 1995) or Rembarrnga (McKay 1975: 287-309). Apart from body parts, the following types of incorporation are attested (the square brackets indicate the apparent thematic role of the incorporated noun).

(71)

gijirriyn	*yul-lik*	*yul-del-bu-m*	*gaji*
bird:species	ground-LOC	ground-open-Aux-PP	that

gut-miyn	*an-gundu*	*gakuy*	*an-wak*	*ngal?-jim*
lay-Aux:PP	III-egg	after	III-little	out-come

'The *gijirriyn* bird burrowed into the ground and laid its eggs. Afterwards the chicks came out.' [locative]

(72)

gaji	*warri-ba-?lul*	*bin-ba-wili-ngiw-a*	*wili–lik*
that	child-PL-pair	3plO-2plS-house-put:in-IMP	house-LOC

'You mob put those two children in the house!' [locative]

(73)

at-dumgiga-gut-miyn	*an-wak*	*ga-yu-yu*	*dumgiga*
1sgS-sleep-put-Aux:PP	III-little	NP-R-lie:NP	asleep

'I put the child to sleep. He is lying asleep.' [?]

(74)

an-mewel	*at-windi*	*at-mirral-lagi-yn*
III-clothes	1sgS-hang:out	1sgS-sun-put-PP

'I hung the clothes out in the sun.' [locative]

(75)

at-wik-lagi-yn	*an-mewel*	*wik-lik*	*gaku*
1sgS-water-toss-PP	III-clothes	water-LOC	later

bat-wurlek-miyn
1sgS:NP-wash-Aux:A

'I tossed the clothes in the water. Later I will wash them.' [locative]

(76)

yow	*at-windi-ny-iyn*	*mirral-lik*	*ga-mirral-nu-n*
yes	1sgS-hang:out-UNM-PI	sun-LOC	NP-sun-burn-NP

'Yes I hung them (the clothes) out in the sun. They are drying in the sun.' [?locative, ?instrumental]

(77)

ngirri	*wang*	*ban-nabat-na-wa-m*	*an-nebe-yang*
dog	meat	1sgO-hand-take:off-Aux-PP	III-hand-ABL

'The dog took the meat out of my hand.' [ablative]

(78) *gimiyn-gurditj-ma*
point-go:around-Aux:IMP
'Go around via the point!' [perlative]

(79) *wik ga-bul-murluk-murluk-marl*
water NP-water-R-wave-Aux:NP
'The water is waving.' [?theme]

The incorporated nouns appear to have a variety of thematic roles. However, as with the independent body part nouns discussed in section 5.3, they are all characterisable in terms of Halliday's (1985: 134) notion of Range. Therefore it would appear that incorporation in Warray can be analysed in a simple and unified manner as the incorporation of Ranges. The majority of the non-body part Ranges appear to have a locative thematic role. In 5.3 we saw that thematic roles are relevant in determining the case marking of independent body part nominals (locatives take Absolutive case, other roles take the appropriate case marking). Thematic roles are also relevant in controlling incorporation.

(80) *ngek amala an-nebe-yi at-gibat-derret-bu-yi-yn*
I no III-hand-INS 1sgS-back-scratch-Aux-refl-PP
'Not me, I scratched my back with my hand.'

In (80) there are two body parts involved in the process: an instrumental (hand), and a locative (back). Both can be interpreted as Ranges. However it is the part in a locative role which is incorporated. A similar situation is found in:

(81) *at-jili-dangʔ-miyn*
1sgS-mouth-click:tongue-Aux.PP
'I clicked my tongue.'

The two body parts involved in the process of clicking one's tongue are the tongue itself and the mouth. The mouth is in a locative role, and the tongue would appear to be interpretable as having either an instrumental or a locative role. As such it would appear that both 'mouth' and 'tongue' should be potentially incorporable. Incorporation structures freely given by my consultants involved only 'mouth'. When presented with incorporated structures involving 'tongue', they stated that these would also be acceptable. However given that presentation of constructed examples is a rather doubtful elicitation process, only a limited credence may be given to these statements. Even if forms with 'tongue' incorporated are acceptable, the evidence still indicates that an unambiguously locative part is preferred for incorporation.

In overall terms incorporation in Warray may be analysed as being the incorporation of Ranges. When there is more than one possible Range, the one with a locative role is the one incorporated. Analysing incorporated body part nouns as Ranges explains why in transitive constructions they are interpreted as being part of the object, rather than of the subject. As they express the domain over which the process operates they must be part of the object, rather than of the subject.

The relationship between the incorporated noun and the verb appears to be one of modification. The incorporated noun delimits the verb in a manner akin to that of adverbs. In this context it is important to note that, apart from the nouns so far discussed, the only other incorporable roots in Warray are certain roots with classical adverbial functions (e.g. *wok* 'heavily'). Therefore noun incorporation in Warray appears to be analysable as a particular type of a more general compounding process of verbs with their modifiers. Compounding of modifiers is the prototypical compounding pattern, and as such noun incorporation in Warray falls naturally within the general parameters of compounding.

The patterning of incorporation in Warray is in general accord with a variety of proposals for a thematic hierarchy controlling the order in which roles compose semantically with the verb (see Bresnan – Moshi 1990: 169).

Agent > Benefactive > Goal > Instrument > theme/patient > Locative

Warray permits incorporation only with the role lowest on the hierarchy (with a set of exceptions involving the next lowest role – see (59)-(61)). The nature of body part incorporation in other languages is an empirical question requiring further investigation. It is however interesting to note that body parts appear universally to be the prototypical incorporable nouns. If body part incorporation is universally that of Ranges with a preference for locative Ranges, then this would accord with the hierarchy proposed by Kiparsky (1989).

Warray while providing some problems for Mithun's analysis of noun incorporation is, as we have indicated, compatible with her thematic role based approach. What is required within Mithun's approach is a re-analysis of the semantics of body part incorporation, and of its status as a secondary form of noun incorporation. The evidence from Warray (and other languages) argues that body part incorporation is the prototypical primary form of incorporation. The evidence from Warray suggests that this is so because the body parts of subcategorised entities are prototypically backgrounded Ranges, which modify the verb.

The organisation of noun incorporation in Warray appears to be fundamentally incompatible with Baker's abstract structure based approach. Baker (1987: 2) sets out the generalisation "A noun can be incorporated into another category in the system of a polysynthetic language only if a noun phrase headed by that noun would be the sister of the category in the phrase structure system of an isolating language." Baker (1987: 6) specifically states "The claim that locatives do not incorporate also needs clarification ... Indeed, there are examples, but to the best of my knowledge they are fairly few and limited".

Baker excludes body part incorporation from his purview (Baker 1987: 3), because "their role in the sentence as a whole is not always clear". This paper has demonstrated that body part incorporation is a regular process in Warray. Nevertheless this does not constitute a direct counter-example to Baker's arguments on the structure of noun incorporation. However the non-body-part locationals exemplified in (71)-(79) are direct counter-examples to both the general and the specific statements that Baker makes about noun incorporation. There would not appear to be any way of generating these examples within Baker's system.

While it is possible to determine the controls on what is potentially incorporable in Warray, it is not possible to set out the controls determining when incorporation will actually occur. It appears that certain incorporations are obligatory (e.g. *dum-bay?-marl* 'to open one's eyes' and *bam-berrngberrng-marl* 'to shake one's head'). However most incorporations are optional and it is probable that discourse factors (independent occurrence as new information vs. incorporation as old information – cf. Merlan 1976) play an important role in controlling the actual occurrence of noun incorporation. However the lack of a suitable textual corpus prevents full comment on this aspect of noun incorporation in Warray.

In Warray the incorporation of a noun is no bar to its also occurring as a free form in the same clause.

(82)	*an-nebe*	*at-nabat-nyim*	*yumbal-lik*
	III-hand	1sgS-hand-enter	log-LOC

'I put my hand in the log.' [Literally: 'Hand, I hand-entered the log.']

(83) *a-wa-rru* *an-jen* *ga-jen-bil?-marl* *bigirring-u*
I-bad-OBL III-tongue NP-tongue-poke-Aux:NP 3pl-DAT
'(He is) bad. He is poking his tongue out at them.' [Literally: 'He is bad. Tongue, he is tongue-poking out at them.']

This type of construction is reasonably common in Warray, but the nature of the contrast between the three possible types of possessor ascension clauses is not entirely clear.

(84) *an-nebe* *at-dayn?-mi-yi-yn*
III-hand 1sgS-cut-Aux-refl-PP
'I cut my hand.'

(85) *an-nebe* *at-nabat-dayn?-mi-yi-yn*
P-hand 1sgS-hand-cut-Aux-refl-PP
'I cut my hand.'

(86) *at-nabat-dayn?-mi-yi-yn*
1sgS-hand-cut-Aux-refl-PP
'I cut my hand.'

(84)-(86) convey the same proposition. It is highly likely that the differences between the three are discourse based. Merlan (1976: 177) in her seminal discussion of the discourse status of noun incorporation states

> In terms of discourse, it is possible to show that incorporated nouns serve to maintain definiteness of discourse reference by functioning as anaphors which maintain co-reference with previously introduced lexical nouns. Because the lexical properties of nouns are preserved under incorporation, noun incorporation functions as a strong reference-maintaining device intermediate between complete repetition of the co-referential adjunct and complete anaphoric pronominalisation.

It is likely that (84)-(86) represent finely graded distinctions in the possibilities for discourse reference maintenance in Warray. From observation it appears that independent occurrence functions in a manner akin to topicalisation, with (84) and (85) representing finely graded distinctions in the degree of prominence. They are normally clause initial and set off by a slight pause. However the relationship between the boundaries of intonation groups, the placement of pauses and putative clause boundaries is complex in Warray. It is definitely not possible to formally distinguish a topicalised clause structure from an ordinary clause structure in terms of the patterning of intonation group boundaries and/or the placement of pauses. Thus the status of independent part nouns in possessor ascension constructions is presently somewhat uncertain.

5.6. The syntax of body part constructions and inalienability

In section 3 we saw that noun class marking divides body parts into an inalienable class, which takes Class III marking, and an alienable class, which takes Class IV marking. Possessor ascension and incorporation also allow us to define alienable and inalienable classes. However these "verbally" defined classes do not correspond exactly to the nominally defined classes. Part nouns which belong to the nominal alienable class take possessor ascension cross-referencing.

(87) *gijewek ban-ba-dayn?-mi*
beard 1sgO-3plS-cut-Aux:PP
'They cut my beard.' [Literally: 'They cut me beard.']

These nouns may also be incorporated.

(88) *bat-du-dum-mila-jup-jup-miyn*
1sgS:NP-PR-eye-tear-R-flow-Aux:A
'My tears are flowing.' [Literally: 'I am tear-flowing.']

Conversely the noun *an-mewel* 'clothes', which belongs to the inalienable nominal class does not fall within the verbal inalienable class. There are no examples of this noun being incorporated and the prefixes invariably refer to it, rather than to the possessor.

(89) *wek at-dirrim-bu-m ngikbayang amala gat-na-n*
fire 1sgS-light-Aux-PP morning neg IRR:1sgS-look-P
wek-yi no-y ngek-u-wu an-mewel
fire-ERG burn-PP 1sg-OBL-DAT III-clothes
'I lit a fire this morning, (but) I did not look out and the fire burnt my clothes.'

It is possible to represent the action as affecting the whole, but only by cross-referencing the whole as an IO, not as a DO.

(90) *an-mewel ban-nat-lala?-mi*
III-clothes 1sgO-IO-tear-Aux:PP
'He tore my clothes.' [Literally: 'He tore the clothes affecting me.']

It would appear that the differences between the nominal and verbal inalienable classes follow from the fact that they are concerned with very different things. Nominal inalienability is concerned with distinguishing the body parts that constitute non-replaceable parts of a person's physical and social identity. Verbal inalienability is concerned with coding the fact that persons have a high discourse status, whereas the parts of persons usually have a low discourse status, irrespective of whether they are non-replaceable parts of a person's identity or not.

It is therefore predictable that body parts treated as alienable by class marking will belong to the verbal inalienable class. The lexeme *an-mewel* 'clothes' presents the converse case. It is class marked as nominally inalienable, but is treated as alienable verbally. Given that it is the only example of its type, any analysis is of necessity somewhat tentative. Nevertheless it is worth noting that 'clothes' as an entity are rather more independent of the individual, than other body parts are. Consequently they do appear to have a higher discourse status than other parts, which would in turn explain why they are verbally marked as alienable.

6. Conclusion

Body parts play an important role in grammatical coding in Warray. The noun class system has as one of its major contrasts the distinction between replaceable and non-replaceble body parts. The most frequent type of nominal compounding is compounding of body parts. There are two classes of body part compounds. One class comprises body part lexemes. The other class comprises compounds of inalienable body parts which describe enduring conditions of the whole. This second class has a wide range of use in its synecdochic function, with many natural species names being compounds belonging to the class.

The system of clause level grammar presents a number of ways of coding part/whole relations. Body parts, normally having a low discourse status, are not usually cross-referenced in the verbal complex, and indeed are frequently incorporated into the verbal complex. They generally function as Ranges in clause level grammar. We have argued that body part incorporation should also be viewed as being Range incorporation, with a preference for locative Ranges. While body parts usually have a low discourse status, and are preferably interpreted as locative Ranges, there are ways of grammatically coding departures from either of these states of affairs.

The alienable/inalienable opposition is an important parameter in discussing the grammar of body parts in Warray. A number of oppositions and coding patterns can be characterised in terms of this opposition, though as we have seen, nominal and verbal inalienability are not the same thing.

The important point to be noted about both types of inalienability, and about the grammar of body parts in general in Warray, is that the prototypical reference whole for the concept of body part is not the 'body' as such. Rather it is a concept akin to that of 'person', involving factors of humanness and social interaction in terms of distinguishing individual characteristics. The grammar of body parts in Warray cannot be properly understood without this central insight.

Notes

1. I would like to thank Hilary Chappell, Bill McGregor, Francesca Merlan and Jane Simpson for comments on earlier versions of this article. Naturally responsibility for any errors or omissions remains my own.
2. The following orthography is used: long and syllable final stops – *p* bilabial, *t* alveolar, *rt* retroflex, *tj* palatal, *k* velar, *ʔ* glottal; other stops – *b* bilabial, *d* alveolar, *rd* retroflex, *j* palatal, *g* velar; nasals – *m* bilabial, *n* alveolar, *rn* retroflex, *yn* syllable final palatal, *ny* other palatal, *ng* velar; laterals – *l* alveolar, *rl* retroflex; rhotics – *rr* alveolar, *r* retroflex. In addition, *n.g* represents the apical nasal followed by the velar stop; *r.n*, the retroflex rhotic followed by the apical nasal; *y.n*, the semivowel *y* followed by the apical nasal.
3. These two forms are dialectal variants.
4. Abbreviations used in the glosses are: A – Set A suffixes (1st person Non-Desiderative/Non-Hortative Future); ABL – Ablative; Aux – Auxiliary; caus – causative; DAT – Dative; detr – detransitiviser; DO – Direct Object; IMP – Imperative; inch – inchoative; INS – instrumental; IO – Indirect Object; IRR – Irrealis; LOC – Locative; Neg – Negative; NP – Non-Past; O – Object; OBL – Oblique; P – Past; PI – Past Imperfective; pl – plural; PP – Past Perfective; PR – Present; R – Reduplication; recip – reciprocal; refl – reflexive; S – Subject; sg – singular; UNM – Unmarked.
5. For one of my consultants this lexeme meant only 'fur' with *mija* meaning both 'head' and 'body hair'. For my other consultant this lexeme had the meaning indicated with *mija* meaning 'head hair'.
6. Mature specimens of the king and western brown snakes have large heads. This particular name for brown snakes (there are other names) refers only to such specimens.
7. This plant has bright red stalks.
8. This tree has rough dark red-grey to black bark to half trunk height, with smooth white bark on the upper trunk and branches.
9. This example was tested twice with both my consultants, and whenever the part occurred as a free noun it took ablative case marking.

References

Baker, Mark C.

1987 Noun incorporation and the nature of linguistic representation. [Unpublished MS presented Wenner-Gren conference, Jamaica.]

1988 *Incorporation: a theory of grammatical function changing.* Chicago: Chicago University Press.

Berndt, Ronald M. – Catherine H. Berndt

1985 *The world of the first Australians.* (4th Edition.) Adelaide: Rigby.

Bresnan, Joan – Lioba Moshi

1990 "Object asymmetries in comparative Bantu syntax", *Linguistic Inquiry* 21: 147-168.

Burzio, Luigi

1986 *Italian syntax: a government binding approach.* Dordrecht: Reidel.

Craig, Collette (ed.)

1986 *Noun classes and categorization.* Amsterdam: John Benjamins.

Dixon, Robert M.W.

1980 *The languages of Australia.* Cambridge: Cambridge University Press.

1982 *Where have all the adjectives gone?* Berlin: Mouton.

Evans, Nicholas R.

1995 "The syntax and semantics of body part incorporation in Mayali", [this volume]

Fox, Barbara

1981 "Body part syntax: towards a universal characterisation", *Studies in Language* 5: 323-342.

Haas, Mary R.

1941 "Noun incorporation in the Muskogean languages", *Language* 17: 311-315.

Hale, Kenneth L.

1981 "Preliminary remarks on the grammar of part-whole relations in Warlpiri", in: John Hollyman – Andrew Pawley (eds.), 333-344.

Halliday, Michael A.K.

1985 *An introduction to functional grammar.* London: Edward Arnold.

Harvey, Mark D.

1990 Warray grammar and dictionary. [Unpublished MS, Sydney University.]

Heath, Jeffrey

1978 *Ngandi grammar, texts, and dictionary.* Canberra: Australian Institute of Aboriginal Studies.

1980 *Basic materials in Warndarang: grammar, texts and dictionary.* (Pacific Linguistics B72.) Canberra: Pacific Linguistics.

1981 *Basic materials in Mara: grammar, texts and dictionary.* (Pacific Linguisitcs C60.) Canberra: Pacific Linguistics.

1984 *Functional grammar of Nunggubuyu.* Canberra: Australian Institute of Aboriginal Studies.

Hollyman, John – Andrew Pawley (eds.)

1981 *Studies in Pacific linguistics in honor of Bruce Biggs.* Auckland: Linguistic Society of New Zealand.

Hopper, Paul J. – Sandra A. Thompson
1984 "The discourse basis for lexical categories in universal grammar", *Language* 60: 703-752.

Jaeger, Jeri *et al* (eds.)
1978 *Proceedings of the Fourth Annual Meeting of the Berkeley Linguistics Society.* Berkeley: University of California.

Kiparsky, Paul
1989 Double objects. [Unpublished MS presented to conference on non-standard case, Tucson, Arizona, July 1989.]

Lakoff, George
1986 "Classifiers as a reflection of mind", in: Colette Craig (ed.), 13-51.

Lieber, Rochelle
1981 *On the organization of the lexicon.* Bloomington: Indiana University Linguistics Club.

McGregor, William B.
1985 "Body parts in Kuniyanti clause grammar", *Australian Journal of Linguistics* 5: 209-232.

McKay, Graham R.
1975 Rembarnga: a language of central Arnhem Land. [Unpublished Ph.D. dissertation, Australian National University.]

Merlan, Francesca C.
1976 "Noun incorporation and discourse reference in modern Nahuatl", *International Journal of American Linguistics* 42: 177-191.
1983 *Ngalakan grammar, texts and vocabulary.* (Pacific Linguistics B29.) Canberra: Pacific Linguistics.

Mithun, Marianne
1984 "The evolution of noun incorporation", *Language* 60: 847-895.
1986 "On the nature of noun incorporation", *Language* 62: 32-37.

Napoli, Donna J.
1988a *Predication theory.* Cambridge: Cambridge University Press.
1988b Review of Burzio 1986. *Language* 64: 130-142.

Nichols, Johanna
1986 "Head marking grammar and dependent marking grammar", *Language* 62: 56-119.

Osborne, Charles R.
1974 *The Tiwi language.* Canberra: Australian Institute of Aboriginal Studies.

Perlmutter, David M.
1978 "Impersonal passives and the unaccusative hypothesis", in: Jeri Jaeger *et al* (eds.), 157-189.

Sapir, Edward
1911 "The problem of noun incorporation in American languages", *American Anthropologist* 13: 250-282.

Stanner, William E.H.
1937 "Aboriginal modes of address and reference in the north-west of the Northern Territory", *Oceania* 7: 300-315.

“My face *am* burning!”: quasi-passive, body parts, and related issues in Yawuru grammar and cultural concepts[1]

Komei Hosokawa

1. Introduction

This paper aims at illustrating modes of incorporation of body-part nouns into the clause grammar of Yawuru, an Aboriginal language spoken in the west Kimberley region, Western Australia. It focuses on the quasi-passive and three other related constructions. The treatment of body parts and body-part-like items in the cultural framework of Yawuru is discussed. It will be claimed that a notion of “identity” is crucial to the understanding of the issue. It will be also shown, in this connection, how this language provides different morpho-syntactic devices according to whether a body part is regarded as an integral part of the whole (that is, possessor) or as a separate entity, a focus of discourse in itself.

2. Language typology

Yawuru (Jauor) is a member of the Nyulnyulan family, one of the non-Pama-Nyungan language families (Wurm 1972: 124-125, McGregor 1988: 94-98). Several characteristics of its morpho-syntactic structure are summarised in this section to provide a basis for the main discussion on body parts in the grammar.[2]

2.1. Nominal morphology and NP case marking

Yawuru nouns have no declension classes or morphologically marked noun class. Encountered however is a fossilised nominal prefix *ni-* (~*n-*),[3] apparently used to mark a particular class of body-part nouns and a few other items. These include:

nimarla	'hand, arm'
niminy	'eye, face'
nimalul	'nose'
nimanya	'neck'
niwal	'foot'
nilawal	'name'
nimarndarl	'shadow'

and many others (also see 2.2 and note 6).

Apart from this prefixation, which is lexical and no longer productive, Yawuru nominal morphology is entirely suffixing.[4] Forms of the case markers are as follows:

-ni	ergative (ERG)
-yi/-dyi	dative (DAT)
-dyina	genitive (GEN)
-gun	locative (LOC)
-ngarn	allative (ALLAT)
-gap	ablative/perlative (ABL)
-barri	instrumental (INST)
-ngany	comitative (COM)
-nyurdany	causal (CAUS)[5]

Other nominal (derivational) suffixes that nouns and pronouns (either ABS or GEN) may carry include, among others:

-ngany	comitative (COM)
-lambu	approximative (APPR)
-gardi	lative ('in/to the side of')
-ngarr	semblative ('like')
-yirr	conjunctive ('and')
-dyunu	intensifier ('really')

When an NP is case-marked, the general principle in Yawuru is "mark the first constituent only", that is, the case marker is attached only to the first word of the NP (cf. Blake 1987: 77ff). For example:

Absolutive:

(1) a. *nyamba wamba*
this man
'this man'

b. *wamba nyamba*
man this
'this man'

Ergative:

(2) a. *nyamba-ni wamba*
this-ERG man
'(by) this man'

b. *wamba-ni nyamba*
man-ERG this
'(by) this man'

Dative:

(3) a. *nyamba-yi wamba*
this-DAT man
'to/for/of this man'

b. *wamba-yi nyamba*
man-DAT this
'to/for/of this man'

2.2. Incorporation

Only very limited traces of verbal incorporation of body parts are encountered in Yawuru, as in the verb *dyinungu* 'disembowel' (*nungu* 'belly, stomach, womb') and the verb *mirdibi* 'run away' (*ni-mirdi* 'shin, lower leg').[6] These instances are lexical and the verb roots, *dyi* and *bi*, appear elesewhere in the language only with meanings that are rather hard to associate with the incorporated verbs (*dyi* occurs as a suppletive root for the verb *ni* 'be'; and *bi* is the verb root for 'drink'). Body-part incorporation is not a productive device in Yawuru grammar.

2.3. Alignment and crossreferencing system

In terms of alignment, Yawuru is an ergative language which, however, has an accusative type verb-agreement system. Verbs are conjugated by a series of prefixes and suffixes/enclitics added to a verb root. Yawuru verbs are either simple or complex. The latter consist of an uninflected preverb and an inflecting auxiliary. All the bound pronominals go onto the auxiliary. The distinction of simple and complex verbs is not relevant to the discussion in this paper.

Details of the inflexional morphology of Yawuru verbs are beyond the scope of this paper. For our purposes, it is enough to notice that subject-marking is by a prefix, while object-marking is by a pronominal enclitic.

Table 3 gives an outline of the crossreference system in Yawuru. I follow Dixon (1987: 2) in employing the following symbols: A for transitive subject; S for intransitive subject; and O for direct object. As Table 3 shows, Yawuru case marking (on the core argument NPs) follows an ergative-absolutive pattern, whereas the verbal agreement follows a subject-object (nominative-accusative) pattern.[7] The first-order verbal prefix agrees in person (1, 2, 3 and 12) and number (minimal vs. augmented) with the syntactic subject (that takes the ergative case in transitive clauses and the absolutive in intranstives).

Table 3. Yawuru crossreference system

	A	S	O
verb:	prefix	prefix	enclitic
NP:	ERG	ABS	ABS

The operation of the Yawuru crossreferencing system may also be formulated as in (4). The NP referring to transitive subject is marked by the ergative, whereas the NPs referring to transitive object and intransitive subject carry zero marker (i.e. absolutive).

(4) transitive: (A-ERG) (O-ø) A-Verb-O
intransitive: (S-ø) S-Verb

The subject-object relation is clearly marked by the verbal inflexion, so that the core NPs are not always overtly expressed in natural discourse. This is why they are in parentheses in (4). The formula is no more than a conceptual schema, showing the core argument NPs only. The schema does not indicate the actual word order, which is highly flexible in Yawuru.

Yawuru actually has another type of transitive structure, namely "semi-transitive", in which the object is given the dative marking (i.e. A-ERG O-DAT A-Verb-O). We will see more about this later.

2.4. Pronominal system

The Yawuru pronominal system is given in Tables 1 and 2. In Table 1, the two left-hand columns are independent pronouns (absolutive and genitive forms); the two right-hand side columns are enclitics, attached to the end of finite verbs. The enclitics are either in accusative or in dative form. The accusative enclitics are mostly identical to the absolutive pronouns, and the dative enclitics are easily identified with the genitive pronouns. Table 2 gives the bound pronominals that occur as verbal prefixes (cross-referencing the subject); these vary according to the tense/mood of the verb.

Table 1. Yawuru pronominal system

		ABS	GEN	ACC	DAT
min	1	*ngay(u)*	*dyanu*	*-ngayu*	*-dyan(u)*
	2	*dyuyu*	*dyiya*	*-dyu(yu)*	*-dyiya*
	3	*ginya(ng)ka*	*dyina*	*-ginya/-dyina*	*-dyina*
	12	*yayu*	*dyaw* (~ *dyayu*)	*-yayu*	*-dyaw*
aug	1	du. *yarrgarda*	du. *dyarra*		
		pl. *yarryirr*	pl. *dyarrayirr*	*-yarrirr*	*-dyarra*
	2	du. *kurrgarda*	du. *dyunggarra*		
		pl. *kurryirr*	pl. *dyunggarrayirr*	*-kurrirr*	*-dyunggarra*
	3	du. *yirrgarda*	*dyirrayirr*		
		pcl. *kangadyunu*[a]			
		pl. *yirrydyurr*	*dyirrayirr*	*-irr*	*-dyirra*
	12	*yadiri*	*dyayrda*	*-yadiri*	*-dyayrda*

a. The 3rd-person paucal (pcl) form *kangadyunu* derives from a demonstrative pronoun.

Table 2. Pronominal prefixes on the verb

		non-future	future	irrealis
min	1	*nga-*	*nga-*	*nga-*
	2	*mi-*	*wal-/nga-*	*mi-*
	3	*i-*	*wa-*	*wa-*
	12	*ya-*	*ya-*	*ya-*
aug	1	*yanga-*	*yang-*	*ya-*
	2	*ku-*	*wa-*	*ku-/mi-*
	3	*inga-*	*wang-*	*wa-*
	12	*yaga-*	*ya-*	*ya-/yaga-*

There are dual/plural distinctions only in the non-12-person augmented-number free-form pronouns (ABS and GEN). The distinction is neutralised in ACC and DAT forms,[8] as well as in the pronominal prefixes on the verb. Paucal/non-paucal distinction is found only in 3aug, and the paucal form is a derivation of distal demonstrative *ka-/~ ~kanga-*.

The pronominal system of Yawuru follows a "four-person" pattern, or minimal-augmented system (Dixon 1980: 351-353). The category which includes both the speaker and the hearer (i.e. 'you and me') behaves grammatically as if it were a singular category (thus termed minimal), since the corresponding verb does not carry the explicit number marker that would be required when the subject is non-minimal (or augmented). The inclusive person in question is indicated by the digits "12" (i.e. 1+2, or 1 and 2 combined).

Case markers (except the dative) are attached only to the absolutive (ABS) forms of the pronouns. For example, the ergative for the 1min and 2min are respectively *ngay(u)-ni* (the stem-final *u* usually drops) and *dyuyu-ni*. Case marking of the independent pronouns follows the same pattern as common nominals.

3. Clause types

Yawuru clauses can be classified into seven types according to the patterns of case marking and crossreference. These are given in Table 4. In this paper our discussion is confined to simple clauses.

Table 4. Types of simple clauses

basic types (I-III)						
I.	intransitive:		ABS_j	*j*-Verb(-k_{DAT})		(DAT_k)
II.	transitive:	ERG_i	ABS_j	*i*-Verb(-j_{ACC}/k_{DAT})		(DAT_k)
III.	semi-transitive:	ERG_i	DAT_j	*i*-Verb-j_{DAT}		(DAT_k)
identity-sensitive types (IV-VII)						
IV.	quasi-passive:	ERG_i	ABS_p	(ABS_w)	*w*-Verb	
V.	double-subject intransitive:		ABS_p	(ABS_w)	*w*-Verb	
VI.	double-subject transitive:	ERG_p	ABS_j	(ABS_w)	*w*-Verb	
VII.	double-object transitive:	ERG_i	ABS_p		*i*-Verb-w_{ACC}	

In Table 4, the core arguments are shown on the left-hand side of the verb, while non-core (or oblique) arguments are shown to the right of the verb (I use the terms "core" and "oblique" as per Foley – Van Valin 1984: 77-80). Here ERG, ABS and DAT stand for argument NPs (free forms, either pronominal or non-pronominal) that take the ergative, absolutive and dative cases, respectively. Indexing letters (*i*, *j*, *k*, *p* and *w*) indicate crossreferencing between bound pronominals on the verb and the free-form NP arguments. The index *p* is applied to NPs referring to a part (typically a body part); while *w* to NPs referring to the whole (typically the possessor of the body part). Letters prefixed to the verb indicate subject agreement: e.g., "*i*-Verb" signifies that the verbal prefix agrees with the NP argument indexed *i*. Letters suffixed to the verb indicate object agreement by a pronominal enclitic, with its case form (either DAT or ACC) indicated by the subscript. Optional agreement (crossreference) is given in parentheses. Core arguments take either absolutive (ABS), ergative (ERG) or dative (DAT) case. Again, the chart here is given as a conceptual formulation and should not be taken as canonical; these formulas do not reflect the actual word order. Instances of each type are given and further analysed in sections 4 and 5.

Types I, II and III (or similar types of constructions respectively with absolutive, ergative-absolutive and ergative-dative case frames on argument NPs)[9] are observed in many Australian Aboriginal languages, although the crossreferencing affixes on the verb such as in Yawuru may be lacking. The reflexive and reciprocal constructions are formally intransitive in Yawuru, conforming to the type I (i.e. absolutive case frame).[10]

In types IV to VII, the NPs indexed *p* are typically body-part nouns. They occur as ABS except in type VI where they are assigned the ergative case (ERG). In types IV, V and VI, the arguments indexed *w* are personal pronouns and are often left out in actual texts.

Type V (the double-subject intransitive construction) involves two absolutive NPs, only one of which is crossreferenced in the verb. Transitive clauses of type VII (double-object construction) take a single absolutive NP, which is not, however, crossreferenced by the "object" enclitic; instead, another entity, typically the whole of which the absolutive NP is a part, is crossreferenced. Double-object constructions are encountered in some other Aboriginal languages, such as Gooniyandi, Warlpiri, Djaru and Nyangumarta. Gooniyandi (and probably others, too) has a double-subject construction (of type V) as well.[11]

Type IV, which I have tentatively termed "quasi-passive", is rather unusual and this is a focal point of the description presented in this paper.[12] In the quasi-passive construction, the pronominal prefix, which normally crossreferences the subject, crossreferences the patient/undergoer. It has further an additional absolutive NP and an ergative NP, neither of which is crossreferenced. The semantic characteristics of the quasi-passive are examined later, in 5.1. and 6.

The clause types IV-VII are jointly called "identity-sensitive" constructions for reasons discussed later (in 7.2). It will be shown in the following sections that the identity-sensitive construction types have a close relationship to each other.

4. Basic construction types

4.1. Intransitive

The formula for the intransitive construction (clause type I) given in Table 4 is repeated below:

(5) ABS_j *j*-Verb(-k_{DAT}) (DAT_k)

The absolutive NP (ABS_j) is the subject of the clause and refers to the actor. The predicate verb carries a bound pronominal prefix crossreferencing the actor. The subject NP may be a pronoun (of ABS form). The postverbal (-k_{DAT}) in the formula stands for an optional dative pronominal enclitic which refers to an undergoer/experiencer, which may occur (with or without the crossreferencing enclitic) in the clause as a datively marked NP. (6) and (7) are examples of intransitive clauses. (6) has a simple verb predicate and (7) has a complex verb predicate.[13]

(6) *wula* *wa-ng-ga-bula(-dyayrda)* *kalbu-gap.*
water(ABS)$_j$ 3_j-EN-FUT-come(-12″ DAT_k) above-ABL
'The rain$_j$ is coming (unfortunately for us$_k$).'

(7) *yardap + i-ny-dyu-rn* *dyalangardi-ngarr* *(kamba*
crawl + 3_j-EN-AUX-IMPF goanna-like that
wamba).
man(ABS)$_j$
'He$_j$ (that man) crawls like a goanna.'

In (6) the subject is 'water' (expressed in the absolutive case), while the verb carries the 12aug dative enclitic, indicating that the involved undergoer is 'all of us' (inclusive). The enclitic may be omitted. In any case, the verb carries the 3min prefix, in agreement with the subject 'water'.

4.2. Transitive

The second type of Yawuru simple clauses is transitive, for which the formula is:

(8) ERG_i ABS_j *i*-Verb(-j_{ACC}/k_{DAT}) (DAT_k)

The verbal prefix agrees with the ergative NP, which is considered the syntactic subject of the clause; (9) and (10) are examples. Notice in (10) that the transitive subject (A) is lower in animacy than the direct object (O).

(9) *nyamba wula nga-m-bula-nda-ngany-dyiya.*
this water(ABS)$_j$ 1_i-EN-come-PF-COM-2(DAT)$_k$
'I_i have brought this $water_j$ for you_k.'

(10) *buru-ni i-na-nya-nda-yarrirr kamba-gun niyamarri.*
sand-ERG_i 3_i-TR-catch-PF-1″(ACC)$_j$ that-LOC hill(LOC)
'We got bogged on the sand hill.' (Literally: 'That $sand_i$ caught us_j.')

In (9), the absolutive NP *nyamba wula* 'this water' is the direct object of the action. The verbal prefix indicates that the agent is 1st person, so that the ergative pronoun *ngay-ni* (1-ERG) can be inserted anywhere in the sentence, although it is left out in (9) as is usually the case in natural utterances in the language.

Transitive verbs may carry an enclitic cross-referencing the object. The object-marking enclitics take accusative forms (given as -j_{ACC} in formula (8) above) when referring to the direct object (patient), as is the case in (10). Dative enclitics (-k_{DAT} in the formula) appear when the reference is to the indirect object (goal, beneficiary, etc), as in the case of (9). Verbs, however, cannot carry more than one object-marker at a time. The general tendency is that non-3rd-person objects are more likely to be cross-referenced. Transitive verbs may carry no object-marker at all, as in:

(11) *dyimbin i-na-ma-nda lindyu-ni.*
inside 3-TR-put-PF police-ERG
'The police locked [him] up.'[14]

(12) *ngay-ni nga-na-ng-ga-makura.*
1-ERG 1-TR-EN-FUT-make
'I can do it.' (Literally: 'I will make [it].')

4.3. Semi-transitive

The third basic clause type is semi-transitive. In formula (13), the clitic pronoun (represented by *-j*) takes the dative, not accusative case.

(13) ERG_i DAT_j *i*-Verb-j_{DAT} (DAT_k)

Examples are as follows:

(14) *nga-na-ngurika-nda-dyirra ngay-ni kamba-rri-yi*
1_i-TR-wait-PF-3″ $(DAT)_j$ 1-ERG_i that-DU-DAT_j
bambi nganka-yi.
long:time word-DAT_k
'I_i have been waiting for the two $fellows_j$ for a long while to have a discussion with them (literally: 'for a $word_k$')'

(15) *gilp + nga-na-ga-dyu-yi-dyina (ngay-ni)*
avoid + 1*i*-TR-FUT-AUX-PURP-3$(DAT)_j$ 1-ERG_i
darlu-yi dyanu.
wife's:mother-DAT_j 1_i(GEN)
'... so that I could avoid facing my mother-in-law.'

Semi-transitive clauses differ from transitive clauses in that the object is assigned dative case marking instead of absolutive. Therefore, the core arguments for a semi-transitive predicate are an ergative NP and a dative NP, and they are respectively crossreferenced by the verbal prefix and the dative enclitic. No other object NP in the absolutive case may occur; nor is an accusative enclitic acceptable.

In addition to the core dative object (DAT_j in the formula), a second dative NP may occur, which is syntactically an oblique constituent (DAT_k in the formula). However, the dative enclitic on the verb always agrees with the core dative, not with the oblique dative. Example (14) shows this.

It should be also noted that the postverbal clitic pronoun is very rarely omitted in semi-transitive verbs, whereas it is optional in transitive and intransitive verbs.

Typical verbs that require the semi-transitive clause structure are those roughly equivalent to 'wait for', 'look for', 'want/like', 'hate/dislike', 'approach', 'watch out for ', 'scold', 'support', 'call', 'sing' (i.e. perform a love magic to somebody), 'infest', etc.. It is presumed that the object of these semi-transitive predicates reflects a semantic relation of goal, rather than patient.[15]

5. Constructions for body parts

The intransitive, transitive and semi-transitive structures are termed "basic clause types", since Yawuru simple clauses are mostly classified into one of these three types. We encounter, however, a relatively limited number of cases where Yawuru clause structure does not quite fit into the threefold typology.

5.1. Quasi-passive

In the quasi-passive (type IV), ABS_w is a free-form personal pronoun of the absolutive form, with which the verbal prefix agrees.

(16) ERG_i ABS_p (ABS_w) *w*-Verb

The quasi-passive is a construction in which the patient/undergoer is marked by the verbal prefix, not by the enclitic object-marker. Like the transitive, the core constituents are an ergative NP and an absolutive NP. Unlike the transitive, however, the verbal prefix does not agree with the ergative NP, but agrees with the undergoer (indexed *w*), which may appear in the clause in the form of the absolutive pronoun (ABS_w).

Examples of quasi-passive sentences are given in (17) and (18). In these sentences, the ergative marker *-ni* cannot be replaced by other case markers, such as locative *-gun*, causal *-nyurdany*, or instrumental *-barri*.

(17) *dyanga-ni gil + nga-ny-dyu-nda niwal (ngayu)*
oyster-ERG_i cut + 1_w-EN-AUX-PF $foot(ABS)_p$ $1(ABS)_w$
buya-gun.
reef-LOC
'My_w $foot_p$ was cut by an oyster $shell_i$ on the reef.' or 'I_w got a cut in my $foot_p$ from an oyster $shell_i$ on the reef.'

(18) *(ngayu) nga-lurra-nda kungkulu-manyan nyamba-ni*
1(ABS)$_w$ 1$_w$-burn-PF hair(ABS)$_p$ -only this-ERG
dyungku.
fire(ERG)$_i$
'The fire$_i$ burned only my$_w$ hair$_p$', or 'I$_w$ burned only my hair$_p$', or 'I$_w$ just had my hair$_p$ burned by the fire$_i$.'

Sentence (18) describes an incident in which a sudden flame of fire licked the face of the speaker, who fortunately didn't get her face burned; she just had her hair burned.

The quasi-passive construction is like a passive, in that the verbal marking of the agent-patient relation (or subject-object relation) is reversed. Unlike passive, however, the agent (or the cause of the process) is not defocused (I use the terms "focusing" and "defocusing" as set out in Shibatani 1985: 832). It is the patient pronoun (rather than the agent ergative NP) that is often omitted in textual examples of the Yawuru quasi-passive. Ergative NPs are frequently left out in Yawuru transitive sentences. The quasi-passive is also different from a passive in that, apart from the reversed agreement, no special changes take place in the form of the verb or in the case marking of the NPs.

Quasi-passive sentences are rather infrequent in the corpus of Yawuru texts, but when they occur, they often involve body-part nouns. The quasi-passive should be syntactically and semantically compared with the remaining three construction types, namely the double-subject intransitive/transitive and the double-object transitive, since these, too, typically involve body parts. The four types of body-part constructions contrast semantically. We shall illustrate later, through a comparison with the double-object transitive and with the double-subject intransitive, that the quasi-passive is a morpho-syntactic device by which the speaker foregrounds the entity with which the mentioned body part has contact, implying the accidental or involuntary nature of that contact.

Ergatively marked quasi-passive agents are always low in animacy (for example, the oyster shell and the fire in (17) and (18)).[16] By contrast, the undergoers (which are cross-referenced by the verbal prefix) of the quasi-passive are usually first or second person – that is, those highest on the animacy hierarchy.[17]

5.2. Double-subject intransitive

The formula for this construction type is:

(19) ABS_p (ABS_w) *w*-Verb

where ABS_w is a free-form personal pronoun, often left out in actual sentences. Examples follow:

(20) *rdardarl nga-ngara-rn niminy (ngayu).*
sick 1_w -become-IMPF eye$(ABS)_p$ $1(ABS)_w$
'I have a sore eye.'

(21) *nungu (ngayu) nga-m-bardika-rn.*
stomach$(ABS)_p$ $1(ABS)_w$ 1_w -EN-be:full-IMPF
'I'm sated.' (Literally: 'Stomach am filled.')

(22) *ngadyi dyuyu/*dyiya marru nyily* +
whether 2(ABS/*GEN)$_k$ head$(ABS)_p$ headache +
mi-ny-dyu-n.
2_k-EN-AUX-IMPF
'Do you have a headache?'

The point here is that the personal pronouns in the double-subject examples above take the absolutive forms, not the genitive, as explicitly indicated in (22) above, despite the underlying genitive relationship (my eye, my stomach, your head). At phrase level, however, the genitive relationship is expressed by using the genitive pronoun, not the absolutive. For similar constructions in Japanese, Mandarin Chinese and certain Southeast Asian languages, see respectively Tsunoda (1995), Chappell (1995) and Clark (1995). The double subject construction in Japanese is also discussed in section 7.1 below and comparisons are made with Yawuru.

(23) *marru dyiya*[18]
head(ABS) 2(GEN)
'your head'

Similarly, (24a) below is an NP, which is well-formed and acceptable by itself. The genitive phrase, however, is unacceptable at clause level, as shown by (24b):

(24) a. *bilyurr* *dyanu*
heart(ABS)$_{P}$ 1$_{w}$(GEN)
'my heart'

b. *[[*bilyurr* *dyanu*] *bidbid-gadya* *i-nga-rn.*]
heart(ABS)$_{P}$ 1$_{w}$(GEN) throb-INTENS 3$_{P}$-be-IMPF
'My heart is throbbing.'

Instead, one has to say something like (25) or (26), where (25) is a double-subject intransitive and (26) is a plain intransitive.

(25) *bidbid-gadya* *nga-nga-rn* *bilyurr* *(ngayu/*dyanu).*
throb-INTENS 1$_{w}$-be-IMPF heart(ABS)$_{P}$ 1(ABS/*GEN)$_{w}$
'My heart is beating hard.'

(26) *bilyurr* *i-nga-rn-dyanu* *bidbid-gadya.*
heart(ABS)$_{P}$ 3$_{P}$-sit-IMPF-1(DAT)$_{w}$ throb-INTENS
dyunku-nyurdany
running-CAUS
'I've got a beating heart because I ran.'

The word order of (25) can be scrambled in all possible ways. But irrespective of the order, the verbal prefix still agrees with the possessor, not with the body part. As far as agreement is concerned, the possessor is the subject. Meaningwise, however, it is the part (heart), not the whole (person), that is throbbing, so that the heart could be regarded as subject of the sentence. A similar remark applies to examples (20)-(22). These are thus called "double-subject" constructions.

Notice also that the personal pronoun (for the possessor) is often omitted; if the possessor is to be denoted by a pronoun, then it should be in the absolutive form (*ngayu* in (25)), not in the genitive (*dyanu*).[19] That means, one has to say in Yawuru something like "I am throbbing heart" or "I heart am throbbing" (though these are ungrammatical in English).

Formula (27) below indicates that the state of affairs expressed by the double-subject construction may also be expressed (with a slight difference in meaning, however) by a usual intransitive construction, if the verb takes a dative enclitic (-*w*) marking the undergoer.

(27) [ABS$_{P}$ (ABS$_{w}$) *w*-Verb] ↔ [ABS$_{P}$ *p*-Verb-w_{DAT}]

The two clause structures seem to reflect different discourse focuses. In the double-subject construction, exemplified by (25), it is the body part rather than the whole that is focal, while in the dative enclitic construction, as in (26) above, the focus seems to be on a relation between the the action of the person and the state of his/her body part (as a result of that action, in this case), rather than what is happenening to the part itself. In other words, the double-subject construction would be more appropriate in a context such as (28), rather than (29).

(28) *My head aches. Have you got any aspirin?*

(29) *I had an headache; that's why I took a day off.*

Thus, when asking somebody about his/her state of health with reference to a particular body part, the double-subject construction is the most suitable way of encoding the question:

(30) *ngadyi niyakan dyuyu mabu mi-nga-rn.*
whether back(ABS)$_{P}$ 2(ABS)$_{W}$ good 2$_{W}$-be-IMPF
'Is your (wounded) back all right?' (Literally: 'Are you$_{W}$ back$_{P}$ good?')

5.3. Double-subject transitive

The formula for the double-subject transitive is unlike the others in that the body-part nouns (indexed *p*) take the ergative case.

(31) ERG$_{P}$ ABS$_{j}$ (ABS$_{W}$) *w*-Verb

Instances of the double-subject transitive are rather few. In (32) and (33) below, the transitive verbs seem to have double subjects: one referring to a body part, the other referring to the possessor of the body part. The verbal prefix agrees with the possessor, as in the double-subject intransitive. Noticc in the examples below that the possessor occurs in the form of an absolutive pronoun, not a genitive one. These pronominal NPs, i.e. ABS$_{W}$ in formula (31), may be omitted.

(32) *buu + wal-dyu (dyungku) nilirr-ni*
blow + 2_wFUT-AUX fire(ABS)$_j$ mouth-ERG$_p$
*dyuyu/*dyiya.*
2(ABS/*GEN)$_w$
'You blow the fire with your mouth (i.e. breath).'

(33) *nilirr-ni-manydyan mi-ny-dyu-nda dyuyu/*dyiya.*
mouth-ERG$_p$-only 2_w-EN-say-PF 2(ABS/*GEN)$_w$
'You replied only with (the movement of) your lips.'[20]

It would be possible to explain the case form of the pronouns in the above examples by regarding the body part noun and the pronoun as forming a single NP, though discontinuous in (33),[21] so that the gloss to *nilirr-ni dyuyu* in (32) would be 'mouth-ERG 2(ERG)'. (Recall the general principle of "mark the first constituent only" outlined in 2.1.) I have chosen, however, to regard them as constituting separate NPs in the clause, considering the phenomenon in the context of body-part syntax.

We have at least one example, in which the ergative marker in the double-subject transitive construction is affixed to the pronominal possessor, rather than to the body part.

(34) *bidyara ngay-ni nga-likarra-rn durrgadya karda.*
ear(ABS) 1-ERG 1-hear-IMPF noise(ABS) yonder
'I heard a distant noise.'

Perhaps related to this is the fact that the word *bidyara* 'ear' refers to hearing ability rather than to the ear (as a visible part of body) itself. The noun *bidyara* in (34) cannot be replaced by the noun *nimalul* 'ear', 'earlobe' which refers to the body part itself.[22]

Double-subject transitive constructions may be paraphrased as the plain (single-subject) transitive. The latter clause type is much more common than the former. Consider (35) and (36):

(35) *kamba yila nga-ny-dyanba-rn niwal-ni (ngayu).*
that dog(ABS) 1_w-EN-kick-IMPF foot-ERG$_p$ 1(ABS)$_w$.
'I kicked the dog.' (Literally: 'My foot I kicked that dog.')

(36) *(ngayu-ni) kamba yila nga-ny-dyanba-rn niwal-gun*
1-ERG$_w$ that dog(ABS) 1$_w$-EN-kick-IMPF foot-LOC$_p$
dyanu.
1$_w$(GEN)
'I kicked the dog with my foot.' (Literally: 'I kicked that dog at my foot.')[23]

Due to paucity of data, the semantic contrast between the two modes of expression is not very clear. It seems, however, that in the double-subject transitive, the body part is foregrounded (as well as in the intransitive double-subject construction), while in the common transitive structure it is the object (patient) that is the main topic. In both cases, the agent (who is the possessor of the body part) is relatively defocused.

5.4. Double-object transitive

The structure of double-object clauses is basically transitive, involving an ergative NP and an absolutive NP, as shown in formula (37):

(37) ERG$_i$ ABS$_p$ *i*-Verb-*w*$_{ACC}$

(37) is, however, different from the usual transitive formula (8), because the accusative enclitic on the verb (-*w*$_{ACC}$) does not agree with the absolutive NP in the clause (ABS$_p$). Thus, it appears as if the sentence has two objects: one referred to by the free-form NP and the other referred to by the bound clitic pronoun. The enclitic cannot take a dative form. Consider the following instances:

(38) *dulbu nga-na-ø-bilka-dyuyu/*dyiya (ngay-ni).*
heart(ABS)$_p$ 1$_i$-TR-IRR-hit-2(ACC/*DAT)$_w$ 1-ERG$_i$
'I might spear you in the heart.' (Literally: 'Heart I might hit you.')

(39) *kirrbadyu-ni ward + i-ny-dyu-n-ngayu*
honey-ERG$_i$ stick + 3$_i$-EN-AUX-IMPF-1(ACC)$_w$
nimarla.
hand(ABS)$_p$
'The honey stuck to my hand.'

Typically, the absolutive noun (ABS$_p$) is a body-part noun and the accusative enclitic (-*w*) indicates the possessor of the body part. As in the double-subject construction examined in the previous subsection, the possessive (or part-whole) relation may not, at clause level, be encoded in the genitive phrase. Consider (40a-c) below.

(40) a. *i-na-bilka-nda-ngayu* *marru.*
3$_i$-TR-hit-PF-1(ACC)$_w$ head(ABS)$_p$
'She$_i$ hit me$_w$ on the head*p*.'

b. *[*i-na-bilka-nda* [*marru* *dyanu.*]]
3*i*-TR-hit-PF head(ABS)$_p$ 1$_w$(GEN)
'She$_i$ hit me$_w$ on the head$_p$.'

c. ?[*i-na-bilka-nda-ngayu* *marru-gun.*]
3$_i$-TR-hit-PF-1(ACC)$_w$ head$_k$-LOC
(?)'She$_i$ hit me$_w$ (perhaps on the head$_p$) with her$_i$ head$_k$.'

What (40a-c) shows is that one has to say something like "he hit me head" (40a), rather than "he hit my head" (40b) or "he hit me in the head" (40c) in Yawuru, in spite of the fact that the phrase *marru dyanu* 'my head' is acceptable by itself.

If the enclitic takes the dative form, as it does in (41) below, then the structure is simply transitive. In that case, however, the head which is to be hit is neither the speaker's nor the hearer's, but that of a third person.

(41) *wal-a-bilka-dyanu* *marru* *dyina.*
2$_i$FUT-TR-hit-1(DAT)$_k$ head(ABS)$_p$ 3$_w$(GEN)
'You$_i$ should hit him$_w$ on the head$_p$ for me$_k$ (i.e. on my behalf).'

5.5. Reflexive actions on body parts

Yawuru verbs may be reflexivised by adding the prefix *ma-* and the suffix *-ndyi* to the verb root.[24] Compare transitive (42) and reflexive (43):

(42) *ngayu-ni* *nyamba* *yila* *nga-na-balu-nda.*
1-ERG$_i$ this dog(ABS)$_j$ 1$_i$-TR-hit-PF
'I$_i$ hit this dog$_j$.'

(43) *ngayu* *nga-ma-balu-ndyi-nda.*
1(ABS)$_i$ 1$_i$-REF-hit-REF-PF
'I$_i$ hit myself$_i$.'

The subject (actor) occurs in the absolutive case (but cf. discussion in 6.4). The verb loses the transitive conjugation marker (TR). In terms of case marking and crossreferencing patterns, the reflexive construction is plain intransitive clause type.

When reflexive actions involve a part of the body, a structure parallel to the double-subject intransitive occurs:

(44) *i-ma-bali-ndyi-n* *ngulyku* *(ginyangka).*
3_w-REF-cut-REF-IMPF whisker(ABS)$_P$ 3(ABS)$_w$
'He$_w$ is shaving.' (Literally: 'He$_w$ cuts himself$_w$ whisker$_P$.' ")

(45) *marlu* *(dyuyu)* *wa-ma-bilka-ndyi* *marru.*
not 2(ABS)$_w$ 2_w-FUT-REF-hit-REF head(ABS)$_P$
'Don't (you$_w$) hit yourself$_w$ in the head$_P$.'

Notice, in the above examples, that both actors (whole) and objects (part) are encoded in the absolutive. The construction may well be interpreted as reflexivised double-object transitive: i.e. $i = w$ in terms of the indices in formula (37).

6. Relations and contrasts between the clause types

We have so far discussed the structures and some characteristics of Yawuru simple clauses. In this section the relationships and contrasts between the different syntactic-semantic constructions are examined.

6.1. Quasi-passive vs. double-subject intransitive

(46) illustrates the syntactic connection of the quasi-passive and double-subject intransitive. (46a) is quasi-passive, while (46b) is intransitive with a double subject. The latter results by taking the ergative NP (*nyamba-ni buru* 'by this sand') away from the former.

(46) a. *dyiladyila* *nga-ngara-nda* *niwal* *(ngayu)*
burningly:hot 1_w-become-PF foot(ABS)$_p$ 1(ABS)$_w$
nyamba-ni *buru.*
this-ERG$_i$ sand(ERG)$_i$
'I$_w$ had my feet$_p$ (i.e. the soles) burned by this hot sand$_i$.'

b. *dyiladyila* *nga-ngara-nda* *niwal* *(ngayu).*
burningly:hot 1_w-become-PF foot(ABS)$_p$ 1(ABS)$_w$
'My feet were burned.'

Whether or not the cause of the state of affairs is mentioned seems to be crucial to the semantic contrast that exists between these two otherwise identical constructions. The quasi-passive implies that the described state of the body part is caused by contact of that body part with the cause (that is, the entity referred to by the ergative NP). The focus is on that contact, and in many cases the unexpected and/or undesirable nature of the contact. Without the ergatively marked NP, on the other hand, the semantic focus shifts to that part of the body which is affected. Thus the double-subject intransitive is used in a context where the speaker's main concern is to specify the affected part.

In either case, the free-form pronoun (as absolutive NP referring to the possessor of the body part) is subject to ellipsis. This, however, does not necessarily mean that the possessor (or the whole) of the body part is defocused, since it is marked by the pronominal prefix on the verb.

Example (46b) above is not a spontaneously uttered sentence, but is an elicited one. We have, however, natural text examples such as (47), which are structurally identical to (46b).

(47) *nilirr* *ribi* *nga-ngara-rn* *(ngayu).*
mouth(ABS)$_p$ toothache 1_w-become-IMPF 1(ABS)$_w$
'I$_w$ have a toothache. (My tooth$_p$ aches).'

6.2. Quasi-passive vs. double-object transitive

The next set of examples, (48) and (49), illustrates that the quasi-passive also has a close relation with the double-object construction. (48a) and (49a) are double-object transitive; (48b) and (49b) are quasi-passive ((49b) is the same as (17) given earlier).

(48) a. *laar + i-ny-dyu-rn-ngayu (munyu)*
crack + 3_i-AUX-IMPF-1(ACC)$_w$ knee(ABS)$_p$
nyamba-ni dila wanangarri.
this-ERG hard stone(ERG)$_i$
'This hard stone wounded me (in the knee).'

b. *laar + nga-ny-dyu-nda munyu (ngayu)*
crack + 1_w-AUX-PF knee(ABS)$_p$ 1(ABS)$_w$
dila-ni wanangarri.
hard-ERG stone(ERG)$_i$
'I_w got my knees$_p$ wounded by the hard stone$_i$ (i.e. the stony surface of the path).'

(49) a. *dyanga-ni gil + i-ny-dyu-nda-ngayu*
oyster-ERG$_i$ cut + 3_i-EN-AUX-PF-1(ACC)$_w$
(niwal) buya-gun.
foot(ABS)$_p$ reef-LOC
'The oyster shell$_i$ cut me$_w$ (in the foot$_p$) on the reef.'

b. *dyanga-ni gil + nga-ny-dyu-nda niwal*
oyster-ERG$_i$ cut + 1_w-EN-AUX-PF foot(ABS)$_p$
(ngayu) buya-gun.
1(ABS)$_w$ reef-LOC
'My$_w$ foot$_p$ was cut by the oyster shell$_i$ on the reef.'

The states of affairs expressed by the double-object transitive can also be encoded by the quasi-passive without changing the logical/referential meaning. It should be noted, however, that it is the body-part absolutive noun which is often defocused and omitted in double-object structures, such as (48a) and (49a). In the quasi-passive structure, on the other hand, the body parts are not defocussed - see (48b) and (49b). The double-object construction focuses on the relation between the agent (or cause) of an incident and the experiencer (i.e. the possessor of the affected body part), relatively defocusing the part itself (which is usually mentioned already in the discourse). The quasi-passive, by contrast, focuses on the relation between the part and the agent/cause. One of the functions of the quasi-passive seems to be relating a whole, part and external causer without defocusing any of them.

6.3. Quasi-passive vs. transitive

Transitive sentences are not always interchangeable with quasi-passive structures, even though they involve body parts as in (51a), the double-object construction; (50b) and (51b) are ungrammatical.

(50) a. *kinykirida-ni i-na-nya-nda-ngayu.*
phlegm-ERG$_i$ 3$_i$-TR-catch-PF-1(ACC)$_j$
'I caught a cold.' (Literally: 'Phlegm$_i$ caught me$_j$.')

b. *[*kinykirida-ni nga-na-nya-nda.*]
phlegm-ERG$_P$ 1$_W$-TR-catch-PF
'I caught a cold.'

(51) a. *i-na-ra-rn-yadiri niminy rumarra-ni.*
3$_i$-TR-stab-12"(ACC)$_W$ eye(ABS)$_P$ sun-ERG$_i$
'The sun dazzled us.'

b. *[*yaga-rr(-a-r)a-rn niminy rumarra-ni.*]
12"$_W$-AGM-TR-stab-IMPF eye(ABS)$_P$ sun-ERG$_i$
'The sun dazzled us.'

The context of discourse is also relevant to the use of the quasi-passive. In (52) below, which is taken from a transcription of a spontaneous narrative, the intransitive structure ('I sit') is followed by the quasi-passive structure ('chill I catch'), with the effect that both finite verbs have the same pronominal prefix, namely the 1st person *nga-*.

(52) *midyala + nga-nga-rn ngayu, kuulbarra-ni*
sitting + 1$_j$-be-IMPF 1(ABS)$_j$ chill-ERG$_i$
nga-na-nya-rn.
1$_j$(=$_W$)-TR-catch-IMPF
'I was sitting (for a long while) and caught a chill.'

The second clause of (52) may be paraphrased as a plain transitive structure:

(53) *kuulbarra-ni i-na-nya-rn-ngayu.*
chill-ERG$_i$ 3$_i$-TR-catch-IMPF-1(ACC)$_j$
'I caught a chill.' (Literally: 'Chill chatches me.')

It seems, however, that the rhythm of story telling (and/or the pattern of topic continuity) made the narrator prefer the verb agreement as in (52). Obviously, the factor of discourse cohesion is relevant.

6.4. Quasi-passive vs. reflexive

Another important function of the quasi-passive becomes clear when we compare it with the reflexive construction. Consider (54)-(57) below. (54) is a double-object transitive, in which the agent (1st person) is the cooker of the meat.[25] (55) is also a double-object transitive, but this time it is the speaker who was burned.

(54) *ngayu-ni warli nga-na-lurra-nda kungkulu-manyan*
1-ERG$_{i}$ meat(ABS)$_{j}$ 1$_{i}$-TR-burn-PF hair(ABS)$_{j}$-only
dyungku-gun.
fire-LOC
'I$_{i}$ singed the game$_{j}$.' (i.e. 'I burned the surface of the skin ready for later steam-cooking in the earth oven.')

(55) *dyungku-ni i-na-lurra-nda-ngayu (kungkulu-manyan).*
fire-ERG 3$_{i}$-TR-burn-PF-1(ACC)$_{w}$ hair(ABS)$_{p}$-only
'The fire$_{i}$ burned me$_{w}$ (but only my hair$_{p}$).'

Now, (56) below is quasi-passive. Contrast it with (57), which has the reflexive form of the verb (*ma-lurra-ndyi*). The implication of the reflexive construction of (57) is that the action of burning the speaker himself/herself was done on purpose – it is wilful (or volitional). Such implication of the actor's intention is absent in the quasi-passive sentence (56):

(56) *(ngayu/-*ni) nga-lurra-nda kungkulu-manyan dyungku-ni.*
1(ABS)$_{w}$/*ERG 1$_{w}$-burn-PF hair(ABS)$_{p}$-only fire-ERG
'The fire$_{i}$ burned only my$_{w}$ hair$_{p}$.' (I$_{w}$ had only my hair$_{p}$ burned by the fire; i.e. I accidentally burned myself with the fire but only my hair got burned.)

(57) *ngayu(-ni) nga-ma-lurra-ndyi-nda nimarla.*
1(ABS)/-ERG 1$_{i}$(=$_{w}$)-REF-burn-REF-PF hand(ABS)$_{j}$(=$_{p}$)
'I have (intentionally/deliberately) burned my (own) hand.' (Literally: 'I$_{w}$ burn myself$_{w}$ hand$_{p}$.')

In the quasi-passive construction (56), the personal pronoun (for the patient or undergoer) is in the absolutive form (*ngayu*) and does not carry the ergative marker *-ni*. Notice, however, that in the reflexive construction (57), the subject pronoun may optionally carry the ergative marker (as given in parentheses; the actual ergative form is *ngay-ni*). It has already been pointed out (in 5.4.) that in Yawuru reflexive clauses the subject generally stands in the absolutive case. However, when the action is something unusual (or at least unexpected) and the willfulness of the action thus needs to be emphasised, which is the case in (57), the ergative marker may appear. Here the subject is a wilful actor; and the fire, which is the direct cause (or instrument) of the incident is rather out of focus in the sentence. In the quasi-passive, the person who suffered the burn is an undergoer, rather than an actor; and the fire as the direct causer (if not a wilful agent) of incident is still in focus.

The following examples, which I assume show the quasi-passive clause structure (though there is no explicit body-part NP), provide further evidence that the notion of non-volitionality (or uncontrollability in this particular example) underlies the syntactic construction in question.

(58) *darra + nga-ny-dyu-n* *wula-ni.*
belch + $1_w(=_i)$-EN-AUX-IMPF water-ERG
'I_w belch because of the $water_i$ (that is, carbonated drink).'

(59) *burrburr + nga-ny-dyu-nda* *durr-gadya-ni.*
frighten + $1_w(=_i)$-EN-AUX-PF roaring-INTENS-ERG_i
'I_w was frightened at the roaring $noise_i$.'

Note, again, that the causal case marker cannot replace the ergative marker in the above sentences.

6.5. Double-object transitive vs. intransitive

(60)-(63) are good examples to sum up the interrelationship of the clause types examined so far: (60) is a double-object transitive; (61) is a quasi-passive, although ungrammatical in this particular case; (62) is a regular transitive; and (63) is a regular intransitive.

(60) *ward + i-na-nya-nda-ngayu* *dyangkurr*
stick + 3_i-TR-AUX-PF-$1(ACC)_w$ $clothing(ABS)_p$
nyamma-ni balu.
this-ERG $tree(ERG)_i$
'My shirt got stuck on this tree (i.e. branch).' (Literally: 'This $tree_i$ caught me_w $cloth_p$.')

(61) *[*ward* + *nga-na-nya-nda* *dyangkurr* *nyamma-ni*
stick + 1_wTR-AUX-PF clothing(ABS)$_p$ this-ERG
balu]
tree(ERG)$_i$
(Literally: 'My shirt *am* stuck by this tree.')

(62) ?[*ward* + *i-na-nya-nda* *dyangkurr* *dyanu*
stick + 3_w-AUX-PF clothing(ABS)$_p$ 1_w(GEN)
nyamma-ni *balu.*]
this-ERG tree(ERG)$_i$
(Literally: 'My shirt got stuck by this tree.')

(63) *dyangkurr* *dyanu* *ward* + *i-ngara-nda*
clothing(ABS)$_j$ 1(GEN) stick + 3_j-become-PF
nyamma-gun *balu.*
this-LOC tree(LOC)
'My shirt got stuck on this tree.'

It has been stated earlier that what is expressed in a quasi-passive can also be expressed in a double-object transitive. Now (60) and (61) above demonstrate that the reverse is not always the case. Incidentally, *dyangkurr* 'clothing' seems to present a borderline case.[26] Whereas this non-body-part noun (although it is very much related to the body) cannot occur in the quasi-passive construction (61), it does not fit into a plain transitive clause, either. As in (62), the possessive phrase *dyangkurr dyanu* 'my shirt' as object NP of the transitive reveals low acceptability (cf. discussion in 5.3). Such a possessive phrase, however, would be more acceptable if it referred to a piece of clothing that is not on the body, as in "they kindly brought my jumper which I had left at their house".

It is important to reiterate that the double-object construction foregrounds the whole rather than the part (see 5.4), but such foregrounding takes effect on the condition that the whole and the part are not alienated. On the other hand, possessive phrases, such as those equivalent to 'my clothes' or 'my head', tend to shift the focus to the objects themselves. Such shifts of focus are more likely to occur when the object is separated from the whole. The case of clothing, discussed above, which can easily be separated from or attached to the body, provides crucial illustration.

To add a further piece of evidence, compare (64) and (65). The genitive modifier 'my (body)' is acceptable in (64) while it is rejected in (65).

(64) *ngay-ni nga-na-ng-ga-dyardi dyangkurr karrikan-gap*
1-ERG 1-TR-EN-FUT-skin clothing(ABS) body-ABL
dyanu.
1(GEN)
'I'll take my clothes off.' (Literally: 'I'll strip clothes from my body.')

(65) *ngurru dyangkurr nga-na-ngama-yi karrikan-gun*
more cloth(ABS) 1-TR-put(REF)-JUSS body-LOC
*ngayu/*dyanu.*
1(ABS/*GEN)
'I have to wear more clothes.' (Literally: 'I have to put more clothes on me body.')[27]

The results of the actions described in the above two sentences are contrastive: in (64) the clothes will be taken off the body, whereas in (65) they are put on.

7. Discussion

We have so far described the structure of Yawuru simple clauses from a morpho-syntactic as well as semantic perspective. It has been illustrated that, when paraphrasing is possible from one clause type to another, there usually exists some semantic contrast between them.

In this section, I first comment briefly on a general linguistic phenomenon which has much to do with the Yawuru quasi-passive and related constructions. I then address a further matter that shows the Yawuru case to be characteristic.

7.1. General linguistic aspect

(66) and (67) below are from Japanese. (66a) is an example of the so-called "double-subject" construction: both the 1st person *watasi* and the body part *sinzoo* could be considered the subject of sentence. There is no need now to go into the debate of which is the "true" subject. It should be noted, instead, that in spite of the well-formedness of the NP *watasi-no sinzoo* 'my heart', such a possessive NP itself does not usually stand as the subject. This is parallel to what has already been observed (in 5.2)

as to the status of the possessive phrase (noun + genitive pronoun) in Yawuru clause grammar. A Japanese sentence like (66b) cannot avoid a special contextualisation, such as "Your heart may not be throbbing, but my heart is (because I feel so nervous!)". Otherwise (66b) would sound quite strange. (66a), by contrast, is quite acceptable as a context-free, independent statement. (See also Tsunoda 1995 on this construction.)

(66) a. *watasi-wa* *sinzoo-ga* *dokidoki-s-ite-iru.*
1-TOP heart-NOM throbbing-do-DUR-PRES
'My heart is throbbing.'
b. [*watasi-no* *sinzoo*]-*wa/ga* *dokidoki-s-ite-iru.*
1-GEN heart-TOP/NOM throbbing-do-DUR-PRES
'My heart is throbbing.'

Now the acceptabilities of the sentences become reversed when we talk of a motorcar instead of the heart. (67a) below is quite unnatural as an independent clause, despite the fact that the construction itself is a structural parallel to that of (66a).[28] On the other hand, (67b), which is parallel to (66b), is completely acceptable.

(67) a. ?[*watasi-wa* *kuruma-ga* *kosyoo-s-ite-iru.*]
1-TOP car-NOM out:of:order-do-DUR-PRES
'I have my car out of order.'
b. [*watasi-no* *kuruma*]-*wa* *kosyoo-s-ite-iru.*
1-GEN car-TOP out:of:order-do-DUR-PRES
'My car is out of order.'

A similar state of affairs can be observed in the English examples given below.

(68) a. *He hit me on the back.*
b. ? *He hit my back.*
c. ?? *He hit me on the watch.*
d. *He hit my watch.*

(68a) and (68d) are quite normal as independent sentences, whereas sentences like (68b) and (68c) are less acceptable unless some special contextualisation is provided.[29]

7.2. Identity-sensitive grammar

One of our common understandings is that human languages often reserve some special kind of morphology and/or syntax for the semantic domain of body parts. The quasi-passive and the related constructions in Yawuru are examples of this. The extent of the semantic domain, however, remains to be specified or modified according to the particular cultural knowledge of the language users.

Kinship terms, for example, may or may not go together with terms for body parts. In Yuman languages (Langdon 1988, Miyaoka 1991), body-part nouns and kin terms can commonly take possessor-marking prefixes, which are identical with the subject markers to verbs. In Jacaltec (Mayan), body-part nouns belong to a noun class different from the one which has kin terms and human terms as its members (Hosokawa 1991b). Kin terms and body terms in this language also differ in the morphology of possessor marking (cf. Day 1973: 29-31).[30] In Australian languages kin and body terms often reveal peculiar morpho-syntactic behaviour (Leeding 1995). This, however, is not the case in Yawuru.

Earlier in this paper it was suggested that the quasi-passive and the double-subject/object constructions in Yawuru typically involve body-part nouns. As a matter of fact, this generalisation is not sufficient. Although these relatively marked clause types do involve a certain class of nouns, the category includes (as we shall see later in this section) entities other than body parts as well. On the other hand, nouns such as *dyidya* 'nail', *kunbulu* 'blood' as well as various kinds of bodily exuviae do not fit into the seemingly body-part-related clause types.[31] Furthermore, in some cases a body-part noun can either be an element of a marked clause structure or of an unmarked structure, depending on the context. Compare, for example, (69) and (70) below. The body-part noun *marru* 'head' occurs as one of the double objects in (69), while it stands as the single object in the plain transitive (70).

(69) *wal-dyurrku-ngayu* *marru.*
2_iFUT-cut:hair-1(ACC)$_w$ head(ABS)$_p$
'Will you cut my hair.' (Literally: 'You$_i$ hair-cut me$_w$ head$_p$")

(70) *rarrp* + *wal-dyu* *marru* *nyamba-yi* *nganydyu,*
comb + 2_iFUT-AUX head(ABS)$_j$ this-DAT$_k$ girl(DAT)$_k$

dyurru	*i-nga-rn*	*manydya.*
lice(ABS)$_l$	3_l-be-IMPF	many.

'Comb her hair (Literally: 'You$_i$ comb hair$_j$ for her$_k$'). She's got many lice.'

Being a part of the body or related to the body is, therefore, not a precise characterisation of the category in question. Then, what is the reason why Yawuru speakers use special syntax for certain kinds of entities (and in certain contexts of discourse)? I consider that the reason is partly cultural, rather than purely linguistic.

Example (71) reveals a double-object transitive construction, in which the noun *nilawal* 'name' behaves just like a term of body part. Here, the person (who is an old lady in the interactive context) was called by her Aboriginal name (or "bush name"), which should not be done without a good reason. The construction involving the possessive NP (*nilawal dyanu* 'my name') as in (72) is inadequate in this particular context.

(71)	*inga-rr-a-kilbira-nda-ngayu*	*nilawal.*
	$3''_i$-PL-TR-sing-PF-1(ACC)$_w$	name(ABS)$_p$

'They called out my name.' (or 'They called me by name.')

(72)	?[	*inga-rr-a-kilbira-nda*	[	*nilawal*	*dyanu.*	]]
		$3''_i$-PL-TR-sing-PF		name(ABS)$_p$	1_w(GEN)	

'They called me by name (but only by an unimportant name).'

Personal names are very important in Aboriginal cultures. For Yawuru people, names are as much a part of themselves as arms and legs, indeed perhaps more so. That is the reason (or at least a part of the reason) why the double-object construction is required in (71). This may seem to have two objects: but they are, in a sense, one and the same.

It should be noted that the possessive structure, as illustrated by (72), becomes acceptable if the old lady were not called by her traditional (proper Aboriginal) name, but simply by her "whitefella name", which is, from the viewpoint of Aborigines, not essential to her identity. In other words, (71) expresses the identity of, and the inseparability between the name and the person, while such close tie is not encoded in (72).[32]

For a similar cultural reason, identity-sensitive constructions are applied to words such as *nimarndarl* 'shadow', *niwal* 'footprint, track' (the same word as 'foot') and *buru* 'native country', as in the double-object transitive (73), reflexive double-subject (74) and double-subject copulative sentence (75).

(73) *wa-y-rr-a-baa-yayu* *niwal.*
3″$_{i}$-IRR-AGM-TR-see-12(ACC)$_{w}$ foot(ABS)$_{p}$
'They$_{i}$ might read our$_{w}$ footprints$_{p}$ (and chase us).'

(74) *i-ma-bura-ndyi-n* *nimarndarl* *wula-gun.*
3$_{i}$-REF-see-REF-IMPF shadow(ABS)$_{p}$ water-LOC
ginyangka
3(ABS)$_{i}$(=$_{w}$)
'He$_{i}$ is looking at himself$_{i}$ in the water.' (Literally: 'He looks himself$_{w}$ shadow$_{p}$.')

(75) *ngayu* *buru* *dyanu* *dyamardagap*
1(ABS)$_{w}$ country(ABS)$_{p}$ 1$_{w}$(GEN) (toponym)
nga-nga-rn.
1$_{w}$-be-IMPF
'My country is Jamardagab.' (I$_{w}$ was born there and I am spiritually associated with that place$_{p}$.)

In the last example above, the identity-sensitive word *buru* is modified by the genitive pronoun but it still triggers the 1st-person verb agreement (cf. the case of the word *dyangkurr* 'clothing' discussed in 6.5).

To further illustrate the contrast between the identity-sensitive syntax and the basic syntactic patterns of the language, compare (76), a double-object transitive, and (77), a normal transitive with an oblique dative argument.

(76) *i-na-kilbira-nda-ngayu* *nilawal.*
3$_{i}$-TR-sing-PF-1(ACC)$_{w}$ name(ABS)$_{p}$
'He called out my (real) name.'

(77) *i-ny-dyu-nd-dyanu* *karrydya* *nganka.*
3$_{i}$-EN-say-PF-1(DAT)$_{k}$ furious word(ABS)$_{j}$
'He called me bad names.' (that is, he abused me.)

The culture-specific notion of identity is also the basis of the seemingly irregular verb agreement such as we observe in examples (78) and (79) below. (78a) is a usual intransitive sentence, in which the verbal prefix takes the 3rd-person singular form (*i-*), agreeing with the subject, 'the snake'. If, however, the snake happens to be the *rayi*, or conceptional dreaming of the addressee, then the bound pronominal agreement is in the 2nd person (*mi-*), as in (78b).[33]

(78) a. *kamba dyurru i-ndira-rn yanban-gadya.*
that snake(ABS)$_j$ 3$_j$-go-IMPF south-INTENS
'The snake went on southwards.'
b. *kamba dyurru mi-ndira-rn yanban-gadya.*
that snake(ABS)$_p$ 2$_w$-go-IMPF south-INTENS
'The snake (which is your dreaming) went on southwards.'

(79) is a similar example of identification by the notion of conceptional dreaming.

(79) *i-na-ra-rn dal + nga-ny-dyu-rn niyakan.*
3$_i$-TR-stab-IMPF clang + 1$_w$-TR -AUX-IMPF back(ABS)$_p$
'My father speared [the turtle, which happened to be my dreaming] and [the spear] hit [the turtle's] back, making a clanging sound.'

The original context of the above sentence is as follows. Before the narrator (female)$_w$ was born, her father went fishing to the coast and saw a green turtle approaching him. He tried to spear it, but missed. He should have hit the turtle's neck, but the spear hit the carapace$_p$. Just after this episode, his wife gave birth to a girl$_w$, the narrator. The baby girl had a natal scar on her shoulder$_{p'}$, just around the place the father's spear had hit the turtle. So the people regarded the green turtle as the newborn baby's *rayi* (conceptional dreaming).

What is common between these instances of the dreaming-wise person agreement and the double-subject intransitive construction is the logic of identification which operates in the context of Yawuru culture – which, I suspect, is probably pan-Aboriginal.

8. Conclusion

Although most Yawuru simple sentences may be classified into one of the three basic clause types (intransitive, transitive and semi-transitive), several other types of syntactic constructions are encountered which do not quite fit into the threefold typology. Four distinct clause types (quasi-passive, double-subject intransitive, double-subject transitive and double-object transitive) are shown to be relevant to the theme of body parts in grammar. All of these are characterised by the fact that the verb crossreferences the whole, rather than the part, even when it is possible to add to the clause a free-form pronominal NP referring to the whole.

This paper has explored the semantic characteristics of these modes of expressing body parts and related phenomena. Contrastive analysis of each clause type has revealed that the double-subject constructions tend to foreground the part, whereas the double-object construction tends to background it (and usually leave it out). The quasi-passive, on the other hand, turns out to be indicative of non-volitionality and uncontrollability, without any of the relevant entities (whole, part and external inanimate agent/causer) being defocused.

The class of nouns which may enter into these marked clause constructions in Yawuru includes not only body parts, but names, shadows, tracks, native places and personal "dreamings" as well. I have attempted to show that both cultural factors and universal (cross-linguistically attested) factors are at work in the formation of the Yawuru quasi-passive and other marked clause types. It has been claimed in this paper that (i) the notion of identity is crucial to the understanding of Yawuru clause grammar; and that (ii) even dealing with body parts there are alternative constructions in which body part and other related items may be encoded.

To conclude, I remark briefly on the notion of inalienability. Body parts are too often referred to by linguists as "inalienable" possessions. Most body-part items, however, can be separated from the body in certain situations: teeth can be taken out, arms or legs can be cut off; and even eyes or head can be alienated from the "possessor", either physically or ideationally/conceptually. An important point here is that languages provide different grammatical devices to indicate whether the body parts referred to remain as integral parts of the body or are separated and alienated from the body. For example, in Mayali of North Australia (Evans 1995 and pers.comm.), although verbs with incorporated body parts usually agree in number with the whole (possessor of the relevant body part), they sometimes show number agreement with the part. This happens precisely when the part is regarded as separated from the whole. Evans also points out that body parts normally incorporate unless there is discourse focus on the part as an entity in and of itself. The case of Yawuru has further clarified that notions such as name can be treated in the same alternative ways as body parts: they may be encoded either as inalienable parts of the person's whole existence or as alienable, and thus not essential to his/her identity.

Notes

1. This is a revision of the paper read at the Australian Linguistics Society conference, August 1988, Armidale NSW. First of all, my deep gratitude goes to Elsie (Nyidyalya) Edgar of Broome, Western Australia, a knowledgeable Yawuru woman, who generously shared her knowledge of the language with me. Fieldwork grants from the Research School of Pacific Studies (of Australian National University) and the Australia-Japan Foundation supported the field research (1986) in Broome. I gratefully acknowledge the detailed comments provided by the volume editors: Hilary Chappell and Bill McGregor. I also thank Felix Ameka, Heather Bowe, Lois Carrington, Bob Dixon, Tom Dutton, Nick Evans, Cliff Goddard, Yukiko Hosokawa, David Nash, Osamu Sakiyama, Tim Shopen, Jane Simpson, Gerda Smith and Tasaku Tsunoda for their comments on earlier versions of the paper. Special acknowledgement goes to the late Don Laycock, who, while himself preparing his paper for the Armidale conference, gave his valuable time and energy to reading the original draft of this paper and discussed with me a number of points. The discussion continued even during our flight to Armidale. Don's tragic death not long after the conference came as a dreadful shock to me, both personally and academically.
2. See Hosokawa 1990 and 1991a (revised version Hosokawa forthcoming) for descriptive details of Yawuru grammar; and Hosokawa 1988 and 1994 for the sociolinguistic situation. Yawuru examples are presented in a spelling convention as follows: *rd*, *rn*, *rl* and *rt* are digraphs for retroflex (apico-postalveolar) consonants (no lengthening of vowel before the digraphic *r* is implied); *dy*, *ny*, *ly* and *ty* are digraphs for lamino-postalveolar consonants; *r* (postalveolar approximant), *rr* (alveolar flap) and *rry* (palatalised flap) are dintinguished; *ng* is the velar nasal; *y* is an unrounded palatal approximant and *w* a slightly rounded labio-velar approximant. Vowels are *a*, *i*, *u* and corresponding long vowels (indicated by the doubling of the vowel letters). Stress is usually on the first syllable. Word-final *p*, *t*, *rt*, *ty* and *q* are tense unreleased stops. All the example of Yawuru sentences in this paper are taken from spontaneous speech of the native speakers unless otherwise noted.
3. This is cognate with the non-personal prefix *ni-* in Nyulnyul described in McGregor (1995), and similar forms in Bardi, Nyikina, and other Nyulnyulan languages.
4. Precisely speaking, Yawuru case markers are enclitics rather than suffixes (see Hosokawa 1991a: sections 2.2.2, 2.5.2 and 6.1). However, the distinction between clitics and affixes does not affect the discussion in this paper, and the term suffix will be used throughout to cover both suffixes and case-marking (and several other) enclitics in Yawuru.
5. Abbreviations used in this paper (mainly in the interlinear glosses) are as follows: 1min – 1st person minimal (similarly 2min, 3min and 12min); 1aug – 1st person augmented (similarly 2aug, 3aug and 12aug); 1 – (in interlinear glosses) verbal prefix or enclitic referring to 1st person minimal-number (similarly 2, 3 and 12); 1″ – (in interlinear gloss) verbal prefix or enclitic referring to 1st person augmented-number (similarly 2″, 3″ and 12″); 12 – the fourth person, or the speaker and the hearer(s) included (as a unit category of grammatical person, not of number); A – transitive subject; ABL – ablative; ABS –absolutive; ACC – accusative; AGM – (in interlinear glosses) augmented number; AUX – auxiliary; CAUS – causal; COM – comitative; DAT – dative; du, DU – dual; DUR – durative; EN – epenthetic nasal; ERG – ergative

(ergative NP, ergative form of a personal pronoun); FUT – future; GEN – genitive; IMPF – imperfective (aspect); INST – instrumental; INTENS – intensifier; IRR – irrealis; JUSS – jussive (mild imperative); LOC – locative; NOM – nominative; O – transitive object; pcl – paucal; pl – non-dual plural (more than three); PF – perfective (aspect); PRES – present; REF – reflexive; S – intransitive subject; TOP – topic; TR – transitive conjugation marker.

6. The prefix element *ni-* of *ni-mirdi* 'shin' was historically the marker of the third person singular possessor. The initial nasal of *nungu* 'belly' probably comes from the same source. See note 3.
7. Such splits in case-marking patterns are not uncommon in Australian Aboriginal languages. See Blake (1987: 13); cf. Dixon (1980: 328-330, 338ff).
8. Issues related to this neutralisation are discussed in chapter 7 of Hosokawa (1991a).
9. See Hale (1982: 248-250) and Nash (1980: 197-200) for the ergative-dative construction in Warlpiri. Semi-transitive is sometimes called "middle" (e.g. Jagst 1982: 46).
10. In other Aboriginal languages, such as Gooniyandi, reflexive/reciprocal verbs take the subject (actor) in the ergative case (McGregor 1990: 318). Cf. example (43) discussed in 5.5.
11. In Gooniyandi, however, ABS_w is often a nominal, and the normal word order is for it to precede ABS_p. Gooniyandi has also a construction of the type ERG_p ERG_w *w*-Verb (Bill McGregor, pers.comm.).
12. A possible parallel is the medio-passive construction reported from Djaabugay (Blake 1977: 48; cf. Hale 1976: 324).
13. See Hosokawa 1990 and 1991a (chapter 5) for details of complex verbs in Yawuru.
14. The word *dyimbin* 'inside' often refers to a jail or temporary lock-up at a police station.
15. The semantic range of the verbs that occur in the Yawuru semi-transitive clauses seems to be more or less shared by other languages which have comparable construction types. As for nearby Australian languages, Gooniyandi (McGregor 1990 and pers.comm.) and Warlpiri (Nash 1982: 190-196) have similar lexical entries for the verbs with the middle (i.e. ERG-DAT) construction type. Bob Dixon (pers.comm.) suggests that the range in question is recurrent outside Australia: it is observed, on the one hand, for a particular group of transitive verbs in Northeastern Caucasian languages that take a DAT-ABS (rather than ERG-ABS) case frame; and, on the other, for a group of intransitive verbs that require an ABS-DAT case frame in Polynesian languages.
16. Actually they are not "agents" in the strict semantic term, but rather causes or causers of the described incidents.
17. In this respect, Yawuru quasi-passive is comparable to the split verb agreement systems reported from Dargwa (Northeastern Caucasian) and Wichita (Caddoan). In Dargwa, when transitive object is 1st or 2nd person and the transitive subject is 3rd person, then the verb tends to agree with the former, rather than with the latter (Anderson 1985: 197). In Wichita, where a 3rd-person subject noun is possessed by 1st or 2nd person, verbal agreement may be with the possessor, rather than with the possessed (Anderson 1985: 197).
18. The word order of (23) can also be *dyiya marru*, with the same meaning. Unlike many other Aboriginal languages, Yawuru does not employ a simple adposition (*marru dyuyu* or *dyuyu marru*) to encode possession (or the whole-part relation). Non-

pronominal possessors (e.g. Dicky's spear) are usually encoded in the dative phrase: spear + Dicky-DAT. See Hosokawa (1991a: section 6.3) for further accounts of dative/genitive expressions in Yawuru.

19. In example sentences, words in parentheses are added after consultation with native speakers. When the original sentences were recorded in natural speech, those words were not spoken.
20. The verb *dyu* 'say' is transitive and requires an ERG-ABS case frame.
21. Cf. the discussion of Blake (1987: 95-98).
22. Example (34) could also be interpreted in such a way that the noun *bidyara* and the pronoun *ngayu* comprise a unified nominal compound and thus case markers do not intervene: i.e. [*bidyara-ngay(u)*]$_N$ *-ni nga-likarra-rn durrgadya karda*. Even though this interpretation holds, the problem of agreement remains: the compound noun is crossreferenced by the 1st-person prefix, not the 3rd person.
23. The phrase *niwal-gun dyanu* in (36) constitutes a single NP. See Hosokawa (1991a: section 7.1) for the encoding of the body-part instrumentality into the locative case (rather than the instrumental).
24. This is a somewhat simplified account. There are verbs that may take only one of the two affixes, as well as those that are reflexivised by some other morphological devices. A comprehensive description of the reflexive verbs in Yawuru is beyond this paper's scope. See Hosokawa (1991a: sections 4.2.7, 4.7 and 7.4.2.6) for detailed accounts of Yawuru verbal and pronominal reflexives.
25. In this particular example of double-object construction, both objects, 'meat' (i.e. animal) and 'hair' (i.e. its skin), are overt. (The word *warli* 'meat' here refers to an entire game animal, not to a part of it; such a use of the term is quite common in Australian Aboriginal languages.) The part noun *kungkulu* 'hair' is not defocused probably because of the limitative suffix *-manyan* attached to it.
26. It is interesting in this regard that in Nyulnyul, a closely related sister language of Yawuru, nouns of clothing items (namely, hats, glasses, bandages, shoes and shoelaces) are found in the class of prefix-taking nominals, as well as terms for body parts and personal representations (McGregor 1995 and pers.comm.).
27. Here the irregular verb *ma* 'put' is inflecting in the 1st-person reflexive jussive (mild imperative).
28. (67a) becomes quite acceptable when it is embedded into a complex clause structure (e.g. *I didn't go out as my car was out of order*).
29. As a matter of fact, native speakers of English (particularly linguists) disagree on the acceptability of (68b) and (68c). Some speakers claim that they would "put a star" on them, whereas others maintain that these are quite acceptable sentences. It is intriguing to note here that the former view was expressed by those working in formal frameworks and the latter by those in systemic grammar. This is perhaps not accidental. I understand that the very ideas of "well-formedness" and "acceptability" unavoidably reflect our theoretical viewpoint.
30. Jacaltec is an ergative language in which a set of bound pronominals functions both as the marker of subject (agent) on transitive verbs and as the marker of possessor on nouns (Craig 1977: 137-138). Attaching the agent set to body-part nouns renders possessed forms. As for kinship terms of reference, however, further morphological changes are required to produce possessed forms (Hosokawa 1991b, Day 1973).

31. Unlike the case for Mayali (Evans 1995), bodily exuviae (and also 'nest') are excluded from the list of the items subject to the body-part constructions in Yawuru.
32. Even with this contextualisation, sentence (72) still sounds somewhat unnatural, since the word *nilawal* 'name' usually refers to the traditional personal names that are inherited through the system of kinship sections.
33. The Yawuru word *rayi* 'dreaming' can refer to various different kinds of "dreamings" (totems) that a person holds, such as personal dreamings (usually conceptional), inherited family dreaming(s), local dreaming (associated with one's birth place) or clan dreaming(s). It most commonly refers to one's conceptional dreaming, usually a plant or an animal that has either come up in the parent's dream, or has made a special kind of encounter with the parent(s), before the person was born. The same word *rayi* also functions as an adjective 'secret, hidden' and as a preverb 'hiding something, making something secret'.

References

Anderson, Stephen R.
1985 "Inflectional morphology", in: Timothy Shopen (ed.), 150-201.

Blake, Barry J.
1977 *Case marking in Australian languages.* Canberra: Australian Institute of Aboriginal Studies.
1987 *Australian Aboriginal grammar.* London: Croom Helm.

Chappell, Hilary
1995 "Inalienability and the personal domain in Mandarin Chinese discourse", [this volume.]

Clark, Marybeth
1995 "Where do you feel? – stative verbs and body part terms in mainland Southeast Asia", [this volume.]

Craig, Collette G.
1977 *The structure of Jacaltec.* Austin: University of Texas Press.

Day, Christophor
1973 *The Jacaltec language.* Blomington: Indiana University Press.

Dixon, Robert M.W.
1980 *The languages of Australia.* Cambridge: Cambridge University Press.
1987 "Studies in ergativity: introduction", in: Robert M.W Dixon (ed.), 1-16.

Dixon, Robert M. W. (ed)
1976 *Grammatical categories in Australian languages.* Canberra: Australian Institute of Aboriginal Studies.
1987 *Studies in ergativity.* Amsterdam: North-Holland.

Evans, Nick
1995 "The syntax and semantics of body part incorporation in Mayali", [this volume]

Foley, William A. – Robert D. Van Valin
1984 *Functional syntax and universal grammar.* Cambridge: Cambridge University Press.

Hale, Kenneth L.
1976 "Dja:bugay", in: Robert M.W. Dixon (ed.), 321-326.
1982 "Some essential features of Warlpiri verbal clauses", in: Stephen Swartz (ed.), 217-315.
Hosokawa, Komei
1988 "Fate of Yawuru in the context of social reorganisation of Broome Aborigines", [Unpublished MS, to appear in: Patrick McConvell (ed.), *Can Aboriginal languages survive?: language shift and maintenance in Aboriginal Australia*. St. Lucia: Queensland University Press.]
1990 "Preverb in Yawuru: a functional slot for predicate formation", in: Osamu Sakiyama – Akihiro Sato (eds.), 248-291.
1991a The Yawuru language of the west Kimberley: a meaning-based description. [Unpublished PhD dissertation, Australian National University.]
1991b "Hakarutekkugo" [The Jacaltec language (of Guatemala)], in: Takashi Kamei – Rokurô Kôno – Eiichi Chino (eds), 103-110.
1994 "Retribalization and language mixing: aspects of identity strategies among the Broome Aboriginis, Western Australia", *Bulletin of the National Museum of Ethnology* 19(3): 491-543.
forthcoming *Meaning in Yawuru: a semantically-oriented description of an indigenous language and culture.* (Pacific Linguistics, Series C.) Canberra: Australian National University.
Jagst, Lothar H.
1982 "A tentative description of Ngardilpa (Warlpiri) verbs", in Stephen Swartz (ed.), 1-68.
Kamei, Takashi – Rokurô Kôno – Eiichi Chino (eds.)
1991 *Gengogaku daiziten III: sekai gengo hen (3)* [Encyclopaedia of Linguistics, volume 3: Languages of the world, fascicule 3A]. Tokyo: Sanseido Publishers.
Langdon, Margaret
1988 Possession and kinship in Yuman languages. [Unpublished MS presented to Linguistics Department, Monash University.]
Leeding, Velma
1995 "Body parts and possession in Anindilyakwa", [this volume]
McGregor, William B.
1984 A grammar of Kuniyanti: an Australian Aboriginal language of the southern Kimberley, Western Australia. [PhD dissertation, University of Sydney.]
1988 *Handbook of Kimberley languages, volume 1.* (Pacific Linguistics, Series C-105.) Canberra: Australian National University.
1990 *A functional grammar of Gooniyandi.* Amsterdam: John Benjamins.
1995 "The grammar of nominal prefixing in Nyulnyul", [this volume.]
Miyaoka, Osahito
1991 "Hokubei Indian syogo" [The languages of North American Indians], in Takashi Kamei – Rokurô Kôno – Eiichi Chino (eds), 1004-1078.
Nash, David G.
1980 Topics in Warlpiri grammar. [Ph.D. dissertation, Massachusetts Institute of Technology.]
1982 "Warlpiri preverbs and verb roots", in: Stephen Swartz (ed.), 165-216.

Sakiyama, Osamu – Akihiro Sato (eds.)
1990 *Azia no syogengo to ippan gengogaku* [Languages of Asia and general linguistics]. Tokyo: Sanseido Publishers.
Shibatani, Masayoshi
1985 "Passive and related constructions: a prototype analysis", *Language* 61(4): 821-848.
Shopen, Timothy (ed.)
1985 *Language typology and syntactic description, volume III: grammatical categories and the lexicon.* Cambridge: Cambridge University Press.
Swartz, Stephen
1982 *Papers in Warlpiri grammar: in memory of Lothar Jagst.* (Work Papers of SIL-AAB, A-6). Darwin: Summer Institute of Linguistics (Australian Aborigines Branch).
Tsunoda, Tasaku
1981 *The Djaru language of Kimberley, Western Australia.* (Pacific Linguistics, Series B-78.) Canberra: Australian National University.
1995 "The possession cline in Japanese and other languages", [this volume.]
Wurm, Stephen A.
1972 *Languages of Australia and Tasmania.* The Hague: Mouton.

Body parts and possession in Anindilyakwa

Velma J. Leeding

1. Introduction

Anindilyakwa [eninhthilyakwa] is a non-Pama-Nyungan language, the sole member of the Groote Eylandt Family (Oates 1970: 15) or Andilyaugwan Family (Wurm 1972: 117). As Yallop (1982: 40) reports, Anindilyakwa and the neighbouring mainland language, Nunggubuyu, "are similar in grammar and possibly share the distinction of being the most grammatically complex Australian languages. They are diverse in basic vocabulary, however, and are therefore allocated to separate families". The language is spoken by just over 1,000 Warnindilyakwa Aborigines on Groote Eylandt, Northern Territory, Australia. The Umbakumba communalect differs from the one at Angurugu in its more strident voice pitch, the use of laminopalatal consonants for some laminodentals and in a few vocabulary items.[1] With intermarriage and job-oriented residency, there is no longer a clear division between the two communities.

Anindilyakwa is a multiple-classifying prefixing language in which all traditional nouns, adjectives, personal and demonstrative pronouns are prefixed for person, number and gender.[2] This prefixation differs from that of many other Aboriginal languages in that the noun classes form part of a gender system that includes first and second as well as third person. There is prefixal agreement between words within the noun phrase and crossreferencing of the subject, object or indirect object of the verb. The deep structure prefix sets are similar but not identical for all word classes, the main difference being for the feminine gender. The different sets of verb prefixes distinguish seven moods (Actual, Negated Actual, Potential, Negated Potential, Imperative, Hortatory and Purposive) which interact with the tense and aspect suffixation. Nominative and accusative cases are marked in the verb prefixation but not on the corresponding noun phrases. Nouns and adjectives, unlike most other Aboriginal languages, have different internal structures (see Dixon 1980: 274-5).

The phonological system for modern Anindilyakwa includes the widest possible range of consonants for Aboriginal languages but, traditionally, the distinction between alveolars and postalveolars was not phonemic. An unusual distinction exists between peripheral and non-peripheral consonants, the former being both rounded and unrounded and the latter only unrounded. The two phonemic vowels (high central and low central) are rounded and backed by rounded consonants or fronted by laminal consonants. The Arandic and Tiwian languages are the only others in Australia known to exhibit a two vowel system. Anindilyakwa words are long, consisting of between three and fourteen syllables. Numerous complex morphophonemic rules reduce the full forms of the deep structure morphemes, thus making the identification of surface structure morphemes extremely difficult. Morpheme-final deep structure vowels are neutralised to the high vowel word-medially and the low vowel word-finally.[3]

The aim of this paper is to describe the way in which body part terms function within the morphological and syntactic structure of Anindilyakwa. Three phenomena are discussed: (a) the semantic classification of body part terms in relation to the overall noun class system; (b) inalienable, semi-alienable, alienable and kinship possession of body parts; and (c) the incorporation of body parts into the morphological and syntactic structure of nominals (nouns and adjectives) and verbs (transitive and intransitive).

2. Semantic classification of body parts

Anindilyakwa nominals are all assigned to a specific noun class. Traditional words are obligatorily prefixed; loanwords are usually not prefixed but have their classification indicated by adjective or demonstrative modifiers. The major body parts are obligatorily classified within this semantic system; the lesser body parts are classified in agreement with one of the major parts.

2.1. Noun classification

Nouns are assigned to five classes in Anindilyakwa and adjectives are marked in agreement with the nouns they modify. Five third person noun classes form part of a gender system which identifies (first, second and third) person, (singular, dual, trial and plural) number and (masculine or feminine) gender. The surface structure gender prefixation for nominals (excluding trial) is as follows:

ngarri-	1st inclusive plural
yi-	1st inclusive dual
yirri-	1st exclusive plural
yirringi-	1st exclusive feminine dual
yini-	1st exclusive masculine dual
ningi-	1st exclusive singular
kwirri-	2nd plural
kwirringi-	2nd feminine dual
kwini-	2nd masculine dual
ningkwi-	2nd singular
wirri-	3rd WIRRA plural
wirringi-	3rd WIRRA feminine dual
wirrini-	3rd WIRRA masculine dual
thi-	3rd THA feminine singular
ni-	3rd NI masculine singular
a-	3rd A (number neutral)
mwi-	3rd MWA (number neutral)

The noun class system can be analysed as a hierarchical network of binary oppositions. The highest rank divides the nouns into personified and nonpersonified. The personified is further subdivided according to singular versus nonsingular number, masculine versus feminine gender and inherently human versus ultimately nonhuman. The nonpersonified oppositions are based on the presence or absence of visibility and lustre. Nonsingular number can be explicated for the nonpersonified but not within the noun class system. The hierarchical system is diagrammed in Figure 1, with the relevant deep structure noun class prefixes listed beneath each category. The taxonomic structure is not derived from a single node but a paired opposition.

The semantic division between of personified and nonpersonified does not correspond with a distinction between animate and inanimate. The personified noun classes include mainly humans and nonhuman animates who acted rationally or in human roles in the traditional "creation" era, e.g., they married, had children, spoke, thought, evaluated, fought with implements or changed their own appearance. These animate beings retain a sibling relationship to each other and to humans even though they are manifested nowadays as animal life. The personified classes do,

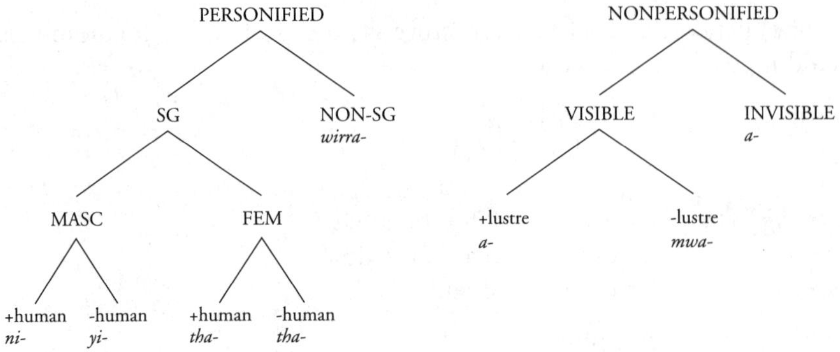

Figure 1. Semantic oppositions for noun classification

however, have a small number of inanimate items which are classified in agreement with these Dreamtime beings and/or have harmful supernatural power. The nonpersonified classes are unmarked categories incorporating the remaining nonhuman animates and the majority of inanimates. The term *lustrous* does not refer to any specific colour but to items that reflect light and colour. The invisible or "never visible" category is restricted to speech and the elements.

The five noun classes and their classifying prefixes are listed below. The paired subclassification for human versus nonhuman for all personified classes is necessary because of morphological restrictions in other parts of the data. Body part prefixation assigns the terms to one of these five noun classes and should not be considered pronominal crossreferencing to a possessor.

Personified classes:

3:1(i)	WIRRA	two or more human animates, spirits
3:1(ii)	WIRRA	pairs/flocks/colonies of non-human animates; clusters of inanimates such as fruit
3:2(i)	THA	human feminine animates
3:2(ii)	THA	non-human animates classified as feminine; inanimates with supernatural power
3:3(i)	NI	human masculine animates
3:3(ii)	YI	non-human animates classified as masculine; inanimates with supernatural power

Non-personified classes:

3:4	A	invisible items and visible items with a lustrous appearance (animate and inanimate)
3:5	MWA	visible items that are lustreless (animate and inanimate)

2.2. Classified body parts

The names of body parts and normal byproducts fall into only three noun classes: 3:3(ii), 3:4 and 3:5. The placement of body parts into the nonhuman masculine 3:3(ii) class, but not the nonhuman feminine 3:2(ii) class, appears to be an arbitrary choice. Both class markers, however, are used when describing abnormalities which are found in or on the human body, e.g. *yiningwinya* 'pimple, fontanelle', *tharririkwirikwa* 'swollen glands in the neck and armpit', *thirimpwa* 'wart, mole', *thinyinya* 'Pterygium (eye disorder)', *thimwingkwirrartha* 'varicose veins (associated with pregnancy)'. Such feminine classification is based on association with the human female or named after THA class items with a similar appearance, e.g., swollen neck glands with the Peaceful Dove; a wart with a tadpole.

Personified body parts are classified as Class 3:3(ii) because of their association with spiritual/supernatural powers and/or the ceremonial world. All sickness or unexpected death is associated with the spirit world. In Table 1, body part 1 is the seat of the human spirit and its emotions; in 2, the greyness may be attributed to the supernatural as a sign of approaching death; fingernail clippings 3 are used for sorcery; and 4-5 are associated with sexual activity and reproduction. Body parts 6 and 7 are of ceremonial significance: blood from opened veins was used as an adhesive for ceremonial body decorations; the tendon from a wallaby's tail was used in making ceremonial spears; and the trachea/oesophagus has the same name as the religiously-important didjeridu (musical pipe). Body part 8 may be simply a loanword from Nunggubuyu because, while some speakers differentiate the meanings of 7 and 8, others insist they are synonyms and both include all three body parts. Reference is usually made to body part 9 when the gland is swollen and its prefixation is probably due to the fact that all sickness is considered to result from supernatural (spiritual) activity. No explanation is available for the classification of 10.

The majority of body parts are classified under 3:4 A or 3:5 MWA noun classes, according to their luminous or nonluminous appearance. All body parts are considered to be visible since they are seen in dissected

Table 1. Personified body parts

1	*yikwithikwitha*	chest
2	*yikwira*	grey hair
3	*yinhanha*	fingernail, toenail
4	*yakika*	scrotum
5	*yilhirra* ~ *yiwarra*	vagina
6	*yayarra*	vein, tendon (Achilles and the one above the elbow)
7	*yiraka*	trachea, oesophagus
8	*yampwiya*	neck
9	*yingampwa*	gland in the groin
10	*yina* ~ *yiwalkwirra*	knee, kneecap

animals (see also Crowley 1995). The body parts in Noun Class A 3:4 are either the internal organs which are inherently shiny or the external areas which catch the light; the Noun Class MWA 3:5 body parts are the internal organs that lack lustre or external areas which are covered by hair or recessed (shadowed) and do not normally reflect the light. For example, on the back of the body from the base of the skull to the leg calf the shoulder and buttocks areas are the only ones which are seen as "catching the light". The most obvious contrast between the two classes is the assignment of the front teeth to A class and the molars to MWA class. One of the least obvious is the concept of head hair as lustreless; this is for cultural reasons – its shininess is regarded as unusual and a sign of pregnancy.

The normal body products fall into Noun Classes YI 3:3(ii), A 3:4 and MWA 3:5 on the same basis as the above. Clear body fluids 56-61 are classified in the same way as fresh water which reflects light and other images; opaque body excreta 55 (MWA not YI class) and 62-64 are in the same MWA class as the nonreflective sea or salty water.

The names of human body parts and natural byproducts are normally used for the corresponding parts of nonhuman animates, e.g., *mwiyatja* 'shoulderblade, scapula' is also used for the wing of a bird or the suprascapula of a fish; *mwatha* 'ear' is one of the synonyms for the horn of an introduced domesticated animal. There are, however, a few body parts which are primarily associated with nonhuman animates and do not refer to human body parts. Body parts 65-68 in Table 4 are classified in the MWA noun class according to their lustreless appearance. The Macassan loanword 69 is categorised as either A or MWA noun class depending upon whether the individual item is lustrous or nonlustrous; the Macassan loanword 70 is classified as 3:2(ii) in agreement with the noun class of the particular turtle.[4]

Table 2. Non-personified body parts

Lustrous, A class:		
11	*aringka*	head
12	*arra*	forehead
13	*amwingkwirra*	cheek (fatty tissue)
14	*aminta*	nose
15	*athirra*	mouth
16	*alyalyikpwa*	lips, tongue
17	*akarrnga*	front teeth
18	*angirnta*	chin, jaw
19	*amwirntha*	shoulder
20	*ayantha*	upper back (shoulderblade area)
21	*angangpwilya*	armpit
22	*angamwilha*	fatty tissue above the breasts
23	*alyikilyingpwarrnga*	lungs
24	*arntirnta*	heart
25	*awa*	liver
26	*ayarrmwirra*	upper arm, biceps muscle
27	*arnta*	elbow
28	*ayarrka*	lower arm (elbow to fingertip)
29	*arrngka* ~ *anka*	hip
30	*akpwalha*	abdomen
31	*akpwa*	buttocks
32	*alhakpwa*	lower leg, shin (knee to ankle)
33	*alhika*	foot
34	*angwirnta*	ankle
35	*ayamwa*	body hair
36	*amwingina*	small joint (e.g., wrist, ankle, finger, toe)
Lustreless, MWA class:		
37	*mwamwingpwa*	head hair
38	*mwangmwa*	brains
39	*mwamwilha*	nostril hair
40	*mwanpwa*	eye
41	*mwarmpwa*	molar teeth
42	*mwatha*	ear
43	*mwamwirrmwa*	nape of the neck
44	*mwiyatja*	shoulderblade
45	*mwirnga*	spine
46	*mwamwirrkwira*	ribs (side of the chest)
47	*mwirirrpwa*	back (from shoulderblade to sacrum)
48	*mwangpwangpwirra*	sacrum, central lower back
49	*mwing(w)ining(w)ina*	hip joint, pelvis
50	*mwilhkwa*	abdominal sac, i.e., stomach, womb, bladder, small intestine
51	*mwikwikwiwarnantha*	large bowel, large colon
52	*mwarrikwa*	penis
53	*mwakarra*	thigh
54	*mwamwarrpwa*	leg calf, calf muscle

Table 3. Body products

55	*yima* ~ *yimi*	faeces
56	*angwatha*	tears
57	*anyirra*	nasal mucous
58	*arrtha*	earwax
59	*alyikwalya*	saliva
60	*angamwina*	breast milk, breast
61	*amwinha*	urine
62	*mwikwana*	perspiration
63	*mwarra*	blood, menstruation
64	*mwamwingwirra*	semen

Table 4. Nonhuman body parts and products

65	*mwarntja*	(bee) brood cells
66	*mwangkarriya*	yellow (turtle) fat, (bird or turtle) egg yolk, stored pollen
67	*mwarapwarra*	(mammal or reptile) tail
68	*mwanintja*	(fish) dorsal fins
69	*karntirra*	(land mammal) horn
70	*karrpwa*	shell (Hawksbill Turtle)

A few additional body parts given in Table 5 appear to have been obligatorily prefixed in the traditional language by the inalienable possession marker *mwi-* which follows the noun classifying and gender prefixes (see section 3.2 below). There appears to be a certain amount of semantic cohesion in that they are either (a) bound nonphysical or spiritual terms; or (b) the body parts referred to are not small sections but cover larger areas in which the parts are individually named, for example, some extend from head to toe. Other similar terms, however, do have free forms. Examples 73, 78 and the loanword 80 only occur as bound forms and, therefore, are not classified semantically within the noun class system. The terms 74, 76 and 77 have been offered as dictionary entries but are rarely used in this form in any current texts. Other roots can be free forms and are also classified as class A 3:4 nouns but only with an extended meaning: 72 *athangkwa* 'certain meat on the back of turtle or fish', 73 *apinga* 'anthill', 75 *athithira* 'bone, shell, nut, hard seed', and 79 *awarrwalya* 'shade of any inanimate object'.

2.3. Incorporated body parts

In Anindilyakwa, there is another set of body part roots which can be incorporated within a verb as subject or object or as part of a compound

Table 5. Bound body parts roots

71	*-kwilyi*	skin
72	*-athangkwi*	flesh, body
73	*-apingi*	torso, (fat) body
74	*-mwalya*	body fat
75	*-athithira*	bone
76	*-mwikwirra*	face
77	*-kirri*	name
78	*-awirrini*	spirit
79	*-warrwalya*	shadow
80	*-tjingwila*	feathers

nominal stem. The incorporated and isolated body parts are not necessarily the same morphemes, as is usually the case in Mayali (Evans 1994) and Ngalakan (Merlan 1983: 143-146). Only about 40% of the incorporated roots show any similarity with corresponding free form nouns. It is easier to define the structure of the free form noun in terms of the incorporated root, that is, the free form body part term usually consists of the first two syllables of the incorporated root (minus any resulting word-final consonant).

The incorporated body part noun roots are not prefixed by nominal classifiers and all end with *(ng)kwi* or *(ng)ka* in their fullest form. This morpheme-final syllable is similar in shape to the nominaliser but cannot be isolated from the morpheme because (a) the morphophonemic rule for neutralisation of morpheme-final word-medial vowels does not operate and (b) morphemes always end in a vowel, not the consonant /*rr*/. The velar consonant /*kw*/ in the final syllable can dissimilate to the labial /*pw*/ following a velar nasal. Many of the incorporated roots have a shortened variant in which the medial syllable is usually deleted. Each incorporated root corresponds with a free root but it is not in a one-to-one correspondence, as shown in the second column of Table 6. In the following list of incorporated body parts, the numbers in the second column crossreference the entries to those in section 2.2.; the asterisk indicates the incorporated and isolated forms are similar in morphemic shape; and the root-initial laminopalatal consonant freely varies with a laminodental.

There are some incorporated noun roots which do not all appear to correlate with a body part. Three such roots do not have free forms: *lharrngkwi* ~ *lyarrngkwi* 'unspecified things', *mwakwi* 'camp, residence' and *ngwirrkwi* which appears to have a primary meaning of 'food'. The incorporated root, *kwikwi* 'water', is a Common Australian term and, in

Table 6. Incorporated body part noun roots

81	1,60	*nyaka*	chest, breast
82	2,37	*mwilhkwi*	head hair
83	3	*rnarrka ~ narrka*	fingertail, toenail
84	4	*arntirrka*	scrotum
85	7*	*raka*	trachea, oesophagus
86	9*	*ngampwaka*	gland in the groin
87	10,49*	*ng(w)inaka*	knee and pelvic joints
88	11*	*lyangarngka ~ lyangka*	head, fontanelle, clitoris
89	11*	*warngka ~ wanka*	head
90	12*	*rraka*	forehead
91	13*	*mwingkwirrkwi ~ mwingkwi*	cheek, temple
92	16*	*lyikpwirrkwi ~ lyikpwi*	lips, tongue, sternum
93	17,41	*warrka*	teeth
94	18*	*ngirntarrka*	chin, jaw
95	19	*ngarngka*	shoulder
96	22,23*	*lyikarrngka ~ lyikarrka*	breast area, lungs
97	24,25,74	*lyingarrka*	liver, heart, kidneys, body fat
98	27*	*arntaka*	elbow
99	28,36	*ngwiyangka ~ ngwangka*	lower arm, hand, joint
100	28	*mwarntakwi ~ mwangpwi*	hand, (turtle) flipper
101	29*	*angkarrka ~ arrka*	hip
102	30*	*akpwalhka ~ pwalhka*	abdomen, body fat
103	32*,52	*lhakpwaka*	leg, shin
104	33*	*lhirraka*	foot
105	34*	*ngwirntirrka*	ankle
106	35,39,65	*lyingkwi ~ lyingpwi*	body hair, nostril hair, bee brood cells
107	38,56,59	*mwilyingka*	brains, saliva, tears
108	40*	*arngarngka ~ a(r)ngka*	eye
109	42	*ngarrkwi*	ear, varicose veins
110	44,80	*alhpwilhka ~ alhka*	shoulderblade, feathers, wing
111	45*	*mwirngka ~ pwirngka*	spine
112	45,46,67 69,75	*lharrka*	spine, ribs, tail, horn, bone
113	47*,72,73	*ripwirrkwi ~ rikwi*	back, body, torso
114	48*	*mwangpwirrkwi*	lower back
115	50	*antjalhka*	abdominal sac
116	51*	*kwiwakwi*	large bowel
117	52,73	*lhingaka ~ lyingaka*	penis, torso
118	53	*ngaka*	thigh
119	57*,60,64	*mwilyirrkwi ~ mwirrkwi*	nasal mucus, breast milk, semen
120	61*,62,63	*mwinharrka ~ mwilharrka*	urine, perspiration, blood
121	63	*pwirrngkwi*	blood, joint fluid
122	71	*mwintjirrka*	skin
123	76*	*lhapwirrkwi*	face
124	78	*mwarngkwi ~ mwirangkwi*	spirit

Anindilyakwa, corresponds with the isolated form *akwingwa* in which the last velar stop has been lenited. There is, however, a possibility that it is connected with *kwiwakwi* 'large bowel' but this body part is defined by the Aboriginal speakers as "containing faeces" (not water).

Anindilyakwa has a second set of incorporated roots which do not substitute for any one specific free form body part term but can refer to the shapes of one or more body parts. The last syllable of these shape adjective roots is *(ng)ki* which correlates with the adjectiviser morpheme. They commonly occur as the first root of a compound stem in contrast with the set of descriptive adjective roots which occur only as the second root. Such roots not only take their shape from human or nonhuman animate body parts but can be incorporated instead of a body part noun root, e.g., the noun root, *ripwirrkwi* 'human back', can be replaced by the shape adjective root, *ramparrki* 'back-shaped, flat'.

The finite set of shape adjective roots is listed in Table 7. The same root can be used for more than one body part. In some instances, the shape adjective root is the only incorporated form for a specific body part but, in others, it freely varies with a body part noun root. The notation following the body part referent corresponds to those in section 2.2. (See also Harvey 1995 for the synedochic use of body part terms in Warray.)

There are some incorporated shape adjective roots for which there is no attributed body part. The root, *akaki* 'hot and burning', refers to 'fire' and has a corresponding shortened form in the noun *aki* 'firewood'; the root, *mwarrki* 'circular', is used of the moon and other circular items but may be associated with *(y)amwarrki* 'leg calf'; and the root, *(ng)arningki* ~ *ningki* 'pointed', primarily refers to a spear and has not as yet been observed used for any body part.

2.4. Extended meanings

For most of the classified body parts and products (section 2.2.), the correspondence between the human and nonhuman animate parts is quite obvious, e.g., *aminta* 'nose, beak'; *mwiyatja* 'wing, shoulderblade'. There are, however, other semantic links between such body parts which are culture specific. The name of a human body part is used when the part of the nonhuman animate or its product is seen as having (a) the same shape, e.g., *ayarrka* 'hand' is used for the 'palmate pectoral and ventral fish fins', *mwamwilha* 'nasal hair' for 'gills of a fish', and *mwatha* 'ear' for the ear-shaped 'sand ring in which the baler shellfish lays its eggs'; (b) the same

Table 7. Incorporated shape adjective roots

125	*warrki* ~ *walhki*	convex Referents: chest 1, emotions
126	*ampwirrki* ~ *(y)amwirrki*	ball-shaped, round solid Referents: scrotum 4, upper arm 26*, little finger 28, eye 40, neck nape 43*, neck glands, wart, mole, pterygium
127	*lyirraki* ~ *lhirraki*	open-mouthed, hole-shaped Referents: vagina 5*, mouth 15*, cupped hand 28, anus opening
128	*rraki*	elongated and flexible Referents: Achilles tendon 6*, tendon above the elbow
129	*(y)arrki*	forked Referent: vein 6*
130	*yalhki*	tubular Referents: neck 8, neck tendon, tendon at the back of the knee or in a wallaby's tail
131	*ngwaningki* ~ *ngwiningpi*	snub-nosed Referents: nose 14, faeces 55
132	*apwi(yi)ngki* ~ *yingki*	hump-backed, round-shouldered Referents: upper back 20, eye 40, body 73*
133	*ngangpilhki*	concave Referent: armpit 21*
134	*rang(pwarr)ki* ~ *rampwarrki*	flat Referents: back of hand 28, palm of hand 28, buttocks 31, sole of foot 33, back 47, breast 60
135	*lyaki*	elongated solid Referent: ribs 46
136	*yamwarrki*	oval solid Referent: leg calf 54*
137	*ampwaki*	curved Referent: U-shaped curve between the upper thighs 53
138	*wilyaki*	? Referents: ear wax 58, soft edge of the turtle shell 70

physical quality, e.g., *athithira* 'bone' is used for the hard, outer casing on shells (not the edible shellfish inside); and (c) the same function, e.g., *alhakpwa* ~ *athakpwa* 'leg' is used for the tail of a fish or ray because it performs the same function of propulsion as the human leg.[5]

Body part names are also extensively used for parts of inanimate objects. The usage is similar to that in such English expressions as: 'table leg', 'bedhead', 'mouth of a river', 'foot of a mountain', etc. In Anindilyakwa, it is particularly productive for describing parts of introduced items such as clothing and vehicles. Even though the name of the new artifact may be a phonologically-adapted loanword, its parts are named in traditional Anindilyakwa terminology. Some correlations with non-human animate and inanimate items are as shown in Table 8.

Table 8. Metaphorical equivalents of free or bound body parts

Body part	*Referent*
head	dress bodice, trouser top
nose	woomera hook, bowl of a Macassan smoking pipe
ear	centipede legs, spear (with pairs of curved hooks) named after the centipede
mouth	well of a dinghy, waterhole
teeth	pandanus thorns, shadow
lower arm, hand	front legs of an mammal, reptile, insect or spider
stomach	tyre tube
leg	end of a spear shaft into which a woomera fits, stern of a boat, back tyre of a bicycle
foot	car or bicycle tyre
heel	butt of a gun, bottom of a billycan
ankle	(raised) scar, knot in a vine or rope
vein	sun's rays, spreading roots, sea wasp tentacles, points of a star
body hair	foliage, grass, leaves, flowers
skin	tree bark, fruit
bone	coconut husk, seed, nut, shell
feathers	wool plucked from a garment
blood, joint fluid	fresh water, river, sea

Incorporated body part noun and shape adjective roots are used extensively in one of the systems for naming flora and fauna, and in cross-referencing a wide variety of animate and inanimate items within the root/stem structure (see section 4 for further discussion). In many cases, only one root describing a salient physical feature can refer to each particular entity but others are seen as having a choice of body parts or other physical features. For example, the bark of the Stringbark tree is crossreferenced by the root, *yalhki* 'neck-shaped or tubular', when first cut from the tree but by the root, *rampwarrki* 'flat', after being softened and flattened in the fire. The body parts shown in Table 9 refer to the items in the second column.

Table 9. Metaphorical equivalents of incorporated roots

Body part	*Referent*
head	largish stone, limestone ridge, spear head
eye	traditional shelter
mouth	cassette player, opened peanut tree pods, waterhole, cave
teeth	hand, foot
trachea	cup, jug, water tank, billycan, petrol drum
elbow	dress, house, root, vine
fingernail	oyster (Ostrea species)
shoulder	wing oyster (Pinctada species)
ribs	tree-covered plain, bed frame, long yam, road
spine	tripod-fish
torso	large fish or tree trunk, dugout canoe, log
feathers	grass, rush-lined swamp, spinifex-covered beach
body hair	leaves, grass, tree and its shade
scrotum	waterlily roots, pandanus fruit, wild figs
flipper	stingray flap, wallaby pouch, human hand
ball-shaped	mango, wild apple, wild prune
open-mouthed	billycan, cup
concave	axehead, knife, razor shell
flat	bark, road, galvanised iron sheet, blade of a shovel

3. Possession

Anindilyakwa distinguishes four types of possession: inalienable, semi-alienable, alienable and kinship. The proprietive and privative are considered to be inalienable and semi-alienable possession, respectively. The distinctions are marked by contrastive morphological features.

Inalienable possession in Anindilyakwa involves an intrinsic whole-to-part relationship in which a body part is seen as an inseparable part of the whole, whether that be a human or nonhuman animate or an inanimate object. Alienable possession involves the possession of something that is physically divorced from its owner and exists as a separate entity in its own right. Semi-alienable possession involves the possession of a separate entity by right of occupancy but the possessor's identity is inextricably linked to that space. The possessor of an alienably possessed item could be said to exert a choice or control over the possession of that item (Lynch 1973:76). No such choice exists within the realm of inalienable or semi-alienable possession.

Proprietive possession is a subclass of inalienable possession and denotes the ownership of either an intrinsic body part or physical attribute. The privative is a subclass of semi-alienable possession and denotes the unnatural divorcing of an item which is seen as the expected part of the whole, e.g., a child without a mother or a body without clothing.

Kinship possession primarily pertains to an intangible inherent union between two human animates, either by birth, marriage or adoption. The inherent nature of the relationship denotes inalienability but there is also an overlap with the alienable in that the possessor and possessed are separate entities. Associative possession expresses ownership by right of occupancy but, like kinship and alienable possession, the possessor and possessed are separable.

3.1. Structural background

The complexity and interdependence of morphophonemic rules in Anindilyakwa cause difficulty in identifying the deep structure morpheme from a variety of surface forms. All morphemes begin with a consonant or the vowel /*a*/. The Vowel Deletion Rule always deletes the first of two vowels, as in (1) in which the 3:4 nominal classifying prefix *a*- preceding a root-initial low vowel is deleted and in (3) and (4); a Vowel Lowering Rule lowers the high vowel between two similar rounded or laminal consonants, as in (3); the Haplology Rule deletes the first of two syllables in which the consonants are the same or similar, as in (3); the Morpheme-final CV Deletion Rule operates on the last syllable of a first order root, as in (2); the Assimilation Rule assimilates a nasal to the same point of articulation as the following stop or nasal, as in (2); the Hardening Rule hardens a morpheme-initial nasal or a lateral following a morpheme-final consonant, as in (2) and (3); and the Vowel Metathesis and Cerebralisation Rules interact and reduce the sequence /*irrani*/ to /*arni*/, as in (4). In all these examples, the deep structure morphemes have been placed below the vernacular so that the reader can identify them. Note the neutralisation of the word-medial morpheme-final vowel to a high vowel and the word-final vowel to a low vowel.[6]

(1) *aringka*
{a}-aringka
{3:4}-head
'head'

(2) *yiningwingwangpwa ~ yiningwingwampwa*
yi-{a}ni-mwi-ngw{iy}ang{ka}-mwi
3:3(ii)-GDR-INALP-hand-PROP
'four-legged mammal or reptile' (Literally: 'nonhuman male with hands')

(3) *nanimwamwalya ~ nanimwalya*
n{i}-ani-mwi-mwalya
3:3(i)-GDR-(INALP)-body:fat
'man' (Literally: 'human male possessing body fat')

(4) *wirranimwawirrina ~ warnimwawirrina*
wirr{a}-ani-mw{i}-awirrini
3:1(ii)-GDR-INALP-shadow
'their [dogs'] shadows'

The inalienable *mwi*- and associative *ngwi*- possession prefixes constitute a paired opposition in pre-stem position. Such possession nouns can be prefixed by the full set of nominal classifiers and the gender morpheme intervenes between the nominal classifier and the possession prefix. The two gender morphemes, *atha*- and *ani*-, distinguish human feminine and non-humanfeminine, respectively. Gender is marked when the nominal classifier signifies a personifed noun class but can also be marked for the non-personified and inanimate classifiers.[7] Variation between the two gender morphemes is found in noun class 3:2(ii) when the difference between the female sex and feminine gender is not always precisely identified. The following examples illustrate prefixation with inalienable possession nouns:

(5) *nanimwikirra*
n{i}-ani-mw{i}-kirri
3:3(i)-GDR-INALP-name
'his [human male's] name'

(6) *yinimwikirra*
yi-{a}ni-mw{i}-kirri
3:3(ii)-GDR-INALP-name
'its [wallaby's] name'

(7) *wirringathimwikirrikiya*
wirring{a}-atha-mwi-kirri-kiyi
3:1(i)FemDu-GDR-INALP-name-DUAL
'their two [human female's] names'

(8)

(8)	*thathimwikirra*	~	*thanimwikwirra*
	th{a}-atha-mwi-kirri		th{a}-ani-mwi-kirri
	3:2(ii)-GDR-INALP-name		3:2(ii)-GDR-INALP-name

'its [frog's] name'

The substitution of nominal classifying prefixes can be seen in the following data in which the compound stem remains constant but the prefixation changes. The Consonant Derounding Rule deletes the rounding on the velar consonants.

(9)	*yinimwikayarra*	*yapwingwirra*
	yi-{a}ni-{mwi}-mwik{wi}-ayarri	y{i}-apwi-ngwirra
	3:3(ii)-GDR-INALP-cheeks-forked	3:3(ii)-?-dangerous

'sea wasp's tentacles'

(10) *thathimwikayarra*
th{a}-atha-{mwi}-mwik{wi}-ayarri
3:2(ii)-GDR-INALP-cheeks-forked
thakwilyingatjanga
th{a}-a{tha}-kwi-lyingatjanga
3:2(ii)-GDR-NSR-pop:out(=star)
'the points of the Pleiades stars or Seven Sisters'

(11)	*mwamwikayarra*	*mwamwawira*
	mw{a}-a{ni}-{mwi}-mwik{wi}-ayarri	mw{a}-a{ni}-mw{i}-awira
	3:5-GDR-INALP-cheeks-forked	3:5-GDR-INALP-alone(=sun)

'sun's rays'

(12)	*amwikayarra*	*anhimwa*
	{a}-a{ni}-{mwi}-mwik{wi}-ayarri	a-nhimwi
	3:4-GDR-INALP-cheeks-forked	3:4-?mangrove

'spreading above-ground roots of a mangrove tree'

3.2. Inalienable possession

Inalienable possession implies that the possessed is intrinsically or "inextricably linked to the possessor who usually has no choice in the matter" (Lynch 1973:76). It can be subdivided into three different semantic notions: whole-to-part, part-to-component and proprietive. Inalienable possession is marked by the prefix *mwi-* which follows the nominal classifiers and precedes the root or stem. Such possession extends beyond the possession of a body part to that of other physical characteristics. The morphological structure of the inalienable possession noun is the same for the whole-to-part and part-to-component relationships but differs for the proprietive.

The inalienable possession prefixation is obligatory for the bound body part terms listed in Table 5 (section 2.2) but can also be used with a few of the free form body parts in Tables 1, 2, 3 and 6. Furthermore, the root or stem can be a descriptive root or a compound stem. The nominal classifying prefix crossreferences the possessor to a referent which may or may not be overtly stated in the same sentence. Within the noun phrase, the inalienable possession noun is always prefixed in agreement with its modifiers.

3.2.1. Whole-to-part relationship

The whole-to-part relationship involves inherent human or nonhuman animate ownership of a specific body part or some other physical characteristic. The same morphological structure is also used to relate inalienable parts of inanimate things such as implements, plants and topographical features to their "wholes". The part of a thing is seen as inalienable because the thing would cease to exist as an entity if one of its parts were removed, e.g., an axe without its head is no longer an axe but an axe handle.

Examples (13)-(18) illustrate the occurrence of a bound body part noun root (Table 5), a free form noun (Table 2) and an incorporated root (Table 6) in the whole-to-part relationship involving a human or nonhuman animate possessor. The referent noun for the "whole" is very rarely juxtaposed to the inalienable possession noun even though it is morphologically possible. No ambiguity arises because the "whole" animate is introduced into the discourse or conversation prior to mention of its parts (see also Hosokawa 1995).

(13) *yi-ni-mw-athithira*
3:3(ii)-GDR-INALP-bone
'its [wallaby's] bone'

(14) *ngarni-mwi-kirra*
1IncPl:GDR-INALP-name
'our names'

(15) *n-ani-mw-awirrina*
3:3(i)-GDR-INALP-spirit
'his spirit'

(16) *karni-mwa-warrwalya*
2Pl:GDR-INALP-tooth:soft(=shadow)
'your shadows'

(17) *a-mwa-mwi-yarrka*
3:4-GDR-INALP-hand
'its [fish's] pectoral fin' (Literally: 'its hand')

(18) *th-athi-mwi-ngarrkwa*
3:2(ii)-GDR-INALP-ear(=spike)
'its [stingray's] spike'

The specific *inanimate* possessor of a body part usually needs to be identified by the overt juxtaposition of a referent noun. As the nominal prefixation refers to any member of the relevant noun class, disambiguation is necessary in most instances. For clarity, the possessor is shown in all of the following data. Examples (19)-(23) illustrate inalienable possession nouns in which the root is a bound body part; examples (24) an incorporated root; and examples (25)-(30) a free form body part.

(19) *yi-ni-mw-akwilya* *kalhkwa*
3:3(ii)-GDR-INALP-skin ø-coconut
'coconut husk' (Literally: 'its skin, the coconut')

(20) *th-athi-mwi-tjingwila* *tjiwata*
3:2(ii)-GDR-INALP-feathers ø-sweater
'wool plucked from a woollen garment' (Literally: 'its feathers, sweater')

(21) *a-mw-athithira* *akwalya*
3:4-INALP-bone 3:4-fish
'shellfish' (Literally: 'its bone, the fish')

(22) *a-mw-athithira* *a-nhinga*
3:4-INALP-bone 3:4-food
'edible nuts' (Literally: 'its bone, the vegetable food')

(23) *a-mwa-mwalya* *a-yakwa*
3:4-INALP-body:fat(=soft) 3:4-speech
'true words' (Literally: 'its body fat, the words')

(24) *a-mwi-lhirrngwa* *tjinapwa*
3:4-INALP-?tongue ø-rifle
'rifle butt' (Literally: 'its tongue, the rifle')

(25) *a-mw-athirra* *a-kwingwa*
3:4-INALP-mouth 3:4-water
'freshwater hole' (Literally: 'its mouth, the freshwater')

(26) *mwa-mwi-walhkwirra* *mwa-mwarra*
3:5-INALP-kneecap 3:5-small:leaved:paperbark
'bark of the small-leaved paperbark' (Literally: 'its kneecap, paperbark')

(27) *th-athi-mw-aringka* *tiritja*
3:2(ii)-GDR-INALP-head ø-dress
'dress bodice' (Literally: 'its head, the dress')

(28) *mwa-mw-akarrnga* *mw-angkwirrkw-pwa*
3:5-INALP-teeth 3:5-?cheek-PROP(=pandanus)
'pandanus spikes' (Literally: 'its teeth, pandanus')

(29) *mwa-mwi-ngwirnta* *mw-arra*
3:5-INALP-ankle 3:5-rope
'knot in a rope' (Literally: 'its ankle, the rope')

(30) *yi-ni-mw-aminta* *yi-karpwa* / *yi-lyarra*
3:3(ii)-GDR-INALP-nose 3:3(ii)-woomera 3:3(ii)-pipe
'woomera hook, bowl of a Macassan pipe' (Literally: 'its nose')

The inalienable possession marker can be prefixed to a descriptive adjective root which describes the salient physical characteristic or shape by which each item is identified (see also Harvey 1995). Such nouns can take more than one nominal classifier, restricted only by the semantics or cultural world view. Some of these words can be used in a generic sense, e.g., the word for 'turtle egg' is an inalienable possession noun meaning 'the round part possessed by a YI Class nonhuman masculine animate' but its meaning has been extended to cover the egg of any creature whether YI class or not. The word, however, still remains part of a paradigm in which the substitution of a classifier indicates a change of referent to another round object such as a large seed, button, bead or marble.

(31) *yi-ni-mw-amwiwa*
3:3(ii)-GDR-INALP-round
'turtle egg'

(32) *th-athi-mw-amwiwa*
3:2(ii)-GDR-INALP-round
'egg of any THA class bird or reptile'

(33) *mwa-mw-amwira*
3:5-INALP-curved
'crab's pincer claw'

(34) *a-mw-amwira*
3:4-INALP-curved
'axe handle, crab's pincer claw'

(35) *yi-ni-mwa-walyiwa*
3:3(ii)-GDR-INALP-ripe
'flower' (e.g., the white gum flower)

(36) *a-mw-amwiwa* *a-nhinga*
3:4-INALP-round 3:4-vegetable:food
'edible fruit' (Literally: 'its round ones, the vegetable food')

Some of the above inalienable possession common nouns appear to have achieved status as the name of a species or a specific thing and, given the particular nominal prefix, cannot refer to several things sharing a common physical feature.

(37) *yi-mwa-pwalya*
3:3(ii)-{GDR}-INALP-soft[8]
'pied goose'

(38) **wirra-ani-mwi-ngarningki-alhikpwa* →
warni-ngw-animpw-alhikpwa
3:1(ii)-GDR-ALP-pointed-lips
'dingo, wild dog'

(39) *a-mwi-lyar-pwirra*
3:4-INALP-?head-wide
'many-lined sweetlips'

(40) *yi-ni-mw-alhilha*
3:3(ii)-GDR-{INALP}-spongy
'cypress pine needles'

(41) *yi-ni-ngwinya*
3:3(ii)-GDR-INALP-soft
'soft ball wattle'

(42) *a-mw-aningk-wa*
3:4-INALP-pointed-SF(=spear)
'stringybark sapling' (Note: It is used for spears)

(43) *mwa-mwi-rikwa*
3:5-INALP-raw
'road'

(44) *a-mwi-yarra*
3:4-INALP-forked
'traditional A-frame bush shelter'

3.2.2. Part-to-component relationship

The component part of another body part is indicated by the use of inalienable possession prefixation. This allows for the naming of what are considered to be the less significant parts of the body but only as inseparable parts of other major body parts, not the body itself. The nominal class of the possessed component is always in agreement with that of the body part. The free form body part noun (as listed in section 2.2.) and the inalienable possession noun always occur in apposition. Examples (45)-(47) and (48)-(53) illustrate the range of components that can be named in relation to one major body part; examples (53)-(56) show the versatility in the use of major body part names with various components.

(45) *mwa-mw-akwilya mw-an-pwa*
3:5-INALP-skin 3:5-eye-PROP
'eyelid' (Literally: 'eye's skin')

(46) *mwa-mw-ayamwa mw-an-pwa*
3:5-INALP-body:hair 3:5-eye-PROP
'eyebrow' (Literally: 'eye's body hair')

(47) *mwa-mwi-tjingwila mw-an-pwa*
3:5-INALP-feathers 3:5-eye-PROP
'eyelash' (Literally: 'eye's plumage')

(48) *a-mwi-ngangpwilya a-lhika*
3:4-INALP-armpit 3:4-foot
'arch of the foot' (Literally: 'foot's armpit')

(49) *a-mwi-kwithikwitha a-lhika*
3:4-INALP-chest 3:4-foot
'ball of the foot' (Literally: 'foot's chest')

(50) *a-mw-arirrpa a-lhika*
3:4-INALP-back 3:4-foot
'instep' (Literally: 'foot's back')

(51) *a-(mwa)-mwalya a-lhika*
3:4-INALP-body:fat 3:4-foot
'sole of the foot' (Literally: 'foot's body fat')

(52) *a-mwingkw-alhiwa a-lhika*
3:4-cheeks-few:and:small 3:4-foot
'toes' (Literally: 'foot's few and small')

(53) *a-mw-amwiwa a-lhika*
3:4-INALP-round 3:4-foot
'toes' (Literally: 'foot's round part')

(54) *a-mw-amwiwa a-ngamwina*
3:4-INALP-round 3:4-breast
'nipple' (Literally: 'breast's round thing')

(55) *a-mwikwirra* *a-ngamwina*
3:4-{INALP}-face 3:4-breast
'nipple' (Literally: 'breast's face')

(56) *mwa-mw-akwilya* *mwilhkwa*
3:5-INALP-skin {3:5}-abdominal:sac
'inside lining of the stomach' (Literally: 'abdominal sac's skin')

Some components appear to have achieved the status of proper nouns and, like examples (37)-(44), refer to a specific item unless otherwise stated. These words have been isolated and given as dictionary entries.

Same noun class as:

(57) *yi-mw-antjantji-nyirra* — knee
3:3(ii)-{GDR}-INALP-REDUP:stomach-soft
'joint fluid or knee cartilage' (Literally: 'its very soft stomach')

(58) *mwa-mw-alyikpwa* — ?abdominal sac
3:5-INALP-lips
'fatty yellow tissue around organs'

(59) *yi-ni-mw-aringka* — vagina
3:3(ii)-GDR-INALP-head
'clitoris'

(60) *mwa-mwirathina* — nape of neck
3:5-{INALP}-became:curved
'occipital area, base of the skull'

(61) *yi-ni-mw-amwira* — neck
3:3(ii)-GDR-INALP-curved
'upper neck and curve of the chin' (Cf:110-1)

(62) *yi-ni-mw-alyalyikpwa* — neck
3:3(ii)-GDR-INALP-tongue
'sternum, cartilaginous bone over the heart' (Literally: 'neck's tongue)

(63) *yi-ni-ngara* — neck
3:3(ii)-GDR-{ASSP}-?
'prominent neck tendon'

(64) *yi-ni-ngwinya* ?
3:3(ii)-GDR-{ASSP}-soft
'fontanelle, pimple'

(65) *yi-ni-ngwirnta* chest
3:3(ii)-GDR-{ASSP}-ankle
'cicatrice and other raised scars'

(66) *yi-ni-mwingkwirra* ?neck
3:3(ii)-GDR-{INALP}-cheek
'cheekbone, jawbone'

There are two examples of inalienable possession, (67) and (68), in which the component of a component is possessed. Until further evidence becomes available, it seems preferable to regard the component of a body part as being raised to body part status, thus maintaining a part-to-component relationship.

(67) *a-mw-alyirra* *a-mw-athangkwa*
3:4-INALP-slippery(=juice) 3:4-INALP-flesh
'blood or juice from red meat' (Literally: 'meat's juice')

(68) *mwa-mwa-mwilhkwa* *mwa-mw-alhika*
3:5-INALP-stomach 3:5-INALP-foot(=tyre)
'tyre tube' (Literally: 'tyre's stomach')

3.2.3. Proprietive relationship

The proprietive or 'having' relationship is realised by the suffix, *-mwirra* ~ *-mwi* and its hardened form *-pwi*. The deep structure shape of the morpheme is *-mwi* which is neutralised to *-mwa* in word-final position; the *-rra* syllable is a stem formative added to morphemes which do not otherwise occur in word-final position. Thus the morphological shape of the proprietive suffix and the inalienable possession prefix is identical. Furthermore, the proprietive suffix and the inalienable possession prefix can cooccur. The semantic concept and the root/stem composition is the same as for inalienable possession. The proprietive suffix is, therefore, redundant in such structures.

(69) *warni-mwa-mwalyi-(mwirra)*
3:1(i):GDR-INALP-body:fat-PROP
'Aborigines, mankind'

(70) *mwa-mwi-lharrngkw-angwiri-mwi(rra)*
3:5-INALP-things-fire-PROP
'pink hibiscus shrub'

(71) *yi-ni-mw-ayarrki-mwirra*
3:3(ii)-GDR-INALP-hand-PROP
'spider's web'

(72) *a-mwi-lyingkwi-ngiri-mwirra*
3:4-INALP-body:hair-white-PROP
'sea anemone species'

The proprietive suffix can occur without the inalienable possession marker. It is interesting to note that some primary body parts (section 2) are formed in this way, as well as the names of flora and fauna. The absence of the inalienable possession marker indicates that focus is not on an intrinsic whole-to-part relationship but simply on the fact that the particular item has a certain physical characteristic or body part as its salient feature. The relationship is stronger than that of the comitative 'with, having' in which the possessed item is external to the possessor (Dixon 1980: 324-326), e.g., the man with the dog. (Note: Some of the body parts in the following examples are incorporated, not free, forms.)

(73) *a-yarr-mwirra*
3:4-thin-PROP
'upper arm'

(74) *mw-amwirr-mwa*
3:5-round-PROP
'nape of the neck'

(75) *mw-amwarr-mwa*
3:5-oval-PROP
'nape of neck'

(76) *apwilhkwi-mwa*
{3:4}-feathers-PROP
'creek lily'

(77) *mw-angkarrk-pwa*
3:5-hips-PROP
'wild plum'

(78) *yi-lyang-mwirra*
3:3(ii)-head-PROP
'ark shell'

3.3. Semi-alienable possession

Semi-alienable possession is indicated by the prefix *ngwi-* which follows the nominal classifier and gender morphemes and precedes the root or stem. There are two categories, associative and privative.

3.3.1. Associative possession

Associative possession involves permanent or temporary ownership by right of occupancy. Its marker is the prefix *ngwi-* which immediately precedes the stem. The possessor is indicated by the nominal prefix which is followed by a gender morpheme. The stem which denotes the position occupied may be a noun (usually without a nominal classifier) or a locative adverb.[9] There is no conclusive evidence of a free form body part noun or noun root occurring in the associative possession nouns (see example (85) for the only exception).

The connotation of ownership by right of occupancy is supported by the common occurrence of the locative case marker (79)-(81). Its absence in examples (82)-(85) is typical of the optional marking of all peripheral case clitics in Anindilyakwa. The possessors in the following examples are the names of human or nonhuman animates which can be considered to have chosen to occupy the particular location as a permanent abode, i.e., they "belong" in that place. There is an additional sense of the place belonging to or reserved for them rather than other members of the same generic family, e.g., only the sea eagle has the right to occupy the "sea space" and only the brown tree snake the right to occupy the place in the sun. The concept of belonging is found in other Aboriginal languages such as Dyirbal (Dixon 1980: 324).

(79) *th-athi-ngwi-mwamwawiri-mwantja*
3:2(ii)-GDR-ASSP-sun-LOC
'brown tree snake'

(80) *ani-ngwi-mwikwimwikwi-mwantja*
{3:4}-GDR-ASSP-ocean-LOC
'octopus'

(81) *n-ani-ngw-akarrngi-mwantja*
3:3(i)-GDR-ASSP-teeth-LOC
'male dentist'

(82) *yi-ni-ngw-akarta*
3:3(ii)-GDR-ASSP-sea
'sea eagle'

(83) *warni-ngw-amwiyarra*
3:1(i):GDR-ASSP-bush:shelter
'ancestors'

(84) *th-athi-ningkwira*
3:2(ii)-GDR-{ASSP}-northwest:wind[10]
'olive shell'

(85) *yi-ni-ngw-ampwaka*
3:3(ii)-GDR-ASSP-?groin
'giant clam'

When an associative possession noun, as in examples (86)-(88), functions as the modifier of a noun phrase, its nominal prefix always cross-references the head.

(86) *ngarni-ngw-aritjilhangwa* (*ngakwirriwa*)
1IncPl:GDR-ASSP-behind(=last:in:time) 1IncPl
'we, the last ones in the line-up, last-born'

(87) *n-ani-ngw-athinipwawiya*
3:3(i)-GDR-ASSP-soon(=first:in:time)
n-ani-ngkw-arpwa
3:3(i)-GDR-NSR-wound(=man)
'he, the first adult male in a row, first-born'

(88) *ani-ngwi-mwakarti-(mwantja)* *a-kw-alya*
{3:4}-GDR-ASSP-sea-LOC 3:4-NSR-eat(=fish)
'saltwater fish'

The associative possession marker also indicates possession in relation to inanimate objects which occupy a specific place or space. In examples (89) and (90), the noun head identifies the possessed item and its modifier, the location of the possessed item, e.g., the teeth belonging to the bottom space or the particular bottom space which is "reserved" for such molars.

In examples (91)-(94), the meaning is more complex because the noun head could be said to name an item and its modifier to identify the purpose for which it is used. It is possible, however, to regard the named item in the noun phrase head as belonging to or located by association with the possessed noun (modifier). The semantic notion in (91) and (92) would thus not concern the net or harpoon capturing and possessing fish or turtle but rather the association with only fish or only turtle, respectively; the notion in (93) and (94) focusses on war vehicles or clothing which are permanently or temporarily associated with a war or located in a war zone.

(89) *mw-ani-ngw-arrawa* *mw-armpwa*
3:5-GDR-ASSP-below 3:5-molars
'bottom teeth'

(90) *th-athi-ngw-arrawa* *thimpwala*
3:2(ii)-GDR-ASSP-below ø-clothing
'underclothes'

(91) *mw-ani-ngw-akwalya* *mwi-kayiwa*
3:5-GDR-ASSP-fish 3:5-dillybag
'fish net' (i.e., net containing or possessing fish)

(92) *ani-ngwi-mwantha* *a-ningkwa*
{3:4}-GDR-ASSP-turtle 3:4-pointed(=harpoon)
'harpoon for spearing turtles' (i.e., harpoon capturing or possessing turtles)

(93) *mw-ani-ngw-angwinyampa* *mw-alhamwikwa*
3:5-GDR-ASSP-a:fight 3:5-canoe
'war canoe' (i.e., a vessel containing fighters)

(94) *th-athi-ngw-angwinyampa* *thimpwala*
3:2(ii)-GDR-ASSP-a:fight ø-clothing
'clothes worn for fighting, army uniform' (i.e., clothes that contain fighters)

The semantic connotation in examples (95)-(97) does not refer to the location of the item. For example, the word for 'bat' (not 'flying fox') is marked for associative possession but the noun root can also be used in isolation for 'peanut tree'. The Aboriginal speakers have explained to me that when the sleeping bats are suspended in a cave they look like the black pods on the peanut tree, not that they live in such trees. In the following examples, the meaning is probably one of resemblance of the physical feature to an important cultural item.

(95) *yi-ni-(ngwi)-mwapwalhpwa*
3:3(ii)-GDR-ASSP-peanut:tree
'bat (generic)' (i.e., resembling the peanut tree's black pods)

(96) *th-athi-ngw-angwira*
3:2(ii)-GDR-ASSP-fire
'omaria cone shell' (i.e., reddish colour resembles fire)

(97) *mw-ani-ngw-alhitha* *mw-amwawira*
3:5-GDR-ASSP-paint(=shininess) 3:5-sun/day
'bright sunshiny day' (i.e., the brightness of the sunshine resembles the shininess of paint)

3.3.2. Privative possession

The privative suffix, *-mi* 'not having, without', always cooccurs with the associative possession prefix *ngwi-*. The semantic notion, however, is one of deprivation of something or someone that is normally possessed. Generally speaking, it corresponds to the concept of *personal domain* defined by Bally (1926 [1995]: 33) as including "objects and beings associated with a person in an habitual, intimate or organic way (e.g. the body and its parts, clothes, the family, etc.) ... [and] is related to organic life since only living beings or objects personified in the imagination can be expressed in this way". In Anindilyakwa, the personal domain is not restricted just to animates, however, but occasionally includes inanimate objects marked by a nonpersonified nominal classifier. The deprivation can be temporary or permanent.

The possessor is indicated by the nominal classifying prefix and the possessed item is named in the stem. The stem can be a word minus its nominal classification. The privative construction is not used for the separation of a body part from a human being but can apparently be used in relation to a nonhuman animate, as in the only available data (102).

(98) *th-athi-ngwi-ningkwarpwi-ma*
3:2(i)-GDR-ASSP-man-PRIV
'widow, husbandless [woman]'

(99) *warni-ngwi-ngarnti-ma*
3:1(i):GDR-ASSP-mother-PRIV
'motherless [children]'

(100) *yi-ni-ngwi-thimpwali-ma*
3:3(ii)-GDR-ASSP-clothing-PRIV
'unspotted [goanna]'

(101) *th-athi-ngwi-thimpwali-ma*
3:2(i)-GDR-ASSP-clothing-PRIV
'naked female'

(102) *th-athi-ngarrkwi-ma*
3:2(ii)-GDR-{ASSP}-spike-PRIV
'spikeless [stingray]'

(103) *n-ani-ngwi-yakwirrarri-ma*
3:3(i)-GDR-ASSP-happy-PRIV
'unhappy male'

(104) *ani-ngwa-kwingwi-ma*
{3:4}-GDR-ASSP-water-PRIV
'waterless [place]'

3.4. Alienable possession

Alienable possession can be said to involve some measure of choice or control (Lynch 1973: 76). The possessor is always a human or nonhuman animate but is not necessarily contiguous to the possessed at all times. Alienable possession is expressed by the genitive suffix, *-lhangwi,* and its hardened form *-thangwi,* which usually occurs on the modifier in the noun phrase but can infrequently be found affixed to the noun head. The semantic function of the genitive case marker does not differ in its most common usage from that in other Aboriginal languages (Dixon 1980: 293) but, in Anindilyakwa, the genitive suffix is not used exclusively for alienable relationships. It can replace or coexist with the inalienable (except proprietive) and kinship possession nouns.

In examples (105)-(109), the most common usage of the genitive suffix for alienable possession is exemplified in noun phrases in which the head noun is not a body part. There is no agreement between the nominal classifiers in the head and modifier(s) of an alienable possession noun phrase. The modifier can be a personal or demonstrative pronoun or a noun.

(105) *ngakwirri-lhangwa wirri-yikwayiwa*
1IncPl-GEN 3:1(i)-small:ones(=children)
'our children'

(106) *mwi-yikwitjiya tjarri-lhangwa mwitjiyanga*
3:5-small ø-Jerry-GEN ø-boat
'Jerry's small boat'

(107) *athirra y-aratji-lhangwa*
{3:4}-mouth(=hole) 3:3(ii)-goanna-GEN
'Gould goanna's hole'

(108) *th-akini-lhangwa thi-kwirirrkwi-lhangwa a-ngwarnta*
3:2(ii)-that-GEN 3:2(ii)-brolga-GEN 3:4-stone
'that brolga's stone'

(109) *a-mwapa a-mwakalyiwakpwi-lhangwa*
3:4-song 3:4-Bickerton:Island-GEN
'Bickerton Island's song'

The possession of a body part or product is expressed by the use of the alienable possession suffix, *-lhangwi* ~ *-thangwi*, adjoined to a pronoun modifier. The possessor can be a human animate or a personified nonhuman animate. The use of the genitive suffix seems to imply that the possession is alienable and therefore considered to be a part that could be separate from the whole. The noun phrase head is a body part noun (from Tables 1-4), modified by a personal or demonstrative pronoun suffixed for genitive case.

(110) *ningkwi-lhangwa mw-arra*
2Sg-GEN 3:5-blood
'your blood'

(111) *ngalhi-lhangwa mw-anpwa-kwiya*
3FemSg-GEN 3:5-eye-DUAL
'her two eyes'

(112) *ani-lhangwa* *arrpwitha* *alhika*
3MascSg-GEN {3:4}-strong 3:4-foot
'his [wallaby's] strong feet'

(113) *yi-pwirathi-lhangwa* *ayarrka*
3:3(ii)-wallaby-GEN {3:4}-hand
'the wallaby's front paw'

(114) *th-akini-lhangwa* *aringka*
3:2(ii)-that-GEN {3:4}-head
'that female's [brolga's] head'

(115) *n-akina* *n-angariya* *akpwilhangwa*
3:3(i)-that 3:3(i)-young(=baby) {3:4}-buttocks-GEN
'that baby boy's buttocks'

(116) *yingalhi-lhangwa* *yi-ni-mw-amwiwa*
3NonhumanMasc-GEN 3:3(ii)-GDR-INALP-round(=egg)
'its [the turtle's] egg' (Literally: 'its egg, its round thing')

Unnamed species in Anindilyakwa can be "named" in relation to a similar named species. In examples (117) and (118), the modifier is a marker with the genitive suffix but the meaning is closer to proprietive than alienable.

(117) *yi-lying-pwirra-thangwa* *yi-ni-mwa-mwirarra*
3:3(ii)-body:hair-wide-GEN 3:3(ii)-GDR-INALP-?(=Grevillea)
'red grevillea' (Literally: 'the grevillea species with wide leaves')

(118) *yi-lyarr-pwirra-thangwa* *yi-ni-kw-ampwa*
3:3(ii)-ribs-wide-GEN 3:3(ii)-GDR-NSR-?(=Paperbark)
'broad-leaved paperbark sp.' (Literally: 'the paperbark species which has wider leaves than the melaleuca viridiflora')

There is some evidence that seems to suggest that external but not internal body parts in Anindilyakwa may have been traditionally marked for inalienable possession (see Chappell – McGregor 1995 and McKay 1995). There are a few inalienable possession forms which are still accepted by the older generation, e.g., *ninganimwalhakpwa* 'my leg', *warnimwiyatjimwa* 'their [birds'] wings'. This conforms to the possession of bound body part noun roots (Table 5) which differ from those immediately above only in that the latter are small sections of the body whose loss would not prevent recognition or identification of the "whole" human being. The possible change from inalienable to alienable possession may be due to the impact of the English structure, which has pronominal modification of body parts. The only common inalienable possession noun referring to humans has a shift in meaning – the term 'foot' is applied to a person's footprints which, to an Aborigine, are as unique as their fingerprints (see also Evans 1995, Harvey 1995, Hosokawa 1995, McKay 1995 and McGregor 1995 for the same phenomenon in other Australian languages). Compare the following pair of words:

(119) *nganyangwa* *a-lhika*
1ExcSg:GEN 3:4-foot
'my foot'

(120) *ning-ani-mw-alhika*
1ExcSg-GDR-INALP-foot
'my footprints, my tracks'

Furthermore, the inalienable possession prefix and genitive suffix can cooccur in the names of nonhuman animates or inanimates, according to their most salient physical feature (121)-(122) or intrinsic quality (123)-(124). In the latter pair of examples, the semantic relationship could best be described as proprietive. (The iregularity in the occurrence of the inalienable possession marker may be due to the random shortening of similar syllables in 'good' to maintain word rhythm/stress by reducing similar syllables.) No explanation is available for the conflict between the two types of possession expressed by different morphological affixation within the same word or phrase but it would seem that semantically inalienability outranks alienability.

(121) *yi-ni-mw-alyi-thangwa*
3:3(ii)-GDR-INALP-lips-GEN
'rainbow bee-eater'

(122) *a-mw-arnin-thangwa*
{3:4}-GDR-pointed-GEN
'spotted eagle-ray'

(123) *mwa-mwi-wirrariya-thangwa* *mw-arntja*
3:5-INALP-bad-GEN 3:5-bee:brood
'inedible portion of bees' brood cells'

(124) **mw-ani-mw-aningapwa → mw-aningapwi-thangwa*
3:5-GDR-INALP-good-GEN
mw-arntja
3:5-bee:brood
'edible portion of the bees' brood cells'

The genitive suffix can be adjoined to a root or stem which has an incorporated body part noun root but is not prefixed for possession. Such combinations of an adjective root and a body part noun root facilitate the naming of animals and other items in accordance with distinctive physical features or characteristics. The morphological structure of the compound stems parallels that for inalienable possession nouns but it has not yet been possible to check whether there are fuller forms of these words which include other prefixation.

(125) *yi-ri-kwitji-lhangwa*
3:3(ii)-body-small-GEN
'brindled bandicoot'

(126) *ya-yarran-thangwa*
3:3(ii)-?forked-GEN
'king brown snake'

(127) *thi-ningkwi-lhangwa*
3:2(ii)-?pointed-GEN
'dugong'

(128) *a-ngkwi-lyingarrk-alyi-thangwa*
3:4-?NSR-liver-wet-GEN
'little blue shark'

(129) *a-mwan-thangwa*
3:4-spirit-GEN
'true words, truth'

The cooccurrence of the associative possessive prefix *ngwi-* with the genitive suffix has been found in only one word. This does not easily fit into the semi-alienable classification because no other roots consist of free form body parts. All flora and fauna terms have been extensively researched for Anindilyakwa and it is unlikely that more data will become available.

(130) *yi-ni-ngw-arntarnti-lhangwa*
3:3(ii)-GDR-heart(=?reddish)-GEN
'mangrove jack'

3.5. Kinship

Kinship terms refer primarily to human blood and affinal relationships. The system extends to classificatory relationships for "outsiders" who are assigned a relationship to a specific member of the community and then addressed by the appropriate kinship terms. Membership in the two moieties is not marked for kinship possession. The suffixation on kinship nouns involves a limited number of person, number and gender distinctions for the propositus (Sheffler 1978) or possessor of the "kin" named in the root; the prefixation denotes the person, number and gender of the possessed relative or "kin".[11]

The kinship roots denote the nature of the relationships. Two-thirds of these are cognates with Nunggubuyu (Heath 1984: 225-228) and three of the remainder are cognates with Yolngu-Matha which is spoken in north-eastern Arnhem Land. Marriage and ceremonial ties exist with both the Nunggubuyu and Yolngu-Matha people. As Anindilyakwa has phonological rules which can account for most of the phonemic irregularities with the Nunggubuyu terms (Leeding 1989), it seems likely that these have been introduced into Anindilyakwa. Such borrowings may also account for the irregularity in what appears to be the gender morphemes.

The set of seven kinship possession suffixes distinguishes the three persons. Only singular and nonsingular number is differentiated; and masculine and feminine gender are contrasted only for the third person singular.

-arrka	1st singular
-ana	2nd singular
-anikpwa	3rd masculine singular
-athikpwa	3rd feminine singular
-anyingwa	1st non-singular
-aningwa	2nd non-singular
-arringpwa	3rd non-singular

Kinship nouns usually consist of a nominal prefix, root and suffix. The full set of nominal gender prefixes, listed in section 2.1, are used for kinship nouns. For a few roots, the gender morphemes, *ani-* ~ *atha-*, follow the prefix but there seems to be no semantic reason for this addition. The following examples illustrate the interrelationship between various nominal prefixes and kinship possession suffixes. Affinal relationships are marked in the same way as the biological.

(131) *ni-ngw-arrka*
3:3(i)-father-KPOSS:1Sg
'he, my father'

(132) *thi-rnt-ana*
3:2(i)-mother-KPOSS:2Sg
'she, your mother'

(133) *wirr-apwirr-anikpwa*
3:1(i)-daughter-KPOSS:3MascSg
'they, his daughters'

(134) *th-angantjamwintj-arringpwa*
3:2(ii)-mother's:father's:sister-KPOSS:3Nsg
'she, their mother's father's sister'

(135) *ning-ani-kwi-mwarntj-ana*
1ExcSg-GDR-MASC-younger:brother-KPOSS:2MascSg
'I, your younger brother'

(136) *warni-kwimwarntj-anyingwa*
3:1(i):GDR-younger:brother-KPOSS:1Nsg
'they, our younger brothers'

(137) *wirring-athathi-ngwiy-aningwa*
3:1FemDu-REDUP:GDR-spouse-KPOSS:2Nsg
'they two women, your wives'

(138) *n-athitj-arrka*
3:3(i)-daughter's:husband-KPOSS:1Sg
'he, my son-in-law'

(139) *th-arrngilh-anikpwa*
3:2(i)-wife's:mother-KPOSS:3MascSg
'she, his mother-in-law'

A noun phrase can consist of a kinship noun head and pronoun or noun modifier that is marked for the genitive. The person, number and gender of the free pronoun correlates with that of the kinship suffix. Such redundancy is used as a stylistic device to emphasise or focus attention on the possessor, rather than the possessed.

(140) *nganyangwa ni-ngw-arrka*
1ExcSg:GEN 3:3(i)-father-KPOSS:1Sg
'my father'

(141) *ngalhi-lhangwa n-ani-ngwiy-athikpwa*
3FemSg-GEN 3:3(i)-GDR-NSR-spouse-KPOSS:3:2(i)
'her husband'

(142) *ni-mwawiri-lhangwa wirring-ampwilyimwa-kiya*
3:3(i)-moon-GEN 3:1(i)FemDu-two-DUAL
wirring-athi-ngwiy-anikpwa-kiya
3:1(i)FemDu-GDR-NSR-spouse-KPOSS:3Sg-DUAL
'Moon's two wives' (Note: The moon and his wives are personified and function as humans)

(143) *wirring-ampwilyimwa-kwiyi-lhangwa*
3:1FemDu-two-DUAL-GEN
thi-rnt-arringpwa
3:2(i)-mother-KPOSS:3Nsg
'the two females' mother'

(144) *thi-warti-lhangwa thi-rnt-athikpwa*
3:2(ii)-dog-GEN 3:2(ii)-mother-KPOSS:3FemSg
'the female dog's mother'

The kinship nouns usually refer to human animates but are also used for nonhuman animates, mainly in traditional Dreamtime (creation) stories where the animal is personified and acts as a human being. The kinship relationship among nonhuman animates (mammals, fish, birds) is restricted to those of their own species, e.g., husband and wife, mother/father and offspring, brothers and sisters. Similar relationships between domesticated dogs and herds of undomesticated animals are nowadays given in accordance with their blood lines, function or appearance, e.g., a male wallaby is termed 'older brother' in relation to the other adult males if his fur has lost its gloss. This division between human and nonhuman kinship differs from the Guurindji custom which, according to my own field notes, puts domesticated pets into a relationship with humans, e.g., 'you call that dog uncle'.

(145) *y-awaw-arringpwa*
3:3(ii)-older:brother-KPOSS:3Nsg
'it [wallaby], their older brother'

(146) *mwi-rnt-athikpwa*
3:5-mother-KPOSS:3FemSg
'it [shark], her mother'

(147) *wirr-apwirr-anikpwa*
3:1(ii)-son-KPOSS:3MascSg
'it [dog], his son'

In Nunggubuyu (Heath 1984: 238-9), a few of the kinship roots are found in other nouns or verbs. A similar phenomenon has only been identified in one set of common nouns, e.g., *amwirarrka* 'my country'. The root is *mwira* 'father's father' and its use here is in accordance with the system for land inheritance.

The suffix *-akpwi* ~ *-kpwi* has a similar phonemic shape to the kinship possession suffixes. It does not indicate a kinship relationship but does involve some sort of possession, the meaning of which cannot be determined because of the paucity of available data. In examples (148)-(150), the suffix indicates that the nonhuman animate lives or belongs to the habitat named in the root. The semantics is similar to that for associative possession (section 3.3.1.) and there is also a corresponding morphological structure in the use of the associate possession prefix and optional locative suffix. In examples (151) and (152), the suffix indicates that the animate or inanimate object possesses an inherent characteristic which is identified in the root. These examples are prefixed for associative possession but the suffix has the same meaning as the proprietive suffix *-mwi* (section 3.2.3). In examples (153)-(155), there is no prefixation for possession and the suffix indicates that the location itself is owned by right of occupation by a nonhuman animate which habitually resides there. Certain animals are always associated with specific types of topographical features. These words are aligned with associative possession nouns only by the semantics.

(148) *yi-ni-ngwi-mwakarti-kpwi-mwantja*
3:3(ii)-GDR-ASSP-sea-?POSS-LOC
'sea snake'

(149) *th-athi-ngw-atjiringk-akpwa*
3:2(ii)-GDR-ASSP-sand-?POSS
'sand frog'

(150) *warni-ngwi-mwiyarraw-akpwa*
3:1(ii):GDR-ASSP-red:kurrajong-?POSS
'beetle that feeds on the red kurrajong shrub'

(151) *warni-ngw-angwir-akpwa*
3:1(ii):GDR-ASSP-fire-?POSS
'glossy ibis' (Note: the bird has a shiny reddish-brown plumage)

(152) *mwi-ngarrki-kpwa*
3:5-ear-?POSS
'spear with hooks along its blade'

(153) *mw-alharr-akpwa*
3:5-stones-?POSS
'its stones' (smaller rocks in which the rock wallaby lives)

(154) *ak-akpwa*
{3:4}-tree-?POSS
'its tree' (the tree which may be inhabited by either a goanna, a snake, or bees)

(155) *a-ngaly-akpwa* ~ *a-ngalyi-kpwa*
3:4-place-?POSS
'its place, its home, its country' (the place associated with the pheasant coucal or the pheasant coucal's home)

Some kin relationships are associated with parts of the body. The touching of these body parts indicates a particular relative and is often combined with other sign language. For example, the touching of the shoulder followed by the backward movement of both hands means that the person's father has gone looking for turtles.

nose	for	father
breast	for	mother
thigh (lap)	for	mother's father, father's father
leg calf	for	brother or sister (walks behind)
abdomen or womb	for	son or daughter
hip	for	husband, wife
knee	for	avoidance relative

4. Incorporation

Body part noun roots (see Table 6, section 2.3) are incorporated in three ways in Anindilyakwa: morphologically, lexically and syntactically (see also Evans 1995 and Harvey 1995). (i) The term, *nominal compounding*, is being used for the combination of a body part noun root and a descriptive or shape adjective root as the stem of a noun or adjective; (ii) *lexical compounding* is the incorporation of a noun root with a verb root or stem to form a new stem with a fixed (usually metaphorical) meaning; and (iii) *syntactic incorporation* is a feature of the verbs in which a body part noun root or a shape adjective root precedes the verb root in agreement with the object of the verb. Lexically-compounded stems always involve metaphor but the meaning in the nominally-compounded and syntactically-incorporated structures is always literal.

An additional two morphophonemic rules reduce the length of compound stems. A Syllable Deletion Rule deletes the morpheme-final *(ng)kwi* syllable preceding a root which begins with a nasal; a Vowel Deletion Rule deletes the vowel in that syllable preceding a root which begins with a peripheral stop. There can also be random reduction and coalescence.

4.1. Nominal compounding

Nominal compounding in Anindilyakwa is one of the systems used for naming. An incorporated body part root often cooccurs with a descriptive adjective root to provide a precise literal description of the salient physical feature or characteristic of an animate or inanimate object. The speaker can focus the attention of an addressee to a specific anatomical part under discussion. For example, a thin woman can be described simply as thin (*thiyarrmiyarrma*), or as being shallow-ribbed (*thilharrkwpwilhalha*) or shallow-bodied (*thirikwpwilhalha*) (see also Harvey 1995 on synechdoche in Warray).

Most of the adjective stems can be compounded morphologically. Some adjective roots are obligatorily bound by a preceding body part noun root, as illustrated in thc following examples. This phenonomen is also found in neighbouring Nunggubuyu, but it does not appear to be as productive or systematic (Heath 1984: 471).

(156) *a-lhingaki-yarrmiyarrma*
3:4-torso-REDUP:thin
'thin' (Literally: 'thin-bodied, e.g. sapling')

(157) *yi-mwintjirrki-mwilhimwilha*
3:3(ii)-skin-REDUP:rough
'very rough-skinned' (e.g., goanna species, tree with rough bark)

(158) *thi-mwirrkw-pwalya*
3:2(i)-breast-soft
'soft' (Literally: 'soft-breasted, e.g., newborn baby')

(159) *a-warrngki-lyikarrk-pwitha*
3:4-DIMINISHER-liver-strong
'slightly unpleasant in taste' (Literally: 'strong-livered, e.g., some fruit and shellfish')

(160) *a-pwirrngkwi-lyimwitha a-kwingwa*
3:4-joint:fluid-completed 3:4-water
'river' (Literally: 'complete freshwater')

(161) *mwa-pwirrngkwi-lyimwitha mw-akarta*
3:5-joint:fluid-completed 3:5-sea
'saltwater lagoon' (Literally: 'complete sea')

The focus on the salient physical feature separates one item from others in the same species. Morphologically, such words could take the full substitution set of nominal classifiers and function as common nouns or adjectives. Most, however, are regarded semantically as the name of one particular item unless otherwise identified by an overt referent. The real world limits the extent of semantic connotations that are possible. Examples (162)-(172) illustrate these in relation to mammals, reptiles, birds, fish and plants.

(162) *yi-ri-kwitji-lhangwa*
3:3(ii)-body-small-GEN
'brindled bandicoot' (Literally: 'one possessing a small body')

(163) *thi-tji-ningk-wa*
3:2(ii)-small-pointed(=spear)-SF
'echidna' (Literally: 'one with small-spears')

(164) *wirr-antjantji-nyirra*
3:1(ii)-REDUP:stomach-soft
'azure kingfisher' (Literally: 'very soft-stomached')

(165) *yi-mwantjimwan-tjirra*
3:3(ii)-REDUP:flipper-long
'mythological bird'

(166) *thi-mwarngkwi-thatha*
3:2(ii)-eye-spotted
'spotted goanna species' (Literally: 'spotty-eyed')

(167) *yi-lyang-mwilhimwilha*
3:3(ii)-head-REDUP:rough
'estuarine stonefish' (Literally: 'very rough-headed')

(168) *yi-ngarr-ampwilya*
3:3(ii)-ear-paired
'centipede' (Literally: 'paired-eared, i.e., pairs of feet')

(169) *wirri-lyangpwi-lyangpwirna*
3:1(ii)-REDUP-head:many
'ant species' (Literally: 'many-heads')

(170) *a-lhingkw-amwarta*
3:4-body:hair-green
'billabong tree' (Literally: 'green-haired, green-foliaged')

(171) *mw-arnta-ningkwiningkwa*
3:5-elbow-REDUP:pointed
'smilax vine' (Literally: 'very pointed elbows')

(172) *a-warr-walya*
3:4-teeth-soft
'shadow, shade' (Literally: 'soft-teethed')

The nominally-compounded stem can be used to name otherwise unnamed body parts. The referent is usually overt as the head of the noun phrase. Only two body parts appear to be named in this way.

(173) *a-lharrki-lyimwitha athithira*
3:4-bone-completed bone
'skeleton' (Literally: 'complete bones')

(174) *mwi-rrak-alhiwa* *mwilhkwa*
3:5-elongated-few:and:small {3:5}-abdominal:sac
'small intestines' (Literally: 'a few small elongated sacs')

Body part noun and shape adjective roots can be used to denote a relationship between a named and an unnamed object that could be called generic/specific. Some of the flora and fauna is not named because it holds no significance in the local culture but these common nouns can be used. Most of the older women, however, point out that these words are not "proper names". Such nouns classify an unnamed item on the basis of its similarity with one named. Morphologically, the body part noun or the shape adjective root precedes a free form noun without its nominal classifier, as in examples (176), (178), (180), (182) and (184). The following pairs of examples contrast the (named) generic and the (unnamed) specific.[12] There is one example (184) in which the body part noun root precedes a Macassan loanword, the shell having originally been used to make a smoking pipe.

(175) *yi-lyang-mwirra*
3:3(ii)-head-PROP
'ark shell'

(176) *a-ngangki-lyangmwirra*
3:4-concave-Ark:Shell
'cardita shell'

(177) *yi-rntirnta*
3:3(ii)-heart(=red)
'strychnine tree'

(178) *a-lyingkwi-rntirnta*
3:4-body:hair-Strychnine:Tree
'poison berry tree'

(179) *awirintja*
ø-orange
'an orange'

(180) *a-mwingkw-awirintja*
3:4-cheek-orange
'grapefruit'

(181) *ani-ngw-akpwitjakpwa*
{3:4}-GDR-ASSP-REDUP:leg
'snub-nosed dart'

(182) *a-lhikwi-ningwakpwilyakpwa*
3:4-foot-snubnosed:dart
'dart or swallowtail'

(183) *thampwakwa*
ø-smoking:pipe
'smoking pipe, tobacco, cigarette'

(184) *arrki-thampwakwa*
{3:4}-hip-smoking:pipe
'screw shell' (which was used as a pipe)

4.2. Lexical compounding

Lexically-compounded stems are very common in Anindilyakwa but are nonproductive. As in Nunggubuyu (Heath 1984: 463), they are learned and not put together spontaneously. Lexical compounding is used to represent customary events or practices that have cultural significance, e.g., body positions in ceremonies or birthing; the darkening of the sky which alerts hunters about the need to return home or to seek shelter from the impending storm. Some of the semantic notions are culture-specific, e.g., the growth of a tree is seen as an increase in foliage (body hair) and the sky as an abdomen. Such events and activities could be said to be "institutionalised" within Warnindilyakwa culture (see Mithun 1984) or restricted simply because of natural phenomena.

Lexical-compounding is found in the stems of verbs, adjectives and nouns. The first root in the compound stem is always selected from the finite sets of incorporated body part nouns and shape adjective roots but the second root can be one of the wide range of descriptive adjective and verb roots.

In verbal constructions involving lexical compounding, the incorporated root that precedes a verb root always refers to the subject of the intransitive verb. The person, number and gender of the possessor of the body part is marked by the referent in the verb prefix and, optionally, by a free form noun, pronoun or demonstrative. As can be seen from examples (191)-(195) which move more closely to a literal meaning, lexical compounding is not clearly distinguished semantically from syntactic (literal) incorporation.

(185) *ni-lyang-mwiri-thi-na*
3:3(i)-head-hollow-INCH-TNS
'He was/became incompetent.'

(186) *ni-nyak-ampwilyimwa*
3:3(i)-chest-was:staying
'He was trusting in the old ways.'

(187) *n-arnti-warrkina*
3:3(i)-elbow-weaved:in:and:out
'He limped.' (Literally: 'His elbow weaved in and out')

(188) *ni-ngangki-wangi-nimwa*
3:3(i)-concave(=chest)-bite-TNS
'He was breathing.' (Literally: 'His chest was biting')

(189) *yingi-yangki-lyikpwi-na*
3:2(i)-speech-eat-TNS
'She spoke but was not heard.' (Literally: 'Her speech was eaten')

(190) *yingi-mwir-arrngalh-ana*
3:2(i)-chin:curve-scratch-TNS
'She was husky.' (Literally: 'Her chin curve scratched')

(191) *na-lhingkwi-warra*
3:4-body:hair-move
'It [the tree] grew taller.' (Literally: 'Its foliage moved')

(192) *n-arra-tjalh-anga*
3:4-forehead-emerge-TNS
'It [the cliff] jutted out.' (Literally: 'Its forehead emerged')

(193) *na-rraki-rntangmwi-na*
3:4-elongated-clap-TNS
'It [the wire] sang.' (Literally: 'The elongated thing clapped')

(194) *ngarri-rikwi-lharri-na*
1IncPl-body-fall-TNS
'We were born.' (Literally: 'Our bodies fell')

(195) *na-mwilyirrkwi-lharri-na*
3:4-breast-fall-TNS
'It was born prematurely.' (Literally: 'Its breast fell')

Lexically-compounded verbs take the same stem formatives and suffixation as for all other verbs. Such verbs take intransitive subject prefixation which does not change for the reflexive and inchoative. The causative verb normally requires transitive prefixation but, in lexically-compounded verbs, is marked for the intransitive. In such cases, the incorporated body part noun roots precede an already established stem. Note the layering in example (199) where the established stem is marked for causative but the verb itself for reflexive. Example (200) exemplifies the rare occurrence of a lexically-compounded stem preceded by another incorporated body part noun root.

(196) *na-lhingkw-apwitja-tjingwi-na*
3:4-hair-jumped-REFL-TNS
'It [the grass] grew.' (Literally: 'Its hair jumped itself up')

(197) *n-apwalh-mwir-thi-nimwa*
3:4-abdomen-brown-INCH-TNS
'It [the sky] was getting darker.' (Literally: 'Its abdomen was becoming brown')

(198) *na-yangki-lhawirra-thi-na*
3:4-speech-return-INCH-TNS
'It [the voice/words] echoed.' (Literally: 'Its speech came back')

(199) *ningkwi-mwarngki-lhawirra-ka-tjingwi-na*
2Sg-spirit-return-CAUS-REFL-TNS
'You remembered.' (Literally: 'Your spirit/mind brought itself back')

(200) *ningi-lyangki-warrpwikwa-tjingwi-na*
1ExcSg-head-chest:carry(=felt)-REFL-TNS
'I thought to myself.' (Literally: 'My head felt itself')

Lexically-compounded verb stems can be nominalised by the preceding nominaliser *(ng)kwi-*, thus indicating that these are fixed collocations. There appears to be two different semantic connotations with some possible overlap. In examples (201)-(203), the form or physical characteristic of the item is in focus but in examples (204)-(205) it is the function that is involved. No explanation is available for the absence of the nominaliser in similar nominal nouns, as in examples (205)-(206).

(201) *a-kwi-kwiwa-lyimwi-tha*
3:4-NSR-water-be:complete-INCH
'swamp, pond, lake' (Literally: 'the one with completed water')

(202) *a-kwi-rrak-artjiyi-nga*
3:4-NSR-forehead-stands:upright-TNS
'chair, seat, stool' (Literally: 'the one with the upstanding forehead')

(203) *th-a-kwi-mwintji-thathi-tjingwa*
3:2(ii)-GDR-NSR-skin-became:cooked-REFL
'spotted monitor species' (Literally: 'the one with self-burnt skin')

(204) *mwa-kwi-lyang-pwarrka*
3:5-NSR-head-weave:in:and:out(=sweep)
'broom' (Literally: 'the one that sweeps')

(205) *arnti-warra*
{3:4}-elbow-shake
'walking stick' (Literally: 'the one that shakes the elbow')

(206) *a-lyingarrki-rrangkwirra*
3:4-liver-swollen
'boil, inflamed sore' (Literally: 'the one with a swollen liver')

Nonphysical human attributes (207)-(209) are realised as adjectives with lexically-compounded stems. There is an additional metaphorical use for such terms when used to describe attributes of inanimate items (210).

(207) *wirri-mwarngkw-ampwilyimwa*
3:1(i)-spirit-two
'two-faced [persons]' (Literally: 'two-spirited or double-minded')

(208) *thi-lyang-pwirna*
3:2(i)-head-many
'ignorant [female]' (Literally: 'many-headed')

(209) *ni-nyak-pwitha*
3:3(i)-chest-strong
'brave [male]' (Literally: 'strong-chested')

(210) *a-lhingaki-nyakpwitha*
3:4-torso-brave
'steady [tree]' (Literally: 'brave-torso-ed, i.e., one that stands firm as an upright post to support a roof')

4.3. Syntactic incorporation

The syntactic incorporation of body part noun and shape adjective roots occurs in transitive and reflexive verbs preceding a root/stem describing an action. Anindilyakwa incorporation, with only two finite sets of verb roots, differs from that for Mayali (Evans 1995) in which nouns other than body parts can be incorporated into the verb structure. The two languages, however, are alike in their capacity for such compounds to be put together spontaneously, without any change in meaning in either constituent. All of the incorporated roots (not the classified body part nouns) can occur as a replacement for the object prefix in a transitive verb, given a suitable semantic context.

(211) *yingi-ngwiyang-pwatj-ana*
3:2(i)-hand-hit-TNS
'She hit [its] hand.'

(212) *ni-mwarngkwi-rringki-na*
3:4-spirit-see-TNS
'He saw the spirit.'

(213) *ni-warrki-rntirrka*
3:4-teeth-take
'He took the teeth.'

(214) *ning-alhki-thaki-na*
1ExcSg-feathers-cook-TNS
'I singed [its] feathers off.'

(215) *wi-lhakpwaki-rith-ana*
2Pl-leg-chop-TNS
'Chop the leg off!'

Body part incorporation is very productive in Anindilyakwa because the roots can be used to refer to any artifact or any natural phenomena, as in examples (216)-(222). The nonagentive possession of the body part is evidenced by the use of an intransitive verb prefix, by the absence of an object prefix, and by the fact that the body part or characteristic cannot possibly refer to the human referent in the subject.

(216) *ni-rraki-rntangmwa-tji-na*
3:3(i)-elongated-clap-TSR-TNS
'He strummed a guitar.'

(217) *ningi-nga-lyipwirrkwi-ripwikwi-na*
1ExcSg-3:2(ii)-lips-fold-TNS
'I hemmed it [the dress].'

(218) *ni-ngakw-pwirrkwa-tji-na*
3:3(i)-thigh-crush-TSR-TNS
'He flattened the mound.'

(219) *ying-arntirrk-pwatj-ana*
3:2(i)-scrotum-hit-TNS
'She knocked the plums off [the tree].'

(220) *yirri-rrak-pwirr-ana*
1ExcPl-forehead-split-TNS
'We split the firewood.'

(221) *ngarri-mwingkwi-lharri-tji-na*
1IncPl-cheeks-fall-TSR-TNS
'We shook the fruit off.'

(222) *warni-ngwa-pwatjakpwa*
3:1(ii):GDR-ALP-?REDUP:leg(=bird)
na-warrwarr-wart-anga *a-lyingpwa*
3:1(ii)-REDUP:teeth-hit-TNS 3:4-clapsticks
'Brush Cuckoo tapped the clapsticks.'

A whole-to-part relationship between the possessor in the subject role and the body part in the object role is expressed by marking the verb with the reflexive suffix *-tjingwi*. Both animates and inanimates can occur as the "whole" possessor in the subject and function as both agents and experiencers. The use of intransitive prefixation for reflexive verbs, however, seems to indicate that the focus is on the subject as experiencer, rather than motivator and agent.

(223) *ningi-yamwirrk-pwatja-tjingwi-na*
1ExcSg-ball:shaped-hit-REFL-TNS
'I hit myself on my upper arm'

(224) *ying-alhpwi-thirrka-tjingwi-na*
3:2(i)-shoulderblade-stretched-REFL-TNS
'She stretched her arms out'

(225) *ni-rikwi-warrkwa-tjingwi-na*
3:3(i)-body-changed:over-REFL-TNS
'He turned himself around.'

(226) *ni-rraki-ripwikwa-tjingwi-na*
3:3(ii)-elongated-folded-REFL-TNS
'It [the snake] coiled itself around.'

(227) *nimwa-mwarrki-warrkwa-tjingwa*
3:5-circular-change:over-REFL
'It [the sun] moved itself to the one o'clock position.'

It is not the purpose of this paper to discuss the problems of possessor ascension (Blake 1984) but the following data is given to clarify the interaction between the object expressed by verb prefixation and the corresponding noun phrase. When the object prefixation in the verb marks the possessor or "whole" person as the object of the verb in the accusative case, then the body part noun or noun phrase occurs externally in the unmarked accusative case or the marked locative case. Thus the status of the whole is raised above the value of the part (as outlined by Fox 1981), the action being seen as directed primarily against the possessor. This provides an alternate system to the one above in which the body part term itself replaces the object prefix. Within the noun phrase, the person, number and gender of the possessor is also often identified by a pronoun or demonstrative modifier; the head of the noun phrase can be a free form noun or an inalienable or associative possession noun.

(228) *ningi-ni-ngatj-amwa ani-lhangwi-mwantja aringka*
1ExcSg-3:3(i)-hit-TNS 3MascSg-GEN-LOC {3:4}-head
'I am hitting him on his head.'

(229) *wirri-warta* *na-ni-rntirrka*
3:1(ii)-longhaired 3:1(ii)-3:3(ii)-grab
n-akina *akwpwi-lhangwa*
3:3(ii)-that {3:4}-buttocks-GEN
'The dog got hold of [the wallaby's] buttocks.'

(230) *kwirri-nga-rringki-namwa* *ngalhi-lhangwa*
2Pl-3:2(ii)-see-TNS 3FemSg-GEN
th-athi-ngapwa *th-athi-mw-athangkwa*
3:2(ii)-GDR-{NSR}-good 3:2(ii)-GDR-INALP-flesh
'You can see her nice body.'

(231) *kampwirra* *n-ani-ngatj-amwa* *n-akina*
therefore 3:3(ii)-3:3(ii)-hit-TNS 3:3(ii)-that
mw-ani-ngwi-wilyarri-mwantja
3:5-GDR-ASSP-middle-LOC
'Therefore he [Seagull] is hitting that [Pheasant] in the middle [of his back].'

A verb stem which includes a syntactically-incorporated body part term can be nominalised but very few examples are available. The meaning conveys the function or behaviour of the item but does not describe its form.

(232) *th-athi-kw-apwalhi-mwingkwa*
3:2(ii)-GDR-NSR-abdomen-partially:cover
'trousers' (Literally: 'the abdomen-covering thing')

(233) *th-athi-kwi-lying-pwitja*
3:2(ii)-GDR-NSR-head:hair-jump
'Flat-pod Bean climbing vine' (Literally: 'the hair-jumping one')

5. Conclusion

Body parts in Anindilyakwa fall into two grammatical categories: independent semantically-classified nouns and the incorporated unclassified roots. The latter have a very productive role, together with shape adjective roots, in forming compound stems with very explicit descriptions of the salient features of flora and fauna. The independent body part nouns are prefixed in the same way as other nominals, i.e., in a gender system which includes first and second person as well as third person noun classes. Unlike Mayali (Evans 1995), only one body part noun or one shape adjective root can occur in a compound stem (except where there is derivational layering). There is no well-established evidence that the possession of external, internal or separable parts or byproducts are treated differently from inseparable human body parts (cf. Chappell – McGregor 1995). The use of incorporated body parts for generic nouns begs the question as to whether the environment is associated with the body parts or vice versa, e.g., the incorporated root, *alyikwi* 'saliva', is in the noun *yalyikwa* 'rain'. To say that these are simply homophonous ignores the significance of the semantic similarities. One of the obvious features of Anindilyakwa is the copious number of recurring morpheme shapes in words which have different nominal classification. For example, the incorporated body part noun root for 'eye' is used to describe anything circular, solid round or oval, such as a traditional paperbark shelter, midnight (eye(=moon):central), tamarind fruit (eye:weak), green plum (eye:small), shelf limpet, snail, small barracuda fish and a jellyfish species. Some Warnidilyakwa Aborigines have identified these roots as body parts, despite the phonological changes.

Proper nouns are generally not incorporated in Anindilyakwa, a restriction attributed by Mithun (1984: 864) to their not being "sufficiently generic to narrow the scope of verbs in a useful way". Anindilyakwa body part noun and shape adjective roots are very "productive" in that they are used for crossreferencing proper nouns which have the same salient feature/s as the body part/s and their shapes. The roots of the few incorporated proper nouns (viz., noun roots for: 'liquid', 'camp', 'things'; and the adjective root for 'fire') have the same meanings as those listed in the closed generic noun sets for Nunggubuyu (Heath 1984: 466), Mayali (Evans 1995) and Ngalakan (Merlan 1983: 143-146).

Kinship possession is identified in Anindilyakwa by suffixation which has only seven distinctions for person, number and gender. There is similarity between the roots/stems of Anindilyakwa and Nunggubuyu kinship nouns in the roots/stems but, unlike Anindilyakwa, Nunggubuyu is said to have a "scrambled" suffixal paradigm for identifying the person, number and gender of the possessor (Heath 1984: 228-229). The kinship relationships in Anindilyakwa correspond with those in Pama-Nyungan Australian languages in that they "are generally classified under inalienable possession" (Dixon 1980: 512).

There are three overlapping definitions for inalienable possession which have been taken into consideration in the analysis of Anindilyakiwa body parts and possession. Fox (1981: 323) defines such possession as "the reflection of the real-world fact that body parts are physically contiguous with their possessors". The notion of choice is added by Lynch (1973: 76) when he states that "inalienable possession implies that the possessed is closely or inextricable linked to the possessor who has no choice in the matter of whether the possession is his or not". A similar definition from Dixon (1980: 512) regards "the possessed [as having] an inherent connection with the possessor [which] cannot be given away". In Anindilyakwa, these three factors can be used to distinguish the four types of possession. Inalienable possession and kinship possession cannot be chosen or given away but contrast with each other because the former is inextricably-linked and contiguous while the latter is inextricably-linked but not contiguous. Kinship possession of affinal and classificatory relationships is, however, an exception because these can be chosen and given away. Alienable possession is said to involve a control or choice in obtaining possession and in opting for subsequent dispossession but, in Anindilyakwa, it now appears to include the possession of body parts which by definition are inalienable. Added to this is the problem that the alienable possession marker is used with inanimates who are incapable of choice. Semi-alienable possession involves a possessor who is inextricably linked to a noncontiguous place or item but who, in the case of animate possession, originally had a choice of location. In Table 10, the four types of possession are contrasted in relation to three semantic features with the brackets indicating that there are minor unsystematic members.

Table 10. Semantic contrasts for possession

	Inherent	Contiguous	Chosen
Inalienable	yes	yes	no
Kinship	yes	no	(no)
Semi-alienable	yes	no	yes
Alienable	(no)	no	yes

Possession is marked in different ways: (i) by two contrastive prefixes and/or proprietive or privative suffixation for inalienable and semi-alienable possession; (ii) by pronominal suffixation for kinship possession; and (iii) by case suffixation for alienable possession. Semantic distinctions are sometimes blurred because nouns can be marked for alienable possession as well as one of the other three types of possession. Alienability in such instances is outranked semantically. Given the additional fact that body parts are now marked for genitive case, it seems advisable to regard such possession as general or unspecified, rather than defining it as distinctly alienable.

Body parts permeate the whole morphological and, to a lesser extent, the syntactic structure of Anindilyakwa. They function as essential building blocks in the naming of the human and nonhuman environment and in its interrelationships. Even though research is not complete, this present study of semantic features in relation to body parts has provided fascinating insights into the Warnindilyakwa classifications for the universe in which they live.

Notes

1. Data was collected during a little over six years residency at Umbakumba between 1974 and 1982 and for about six months at Angurugu whilst employed in the bilingual education program, Northern Territory Department of Education. I encouraged the Umbakumba Aborigines to discuss their own views about the phonological, grammatical and semantic features of their language, not just to supply data. Many fascinating and profitable conversations helped me to gain insights into their world view and values. I am indebted to many people in the Umbakumba community but my special thanks go to my friends and colleagues: Katie Yantarrnga, Stella Yantarrnga, Tapinkanga Yantarrnga, Greta Yantarrnga, Alec BaraBara, Arthur Bara and Wesley Bara. I am indebted to Hilary Chappell, Bill McGregor and Rosemary Young (Summer Institute of Linguistics) for their comments on earlier drafts of this paper; and to Dulcie Levitt (1981) for the technical information and popular names for plants.
2. Abbreviations: * not an acceptable form for Anindilyakwa, or cross-referenced to the preceding number; [ ] enclosing a possible referent within that noun class; ( ) optional when enclosing Anindilyakwa phonemes; { } deletion of the deep structure morpheme or segment/s; ? meaning of the gloss is unknown or uncertain; = extended meaning; ø zero morpheme; ASSP associative possession; ASR adjectiviser; CAUS causative; Du dual; Exc exclusive; Fem feminine; GDR gender; GEN genitive; INALP inalienable possession; Inc inclusive; INCH inchoative; LOC locative case; Masc masculine; Nsg/nonsg nonsingular; NSR nominaliser; Pl plural; PRIV privative; PROP proprietive; REDUP reduplicated; REFL reflexive; SF stem formative; Sg singular; POSS possessive; TNS tense and/or aspect; TSR transitiviser. In the interlinear glosses, numerical notation refers to person, i.e, 1(st), 2(nd) and 3(rd) persons. For the third person, the numeral 3 is followed by a colon and the number of the noun class.
3. A word-medial morpheme-final low vowel can occur between two consonants with the same manner of articulation, e.g., two rounded consonants or two laminals. This prevents the operation of the Haplology Rule.

4. The noun class of an unprefixed loanword is identified by the prefixation of the adjective or demonstrative modifier within the noun phrase or by the subject or object crossreferencing in the verb prefixation.
5. The action of swimming is seen as movement produced by the hands of humans or the two front legs (literally: 'hands') of four-footed mammals and reptiles. Other animals go (literally: 'foot-walk') or run. The pectoral fish fin is referred to as a hand but not regarded as the prime mover.
6. These phonological rules cover most but not all allomorphic variation or morpheme deletion. Anindilyakwa has short forms of words which are unpredictable, e.g., the full of correct form of the word for 'sea snake' is *yinikapwangwa* but the Aboriginal speakers say that "its nickname is *yapwangwa*". Full forms have been used wherever they are known but some irregular shorter forms are included. The vernacular is only segmented into the morphemes relevant to the discussion.
7. The use of the gender morpheme *ani-* with the nonpersonified A and MWA nominal prefix is irregular. The most likely solution is that the Haplology Rule deletes their final CV syllables and the morpheme is represented by only the vowel /a/. In this paper, the gender morphemes are not listed in the glosses unless the *full* form is in the surface structure.
8. The root, *mwalya* and its hardened counterpart *pwalya*, has two extended meanings of 'body fat' and 'soft'. The latter seems to be more applicable in these words.
9. There is one word with a stem consisting of two adjective roots which has not been recorded as a free form noun: *yiningwaningpwirna* 'blue tusk-fish' (Literally: 'pointed:many'). Semantically the word fits with the inalienable possession nouns. The prefix *mwi-* is best considered to be assimilated by the following *ngpwi* even though such remote assimilation is unusual.
10. Anindilyakwa, unlike most other Aboriginal languages, does not name the compass points. The names of the winds are used in certain contexts to refer to these directions.
11. The locative suffix *-mantja* can also be used to denote a kinship affiliation but does not usually occur with a kinship root. e.g., *winathangmwantja* 'two marriageable cousins', *wirringathiyarrimwantja* 'two sisters' (literally: 'two similar females'). The relationship between siblings can, however, be expressed by replacing the kinship suffix with the locative marker, e.g., *kwirripwikwiyapwimwantja* 'you three, her older sisters'.
12. The nominal classifier for the personified which pertains to the Dreamtime is not carried across to the unnamed shell.

References

Bally, Charles
1926 "L'expression des idées de sphère personnelle et de solidarité dans les langues indo-européennes", in: Franz Fankhauser – Jakob Jud (eds.), 68-78.
[1995] [Translated and reprinted in this volume.]

Blake, Barry
1984 "Problems for possessor ascension: some Australian examples", *Linguistics* 22: 437-453.

Chappell, Hilary – William McGregor
1995 "Prolegomena to a theory of inalienability", [this volume.]

Crowley, Terry
1995 "Inalienable possession in Paamese grammar", [this volume.]
Dixon, Robert M.W.
1980 *The languages of Australia.* Cambridge: Cambridge University Press.
Evans, Nicholas
1995 "The syntax and semantics of body part incorporation in Mayali", [this volume.]
Fox, Barbara
1981 "Body part syntax: towards a universal characterisation", *Studies in Language* 5: 323-342.
Harvey, Mark
1995 "Body parts in Warray", [this volume.]
Heath, Jeffrey
1984 *Functional grammar of Nunggubuyu.* Canberra: Australian Institute of Aboriginal Studies.
Hosokawa, Komei
1995 " 'My face *am* burning!': quasi-passive, body parts, and related issues in Yawuru grammar and culture concepts", [this volume.]
Leeding, Velma J.
1989 Anindilyakwa phonology and morphology. [Unpublished Ph.D. dissertation, University of Sydney.]
Levitt, Dulcie
1981 *Plants and people–Aboriginal uses of plants on Groote Eylandt.* Canberra: Australian Institute of Aboriginal Studies.
Lynch, John
1973 "Verbal aspects of possession in Melanesian languages", *Oceanic Linguistics* 12: 69-102.
McGregor, William
1995 "The grammar of nominal prefixing in Nyulnyul", [this volume.]
McKay, Graham R.
1995 "Body parts, possession marking and nominal classes in Ndjebbana", [this volume.]
Merlan, Francesca
1976 *Ngalakan grammar, texts and vocabulary.* (Pacific Linguistics B 89) Canberra: Pacific Linguistics.
Mithun, Marianne
1984 "The evolution of noun incorporation", *Language* 60: 847-893.
Oates, William J and Lynette F.
1970 *A revised linguistic survey of Australia.* Canberra: Australian Institute of Aboriginal Studies.
Sheffler, Harold
1978 *Australian kinship classifications.* Cambridge: Cambridge University Press.
Wurm, Stephen A.
1972 *Languages of Australia and Tasmania.* The Hague: Mouton.
Yallop, Colin
1982 *Australian Aboriginal languages.* London: Andre Deutsch.

The grammar of nominal prefixing in Nyulnyul

William McGregor

1. Introduction[1]

1.1. Aims and scope

This paper investigates the phenomenon of pronominal prefixing to nominals in Nyulnyul, a language traditionally spoken in the vicinity of Beagle Bay, towards the northern extremity of the Dampier Land peninsula in the North West of Western Australia. In this language a small set of about two score or so nominal roots and stems are prefixed according to the person and number of their 'owner', 'possessor', or the whole of which they are a part. The majority of these prefix-taking (or prefixing) nominals are terms for parts of the human body; there is also a (much smaller) number of other semantic domains covered, including personal representations and items of clothing. However, most nominals from each of these three semantic domains do not take prefixes. The main point of the paper is to argue that whether or not a nominal is prefixing is not an accidental fact of morphology or (morpho-) phonology, but is semantically motivated. It is suggested that the relevant semantic factor is the notion of the *personal sphere* or *personal domain* – that is, roughly, whether or not the item is represented in the language as an aspect of – or part of – the persona of a human being, or, rarely of an animal (Bally [1995], and Chappell – McGregor 1995).

It is proposed, moreover, that the phenomenon of prefixation may have arisen historically through the morphologisation of a clause-level syntactic construction. The crucial property of this construction is that the whole is typically realised by a pronominal which immediately precedes the nominal referring to the part, the two being in the morphologically unmarked (absolutive) case. This is a special instance of the construction type which has been referred to in recent literature as the "favourite" construction for the expression of body parts in Australian languages (Hale 1981, McGregor 1985), which as the name suggests is perhaps the most common means of integrating parts of the human body (and more generally parts of non-human wholes) into clauses.

1.2. Typological sketch

Nyulnyul is a non-Pama-Nyungan language belonging to the Nyulnyulan family, a small family consisting of ten fairly closely genetically related languages spoken on the Dampier Land peninsula and adjoining areas of the Kimberley region (McGregor 1988b: 49, Stokes – McGregor 1989). These languages have been classified typologically as "prefixing languages without noun classification" (Capell 1940: 244). That is, they have pronominal prefixes to the verb indicating the person and number of the "subject"; furthermore, many – all of the Western Nyulnyulan languages, which include Nyulnyul, (see Stokes – McGregor 1989) – also take pronominal prefixes to some nominals indicating the person and number of the "possessor". However, these languages do not distinguish noun classes, as do the Worrorran languages spoken in the northern Kimberley.

The significance of Capell's (1940) prefixing parameter is that in a considerable number of languages of northern Australia there are both prefixes and suffixes, whereas in most other Australian languages (in particular, in the Pama-Nyungan languages) there are suffixes only, and no prefixes. And like most other prefixing languages, the Nyulnyulan languages all have a fairly small set of around a dozen postpositions, which indicate "case" relationships. One of these, the ergative postposition – some reflex of **-ni* (see Stokes – McGregor 1989) – marks the "subject" of a transitive clause (irrespective of its person or number), but not the "subject" of an intransitive clause, which, like the "object" of a transitive clause, takes no postposition.

Like almost all Kimberley languages, Nyulnyul does not distinguish alienable from inalienable possession at the rank of phrase. However, the mode of representation of possession is rather unusual for an Australian language: it is by means of a free pronominal denoting the possessor, rather than by the genitive or dative form of the NP referring to the possessor (as in most Australian Aboriginal languages, including the closely related Yawuru (Hosokawa 1995)). The pronominal referring to the possessor is an oblique (OBL) or emphatic (EMP) form of the appropriate free pronominal (see section 2 for the forms): for instance, 'my dog' can be expressed as either *jan yil* (1:sg:OBL dog) or *janijirr yil* (1:sg:EMP dog). Furthermore, it may either precede or follow the nominal referring to the possession. Thus, both *jan wunyjub* (1:sg:OBL mother) and *wunyjub jan* (mother 1:sg:OBL) are equally acceptable, and mean 'my mother'. The possessor may also be referred to by a distinct NP, usually either for reasons of referential specificity, or focus. Examples (1) and (2) show this type of possessive construction for a body part and an owned possession respectively.[2]

(1) *bin wamb nimal jin*
this man his:hand 3:sg:OBL
'this man's hand'

(2) *bin wamb yil jin*
this man dog 3:sg:OBL
'this man's dog'

There are two types of verb in Nyulnyul (and other Nyulnyulan languages): preverbs and verb roots. Verb roots are inflecting, and take prefixes and/or suffixes for tense and person and number of the "subject" and "object" (in transitive clauses). Preverbs are uninflecting. Verb roots and preverbs may occur alone or in compounds, in which case the preverb is almost always followed by the verb root, which functions as a type of auxiliary. (See McGregor 1995 for further details.)

1.3. Data on Nyulnyul used in this study

Today Nyulnyul is almost dead. It has a single fluent full speaker, and perhaps ten or so part speakers (McGregor 1988c). The primary corpus for this study is the material elicited from the remaining full speaker during the course of three field trips (1985-1986, 1988 and 1990). During this time somewhat over 120 hours of elicited Nyulnyul words and sentences were recorded, as well as a few short texts.

A secondary source of data is previous investigations of the language. Nyulnyul has been studied since at least the turn of the century, and a considerable number of scholars and amateurs have collected material in it. These include the well known amateur anthropologist Daisy Bates (who collected a short wordlist in the late 1890s – Bates n.d.); a number of Pallottine Missionary Fathers from Beagle Bay, including Joseph Bischofs (Bischofs n.d.a and n.d.b), Herman Nekes and Ernest Worms (e.g. Nekes 1939 and Nekes – Worms 1953); the anthropologist Norman Tindale (in the early 1950s); and the linguists Nora Kerr (of the Summer School of Linguistics) in the 1960s and Bronwyn Stokes in the late 1970s. Although there are inadequacies in the earlier records in terms of both the comprehensiveness and accuracy of the transcriptions and analyses, they are on the whole reasonably reliable, and have proved invaluable sources of information in various instances when I was unable to elicit particular forms.[3]

1.4. Organisation of the argument

I begin, in section 2, by describing the morphology of pronominals, including both free and bound forms. Having done this, I then examine (in section 3) the range of nominals which take pronominal prefixes, and show that the class of such items is not a random one, but is strongly motivated semantically. In this section I argue for my main thesis, that the set of prefixing nominals consists precisely of those nominals which refer to items belonging to the personal sphere of a human being. Section 4 proposes a possible historical origin for nominal prefixing in a clause level grammatical construction. Then in the following section I make some comparative remarks, looking at nearby languages which also have nominal prefixes. The set of nominals which take prefixes in Nyulnyul is compared with the sets that take prefixes in other languages, including some which are at best distantly related to Nyulnyul; it is shown that there is considerable agreement amongst these languages. Indeed, there is too much agreement for prefixing to be purely a "customary" or accidental feature of morphology, and this provides indirect evidence shoring up the argument of section 3, which was based on an intuitive semantics. The paper concludes with a brief summary and remarks on possible directions for future research.

2. Pronominal categories in Nyulnyul

The Nyulnyulan languages, including Nyulnyul, distinguish a four person and two "number" system in their free pronouns (Stokes – McGregor 1989; see also Hosokawa 1995). The person system distinguishes between 1 (that is, a group including the speaker and optionally one or more others, excluding the hearer), 1+2 (a group including the speaker and hearer and optionally one or more others), 2 (the hearer, possibly with others (not the speaker)), and 3 (one or more others, in a non-speech role). In the minimal/augmented number system (McKay 1978, McGregor 1989a), the minimal category refers to the smallest possible class within the particular person category, whilst the augmented category consists of minimal plus one or more others. Except in the case of the 1+2 person, of course, the minimal category consists of a single member; for 1+2, the minimal consists of the speaker and the hearer – that is, two individuals.[4]

The free pronouns of Nyulnyul are as laid out in Table 1. Three case forms are given for each pronoun: nominative, oblique and emphatic. The first occurs when the pronominal fulfils a role that is cross-referenced in the verb, either by nominative prefixes or accusative suffixes; the oblique occurs in most other environments – that is, when the pronominal realises any other clausal or sub-clausal role; and the emphatic form appears to be used exclusively to stress ownership.

Table 1. Nyulnyul pronouns

		Minimal	Augmented
1	NOM	*ngay*[a]	*yarrad*
	OBL	*jan*	*jarrad*
	EMP	*janijirr*	*jarrajirr*
1+2	NOM	*yay*	*#yadir*[b]
	OBL	*jay*	*#jadir*
	EMP	*#jajirr*	*#jadirjirr*
2	NOM	*juy*	*kurr*
	OBL	*jiy (jii)*[c]	*jungkarr*
	EMP	*jijirr*	*#jungkarrjirr*
3	NOM	*in*[d]	*irr* ~ *yirr*
	OBL	*jin*	*jirr*
	EMP	*jinijirr*	*jirrijirr*

a. The form *ngayu* is sometimes heard instead of *ngay*; this is almost certainly a borrowing from languages spoken to the south of Nyulnyul.

b. The forms for the 1+2 augmented are not certain, as Nekes (1939) does not distinguish between the tap /rr/ and the glide /r/. My guess that it is a final glide is based on the cognate form in Yawuru, *yadiri*, etc. (Hosokawa 1991).

c. Phonetically this is [ji.], with at least a half long vowel. Phonemically it could be any of /ji/, /jii/ or /jiy/, depending on the analysis.

d. Nekes (1939) gives the form *kinyingk* for this pronoun. However, there is reason to believe that this is a demonstrative in Nyulnyul (see e.g. McGregor 1989b).

The paradigm of pronominal prefixes to nominals in Nyulnyul is set out in Table 2.[5] The system is somewhat different from the system for free pronominals. There is a single form *yarr-* which covers both 1 augmented and 1+2 augmented, and a form *ya-* which covers just 1+2 minimal (see also e.g. Bischofs n.d.a: 1, Nekes 1939, and Nekes – Worms 1953: 16).

This gives us the so called Assiniboine system (Greenberg 1988, McGregor 1989).[6] (It should be noted that although the same person features are used here as in Table 1, they have a different meaning: 1 in Table 2 includes all first persons other than the speaker-hearer dyad.)

Table 2. Pronominal prefixes to Nyulnyul nominals

	Singular	Dual	Plural
1	*nga- ~ ngi-*		*yarr-*
1+2		*ya-*	
2	*nyi- ~ nya- ~ nyu-*		*kirr- ~ kurr-*
3	*ni- ~ na- ~ nu-*		*(y)irr-*

Clearly the pronominal prefixes are quite similar to the nominative case forms of the free pronouns. There are also striking similarities between the pronominal prefixes to nominals and the pronominal prefixes to verbs, as illustrated by the forms provided in Table 3.[7]

Table 3. Nominative prefixes to Nyulnyul verbs

	Set	Minimal	Augmented
1	I	*ngan(a)- ~ ngal(a)-*	*ya- -rra*
	II	*nga-*	*ya- -rr*
1+2	I	*yan(a)- ~ yal(a)-*	*?ya- -rra ~ yi- -rra ~ yu–rra*
	II	*ya-*	*?ya- -rr ~ yi- -rr ~ yu- -rr*
2	I	*min(a)- ~ wana- ~ mil(a)-*	*ku- -rra ~ wa- -rra*
	II	*mi-*	*ku- -rr(a) ~ wa- -rr(a)*
3	I	*in(a)- ~ guna- ~ ila-*	*(y)i- -rra ~ yu- -rra*
	II	*i- ~ yu-*	*(y)i- -rr(a) ~ yu- -rra*

Similarities between nominative prefixes to verbs and "possessive" prefixes to nominals are not uncommon in the world's languages – see for example, Allen ([1972]: 83-83), and the references therein. According to Allen, the connection is usually between "possessive" markers and the transitive subject, rather than between these markers and the intransitive subject (Allen [1972]: 83). However, it is clear that in Nyulnyul the closest similarity is between the nominal prefixes and the set II prefixes, which are by and large restricted to intransitive clauses. We will return to this point later, at the end of section 4.

3. Nominal prefixing in Nyulnyul

As in other Western Nyulnyulan languages there are no more than forty or so lexical items in Nyulnyul which take the nominal series of pronominal prefixes from Table 2. (For convenience we will refer to this series of prefixes as nominal prefixes.) The vast majority (at least three-quarters) of these are clearly unanalysable nominal roots. Of the remainder, about half a dozen are complex nominal stems, and approximately the same number are, at least notionally, not nominals (that is, they refer neither to entities nor to qualities). Despite their small numerical size, however, the prefix-takers are important lexemes, as the argument of this paper will demonstrate. We discuss the three types in order in the following subsections.

3.1. Prefixing nominal roots in Nyulnyul

The majority of prefixing nominal roots in Nyulnyul are terms for parts of the human body. However, not all body part nominals take pronominal prefixes. For instance, the word for 'hand', *-marl*, takes a prefix, but the words for 'hair' (*mugurn*) and 'fingernail' (*wurrul*) do not. In fact, only about one third of the nominal roots referring to body parts which I have so far collected are prefixing. On the other hand, in addition to words referring to body parts, there is a small number of other nominal roots which also take prefixes.

These two observations raise the question, the main issue we set out to investigate in this paper: Why are some nominals prefixed and others not prefixed? Is there some consistent principle whereby we can predict whether a particular nominal will take a pronominal prefix, or is there at least an explanation which will account for the known facts? Or, alternatively, is there no systematic explanation, and the distribution of prefix taking nominals merely haphazard within this word class? As indicated in section 1.1, I will be arguing that there is an explanation for the differential treatment of nominals, and that, more particularly, whether or not a nominal root is prefixing is semantically motivated.

Prefixing nominal roots fall into two semantic classes: (i) terms for parts of the human body; and (ii) terms for personal representations – that is, things which represent or stand for persons they are associated with (Chappell – McGregor 1995). Table 4 lists all of the known prefix

Table 4. Prefixing body part nominals in Nyulnyul

-alm	head	*-k*	back
-arnkarr	forehead	*-migil*	small of back, loins, anus[c]
-lirr	lips, mouth	*-wink*	chest, breast[d]
-m	eye	*-ng*	stomach, belly
-mirl	nose[a]	*-kurrinykurriny*	navel
-ilabab	ear	*-jirrjirr*	navel (*)[e]
-angal	tongue	*-murr*	buttocks, bum
-ward	chin (*)[b]	*-ur*	anus (*)
-many	(front part of) neck	*-rnmurr*	thigh, lap
-ngkurn	nape of neck	*-mird*	leg, knee, shin, calf
-marl	hand, arm, upper arm	*-imbarl*	foot
-mbarrm	armpit	*-kard*	body
-yalangkun	elbow (*)	*-wal*	tail[f]
-marrangk	fingers		

a. Bischofs n.d.a gives the declinable form *-mel* 'beak', which is presumably a mishearing and misrepresentation of *-mirl* 'nose'.

b. According to data elicited in 1990, this form is non-prefixing.

c. Nekes – Worms (1953: 437) give this term as meaning 'small of back, loins'; according to my fieldwork, however, it means 'anus'.

d. Bishofs n.d.b gives *-wink* as meaning 'breast', whilst Nekes – Worms (1953: 441) cite it as meaning 'chest'. The latter would seem to be the better rendition into modern English.

e. This form is given in Bischofs n.d.a and n.d.b only; both my data and Nekes – Worms 1953 have *-kurrinykurriny* as the only form for 'navel'. (The two forms may perhaps be dialectal variants.)

f. Both Bischofs n.d.a and Nekes – Worms 1953: 441 cite extensive paradigms for this nominal. Lest the reader think forms such as *ngawal* 'my tail' (Nekes – Worms 1953: 441) dubious – thus casting doubt on the reliability of this source – it might be pointed out that in some nearby languages (including Gooniyandi – see McGregor 1990), the term for 'tail' is used in reference to the foreskin. I have been unable to verify this possibility with the remaining speakers of Nyulnyul, and in my data the only forms are for third person singular and non-singular. (Nekes 1939 also gives the non-prefixing form *juburr* 'tail'.)

Table 5. Prefixing nominals for personal representations in Nyulnyul

-marraj	shadow, reflection, soul
-jimbarl	footprint, tracks (*)
-lawirl	name
-ginbal	appearance

taking nominals in Nyulnyul which refer to parts of the body, and Table 5 lists all known prefixing terms for personal representations.

These tables are based on my own corpus, as well as the material cited in Bischofs (n.d.a and n.d.b) and Nekes – Worms 1953. There is a very high level of agreement between these earlier sources and my own data; where there is disagreement, this is marked in the tables by an asterisk.

Despite extensive checking with the remaining speaker, it proved impossible to identify the starred forms for 'elbow' and 'navel' in Table 4; and for the remaining two starred items, the terms for 'chin' and 'anus', only *ni-* initial forms have been elicited, even for a first person singular possessor. Since in all instances a partial paradigm for the nominal is provided in the earlier works, I am inclined to consider the classification of these nominals as prefixing to be reliable, and that the corresponding gaps in my data reflect language attrition (or perhaps dialectal differences). On the other hand, except for the term *-jimbarl* 'footprint, track' in Table 5, there are paradigms in Nekes – Worms 1953 for each of the items which show up as prefixing in my own data.[8]

There remains a handful of nominal roots which may or may not be prefixing. These all have initial *ni*, but it is uncertain whether or not this is the third person singular prefix, as no forms for possessors other than the third person singular are known. They are:

ninggaal	ankle
ninngirr	vagina
nimidijin	anus of an animal
nimbirr	flipper(s)
nimandar(r)	shadow, reflected image

Only *nimbirr* 'flipper(s)' comes from my own corpus. Of the others, the first comes from Tindale 1952-1954; the second and third from Nekes – Worms 1953: 439 and 437 (respectively), and the final one from Nekes – Worms 1953: 435.

For the purposes of comparison, Tables 6-10 list all known non-prefixing body part terms and other nominals which one might expect to fall into the same, or nearby, semantic domains. These tables also contain items from my own and previous investigators' corpora, and when forms from the latter sources are not represented in my own corpus, they are marked with an asterisk. (It will be noticed that in many instances more than one Nyulnyul term is given the same English gloss – see, for instance, 'shoulder' and 'urine' below. It is not known in what way these alternative designations differ from one another semantically; it is quite possible that some of them are borrowings from nearby languages, and may have been given when the interviewee had forgotten the Nyulnyul term.)

Comparison of Tables 4 and 5 with Tables 6-10 reveals that terms for bodily parts, bodily products, personal representations, human propensities, and so on are not randomly distributed across the classes of prefixing

Table 6. Non-prefixing body part nominals: external

mugurn	hair	*bud*	neck, nape of neck (*)
jabuly	grey hair	*bulgumara*	nape of neck (*)
rdirdird	curly hair	*(r)langarn*	shoulder
maanda	eyebrow	*kundijin*	shoulder
muljurrjurr	eyelash (*)	*kurlujunu*	shoulder
rlabirl	body hair	*ngamarn*	breast
jiidi	beard	*namini*	chest, breast (*)
bardin	skin	*ngurrngk*	knee cap, elbow
birrirr	skin, foreskin	*milg*	ankle (*)
wurrul	fingernail	*kulul*	penis
barndal	feather, body hair	*narndi*	testicles[a]
bulgumara	occiput (*)	*jaway*	vagina
jimindi	fontanelle	*bara*	groin (*)
rlamard	cheek	*manygarr*	gills

a. I was given this as one of two terms for a man's "private parts" by the remaining full speaker of the language, who would not elaborate. Fortunately, Nekes – Worms (1953: 422) also include the word, and provide a more precise meaning, namely 'testicles'.

Table 7. Non-prefixing body part nominals: internal

Soft			
mabaar	flesh	*kaabir*	liver
julawu	nerve	*dilb*	kidney (*)
kurrburl	throat (hollow)	*kaburr*	guts, stomach, bowels
langirr	oesophagus (*)	*jilbid*	stomach (*)
nawang	neck, throat	*bulg*	bowels (*)
kudurd	heart	*namidijin*	duodenum (*)
liyan	heart, like	*nayirrmala*	bladder (*)
barrij	lungs		
Hard			
kinyji	bone, scales, shell	*murdumurd*	backbone
jiwarr	skeleton, dead body	*jaarringk*	tooth
karrabard	ribs	*daangga*	jawbone
wiirri	ribs	*bulurr*	skull (*)

and non-prefixing nominals. Thus, no terms for bodily products, items of clothing or adornment, or human propensity are prefixing. On the other hand, virtually all terms for personal representations are prefixing. There are two possible exceptions: the term *nimandar(r)* 'shadow, reflected image' (which Nekes – Worms 1953: 435 treat as non-prefixing), and *jaarlngk* 'spirit'. The first makes its primary reference to shade, rather than to a person's shadow, and so is not really an exception. It remains unclear how to account for the second term, however, unless it makes

Table 8. Bodily products and exuviae (non-prefixing)

Attached			
baaburr	scar	*mirniny*	bruise?
barndird	boil	*mugurdarl*	cicatrice
kunburl	blood	*yig*	sore
kundany	wart	*budungkurr*	big pimples, pock marks
liird	fat		
Not attached			
baarrj	spittle	*kurdawi*	urine
jiwil	spittle (*)	*ngardir*	urine
karirr	spittle, expectoration (*)	*ngurnd*	urine
kurnb	mucus	*yabiyab*	urine
kurnbukurt	mucus	*mili*	sperm (*)
buwa	pus, bad smell	*nundurr*	sweat
lakurr	egg	*wangkirr*	tears
burdi	faeces		

Table 9. Clothing and adornment (non-prefixing)

baal	belt
binyjibinyji	armbands, long pearlshell pendant
riij	pearshell pendant, hair string belt
guwarn	pearl shell
jad	clothes
jangkurr	hat
jardi	shirt
jinaburd	shoe

Table 10. Human propensity (non-prefixing)

balbirr, blurru	bald	*marrgin*	hunger
bamburl	blind	*burlji*	tired
kuuk	dumb	*ngangimarr*	thirsty
rdakardak	deaf	*ninyji*	alive
maad	crazy	*nyun*	ache
mukumuk	cripple	*nyunnyun*	ache, headache
biri	naked	*wararr*	pain
kurda	naked	*yuburl*	sick
kurdabirl	naked	*warrwal*	fit
jiwarr	dead	*warrwarr*	cramp
kurrii	greedy	*jibiljbilj*	smallpox
kunyurrji	sleepy	*rarrjin*	shame
kunyurrk	asleep	*rinyiriny*	sensible
manjurn	alone	*jimbijimb*	tip toes

reference to a disembodied spirit, which is not associated with a human being.

Within the terms for body parts themselves, things may at first glance appear rather random. However, more careful consideration reveals a number of systematic differences. All terms for hard internal organs (bones and teeth) and almost all soft internal organs are non-prefixing. There are just a few apparent exceptions, which can be readily explained. The prefixing term *-ng* 'stomach' refers not just to the stomach as an internal organ, but also to the external area of the torso (as in English), and is therefore not really an exception. The other exceptions, the tongue and anus, might also be reasonably regarded as external organs (see also below for another explanation).

Finally, we come to external parts of the body, of which roughly the same number (around 30) are prefixing as non-prefixing. Here we also find some important differences. About a third of the non-prefixing nominals refer to outer coverings of the body, including skin, hair and fingernails; the gills of a fish might also be regarded in the same way as a covering. Conversely, none of the prefixing nominals refer to outer coverings. Secondly, of the half a dozen nominals which refer to genitals, all are non-prefixing. (*Ninngirr* 'vagina' is a possible, though unlikely, exception – see above.)

Distinguishing the remaining non-prefixing nominals from the prefixing nominals is more problematic, although not impossible. Thus the tables show that there are two terms for 'elbow', one prefixing and one non-prefixing. However, the non-prefixing term, *ngurrngk* also refers to the knee-cap. This suggests that *ngurrngk* may be better glossed 'joint', and that it refers to the bone of, or protecting a joint. Thus the gloss 'elbow' may be inaccurate: 'elbow bone' might well be nearer the mark. This hypothesis is supported by the observation that *ngurrngk* does not refer to the external part of the knee, for which I have been unable to obtain a distinct term – the prefixing term *-mird* 'leg' was always given in response to my prompts to elicit the term for this part of the body. A similar explanation may account for some of the other apparent irregularities in the treatment of erstwhile external parts of the body. Thus, there is one prefixing and two non-prefixing terms meaning 'nape of the neck'. Possibly the two non-prefixing nominals given in Table 6 for 'nape of neck' refer instead to the neck bones (and note that there is no term for neck bone given in Table 7), rather than the nape. (Such a 'mistake' is not unlikely in field elicitation, particularly if, for example, Fr. Worms – who collected these words – had indicated the region by touch or pointing.)

Secondly, Table 6 gives a non-prefixing term for 'ankle' (from Nekes – Worms 1953), whereas the form given by Tindale 1952-1954 could well be prefix taking. It is possible that these terms differ in that the former refers to the bones of the ankle, the latter to the visible region.

Thirdly, the Nyulnyul terms for 'occiput', 'cheek', and 'shoulder' are all non-prefixing. These terms might refer primarily to bones: the bones at the back of the head, the cheek bone, and shoulder bones, respectively. It may then be that by extension they refer to the corresponding regions on the outside of the body.

We are thus left with three non-prefixing nominals to account for: *jimindi* 'fontanelle', *ngamarn* 'breast', and *namini* 'chest, breast'. The first of these, unlike the prefixing nominals of Table 4, refers to a temporary part of the body, one which disappears naturally shortly after birth (see also Crowley 1995.) It is not clear why *ngamarn* 'breast' and *namini* 'breast, chest' should be non-prefixing. It could be hypothesised that *ngaman* refers in the first place to a bodily fluid, breast milk, and by extension to 'breast' – in the majority of Australian Aboriginal languages there is a single nominal covering both meanings (see also Crowley 1995). Similar remarks might also apply to *namini* 'chest, breast'. However, there is no language internal support for this supposition. (A more satisfactory explanation will emerge shortly.)

It seems reasonable to conclude that the contrast between prefixing and non-prefixing nominals is by and large semantically motivated. We have suggested that nominals denoting personal representations and external parts of the body other than coverings and genitals are prefixing; all others (including those denoting bodily products, internal organs, human propensities, and so on) are non-prefixing. We have been able to account for most apparent exceptions. However, these explanations are in many instances speculative, and further intensive lexical research would be required to test them.

A more satisfactory, exceptionless, and less *ad hoc* characterisation of the class of prefixing nominals is provided by the notion of the *personal sphere* or *domain* – the domain of a human being's persona (Bally 1926 [1995]). It is the person, rather than their body that is pertinent to nominal prefixation, and those parts of the human body that are most important, obvious, most visible and central to the human being, tend to be prefixed. And interestingly, despite the fact that only about half of the known terms for external parts of the body are prefixing, they actually cover the entire body from head to foot, and both front and back. On the other hand, the less obvious, internal, and more peripheral parts are not prefixed, not being central to a human being's persona; nor, for the same reason, are nominals referring to bodily products, outer coverings and clothing prefixed. Human propensity nominals are not prefixed because they typically refer to non-essential attributes of the person – attributes which do not define the person as a person, and which are temporary (see also next paragraph).

It may seem surprising that the genitals are not prefixed (but cf. Harvey 1995 who remarks that genitals are treated in unexpected ways in the grammars of many languages). However, this fact can be accounted for under the more restrictive hypothesis that (in Nyulnyul) the domain of the person includes only those attributes which are common to every normal human being. What is involved then is a generalised notion of person, which does not refer to characteristics (no matter how personal) peculiar to an individual. The advantage of this hypothesis is that it explains why human propensity terms are never prefixed – thus improving on the explanation of the previous paragraph; in addition, it accounts for the two problematic non-prefixing terms commented on above: *jimindi* 'fontanelle' and *ngamarn* 'breast'. We will henceforth employ the notion of personal sphere in this restricted sense in Nyulnyul.

Personal representations are not, of course, parts of the body. However, they are closely associated with the person, and obviously fall within the personal sphere: that is, within the domain of the persona of any individual person. They include, in Nyulnyul, a person's name, footprints, shadow, reflection, soul, and appearance – all of which are denoted by prefixing nominals. These linguistic facts correlate with cultural facts. In Aboriginal cultures generally (as in many other cultures) names are regarded as inherent to, rather than an arbitrary adjunct of, a human being; the same holds true for a person's shadow, reflection, spirit, and footprint (see also Hale 1981, McGregor 1985, Hosokawa 1995, and Harvey 1995).

To conclude this section, I briefly discuss an alternative explanation which has been proposed to account for the difference between prefixing and non-prefixing nominals in some Kimberley languages. Capell (1972) proposes a purely phonological explanation of the difference between the two classes of nominal, specifically in the Worrorran languages (although his wording does suggest that he regards it as being more widely applicable). His explanation is encapsulated in the aphorism "no prefixation without initial vowel"; in other words, only vowel initial nominals are prefix taking; this explanation has been repeated more recently in Capell – Coate (1984). Clearly this does not account for the Nyulnyul facts: only about 16% of the prefix taking nominals listed in Tables 4 and 5 are vowel initial. Of course, it is always possible to suggest that the roots listed in Tables 4 and 5 are in fact all vowel initial underlyingly, and set up different morphophonological rules in order to derive the occurring forms. However, such an analysis would be much more complex and abstract than the one I have proposed here, and would involve numerous arbitrary decisions as to the quality of the initial vowels in the underlying forms. Furthermore, it would be descriptively uneconomical, since the initial vowel would have to be deleted in 84% of the roots when following a non-singular *rr*-final prefix – a rule which would be quite unnatural phonetically, in as much as its effect would be to give rise to a consonant cluster.

A similar (if less pronounced) state of affairs obtains in Ungarinyin, for which Rumsey (1982: 42-45) proposes a modification of Capell's principle. He points out that in Ungarinyin the reverse is more nearly the case: no initial vowel without prefixation – that is, only prefixing nominals are vowel initial. An examination of Tables 4-10 above attests to the validity of the latter hypothesis in Nyulnyul. The only nominal roots (from these semantic domains) with initial vowels are to be found amongst the prefix taking roots. Moreover, this would seem to be a very strong regularity even when the entire nominal lexicon is taken into account. The only certain exception I know of is an *a*-initial nominal root, *angga* 'what'. (There is a small number of nominal roots with initial [i] and [u], but it is not yet clear whether they should be analysed phonemically as having initial /i/ or /u/, rather than /yi/ or /wu/ – see footnote 3.) The main problem with this proposal is that – even granted its validity – it does not provide an explanation for why some nominals are prefixing and others are not.

Summing up, it seems clear that whether or not a nominal is prefixing is better accounted for semantically than phonologically (pace Capell 1972). A modification of Capell's phonological hypothesis along the lines suggested by Rumsey 1982 appears to be valid, but lacks explanatory power.

3.2. Prefixing of morphologically complex stems

In the previous subsection we proposed an explanation for why some, but not all, nominal roots take prefixes in Nyulnyul. We now turn to the rather small set of morphologically complex nominal stems which are prefix taking. These are listed in Table 11.

Table 11. Prefixing nominal stems

-alm-ird	head-ASSOC	hat
-m-ingid	eye-ASSOC	glasses, mirror
-marl-ingid	arm-ASSOC	bandage[a]
-imbal-ingid	foot-ASSOC	boot, shoelace
-imbal-ird	foot-ASSOC	shoelace
-mird-igurd	leg/knee-DYS	lame
-m-ukun	eye-ABL	eyebrow

a. It is not known whether this lexical item is restricted in meaning to bandages of the arm, or refers more generally to any bandage. In each of the contexts in which it occurs in the corpus, the bandage is attached to a person's arm.

Each of the stems in this table consists of a prefixing nominal root followed by a stem-forming suffix or postposition. The two morphs *-ingid* (the status of the final segment is uncertain, and may possibly be the postalveolar stop *rd*) and *-ird* each indicate that the thing to which the new stem refers is closely associated with the entity referred to by the nominal to which they are attached. Accordingly, they have been glossed ASSOC, that is, 'associated with', although it is likely that there is a subtle meaning difference between them. Thus, for example, *-imbalingid* (foot-ASSOC) and *-imbalird* (foot-ASSOC) both occur: the first appears to refer to both 'shoes' and 'shoelaces', while the second apparently refers just to 'shoelaces'. However, as there are only a few instances of each stem, the precise nature of the semantic difference (if any) between them remains uncertain. The remaining bound morphemes are the suffix *-igurd*, indicating an association with negative connotations, here dysfunctionality; and *-ukun*, the ablative postposition.

The claim that the stems listed in Table 11 are prefixing requires some qualification. First, all of them are bound forms, and do not have privileges of independent occurrence. Second, at least two of them, namely *-marl-ingid* 'bandage' and *-imbal-ingid* 'shoe, shoelace', frequently occur in the third person singular form – that is, as *ni-marlingid* 'bandage', and *ni-mbalingid* 'shoe, shoelace' respectively, irrespective of the person and number of the 'owner' or human being they are attached to (cf. remarks below concerning cognates of prefixing nominals in the non-prefixing Nyulnyulan languages). For instance, in example (3) the owner of the shoe, the speaker, is not cross-referenced in the prefix.

(3) *nganambarrgand jan ni -mbal -ingid*
I:tied:it my 3:sg -foot -ASSOC
'I tied up my shoe.'

However, as the following example illustrates, at least the stem form *-mbalird* 'shoelace' can take other person-marking prefixes:

(4)	*nga*	*-mbal*	*-ird*	*ruk*	*injid*	*gard*	*marriny*	*nganjid*
	1:sg	-foot	-ASSOC	undo	it:went	still	move	I:went

'My shoelace came undone as I was walking along.'

The phenomena exemplified in (3) and (4) may be referred to as *impersonal* and *personal* prefixing respectively. All other prefixing nominal stems (and roots) seem to invariably carry the appropriate prefix cross-referencing the owner of the item, and thus take personal prefixing.

Evidently a semantic explanation for the fact that these stems are prefixing along the lines suggested in the previous section would run into difficulties. For these stems fall into semantic domains associated with non-prefixing nominal roots: items of clothing, human propensity, and bodily coverings (eyebrows). It would seem strange to suggest that these items belong to the personal sphere of a human being whilst other like items are not (we return to this point later).

A number of alternative explanations suggest themselves. The most obvious is that these particular stems are prefixing because the roots on which they are based have no privileges of occurrence without a prefix. The problem with this is that it does not explain why the third person singular form, with initial *ni-*, does not invariably occur. Nor does it account for the fact that some of the stems are invariably person-marked, while two are only sometimes person marked.

A second hypothesis might go along the lines that the appropriate constituency analysis for these stems is as illustrated in Figure 1 rather than Figure 2 (see also Crowley 1995). That is, the prefix should be bracketed with the prefixing root, rather than with the root plus suffix.[9] This is supported by the closer connection which obtains between the prefix and root than the suffix and root.

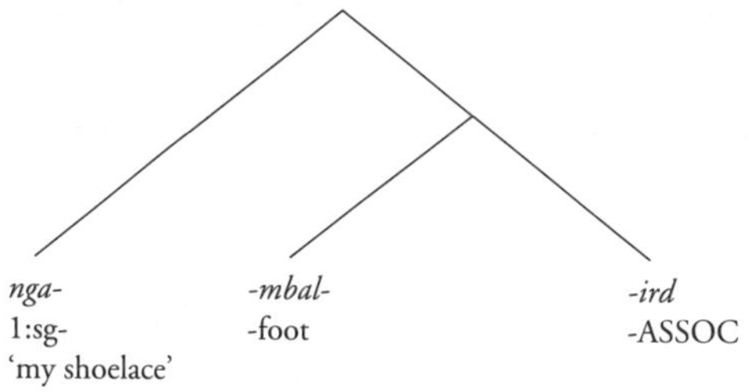

Figure 1. Proposed constituency analysis for a prefixing stem

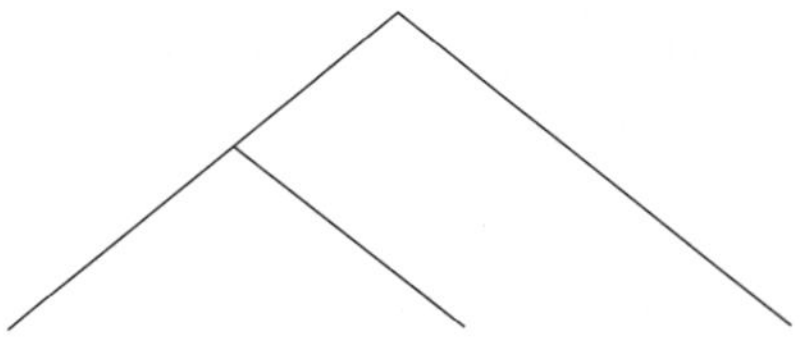

Figure 2. Alternative constituency analysis for a prefixing stem

Even if it is accepted that Figure 1 rather than Figure 2 represents the appropriate structure for prefixing stems, semantic problems emerge, albeit in a somewhat different guise. Thus, why should it be that the items in Table 11 are morphologically complex stems rather than lexically simple, unanalysable roots? Why aren't other items of clothing morphologically complex – e.g. why isn't the word for 'shirt' constructed in the same way as that for 'shoelace' – and why are there alternative non-prefixing terms for most of the items in Table 11? Are these purely accidental lexical facts of Nyulnyul? The answer to the last question, it seems to me, is a qualified no. In fact, if we examine the items from Table 11 carefully, and contrast them with non-prefixing nominal roots from the same semantic domains, we find that there are some subtle differences.

Hats and boots are closely associated with parts of the body which fall into the personal sphere: the head and feet respectively. In a similar way, glasses are associated closely with the eyes. It might be possible to account for the term for 'bandage' in a similar way, if, as per the footnote to Table 11, this term refers just to bandages around the arm. By contrast, the items of clothing listed in Table 9 above are not so specific to aspects of the personal sphere: shirts are associated with the upper part of the trunk, but there is no distinct term for this in Nyulnyul. Nor is there (as far as I know) a distinct term for the waist, which may explain why the terms for belt are not prefixed. For other items such as pearlshell pendants and items of adornment, it is easy to appreciate why these are not associated with prefixing nominals – they are gender-specific items in traditional Nyulnyul culture. There remains, however, an as yet unresolved problem: why are there also non-prefixing nominals referring to hats and boots?

A few observations lend further credence to the suggestions of the previous paragraph. It is these prefixing items of apparel that are most closely related to the person. They are amongst the physically closest to the relevant part – they are typically tight rather than loose on that body part, and certainly more so than most other items of apparel. Secondly, they are typically more protective and/or functional than other items of apparel: hats protect from the sun's rays (Aborigines in the region, like whites, use hats not just for fashion, but also for protection); bandages protect an infected or sore part; glasses are used for vision, and also protect the eyes from dust and grit; and shoes and shoelaces protect the feet (even traditionally, some sort of footwear was occasionally used for protective purposes in difficult terrain).

Before discussing the remaining terms from Table 11, we sum up our findings so far: prefixing stems refer to those items which are very closely associated with entities belonging to the personal sphere of a human being. In fact they may be so closely associated as to constitute parts of the personal sphere themselves – we will return to this matter in section 4 below. Items from the same general semantic domains which are not closely associated with aspects of the personal sphere are not prefixed.

Our explanation may perhaps be extended to account for the exceptional prefixing term for a body covering, *-m-ukun* 'eyebrow', as well as the prefixing term *-m-ingid* 'mirror'. The eyebrow protects the eye and is obviously very closely associated with it. The eye belongs (in Nyulnyul) to the personal sphere. Moreover, it is not unreasonable to regard the eyebrows, as distinct from other coverings or types of hair, as belonging to the personal sphere: for instance, they, as distinct from all other types of hair, are used in interpersonal communication. To suggest a similar explanation for 'mirror' might seem to be stretching things a little too far: surely this is no more than a semantic extension from 'glasses'. Nevertheless, it is interesting to observe that the only occasion on which a prefixing term was used to refer to a mirror was in (5), which described a real event in which the speaker was temporarily blinded by sunlight reflected in a motor car's rear-vision mirror. Mirrors are otherwise – e.g. when being used to look at the self – referred to by the loan-word *kilaj* (from English *glass*). It is perhaps reasonable then to consider the mirror in (5) as being very closely related to the eye, and ultimately to the speaker's personal domain.

(5) *bin ngiim -ingird dibirr nganggam ngiim inmarrin*
this my:eye -ASSOC turn I:will:do my:eye it:burns
'I want to turn the mirror around, as my eye is burning (from the sun).'

This leaves us with the stem *-mird-igurd* 'lame', the only known prefixing human propensity nominal. Although it may seem reasonable for this stem to be prefixing in terms of the explanation of the previous paragraph – it indicates dysfunctionality of something in the personal sphere – it is difficult to understand why the terms for dumbness, blindness and deafness are not also prefixing. Perhaps a culturally-specific explanation is possible; but not on present knowledge.

Finally, it remains to raise the issue of the difference between personal and impersonal prefixation of prefixing nominal stems. I cannot provide a satisfactory explanation at this stage. I can do no more than draw attention to the obvious fact that it is only where there is no other indication in the clause of the owner/possessor of the item – no oblique pronoun or pronominal prefix or suffix in the verbal complex referring to the possessor – that the person and number is indicated in the nominal prefix (see example (4)). Where the possessor is indicated by the oblique pronoun and/or the prefixes to the verb (as in (3)), the nominal prefix does not indicate the person of the possessor. The difficulty of this purely formal explanation is that although it accounts for the known facts concerning morphologically complex nominal stems, it does not hold for nominal roots, where the person and number is always indicated by the prefix to prefix-taking roots, regardless of whether reference is made elsewhere in the clause to the 'owner' of the part. This matter will clearly repay close investigation – although it may well never be resolved!

3.3. Nominal prefixes to words of other classes

Nominal prefixes are not restricted to occurring on nominal lexemes. They also occur on a small number of items from other word classes, in particular, what would seem to be particles and adverbials. An analysis of the parts of speech in Nyulnyul has not yet been undertaken, so this statement must be accepted with some caution. However, these lexical items do show syntactic behaviour which distinguishes them from the nominals discussed in the two previous subsections; this is not the place to go into details.

Why is it that these particular non-nominals are prefixed, and why with the nominal rather than the verbal set of prefixes? These are important questions, albeit questions which we cannot do justice to at this stage of the investigation. We could begin by alluding to morphological criteria in an initial attempt at explanation: the nominal prefixes are used

Table 12. Prefixing roots from other word classes

-mungk	believe, think
-malkang	self, alone[a]
-yam	abstain (tabooed?)[b]
-malugul	lack, want
-nangan	?[c]

a. Nekes (1939) cites this as the Bardi form, giving *-malg* as the corresponding Nyulnyul form. However, in my corpus only the longer form occurs.

b. This word occurs in Nekes (1939), but not in my database. Attempts to elicit it failed. A cognate form occurs in Bardi with the same meaning (see section 5.1).

c. Bischofs (n.d. a: 11) cites this item as a verb which declines as a noun; however, he omits to gloss it. Unfortunately, it has proved impossible to find cognates in related languages which might throw light on this, primarily because the form of the root itself varies with each pronominal prefix in Bischofs's manuscript.

because the verbal ones must occur with tense prefixes, and these do not occur on the above non-nominal roots. This describes, but does not explain the facts – why isn't tense relevant to (some of) these terms? For instance, why isn't it relevant to *-mungk* 'believe, think', *-yam* 'abstain' and *-malugul* 'lack, want', since these would all appear to be (at least in the normal interpretations of the glosses) attributive states which could have occurred in the past and no longer obtain, or might occur in the future. (Interestingly, however, in certain other languages (e.g. English and Mandarin) similar stative verbs show restrictions in terms of the tense and aspectual marking they permit.) On the other hand, a semantic explanation along the lines suggested for nominals would seem improbable: these terms do not seem to refer to anything which one would normally consider as part of one's personal sphere, and moreover, as qualities they are normally temporary and associated with particular individuals rather than people in general (see also above pages 263-264).

Finally, we mention that there is a stem which is apparently an adverbial, which consists of the reduplication of a prefixing nominal root. It is *PF-k-ung-PF-k-ung* (PF-back-ALL (reduplicated)), meaning 'backwards'. As for the other lexemes discussed in this section, there is no obvious semantic explanation for the fact that this term is prefixing whilst, for instance, the adverbials for 'forwards' and 'sideways' are non-prefixing roots.

4. Historical development of nominal prefixation in Nyulnyul: a hypothesis

It is a commonly accepted principle of historical linguistics that morphological elements frequently derive historically from what were previously independent words (see e.g. Hock [1988]: 616, Anderson 1988: 336). That is, the morphological structure of words reflects and derives from some earlier syntactic construction, and is thus a fossilised representation of the latter. Givón (1971) articulates a very strong version of this position in his dictum that "today's morphology is yesterday's syntax". In this section I propose a possible historical origin for nominal prefixation in Nyulnyul in a (part of a) previous clause-level syntactic construction.

It is not difficult to find a good contender for a construction type which could have given rise to nominal prefixation. Many Australian languages display a favourite construction for the expression of body parts and some other inalienable possessions within clauses. The defining characteristics of this construction are that: (i) the part and whole are accorded identical morphological marking; and (ii) the whole, but not the part is cross-referenced by a bound pronoun either on the verb, or encliticised to an auxiliary or to the first constituent of the clause, depending on the language (McGregor 1985: 209, Hale 1981). It is furthermore typically the case in this construction that if both the part and the whole are represented by overt nominal expressions, then the expression for the part follows – frequently immediately – the expression referring to the whole (see also Chappell 1986, Chappell – McGregor 1995). These properties are illustrated in the following example from Gooniyandi, a language spoken about three hundred kilometres to the east of Nyulnyul, and belonging to a different family:

(6) *nganyi marla diribloondi langgagoorloo -ya*
I hand I:entered hollow:log -LOC
'I put my hand into the hollow log.'

Both McGregor 1985 and Hale 1981 argue that the expression for the part and the expression for the whole – in Gooniyandi and Warlpiri (central Australia) respectively – do not together constitute a single NP. Rather, each expression separately constitutes a single NP, which realises some distinct role in the clause. According to McGregor (1985: 217), in the favourite construction in Gooniyandi the NP referring to the whole fulfils a participant role (roughly equivalent to an argument role in formal grammar) in the clause, whilst the part realises an *inner role*. Briefly, inner roles are non-participant roles which correspond to participant roles in a one-to-one manner, the corresponding roles being semantically identical except that the former do not participate in the transitivity of the clause – i.e. in the 'direction' of the action (McGregor 1985: 215). The two most important inner roles are Instrument and Range, the former corresponding to the Agent participant role, the latter to the Medium participant role, the thing through which the situation comes into being or is enacted – in short, a grammatical category embracing the intransitive subject and transitive object (McGregor 1985: 213, 1990: 323 and Halliday 1985: 144). Parts of the body most frequently (in Gooniyandi) realise the inner role of Range, and usually these are Ranges corresponding to Goals – i.e., roughly, they correspond to what are sometimes called cognate objects (as in *sing a song*), except that the nominal and verbal are not cognates.

My proposal is that at some earlier stage in the development of the Nyulnyulan languages, no nominals were prefix taking, and that, moreover, body part nominals typically occurred in a construction resembling the favourite construction for body parts in Gooniyandi – that is, one in which the whole was treated as the participant (or argument), the part as a non-participant inner role. It is reasonable to presume that the body part would most commonly have been a Goal/Range (or cognate object), as is normally the case in modern languages displaying the construction. Let us assume furthermore that at this time only the "subject" participant was cross-referenced in the verb.[10] Then it is reasonable to expect that an NP referring to the Goal would appear relatively frequently in clauses, and at least more frequently than they would in a language in which the Goal was obligatorily cross-referenced (where ellipsis might be more frequently expected). The meaning 'he hit me in the head' would be most naturally expressed in proto-Nyulnyulan as per example (7), which would illustrate the predominant construction type in which body parts were found, the favourite construction:

(7) **indam ngay alm*
he:hit 1:sg:NOM head
'He hit me in the head.'

where *ngay* 'I/me' is the Goal, and *alm* 'head' is a Range.

A construction such as this could readily be reanalysed in such a way that *ngay alm* (1:sg:NOM head) is treated as a single constituent. Ultimately, the two words might coalesce into a single morphologically complex word. Recall also the very close formal similarity between the nominative cases of the free form pronouns and the nominal prefixes. The scenario we have elaborated also accounts for the fact that the possessor, the whole, is not cross-referenced in the verb: for in the process of reanalysis of the construction represented by (7), the possessor/whole would no longer be a Goal participant.

That this proposal – based on a dominant subpattern within the favourite construction for expression of body parts within clauses – is not too unreasonable as an explanation of the general phenomenon of nominal prefixation (not restricted to this particular domain) finds some support in the well known fact that syntactic change frequently starts in a relatively restricted domain (Hock [1988]: 378).

But there is another construction type commonly found in Australian languages which might at first glance appear to be an even better candidate for the historical source of nominal prefixation. This is the NP-level construction in which inalienable possession is shown by apposition of the possessor and the possessed, in that order, as e.g. in the Warrgamay NP *ngulmburu bingany* (woman foot) 'woman's foot' (Dixon 1980: 293). If this were the source of pronominal prefixes, the only historical change we would need to postulate would be the morphologisation of free forms; no constituency reanalysis would be involved. However, there are two reasons why I prefer the more complex suggestion elaborated above over this apparently simpler one. Firstly, none of the non-Pama-Nyungan languages in the vicinity express inalienable possession by apposition, or indeed make a distinction between alienable and inalienable possession at phrase level. This makes it less certain that juxtaposition was ever employed in the expression of inalienability in some ancestor of Nyulnyul. This is not, of course, a strong argument, and the modern languages may all reflect the results of widespread diffusion of the typological parameter, lack of distinction between alienable and inalienable possession, in the Kimberley and neighbouring regions. However, there is a second, much more important reason: the historical source construction I have proposed shows semantic features which are quite consistent with the semantic features attributed to prefixation, and significantly more so than the inalienable possession construction in the NP. Let us review the evidence.

As we have seen, it is only a small subset of the set of body parts which are actually prefixed in Nyulnyul. This would be difficult to explain if the historical source of prefixation was in the appositive inalienable construction mentioned in the previous paragraph, for in Australian Aboriginal languages which display this latter construction type it is typically the case that all body parts are treated as inalienable possessions. On the other hand, the facts of prefixation are quite consistent with the semantics of the favourite construction. I will attempt to demonstrate this claim by a brief discussion of the favourite construction in Gooniyandi.

In a paper on body parts in Gooniyandi (McGregor 1985), I suggested that the semantic factors of *separability* and *individuation* are relevant to the way in which body parts (and a small set of other items) are incorporated into the Gooniyandi clause. Parts which are separable, and individuated tend to be treated as participants, and are cross-referenced in the verbal complex (McGregor 1985: 226). Inseparable parts on the other hand are more likely not to be treated as participants; the owner of the part is usually cross-referenced instead of the part. In fact, my claim was really the reverse of this: participants are individuated, while entities in the corresponding non-participant roles lack individual identity. Thus the favourite construction in Gooniyandi (see above) expresses the inseparability of a body part and the whole of which it is a part.

Some comment on the notions of separability and individuation is called for (see also Chappell – McGregor 1995). These are abstractions which encompass a number of quite different (but interrelated) particularisations in different contexts. In Gooniyandi, separability and individuation relate strongly to the referent process or action, and the way in which the part and the whole interact with respect to that process; they are not merely properties of the parts of the body themselves. In particular, separability does not coincide with the property of being able to be removed or cut off from an animate being, as the term might seem to suggest. At least the following parameters are relevant: whether the 'entity' has independent existence (either independent of some other entity (as is the case of say bodily products once they have been excreted by the person), or independent of a process (as in the case of say a song, which may be presumed to have some existence in abstraction from particular performances)); whether the part can be physically isolated from the whole without hurting the whole (as in 'they cut his hair'); whether action may be represented as directed at the part to the exclusion of the whole (as in 'they cut off his toe'). Inseparability, on the other hand, relates to the property whereby action involving the part (or whatever) automatically involves the whole in the same semantic role relationship: it cannot be seen as being restricted to the part to the exclusion of the whole. Putting it another way, in a given language, inseparable parts of the body are those parts which belong to the personal sphere of a human being, within a particular referent process or action.

Significantly, Gooniyandi words corresponding to the nominals which are prefixed in Nyulnyul tend to be treated as non-participants in the favourite construction in the former language – that is, they tend to be treated as inseparable from the whole of which they are a part. This observation holds for both body parts and personal representations. Conversely, if we turn to the body parts that are not prefixed in Nyulnyul, we find that they are usually treated as participants in Gooniyandi clauses, and do not usually occur in the favourite construction (McGregor 1985: 222ff). This holds for the external parts – with the exception of the second group of more "personal" parts on Table 6 (neck, shoulder, breast, knee, ankle, etc.) – as well as the soft and hard internal body parts of Table 7. The case is even stronger for bodily products, where in Gooniyandi they are always treated as participants in their own right, and never occur in the favourite construction (McGregor 1985: 229). Finally, human propensity nominals in Gooniyandi usually occur as qualifiers within NPs, or as secondary predicates, occasionally as verbs; they are not treated as inseparable parts of the personal domain. Similar remarks apply to the corresponding construction in Yawuru (see Hosokawa 1995).

My proposal is, then, that Nyulnyul has grammaticalised this separable/inseparable distinction, so that the class of nominals falls into two disjoint subclasses, one referring to separable and individuated things, the other to inseparable or non-individuated things which belong to the personal sphere of a human being. (In Gooniyandi this distinction is not grammaticalised, of course, and a given part of the body may (normally) be treated as either separable or inseparable, depending on the situation in which it occurs.) In this way we are able to account historically for the restricted subset of inalienable possessions which are prefixed in Nyulnyul.

By comparing Nyulnyul and Gooniyandi, it is possible to appreciate and begin to understand some of the apparent exceptions in nominal prefixation. Most notable among these are the prefix taking nominal stems. We proposed that the referent items are very closely associated with some item which falls within a human being's personal sphere, to the extent that they could (almost) be regarded as part of this sphere themselves. This suggestion finds additional support in facts of Gooniyandi grammar. Thus, the eyebrow can be treated as inseparable in Gooniyandi, for action on the eyebrow usually affects the person, and the eyebrows are also used instrumentally in communication (as mentioned above). Secondly, it is interesting that in Gooniyandi *magarda* 'hat' may be treated as inseparable – one way of expressing the notion 'I put on my hat' is as shown in (8) below:

(8)	*nganyi*	*-ngga*	*magarda*	*yoodjawilimarni*
	I	-ERG	hat	I:might:put:on:myself

'I might put on the hat.'

This Gooniyandi clause is a favourite construction: a reflexive/reciprocal in which *magarda* 'hat' discharges the inner role of Range. This makes the case for treating this item of clothing as inseparable somewhat more reasonable. Here, the inseparability relates to the fact that action done to the hat is done to the owner – thus, putting on the hat affects the appearance of the owner more than it affects the appearance of the hat. Thirdly, in Gooniyandi one could treat bandages and shoelaces as inseparable in clauses referring to tying up such items: these notions could be expressed as reflexive clauses of the type 'I tied myself up bandage'. This suggests that it is not quite so unreasonable that *-marl-ingid* 'bandage', and *-imbal-ird* 'shoelace' should be prefixing. In each of these cases we have long, thin, rope-like, items which are tied up, and in so doing, the person is, in a sense, also tied up. Other items of clothing or apparel which are rope like – belts, necklaces and so on are not normally tied up. This explanation does not account for 'hat'. But it might if this term referred to the ceremonial head-dress consisting of a long piece of string tying the hair up into a high coiffure. Here is perhaps an alternative explanation for the fact that 'hat' is prefixed, although it is not possible to decide at this stage which explanation is to be preferred.[11]

It might also be possible to account for prefixation of some of the non-nominals in terms of the historical hypothesis outlined in this section. For instance, let us briefly consider *-mungk* 'know, believe, believe mistakenly, think'. It is notable that in Gooniyandi there is a somewhat similar term, albeit a nominal, *binarri* 'know, knowledge'. This lexeme frequently occurs in constructions like *nganyi binarri* (I know) 'I know', which may be followed by a clause expressing a proposition that is known, in a manner reminiscent of the way in which the particle *-mungk* is used in Nyulnyul. This suggests that the prefixing of *-mungk* 'believe' might have arisen historically from a similar configuration of personal pronominal and (possibly nominal) root, as in **ngay mungk* (I think) 'I think'. If this is the case, then the fact that the semantic features suggested for Nyulnyul nominal prefixation do not account for the prefixing of this particle becomes explicable. A similar explanation might be adduced to explain the prefixing of *-malkang* 'self, alone', *-yam* 'abstain' and *-malugul* 'lack, want'; however this is beyond the scope of the present investigation.

It is well to bear in mind Anderson's caution that "[u]nfortunately, it is impossible to identify all of 'today's morphology' with 'yesterday's syntax'. For one thing, not all affixes have a (relevant) source in syntactic material" (Anderson 1988: 338). (See also Hock [1988]: 617 and Austin 1989: 69.) I believe, however, that the arguments I have presented in this section do lend support to the historical hypothesis. For not only do the relevant affixes have reasonable sources in syntactic material – in the form of free pronominals which are reconstructable for proto-Nyulnyulan (see Stokes – McGregor 1989), but also the construction which constitutes the putative source for the morphology of today is one which recurs in languages of the area, and may reasonably be supposed to have existed in the parent language. Moreover, we have been able to demonstrate a close semantic connection between the morphological and the syntactic constructions, one which would seem to be too striking to be attributable to chance or to areal diffusion.

I conclude by remarking that this discussion points to an important limitation in Allen's suggestion that there is a strong cross-linguistic connection between possessive prefixing of nominals and nominative prefixes to verbs – indeed, more particularly to transitive subject prefixes (Allen [1972]). It seems likely that this generalisation would apply primarily in the case of prefixing of alienably possessed nominals, via the notion of ownership. For inalienable possession, my expectation is that the relationship would be between possession and stativity or attribution/identification, as is found in Nyulnyul. Further justification for this hypothesis is provided by Amerindian language data cited in Langdon 1988; the issue, however, begs further investigation.

5. Comparison with other nominal prefixing languages

Pronominal prefixation of (a subset of) nominals occurs in a number of non-Pama-Nyungan languages of the Kimberley and Arnhem Land. These include all known Worrorran languages, as well as all western Nyulnyulan languages. The Eastern Nyulnyulan languages Yawuru

(Hosokawa 1995) and Nyikina (Stokes 1982) do not show this system and it is present in the speech of only one of the two remaining speakers of Warrwa (McGregor 1994: 15–16). These languages form two discontinuous blocks, one located in the northern Kimberley region, and the other in Dampier Land. There is evidence that historically the Nyulnyulan languages all had pronominal prefixes to nouns, and that they were lost relatively recently in Nyikina, Yawuru and Warrwa. This is because cognates of the prefixing nominals in these three languages correspond to the third person singular forms in the prefix taking languages. Thus, the Nyikina word for 'head', *nalma*, is cognate with the Bardi word *nalma* 'his/her/its head' and Nyulnyul *nalm* 'his/her/its head' (see also Stokes 1982: 409-412). There are no correspondences involving the bare root forms of the prefix-taking nominals, as would be expected if the prefixing languages were the innovators. It would appear then that at some point in the history of Kimberley languages, noun-prefixing languages formed a continuous geographical region.

In this section we examine the range of nominals in these other Kimberley languages which take prefixes, showing that they overlap to a considerable degree. This suggests that there is similar motivation for the systems in these other languages, and this lends further support for my hypothesis that prefixing is semantically motivated.

5.1. Prefix taking nominals in other Nyulnyulan languages

We begin by comparing Nyulnyul with its Western Nyulnyulan neighbours, Bardi, Jawi, Jabirrjabirr and Nimanbur. Tables 13 and 14 list the available data on these languages. Table 13 provides, where known, the forms of prefixing nominals, whilst Table 14 lists the relevant non-prefixing nominals. However, as there was insufficient data available for bodily products and human propensity nominals, these have not been included in the table.[12]

It will be clear from an inspection of these tables and comparison with Tables 4-10 above that there is an extraordinarily high level of agreement among the five languages as to which parts are prefixed, and which are not. Among the terms for parts of the body, there are just a couple of points of difference. A prefixing nominal has been given for 'calf' in Bardi, for which no corresponding Nyulnyul word has been elicited; my suspicion is that this is the Bardi word for 'leg' (see two lines previous on Table 13). Similarly, the Bardi term for 'throat' on this table is possibly the same word as for 'front of the neck'. There are a few more differences in the last two groups of rows in Table 13 as compared to Table 12 above, but since the evidence here is inadequate and corresponding terms have not been elicited for 'sweetness', 'right' and 'meaning' in Nyulnyul, it is pointless attempting to account for these possible differences. Finally, inspection of Table 14 reveals that the only non-prefixing nominal that does not correspond to a non-prefixing nominal in Nyulnyul is *baara* 'bum'; clearly, however, this is cognate with the Nyulnyul term for 'groin', which may account for this difference.

Table 13. Prefixing nominals in some Nyulnyulan languages

	Bardi	Jawi	Jabirrjabirr	Nimanbur
Body parts				
head	*-alma*	*-alm*		
forehead	*-nangara* (*-rnkarr*?)			
lips	*-lirr*			
mouth	*-lirr*	*-lerr*	*-lerr*	*-lerr*
eye	*-mi*	*-miny*	*-m*	*-m*
nose	*-mala*	*-marl*		
ear	*-lamarr*			
tongue	*-yangala* ~ *-jangala*[a]	*-(y)angarl*		
chin	*-yoda* ~ *-joda*			
neck, front		*-many*		
hand	*-marla*	*-milirriny*		
arm	*-marla*			
armpit		*-nganybu*		
back	*-ya* ~ *-ga*	*-ga* ~ *-ya*		
stomach	*-ngu*	*-ngu*	*-ng*	*-ng*
navel	*-runggurunggu*			
anus	*-munggulu*			
leg	*-lara*	*-langka*		
knee	*-mirdi*	*-mird*		
calf	*-lari* (McG)[b]			
foot	*-yambala* ~ *-jambala*	*-ambal*	*-mbal*	*-mbal*
thigh	*-rnmurru* (McG)			
body	*-yarda* ~ *-karda*			
throat	*-manyi*			
chest	*-mara*			
tail	*-yala* ~ *-jala*	*-yal*		
elbow	*-yalanggun* ~ *-jalanggun*			

Table 13. (cont.)

	Bardi	Jawi	Jabirrjabirr	Nimanbur
Personal representation				
shadow	*-minggarra*			
name	*-nga*	*-nga*		
Others (including non-nominals)				
believe,	*-munggun*	*-minggon*	*-mongg*	*-mongg*
self, alone	*-malgang*		*-malg*	*-malg*
tabooed	*-yamu* ~ *-jamu*		*-yam*	*-yam*
sweetness	*-yarra* ~ *-jarra*			
right,	*-molon*			
Possible prefixing roots				
back of head	*-nggan*			
palm of hand	*-ngonyi*			
penis	*-rnda*			
vagina	*-wa*			
meaning	*-nmurru* (maybe)			

a. In Bardi *y* occurs following a vowel, *j* or *g* following a consonant.

b. McG in brackets following a nominal indicates that the only source of this nominal is in my own field notes.

Table 14. Non-prefixing body part nominals in some Nyulnyulan languages

	Bardi	Jawi	Jabirrjabirr	Nimanbur
External				
hair	*muwarn*	*muwarn*		
eyebrow	*maandi*			
nape of neck		*buuda*		
skin		*janggala*		
fingernail	*orol*	*orol*		
breast	*ngamarn*	*ngamarn*	*ngamarn*	*ngamarn*
bum	*baara*			
ankle	*milkal*			
shoulder		*laangan*		
Soft internal				
throat (hollow)	*korrbal*			
heart		*liyan*		
liver		*kawir*		
Hard internal				
bone, scales		*gaanyji*		
rib (bone)	*iirri*			
tooth	*jarrangku*	*jarrungk*		

Much of the agreement on Table 13 can be accounted for as retention from a common ancestor language, proto-Western-Nyulnyulan: the vast majority of both prefixing and non-prefixing nominals in each of the languages are cognates. Nevertheless there are eight prefixing nominals on Table 13 which appear not to be cognate with the corresponding Nyulnyul nominals – the terms for 'ear', 'chin', 'navel', 'anus', 'leg', and 'chest' in Bardi, and 'hand' and 'armpit' in Jawi. This agreement cannot be accounted for in terms of retention of lexemes from a proto-language, and lends at least some measure of support for the hypothesis that whether or not a nominal is prefixing is governed by semantic factors – although obviously no firm conclusions can be drawn. More importantly, the data from these other languages strengthens the case for Nyulnyul, particularly since the Nyulnyul data may not be fully reliable, based as it is on the recall of one surviving speaker and data collected over half a century ago by investigators who in most cases had little or no formal linguistic training.

5.2. Nominal prefixing in some Worrorran languages

The Worrorran languages – including Ungarinyin, Worrorra, Yawijibaya, Unggumi and Wunambal – seem to have a slightly larger set of prefix taking nominals than do the Nyulnyulan languages.[13] Vasse (1991: 29), for instance, says that there are about fifty prefixing nominals in Wunambal. (As we have already seen in section 3, there are about forty in Nyulnyul.) They fall, in Ungarinyin, Worrorra, Wunambal and Unggumi at least, into two major groups: prefixing body parts and prefixing adjectives; as for Yawijibaya and Gwnin/Kwini (both Worrorran languages I have gathered some data on), I have as yet found no evidence in the available data of prefixing of adjectives. Next to these prefixing nominals there are also non-prefixing body part nominals and non-prefixing adjectives, as is the case in Nyulnyul. For the purposes of comparison, I have listed in Table 15 the body part and personal representation nominals in the Worrorran languages which correspond to prefixing nominals in Nyulnyul. Then in Table 16 are listed terms in the northern languages for those parts of the body and nearby semantic domains which are not prefixed in Nyulnyul. In each table, a *y* in a column indicates that the corresponding nominal in that language is prefixing; an *n* indicates that it is not. It should be borne in mind, however, that there are many gaps in the information, and that in some instances information is not particularly reliable.

Table 15. Prefixing nominals in some Worrorran languages which are also prefixing in Nyulnyul

	Yawi-jibaya	Unggumi	Worrorra	Ngarinyin	Wunambal	Kwini
Body parts						
head	y	y	y	y	n	n
forehead		y	y	y	n	n
lips, mouth		y	y		n	y
eye	y	y	y	y	n	n
nose	y	y	y			
ear		y		y		n
face	y	y	y		y	n[a]
tongue	y	n				n
chin	y	n	y			
armpit	y	y				n
neck	y	y	y			n
hand	y	y	y	y		n
arm		y	y		y	y
upper arm			y		y	
elbow	y		y			n
back	y	y				n
chest		n	y		y	y
stomach		y	y			
navel	y					n
bum		y			y	n
leg		y				n
knee		y	y			n
shin			y			
calf		n				n
foot	y		y	y	y	y
thigh	y	y	y			
body					y	y
tail		n		y		n
ankle						n
Personal representations						
shadow						y
name				y	y	
soul, spirit			y	y	y	

a. The term for face in Nyulnyul is the two word construction *PF-lerr-PF-mirl* 'mouth nose', each word of which is prefixed according to the person and number of the "owner" of the face. For this reason, we have admitted the term for 'face' on this table.

Table 16. Prefixing nominals in some Worrorran languages which are not prefixing in Nyulnyul

	Yawi-jibaya	Unggumi	Worrorra	Ngarinyin	Wunambal	Kwini
Body parts						
External						
hair		n			y	n
beard	y	n	y			n
skin	n	n				n
fingernail	n	n				n
moustache						n
pubic hair	y					
eyelid						n
cheek				y		
shoulder	y	y	y		y	y
breast	n					n
penis	y			y		
testicles	y					
big toe						n
wrist		y	y	y		n
heel						n
Internal						
Soft						
brain						n
throat	y	y	y			
flesh					y	
heart	n	y				n
liver	n	?				n
guts						n
vagina	n					
vulva	y					
clitoris	n					
kidney	y	y			y	
vein	n					
tendon	n					n
intestines	n					n
muscle					y	y

What is striking about the data presented in these tables is the level of consistency with the Nyulnyulan languages, given the distance of their genetic relationship and allowing for the numerous gaps in the data. The body parts listed in Table 15, which are all prefixed in Nyulnyul, are with just two exceptions prefixed in at least one Worrorran language. The exceptions are the terms for 'ankle' and 'calf'; which are not prefixing in any Worrorran language for which I have data. However, it is not certain that

Table 16. (cont.)

	Yawi-jibaya	Unggumi	Worrorra	Ngarinyin	Wunambal	Kwini
Hard						
bone	y	y		y	y	n
scales		n				
ribs	y				y	n
tooth	y		y			n
skull						n
cheekbone					y	
hip		n			y	n
pelvis					y	
jaw		y				n
knuckles						n
Body products						
Attached						
blood	n	n				n
fat	n	y			y	n
cicatrice						n
sore	n	n				n
Miscellaneous (unclassified)						
ritual meatfood				y		

the Nyulnyul word for 'ankle' is prefixing either (see page 259 above), and there is no corresponding particular term for 'calf' (the term *-mird* 'leg' covers this meaning as well). If we consider the full quota of nominals in Table 15 on which there is data, nearly three-quarters of them are prefixing, in perfect agreement with Nyulnyul. Moreover, if the more divergent Kwini is ignored, the agreement is more striking – 88% of the nominals listed in Table 15 are prefixing, as are the corresponding nominals in Nyulnyul. This level of agreement would seem to be too high to be accounted for either by chance, or by the hypothesis that they reflect retentions from a single Worrorra-Nyulnyulan proto-language (presuming that there was one), especially since there are few apparent cognates between the Nyulnyulan and Worrorran terms. Instead, it lends strong support to the semantic explanation proposed in section 3.

There is less consistency between the Nyulnyulan and Worrorran languages in terms of the nominals listed in Table 16. In the first group of external body parts on Table 16 – which are non-prefixing parts in Nyulnyul – significantly more parts seem to be prefixed. There seems to be a strong tendency for nominals referring to outer coverings not to be prefixed, just as in Nyulnyul (but note that the more "personal" sorts of hair, including pubic hair and beards, are prefixed in two languages). However, joints and male – but not female – genitalia tend to be prefixed; interestingly, it is this group of nominals which constitute the main differences between Nyulnyul and Gooniyandi in terms of what is generally treated as – and perceived of as being – inseparable. The other obvious point of difference from Nyulnyul lies in the hard internal parts, where unlike Nyulnyul the Worrorran languages show a definite tendency towards prefixing: bones, teeth and so on generally take prefixes in these languages. (These are, as one might expect, usually treated as separable in Gooniyandi.) Finally, just a few of the soft internal organs, none of which are prefixed in Nyulnyulan languages, are prefixed in Worrorran languages.

Clearly, if there is anything in my suggestion that there is a semantic principle in operation, it must be somewhat different in the two language families (and even amongst the languages of each family). Alternatively, some differences may be attributed at least in part to differences in the conceptualisation of particular body parts. For instance, in Ungarinyin, the kidney is regarded as the seat of the emotions (Rumsey 1982: 65), and the meaning 'I am getting happy' is expressed metaphorically as *ngiyamad-ju barij wi* (my:kidney-lative rise it:is) 'my kidney rises'. If such cultural and linguistic factors are taken into account it may be possible to explain some at least of the unexpected prefixed body parts (cf. Dixon 1972: 308ff who uses ethnographic information to explain unexpected noun class membership in Dyirbal). To adequately account for the differences between the various languages would take us well beyond the scope of the present article, however.

6. Conclusion

In this paper I have argued that nominal prefixation in Nyulnyul partitions the nominals into two classes on a semantic basis. Prefixing nominals indicate those items which are conceived of as belonging to the personal sphere of a human being, those items which are viewed of as inseparable from the individual. Non-prefixing nominals indicate those entities which do not belong to a human being's personal sphere, and which have independent status as things.[14] We were able to account for a number of apparent exceptions by a careful formulation of the semantic principle involved. However, it must also be acknowledged that there remain some problems. In particular, I have been unable to account in the same way for the (nominal) prefixing of some non-nominals, for which the personal sphere can hardly be relevant. I have suggested that these exceptions might be explained through the postulation of development along similar historical lines from two quite different construction types which accidentally shared the formal property that a pronominal and nominal were juxtaposed, in that order.

More generally, this paper can be understood as an attack on the structuralist dogma that form-classes are semantically arbitrary. It presents further evidence in support of a view expressed particularly forcefully by Anna Wierzbicka: "that form classes are semantically motivated, and that differences in grammatical behaviour reflect iconically differences in meaning" (Wierzbicka 1985: 313).

We have devoted the bulk of this paper to the first claim. The second is rather more obvious. For, clearly the fact that a nominal is prefixing can be seen as an iconic reflection of the fact that the referent entity is so closely connected with its "possessor" that it can have no existence independent of the latter. And conversely, that a nominal is not prefix taking reflects the fact that the referent is an independent entity. Grammatical distance iconically reflects conceptual distance (see Haiman 1983: 782). This observation then suggests why it is that – as against expectations – the most important, obvious, and possibly most frequently referred to (see footnote 14) body parts and related terms are the morphologically most complex.

In conclusion, it must be emphasised that as far as I have been able to determine it is only the one remaining full speaker of the language who uses the system of nominal prefixing consistently. Semi-speakers do not appear to control it at all; they appear to use the third person singular form as the invariant shape of erstwhile prefixing nouns. (Compare my remarks above regarding cognates in the non-prefixing Eastern Nyulnyulan languages.) Interestingly, it appears that the one other full speaker who Bronwyn Stokes interviewed in 1979 (now deceased) did not use the system, or at least did not use it consistently, normally choosing the third person singular forms, even in reference to parts of the hearer and speaker (Bronwyn Stokes, pers.comm.). It would thus appear that the system of nominal prefixing has been in the process of dying for some considerable time.

Notes

1. This is a revised version of a seminar presented at Melbourne University in 1987, and subsequently (in considerably revised form) to the *Body Parts in Grammar* workshop at the Australian Linguistics Society Conference, Armidale, 1988. I am grateful to Peter Austin, Kate Burridge, Hilary Chappell, Mark Durie, and Bronwyn Stokes for their comments on earlier drafts. I would like to express my gratitude to Kimberley Language Resource Center, Australian Institute of Aboriginal and Torres Strait Islander Studies, Australian Research Council and National Aboriginal Languages Program for financial and/or other assistance. I am also indebted to the Pallottine Missionary Organisation for permission of access to their archives and library; to Howard Coate for access to his unpublished materials on many Worrorran languages; and to La Trobe University for funding my library research. Nick Thieberger and Linda De Veer are to be thanked for their assistance in cataloguing the manuscript material and inputting it onto computer. My greatest debt is, however, to my Nyulnyul teacher, Mary Carmel Charles, who, over the past ten years, has spent so many hours of her time teaching me her language.
2. The orthography I use here for transcribing Nyulnyul words has been chosen for convenience, and is a variant of the North Kimberley orthography (McGregor 1988a). It is almost identical with the practical orthography used in Torres – Williams 1987 and Charles 1993 (the only publications in the language). However, this latter system uses *oo* for the high back vowel, and over-differentiates *g* and *k* (which are allophones of a single phoneme), and (probably) *o* and *oo* (= *u*) (see footnote 3 below). For convenience, I employ *u* and *k* respectively to represent these two phonemes.

   The following abbreviations are used: ASSOC – associated with; DYS – dysfunctional; EMP – emphatic; ERG – ergative; NOM – nominative case; OBL – oblique case; PF – prefix; sg – singular; 1 – first person; 1+2 – the person category consisting of the speaker and one or more addressees; 2 – second person; 3 – third person. The symbol # preceding a form indicates that it is not represented in my corpus of elicited material, but is found in an earlier source. However, I spell the forms according to my own orthographic conventions, rather than those of the source. In some instances this means that I have had to make a guess at the phonemic representation of a word; some comment will normally be provided as to why that particular phonemicisation has been chosen over other possibilities.
3. Some important qualifications should be borne in mind. Firstly, the available data is incomplete and in places fragmentary. Secondly, I have not completed my analysis of Nyulnyul phonology, and there may be some inaccuracies in the present transcription. Unresolved problems include: status of *o* as a distinct phoneme (here I assume it to be an allophone of *u*); the status of the length contrast in the high vowels (whether it is to be accounted for in terms of phonemic length, or as the sequences *iy* and *uw* sequences (following Stokes' analysis of Nyikina – Stokes 1982); and which words have initial vowels *i* and *u*, rather than initial *yi* and *wu* respectively. These limitations on both the size of the corpus and analysis should be borne in mind.
4. This traditional system is no longer operative, even in the speech of the remaining full speaker: the distinction between 1 and 1+2 has all but disappeared. As indicated in Table 1, the forms for 1+2 augmented do not occur at all in my corpus; and the 1+2 minimal occurred only sporadically. (See McGregor 1988c for further details.)

5. It should be noted that the forms given here are somewhat tentative, and the details of Nyulnyul morphophonology have yet to be worked out. As the reader will see from an inspection of the Nyulnyul words cited here, the phonological forms of prefixes to particular nominals do not always relate in a straightforward way to the allomorphs shown in this table. This article is not the place to delve too deeply into these matters.
6. This same collapsing appears also to occur in Bardi (Metcalfe n.d.: 1), a comparatively strong language with some hundreds of speakers. Thus it seems reasonable to suppose that it is not merely the result of grammatical levelling attendant to language attrition, but a characteristic of the traditional pronominal prefix system.
7. The forms given in the following table are tentative, and details of both the underlying forms and morphophonemic changes have yet to be worked out (see also footnote 5). The forms preceded by the question mark are uncertain – there are possible, but not certain instances of them in my data, and Stokes (1988) does not fill in these gaps. Thus, it would seem that 1 and 1+2 are distinct persons, and there is a minimal/augmented "number" distinction, although these days the system has collapsed somewhat, and the 1 augmented appears to be the unmarked form, which can be used to refer to any non-singular 1 or 1+2 person.

   The two rows labelled I and II correspond to two distinct sets of allomorphs, which, following Stokes (1982 and 1988), I designate sets I and II. Set I prefixes normally occur in transitive clauses, whilst set II prefixes normally occur in intransitive clauses.
8. It might seem unlikely to Australianists that there would be a distinct term for 'footprint' – in most Aboriginal languages the term for 'foot' covers both the body part and the mark it leaves on the ground. However, there can be no doubt that in the elicited Nyulnyul of the last remaining speaker, there are two distinct terms – which differ semantically in this respect.
9. The two exceptional stems mentioned previously might permit both constituency analyses, and might perhaps be at the leading edge of a process of change (some might prefer "attrition"), in which prefix taking stems are being reanalysed as in Figure 2. The presence of the third person singular prefix might be accounted for morphologically, as a consequence of the fact that the nominal root is not a free form, and requires a prefix.
10. This is a reasonable hypothesis given that: (i) the Goal in modern Nyulnyul is cross-referenced by one of a set of enclitic bound pronominals attached to the verb, which are clearly closely related to and cognate with the nominative forms of the free pronominals; and (ii) the nominative prefixes cross-referencing the Actor are clearly older, in as much as they are more closely fused with the verb root, show more allomorphy and irregularities, etc..
11. There is an alternative term for 'hat', *jangkurr*. Why this is not prefixing while *-almid* is is not clear. One possibility, along the lines of the argument of this paragraph, is that the former term refers primarily to modern hats, whilst the latter refers to traditional hats. There is however no evidence in support of this hypothesis. It is more likely that an explanation is to be found in whether or not the item is considered as a part of someone's personal sphere – see discussion of (5) above.
12. Sources of data are Metcalfe (n.d.), McGregor's field notes, and Nekes (1939) for Bardi; and McGregor's fieldnotes for Jawi.

13. However, it seems that Gwnin/Kwini, a fairly close relative of Wunambal, is an exception in that it has few prefix taking nominals – indeed, fewer than are found in the Nyulnyulan languages.
14. As Mark Durie has suggested to me, it may be that these are also the most frequently referred to body parts in natural discourse. This might have been a fruitful direction for future investigation, were it not for the unfortunate fact that the language is no longer used in normal situations, and is remembered only.

References

Allen, W. Sidney
1964 "Transitivity and possession", *Language* 40: 337-343.
[1972] [Reprinted in: Fred W. Householder (ed.), 82-90.]
Anderson, Stephen R.
1988 "Morphological change", in: Frederick J. Newmeyer (ed.), 324-362.
Austin, Peter
1989 "Verb compounding in central Australian languages", *La Trobe Working Papers in Linguistics* 2: 43-71.
Bally, Charles
1926 "L'expression des idées de sphère personnelle et de solidarité dans les langues indo-européennes", in: Franz Fankhauser – Jakob Jud (eds.), 68-78.
[1995] [Translated and reprinted in this volume.]
Bates, Daisy
n.d. Native vocabularies – Broome Magisterial District. [Unpublished MS.]
Bischofs, Joseph
n.d.a Premiers éleménts de la langue Niol/niol. [Unpublished MS.]
n.d.b Vocabulary of the Niol Niol language. [Unpublished MS.]
Capell, Arthur
1940 "The classification of languages in north and north-west Australia", *Oceania* 10: 241-272, 404-433.
1972 "The languages of the northern Kimberley, W.A.: some structural principles", *Oceania* 43: 54-65.
Capell, Arthur – Howard Coate
1984 *Comparative studies in northern Kimberley languages*. Canberra: Pacific Linguistics.
Chappell, Hilary
1986 "The passive of bodily effect in standard Chinese", *Studies in Language* 10: 271-296.
Chappell, Hilary – William McGregor
1995 "Prolegomena to a theory of inalienability", [this volume.]
Charles, Mary
1993 *Winin: Why the emu cannot fly*. Broome: Magabala Books.
Crowley, Terry
1995 "Inalienable possession in Paamese grammar", [this volume.]

Dixon, Robert M.W.
1972 *The Dyirbal language of north Queensland.* Cambridge: Cambridge University Press.
1980 *The languages of Australia.* Cambridge: Cambridge University Press.

Elkin, Adolphus P. (ed.)
1939 *Studies in Australian Linguistics.* (Oceania Monograph 3.) Sydney: University of Sydney.

Fankhauser, Franz – Jakob Jud (eds.)
1926 *Festschrift Louis Gauchat.* Aarau: H.R. Sauerlander.

Givón, Talmy
1971 "Historical syntax and synchronic morphology: an archaeologist's field trip", *CLS* 17: 394-415.

Greenberg, Joseph H.
1988 "The first person inclusive dual as an ambiguous category", *Studies in Language* 12: 1-18.

Haiman, John
1983 "Iconic and economic motivation", *Language* 59: 781-819.

Haiman, John (ed.)
1985 *Iconicity in syntax.* Amsterdam: John Benjamins.

Hale, Ken
1981 "Preliminary remarks on the grammar of part-whole relations in Warlpiri", in: John Hollyman – Andrew Pawley (eds.), 333-344.

Halliday, Michael A.K.
1985 *An introduction to functional grammar.* London: Arnold.

Harvey, Mark
1995 "Body parts in Warray", [this volume.]

Hock, Hans H.
1986 *Principles of historical linguistics.* Berlin: Mouton de Gruyter.
[1988] [Reprinted Berlin: Mouton de Gruyter.]

Hollyman, John – Andrew Pawley (eds.)
1981 *Studies in Pacific linguistics in honor of Bruce Biggs.* Auckland: Linguistic Society of New Zealand.

Hosokawa, Komei
1995 "'My face *am* burning!': quasi-passive, body parts, and related issues in Yawuru grammar and culture concepts", [this volume.]

Householder, Fred W. (ed.)
1972 *Syntactic theory 1, structuralist: selected readings.* Harmondsworth: Penguin.

Langdon, Margaret
1988 Possession and kinship in Yuman languages. Seminar presented to Linguistics Department, Monash University, August 1988, and subsequently to the *Body Parts in Grammar* Workshop at Australian Linguistics Society Conference, Armidale, August 1988.

Malcolm, Ian (ed.)
1991 *Lingustics in the service of society: essays to honour Susan Kaldor.* Perth: Institute of Applied Language Studies

McGregor, William
1985 “Body parts in Kuniyanti clause grammar”, *Australian Journal of Linguistics* 5: 209-232.
1988a “A survey of the languages of the Kimberley region”, *Australian Aboriginal Studies* 1988/2: 90-102.
1988b *Handbook of Kimberley languages, Volume 1: General information.* Canberra: Pacific Linguistics.
1988c Language shift among the Nyulnyul of Dampier Land. [Unpublished MS.]
1989 “Greenberg on the first person inclusive dual: evidence from some Australia languages”, *Studies in Language* 13: 437-451.
1990 *A functional grammar of Gooniyandi.* Amsterdam: John Benjamins.
1995 *Nyulnyul.* München and Newcastle: Lincom Europa.

McKay, Graham
1978 “Pronominal person and number categories in Rembarrnga and Djeebbana”, *Oceanic Linguistics* 17/1: 27-37.

Metcalfe, Christopher D.
n.d. Prefixing nouns in Bardi. [Unpublished MS.]

Nekes, Herman
1939 “The pronoun in Nyol-Nyol (Nyul-Nyul) and related dialects”, in Adolphus P. Elkin (ed.), 139-163.

Nekes, Herman – Ernst A. Worms
1953 *Australian languages.* (Micro-Bibliotheca Anthropos 10.) Fribourg: Anthropos-Institut.

Newmeyer, Frederick J. (ed.)
1988 *Linguistics: the Cambridge survey. Volume 1, Linguistic theory: foundations.* Cambridge: Cambridge University Press.

Rumsey, Alan
1982 *An intrasentence grammar of Ungarinyin.* Canberra: Pacific Linguistics.

Stokes, Bronwyn
1982 A description of Nyigina, a language of the west Kimberley, Western Australia. [Unpublished Ph.D. thesis, Australian National University.]
1988 Alternate pronominal prefixing in Dampier Land. [Unpublished MS.]

Stokes, Bronwyn – William McGregor
1989 Classifying the Nyulnyulan languages. [Unpublished MS.]

Tindale, Norman
1952-1954 Field journal of Norman B. Tindale: 18th expedition under the auspices of Board for Anthropological Research, University of Adelaide and University of California at Los Angeles, 1952-1954. [Unpublished MS.]

Torres, Pat – Magdalene Williams
1987 *The story of crow.* Broome: Magabala Books.

Vasse, Eric
1991 “Nouns and nominals in Wunambol”, in Ian Malcolm (ed.), 27-34.

Wierzbicka, Anna
1985 “Oats and wheat: the fallacy of arbitrariness”, in: John Haiman (ed.), 311-342.

Body parts, possession marking and nominal classes in Ndjébbana

Graham R. McKay

1. Introduction

Ndjébbana or Kunibidji is a prefixing non-Pama-Nyungan language perhaps best known in the earlier literature as Gunavidji. The traditional territory of its 150 or so remaining speakers includes the settlement or town of Maningrida (in Ndjébbana *Manayingkarírra*) at the mouth of the Liverpool River on the north coast of central Arnhem Land, in the Northern Territory of Australia. The language, once classified as the sole member of the Gunavidjian language family (Wurm 1972: 114), now appears to me, on the basis of my own observations, to be related to the nearby languages Nakkara and Burarra and more distantly to the Gunwinjguan languages, though this has not been fully established. Ndjébbana is a polysynthetic language with relatively free word order. The verb complex is the main element of a sentence, containing, as it does, pronominal reference to subject and object or indirect object in a prefix, as well as tense and aspect suffixes. There is no case marking on independent noun phrases except for an ablative marker. My own fieldwork in the language spanned the period 1975 to 1982, and was directed towards the development of a bilingual education program in the language for the Northern Territory Department of Education.

Over forty five years ago in his survey "The languages of Arnhem Land", Arthur Capell remarked on the complexity of the system of possession marking in Ndjébbana (Kunibidji). He wrote: "Different parts of the body take different possessive suffixes, the plan of which does not yet appear." (Capell 1943: 28) While my own subsequent work on the language has not yet provided a definitive analysis or "plan" of the area of possession, some patterns have become apparent.[1]

Four basic methods of possessor marking are employed in Ndjébbana. These involve respectively a juxtaposed free cardinal pronoun, a pronominal prefix on a form of the verb *réndjeyi* 'to stand, be', a suffixed possessive pronoun and a pronominal prefix attached directly to the possessed nominal. It is possible to classify nominal roots in the language according to which of these means of possessor marking they use. Particularly with body part nominals, semantic differences between the methods of possessor marking are apparent from cases in which the same lexical form or semantically related forms are used with two different methods of possessor marking. Such evidence shows the relevance of three main features in the description of the differences between the three methods of possessor marking: alienability, animacy, and relative external visibility. In addition relative inalienability correlates with relative morphological tightness or inseparability in the marking of possessor.

Thus the four types of possession marking used in Ndjébbana appear to form a continuum of inalienability correlated with the relative closeness of the morphological linkage between possessor and possessed. Complete identity between possessor and possessed stands at one extreme, with the possessor marked by an affix (in one type by prefix and in another type by suffix), while separability of possessor and possessed stands at the other, marked by separate word forms for possessor and possessed. There is an intermediate type, in which the possessor is marked as a prefix to the accompanying positional verb. External visibility appears to be a contributing factor in distinguishing this intermediate type, while animacy also plays some role. Furthermore this intermediate type of possessor marking, which makes use of one of three positional verbs, *réndjeyi* 'to stand', provides some support for claims of a universal link between possessive, locational and existential constructions (Lyons 1967, Clark 1978).

2. Nominal possession classes and gender classes

Nominals in Ndjébbana form a large class of words which covers most of the range of the classes of nouns and adjectives in English. Nominals in Ndjébbana can be classified according to two intersecting sets of criteria, the first involving the method of marking possession, the second involving gender.

2.1. Possession classes

Nominals can be classified in Ndjébbana on the basis of four mutually exclusive means of marking the possessor, as follows:

A The possessor is marked by means of the appropriate free form cardinal pronoun juxtaposed to the noun which marks the possessed item.
B The possessor is marked using the subject prefix to the PAST 1 (which includes present)[2] tense form of the verb *réndjeyi* 'stand, be', which always follows the possessed noun.
C The possessor is marked using the possessive pronoun suffix on the possessed noun.
D The possessor is marked using a pronominal prefix directly attached to the possessed noun.

Each of these means of marking possessor defines a single possession class of nominals, with the exception of D which is common to six possession sub-classes. The sole criterion for separating these six possession sub-classes (D1 to D6) is the differing form of the third person minimal masculine prefix used in each. The possession classes are set out in detail in section 4 below. Details of the pronominal categories of Ndjébbana and paradigms of the pronominal forms referred to in this paper are set out in section 3.

2.2. Gender classes

As noted by Capell (1943: 24, 28) Ndjébbana is a "dual classifying" language. That is, it has two grammatical gender classes which can be labelled masculine (M) and feminine (F). Most (if not all) nominals of possession class A have inherent gender which is morphologically marked only by means of agreement on the verb and on associated nominals of possession classes B, C and D. One nominal of possession class C also appears to have inherent gender. This is *djakí* 'speed(iness)'. See example (25) and note 4.

Gender agreement, marked both on verbs and on modifying nominals, is exemplified in (1) and (2). For details of the pronominal categories used in the examples see section 3.1 below.

(1)	*Makéddja*	*ya-rrikka*	*nja-barlámbarla*
	long:necked:turtle:(F)	3:MIN:F-crawl	3:MIN:F-freshwater

nja-barrábarra (XXV/15-16)
3:MIN:F-big

'(We saw) a long-necked turtle crawling along – the/a freshwater one, a big one.'

(2) *Balawúrrwurr ka-rakarawéra yinjírra*
wind:(M) 3:MIN:M-moved 1:AUG:DAT
nga-namánda (XVIII/6)
3:MIN:M-small

'A little bit of wind blew for us.'

The gender distinction is neutralised in the augmented number for nominals with human referents, while for nominals with non-human referents it is number which is neutralised.

(3) *yídja barra-karrówa*
man 3:AUG-many

'many men'

(4) *barra-ngarrámi-ba*[3] *barra-karrówa*
3:AUG-woman-PL 3:AUG-many

'many women'

(5) *yókkarra n-karrówa*
fish:(M) 3:MIN:M-many

'many fish'

(6) *bíbbo nja-karrówa*
crab:(F) 3:MIN:F-many

'many crabs'

(7) *Njirri-nána nja-múlbbum*
1:UAUG:A+3:MIN:O:(M or F)-saw 3:MIN:F-several
karrddjúnja yaka-yóra (XXIII/5-6)
stingray 3:MIN:F:S-lie

'We saw several stingrays there.'

In discussing gender classes it is pertinent to mention that there are homophonous word pairs with distinct genders and distinct, though related, meanings. The different gender is marked, as noted above, only by means of agreement markers on other words. Thus the word *makéddja* in example (1) shows by its agreements that it is clearly feminine and refers to a specific species of turtle, the freshwater long-necked turtle. There is also a masculine word *makéddja* and this is the generic term for all turtles, including members of all marine and freshwater turtle species. Compare (8) with (1).

(8) *makéddja* *n-barrábarra*
turtle:(generic):(M) 3:MIN:M-big
'a big turtle (generic, any unspecified species)'

Another word pair of this type is the word *ngalidjbínja* 'didgeridoo or drone pipe' (M) and 'shotgun' (F), which is mentioned below in sections 4.1 and 5.1.

3. Ndjébbana pronouns and pronominal categories

In this section we will outline the pronominal categories and the pronoun forms, both free and bound, of Ndjébbana. This will clarify the examples quoted throughout the paper and provide brief information on other uses of the pronoun forms. More detailed discussion of the free form pronouns and further examples of their use are provided elsewhere (McKay 1984a: 134-141). Bound pronoun forms have also been partially analysed elsewhere (McKay 1989). In some paradigms given in the following sections the unit augmented feminine forms have been omitted as not essential to the present discussion. This category is there indicated by the regular augmented form with the addition of the suffix *-nja*. Further details of the meaning of this category are already available (McKay 1979).

3.1. Pronominal categories

As in some of the neighbouring languages of Arnhem Land, the traditional pronominal categories are not entirely suitable for Ndjébbana (McKay 1978). The analysis adopted here better suits the formal patterns evident in the Ndjébbana pronoun paradigms. Person categories used are *1 person* (first person exclusive), *1/2 person* (first person inclusive), *2 person* (second person) and *3 person* (third person). Number categories are *minimal* (MIN), *unit augmented* (UAUG) and *augmented* (AUG). As its name suggests, minimal marks the minimum number for the particular person category. That is, in each of 1, 2 and 3 person minimal number indicates a single referent (or singular), but in 1/2 person minimal number indicates two referents (or dual) – "you and me". Unit augmented number is "minimal plus one" number, that is two or three referents (dual or trial), depending on person. Augmented indicates any number in excess of unit augmented, that is three-or-more or four-or-more, depending on person.

3.2. Cardinal pronoun

The cardinal pronoun is clearly a free pronoun form. Table 1 sets out the paradigm of cardinal pronoun forms.

Table 1. Cardinal pronoun forms

	Minimal M	Minimal F	Unit augmented M	Unit augmented F	Augmented
1	*ngáyabba*		*njirrikébba*	*njarrayábbanja*	*njírrabba*
1/2	*ngárrabba*		*ngirrikébba*	*ngarrayábbanja*	*ngúrrabba*
2	*njínjdjabba*		*nirrikébba*	*narrayábbanja*	*núrrabba*
3	*nakébba*	*ngayábba*	*birrikébba*	*barrayábbanja*	*barrayabba*

An emphatic pronoun is formed by replacing the cardinal pronoun suffix *-bba* with the emphatic pronoun suffix *-mala*.

Apart from its use in marking possession, which is the subject of this paper, the cardinal pronoun may be used, without case marking, to provide free form marking of subject or object in the sentence. Since major noun phrase constituents are obligatorily marked by means of prefixes on the verb, a free form pronoun in apposition to this is not necessary in the majority of sentences. As a result, where the cardinal pronoun (or the emphatic pronoun) is used it has contrastive or emphatic force, as in (9).

(9) *Barrayabba yaláwa barra-balayirríya yawúyakka njírrabba*
3:AUG then 3:AUG-came first/ahead 1:AUG
njarra-yóra bakkándja njarra-wolakkamíya.
1:AUG-camped later 1:AUG-came
(XXV/91-94)
'They came on ahead. We camped there and came on later.'

Table 2. Possessive pronoun forms

	Minimal M	Minimal F	Unit augmented M	Unit augmented F	Augmented
1	*njabba*		*njaddana*	*njaddayúnja*	*njaddabirra*
1/2	*ngadda*		*ngaddana*	*ngaddayúnja*	*ngaddabirra*
2	*ngka*		*naddana*	*naddayúnja*	*naddabirra*
3	*na*	*ngaya*	*baddana*	*baddayúnja*	*baddabirra*

3.3. Possessive pronoun

The available evidence suggests that the possessive pronoun is best analysed as a suffix (McKay 1984a: 140-141). The possessive pronoun paradigm is set out in Table 2.

Apart from its use in marking possession when suffixed to a nominal, which is the subject of this paper (see section 4.3 below), the possessive pronoun may be suffixed to other parts of speech and then functions as a "dative of interest" as in (10). The dative pronoun (section 3.4) is, however, more usual in this function.

(10) *Bárriya birri-mérbara-njabba birri-nóra.*
3:UAUG 3:UAUG:A-hide-1:MIN:POSS 3:UAUG-sit
(VII/64-65)
'Those two are hiding it from me.'

3.4. Dative pronoun

Though not featuring in the possession marking of nominals, the dative pronoun is included here to show the full range of pronominal types available within the language. The paradigm is set out in Table 3. Evidence suggests that the dative pronoun may be best considered a suffix (McKay 1984a: 137).

Table 3. Dative pronoun forms

	Minimal M	Minimal F	Unit augmented M	Unit augmented F	Augmented
1	*ngabúyanga*		*yinjerrekéyanga*	*yinjerreyánja*	*yinjírra*
1/2	*yingárra*		*yingerrekéyanga*	*yingarrayánja*	*yingúrra*
2	*yikkóyanga*		*yinerrekéyanga*	*yinerreyánja*	*yinúrra*
3	*-yana*	*yangaya*	*yiberrekéyanga*	*yiberreyánja*	*yibérra*

The dative pronoun has two specific functions. Firstly it is used to mark an "interested party", often an indirect object in sentences with a transitive or an intransitive (but not a ditransitive) verb. See (11) and (12).

(11) *Nga-bbúdjeya-yana* *ngi-yirríya.* (XXV/36)
1:MIN:S-shouted-3:MIN:M:POSS 1:MIN:S-went/walked
'I called out to him as I walked.'

(12) *Lúrra* *ka-bbándjanga* *yinjírra* *Ndjébbana*
ceremony 3:MIN:M:A+3:MIN:M:O-put 1:AUG Ndjébbana
njarra-ngúdjeya. (VII/122)
1:AUG-speak
'He established the *Lúrra* ceremony (or fishing place) for us Ndjébbana speakers.'

Secondly, the third person minimal dative pronouns are used to mark gender of third person agent (A) and to distinguish third person from non-third person agents with certain transitive pronominal prefixes as in (13), (14) and (15).

(13) *ngana-wúna*
MIN:A+2:MIN:IO-gave
'I gave it to you.'

(14) *ngana-wúna-yana*
MIN:A+2:MIN:IO-gave-3:MIN:M:POSS
'He gave it to you.'

(15) *ngana-wúna-yángaya*
MIN:A+2:MIN:IO-gave-3:MIN:F:POSS
'She gave it to you.'

3.5. Pronominal prefix forms

Each Ndjébbana verb bears a pronominal prefix form. In the case of intransitive verbs, this indicates the intransitive subject (S) of the sentence. In the case of transitive verbs the prefix indicates both the agent (A) and patient or direct object (O) of the sentence, while with ditransitive verbs the prefix form indicates both the agent (A) and the indirect object (IO) of the sentence. Thus with transitive and ditransitive verbs (hereafter simply termed "transitive") the prefix forms normally (but not always) overtly mark two pronominal elements. Only the element which stands first may be fully specified for person and number, while the element which stands second must undergo some neutralisation and may be partially specified or unspecified. Ordering of the agent and patient (or agent and indirect object) within the prefix form varies with different combinations, though this is not relevant here – see, however, McKay 1989.

For our present purposes it will be sufficient to present a partial pronominal prefix paradigm, omitting all transitive prefix forms with non-third person patient (O) and omitting all feminine unit augmented forms. Thus Table 4 sets out intransitive subject (S) forms and transitive agent (A) forms where patient or direct object (O) is third person. Where this patient (O) is third person minimal it has zero realisation except in the few forms on the table where it is marked as feminine.

Table 4. Intransitive (S) prefix forms and transitive (A and O) prefix forms with third person patient (O)

	S	O		
A/S		Minimal	Unit augmented	Augmented
1:MIN	*nga-*	*nga-*	*ngabindi-*	*ngabana-/ ngabanda-*
1:UA:M	*njirri-*	*njirri-*	*njanbirri-*	*njanbirri-*
1:AUG	*njarra-*	*njarra-*	*njanbirri-*	*njanbirri-*
2:MIN	*nja-/ngana-*[a]	*dja-*	*djabindi-*	*djabana-/ djabanda-*
2:UA:M	*nirri-*	*nirri-*	*nanbirri-*	*nanbirri-*
2:AUG	*narra-*	*narra-*	*nanbirri-*	*nanbirri-*
1/2:MIN	*yV-/ka-*[a]	*ka-/karrka-*	*nganbirri-*	*nganbirri-*
1/2:UA:M	*ngirri-*	*ngirri-*	*nganbirri-*	*nganbirri-*
1/2:AUG	*ngabarra-*	*ngabarra-*	*nganbirri-*	*nganbirri-*
3:MIN:M	*ka-*	*ka-* (O=M) *yaka-* (O=F)	*bindi-*	*bana-/banda-*
3:MIN:F	*nja-/yV-*[a]	*ka-* (O=M) *yaka-* (O=F)	*bindi-*	*bana-/banda-*
3:UA:M	*birri-*	*birri-* (O=M) *yibirri-* (O=F)	*banbirri-*	*banbirri-*
3:AUG	*barra-*	*barra-* (O=M) *yabarra-* (O=F)	*banbirri-*	*banbirri-*

a. Left hand form is used in future and irrealis; right hand form, in non-future realis.

The prefix form used with nominals is identical to the intransitive subject (S) set in Table 4 except that the third person minimal feminine form on nominals does not vary with tense but is always *nja-* and the third person minimal masculine form varies with possession subclass (D1 to D6 – see 2.1 above and 4.4 below): *n-* D1; *nga-* D2; *ka-* D3; *ø-* D4; *ma-* D5; *na-* D6.

4. Nominal possession classes

4.1. Possession class A

With nominals of this class the cardinal pronoun (a free form) marks the possessor, being juxtaposed in either order to the noun which indicates the thing possessed. (For forms, examples and some discussion of this pronoun see 3.2 above and McKay 1984a: 134-136.) Capell (1943: 28) noted that when the possessed item is feminine, the feminine prefix *nja-* marks agreement on the cardinal pronoun itself. In fact agreement marked by a prefix appears to extend to the masculine and to the non-minimal numbers, at least for the first person minimal possessor, as exemplified in (16) to (20). Many speakers, especially younger speakers, omit the third person minimal masculine prefix *n-*.

(16) *nja-ngáyabba* *kíkka*
3:MIN:F-1:MIN mother
'my mother'

(17) *n-ngáyabba* *bábba*
3:MIN:M-1:MIN father
'my father'

(18) *barra-ngáyabba* *barra-rókaddji-ba*
3:AUG 3:AUG-child-PL
'my children'

(19) *marddúrddiba* *ngáyabba*
heart 1:MIN
'my heart'

(20) *njíndjabba wíba*
2:MIN camp/place/home
'your place'

Possession class A is an open class of nominals, a selection of which is listed in Table 5, in three groups: body part nominals, kin terms and a selection of other nominals. Where known, the intrinsic gender class of each word is indicated (M = masculine, F = feminine – see 2.2 above).

Table 5. Possession class A – selected terms

Body parts and products	
bálbbal M/F 'wing of bird'	*kélbarla* 'shoulder blade'
barrúra M/F 'egg'	*kurlarrabarrábarra* 'backbone'
bórdanja 'urine, piss'	*mankimánki* M 'spines, needles'
djarnarrába M 'cough' (deverbal)	*marddúrddiba* M 'heart'
djúrddjurd 'heart'	*ngánjbarl* 'sneeze'
karnbilíbala 'blood'	*ngárnba* 'perspiration, sweat'
Kin terms	
bábba M 'father'	*kíkka* F 'mother'
djábba M/F 'big brother/sister'	*kúdjala* 'grannie'
djóngok M/F 'cousin'	*ngánja* 'aunt (father's sister)'
Other terms	
bábburr 'patrilineal clan'	*kakárra* M 'shade'
badjubádju 'shirt' (Macassarese)	*mabbúrla* 'blunt'
bárrbaya M/F 'dog/bitch'	*nmáma* M 'harpoon shaft'
barángka 'lazy, tired'	*namarnakkúrrkka* M 'creek'
bírraddja M 'rice' (<Macassarese *berasa'*)	*módj* 'dreaming, totem'
njamánja F 'mussel'	*narrúrra* 'open (area), ?naked'
budmánda 'suitcase' (<English *portmanteau*)	*ngalidjbínja* F 'shotgun, rifle, gun'
budjúlung M 'bottle'	*ngalidjbínja* M 'didgeridoo'
dílkarra M 'moon, month'	*wíba* M 'place, home'
djábbarda M 'fullness'	

4.2. Possession class B

Nominals of this possession class have their possessor marked using the PAST 1 (covering present and recent past) tense form of the verb *réndjeyi* 'stand, be', with the subject (S) prefix of the verb indicating possessor (POSS), as in (21) and (22).

(21) *márnabba nga-réndjeya*
chest 1:MIN:S-stand/be
'my chest'

(22) *kála ka-réndjeya*
ear 3:MIN:M S-stand/be
'his ear'

Possession class B appears to be a closed class of nominals containing only body part terms. All known nominals of this class are listed in Table 6.

Table 6. Possession class B – all known terms

Body parts	
bárnka 'elbow'	*márdba* 'neck, new shoot of palm'
barrábarra 'back, backbone'	*wánba* 'arm'
djábbarda 'lip'	*márnabba* 'chest'
djánama 'belly (external)'	*marnakkúrrkka* 'arm, wing'
djangórna 'tongue'	*mémarla* 'thigh, upper leg'
djaréya 'nape of neck'	*míbba* 'forehead'
kála 'ear'	*ramalówandja* 'jawbone'
kúdja 'base of tree'	*rénjmarla* 'cheekbone'
lékka 'testicles'	*rídda* 'tooth, (axe) blade'
mangúya 'nose; hook of woomera/spear thrower'	*waláya* 'lower leg, tail'

4.3. Possession class C

For nominals of this class the possessor is marked using the possessive pronoun suffix attached to the nominal, as exemplified in (23), (24) and (25). For forms, further examples and some discussion of this pronoun see section 3.3 above and McKay 1984a: 139-141.

(23) *marnákarna-njabba*
rib:bone-1:MIN:POSS
'my rib bone'

(24) *rakáya-njabba*
fingernail/toenail-1:MIN:POSS
'my fingernail/toenail'

(25) *bíbbo djakí-ngaya n-bókka*[4]
crab speed-3:MIN:F:POSS M-bad
'The crab goes slowly/is slow.'

While most nominals in this possession class use the possessive pronoun to mark a possessor (e.g. *lárla* 'penis'), a small number use it to mark agreement (e.g. *lárla* 'male', *warábba* 'one, alone'). Contrast the two meanings of the respective nominals in the example pairs (26) and (27), (28) and (29). Compare (26) with the form of English *my-self*. This distinction between possession and agreement appears not to be formally marked in Ndjébbana and may be an artefact of the English translation. In fact it seems to me that the best way to understand this is to interpret the relationship between the nominal and the pronominal suffix as expressing identity of the referents of both in all cases. When the nominal refers to a part of a whole, then in English the relationship is expressed as "possessed-possessor". Otherwise other realisations are more natural in English (adjective, subject of predication).

(26) *Nga-nóra warábba-njabba*
1:MIN-sit alone-1:MIN:POSS
'I am sitting by myself', 'I am alone.'

(27) *karnayédjabba warábba-baddabirra*
two one-3:AUG:POSS
'three (3:AUG:HUM)'

(28) *Lárla-baddabirra*
penis-3:AUG:POSS
'their penes'

(29) *Lárla-baddabirra birri-búlanj.*
male-3:AUG:POSS 3:UAUG-(subsection)
'The two boys are both Búlanj/both belong to the Búlanj subsection.'[5]

The number agreement in (29) is not complete (AUG-UAUG) but no explanation for this has been found. The translation given was provided in explanation by thc speaker.

It appears that possession class C is a closed class, containing mainly body part nominals, as listed in Table 7.

Table 7. Possession class C – all known terms

Body parts	
díla 'eye; seed; cartridge (gun)'	*marnákarna* 'rib bone'
njamanja 'knee'	*marnawarrínjba* 'kidney'
djamíbbala 'eyebrow'	*njakkárla* 'lower leg'
djúya 'buttocks; butt/stock (gun)'	*njakkórdba* 'brain'
kalórla 'nipple, breast'	*njayára* 'shoulder'
kárla 'Adam's apple, throat'	*ngalidjbínja* 'throat, windpipe'
kúdja 'hand, finger, foot, toe; tracks'[a]	*rakáya* 'nail of finger/toe'
yúrnka 'head'	*rárra* 'vagina'
lárla 'penis, male'	*yalangaló* 'pubic hair'
manbáma 'side'	*yakára* 'shoulder'
mandjamándja 'whiskers, beard'	*yírda* 'bridge/bone (of nose)'
marléwa 'shoulder blade, collar bone'	
Other terms	
búrrbba 'strong'[b]	*warábba* 'one, alone'
djakí 'speed'	*wíba* 'country (?of origin)'[b]

a. The word *kúdja* provides counterevidence to Andersen's claim (1978: 352) that it is a universal of language that *hand* and *foot* never share the same label.

b. See section 5.4.

4.4. Possession class D

Nominals of this possession class mark their possessor in the form of a prefix to the nominal – see (30) and (31). Minimal number nominal prefix forms are the same as intransitive subject (S) verbal prefixes in all persons and numbers with the exception of third person minimal forms. For these nominal forms the feminine is always *nja-*, while six different masculine forms occur, providing the basis for a division into six distinct sub-classes: *n-* D1; *nga-* D2; *ka-* D3; *ø-* D4; *ma-* D5; *na-* D6. In other words these six subclasses differ from one another only in the third person minimal masculine form. In examples in this section the possession sub-class of each nominal will be indicated in brackets to facilitate location of the words in the lists provided below.

(30) *nga-ngardabbámba* (D4)
1:MIN-liver
'my liver'

(31) *nga-méyameya* (D4)
1:MIN-head:hair
'my hair'

As with possession class C, the prefixing possession class of nominals (D1 to D6) contains some nominals for which the prefix marks the possessor and others for which the prefix seems to mark agreement or even person and number of the subject of a predication, as in (32) to (36). Again, as with possession class C, there is no known formal feature distinguishing these. Meaning and the English translations offered by Ndjébbana speakers provide the only guide, but the distinction may well be an artefact of the translation.

Once more it seems to me that the best way to capture this is to interpret the relationship between the nominal and the pronominal prefix as expressing identity of the referents in all cases. When the nominal refers to a body part, English will translate the relationship as one of possession of the part by the whole. When the nominal refers to a quality, English is likely to translate using an adjective. Note that (32), (33) and (36) are verbless sentences.

(32) *(njíndjabba) ngana-kóyok* (D1)
2:MIN 2:MIN-(subsection)
'You are in the *Kóyok* subsection', 'Your skin is *n-Kóyok.*'

(33) *nga-nawarrámaya* (D2)
1:MIN-skinny
'I am skinny.'

(34) *Bárri-na birri-yáwarlbba.* (D1) (XVII/12)
3:UAUG-stayed 3:UAUG-old:person
'The two old people stayed.'

(35) *ngalidjbínja nja-mádjarna* (D1)
gun:(F) 3:MIN:F-proper/real
'proper guns'

(36) *Nga-ndakabbírda.* (D5)
1:MIN-right:handed
'I'm right handed.'

The only nominals in this possession class which refer to individuals or could be taken as kin terms use the prefix not to express a relationship to a possessor but to indicate person, number and/or sex of the one individual as in (37), (38) and (39).

(37)	*n-karókaddja* (D1) 3:MIN:M-child 'boy' (not 'his child')	*nja-karókaddja* 3:MIN:F-child 'girl' (not 'her child')
(38)	*nga-ndjórrkkabba* (D2) 3:MIN:M-spouse 'husband' (not 'his spouse')	*nja-ndjórrkkabba* 3:MIN:F-spouse 'wife' (not 'her spouse')
(39)	*barra-múyi-ba* (D4) 3:AUG-dead:person/spirit-PL 'dead people/spirits' (not 'their spirits')	

Possession subclasses D1 to D6 appear to be closed classes and in some cases contain very few nominals. For instance D5 and D6 contain only one and two nominals respectively known to me. Nominals of the various subclasses are listed in their third person minimal masculine forms in Tables 8, 9, 10 and 11.

Table 8. Possession class D1 – all known nominals

Body parts	
n-barélabba 'little one'	*n-kánkarra* 'meat, flesh'
n-barrómaya 'shadow'	*n-káro*[a] 'fat (noun)'
n-búrrbba 'guts, innards'	*n-kódda* 'skin'
n-djarramáya[a] 'thoracic spine'	*n-mardárda* 'bone, shell'
-wála[a,b] 'name'	*n-mangalówarra* 'jawbone'
Other nominals	
n-bárdbana 'dry, shallow'	*n-káma* 'not cooked (fish), without food (people)'
n-barlánga 'raw; unripe'	*n-kárddja* 'cooked; ripe'
n-barrábarra 'big'	*n-kamárrang* (subsection)
n-báymala 'left handed, left'	*n-kángila* (subsection)
n-bókka 'bad'	*n-karírra* 'empty'
n-bórrmanga 'countryman, friend'	*n-kárla* 'wet, green, living'
n-karókaddja 'child, juvenile'	*n-kálakarra* 'Aboriginal'
n-búlanj (subsection)	*n-karrakárramardba* 'long, tall'
n-djaráma 'heavy'	*n-karrówa* 'many, much'
n-djárwarra[a] 'young person, adolescent'	*n-karrúmakkarra* 'smooth, slippery'
n-djáwarlbba[c] 'old person'	*n-kékkaka* 'new'

Table 8. (cont.)

Other nominals (cont.)	
n-djawukkaléla 'light (weight)'	*n-kóyawa* 'crooked'
n-djídjabba 'same'	*n-kóyok* (subsection)
n-djídja 'old'	*n-mádjarna* 'proper, real, authentic, good'
n-djínjawa 'alive'	*n-nganjónganja* 'black, dirty'
n-kábburla 'old person'	*n-ngówolk* ' "quiet", harmless, placid'

a. These nominals shift the main word stress to the initial syllable of the prefix when this syllable has a vowel. In such cases root initial stop *dj* lenites to semivowel *y* (McKay 1984b: 111-112)

b. 3:MIN:MASC irregular form is *nála* (<**ná-wala*).

c. Feminine singular form is *nja-yáwarlbba* (with *dj* lenition but without stress shifting) (McKay 1984b: 109, 112).

Table 9. Possession class D2 – all known nominals

Body part	
nga-rakkíla 'meat (shellfish)'	
Other nominals	
nga-nabarlámbarla 'from freshwater'	*nga-narárrma*[a] 'white, clean, light (colour)'
nga-nabbórla 'tame, harmless'	*nga-náwarla* 'different, another'
nga-nabbórnba 'stinking, rotten'	*nga-náworrkala* 'different, another'
nga-nabélbala 'little one, young person'	*nga-nawarrámaya*[a] 'skinny'
nga-namánda 'small'	*nga-nawarréya* 'greedy'
nga-nakkúrrngarnawa 'straight'	*nga-ndjórrkkabba* 'spouse'
nga-rárrangala 'sharp (blade)'	

a. These nominals are known to be deverbal derivatives. For example,

nga-na-rarrma
3:MIN:MASC-INFIN-be:white/clean

Table 10. Possession classes D3, D5 and D6 – all known nominals

D3	
ka-bbarráma 'sore, aching'	*ka-múkkabba* 'together, "level" '
ka-ndarládja 'short'	
D5	
ma-ndakabbírda 'right handed'	
D6	
na-bangárda (subsection)	*na-ngórdo* 'leper'

Table 11. Possession class D4 – all known nominals

Body parts	
marlúkarla 'hip, thigh'	*múraka* 'hair on body or limb'
méyameya 'head hair; feather, plumage; palm leaf'	*múya* 'dead person, ghost, spirit'
móya 'sore (noun), scar'	*ngardabbámba* 'liver'
Other nominals	
kómdudj 'initiand'	*ngarráma* 'woman, female'
malamalayínba 'clean, clear (of water)'	*ngárridj* (subsection)
ngurrarákka 'big'	*njárlanga* 'exposed, naked, uncovered, not in bag'
marndálangurrnga 'hard'	*wákadj* (subsection)
múlbbum 'several'	*wámud* (subsection)
njálkkidj 'soft'	*warríwarra* 'dangerous, "cheeky"'

4.5. Nominal and verbal prefixes

We have already noted that Ndjébbana nominals of possession class D mark possessor, agreement and subject of predication, here all taken as forms of identity marking, using a set of pronominal prefixes substantially identical with those used on intransitive verbs (see 3.5 above). This set of prefixes is also substantially the same as the almost identical set of prefixes used to mark the agent of transitive verbs when there is a third person minimal object. The major exceptions to this are:

(a) in the case of third person minimal possessor/subject where the verbal and nominal prefixes are distinct in the masculine except for possession class D3 nominals, and distinct in the feminine except for future tense verbs.

(b) in the case of second person minimal where the nominal prefix form *ngana-* is identical to the non-future realis allomorph of the intransitive subject (S) and transitive object (O) prefixes, but different from the agent (A) form *dja-* and from the future and irrealis allomorph of S and O, which also has the form *dja-*.

In fact in the Ndjébbana pronominal prefix forms (as outlined in McKay 1989) it is only first person minimal and second person minimal which have distinct forms for agent (A) and patient (O). In each case one of these is found to pattern with the intransitive subject (S) as shown in Table 12. Clearly the nominal prefixes match A and S in the first person minimal but match (the distinctive allomorph of) O and S in the second person minimal.

Table 12. Separate minimal A, O and S prefix forms

	A	O	S	Nominal
1 Minimal	*nga-*	*nja-*	*nga-*	*nga-*
2 Minimal	*dja-*	*dja-/ngana-*	*dja-/ngana-*	*ngana-*

Allen (1964: 338-339) draws attention to Hungarian, Abaza and a number of Amerindian examples which show that in these languages the same prefixes or suffixes mark possessors on nouns but agents (and in some cases intransitive subjects) on verbs, as part of his wider establishment of links between agency and possession, especially in connection with the perfect aspect. In this connection it is interesting to note that Ndjébbana nominal prefixes show an unambiguous link with the agent prefixes only in the first person minimal. With second person minimal, nominal prefixes are linked more closely with the patient or direct object prefix. With other persons and numbers there is a clear link not only with agent, but also with patient and intransitive subject forms.

4.6. The special case of *-wála* 'name'

The nominal *-wála* represents a special case. This word belongs to possession class D1 and marks the possessor using a prefix, as is usual for that class:

(40) *Barra-njúngkowara bárra-wala?*
3:AUG-who 3:AUG-name
'What are their names?'

In sentences citing the actual name, where English would use a copula, Ndjébbana uses the verb *réndjeyi* 'stand, be' as in (41) and (42).

(41) *Nála ka-réndjeya Kyle-Little.* (V/212-213)
name:(3:MIN:M) 3:MIN:S-stand Kyle-Little
'His name is Kyle-Little.'

(42) *Ngá-wala nga-réndjeya Graham.*
1:MIN-name 1:MIN:S-stand Graham
'My name is Graham.'

Note that the verb prefix agrees in person and number with the possessor of the name.

It is plausible to suggest, as per section 4.4 above for other possession class D nominals, that the use of a pronominal prefix on the nominal signifies an identity relationship between "possessor" and "possessed" (name). At the same time the use of the verb *réndjeyi* in these predicative examples is suggestive of the patterns associated with possession class B (section 4.2).

Alternatively, the reflexive *ngódjiya* 'call oneself' may be used, as in (43).

(43) *Márru* *nála* *ka-ngódjiya-na.* (II/81)
(name) name:(3:MIN:M) 3:MIN:S-call:oneself-PAST:2
'people who called themselves by the name *Márru*', or 'people who were named *Márru* people'

5. Semantic correlates of possession marking types

It will be clear from the lists and examples given of each possession class that each of the four methods of possessor marking is used with a range of different nominals, including some body part terms in each class. Furthermore it is clear that possession classes A to D form a continuum from least to most inalienable. Possession class A includes the highest proportion of nominals referring to independent, non-possessed items, while possession class D, at the other extreme, contains a very high proportion of nominals referring to attributed qualities, which would be translated into English as adjectives.

We will now examine a number of insights which can be gained into the semantic factors which correlate with the different types of possessor marking. These insights are provided by cases of two types:

(a) where a single lexical root form is used with more than one of the types of possessor marking. That is, the same root belongs to two of the possession classes outlined above. In these cases the root has a different meaning in each possession class.
(b) where two different lexical roots which have some semantic relationship or similarity each belong to different classes with respect to possessor marking.

5.1. Cardinal pronoun as marker of relatively alienable possession

That the A pattern of possessor marking using the cardinal pronoun is a fairly general one is shown firstly by the fact that possession class A appears to be the only open possession class – it appears to be the only class to receive borrowings.

Furthermore the relatively alienable nature of the possession marked by this means is shown by a comparison of several contrasting pairs using the same nominal form in two different possessor marking classes with two different meanings. The more alienable possession is always marked using the free form cardinal pronoun, as in (44) to (47)[6] (possession class is indicated in brackets):

	Cardinal pronoun possessor	Prefix/suffix possessor
(44)	*ngáyabba ngalidjbínja* M-my didgeridoo 'my didgeridoo/drone pipe'(A)	*ngalidjbínja-njabba* throat/windpipe-my 'my throat/windpipe'(C)
	nja-ngáyabba ngalidjbínja F-my shotgun 'my shotgun'(A)	
(45)	*njamánja nja-ngáyabba* mussels F-my 'my mussels'(A)	*njamánja-njabba* knee-my 'my knee'(C)
(46)	*marnawarrínjba ngáyabba* yams M-my 'my yams'(A)	*marnawarrínjba-njabba* kidney-my 'my kidney'(C)
(47)	*kánkarra ngáyabba* meat my 'my meat (for eating)'(A)	*nga-kánkarra* my-meat 'my meat (part of my body), my flesh'(D)

Compare a number of further possession class A nominals with identical nominals of different meaning in possession class C.

	Cardinal pronoun possessor	Suffix possessor
(48)	*kalórla* flower 'a flower'(A)	*kalórla-ngaya* nipple/breast-her 'her nipple/breast'(C)

(49)	*marnákarna* stick:frame:for:scoop:net 'a net frame'(A)	*marnákarna-njabba* rib(s)-my 'my ribs'(C)
(50)	*lárla* sea:creature:like:trepang 'a sea creature like trepang' (impossible to mark possession)	*lárla-na* penis-his 'his penis'(C)

On the other hand a number of apparently inalienable body parts, and, it seems, most, if not all, kin terms[7] follow the possession class A pattern and mark possessor using the free form cardinal pronoun, as in (51) and (52) respectively.

(51)	*marddúrddiba* heart 'my heart'	*ngáyabba* my	*karnbilíbala* blood 'my blood'	*ngáyabba* my
(52)	*bábba* father:(M-) 'my father'	*(n-)ngáyabba* my	*nja-ngáyabba* F-my 'my mother'	*kíkka* mother

Examples such as (51) show that while the "core" significance of class A possessor marking is alienability, there are some exceptional cases on the periphery of the class. Examples such as (52) show clearly that kin terms are classified as fundamentally alienable. Further, in the word pairs (44) to (50), the more alienable term has an inanimate (or non-human) referent, while the more inalienable refers to a human body part.

5.2. Verb *réndjeyi* 'stand, be' as marker of external body parts

A comparison of the nominals in possession classes B, C and D (that is the relatively closed classes marking relatively inalienable possession) reveals a number of sets of different words, each relating to similar body parts. There is a pattern apparent in which the externally visible aspect of the body part is in possession class B, in which the possessor is marked by a verbal prefix as the subject (S) of the verb *réndjeyi* 'stand, be'. This possession class, as noted above (4.2) contains only body part nominals. In each set, (53) to (56), a different lexical item refers to the internal aspect and marks possession directly on the nominal itself using a suffix (possession class C) or a prefix (possession class D). Nominal possession class is again indicated in brackets in the examples.

	'Stand' possessor	Prefix/suffix possessor
(53)	*márdba nga-réndjeya* neck I-stand 'my neck'(B)	*ngalidjbínja-njabba* throat/windpipe-my 'my throat/windpipe'(C)
(54)	*djánama nga-réndjeya* belly I-stand 'my belly (external)'(B)	*nga-bbúrrbba* my-guts/innards 'my guts/innards'(D1)
(55)	*mangúya nga-réndjeya* nose I-stand 'my nose'(B)	*yírda-njabba* bridge/bone:of:nose-my 'the bridge of my nose'(C)
(56)	*márnabba nga-réndjeya* chest I-stand 'my chest'(B)	*marnákarna-njabba* rib(s)-my 'my ribs'(C)
		kardíddja-njabba collarbone-my 'my collarbone'(C)
		nga-yarramáya[8] my-back:(muscle/bone) 'my back'(D1)
		n-djarramáya his-back 'his back'(D1)

The pattern of using the pronominally marked verb *réndjeyi* to mark the possessor of external body parts is continued with a further group of nominals including those listed in (57), but there are several others, similarly referring to external body parts, which do not use this method of possessor marking, as set out in (58). Possession class is indicated in brackets.

(57)	*míbba nga-réndjeya*	(B)	'my forehead'
	kála nga-réndjeya	(B)	'my ear'
	djangórna nga-réndjeya	(B)	'my tongue'
	rídda nga-réndjeya	(B)	'my teeth'
	bárnka nga-réndjeya	(B)	'my elbow'
	marnakkúrrkka nga-réndjeya	(B)	'my (lower) arm'

(58)	*kalórla-njabba*	(C)	'my breast, nipple'
	yúrnka-njabba	(C)	'my head'
	díla-njabba	(C)	'my eye'
	rakáya-njabba	(C)	'my finger-/toenail'
	nga-móya	(D4)	'my sore (on skin)'
	nga-kkódda[9]	(D1)	'my skin'

Alternatively it could be suggested (Hilary Chappell and William McGregor, pers. comm.) that the body parts marking possession with the verb *réndjeyi* 'stand' are more "locational" than "actively used" or "capable of being independently used or directed". While this may be true for many of these nominals, it does not apply so readily to a small number of such words, such as *marnakkúrrkka* 'arm, wing' or *djábbarda* 'lip'. Thus neither suggested semantic basis for the possessor marking with the verb is fully convincing, but both have some plausibility.

One interesting word pair comparison between possession class B and possession class C points to the more personal, inalienable significance of possession when marked with the possessive pronoun rather than the verb *réndjeyi* as in (59). Once again the more inalienable possession is associated with animacy or humanness.

(59)	*kúdja-njabba*	*kúdja*	*ka-réndjeya*
	hand/foot-my	base	it-stands
	'my hand/foot'(C)	'its base (tree)'; 'its big end (didgeridoo or dronepipe)'(B)	

The lack of parallel examples, however, renders this no more than suggestive.

5.3. Three problem sets

Of some interest in this comparison of methods of possessor marking are three sets of nouns, one referring to types of hair (60), the second to the leg (61), and the third to male sex organs (62). The claimed patterns of difference in possessor marking appear not to hold in these cases. In (60) and (61) there is no apparent reason for the differences and in (62) the pattern is the opposite of what one would expect on the "internal-external" basis discussed in 5.2 above, though it supports the "actively used" basis. Possession class is indicated in brackets.

(60)	*djamíbbala-njabba* eyebrow-my 'my eyebrow'(C)	*nga-méyameya* my-head:hair 'my hair (on head)'(D1)
	mandjamándja-njabba whiskers/beard-my 'my beard'(C)	*nga-múraka* my-body/limb hair 'my body/limb hair'(D1)
	yalangaló-njabba pubic:hair-my 'my pubic hair'(C)	
(61)	*mémarla nga-réndjeya* upper:leg/thigh I-stand 'my thigh'(B)	*nga-marlúkarla* my-upper:leg/hip 'my hip'(D4)
	waláya nga-réndjeya lower:leg I-stand 'my lower leg'(B)	*njakkárla-njabba* lower:leg-my 'my lower leg'(C)
		njamánja-njabba knee-my 'my knee'(C)
(62)	*lékka nga-réndjeya* testicles I-stand 'my testicles'(B)	*lárla-njabba* penis-my 'my penis'(C)

5.4. Possession versus predication

We noted in 4.3 and 4.4 above that some nominals in possession classes C and D use the pronominal affix to the nominal itself, not to mark possessor but rather to mark the bearer of an attribute or even the subject of a predication – see examples (30) to (33) and (38) to (42). Of particular interest in this connection are examples (32) and (33), in which the form *lárla* 'penis' or 'male' uses a possessive pronoun suffix to mark possessor in the former meaning (32) and attribution in the latter meaning (33).

Two nominal forms have been found ((63) and (64)) for which the possessive pronoun suffix appears to mark the subject of a predication (compare possession class C), while another means of marking possession is used to indicate possessor. Conclusions are limited by lack of parallel cases. Data on these two distinctions is consistent but rare. In each case the "possessed" or non-predication example is the only common one.

(63)	*búrrbba-njabba* strong-1:MIN 'I am strong'(C)	*nga-bbúrrbba*[10] 1:MIN-guts 'my guts'(D1)
(64)	*wíba-njabba* country-1:MIN 'I belong to this country'(C)	*ngáyabba wíba* 1:MIN country 'my country'(A)

6. Possession and location

In a now classic paper Lyons (1967) explores semantic and syntactic links between locational, existential and possessive constructions such as (65) to (69).

(65)	Existential	*There are lions (in Africa).* *Lions exist.*
(66)	Locative	*The book is on the table.*
(67)	Locative	*There is a book on the table.*
(68)	Possessive	*The book is John's.*
(69)	Possessive	*John has a book.*

Lyons argues that, rather than deriving John's book transformationally from a source like (69) in which John is the surface subject, it should be derived from an underlying sentence more similar to (68) in which book is the subject and John forms part of the predicate, as a sort of "location". The *have* in (69) is transformationally introduced when John is moved to subject position under a general principle that animates are preferred to inanimates as surface subject unless the inanimate is marked as topic. He also comments on Latin (70), in which possession can be marked using the dative case, a case which covers a broad range of "implication" relationships.

(70) *Est Johanni liber.*
is John:DATIVE book
'John has a book.'

The general thrust of Lyons' argument (1967: 392) is that the "distinction of locatives and possessives ... [rests] largely upon the syntactic recognition of a distinction between animate and inanimate nouns". The essential claim he makes is that possessors should be seen as animate "locations". Lyons claims further (1967: 393) that the parallelism between existential, locative and possessive sentences is even more obvious in Mandarin Chinese and Russian than in English or Latin. Compare Comrie's (1981: 212-218) discussion of Maltese possessives and locationals for further parallels.

Eve Clark (1978) carried out a cross-linguistic comparison of the types of sentences discussed by Lyons – existential, locational and possessive – in an attempt to establish that possession is a fundamentally locational notion which differs from location per se primarily in the animacy of the possessor. Word order differences between the members of each pair – locative ((66) and (67)) and possessive ((68) and (69)) – relate to definiteness, or rather referentiality (or specificness), as is apparent in Lyons' examples quoted above (66)-(69). Clark's claims are borne out by predominant word order patterns and verb use in the languages studied, while additional support comes from what Clark (1978:113-117) calls locative properties of the possession relationship. These "locative properties" are found in the non-possessive uses of the cases, prepositions, postpositions, copulas and so on used to mark the possessor or the possession relationship. A very high proportion of these "other uses" is locational.

Ndjébbana provides some support for the proposal that possession is, at bottom, a locational concept in its use of a positional/existential/copulative verb to mark possession with nominals of possession class B.

Ndjébbana has three of these positional/existential/copulative verbs. They are *réndjeyi* 'stand', *yó* 'lie' and *nó* 'sit'.

The first two are used in locative and existential sentences like those in (71)-(74).

(71) *Nganéyabba yeláwa djarríbbang ya-kkóya.* (V/36-37)
there then/so trepang:(F) 3:MIN:F:S-lay/was
'That was where the trepang was', 'There was trepang there.'

(72) *Djawalárra ka-yóra.* (IV/227-228)
lilies:(M) 3:MIN:M:S-lie/be
'There are lilies there.'

(73) *Warábba-na middjin ka-rendjína*
one-3:MIN:M mission 3:MIN:M-stood/was

Márdbalk. (V/132)
Goulburn:Island
'There was a/one mission at Goulburn Island.'

(74) *Njanakkábba ka-rendjína ka-rakarawéra ngána*
dry:land 3:MIN:S-stood/was 3:MIN:S-moved there
Kabálku. (VII/70)
Entrance:Island
'There was dry land right out to Entrance Island.'

The third verb, *nó* 'sit', is used with meanings like 'sit', 'stay', 'live/reside' as well as functioning as a copula in non-present tenses. In the present tense no overt copula is used ((78) and (79)). For "copulative" (or attributive) sentences in present and past see (77)-(79), in all of which English would use a copula, while Ndjébbana uses a copula only in the non-present.

(75) *Márdbalk ka-nóra.* (V/216-217)
Goulburn:Island 3:MIN:M:S-sit/live:(PRES)
'He's living at Goulburn Island.'

(76) *Ngurrarákka kóma ka-kóna*
big:brother:(M) NEG 3:MIN:M:S-sit:(PAST)
pass 'way. (V/310)
'The big brother died (literally: didn't sit), he passed away.'

(77) *Nga-namánda ka-kóna,*
3:MIN:M-small 3:MIN:M-sit:(PAST)
dja-nána ka-yarrárlmanga. (XVII/53-55)
2:MIN:A+3:MIN:O-see:(PAST) 3:MIN:M:S-grow:up:(PAST)
'He was small (then) but you've seen he's grown up now.'

(78) *Nga-yarrárlmanga yaláwa, nga-ngurrarákka*
1:MIN:S-grow:up:(PAST) so 1:MIN-big
yaláwa. (II/71-72)
so
'I grew up to be a man.' (Literally: 'I grew up, so I'm an adult/big now'.)

(79) *Njanabbardákka Yírriddjanga.* (XXI/3)
trevally:(fish) (moiety)[11]
'Trevally is Yírriddjanga.'

The use of *nó* as a copula suggests that this verb is in some sense the unmarked member of this set. This receives further support from the fact that, although all three positional verbs are used as auxiliaries, *no* appears there, too, to be the unmarked auxiliary (McKay 1984a: 123).

One nominal root can occur with all three positional verbs, having a different meaning with each. The significance of this is difficult to determine at present because of the lack of parallels involving other nominals. Compare the uses of *rénjmarla* in (80) to (82). It is very unclear how *rénjmarla* should be glossed in (81) and (82) in the sense that there is no evidence on whether or not there are three homophonous forms or three separate but connected uses of a single term. Only the meaning of the phrase or sentence as a whole was clear from the provider of the text.

(80) *rénjmarla ka-réndjeya*
cheekbone 3:MIN:M-stands
'his cheekbone'

(81) *Dílkarra rénjmarla ka-kkóya.* (XXI/9)
moon:(M) ?new:moon 3:MIN:M:S-lies
'It was new moon time.'

(82) *Mángkaddjarra bókka rénjmarla ka-kóna.* (V/79)
Macassans bad ?mood 3:MIN:M:S-sat
'The Macassans were wild/jealous.'

The special case of *-wála* 'name' was mentioned above (section 4.6). This provides a further relevant example of *réndjeyi* being used as a copula, parallel to the *have* possessive of (69) in that the possessor is marked as the subject (S) of the sentence. Interestingly, however, the nominal also retains possessor marking in the form of a pronominal prefix.

In summary, in the light of the locative and existential functions of the verb *réndjeyi* in particular, it is significant that the same verb provides one of the means of marking possession of body parts in Ndjébbana. This fact gives further support to the suggestion of a formal relationship between existential, locative and possessive constructions, as proposed by Lyons, Clark and others. The Ndjébbana possessive construction used in possession class B is parallel to the *have* possessive of (69) in that the possessor takes the form of the subject (S) of the verb *réndjeyi*, and in addition appears to be referential in all known examples.

7. Conclusion

Ndjébbana provides its speakers with four different means of marking the possessor of a body part (or other) nominal. These involve, in ascending order of inalienability, juxtaposition of a free form pronoun, use of a pronominal verb prefix on a form of the verb 'to stand' associated with the nominal, and use of a pronominal affix attached directly to the nominal itself. In some cases this pronominal affix is a suffix and in others a prefix. Each nominal root in a given meaning is used with just one of these specific types of possessor marking, thus providing a basis for the classification of nominals into four major possession classes (A, B, C, D). The four possession classes defined in this way intersect with the two classes defined in terms of gender (masculine, feminine), fixed or inherent gender being characteristic only of nominals in possession class A (maximally alienable). Nominals in the other possession classes, in the main, have variable gender, depending on the gender of their "possessor".

For the two most inalienable possession classes of nominals, C and D, both marking possessor with an affix directly attached to the nominal, it is argued that the relationship signified by this affix is best described as one of identity, rather than of possession.

Comparison of the four patterns of possessor marking, of the nominals in the four possession classes and, most significantly, of the cases where semantically or formally related word pairs occur in different possession classes points to a number of significant conclusions:

(a) There is a continuum of increasing inalienability (or identity) from possession class A through to possession class D, which is matched by a continuum of increasing closeness of morphological link between possessor and possessed. The loosest morphological link occurs in possession class A where the possessor is a free form pronoun, while the tightest morphological link occurs in possession classes C and D where the possessor pronoun is affixed directly to the nominal as suffix and prefix respectively.

(b) In homophonous word pairs in which one word belongs to a more alienable possession class than the other and both have distinct (though often related) meanings, there is a clear tendency for the more inalienable nominal to have a referent higher on the scale of animacy or humanness than the more alienable term.

(c) Many of the body part terms in possession class B, which mark possession using a form of the positional verb *réndjeyi* 'stand', have semantically related counterparts in the more inalienable possession classes C and D. In these cases there is a tendency for the nominals in possession class B to refer to more externally visible (and perhaps less actively used or independently controllable) body parts than those in the more inalienable classes.

(d) The fact that the known kin terms fall into possession class A and mark possessor with a free form pronoun shows that these nominals are treated as alienable.

Finally, in the use of a positional/existential/copulative verb, *réndjeyi* 'to stand', in the marking of possession in possession class B, Ndjébbana provides some support for claims that there is a universal formal and semantic link between possessive and locational constructions.

Notes

1. The present paper expands two previous unpublished versions (McKay 1980, 1981). For Ndjébbana language data I gratefully acknowledge the help of Jockey Bundubúndu, Joseph Mangkúdja (deceased), Johnny Karddáwarr and others, of Maningrida. Carolyn Coleman assisted by checking a number of points with Ndjébbana speakers for me. Hilary Chappell and Bill McGregor's comments on the version prepared for the Body Parts in Grammar Workshop at the 1988 Australian Linguistic Society Conference have led to significant improvements in the paper. The final version of the paper was written during a visiting fellowship in the Department of Linguistics, Arts, Australian National University, following some Rembarrnga dictionary research which brought me to Canberra with the support of the Institute of Applied Language Studies (WACAE).

   All Ndjébbana material is presented in the practical orthography approved for use in the school program. Stops *b*, *d*, *rd*, *dj*, *k*; nasals *m*, *n*, *rn*, *nj*, *ng*; laterals *l*, *rl*; rhotics *r*, *rr*; semivowels *w*, *y*; vowels *i*, *e*, *a*, *o*, *u*; vowel length coupled with major stress is marked by an acute accent. Stop, nasal and lateral digraphs with *r* (*rd*, *rn*, *rl*) represent retroflexes; stop and nasal digraphs with *j* (*dj*, *nj*) represent laminal articulations; *rr* represents a trill and *r* a retroflex continuant. In consonant clusters where two of the retroflex or lateral digraphs form a sequence the *r* or *j* is written only once for normal purposes though in some technical materials the full form may be found to avoid any ambiguity. For example the sequence fully specified as *njdj* is normally written *ndj*, the sequence *rnrd* is normally written *rnd*. Syllable initial stops are lenis (usually voiced) while syllable final stops are fortis (usually voiceless). Where two stops occur together at a syllable boundary within a word the second (syllable initial) stop assimilates to the first (syllable final) becoming fortis and voiceless. Single peripheral stops *b* and *k* have optional voiced fricative allophones between vowels.

The following abbreviations are used: A – Agent, Transitive subject; AUG – Augmented number (see 3.1); F – Feminine; HUM – Human; INFIN – Infinitive; IO – Indirect object; M – Masculine; MIN – Minimal number (see 3.1); NEG – Negative; O – Transitive object or patient; PAST 1 – Recent past and present in progress; PAST 2 – Distant past and earlier today; PL – Plural; POSS – Possessive; PRES – Present; S – Intransitive subject; SG – Singular; UAUG – Unit augmented number (see 3.1)

Many of the examples are taken from transcriptions of tape recorded texts. References are given to the original manuscript in the form (Text Number/Line Number(s)), for example, (XVII/15-18).

2. PAST 1 tense denotes recent past before today (e.g. "yesterday") as well as present (in progress). PAST 2 tense denotes "earlier today" as well as more distant past.
3. *-ba* appears to be a general suffix of extent which always causes the final vowel of the stem to which it is affixed to be realised as *-i-*. It is involved in pluralising two common nominals *ngarráma* 'woman' (plural *barra-ngarrámi-ba*, possession class D4) and *karókaddja* 'child' (plural *barra-(ka)rókaddji-ba*, possession class D1). It is also used with other nominals with a meaning something like 'in the area of X' e.g. *narrúrri-ba* '(in) the open area' (<*narrúrra* 'open (area)'), *bana-kkábbi-ba* 'littoral' (<*kábba* 'water').
4. In example (25) the predicative nominal *nbókka* 'bad' (possession class D1) agrees in person, number and gender (3 minimal masculine) with the noun *djakí* 'speed(iness)' (possession class C). This "bad speed" is possessed by the *bíbbo* 'crab' (possession class A, 3 minimal feminine) as marked using the possessive pronoun suffix *-ngaya* (3 minimal feminine).
5. Aboriginal society in this area is classified into two exogamous moieties, each further subdivided into four subsections. A person's subsection membership is fully determined by that of his mother, though in the opposite moiety to his mother. Ndjébbana terms for the moieties are *Djówanga* and *Yírriddjanga* (in other parts of Arnhem Land often known as *Duwa* and *Yirritja* respectively). The eight subsection terms, listed by moiety (and following the Western Arnhem Land pattern) are as follows:

   Djówanga: *n-Búlanj, Ngárridj, n-Kóyok, na-Bangárda*
   Yírriddjanga: *n-Kangíla, Wákadj, Wámud, n-Kamárrang*

   Only the masculine forms are given here. Third person minimal feminine forms involve *nja-* prefixed to the root in all cases, replacing the masculine prefix where necessary. All subsection terms belong to possession class D.
6. A further pair of some interest here is *namarnakkúrrkka* 'creek' (possession class A) and *marnakkúrrkka* 'arm, wing' (possession class B). In this case the match is not perfect but the resemblances, both formal and semantic are strong. This is pehaps the only evidence that possession class B marks possession which is more inalienable than that of possession class A.
7. Though not of clear relevance here, it is interesting to note in passing that some of the kin terms quoted by Capell (1943: 28) are actually verb forms using the PAST 2 tense (meaning 'earlier today' or 'distant past') of the verb *réndjeyi* 'stand/be'. For example Capell's *njirarindjina* 'my brothers' represents:

njarra-rendjína	or	*njirri-rendjína*
1:AUG-stood		1:UAUG-stood
'my brothers (plural)'		'my brother (singular)'
(Literally, 'we (plural) stood')		(Literally, 'we (dual) stood')

8. Lenition of the root initial stop *dj* to *y* before an unstressed vowel and after a vowel occurs by a regular phonological rule outlined by McKay (1984b).
9. The doubling of the root initial *k* of *n-kódda* occurs by a regular phonological rule when the stop follows a vowel and precedes a long, stressed vowel. See McKay 1984b.
10. The doubling of the root initial *b* occurs by a regular phonological rule when the stop both follows a vowel and precedes a long stressed vowel within a word (McKay 1984b).
11. See example (29) above and the explanation in footnote 5.

References

Allen, W. Sidney

1964 "Transitivity and possession", *Language* 40: 337-343.

Andersen, Elaine S.

1978 "Lexical universals of body-part terminology", in: Joseph H. Greenberg (ed.), 1978a: 335-368.

Capell, Arthur

1943 "Languages of Arnhem Land, North Australia", *Oceania* 13: 24-50. (Continued from *Oceania* 12: 364-392.)

Clark, Eve V.

1978 "Locationals: existential, locative and possessive constructions", in Joseph H. Greenberg (ed.), 1978b: 85-126.

Comrie, Bernard

1981 *Language universals and linguistic typology: syntax and morphology*. Oxford: Blackwell.

Greenberg, Joseph H. (ed.)

1978a *Universals of human language, Volume 3: Word structure*. Stanford: Stanford University Press.

1978b *Universals of human language, Volume 4: Syntax*. Stanford: Stanford University Press.

Lyons, John

1967 "A note on possessive, existential and locative sentences", *Foundations of Language* 3: 390-396.

McKay, Graham R.

1978 "Pronominal person and number categories in Rembarrnga and Djeebbana", *Oceanic Linguistics* 17: 27-37.

1979 "Gender and the category *unit augmented*", *Oceanic linguistics* 18: 203-210.

1980 Possession marking in Ndjébbana (Kunibidji). [Unpublished MS. Department of Education, Maningrida, Northern Territory, Australia.]

1981 "Nominal classes and possession marking in Ndjébbana (Kunibidji)." [Paper presented to Northern Territory Linguists' Group. Batchelor, Northern Territory, Australia.]

1984a "Ndjébbana (Kunibidji) grammar: miscellaneous morphological and syntactic notes", *Papers in Australian linguistics No. 16.* Canberra: Pacific Linguistics, 1984, 119-151.

1984b "Stop alternations in Ndjébbana (Kunibidji)", *Papers in Australian linguistics No. 16.* Canberra: Pacific Linguistics, 107-117.

1989 Agent and patient in Ndjébbana bound pronominal forms. [Paper presented to Australian Linguistic Society Conference, Monash University, Melbourne.]

Wurm, Stephen A.

1972 *Languages of Australia and Tasmania.* The Hague: Mouton.

Body parts in Murrinh-Patha: incorporation, grammar and metaphor [1]

Michael Walsh

1. Introduction

This account focusses on body parts and their metaphorical extensions in Murrinh-Patha,[2] a non-Pama-Nyungan language spoken by around 1000 people, mainly at Wadeye on the west coast of the Northern Territory of Australia. There appears to be a continuum from purely concrete senses of body parts to metaphorical extensions. So an attempt is made to characterise these extensions in terms of "metaphorical themes". The intention is to develop these "metaphorical themes" as a device to further explore body part metaphor in Australian languages and, more generally, to consider universals in metaphor formation.

One purpose in presenting so much detail on metaphor formation in Murrinh-Patha is to encourage fuller treatments of metaphor in so-called 'exotic' languages. The pervasiveness of body part metaphors is surely not restricted to Murrinh-Patha. So it is reasonable to ask how should a grammatical description embrace these phenomena. The preferred treatment seems to have been to relegate this kind of material to a small section somewhere in the "background" of the grammar. One example of this approach can be seen in Dixon's account of another Australian language, Yidiny (from north Queensland) in which metaphor receives a brief account within the section on semantics (Dixon 1977). This should not be taken as a shortcoming of Dixon's grammar: on the contrary, one has to be thankful that material of this kind rates a mention at all. More typically, such material is simply invisible.

This invisibility of metaphor is not restricted to the grammars of Australian languages. It is usual to find little or no discussion of the topic in substantial accounts of so-called 'exotic' languages. Let us consider a few examples. Derbyshire's grammar of Hixkaryana (1979), a language of South America, devotes scant attention to this subject. Haiman's comprehensive treatment of Hua, a language of Papua New Guinea, (1980) refers to some metaphorical expressions in passing. Matisoff's detailed account of Lahu, a language spoken in northern Thailand (1973) has no separate section devoted to any kind of semantic issues (let alone metaphor) even though other publications of his (for example, Matisoff 1978) examine metaphor and body parts, clearly indicating he is not merely a "morpheme-cruncher". These examples have been chosen because, like Dixon's treatment of Yidiny, they can be regarded as exemplars of descriptive grammars. Nevertheless in each case the treatment of metaphor is relatively perfunctory.

Why should metaphor receive so little attention? Perhaps one reason is that it can be hard to see. Even the most perfunctory treatment of a language can scarcely miss seeing some of the more obvious features like the broad outlines of the phonological system, the pronominal system and the case system. Such phenomena are felt to be obvious precisely because they are so pervasive and because they form closed, and therefore relatively tractable, systems. Metaphor, on the other hand, is open-ended and there are subtleties about the formation of metaphor which only become obvious after a long apprenticeship in the language. Witherspoon's account of Navajo (1977) is one exception to the invisibility of metaphor in an account of an "exotic" language. Predictably the praise justifiably heaped on this work is tempered by references to Witherspoon's long acquaintance with Navajo and the inherent difficulties involved for everyone else.

What then of metaphor in less 'exotic' languages? It seems then that metaphor might not so much be invisible as something that grammarians choose not to focus on. This is certainly the view of a number of linguists including Lakoff (1987) and Langacker (1987). As Langacker puts it:

> Figurative language [including idiomaticity, metaphor, and semantic extension] is generally ignored in current theories; at best it is handled by special, ad hoc descriptive devices. Yet it would be hard to find anything more pervasive and fundamental in language ... We therefore need a way of conceiving and describing grammatical structure that accommodates figurative language as a natural, expected phenomenon rather than a special, problematic one. An adequate conceptual framework for linguistic analysis should view figurative language not as a problem but as part of the solution. (Langacker 1987: 1)

With regard to metaphor, Australian languages are pervasive in their use of body part metaphors. This observation is based not so much on published work as on impressions drawn from the almost universal agreement among specialists in the study of Australian languages that body part metaphors are important in the languages they are familiar with. These personal communications are supported by my own studies of Australian languages. The material presented here from Murrinh-Patha seeks to demonstrate this pervasiveness.

Nominal incorporation in Murrinh-Patha is restricted to body parts, unlike other Australian languages such as Mayali (Evans 1995), Rembarnga (McKay 1975) and Warray (Harvey 1995). An incorporated body part is prefixed immediately before the verb root. It is possible to have an unincorporated body part expression outside the verbal word *as well*. In such instances the "external" body part expression is typically redundant. The following examples show that the relationship between the external body part nominal and its incorporated counterpart varies from identity through similarity to suppletion:

(1) a. *mam-nhi-me-purl*
1sgS-2sgO-foot-wash[3]
'I washed your foot/feet.'

b. *me (nhinhi) mam-nhi-me-purl*
foot (2sg) 1sgS-2sgO-foot-wash
'I washed your foot/feet.'

(2) a. *mam-nhi-ma-purl*
1sgS-2sgO-hand-wash
'I washed your hand(s).'

b. *mange (nhinhi) mam-nhi-ma-purl*
hand (2sg) 1sgS-2sgO-hand-wash
'I washed your hand(s).'

(3) a. *mam-nhi-we-purl*
1sgS-2sgO-head(hair)-wash
'I washed your head(hair).'

b. *pelpith/pemarr (nhinhi) mam-nhi-we-purl*
head/head hair (2sg) 1sgS-2sgO-head(hair)-wash
'I washed your head/head hair.'

(4) a. *mam-nhi-rdarri-purl*
1sgS-2sgO-back-wash
'I washed your back.'

b. *pangkin (nhinhi) mam-nhi-rdarri-purl*
back (2sg) 1sgS-2sgO-back-wash
'I washed your back.'

This account focuses on body part incorporation because that is the most revealing area for a study of body part metaphor in this language. In section 2 I provide some typological background and outline some of the complexity involved in Murrinh-Patha verbs. In section 3 I introduce the system of nominal classification in Murrinh-Patha, indicate the membership of body parts in this system and first mention some metaphorical extensions of body parts. The basic facts of body part incorporation are laid out in sections 4 while 5 discusses the continuum from concrete to metaphorical senses of body parts. In section 6 I survey the uses of body parts which are not incorporated.

2. Typological background

Murrinh-Patha is a highly agglutinating language showing nominal classification but relatively little nominal case morphology. Verbs in Murrinh-Patha are complex expressions marking enough information to stand often as independent utterances. In Street's analysis (1987) thirty-five verb classes are distinguished (see also Walsh 1989, 1993). Each verb class is characterised by the (typically) verb-initial portmanteau morpheme here called AUX which marks subject (pronominally) and other features such as mood, tense and 'mode of action'.

Murrinh-Patha verbs are dense in information marking the major participants of the action of the verb, incorporating body part expressions and adverbs as well as optionally using a 'secondary' auxiliary to modify the action of the verb. The 'secondary' auxiliary can be drawn from one of seven of the total of 35 verb classes found in Murrinh-Patha (Street 1987; Walsh 1989). It is identical in form to the verb-initial portmanteau morpheme (here called an auxiliary) but is suffixed. Its function is to mark aspect and/or the 'orientation' of the action of the verb – whether carried out while sitting, standing, or moving.

The order of elements in the Murrinh-Patha verb is as follows:

AUX_1-(OBJ)-(BP)-VERB ROOT(x2)-(ADV)-TM-NI-(AUX_2)

where: AUX_1 = (primary) auxiliary; OBJ = bound pronominal object form (whether direct or indirect); BP = incorporated body part; (x2) = reduplication; ADV = incorporated adverb; NI = number indicator; TM = 'tense' marker; AUX_2 = secondary auxiliary.

In brief, this means that each Murrinh-Patha verb must have an initial auxiliary from one of the 35 verb classes. That is, the form of this obligatory auxiliary is to be chosen from one of 35 sets of paradigms. The verb root may be reduplicated to indicate such notions as habitual action or iterativity. Immediately before the verb root an incorporated body part may occur. In the next structural position is a cross-referencing bound pronoun to mark certain non-subject functions, as illustrated in examples (5) and (6) and (9) and (10). After the verb root an adverb may appear. The next structural position is occupied by markers of tense/aspect/mood:

nu	'Future'
ø	'Perfect'
dha	'Imperfect'
nukun	'Subjunctive'

In Murrinh-Patha pronouns four numbers are distinguished: singular; dual; paucal (from 3 to about 10) and plural. A number indicator discontinuously indicates whether a pronoun form prefixed to the verb is dual or paucal. Within these two number categories there is further specification of feminine and masculine. Finally a "secondary" auxiliary may appear to modify the action of the verb.

Basic transitive and intransitive patterns (S=Subject; O=Object) are illustrated in (5)-(8):

(5) *mam-nhi-purl*
1sgS-2sgO-wash
'I washed you.'

(6) *ma-nhi-purl-nu*
1sgS-2sgO-wash-FUT
'I will wash you.'

(7) *ngem-kampa*
1sgS-laugh
'I laughed.'

(8) *ngi-kampa-nu*
1sgS-laugh-FUT
'I will laugh.'

Note that the shape of the verb-initial portmanteau morpheme changes with tense (*mam/ma* for PAST/FUT of 'wash' and *ngem/ngi* for PAST/FUT of 'laugh') and according to the verb root ('wash' belongs to Verb Class 8 and thus takes *mam/ma* while 'laugh' belongs to Verb Class 1 and thus takes *ngem/ngi*).

Just two participants can be marked as pronominal prefixes on these complex verbal words. The "subject" is always marked as part of the verb-initial portmanteau morpheme. The next participant will be direct object, if there is one, or otherwise some kind of indirect object (Benefactive or Source):

$$\text{Direct object} > \left\{ \begin{matrix} \text{Benefactive} \\ \text{Source} \end{matrix} \right\}$$

The one exception is when the direct object is morphologically zero, as it is for one of the more common participants: third person singular. In this one instance the cross-referencing indirect object pronominal prefix "displaces" the direct object from its position in the verb.

(9) *mam-ø-purl*
1sgS-3sgO-wash
'I washed her/him/it.'

(10) *mam-mpa-purl*
1sgS-2sgBEN-wash
'I washed her/him/it for you (singular).'

For each pronominally prefixed participant there can be an "unincorporated" form external to the verbal word. This is exemplified in these "fuller" counterparts of (5) and (9):

(5′) *ngay-re nhinhi-ø mam-nhi-purl*
Isg-ERG 2sg-ABS 1sgS-2sgO-wash
'I washed you.'

(9′) *ngay-re nigunu/nukunu/nanthi pangu-ø mam-ø-purl*
Isg-ERG 3sgFem/3sgMasc/NC that-ABS 1sgS-3sgO-wash
'I washed her/him/it.'

In the case of (5′) the "fuller" version is redundant but with (9′) the free form (pro)nominal expression further specifies the participant. For the most part though, the pronominal forms are so specific that there is no need to include a (semantically redundant) copy of the prefixed pronominal form.

Pronouns distinguish four numbers: singular; dual; paucal, and, plural. Dual and paucal pronouns further distinguish between groups consisting entirely of siblings (actual or classificatory) and other groups.

3. Body parts and noun class membership

In Murrinh-Patha each noun can be associated with a noun class marker. The noun class marker appears first in a nominal expression. Membership in noun classes is semantically based (Stanner 1964; Street 1987: 41-44; Walsh 1976c; 1991), for example:

(11) *(ku)* *baybayye*
NC kangaroo

(12) *(thu)* *kuragadha*
NC boomerang

(13) *(mi)* *lawam*
NC flour

Members of the *ku*-class include animals, birds, fish etc. and their products (honey, eggs, meat etc.); the *thu*-class refers to offensive weapons and the *mi*-class relates to fruit and vegetable food.

In all, around ten noun classes might be distinguished for Murrinh-Patha. Street (1987: 41-44) sets up ten noun classes which are semantically based:

1. *kardu* Aboriginal people and human classification (including human spirits):

kardu thipmam	'black person [i.e. Aboriginal]'
kardu pule	'old man'; 'husband'; boss'
kardu ngepan	'spirit/soul of a living person'
kardu warnangkal	'clever man; 'witch doctor''
kardu kawu	'mother's mother'

2. *ku* Non-Aboriginal people and all other animates and their products, including flesh foods:

ku thipmam	'black person [non-Aboriginal]'
ku pule	'old man'; 'husband'; boss [non-Aboriginal]'
ku warnangkal	'clever man'; 'doctor [non-Aboriginal]'
ku kulerrkkurrk	'brolga'
ku lawarnka	'wallaby'
ku murl	'fly'
ku thitay	'(wild) honey, sugarbag'

3. *kura* Fresh water and associated concepts, such as potable fluids (except 'milk' which is in Class 5) and different collective terms for fresh water like 'rain' or 'river':

kura thurrulk	'beer (=foam water)'
kura ngipilinh	'creek, river'
kura yelyel	'rain'

4. *mi* Flowers and fruits of plants and any vegetable foods and also faeces:

mi thathangadhay	'flower of certain trees'
mi marrarl	'fruit of native tree (Terminalia ferdinandiana)'
mi lawam	'flour'
mi ngukin	'faeces'

5. *nanthi* Essentially a residue category in that a noun can usually be assigned to the *nanthi*-class if it does not obviously fit into any of the other nine classes which are relatively well-defined. This class includes most inanimates – objects, implements and natural phenomena of all kinds. Examples are:

nanthi thelput	'house'
nanthi wirrirr	'wind'
nanthi thay	'stick'

6. *thamul* Spears.

thamul nguni	'short light spear'
thamul menek	'ironwood spear'
thamul waya	'fishing spear; 'fish wire'

7. *thu* Offensive weapons (defensive weapons like shields predictably go into the *nanthi*-class), also thunder and lightning, and playing cards:

thu kuragadha	'boomerang'
thu paku	'large club'
thu malarntath	'thunder; lightning'
thu kat	'playing cards'

8. *thungku* Fire and things associated with fire.

thungku thay	'firestick'; firewood'
thungku len	'hot coals'
thungku methith	'matches'

9. *da* Time and space: names for localities, seasons etc.

da pemanhay	'sandhill'
da therri	'dry grass time'; 'latter part of the dry season'
da yidiyi	'Yidiyi(place name)'

10. *murrinh* Speech and language and associated concepts such as news, songs, school.

murrinh mamay	'baby talk'; 'children's talk'
murrinh thelerrdhe	'news'
murrinh school	'school'

I have said about ten classes might be distinguished because there does not seem to be a clear distinction in Murrinh-Patha between nominal classification and compounding. For the first eight classes the morphosyntax provides a criterion: to indicate existential negation ('There is/are no ...), a prefix is attached to the noun class marker in this way:

(14) *banana ma-mi*
banana NEG-NC_{veg}
'[There are] no bananas.'

(15) *thurrulk ma-kura*
foam NEG-NC_{aqua}
'[There is] no beer.'

For the other two classes there are no syntactic characteristics which would set them off from (?other) generic-specific combinations like:

(16) *thay karnmurrirn*
tree woolybutt
'woolybutt (tree)'

where, as in English, the specific term, *karnmurrirn* 'woolybutt', can stand alone or in combination with the generic term, *thay* 'tree'. Looking at nominal classification in a typological perspective Dixon (1986) has distinguished between two quite separate types: noun classes and nominal classifiers. In fact Murrinh-Patha seems to stand somewhere between these ideal types (for further details, see Walsh 1991) but for our purposes here it will suffice to refer to them as noun classes.

A particular noun may occur in a number of noun classes according to its function or according to the way it is viewed in the culture. When a boomerang is considered merely as an object and not as an offensive weapon it will be assigned to the *nanthi*-class. So *nanthi kuragadha* might be said of a boomerang employed as a doorstop or as an improvised back-scratcher. Although *lawam* 'flour' can be assigned to the *mi*-class because it is a foodstuff derived from plants, *nanthi lawam* is used to refer to flour bought in a shop.

Most body parts belong to the *nanthi*-class, the three exceptions being (see also Evans 1994 and Harvey 1994):

(17) a. *ku thimu* 'clitoris'
b. *ku midhut* 'vulva'
c. *ku weyi* 'vagina'

Otherwise *thimu* belongs to the *nanthi*-class wherein it refers to 'nose' while *weyi* belongs to the *nanthi*-class wherein it refers to any kind of hole. It is difficult to give any satisfactory explanation for this division of the body part domain, given the reticence of language teachers to enter into discussion. One can observe that other languages including English have overlaps between words referring to male genitalia and food e.g. eggs, nuts; cucumber, sausage etc.. In Murrinh-Patha these three body part terms have been assigned to that noun class which includes flesh food. One category of body products is assigned to the *mi*-class: *mi ngukin* '(solid) faeces'; *mi yilulul* 'liquid faeces'. Presumably the association is between food and that body *product* which is itself the *product* of food (*mi* can be used generically to refer to 'food'). All other body products whether blood, sputum or urine are assigned to the *nanthi*-class.

For some body parts it is possible to switch noun classes with an attendant change in meaning:

(18)	a.	*nanthi kamarl*	'eye/face'
	b.	*kura kamarl*	'water-hole'
	c.	*kardu kamarl*	'sweetheart'
	d.	*mi kamarl*	'seed'
(19)	a.	*nanthi ngi*	'penis'
	b.	*ku ngi*	'death adder'
(20)	a.	*nanthi nginipunh*	'body'
	b.	*murrinh nginipunh*	'skin(-name)'[4]
(21)	a.	*nanthi pangkin*	'back'
	b.	*da pangkin*	'ridge'

In these sets of expressions it is certainly the body part meaning which is basic. When a term in one of these sets occurs without a noun class marker the unmarked reading for that term is the body part sense. In addition native speakers spontaneously volunteer glosses for the "other" meanings in a way that suggests that the body part sense is seen as being basic: "the seed is round like your eye"; "that snake looks like that "man-thing" ".

4. Incorporation of body parts

4.1. Incorporated forms

Apart from their occurrence in verbs, incorporated body parts commonly appear in a range of other construction types (see 4.2). The list of incorporated forms given in Table 1 is based in part on Street (1987: 103-4) but some forms appear to have variants. These appear to be the only body part terms that can be incorporated.

The relatedness of the free forms and their incorporated counterparts varies (see also Leeding 1995). Some, like *marda*, *me*, *ngurru*, and *walmpu* show no change. For some others the the bound form is a truncated version of the free form:

lamala	*mala*	shoulder
mange	*ma*	hand
manthingapa	*manthi*	collarbone ("yoke-area")
mardayidi	*marda*	chest

nginipunh	*ngini*	body (internal)
wulumu	*lu*	forehead

Of the same kind but not quite so obvious are the following:

kamarl	*ngka*	eye(s)/face
parnta	*warnta*	groin
pelpith	*we*	head
pemarr	*we*	hair of head
pingkarl	*wingka*	knee
thamamay	*dhamurru*	beard
thamarl	*dhamurru*	side of face
tharrmu	*dharrmu*	(lower) leg, shin
thathpi	*dhawi*	mouth, lips
thathpi	*dhayi*	mouth, lips
the	*ye*	ear
themen	*dhanme*	tongue
thimu	*yi*	nose
thirrimeme	*yirri*	navel
thunmu	*dhanmu*	nape of neck

The bound forms beginning with the semivowels /w/ and /y/ also appear as forms starting with /p/ and /th/ and thereby become much more similar to their free counterparts:

(22) *mam-nhi-we-purl*
1sgS-2sgO-head-wash
'I washed your head (hair).'

(23) *mam-ø-we-purl > mampepurl*
1sgS-3sgO-head-wash
'I washed his head (hair).'

(24) *mam-nhi-ye-purl*
1sgS-2sgO-ear-wash
'I washed your ear(s).'

(25) *mam-ø-ye-purl > manthepurl*
1sgS-3sgO-ear-wash
'I washed his ear(s).'

This alternation is treated synchronically as hardening because this analysis is the simpler one. For that reason the underlying forms are shown

Table 1. Incorporated body part forms in Murrinh-Patha.

FREE FORM	INCORPORATED FORM	GLOSS
dimu	*rdi*	tooth/teeth
kamarl	*ngka*	eye(s)/face
kanga	*kanga/dhanga*	small of back
kardirrak	*ngkardarri*	eyebrow
lamala	*mala*	shoulder
lawali	*bu*	upper leg/thigh
lumpu	*ri*	buttocks
mange	*ma*	hand
manthingapa	*manthi*	collarbone ("yoke-area")
marda	*marda*	belly
mardayidi	*marda*	chest
me	*me*	foot/feet
nginipunh	[unmarked]	body (external)
nginipunh	*ngini*	body (internal)
ngurdengke	*ngurdengke*	calf of leg
ngurru	*ngurru*	side
nithi	*be*	arm
pangkin	*rdarri*	back
parnta	*warnta*	groin
pelpith	*we*	head
pemarr	*we*	hair of head
pingkarl	*wingka*	knee
thamamay	*dhamurru*	beard
thamarl	*dhamurru*	side of face
thamarru	*dhamurru*	side of face
tharrmu	*dharrmu*	(lower) leg; shin
thathpi	*dhawi*	mouth/lips
thathpi	*dhayi*	mouth /lips
the	*ye*	ear
themen	*dhanme*	tongue
thimu	*yi*	nose
thirrimeme	*yirri*	navel
thunmu	*dhanmu*	nape of neck
walmpu	*walmpu*	testicles
wulumu	*lu*	forehead

as above. There is also a hardening of /dh/, a voiced interdental fricative, to /th/, a voiceless interdental stop:

(26) *mam-nhi-dharrmu-purl*
1sgS-2sgO-lower:leg-wash
'I washed your lower leg(s).'

(27) *mam-ø-dharrmu-purl > mantharrmupurl*
1sgS-3sgO-lower:leg-wash
'I washed his lower leg(s).'

The connection between *kamarl* and *-ngka-* becomes transparent by considering certain phonotactic constraints:

(28) *mam-nhi-ngka-purl*
1sgS-2sgO-eye/face-wash
'I washed your eye(s)/face.'

(29) *mam-ø-ngka-purl > mamkapurl*
1sgS-3sgO-eye/face-wash
'I washed his eye(s)/face.'

Across morpheme boundaries there is an obligatory consonant cluster reduction:

$$C_1C_2C_3 \rightarrow C_1C_3$$

which deletes the second of a three member consonant cluster. Underlyingly *kamarl* might be regarded as *ngkamarl* in which case the similarity of *kamarl* to *ngka* becomes more apparent. In this scenario underlying *ngkamarl* is reduced to *kamarl* by a very general constraint in Australian languages which prohibits initial consonant clusters.

Most of the other incorporated forms are suppletive:

lawali	*bu*	'upper leg/thigh'
lumpu	*ri*	'buttocks'
mikmu	*mili*	'hips'
miyernu	*gumu*	'tail'
ngapurlu	*yi*	'breast'
pangkin	*rdarri*	'back'

There is just one incorporated form which might be treated as a compound, at least historically:

kardirrak *ngkardarri* 'eyebrow'

The incorporated form, *ngkardarri*, superficially resembles the concatenation of *ngka* 'eye' and *rdarri* 'back', and is semantically not altogether inappropriate given what an eyebrow is in relation to the eye.

Of course the glosses given for body part expressions above can only be regarded as approximate. For instance, *manthingapa* has been glossed as 'collar-bone' and also as 'yoke-area'. One language teacher explained it to me by saying that it was the part which a small child touches when sitting on your shoulders. This means that 'collarbone' is only a part of the story: the collarbone is involved but also the area 'around one's neck' and I have tried to capture this by the expression "yoke-area".

Another area which is vague in its gloss is:

parnta *warnta* 'groin'

Even the term 'groin' is rather vague in English: the Macquarie Dictionary (Delbridge et al. 1985) gives this definition:

> the fold or hollow on either side of the body where the thigh joins the abdomen

In Murrinh-Patha *parnta* is sometimes glossed as 'lap' which is likewise rather hard to pin down. To further complicate matters Murrinh-Patha has an expression *lawali maparr* which can be glossed as 'groin'. In fact *lawali maparr* seems to refer to the 'join of the thigh to the lower midsection'.

4.2. Construction types involving incorporated body parts

The most common type of incorporation is in verbs. This construction type has already been illustrated in examples (1)-(4). Body parts may also be incorporated into adjectival and nominal expressions. These will be outlined below after a brief account is given of a particular subset of verbs.

4.2.1. Impersonal verbs

In Murrinh-Patha there is a quasi-transitive construction used mainly to encode bodily and emotional states in which the experiencer is treated syntactically as the direct object:

(30) *dem-ngi-rurr*
3sgS-1sgO-heat
'I am hot.'

Such expressions may include an incorporated body part which is clearly segmentable and is one member of a substitution set:

(31) *dam-ngi-mala-yiyip*
3sgS-1sgO-shoulder-black(REDUP)
'I've got black spots on my shoulder(s).'

(32) *dam-ngi-ma-yiyip*
3sgS-1sgO-hand-black(REDUP)
'I've got black spots on my hand(s).'

or they may involve a verb root which can be interpreted as containing an incorporated body part (see also 5.5.):

(33) *dam-ngi-mardathin*
3sgS-1sgO-sad
'I'm sad.'

where the first part of the verb root is homophonous with the incorporated body part, *marda* 'belly/chest'. For further details on the impersonal verb construction see Walsh (1987).

4.2.2. Adjectival expressions

Adjectives occur unremarkably in predicative and attributive uses as follows:

(34) *ngay-ka bamam*
1sg-TOP white
'I am white.'

(35) *ngay-ka kardu wardi*
1sgS-TOP person thin
'I'm a thin man.'

But adjectives also appear in constructions in which the person being modified by the adjectival root is suffixed to that root as a cross-referencing bound pronoun (cf. McKay's analysis (1995) of possession classes in Njébbana):

(36) *wardi-ma-nhi*
thin-hand-2sgO
'You have thin hand(s)/finger(s).'

(37) *wardi-be-nhi*
thin-arm-2sgO
'You have thin arm(s).'

(38) *pangki-ngi-ma*
long-1sgO-hand(s)/finger(s)
'I have long hand(s)/finger(s).'

(39) *ngaliwe-nhi-ma*
short-1sgO-hand(s)/finger(s)
'You've short fingers.'

(40) *pangki-ngi-dhayi*
long-1sgO-lip
'I'm long-lipped.'

(41) *pangki-ngi-yi*
long-1sgO-nose
'I'm long-nosed.'

(42) *pangki-ngi-ngka*
long-1sgO-face
'I'm long-faced.'

(43) *pangki-ngi-we*
long-1sgO-head
'I'm long-headed.'

Other examples of this construction type using a substitution set of incorporated body parts are as follows:

(44) a. *lurruwith*
'quick'
b. *lurruwith-ngi*
quick-1sgO
'I'm quick.'
c. *lurruwith-ngi-bu*
quick-1sgO-(lower):leg
'I run fast; I'm quick-(lower-)legged.'
d. *lurruwith-ngi-be*
quick-1sgO-arm
'I throw fast/further; I'm quick-armed.'

e. *lurruwithngima*
quick-1sgO-hand
'I drive fast; I'm quick-handed.'

f. *lurruwith-ngi-we*
quick-1sgO-head
'I've a strong head (as when carrying a weight on my head).'

(45) a. *dutmarnarri*
'crooked'

b. *dutmam-ngi-rdarri*
crooked-1sgO-back
'I'm crooked in the back.'

c. *dutmam-ngi-be*
crooked-1sgO-arm
'I'm crooked in the arm.'

Such constructions can be used as whole utterances and even be modified so as to have an irrealis sense:

(46) *burrekburrek-ngi-me-ya*
sticky-1sgO-foot-DUB
'Might be I've got something sticky on my foot.'

or be modified with adverbs (incorporated) or secondary auxiliaries like other predicates which are more obviously verbs:

(47) a. *pakpak*
'cramped'

b. *pakpakmamme*
pakpak-mam-ø-be
cramped-AUX-3sgO-arm
'His arm is cramped.'

c. *pakpakmamngibengurran*
pakpak-mam-ngi-be-ngurran
cramped-AUX-1sgO-arm-AUX_2
'My arm is cramped.'

d. *pakpakmamngibedangathangurran*
pakpak-mam-ngi-be-dangatha-ngurran
cramped-AUX-1sgO-arm-still(ADV)-AUX2
'My arm is still cramped.'

e. *pakpakmamngibedeyidangurran*
pakpak-mam-ngi-be-deyida-ngurran
cramped-AUX-1sgO-arm-again(ADV)-AUX_2
'My arm is cramped again.'

Some combinations are much more limited presumably because of limited semantic compatibility:

(48) a. *mutmutthe*
mutmut-ø-ye
deaf-3sgO-ear
'deaf'

b. *mutmut-ngi-ye*
deaf-1sgO-ear
'I'm deaf/inattentive to sound'

(49) a. *dhepdhep*
'greedy'

b. *dhepdhepwunkudhayinintha*
dhepdhep-wunku-dhayi-nintha
greedy-3du/pcO-lip-du.m
'They two men are greedy.'

The Murrinh-Patha word for 'policeman' relies on this construction type, as illustrated by (50):

(50) a. *birlmam*
'shining'

b. *birl-birl-mam-ngka* *pilinh*
shine-shine-AUX-eye/face star
'shining star'

c. *ku* *birl-mam-ø-lu*
NC:meat shine-AUX-3sgO-forehead
'policeman [referring to something shining on his forehead]'

(Compare *nanthi birlmalu* 'medal (on hat)'.) Here the noun is assigned to the *ku*-class because policemen were exclusively, and still usually are, non-Aboriginal people (see section 3). Thus *birlmalu* derives from the distinctive (and usually shiny) badge featuring a kangaroo that appears on the hats worn by police in the Northern Territory of Australia.

Any of these expressions can also function as nominals like participial expressions or relative clauses in English:

(51) *pangki-ngi-ngka-re* *ngaliwe-nhi-ma-ø*
long-1sgO-face-ERG short-1sgO-hand(s)-ABS
mam-nhi-purl
1sgS-2sgO-wash
'I, who am long-faced, washed you, who are short-fingered.'

Perhaps these incorporated expressions perform a useful function in the language in keeping mere bodily descriptions separate from cursing or teasing which can focus on purported physical deformities coded by free form nouns (cf. Crowley's analysis (1995) of free form alienable constructions in Paamese with a similar function):

(52) *nhinhi-ka pelpith pangkuy*
2sg-TOP head long
'You have got a long head!!' (taunt)

(53) *pangki-nhi-we*
long-2sgO-head
'You are long-headed.' (statement of fact)

(54) *nhinhi-ka thimu pangkuy*
2sg-TOP nose long
'You have a long nose!!' (taunt)

This insult gains added force because of the double entendre on *thimu*, which can also mean 'clitoris'.

(55) *pangki-nhi-yi*
long-2sgO-nose
'You are long-nosed.' (statement of fact)

The harmlessness of the 'compacted body-part descriptions' is evidenced by the fact that such expressions can be used as personal names. For example, one woman is called *ngadadama* 'swollen/sore-ankled'.

Some of these adjectival expressions appear to have yielded 'frozen' nominal expressions which contain incorporated body parts:

(56) a. *pekpek*
'soft'
b. *pekpekpe*
'comb'

from:

c. *pekpek-we*
'soft-head hair'

(57) a. *pangkuy*
'long, tall'
b. *pangkima*
'sugarbag species'

from:

c. *pangki-ma*
long-finger

(58) a. *pangkuy*
'long, tall'
b. *pangkidhayi*
'long knife'

from:

c. *pangki-dhayi*
long-lip

(59) a. *werlpen*
'flat'
b. *werlpenthayi*
'chisel'

from:

c. *werlpen-dhayi*
flat-lip

4.2.3. Inflected nouns (with verbal affixes)

Some nouns occur with verbal affixes (see section 2), that is, with cross-referencing bound pronouns and incorporated body parts.

(60) *murrinh-nhi-dhayi*
language-2sg-mouth/lip
'You're a chatterbox.'

(61) *nanthi-ø-dhayi*
NC:thing-3sg-lip/mouth
'acquisitive'

(62) *nanthi-nhi-dhayi*
NC:thing-2sg-lip/mouth
'You want everything.'

(63) *ku-ø-ngka*
meat-3sgO-eye/face
'hunter'

5. The continuum from concrete to metaphorical

It appears that body part metaphors have a central role in the grammar of Murrinh-Patha. Moreover verbs are central in that texts consist largely of complex verb forms. Often verb roots contain an incorporated body part or even consist entirely of an incorporated body part form – e.g. *manthi* 'carry over one's shoulder'. It will become clear that there are many verb roots which involve an incorporated body part while others look like they might have historically. The confidence one may have about identifying a particular form as an incorporated body part form will vary.

5.1. Incorporated body part forms as verbal affixes

The most unequivocal cases appear in frames like the following:

(64) *mam-nhi-purl*
1sgS-2sgO-wash
'I washed you.'

with incorporated body parts:

(65) *mam-nhi-ma-purl*
1sgS-2sgO-hand-wash
'I washed your hand(s).'

(66) *mam-nhi-rdarri-purl*
1sgS-2sgO-back-wash
'I washed your back.'

(67) *mam-nhi-me-purl*
1sgS-2sgO-foot-wash
'I washed your foot/feet.'

(68) *mam-nhi-ye-purl*
1sgS-2sgO-ear-wash
'I washed your ear(s).'

where a substitution set of incorporated body part forms with each expression is readily segmentable as shown.

This kind of substitutability extends to incorporated body part forms used to refer to spatial orientation. In Murrinh-Patha there is a general purpose locational preposition *ngarra* which can be construed more specifically by contextual cues or be made more specific by using an incorporated body part term to indicate spatial orientation:

(69) *lithpurr thu-pak ngarra palyirr*
axe 2sgS-put LOC stone
'Put your axe under/on top of/beside the stone.'

(70) *lithpurr thu-rdarri-pak ngarra palyirr*
axe 2sgS-back-put LOC stone
'Put your axe on top of the stone.'

(71) *lithpurr thu-ngurru-pak ngarra palyirr*
axe 2sgS-side-put LOC stone
'Put your axe beside the stone.'

And there are other constructions which do not employ an incorporated body part term:

(72) *lithpurr thu-pak ngarra palyirr pepe*
axe 2sgS-put LOC stone below
'Put your axe under the stone.'

The next set of examples illustrates the imprecision inherent in glossing body part terms from another language:

(73) *bangam-manthi-rtal*
1sgS-collarbone-cut
'I cut the top off something [with an axe/knife].'

(74) *bangam-ri-rtal* (< *bangarntirtal*)
1sgS-arse-cut
'I cut the bottom off something [with an axe/knife].'

(75) *bangam-me-rtal*
1sgS-foot-cut
'I cut the leg off something (e.g. chair) [with an axe/knife].'

5.2. Incorporated body part forms as verb roots

A number of verb roots are identical in form to incorporated body parts and there is shared meaning. Consider the verb class 21 root *ma* 'make string designs with hands, pass on/along':

(76) a. *nanthi perrene-ma-dha-neme-pirrine*
NC: thing 3pcS-pass:along-IMPF-pc.masc-AUX_2
'They were passing the thing on/along.'

b. *murrinh wangu ne-na-ma-nu*
word towards 2plS-3sg.m.BEN-pass-FUT
nukunu-nu
3sg.m.-DAT
'Pass the word along to him.'

c. *nanthi pi dem-ma-dim*
NC: thing string 3sgS-make-AUX_2
'She is making a string design.'

The incorporated body part form, *ma* 'hand', is identical to this verb root and the two senses of the verb, 'make string designs with hands' and 'pass on/along', relate semantically to the body part involved quite transparently. Other such roots include the VC 24 *manthi* 'carry over one's shoulder' (cf. the incorporated body part *manthi* 'collarbone') and the VC 8 *ngka* 'to gaze, to look on' (cf. the incorporated body part *ngka* 'eye/face'):

(77) *nguy-manthi-nu*
1sgS-carry:over:shoulder-FUT
'I will carry it over my shoulder.'

(78) *kangkarl ma-ngka-nu*
above 1sgS-gaze-FUT
'I will gaze above.'

In (77) and (78) the equation of an incorporated body part with the verb root is relatively easy to motivate. The semantic range of the incorporated body part, *manthi*, closely ties in with the area of the body involved in the meaning of the verb root, *manthi*. Similarly, the sense of the verb root, *ngka* 'gaze', dovetails with the semantic range of the incorporated body part, *ngka* 'eye/face'. Rather less obvious is *ngurru* 'to provoke' (VC 19?) (cf. the incorporated body part *ngurru* 'side'):

(79) *dam-ngi-ngurru*
3sgS-1sgO-provoke
'He provoked me.'

An added complication arises in that a given verb root may belong to a number of separate verb classes with a resulting modification in meaning (see Walsh 1989, for further details). In examples (80)-(84) a single verb root, *warnta*, appears in four different verb classes, each occurrence being signalled by the change in form of the initial portmanteau morpheme (AUX). Consider, for example:

VC 1 *warnta* 'to split open (intransitive)' (cf. *parnta* 'groin')

(80) *ki-warnta-nukun*
3sgS-split-IRREALIS
'It might split open.'

VC 14 *warnta* 'to split open (with a stone or axe)' (transitive)

(81) a. *ba-warnta-nu*
1sgS-split-FUT
'I'll split it open.'

b. *bangam-warnta* (> *bangamparnta*)
1sgS-split
'I split it open.'

VC 17 *warnta* 'to split open (by hitting down hard)' (transitive)

(82) a. *bu-warnta-nu*
1sgS-split-FUT
'I'll split it open'

b. *ban-warntarta-wurran* (> *banparntartawurran*)
3sgS-split REDUP-AUX_2
'He continually splits it open.'

VC 19 *warnta* 'to split open (with a gun)' (transitive)

(83) a. *nga-warnta-nu*
1sgS-split-FUT
'I'll split it open.'

b. *ngam-warnta* (> *ngamparnta*)
1sgS-split
'I split it open.'

VC 23 *warnta* 'to split open (with a stick)' (transitive)

(84) a. *ngu-warnta-nu*
1sgS-split-FUT
'I'll split it open'

b. *pan-warnta* (> *panparnta*)
1sgS-split
'I split it open'

where the core meaning, 'split open', is shared, being a reflection of one aspect of the body part term, *parnta*: the major bifurcation of the body.

When we turn to *ngka* the connection with the concrete body part sense 'eye/face' is less obvious:

VC 10 *ngka* ‘to be ready to go or to take off (only used of birds, animals, things)’

(85) *tharntu nhini me-ngka-dha-pirri-dha ngamengka-nu*
boat that 3sgS-ready-IMPF-AUX_2 foreign-DAT
wangu
to
‘That boat was standing there ready to go to a foreign place.’

VC 19 *ngka* ‘to scrounge for scraps of food’

(86) *mi nga-ngka-nu*
NC: food 1sgS-scrounge-FUT
‘I will scrounge for food.’

VC 28 *ngka* ‘to watch over, to guard (to police someone)’

(87) *ngirra-ngka-nu*
1sgS-watch:over-FUT
‘I’ll watch over it.’

VC 8 *bath* + *ngka* ‘to wait for’

(88) *bath ma-ngka-nu*
PREVB 1sgS-wait:for-FUT
‘I’ll wait for.’

Probably the most obvious connection here, at least from the perspective of English, is between ‘eye’ and the two senses given in (87) and (88) which have near synonyms in English: keep an *eye* on; keep an *eye* out for.

Again in the case of *marda* ‘belly, stomach, chest’ some senses are quite obvious while others are not so clear:

marda:	VC 14	‘to give’
marda:	VC 15	‘to give oneself (reflex.)’
marda:	VC 22	‘to hold/take to one’s chest’
marda:	VC 23	‘to skirt, to move along the edge; to request’
marda:	VC 24	‘to rejoice (intransitive)’
marda:	VC 19/23	‘to rejoice (transitive)’

The last two uses, connected with rejoicing, relate to the Murrinh-Patha view that *marda* ‘belly’ is the seat of the emotions. When *marda* occurs in VC 22, which imparts the general idea of having and holding, it refers to *mardayidi* 'chest'. When *marda* appears in VC 23 'to skirt, to move along the edge' the container sense of *marda* is probably being invoked. I have no convincing explanation for the two senses which involve 'giving'.

Another puzzle is the verb root, *ye*, which is identical to the incorporated body part, *ye* 'ear', which also relates to knowledge and intelligence as evidenced by expressions such as *thema/thama* 'person who knows everything'. The next two examples involve the body part appearing as a preverb, forming a single syntactic unit with the verb:

PREVERB (*the*) +VC 22 'know' [literally: to have ear]

(89) *the* *nganthin-pun*
PREVB 1sgS:have-3plO
'I know them(pl).'

PREVERB (*the*) +VC 8 'to know (literally 'to take ear'), to understand, to remember'

(90) a. *the* *ma-bath-nu*
PREVB 1sgS-take-FUT
'I will know.'
b. *the* *ma-nhi-ma-bath-nu*
PREVB 1sgS-2sgO-from-take-FUT
'I'll know your(sg) actions/ways.'

There seems to be no immediately apparent source for the sense of the VC 8 *ye* 'to discipline by talking (for wrong that has been done)'

More transparent is the connection between the verb roots, *thirri* and *yirri* (glossed immediately below), and the body part term *thirrimeme* (incorporated form *-yirri-*) roughly glossed as 'navel':

VC 10 *thirri* 'to wear male pubic covering'

where the form, *thirrimeme*, appears elsewhere to refer not just to 'navel' but also to the area near it:

VC 19 *yirri* 'to put something in someone's belt' VC 1 *yirri* 'to put something in one's own belt'

Another related form, this time a noun, is *thirriwup* (NC: *ku*) 'male pubic covering', which is made from *thay palathi* or *thay yipi* (types of trees)' where the form, *wup*, is transparently allied to the verb root, *wup*, 'sit down'.

There is also a euphemistic expression involving *thirrimeme* which suggests that the term involves the area *around* the navel:

(91) *thirrimeme ngurru-nu*
'navel' 1sg:move-FUT
'I'll go to the toilet.'

(Cf. *ngay ngirrayerryerrnu* 'I'll urinate' [less polite].)

Recalling examples (81)-(83) where an incorporated body part term is used to indicate spatial orientation, we have:

(92) *ba-yirri-rtal-nu*
1sgS-navel-cut-FUT
'I'll cut off one branch 'halfway' [up the tree],' 'I'll cut a piece out of [biscuit].'

where one sense of *yirri* relates to spatial orientation and the other, presumably, connects with the shape of the excision (cf. 'I'll cut the *eye* out of the potato').

5.3. Incorporated body part forms in compound verb roots

There are numerous verb roots in Murrinh-Patha in which a portion of the verb root bears a resemblance or is identical to an incorporated body part form. First a few clear examples are presented:

- incorporated body part *me* 'foot':
  'to *trip* another' *ME*NGKAT
  'to be knocked off one's *feet*' (by waves or strong current) *ME*THARDAY
- incorporated body part *ye* 'ear':
  'to *hear*, to *listen*' *YE*PUP (VC 16)
  'to *hear* about' *YE*PUP (VC 24)

(Here the different senses of the verb root, *yepup*, depend on the occurrence of that verb root in a particular verb class.)

- incorporated body part *yi* 'nose':
  'to *smell*' *YI*NGAP:
  'to drown, to submerge [up to the *nose*]' *YI*WIK:

These forms are all fairly transparent. The first two involve *me* 'foot' and refer to feet as part of the meaning of the verb. The next two have *ye* 'ear' in combination with *pup* which refers to certain kinds of stance. The last pair include *yi* 'nose' and while the first of this pair 'to smell' obviously relates to 'nose' the second 'to drown, to submerge' is not so obvious. However this gloss can be 'unpacked' in such a way as to make the connection with 'nose' explicit. Consider the following:

(93) *wakay dam-ngi-yi-wik*
finished 3sgS-1sgO-nose-'submerge'
'[I'm] finished; [water] is up to my nose.'

(94) *dam-ngi-ngka-wik*
3sgS-1sgO-eye-'submerge'
'Water is up to my eyes.'

(95) *dam-ngi-we-wik*
3sgS-1sgO-head-'submerge'
'Water is up to my head.'

It now becomes clear that *yiwik* actually includes 'nose' as part of its meaning and that other incorporated body part terms like *ngka* 'eye/face' and *we* 'head' can contrast with it paradigmatically. So the use of the body part 'nose' reflects that this level is crucial for survival (as the water rises!). I do not believe the pattern is productive in the way that (virtually) any body part can be combined with *purl* 'wash', as in examples (64)-(68). It is reminiscent of the usage in English:

(96) *I'm up to my nose in work.*

which allows some other variants:

(97) *I'm up to my ears in work.*
I'm up to my eyes in work.
I'm up to my neck in work.

but scarcely:

(98) *????I'm up to my ankles in work.*
???I'm up to my knees in work.
??I'm up to my waist in work.
?I'm up to my armpits in work.

That is to say, the force of these expressions depends on a (culturally conceived) crucial level in the body. In English I believe acceptability improves the higher one goes.

The incorporated body part form (or something homophonous with it) need not be the first element in the verb root. For example, *thurrulk-thay* and *dharryirr* both translatable as 'foam at the mouth'; they differ in that the first can be analysed as *thurrulk-thay* 'foam-lip/mouth', where the incorporated body part form, *dhay/thay*, occurs at the end of the verb root. The other verb root, *dharryirr*, has no readily separable incorporated body part form although the initial sequence, *dha*, is suggestive of the initial sequences in these semantically compatible incorporated body part forms:

thamamay	*dhamurru*	'beard'
thamarl	*dhamurru*	'side of face'
thathpi	*dhawi*	'mouth/lips'
thathpi	*dhayi*	'mouth/lips'
themen	*dhanme*	'tongue'

The following two verb roots contain the element, *ka*, which is presumably related to the incorporated body part form, *ngka* 'eye': *minminka* 'to blink one's (*eyes*)', and *thapka* 'to be glamorous, to be pleasing to look upon' – the latter refers to *face* only (cf. *thapngini* – for body, where the incorporated body part for 'body' is *ngini*).

In addition there are verb roots *within* which it is tempting to see an incorporated body part form. Perhaps the most clear-cut of these are *winhimardapak* (VC 19) 'to pour into' and *winhimardapak* (VC 21) 'to pour into (intransitive)', which can be compared to *winhipak* (VC 19) 'to spill (transitive)' and *winhipak* (VC 21) 'to spill (intransitive)'.

Given that *marda* is often used with the sense of 'container', as in the following example:

(99) *bamardakangararanu* 'I'll hollow out [a log for a canoe]'
bamardakatnu 'I'll cut open [a drum]'

it seems reasonable to analyse *winhimardapak* 'pour into' as containing the incorporated body part form, *marda* 'belly'. Less obvious are *winhimekpak* (VC 19) 'to delay another' and *winhimekpak* (VC 21) 'to delay (intransitive)', which might be viewed as containing the incorporated body part form, *me*.

The verb root *winhimune* 'to recline, leaning on one's elbow' could well relate to the incorporated body part form, *bu/mu* 'thigh/upper leg'.

Moreover one might speculate that the incorporated body part form, *ngka* 'face/eye' is contained in *winhingkatthat* 'to imitate, mimic (e.g. used for driving a car, riding a horse etc.)' and *winhingkayit* 'to instruct'.

5.4. Two incorporated body part forms in a verb root

Although fairly limited some verb roots suggest that two body part forms have been incorporated: *mardangkardu* 'to know another's thoughts (mind)' (literally, 'to see one's belly, the seat of emotions') involves the root, *mardangkardu*, which might be analysed as *marda-ngkardu* 'belly/chest-see', with *ngkardu* itself being a compound involving *ngka* 'eye/face'.

More tenuous is this verb root *ringkapurl* 'moon rising, last quarter (around midnight)', in which *ringkapurl* might be analysed as *ri-ngka-purl* 'buttocks-eye/face-wash'. Such an analysis may seem far-fetched at first blush but becomes more plausible when (100) is considered:

(100) *(da)* *pa-ngka-purl* *warda-ya*
(NC: time/place) 3sgS-eye-wash now-DUB
'dawn'

If dawn is conceived as as being prior to moon-rise then the posteriority of moon-rise with respect to dawn is aptly expressed by the addition of *ri* 'buttocks'.

In *wengkawuy* 'to be confused' the body part *we-* (head) VC 23/VC 30 has been used, it being the most common form. However the following body parts can also be used, in each case slightly changing the shade of meaning: *ye-* 'ear', *dhawi-* 'mouth/speech', *marda-* 'belly', *rdarri-* 'back', and *me-* 'foot'. So the first portion of the verb root amounts to some kind of substitution set of incorporated body parts while the next part may be seen as the incorporated body part for 'eye/face': *we-ngka-wuy*. Consider also *weyi* (VC 13) 'make quiet (as, of child)', where the verb root, *weyi*, might be seen as *-we-* 'head' + *-yi-* 'nose/breast', as head to breast, and *yengkarnirn~ngkarnirn* (VC 15) 'to think', where it has already been noted in connection with (97) that *ye* is associated with knowledge, intelligence and the like and for this verb root may be added onto the simpler verb root, *ngkarnirn*, which itself might break up into *ngka* 'eye/face' plus an element, *rnirn* (so far unanalysable). Finally, consider *yengkawerr* (VC 23) 'to be terrified from what is *seen* and *heard*', where the seeing and hearing may be directly encoded into the verb root as *ye* 'ear' and *ngka* 'eye'.

5.5. The continuum from concrete to metaphorical in detail

Once the more obvious connections between verb roots and incorporated body part forms have been observed it becomes tempting to look for more tenuous connections. In some instances it will not be clear whether there is a connection but one just cannot see it or that there is simply no connection at all. Connections may be undetectable because the current forms are merely a relic of a previously productive (and transparent) pattern. Otherwise the connection might not be noticed because it derives from the metaphorical resources of the language and may be beyond the limits of awareness even of native speakers. However by looking at a range of examples, some of the more obvious connections can be seen along the continuum from concrete to metaphorical.

(i) *-rdarri-* 'back'. The most concrete use of the incorporated body part obviously relates to the body part itself as in *rdarriraptap* 'to thump someone's back (to remove congestion etc.)'. Slightly less concrete senses relate to the shape of the back or the orientation of the back with regard to the rest of the body. As an unincorporated form, *pangkin* 'back', occurs in such expressions as *pangkin* (NC: *da*) 'ridge' and *mayern pangkin wangu* 'overland road'. As an incorporated body part *rdarri* 'back' appears in forms like:

(102) *bangam-nhi-rdarri-gurduk*
1sgS-2sgO-back-drink
'I'm drinking behind your back.'

Otherwise there are a number of metaphorical extensions:

(103) *nga-rdarri-riwak-nu*
1sgS-back-follow-FUT
'I'll imitate him.'

(104) *dam-nga-rdarri-mit* *kardu-dha*
3sgS-1sgBEN-back-keep NC:person-?from
'Mob sits with me and keeps me company.'

(105) *dam-nga-rdarri-mit* *mi-dha*
3sgS-1sgBEN-back-keep NC:food-?from
'There's a lot of food for me/I have had a lot of food.'

(106) *ngem-luk-rdarri*
1sgS-??-back
'I chuckled to myself.'

Some of the metaphorical themes raised here have direct counterparts in English metaphorical extensions of 'back':

BEHINDNESS	*Biloela is the back of beyond; I'll wait back here*
TEMPORAL ANTERIORITY	*Prices were low back then*
PLENTY	*He has the backing of millions; We've got orders backed up for months*
SOLIDARITY	*You have to back your mates*

Significantly the verb root in (106) refers to chuckling about events that took place in the past and thus invokes the metaphorical theme of temporal anteriority.

Many verb roots can be classified according to these 'metaphorical themes' while a few must at this stage be relegated to a residue category which awaits explanation or is simply a case of chance resemblance:

BEHINDNESS
rdarri 'to hide(behind)'
rdarriyet 'to talk behind someone's back'
rdarrup (VC 19) 'to cover (with blanket, etc.)'
rdarrup (VC 21) 'to cover (reflexive) (with blanket, etc.)'
winhirdarriparl 'to deviate, change direction'
rdarrirereth 'to sense someone's presence behind'
rdarririwak 'to imitate' [i.e. 'to follow behind']
rdarriral 'to go too far (regarding anything), to overstep'
rdarrithuk 'to avoid; keep away'

TEMPORAL ANTERIORITY
rdarrirda 'to report, confess, reveal to someone something one already knew'
lukdarri 'to chuckle to oneself when thinking about previous funny event'

PLENTY
rdarrirdurdu (VC 8) 'to heap up'
rdarrirdurdu (VC 10) 'to heap up on oneself'

SOLIDARITY
rdarrimit 'to keep company'
rdarrirtirt 'to observe, to keep a lookout'
rdarrithak 'to amuse oneself'
rdarrirereth 'to sense someone's presence behind'
rdarririwak 'to imitate' [i.e. 'to follow behind']

RESIDUE
rdarrimepu 'to be overcome with awe, to be greatly astonished, to be spellbound, to be preoccupied'
murrinh rdarrithak 'to slander another' (*murrinh* 'language, speech')
rdarrimirrkat 'to be burdened with many things on one's mind'

Verbs in the residue category all relate to some kind of mental state but it is not clear how the incorporated body part, *rdarri*, contributes to these senses.

(ii) *-dha/dhawi/dhay-* 'lip, mouth'. At least three metaphorical themes can be identified for the incorporated body part set, *-dha/dhawi/dhay-*:

EDGE
LANGUAGE/SPEECH
OPENING

Each of these has counterparts in English where the lip is associated with an edge (lip of a drum), the mouth with an opening (mouth of a cave/gun etc.) and with language or speech. (English also associates the tongue with language and speech whereas Murrinh-Patha appears not to.) In addition, Murrinh-Patha has some examples of the incorporated body part for 'tongue' being used for its shape as in *dhanmeparl* 'to break a sharp pointed object' (cf. *parl* 'fall').

EDGE
dhawirlerr 'to join ends together'

LANGUAGE/SPEECH
dhap (VC 8) 'to close one's mouth, to be quiet, to shut up'
dhap (VC 17/VC 14) 'to stop another talking by giving them something'
dhapep 'to stop another talking, by words'

dharrpu 'to ask'
dhawirtirt 'to watch one's mouth (e.g. for how to pronounce a word, etc.)'
dhaykay 'to yell out'

OPENING

(107) a. *nganthawidum* 'I enclose object with my hand.'
b. *nganthawidum* 'I put [candle in bottle].'
c. *nganthawidum* 'I put [cork in bottle].'
d. *nganthawidhap* 'I put [the top on pot].'
e. *nganthawimnum* 'I keep pushing [tobacco into pipe].'

(iii) *-ma-* 'finger/hand'. The metaphorical themes associated with *-ma-* 'finger/hand' are fairly predictable:

ACQUISITION
CONTROL
EXCHANGE
FRONTNESS

The latter sense presumably derives from the fact that the hands are more often than not at the front of the body. Just a few examples should suffice:

FRONTNESS

(108) *da-ngi-ma-gurduk* *(kumparra)*
2sgS-1sgO-hand-drink (first)
'Drink in front of me!'

EXCHANGE

(109) *ngerramma*
'We plural exclusive exchange with each other.'

ACQUISITION, CONTROL

(110) *kardu* *wakal* *mangan-ma-art*
NC:person baby 1sgS-hand-take
'I passed over the baby.'

(iv) *-marda-* 'belly/chest'. At least these three metaphorical themes can be identified for *-marda-* 'belly/chest':

CONTAINER
EMOTION
FRONTNESS

One of the clearest cases of the distinction between the sense of *marda* for EMOTION and as a CONTAINER is found in the following examples, the first of which uses *marda* as a pre-verbal element (see section 6.4):

(111) *marda ma-art-nu*
belly 1sgS-take-FUT
'I will desire [it/him/etc.].'

(112) *na-marda-art*
2sgS-belly-take
'Get it out [of a container, such as a billy, cardcase, etc.].'

CONTAINER
mardardu 'to be puffed out, to be out of breath'
mardarl 'to plough, to grade with an instrument, either dirt or water (i.e. boat)'
mardapup 'to lay (something) down in an object or in the middle of an object'
mardarartal 'to have a stomach ache'
mardarr 'to save up, to hoard'
mardawal 'to move right through the middle (of place, people); to tell another off; to reprove' (cf. *wal* 'to jump; strike')
mardayarrarr 'to have hunger pains'
mardawurl 'to fill a container'
mardap 'to hop into something (e.g. truck)'

EMOTION
mardabay 'to be disappointed with something, to be disillusioned'
mardabi 'to meet'
mardarde 'to know another's thoughts (mind)' (cf. *rde* 'to hit')
mardakat 'to be angry'
mardarlart 'to be angry' [Literally: 'to be bitten in the belly'].
mardamarda 'to wait for'
mardan 'to be satisfied'
mardangkardu 'to know another's thoughts (mind)' [Literally: 'to see one's belly (the seat of emotions)' (cf. *ngkardu* 'to see')]

mardangurrk ‘to mumble’
mardarraki ‘to make angry by talking’
mardat (VC 8) ‘to amaze another (transitive)’
mardat (VC 18) ‘to be amazed (intransitive), to marvel’
mardarta ‘to be upset, to be troubled’ (cf. *rta* ‘to catch’)

FRONTNESS
mardabi ‘to meet’
mardalili ‘to sway’ (cf. *lili* ‘to walk’)
mardamarda ‘to wait for’

(v) *-ngka-* ‘eye/face’. The incorporated body part, *-ngka-* ‘eye/face’, presents the greatest problems for analysis in part because the greatest number of Murrinh-Patha verb roots begin with the sequence, *-ngka-*. Tentatively the following metaphorical themes are proposed:

EMOTION
FRONTNESS
ROUNDNESS
VISION

VISION is probably the most obvious. EMOTION is problematic in that my intuitions about Murrinh-Patha suggest to me that *-ngka-* ‘eye/face’ relates to a different emotional range than *marda* ‘belly’ but I do not know how to characterise the difference. FRONTNESS is fairly straightforward particularly when the ‘face’ sense is foremost since it mirrors the metaphorical extension to be found in English. More problematic is ROUNDNESS which relates to the shape of the eye. In Murrinh-Patha the extension relates to anything round whether it be flat, convex like a bead or concave like a depression. Overlaid on or in some way connected with this sense, there may be some idea of selection/choice which in turn connects with the other metaphorical themes of VISION and EMOTION. Again there are counterparts in English in that one primary method for reaching a selection is through sight, which can then be metaphorically extended, as in: *What caught your eye? What have you been eyeing off?* Some of the rather large residue category would be captured with an additional metaphorical theme of SELECTION. Even some of those verb roots already assigned could well fit in this category.

EMOTION
ngkabat ‘to be surprised, to surprise’ (cf. *bat* ‘to strike’)
ngkamardathin ‘to frown’ (cf. *mardathin* ‘to be sorry’)

ngkawarrang 'to be silly, to be foolish'
ngkawerr 'to be terrified from what is seen'
ngkawuy (VC 29) 'to have many things on one's mind'

FRONTNESS
ngkadhath 'to face something towards someone'
ngkayi 'to lead another to a place they don't know, to guide'
ngkayith 'to accompany'

ROUNDNESS
ngkardal (VC 30) 'to lay one's head on a pillow'
ngkardal (VC 35) 'to balance people on a floating log [*thay yempi*] in the water, so the front is a little higher than the back'
ngkapurl 'to sharpen' [Possibly this involves an incorporated body part -*ngka*- 'eye' plus -*purl* 'wash'.]
ngkaputh 'to dig out sand in order to get water'
ngkay (VC 14) 'to collect, to pick up nerite shells from the water'
ngkay (VC 19) 'to poke/probe in a hole using an instrument'
ngkarderde 'to burst pimples' (cf. *rde* 'to hit')
ngkarart 'to choose (more than one item), to select (more than one item)'
ngkare 'to carry fish on a forked stick'

VISION
ngkabath (VC 19) 'to aim a spear at'
ngkabath (VC 21) 'to aim a gun at'
ngkabing 'to lift an eyelid, to turn over a page'
ngkarda 'to point out'
ngkardarrath 'to lose track of, to misplace'
ngkardarrup (VC 19?) 'to cover one's face'
ngkardarrup (VC 21) 'to cover one's face (reflexive)'
ngkardith 'to wink'
ngkardu (VC 13) 'to see, to look'
ngkardu (VC 15) 'to look, to see oneself (reflexive)'
ngkarnerr 'to look through cupped hand or cylinder'
ngkanhinhnhinh 'to be picturesque to one's eyes, to have itchy eyes'
ngkangath 'to play hide-and-seek'
ngkapi 'to not recognise, to be ignorant of'
ngkamumuy 'to be blind'
ngkamut (VC 19?) 'to miss writing something down'
ngkamut (VC 23) 'to miss event or person'
ngkarr (VC 19) 'to show'

ngkarr (VC 8) 'to be dizzy'
ngkatum 'to have a vision, trance (with one's eyes open)' (cf. *tum* 'to dry')
ngkawerr 'to be terrified from what is seen'
walngka 'to keep one's eyes open'
ngingkalak 'to be good-looking'
ngkamurruwul 'to be good-looking'
ngkamurruwulngka 'to be good-looking'
ngkawip 'to shine (of moon or torch)'
ngkawurl 'to have a headache'
ngkawuy (VC 23) 'to have blurred vision'
ngkarnern 'to have a headache, in the temple area'

RESIDUE
ngkabang 'to recede, (fresh or salt-water) until completely gone'
ngkardap (VC 8) 'to steal, to cheat'
ngkardap (VC 23) 'to sprinkle fine matter (e.g. salt, dirt – but not sand, sugar)'
ngkardarrarart 'to steal'
ngkardurr (VC 32) 'to leave/depart without another'
ngkardurr (VC 33) 'to be left by another'
ngkadhap 'to sin'
ngkala 'to climb onto' (cf. *la* 'to climb up')
ngkale (VC 8?) 'to hang up'
ngkale (VC 10) 'to hang oneself (reflexive)'
ngkalil 'to fan a fire (with something)'
ngkarl 'to return/replace (transitive), to put back (transitive)'
ngkarlay (VC 8) 'to wave'
ngkarlay (VC 10) 'to wave (reflexive)'
ngkamit (VC 8) 'to be blocked off, to be obstructed from moving, to be hindered from moving (by river, people, fire, tide, etc.)'
ngkamit (VC 19) 'to refuse, to say no'
ngkan (VC 1) 'to extinguish by itself (reflexive) (e.g. of fire)'
ngkan (VC 8) 'to extinguish fire or light (by hand or sand)'
ngkan (VC 23?) 'to extinguish fire (with branch, etc.)'
ngkarara 'to clean up water debris, by the dry season tides'
ngkarel 'to expel, to send off because of bad behaviour'
ngkathan (VC 8) 'to tame; to domesticate, by hand (e.g. dog)'
ngkathan (VC 17) 'to tame, to break in, by saddle etc. (e.g. horse)'
ngkathan (VC 23) 'to tame, to break in, by whip, stick etc. (e.g. horse)'

ngkathan (VC 19) 'to learn'
ngkawadha (VC 8) 'to call by name, to say one's name'
ngkayeth 'to elope with a (married or single) woman (with her consent), to hit lightly (e.g. with a cane, whip)'

(vi) *-ri-* 'buttocks'. The two most immediate metaphorical themes are:

BEHINDNESS
UNDESIRABILITY

exemplified in English expressions such as *He's a pain in the arse* or *Grab the arse end of it!*.

BEHINDNESS
rirda (VC 15) 'to back off, to move backwards'
rikat (VC 14/VC 23) 'to circuit, to go around, to avoid, to bypass danger (etc.)'
rimut 'to leave over' [cf. *mut* 'give']
ripak 'to catch up to another'
ripurl (VC 8) 'to be behind, to be late (in time)' [also means 'to wash another's buttocks']
riwak 'to follow'

UNDESIRABILITY
ribay 'to commit adultery' (cf. *bay* 'to be sated, to be full of food')
rikat (VC 14/VC 23) 'to circuit, to go around, to avoid, to bypass danger (etc.)'
rirda (VC 23) 'to blame another, to accuse another'
rirda (VC 24) 'to blame oneself, to accuse oneself'
ripurl (VC 8) 'to be behind, to be late (in time)' [also means 'to wash another's buttocks']
ripurl (VC 26) 'to wipe clean another's or one's own bottom'
thithimarnti [i.e. *thithimam+ri*] 'bitter (for food and water), sour/salty/spicy' [of chillies, brinjal pickles, very strong coffee, curry, etc.]' (cf. *thithimampe* [i.e. *thithimam+we]* 'fan fern')
wertmarnti [*wertmam+ri*] 'greedy'

But this leaves a substantial residue category only a few of which can be explained at this stage. For instance, it is probable that *rikat* 'to cut' refers specifically to cutting the end of something and so could be related to the incorporated body part *-ri-* 'buttocks'.

RESIDUE
ribap 'to aim a small spear at'
rirda (VC 7) 'to push with foot, to kick'
rirda (VC 8) 'to push with hands'
rirda (VC 14) 'to push over by one's body or by means of an implement, to knock over by one's body, or by means of an implement (e.g. stone, or large spear)'
rirda (VC 19) 'to push by means of an implement (e.g. truck)'
rirda (VC 23) 'to push over by means of an implement, to knock over by means of an implement (e.g. stick)'
ridharl (VC 23?) 'to slip walking, to go/move without stopping'
ridharl (VC 8) 'to slip (i.e. one's hand)'
riguluk 'to tickle on the chest'
rikat (VC 23) 'to cut, with a knife (e.g. bread)'
rikat (VC 19) 'to bite in half'
rilil 'to write, to strike (as of matches)'
ringeng 'to talk (transitive)' (used also for 'to play guitar, to make an engine go, etc.')
ringkarlpurl 'moon rising, laster quarter (around midnight)'
ringyung 'to blow bubbles under water'
riparl (VC 19) 'to praise another' (cf. *parl* 'to fall')
riparl (VC 21) 'to praise oneself'
rirta (VC 1) 'to puncture (intransitive), to pierce (intransitive)' (cf. *rta* 'to catch')
rirta (VC 8) 'to puncture by hand, to pierce by hand'
rirta (VC 7) 'to puncture by foot, to pierce by foot'
rirturt 'to light/ignite dry grass, leaves, paper, etc.'
ritharrame 'to speed up an object (e.g. clapsticks, a child, etc.)'
riwu 'to count, to admire, to pinpoint position where a person is'
riyith 'to explain'

(vii) *-we-* 'head (hair)'. Once again the principal metaphorical themes relate to the relative position of the body part and the perceptual or cognitive properties culturally associated with that body part. For the incorporated body part, *-we-* 'head (hair)', the two main themes seem to be:

ABOVENESS
EMOTION/THOUGHT

which have obvious counterparts in English expressions such as *He/she was the head of the class*, *Don't worry your head about it*, etc..

ABOVENESS
wengkay 'to put [handle] on '
wepek (VC 24) 'to comb hair (reflexive)'
wepek (VC 23) 'to comb hair (transitive)'
wepith 'to peel off stringy part of fanfern leaf'
wet 'to go over one's head'
wertirt 'to delouse one's hair'
weway 'to glide (as of a bird)'
wewu 'to wet, to moisten'

EMOTION/THOUGHT
werdi 'to bring shame upon, to disgrace'
werduy 'to be made cross by another's actions'
wengkawuy (VC 23) 'to confuse'
wengkawuy (VC 30) 'to be confused' [Note: the body part *we-* (head) has been used here, being the most common form, however the following body parts can also be used, in each case slightly changing the shade of meaning: *ye-* 'ear', *dhawi-* 'mouth/speech', *marda-* 'belly', *rdarri-* 'back', and *me-* 'foot'.]
werr (VC 19?) 'to be anxious (transitive), to not give opportunity, to not give one a chance'
werr (VC 21) 'to be anxious (intransitive)'
wererr 'to shake, to shiver with fear or cold'
wert (VC 19?) 'to illustrate by story (sometimes in a hidden way, e.g. parable), to allude by story'
weth 'to be skillful'
wethin 'to worry' (cf. *mardathin* 'to be worried')

RESIDUE
werdum 'to open up house of deceased ceremonially'
wekarl (VC 23) 'to roll up'
wekarl (VC 24) 'to roll up, to curl up (reflexive)'
wekurrk 'to bequeath, to will'
werr (VC 23?) 'to gather together'
werr (VC 24) 'to gather together (reflexive)'
wetuth 'to roll (waves)'
wert (VC 7) 'to creep silently'
wertal 'to recede, of water'

(viii) *-ye-* 'ear'. Murrinh-Patha relates the ear to knowledge or intelligence (cf. English *He has the ear of the President* etc.) and also has a metaphorical extension connected with the relative position of the ear on the body:

ABOVENESS
KNOWLEDGE/INTELLIGENCE

A few examples follow:

ABOVENESS

(113) *kura kangkarl-ye* [> *kangkarlthe*]
water above-ear
'There's a little water lying on top'.

KNOWLEDGE/INTELLIGENCE
yekum 'to forget' (cf. *kum* 'to swim')
yelel 'to be bitter'[5]
yelerr 'to remind'
yerr 'to remind, to let one know, to promise'
yengkawerr 'to be terrified' (cf. *ngkawerr* 'to be terrified')
ngulyengthe/the ngulyeng 'someone does not listen'

(ix) *-yi-* 'breast/nose'. This incorporated body part has a relative body position sense of ABOVENESS in verbs such as *yirara* 'to cut branches off from top [nose] [of tree]'. There is also the verb root *yibirnu* 'to save life' (cf. *birnu* 'alive'), which might suggest the feature LIFE and might be related to the sense of *-yi-* as 'breast' with associated life-giving properties.

There is also an association of *-yi-* in the sense of 'nose' with INTELLIGENCE/SKILL (cf. *He has a good nose for business*) as in *putputthi* 'stubborn' (< *putput-yi*).

(x) *-rdi-* 'tooth'. Finally, the incorporated body part, *-rdi-* 'tooth', might be associated with the relative body position of the teeth or their shape, as in *rdirryerrl* 'to ascend a mountain (i.e. a very high object)'.

5.6. Metaphorical themes as a heuristic device

Within a language the identification of metaphorical themes, however crude initially, can suggest explanations for other parts of the lexicon. In section 6.3 a few examples of nominal compounds are provided. Some of these nominal compounds may be further analysed in terms of metaphorical themes that have evolved in the analysis of complex verbs. For example, (114) might remain no more analysed than (115) or (116).

(114) *mange mardanu*
hand ?
'palm of hand'

(115) *mange putung*
hand knuckle
'knuckle'

(116) *pingkarl kurnern*
knee cap?
'kneecap'

But one way of looking at *mange mardanu* (example (114)) is to see the second element as per (117):

(117) *marda-nu* OR *marda-nu*
belly-DAT 'CONTAINER'-DAT

In this case the composite expression is relying on the CONTAINER sense of *marda*, and is reminiscent of 'cupped hand' in English in which the palm of the hand is used as a container. What differs in English is the metaphorical strategy employed. More speculatively, one of the hitherto unexplained variants for 'clitoris' is as shown in (118a) (the other variant is given in section 1). Applying the same reasoning as for *mange mardanu* we would have the analysis shown in (118b).

(118) a. *ku thinu*
NC:meat ?
'clitoris'
b. *ku thinu*
NC:meat ?LIFE-DAT
'clitoris'

This implies that *ku thinu* 'clitoris' is seen to be associated with LIFE, given its proximity to the generative process.

6. Remarks on unincorporated body parts

6.1. Possession

Possession is not formally distinguished between alienable and inalienable in Murrinh-Patha:

(119) *palngun ngay*
woman 1sg
'my woman/wife'

(120) *mange ngay*
hand(s) 1sg
'my hand(s)'

(121) *kuragadha ngay*
boomerang 1sg
'my boomerang'

6.2. Inflected body part nouns (with nominal affixes)

6.2.1. -ma

The proprietive suffix *-ma* signals that some entity is particularly associated with the noun to which it is suffixed (see also Leeding 1995):

(122) *kardu thamul-ma*
person spear-PROP
'person associated with a spear, a good spearsman'

(123) *kardu palngun-ma*
person woman-PROP
'person associated with a woman, a married man'

(124) *palngun ma-kardu-ma*
woman NEG-person-PROP
'woman not associated with a person, an unmarried woman'

(125) *kardu ma-nanthi-ma*
person NEG-thing-PROP
'man with no things'

(126) *ma-mi-nhi-ma*
NEG-food-2sg.-PROP
'You've got no food.'

(127) *pirtpirt-ngi-ma*
fast-1sg-PROP
'I can make it fast.'

The same construction is used with body parts:

(128) *palngun ngardada-ma*
woman ankle-PROP
'woman associated with ankle(s), a woman with sore ankle(s)'

(129) *kardu ngi-ma*
person penis-PROP
'person associated with penis, a surgeon [i.e. one who performs a circumcision]'

(130) *mardinhpuy ngapurlu-ma*
girl breast(s)-PROP
'girl associated with breast(s), girl with budding breast(s)'

(131) *nanthi nithi-ma*
NC: thing arm-PROP
'thing associated with arm(s), shirt'

(132) *nanthi lawali-ma*
NC: thing thigh-PROP
'thing associated with thigh(s), trousers'

Other examples include: *kamarlma* 'blind' (eye-PROP); *pelpithma* 'silly man' (head-PROP); *thimuma* 'sore nosed person' (nose-PROP); *thema/thama* 'person who knows everything' (ear-PROP); *thema* 'person with sore ear(s)' (ear-PROP). For further discussion, see Walsh 1976b.

6.2.2. -nu

The nominal suffix *-nu* DAT also occurs with body part nouns:

(133) *nanthi* *nithi-nu*
NC: thing arm-DAT
'thing for arm(s), artificial arm(s)'

(134) *nanthi* *lawali-nu*
NC: thing thigh-DAT
'thing for thigh(s), artificial leg(s)'

6.3. Nominal compounds

Nominal compounds are common in Australian languages and quite a few body parts are composite terms:

(135) *me* *ngardarda*
foot ankle
'ankle'

(136) *thathpi* *yuthpan*
lip/mouth straight
'chin'

(137) *ngapurlu* *thimu*
breast nose
'nipple'

(138) *ngukin* *weyi*
faeces hole
'anus'

(139) *ngi* *pelpith*
penis head
'glans'

(140) *mange* *pangkin*
hand back
'back of hand'

6.4. Body part nominals as pre-verbal elements

One construction involves an unincorporated body part nominal preposed before the verb but contributing to the meaning of the verb:

(141) *marda* *mangan-nhe-art* [> *mangannhert]*
belly/chest 3sgS-1du.incl.O-take
'He likes you (singular) and me.'

(142) *the* *ma-nhi-ma-bath-nu*
ear 1sgS-2sgO-from-have-FUT
'I'll know you.'

6.5. Body parts as metaphor

As with incorporated body parts in verbs in nominal expressions, there are differing degrees of metaphorical extension (see also section 5). That is to say some extensions of the core senses are quite obvious like the following:

(143) *payp* *pelpith*
pipe head
'bowl of (smoking) pipe'

(144) *payp* *ngi*
pipe penis
'shaft of (smoking) pipe'

(145) *tharntu* *thimu*
boat nose
'prow/nose of boat'

(146) *thelput* *ngurru*
house side
'side of house; wall'

However some metaphorical extensions are a little less obvious:

(147) *tharntu* *dimu*
boat teeth
'propellor of boat'

(148) *ngunga pelpith*
sun head
'midday, noon'

The following body part has been extended even further:

(149) *mange* 'hand' → action
→ Law
→ REFLEXIVE
→ similar

Here Law refers to the body of custom and practices that the Murrinh-Patha adhere to. In this sense *mange* is often used in the collocation *mange murntak* literally 'old hand' and sometimes, with a hint of nostalgia for those times when adherence to the Law was maintained through harsh sanctions: *mange tetemam*, literally 'hard hand'. Why should *mange* 'hand' be used for this sense 'Law, custom, traditional practice'? We can look to body part metaphors in English for at least part of the answer. In a range of English expressions notions such as CONTROL and EXCHANGE are expressed using 'hand':

(150) *Beware the hand of God*
She's got him in the palm of her hand.
Hands across the ocean.

In Murrinh-Patha both these notions can be seen in *mange* 'the Law' in that it is the Law that is regarded as the seat of control/authority and at the same time it is the Law which is held by the older generation and then passed on to the young – that is, through a process of exchange.

The same notion of CONTROL is at work in the use of *mange* in connection with reflexives. Like many Australian languages there is no formal distinction between verbal expressions which denote reflexives and reciprocals:

(151) *ngumemka-purl-neme*
1pc.excl.S-wash-pc.m.
'We few males washed ourselves', or 'We few males washed each other.'

But if it is necessary to make it clear that the verb is to be interpreted as reflexive rather than as reciprocal a free-standing pronoun is postposed after *mange* 'hand':

(152) *mange nganku neme ngumemkapurlneme*
hand 1pc.excl.m. 1pc.excl.S-wash-pc.m.
'We few males washed ourselves.'

The force of this expression is that the subject group has carried out its *own* actions. The same kind of force is exemplified in *mange ngay* 'by myself' (literally 'my hand'):

(153) *mange ngay ngunungam-ngem*
hand 1sg 1sgS:go-AUX_2
'I'm going by myself.'

This construction type can be used with any pronoun.

Finally there is the sense of *mange* to mean 'like, similar':

(154) *mange kanhi mani*
hand this ??
'like this'

Another way of expressing this notion is to use *nginipunh* 'body':

(155) *thamul ngay nginipunh thamul nhinhi*
spear 1sg similar spear 2sg
'My spear is similar to yours.'

The relationship between *body* and *like* or *similar* is familiar from English: Old English *lic* denoting 'body' gave rise to Modern English *like*.

7. Conclusion

Body part incorporation provides an avenue for the investigation of metaphor formation in Murrinh-Patha. More broadly one might expect that an investigation of other Australian languages will yield metaphorical themes of the kind identified here. The coverage given here is very preliminary but suggests that some of the recurring metaphorical extensions to be found in Murrinh-Patha will have counterparts in less exotic languages such as English. It is by the careful and detailed examination of metaphor in these exotic languages that we can expect to learn more about metaphor in general.

Notes

1. The data on Murrinh-Patha are drawn from my own fieldwork and from the very extensive knowledge of Chester Street of the Summer Institute of Linguistics, Australian Aborigines Branch. I have relied heavily on Street's detailed knowledge of the lexicon of Murrinh-Patha which is partly documented in Street's (1983) dictionary. Otherwise I owe a debt to my Murrinh-Patha language instructors especially the late Kevin Bunduck, the late Harry Kulampurrurt and the late Jumbo Dala. The fieldwork was supported by grants from the Australian Institute of Aboriginal Studies: my thanks go to this body for making the fieldwork possible and to its staff for their considerable assistance. I also thank the editors of this volume for their encouragement, their very useful suggestions for improvement and for their forebearance.
2. The orthography adopted here is identical to that developed by the Summer Institute of Linguistics literacy team, Chester and Lyn Street, and now in use by the literate speakers of Murrinh-Patha. Stops are as follows: *p*, *b*; *t*, *d*; and *k*, *g* are essentially as in English; *rt*, *rd* are voiceless and voiced retroflex; *th* is voiceless and laminal, being dental before /a, u/ and palatal before /i, e/; *dh* is voiced and laminal, realised as a voiced laminointerdental fricative before /a, u/ and as a voiced laminopalatal stop before /i, e/. Nasals are bilabial *m*, laminal *nh* (dental before /a, u/ and palatal before /i, e/), alveolar *n*, retroflex *rn* and velar *ng*. Laterals are alveolar *l* and retroflex *rl*. Rhotics are semi-retroflex continuant *r* and alveolar flap/trill *rr*. There are four vowels: *i*, *e*, *a*, *u*.
3. Abbreviations in glosses are: sg – singular; du – dual; pc – paucal (number i.e. 3 to about 10); pl – plural; m – masculine (i.e. the group is exclusively male); f – feminine (i.e. at least one member of the group is female); S – subject; O – (direct) object; BEN – benefactive; DAT – dative; ERG – ergative; FUT – future; IMPF – imperfect; NC – noun class; NEG – negative; TOP – topicaliser; DUB – dubitative; AUX – auxiliary; AUX_2 – secondary auxiliary LOC – locative; REDUP – reduplicated form; PREVB – preverb; incorporated body part – incorporated body part; VC – verb class.
4. The term, 'skin' or 'skin-name', refers to a system of social classification, common in Aboriginal Australia, whereby each individual is assigned to a particular 'skin' on the basis of which all that person's kin connections can be plotted.
5. However the verb-stem initial /y/ hardens to /nh/ as:

   damnhelelel 'he's bitter'
   damngiyelelel 'I'm bitter'

   whereas the /y/ of *-ye-* incorporated body part 'ear' would be expected to harden to /th/.

References

Blake, Barry
1987 *Australian Aboriginal grammar*. London: Croom Helm.

Craig, Colette (ed.)
1986 *Noun classes and categorization*. Amsterdam: John Benjamins.

Crowley, Terry
1995 "Inalienable possession in Paamese grammar", [this volume.]

Delbridge, Arthur *et al.* (eds.)
1985 *The Macquarie dictionary.* (Revised edition.) Sydney: Macquarie Library.

Derbyshire, Desmond
1979 *Hixkaryana.* (Lingua Descriptive Studies 1.) Amsterdam: North Holland.

Dixon, Robert M.W
1977 *A grammar of Yidiny.* Cambridge: Cambridge University Press.
1986 "Noun classes and noun classification in typological perspective", in: Colette Craig (ed.), 105-112.

Dixon, Robert M.W (ed.)
1976 *Grammatical categories in Australian languages.* Canberra: Australian Institute of Aboriginal Studies.

Evans, Nicholas
1995 "The syntax and semantics of body part incorporation in Mayali", [this volume].

Haiman, John
1980 *Hua: a Papuan language of the Eastern Highlands of New Guinea.* Amsterdam: John Benjamins.

Harvey, Mark
1995 "Body parts in Warray", [this volume.]

Lakoff, George
1987 *Women, fire and dangerous things.* Chicago: University of Chicago Press.

Langacker, Ronald W.
1987 *Foundations of cognitive grammar, Volume 1.* Stanford: Stanford University Press.

Leeding, Velma
1995 "Body parts and possession in Anindilyakwa", [this volume.]

McKay, Graham
1975 Rembarnga: a language of central Arnhem land. [Unpublished Ph.D. thesis, Australian National University.]
1995 "Body parts, possession marking and nominal classes in Njébbana", [this volume.]

Matisoff, James A.
1973 *The Grammar of Lahu.* (University of California Publications in Linguistics 75.) Berkeley and Los Angeles: University of California Press.
1978 *Variational semantics in Tibeto-Burman.* Philadelphia: Institute for the Study of Human Issues.

Stanner, William E.H
1964 *On Aboriginal religion.* (Oceania Monographs 11.) Sydney: Oceania.

Steele, Ross – Terry Threadgold (eds.)
1987 *Language topics. Papers in honour of Michael Halliday.* Amsterdam: John Benjamins.

Street, Chester
1983 *English-Murrinh-Patha, Murrinh-Patha dictionary* (with the assistance of Gregory Panpawa Mollingin). Wadeye: Wadeye Press.
1987 *An introduction to the language and culture of the Murrinh-Patha.* Darwin: Summer Institute of Linguistics, Australian Aborigines Branch.

Walsh, Michael
1976a "The bivalent suffix -ku in Murinjpata", in: Robert M.W. Dixon (ed.), 441-444.
1976b "The having suffix in Murinjpata", in: Robert M.W.Dixon (ed.), 287-290.
1976c The Murinypata language of north-west Australia. [Unpublished Ph.D. thesis, Australian National University.]
1987 "The impersonal verb construction in Murrinh-Patha", in: Ross Steele – Terry Threadgold (eds.), 425-438.
1989 Murrinh-Patha verb classes: past, present and future. [Paper presented at the Non-Pama-Nyungan Comparative Workshop, Australian Linguistics Society Conference, Monash University, September 1989.]
1991 Nominal classification in Murrinh-Patha. [Unpublished MS.]
1993 Vouns and nerbs in Murrinh-Patha: a category squish in a north Australian language. [Unpublished MS.]

Witherspoon, Gary
1977 *Language and art in the Navajo universe.* Ann Arbor: University of Michigan press.

Part III

Languages of the Pacific

Inalienable possession in Paamese grammar

Terry Crowley

1. Introduction[1]

1.1. The language and its type

Paamese is an Austronesian language spoken by about 4,000 Melanesian people, mostly on the island of Paama in the Republic of Vanuatu (formerly the Condominium of the New Hebrides), and in the towns of Vila and Santo, as well as in Nouméa in New Caledonia. All Paamese adults are at least bilingual, speaking, in addition to their own language, also Bislama. This is an English-lexifier pidgin which is spoken as a second language by most speakers of all 105 vernaculars in Vanuatu. Small numbers of Paamese also speak the neighbouring languages of Southeast Ambrym and the northern part of Epi. Younger Paamese who have been to school also have a working knowledge of either English or French. These two languages are for the most part only used in reading and writing, and are seldom used orally.

Paamese is structurally fairly typical of the languages of northern and central Vanuatu, which together constitute a fairly cohesive subgroup within the larger Eastern Oceanic subgroup, within the yet larger Oceanic subgroup of Austronesian languages (Clark nd). It is an SVO language, in which nominal case roles are marked predominantly by prepositions. It has fairly extensive inflectional morphology on verbs, distinguishing morphologically between a number of different modal categories that are expressed as prefixes. There are separate prefixes in each modal category for each of the fifteen subject categories, distinguishing between first, second and third persons, singular, dual, paucal and plural number, and in the first person non-singular categories, also inclusive and exclusive.

This paper concentrates on the expression of possession in Paamese. Oceanic languages in general exhibit a wealth of different constructions in the area of possession, in which the distinction between alienable and inalienable possession represents a recurring theme. Lynch (1973) and (1982) and Lichtenberk (1985) present wide coverage of data on possessive constructions in Oceanic languages, concentrating on the structurally more complex (and diverse) languages of Melanesia and Micronesia. Osumi (1995) discusses possession in Tinrin, a fairly distantly related Oceanic language spoken in New Caledonia. The Polynesian languages represent a rather different range of possessive constructions, which are discussed in Clark (1976: 115-119) and Wilson (1982).

1.2. Overview

This paper describes a broad range of grammatical constructions that are associated with the expression of inalienable possession in Paamese. Most of these constructions are referred to in one way or another in Crowley (1982), but specific references to the behaviour of body part nouns in this work are scattered throughout the volume. At the same time, significant new facts and interpretations have come to light since that work appeared. This paper includes all new data and updated analyses, as well as exhaustive listings of subclass membership for nouns with regard to each construction type. This paper is based on a total body part corpus in Paamese of about 230 lexical items, out of a total published lexicon that probably runs to around 6000 items (Crowley 1992).

There are basically two types of nouns in Paamese, those which can occur with pronominal possessive suffixes directly attached to them, and those in which the pronominal possessive suffixes attach instead to an adjacent "possessive constituent". Thus, compare the possessive constructions below in Paamese, involving the first person pronominal suffix, which appears on the surface as *-k*:[2]

(1) NOUN-SUFFIX
vatu-k
head-1SG[3]
'my head'

(2) NOUN POSS-SUFFIX
vakili ona-k[4]
canoe POSS-1SG
'my canoe'

With regard to the question of the correspondence between semantic and formal categories in language, there are two possible extreme positions. On the one hand, it could be argued that grammar is totally independent of semantics and that we should not expect that formal and semantic categories should coincide. The opposite position would be to argue that all grammar is derived from, and directly expresses, semantic categories, and that there is no real distinction between grammar and semantics.

A proponent of the first of these positions, when faced with the grammatical category of directly suffixed nouns and the semantic category of inalienable possession in Paamese, would not be unduly disturbed by the fact that there are exceptions to any semantic generalisations about the distribution of these two constructions in the lexicon. This was, in fact, my own approach in Crowley (1982: 67-69), and it is also the approach taken for similar constructions in Fijian by Schütz (1985: 463). A proponent of the second position, however, would seek to find a semantic explanation for every grammatical fact in the language. Thus, apparent irregularities in the distribution of directly suffixed nouns with respect to the category of inalienable possession should, after sufficiently detailed investigation, prove to have semantic explanations.

The approach in this paper will be to follow a course somewhere between these two extremes, and is basically that followed in Lichtenberk (1985: 125-126). It will be seen that while the distribution of nouns between these two possessive constructions at first glance appears only to correspond partly to the semantic distinction between alienable versus inalienable possession, with a fair degree of fine-tuning to the understanding of what constitutes inalienability in Paamese, it is in fact possible to a large extent to account for which nouns will behave "inalienably". This also requires the inclusion of some culture-specific knowledge in our account of linguistic facts.

Inalienability in Paamese can be defined roughly as the relationship that holds between the referents of two nouns in which one normally has animate reference and the other is an aspect of this animate being which cannot exist independently of the being. Thus, body parts which would be life-threatening (or at least unbearably painful) if removed (under non-medical conditions) are possessed inalienably in Paamese, as well as a variety of body products that are exuded as part of normal rather than transient bodily functions (even though these all eventually become separated from our bodies). Internal body parts are generally treated as inalienable when they are perceived to be central to our emotions and our sense of individuality and to the maintenance of life itself. However, when an internal organ is one that is typically removed separately when butchering or cooking an animal for eating, it is considered to be alienable.

This summary of what inalienability entails in Paamese also encompasses the kind of relationship that holds between animate referents and some nouns not referring to body parts as well. Thus, nouns referring to blood relations, rather than relations by marriage, often also behave in possessive constructions like body parts, as these represent relationships in which people are inextricably bound, and the relationships cannot be terminated. There are several other smaller categories of nouns which enter into this kind of possessive relationship about which it is more difficult to generalise, though there is still clearly an element of inseparability involved in their relationship with their "possessor" nouns. All of these categories of nouns will be discussed separately below.

There are only limited possibilities for structural interchange in Paamese between nouns which behave inalienably and those which do not. Although it is possible semantically to "alienate" the referents of many directly suffixed nouns, in most cases, semantic alienability still has to be expressed with a dummy inalienable possessor. Even with the minority of directly suffixed nouns which do have separate forms that are not directly suffixed, when an alienable possessor is to be expressed, this can normally only be achieved on the basis of the inalienably possessed form (thus requiring that two possessors be formally expressed).

1.3. Possessive constructions

Paamese is typical of North-Central Vanuatu languages in that many nouns referring to body parts, kinship relations and some other categories belong to a morphologically distinct subclass of nouns, which is distinguished primarily by the fact that the possessor is expressed by means of a pronominal suffix that is attached directly to the noun itself. Possession with other nouns, however, is indicated by using the same set of pronominal suffixes, but by attaching them instead to a special possessive constituent, which normally follows the noun (Lynch 1973), as illustrated above in (1) and (2). Possessive constituents carrying a pronominal suffix can, as an occasional stylistic variant, precede the possessor noun rather than follow it. Thus compare (2) with:

(3) *ono-m* *vakili*
POSS-2SG canoe
'your canoe'

Table 1. Possessive pronominal suffixes[a]

	SG		DL	PCL	PL
1	*-k*	INCL	*-ralu*	*-ratel*	*-r*
		EXCL	*-mal*	*-maitel*	*-mai*
2	*-m*		*-mil*	*-mītel*	*-mi*
3	*-n*	After *-e*	*-alu*	*-atel*	*-ø*
		After *-a/-o*	*-ialu*	*-iatel*	*-i*
		After *-i/-u*	*-lalu*	*-latel*	*-l*

a. Note that the forms in this paper are presented in the northern dialect as described in Crowley (1992), and not in the southern dialect as in Crowley (1982).

Table 1 sets out the surface forms of the full set of pronominal possessive suffixes which are attached either directly to nouns in Paamese or to these special possessive constituents.

The non-singular suffixes show a partial similarity to free form pronouns, as can be seen by a comparison with the paradigm set out in Table 2.

Table 2. Free form pronouns

	SG		DL	PCL	PL
1	*inau*	INCL	*ialu*	*iatel*	*īr*
		EXCL	*komal*	*komaitel*	*komai*
2	*keik*		*kamil*	*kamītel*	*kami*
3	*kei*		*keilu*	*keitel*	*keil*

Transitive verbs take suffixes to mark verbal objects when the object is a singular pronoun, the forms of which are underlyingly *-nau* '1SG', *-ko* '2SG' and *-e* '3SG'. It can be seen that these bear no formal resemblance to the possessive suffixes.

There are four different possessive constituents which can carry pronominal suffixes in Paamese, of which *ona-k* 'my' and *ono-m* 'your' in examples (2) and (3) represent just one. These different possessive constituents indicate a number of different kinds of semantic relations that hold between the possessor noun and the possessed noun. The forms of these different possessive constituents, and the semantic relationships that they express, are summarised in Table 3.

Thus, compare the following examples of the word *ani* 'coconut' which can occur with a possessive suffix attached to each of these different possessive forms:[5]

Table 3. Possessive constituents (with third person singular suffixes)

ā-n	'his/her/its (to eat)'; 'intended specially for him/her/it'; 'specially characteristic of him/her/it' [a]
emo-n	'his/her/its (to drink, to wear, or to use domesitically)'
ese-n	'his/her (owned as something that one has planted, as an animal one has reared, or as something kept on one's own land)'
one-n	'his/her/its (in all other kinds of possession)'

a. The latter two of this possessive constituent reflect a widespread Oceanic use of the possessive constituent in edible constructions to express benefactive and characterising possession. There will be further discussion of these uses of this form in section 2.2.1.

(4) *ani ā-n* 'his/her coconut (which he/she plans to eat the flesh of)'

(5) *ani emo-n* 'his/her coconut (which he/she plans to drink the water of)'

(6) *ani ese-n* 'his/her coconut (which is growing on his/her land)'

(7) *ani one-n* 'his/her coconut (which he/she plans to use in any other way, e.g. to stop the car rolling while changing a tyre, to knock her husband unconscious with while sleeping, to put on top of a pile of papers to stop the wind blowing them away, etc.)'

A very similar kind of construction in closely related Fijian has formerly been referred to as a noun class system, e.g. Milner (1972: 65-66). Lichtenberk (1985) also argues that these possessive suffixes should be analysed as nominal classifiers. However, Schütz (1985: 465) and Lynch (1973: 90) contest this analysis, noting that the function of these possessive constituents is primarily to express a particular semantic relationship between the possessor and the thing possessed. For Paamese at least, I would favour the latter solution as it is possible for these possessive constituents to express a benefactive relationship in which there is no possessed noun phrase expressed, for example:

(8) *Kei* *mūm* *ona-k*
3SG 3SG:REAL:work POSS-1SG
'He/she works for me.'

When the possessor is a noun rather than a pronoun, the possessive constituent with the third person singular pronominal suffix continues to directly follow the possessed noun. This inflected constituent is then followed by the possessor noun. As the noun *ani* 'coconut' does not take a pronominal suffix, nominal possession with this form can only be expressed as follows:

(9) *ani* *emo-n* *ēhon*
coconut POSS:POT-3SG child
'child's drinking coconut'

Although it was mentioned above that a third person singular pronominal possessor can be expressed by placing the possessive constituent either before or after the noun, this variation in order is not possible when the possessor is a noun. Thus:

(10) **emo-n* *ēhon* *ani*
POSS:POT-3SG child coconut

When the possessed noun is one of those that is directly suffixed (that is, typically, an inalienably possessed noun), the possessor comes immediately after the possessed noun itself, which again carries the third person singular possessive suffix. Thus, with a pronominal possessor, we find a construction of the following type:

(11) *vati-n*[6]
head-3SG
'his/her head'

When the possessor is a noun, the construction appears as follows:

(12) *vati-n* *ēhon*
head-3SG child
'child's head'

No morpheme of any kind can intervene between a nominal possessor and the possessed form in such constructions. The suffix *-s* 'only' can be attached to the possessive suffix in the following example:

(13) *vati-ne-s*[7]
head-3SG-only
'only his/her head'

However, when there is a nominal possessor, suffixes such as this can only be attached at the end of the possessive phrase. Thus, compare the following:

(14) *vati-n ēhono-s*
head-3SG child-only
'only the child's head'

(15) **vati-ne-s ēhon*
head-3SG-only child

The pronominal suffix that is attached to the possessed noun in such constructions is invariably the third person singular suffix *-n*, regardless of the number of the possessor noun. This means that the suffix on the possessed noun does not show number agreement with the number category of the possessor noun.[8] Thus:

(16) *vati-n huli keilu*
head-3SG dog DL
'two dogs' heads'

(17) **vati-lalu huli keilu*
head-3DL dog DL
'two dogs' heads'

(18) *vati-n huli keitel*
head-3SG dog PCL
'few dogs' heads'

(19) **vati-latel huli keitel*
head-3PCL dog PCL
'few dogs' heads'

(20) *vati-n huli keil*
head-3SG dog PL
'dogs' heads'

(21) **vati-l huli keil*
dog-3PL dog PL
'dogs' heads'

2. Body part nouns

In this section, and in the section which follows, I will detail the morphological and semantic characteristics of those nouns in Paamese acting as "possessed" objects which accept direct pronominal suffixes. I will contrast the behaviour of these forms with that of those semantically related nouns which accept no pronominal suffixes. I will begin by discussing body part nouns, which constitute the largest group of directly suffixed nouns in Paamese. These formal categories will then be matched with various semantic criteria.

2.1. Directly suffixed body parts

The first set of body part nouns that we can consider are all morphologically simple (apart, of course, from the fact that there is an obligatory pronominal suffix). I will demonstrate that there is a largely consistent relationship between the grammatical class of directly suffixed nouns and the semantic notion of separability from the "owner" of the body part. All attested directly suffixed body part nouns are listed in Table 4 (with all examples given in their third person singular possessed forms). Many of these body parts can be used equally with respect to humans and to the analogous parts of animals. However, this list also includes some morphologically simple suffixed body part nouns which are only used with respect to the body parts of animals.

Many directly suffixed nouns referring to body parts are also expressed by nouns that involve two bound nominal roots. The first of these nouns belongs to what is called the class of "linked nouns", which are obligatorily attached to some other noun, as described in Crowley (1982: 91). Nouns of this type that are included in my corpus are set out in Table 5.

To the list in Table 5, we can also add the forms *hidihē-n* 'his/her nails', *kalihē-n* 'his/her finger/toe; its claw' and *uruvote-n* 'his/her anus'. These appear to be constructed on the same pattern, but the initial morpheme does not occur independently of the suffixed body part to which it is bound. Thus, while *hidihē-n* and *kalihē-n* are derived from *hē-n* 'limb', the initial elements *hidi-* and *kali-* are not otherwise attested in the language. The same is true of the initial element *uru-* in the word *uruvote-n* 'his/her anus', although the word *vote-n* 'his/her buttocks' is recognisable within this word.

Table 4. Underived human and animal body part nouns with directly attached pronominal suffixes

ave-n	'his/her body/body part'	*vati-m*	'his/her head'
ahā-n	'his/her brain'	*hili-n*	'his/her hair of head'
vīvi-n	'his/her cheek'	*mete-n*	'his/her eye'
vulingasi-n	'his/her nose'	*ralinge-n*	'his/her ear'
veāse-n	'his/her cheek/jaw'	*ongo-n*	'his/her mouth'
hirē-n	'his/her neck/throat'	*nā-n*	'his/her face'
ālue-n	'his/her nape'	*loho-n*	'his/her tooth'
vatisī-n	'his/her shoulder'	*mē-n*	'his/her tongue'
vēvē-n	'his/her chest'	*sī-n*	'his/her bone'
voimase-n	'his/her chest'	*hē-n*	'his/her limb'
kou-n	'his/her rib'	*ale-n*	'his/her lower leg'
hinge-n	'his/her armpit'	*vango-n*	'his/her belly'
urili-n	'his/her knee/elbow'	*vite-n*	'his/her navel'
eisili-n	'his/her back/backbone'	*vile-n*	'her vagina'
mari-n	'his/her upper back'	*avoue-n*	'its flipper/wing'
levi-n	'his/her trunk'	*kē-n*	'its tail'
sange-n	'his/her lower belly'	*asupote-n*	'its tail'
ti-n	'his/her intestines'	*kasipote-n*	'its tail'
eate-n	'his/her liver'	*hī-n*	'its dorsal fin'
horā-n	'his/her thigh'	*pusi-n*	'its spur'
rī-n	'his/her hip'	*range-n*	'its comb (of rooster)'
vote-n	'his/her buttocks'	*maue-n*	'its tail (of rooster)'
ō-n	'his penis'	*horate-n*	'its crop (of chicken)'
ase-n	'his external genitalia'	*vatupi-n*	'its side (of bird, where wing joins onto body'

The "linked" nouns just mentioned have as one of their defining characteristics the fact that they must be attached directly to a nominal form and cannot be followed directly, by either a free form or suffixed pronoun. If the linked noun *hili-* 'skin' is to be expressed instead with a pronominal possessor without referring specifically to the skin of any specific body part, it must first of all be bound to a nominal form as a "carrier" for the suffix. The noun that fulfils this function is *ave-n* 'his/her/its body'. Table 6 sets out a number of linked nouns with *ave-n* as a carrier for a pronominal possessor.

There are also some examples in which pronominal possessive suffixes are carried by *ave-n* and the linked noun is metaphorically extended to refer to a body part, as shown in Table 7.

There is a final semantic category of nouns referring to body products and exusions which accept direct pronominal suffixes. These are listed in Table 8.

Table 5. Morphologically complex human and animal body part nouns carrying pronominal suffixes

ami-veāse-n	body:hair-chin-3SG	'his beard'
ami-mete-n	body:hair-eye-3SG	'his/her eyelash/eyebrow'
ami-ongo-n	body:hair-mouth-3SG	'his moustache'
ami-hinge-n	body:hair-armpit-3SG	'his/her underarm hair'
ami-ō-n	body:hair-penis-3SG	'his pubic hair'
ami-vile-n	body:hair-vagina-3SG	'her pubic hair'
ami-tā-n	body:hair-excrement-3SG	'his/her anal hair'
oreli-mete-n	egg-eye-3SG	'his/her eyeball'
oreli-ō-n	egg-penis-3SG	'his testicles'
oreli-ase-n	egg-male:genitalia-3SG	'his testicles'
hili-ongo-n	skin-mouth-3SG	'his/her lip'
hili-ō-n	skin-penis-3SG	'his foreskin'
hei-hē-n	fruit-limb-3SG	'his/her finger/toe'
hei-ase-n	fruit-male:genitalia-3SG	'his testicles'
hei-nā-n	fruit-face-3SG	'his/her forehead'
vati-hē-n	joint-limb-3SG	'his/her wrist'
vati-urili-n	joint-elbow/knee-3SG	'his/her elbow/knee joint'
vati-horā-n	joint-thigh-3SG	'his/her hip'
vati-loho-n	base-tooth-3SG	'his/her molar'
vati-mēn	base-tongue-3SG	'base of his/her tongue'
vati-ō-n	base-penis-3SG	'base of his penis'
vati-sange-n	base-lower:belly-3SG	'his/her lower belly'
kahi-mete-n	covering-eye-3SG	'his/her eyelid'
kele-ralinge-n	end-ear-3SG	'his/her ear lobe'
kele-ō-n	end-penis-3SG	'tip of his penis'
valenge-ō-n	hole-penis-3SG	'his urethral opening'
vāse-hē-n	underside-limb-3SG	'sole of his/her foot'
vuli-tā-n	hole-excrement-3SG	'his/her anus'
tine-hēn	inside-limb-3SG	'his/her palm/sole'
tine-ale-n	inside-lower:limb-3SG	'his/her sole'
uti-hē-n	seed-limb-3SG	'his/her nails'
uti-vile-n	seed-vagina-3SG	'her clitoris'
tavoi-ō-n	trunk-penis-3SG	'shaft of penis'
voi-range-n	container-cock's:comb-3SG	'his/her gum'
voi-nā-n	container-face-3SG	'his/her forehead'
ue-hirē-n	pipe-neck-3SG	'his/her trachea'
vati-kē-n	base-tail-3SG	'its tail'
vati-veāse-n	base-chin-3SG	'its external gill'

The morphologically complex nouns with pronominal suffixes listed in Table 9 also refer to exusions produced by body parts.

The body parts listed in Tables 1-5 are almost all items which are normally considered to be inseparable parts of the individual to which they are attached. Their removal would generally involve either death,

Table 6. Pronominally possessed linked nouns with *ave-n* 'his/her body' as a carrier for suffixes

Linked noun	Pronominally possessed linked noun		
ami-	'body hair'	*amīvan*[a]	'his/her body hair (in general)'
enge-	'dry skin'	*engeiven*	'his/her dry skin'
hili-	'skin'	*hilīven*	'his/her skin'
sine-	'fur/scales'	*sineiven*	'its fur/scales'

a. These forms are difficult to segment in their surface realisations. The noun *ave-n* 'his/her/its body' unpredictably loses its initial *a-* when it is attached to a linked noun. This reduced form is then predictably bound to the preceding linked noun by means of an intervening *-i-* (Crowley 1982: 91, 103-104). Thus, the constructions in Table 6 can all be represented underlyingly as: [LINKED NOUN-*i-ve*-[PRONOMINAL SUFFIX].

Table 7. Non-body part linked nouns with *ave-n* carrier for pronominal suffixes

Linked noun	Pronominally possessed linked noun		
vī-	'smell'	*vīven*	'his/her body odour'
kokohi-	'bubble/foam'	*kokohīven*	'his/her saliva'
toe-	'sap'	*toeiven*	'his/her perspiration'
usi-	'side'	*usīven*	'his/her side'
hati-	'piece'	*hatīven*	'his/her body part'

Table 8. Morphologically simple body products carrying pronominal suffixes

mē-n	'his/her urine'	*luo-n*	'his/her vomit'
tive-n	'his/her saliva'	*rā-n*	'his/her blood'
mease-n	'his/her bile'	*tā-n*	'his/her excrement'
rahi-n	'his/her nasal mucus'		

Table 9. Morphologically complex body products carrying pronominal suffixes

sī-mete-n	juice-eye-3SG	'his/her tears'
sī-me-n	juice-urine-3SG	'his/her urine'
sī-ō-n	juice-penis-3SG	'his semen'
sī-vile-n	juice-vagina-3SG	'her vaginal mucus'
soko-tā-n	remains-excrement-3SG	'remains of excrement left on anus after defecation'

risk of death or unacceptable levels of pain. There is a single exception to this generalisation, i.e. *hili-n* 'his/her hair'. This is clearly removable without pain or risk to life, yet it belongs to the set of directly suffixed nouns. I have no satisfactory explanation for this apparent anomaly.

Those removable suffixed body parts that do not fall into this semantic generalisation in Paamese, such as 'beard', 'fingernail/toenail', 'body hair' and 'foreskin'[9] and so on can still occur as directly suffixed nouns because these meanings are expressed by morphologically complex nouns in which the referent of the final constituent is itself inseparable. Bodily exusions are by definition separable, but some such body products in Paamese are expressed as directly suffixed nouns. It will be seen below that there are other bodily exusions which are expressed without direct pronominal suffixes. The major conditioning factor in this case is whether or not the exusion arises as a result of normal or abnormal (i.e. transient) body functioning. If the function is one that is always present and not some temporary manifestation, then the possessor is expressed by means of direct pronominal suffixation. All of the body products listed in this section can be considered to be permanent rather than transient, in the sense that they do not reflect passing illnesses or reparable disfigurements, with the exception of *luo-n* 'his/her vomit'.

2.2. Indirectly suffixed body parts

Although there are a great many body part nouns in Paamese which express possession by directly taking pronominal suffixes, there is, at the same time, quite a number of such nouns which cannot occur in this kind of construction, and which can only express possession by attaching these suffixes to an adjacent possessive constituent, and can therefore be considered to be only indirectly suffixed. In this section, I will group these indirectly suffixed nouns into semantic categories, demonstrating how each of these can be considered in various ways to be less closely bound to the "possessor" than those nouns that were discussed in the preceding section.

2.2.1. Temporary manifestations

One very significant observation that can be made is that nouns referring to temporary surface manifestations of the body are only expressed as indirectly suffixed nouns. This includes any kind of sore, skin infection or irritation present on the body (as well as any resident creatures), as set out in Table 10.

This category can also include temporary swellings and other symptoms of illness: *hinahin one-n* 'his/her swollen groin glands' and *anivul*

Table 10. Indirectly suffixed nouns referring to temporary manifestations of the body

ahis one-n	'his/her ringworm'	*ingisu one-n*	'his/her cold sore'
heipus one-n	'his/her wart'	*metat one-n*	'his/her pimple'
āhon one-n	'his/her plantar wart'	*utua one-n*	'his/her boil'
akol one-n	'his/her head sores'	*out one-n*	'his/her lice'
amuker one-n	'his/her scabies'	*vokavok one-n*	'his/her white spot tinea'
manu one-n	'his/her sore/wound'	*mange one-n*	'white coating on his/her tongue'
vulihesihes one-n	'his/her tinea of the crotch'		

one-n 'his/her shivering spell'. Finally, we can add the nouns *hanhanoai one-n* 'his/her fontanelle' and *mung one-n* 'his uncircumcised penis' to this category. *Hanhanoai* refers to the soft spot on a baby's head which is only a temporary feature of its body, as it eventually grows over and disappears. *Mung* 'uncircumcised penis' contrasts with the directly suffixed noun *ō-n* 'his penis'. As mentioned in note 9 above, circumcision is the norm for males in Paamese culture and so the possession of an uncircumcised penis is also only a childhood "aberration".

Although the nouns listed in Table 10 do not express possession by means of direct suffixation, they do not fall completely in line with ordinary alienable possessive constructions. Such nouns are grammatically different from other alienably possessed nouns that do not refer to body parts in that there is an additional structural option available to them that is not available with most alienably possessed nouns. As was mentioned in Section 1.3 above, in addition to marking edible possession, the possessive constituent *ān* can also be used to mark a "benefactive" relationship between the possessor and a possessed noun (Crowley 1982: 216-218). This means that the referent of the possessed noun is in some way specially reserved just for referent of the possessor noun. Thus, contrast the following forms:

(22) *āi one-n* 'his/her stick (which he/she is going to hit someone with)'

(23) *āi ā-n* 'his/her stick (which someone is going to hit him/her with)'

The possessive constituent *ān* has also been attested in association with many of the category of nouns listed in Table 10, in alternation with *onen*. However, when the possessor with these nouns is indicated by means of *ān* rather than *onen*, this indicates the idea that the affliction or bodily exusion is particularly noticeable, or that it is something which is characteristic of a particular individual. So, contrast:

(24) *manu one-n* 'his/her sore'

(25) *manu ā-n* 'his/her (unusually large or noticeable) sore' or 'his/her (particularly numerous) sores'

(26) *out one-n* 'his/her lice (which are resident in his/her hair)'

(27) *out ā-n* 'his/her lice (referring to the typically lousy person from whom the lice were caught)'

Nouns outside those listed in Table 10 are not normally able to occur in association with *ān* with this kind of particularising reading, but must instead be given the "benefactive" reading described earlier.

Table 11 lists five other nouns expressing body products found in my corpus which never attach possessive suffixes directly to the noun.

Table 11. Body products which do not carry pronominal affixes

mād one-n	'his/her perspiration'	*sūvo one-n*	'his/her phlegm/pus'
arul one-n	'his/her ear wax'	*metave one-n*	'sleep in his/her eyes'
batīra one-n	'his/her blood clot'		

The set of nouns in Table 8 also referring to bodily exusions and products contrast with these nouns in Table 11 in that the former involve directly attached possessive suffixes. All of the indirectly suffixed nouns in Table 11 refer to body products that arise as a result of some particular short-term activity, or because of a temporary illness or injury. The nouns in Table 8 on the other hand refer to bodily products and exusions which are permanently with us (i.e. blood, saliva, bile, nasal mucus, excrement and urine), and which are not produced by any abnormal bodily activity. There is a single exception to this generalisation with the form *luo-n* 'his/her vomit' in Table 8, which is as temporary a manifestation as those items listed in Table 11, yet it happens to take direct suffixation.

2.2.2. *Internal organs*

In addition to these nouns referring to body products, there are also some nouns referring to actual body parts which do not directly accept pronominal suffixes in Paamese. There is one such semantic set which refers to internal organs, and its membership is listed in Table 12.

Table 12. Indirectly suffixed human and animal body part nouns

heihus one-n	'his/her heart'	*atu one-n*	'her womb'
heihūh one-n	'his/her bladder'	*maleles one-n*	'his/her lungs'
heivangvang one-n	'his/her stomach'	*avoi one-n*	'his/her spleen'
heias one-n	'his/her kidney'	*uikōko one-n*	'his/her larynx'
heinīnias one-n	'his/her kidney'	*hungaiel one-n*	'its internal gill'
hashas one-n	'her uterus'	*ulkahkah one-n*	'membrane around his/her internal organ'

The fact that such nouns should not take direct suffixation is not surprising. Internal organs are the kinds of things that would normally only be directly observed when there is a dead body that has been opened up. The possession of these items by the butcherer of the animal is clearly transferrable and so there is an alienable relationship between him and the body parts, in contrast to the formerly inalienable relationship between the animal and the body parts when it was still alive. It is also worth noting that the term *hungaiel* refers to the inside part of the gill that is removed when cleaning a fish for cooking. The external gill however is expressed as *vativeāse-n*, which belongs in the set listed in Table 5 above.

This generalisation holds for all but four of the nouns in my corpus referring to internal organs. These exceptional forms come from Table 4 and are set out separately in Table 13. Although these four nouns all refer to internal organs and might have been expected to fall into the set listed in Table 12, they accept direct pronominal suffixes.

Table 13. Exceptional internal organs accepting direct pronominal suffixes

ti-n	'his/her intestines'	*ahā-n*	'his/her brain'
eate-n	'his/her liver'	*horate-n*	'its crop (of bird)'

A possible explanation for the irregularity of *ti-n* 'his/her intestines' could be that this is the only specific internal organ that is used in Paamese idioms for the expression of emotions, in which case the body part would be considered to be inseparable from the individual when alive rather than separable. We find examples such as the following idioms:

(28) *Ti-n* *mīs.*
intestines-3SG 3SG:REAL:cry
'He/she feels sorry.'

(29) *Ti-n* *katinau.*
intestines-3SG 3SG:REAL:bite:1SG
'He/she is angry with me.'

(30) *Ti-n* *tīsa.*
intestines-3SG 3SG:REAL:bad
'He/she is angry.'

The noun *eate-n* 'his/her liver' never occurs in such idiomatic constructions in Paamese. However, in many Oceanic languages, the liver is regarded as the seat of the emotions.[10] Perhaps *eate-n* 'his/her liver', in directly attaching pronominal suffixes, reflects an earlier situation in Paamese in which the liver, as well as (or perhaps instead of) the intestines was used to express emotions and feelings. The perceived crucial importance of the brain over any other organ in staying alive is perhaps what is responsible for the suffixed rather than free form status of this form. This would leave *horate-n* 'its crop' as the only completely unexplained item.

2.2.3. *Body parts as potential food items*

The nouns in Table 14 also all clearly have body part reference, but they do not directly take pronominal suffixes.

Table 14. Indirectly suffixed body parts which can also be considered as food

husi one-n	'his/her muscle'	*amur one-n*	'his/her fat'
rengareng one-n	'his/her calf'	*sumor one-n*	'its eggs inside body without shells (of chicken)'
hatesim one-n	'his/her breast muscle'		

While these nouns express inseparable parts of a live individual's body when the possessor is expressed by means of suffixes attached to the possessive constituent *onen*, with the possessive constituent *ān*, they can all be viewed as cuts of pork, chicken or beef (remembering that animal body part names are generally parallel to human body names anyway), or other edible parts of an animal.[11] As edible body parts, they would presumably be no longer viewed as an inherent part of the body, but as being separated from it.

One could also include in the set listed in Table 14 the noun *sūsu* 'breast', which is a free form noun that does not take direct pronominal suffixes. Although one would not consider breasts in the same category as cuts of meat, I have included this as an additional member of this subset because this word is also used with the meaning of 'breast milk', thus representing a drinkable body fluid which passes from one individual to another. That this is the correct explanation for the seemingly anomalous behaviour of this noun is suggested by the fact that *sūsu* normally expresses the possessor by attaching the pronominal suffixes to the possessive constituent for drinkable nouns, i.e. *emon*. This is so, even when it is clearly the body part itself that is being referred to, rather than its drinkable product. In fact, even the breast of a male would be referred to as:

(31) *sūsu* *emo-n*
breast POSS:POT-3SG
'his (drinkable) breast'

Clearly, there is no question here of there being any literal interpretation of drinkability.

There are two other candidates for inclusion in the same set, i.e. *hou one-n* 'his/her tendon/vein' and *angol one-n* 'his/her cartilage'. While these do not represent cuts of meat, they are certainly impediments to the enjoyment of meat which must be removed, and so behave as nouns that are considered as separable body parts.

One problem noun with respect to these generalisations is the word for 'egg'. When a chicken is being cleaned for cooking, the *sumor* 'eggs inside body without shells' are generally carefully removed and cooked separately from the bird itself, so it is not surprising that this noun should be indirectly suffixed. However, it is difficult to explain why the ordinary word for a fully formed egg that has been laid should be expressed as the linked noun *oreli-*, which must be followed by the name of the animal that produced it, e.g. *oreli-ahu* 'turtle egg', *oreli-aman* 'bird egg', *oreli-ato* 'chicken egg'. However, this linked noun construction is one that canonically expresses a part-whole relationship (rather than a relationship of inalienable possession) (Crowley 1991).

2.3. Interaction between semantic and formal criteria

In addition to the semantic criteria for the classification of body part nouns into suffixing and non-suffixing categories, there is also an important generalisation to be made that involves formal criteria. For instance, we can argue that *sī-mete-n* 'his/her tears' should be indirectly suffixed rather than directly suffixed because tears arise as a result of some non-permanent body action (that is, crying). However, because this word is morphologically derived from *mete-n* 'his/her eye' which does accept pronominal suffixes, it can be argued that *sī-mete-n* accepts direct suffixation on this basis. Thus, it could be argued that the structure of this word is that represented in Figure 1 rather than Figure 2. The word expressing this meaning does not belong in the morphological class that we would expect on the basis of its semantics because morphological criteria take precedence.

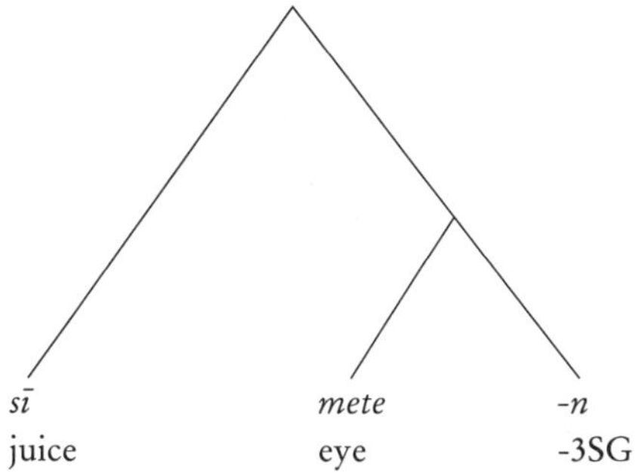

Figure 1. Preferred constituent analysis of *sī-mete-n*

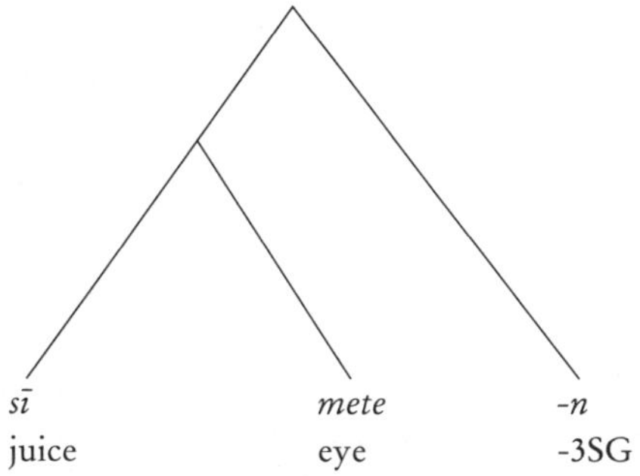

Figure 2. Alternative constituent analysis of *sī-mete-n*

Nouns referring metaphorically to body parts can also involve nominals that are themselves not directly suffixed. By virtue of their primary or literal meanings, such nouns do not accept pronominal suffixes either when they refer to body parts. Table 15 lists some examples of this type.

Table 15. Indirectly suffixed body part nouns involving compounding and metaphorical extension

		Body part reference
tei-ai	log-tree	body
vuli-rais	hole-rice	anus
vuli-tāro	hole-taro	anus
vone-manu	mark-sore	scar
aroe-sūsu	handle-breast	nipple
tīvāva	newborn baby	pupil of eye
sūvon	mat	placenta
avati	moon	kneecap
vuluai emon talo	pool POSS:POT:3SG bird:type	hollow above collar bone

To this set, we can also add a number of metaphorical colloquial and avoidance terms relating to male and female genitalia, which are also not directly suffixed, even though the ordinary words *ō-n* 'his penis' and *vile-n* 'her vagina' do take pronominal suffixes. Thus, *āi* 'stick', *hopu* 'arrow' and *maripu* 'stub' are all used colloquially to refer to 'penis' and *tivirit* 'bat' is used as a euphemism with the same meaning. *Ruruvek* 'puffer fish' is used colloquially to mean 'accidentally exposed penis'. *Meneas* 'ripe chestnut' and *āi kān* 'weapon' (literally: 'sharp stick') are colloquialisms for 'vagina', whereas *votelu* 'shellfish type' is a euphemism for the same meaning. *Oai* 'water' is also used as a euphemism for *sī-ō-n* 'his semen'.

We could also include in this category of nouns used metaphorically the forms *engeoha* 'scaly skin on baby's head' (literally: 'skin of bread-fruit seed') and *utīlau* 'swollen neck glands' (literally: 'seed of the dragon plum'). However, these could equally be considered to be indirectly suffixed because they represent transient rather than permanent features. The behaviour of these morphologically complex nouns should be compared with that of those nouns listed in Table 5 above. It will be remembered that while some of these morphologically complex nouns refer to physically removable body parts, they still accept direct pronominal suffixes because they are morphologically derived from nouns which themselves accept direct suffixes.

There are two examples of nouns in Paamese derived from verbs by means of reduplication, which do not accept pronominal suffixes,i.e. *mēme onen* 'his/her urethral opening' (from *mē* 'urinate') and *uererēm onen* 'his/her oesophagus (from *rēmi* 'swallow', literally: 'pipe-swallow-RED'). Verbs in Paamese take suffixes marking only singular pronominal objects, and the forms of these suffixes bear no relation to the possessive suffixes found on nouns. We could argue, therefore, that these two permanent body parts do not accept direct pronominal suffixes because they are derived from verbal forms, which, as part of their defining characteristics in Paamese are basically non-suffixing. However, we are faced with the fact that there are still a few suffixed nouns in Paamese which are clearly related formally to verbs. Table 16 sets out the forms of this type that are included in my corpus.

Table 16. Related verbs and suffixed nouns

Verb		Suffixed nouns	
mē	'urinate'	*mē-n*	'his/her urine'
pusi	'kick'	*pusi-n*	'its spur (of rooster)'
lū	'vomit'	*luo-n*[a]	'his/her vomitus'
tiv	'spit'	*tive-n*	'his/her saliva'
sū	'dig up ground with snout (of pig)'	*sue-n*	'ground dug up by its snout (of pig)'
rah	'blow nose'	*rahi-n*	'his/her nasal mucus'

a. The appearance of the vowels in these examples is predictable according to the general morphophonemic rules of the language.

We could perhaps argue that the forms in Table 16 are fundamentally nouns which can be used as verbs, and are therefore amenable to direct suffixation. Forms such as *rēmi* 'swallow', however, could be analysed fundamentally as a verb which can be used nominally in *uererēm* 'oesophagus', hence its lack of direct pronominal suffixation. However, the occurrence of derivatives of *mē* 'urinate' in both constructions represents an argument against this. Thus, the behaviour of these forms remains unaccountable.

Finally, nouns borrowed from Bislama, in which there is no pronominal suffixation, are invariably indirectly suffixed in Paamese, regardless of their semantic category. The only body part nouns of Bislama origin that are commonly used in ordinary registers of Paamese are *mustas* 'beard/moustache', *kirās* 'body hair' (ultimately from English "grass", though *gras* has acquired the meaning of 'body hair' in Bislama) and *rīng* 'corona of glans penis' (again, with this meaning in Bislama). There are some other indirectly suffixed body part nouns that have been borrowed from Bislama, which tend to be used as avoidance terms to existing Paamese forms. Thus, we find *melek* 'semen' (from Bislama *melek* 'milk/semen') as a euphemism for *sī-ō-n*, *pōl* 'external male genitalia' (ultimately from English "balls") as a euphemism for *ase-n* and *hōl* 'anus/vagina' (ultimately from English "hole") as a euphemism for both *vuli-tā-n* 'his/her anus' and *vile-n* 'her vagina'.

2.4. Residual indirectly suffixed body parts

The discussion in sections 2.1 to 2.3 leaves a number of nouns in Paamese referring to body parts which do not accept direct suffixation even though they are all permanent features of the body, and they are external. These apparently exceptional forms are set out in Table 17.

Table 17. Indirectly suffixed inseparable body part nouns

hosetau one-n	'his/her canine tooth'	*heivoul one-n*	'his/her ankle'
kurkurumu one-n	'his/her incisor teeth'	*heiali one-n*	'its sucker (of octopus)'
hāsuaev one-n	'his/her thumb/big toe'	*luheavati one-n*	'his/her kneecap'
vāsuaev one-n	'his/her thumb/big toe'	*uniavati one-n*	'his/her kneecap'
kiskisvot one-n	'his/her index finger'	*haroma one-n*	'his/her shin'
salauvul one-n	'his/her middle finger/toe		

There are two nouns in Table 17 referring to different kinds of teeth, i.e. *hosetau* 'canine tooth' and *kurkurumu* 'incisor'. The generic word for 'tooth' is the suffixed form *loho-n* 'his/her tooth', and there is a form derived from this which also refers to a specific kind of tooth, i.e. *vati-loho-n* 'his/her molar'. The two exceptional words for teeth in Table 17 are synchronically not further analysable, but historically at least, they both appear to represent compound constructions in which the final constituent is not a nominal form at all. The various examples in section 2.3 above suggest that a form can be indirectly suffixed for purely formal rather than semantic reasons, such as when it is used metaphorically on the basis of a form that is itself indirectly suffixed, when it is borrowed from Bislama, or when the final constituent belongs to a word class whose members do not normally accept pronominal suffixes. The following paragraphs discuss a number of examples of this type.

Clark's (nd) list of Proto-North Central Vanuatu reconstructions includes the adverbial form **muʔa* 'before/in front', and the final syllable of *kurkurumu* 'incisor' would derive regularly from this. The initial element of this form could conceivably be a reduplicated intransitive derivative of *kur* 'take' (even though the suggested intransitive form does not otherwise occur in the language). Thus, this could historically have been a compound meaning 'take first', referring to the biting off of pieces of food.

Hosetau 'canine tooth' possibly involves a second original unsuffixed morpheme *tau*. This is still retained in Paamese with the meaning 'next'. There is no indication of what the initial morpheme may be derived from, but the canine teeth are the teeth found adjacent to the incisors.

There is also a set of nouns above which refer to different fingers, i.e. *hāsuaev/vāsuaev* 'thumb/big toe', *kiskisvot* 'index finger' and *salauvul* 'middle finger/toe'. *Hāsuaev* (and its variant *vāsuaev*) 'thumb/big toe' appears to derive historically also from a compound, in which the second part would not be expected to have carried pronominal suffixes. Clark's Proto-North Central Vanuatu reconstructions include **bisu* 'finger' and **laba* 'big', which could represent constituent parts of an original compound.

The noun *kiskisvot* 'index finger' is also a compound, derived regularly by the same derivational process suggested above for the detransitivisation of *kur* 'take'. The transitive verb *kis* 'poke' has a reduplicated intransitive derivative in Paamese, that is, *kiskis* 'poke'. The suffixed noun *vote-n* 'his/her buttocks' is one of the minority of suffixed body part nouns in Paamese that has an unsuffixed equivalent (as discussed in section 4 below). Thus, *kiskisvot* literally means 'poke buttocks' (as it is this finger that pokes the anus when wiping after defecation). The fact that this compound is formed on the basis of the otherwise rare unsuffixed form of *vote-n* 'his/her buttocks' rather than the suffixed form is anomalous, however.

I have no suggestion on historical grounds as to why the remaining form referring to digits, i.e. *salauvul* 'middle finger/toe', should belong in this set rather than in the set of forms which accepts direct pronominal suffixes. One possibility is that the middle finger, being longer and more prominent, is in a sense more "individualised", and hence is considered separately from all of the other fingers. However, the middle toe, which is also referred to as *salauvul*, is not the longest toe. This is therefore not consistent with this suggestion, unless *salauvul* refers to 'middle finger' in its primary sense, and only secondarily to 'middle toe'.

The alternative forms *uniavati* and *luheavati* for 'kneecap' are also morphologically derived from the unsuffixed noun *avati*, which in its primary sense means 'moon'. I have no information as to what the original meanings for the initial *uni-* and *luhe-* might have been. These initial morphemes can be compared with the otherwise non-occurring initial morphemes *hidi-*, *uru-* and *kali-* that were mentioned in the supplement to forms set out in Table 5 above.

Of the remaining nouns in Table 17, *heiali* 'sucker of octopus' is formally identical to a regular derivation of *hei-* 'fruit' and *ali* 'kind of tree', meaning 'fruit of the *ali* tree'. Similarly, *heivoul* 'ankle' is formally identical to the word regularly derived from *hei-* 'fruit' and *voul* 'tamanu tree' meaning 'fruit of the tamanu tree'. If this explanation for the behaviour of these two body part nouns is correct, I am not aware of any semantic connection between these fruits and the body parts in question. (However, some of the examples of body parts expressed as extended meanings of other nouns in Table 16 indicate that wide semantic extensions are possible.) The word *heias* 'kidney' from Table 12 also means 'fruit of Tahitian chestnut', and there is a clear resemblance in shape to the body part in question. Thus, this body part may be unsuffixed for formal rather than semantic reasons.

This leaves the form *haroma* 'shin', which also appears to be anomalous. There is no obvious explanation as to why this should be expressed as a free form noun rather than as a noun which accepts direct pronominal suffixes as it is a permanent non-removable body part. The semantically very close term *ale-n* 'his/her lower leg' is suffixed, while *haroma* is not.

3. Suffixed nouns not referring to body parts

The possessive construction in which pronominal suffixes are directly attached to the possessed noun is not restricted just to body part nouns in Paamese. There are some other smaller semantic categories of nouns relating to a person's social and psychological self which also fall into the same morphological category. Again, in most of these cases, we can argue that there is an especially close, inseparable semantic relationship between the "possessor" and the "possessed" noun phrases which would account for the morphological behaviour of a large proportion of these nouns.

3.1. Aspects of individuality

There is a small number of abstract nouns in Paamese that refer to different aspects of an individual's psychological and spiritual self. Some of these accept direct pronominal suffixes while others do not. Generally, transient aspects of our individuality such as feelings and sensations are not directly suffixed while more permanent aspects of our individuality accept direct pronominal suffixes.

3.1.1. Suffixed nouns

Table 18 sets out those nouns of this type which are directly suffixed for possession. It can be seen that these nouns all refer to non-transient aspects of a person's psychological or spiritual individuality.

Table 18. Aspects of individuality carrying pronominal suffixes

ise-n[a]	'his/her name'	*rē-n*	'his/her voice/opinion'
nini-n	'his/her soul/spirit'	*uli-n*	'his/her seat of emotions'
sāse-n[b]	'his/her self'		

a. The noun *ise-n* 'his/her/its name' can have either human or non-human reference, depending on whether the thing named is itself human or non-human. Thus, note the difference between:

Isen	*asa?*
name-3SG	what

'What is its name?'

Isen	*isei?*
name-3SG	who

'What is his/her name?' [Literally: 'Who is his/her name?']

b. The noun *sāse-n* 'his/her self' is usually used as a verbal modifier, in constructions such as the following:

Inau	*nakanian*	*sāsok.*
1SG	1SG:REAL:eat	self:1SG

'I ate by myself.'

It can also be transitivised with the addition of *-ni* (Crowley 1982: 192-194):

Inau	*nakan*	*sāsokun*	*ouh.*
1SG	1SG:REAL:eat	self:TR	yam

'I ate the yam by myself.'

When the object is the same as the subject in such a construction, this gives a reflexive reading:

Inau	*naruv*	*sāsokunau.*
1SG	1SG:REAL:shoot	self:1SG:TR:1SG

'I shot myself.'

3.1.2. Aspects of individuality expressed by unsuffixed nouns

At the same time, however, there is one abstract noun of this semantic type which does not fall into this subclass, as it does not take direct pronominal suffixes, i.e. *alual ā-n* 'his/her life force'. The possessive constituent that is used with this single exceptional noun is the particularising *ān* rather than the neutral possessive constituent *onen*, as described in sections 1.3 and 2.2.1 above.

3.1.3. Transient sensations

There are some abstract nouns referring to feelings and sensations associated with our bodies which also do not take direct pronominal suffixes. This subclass of nouns includes forms such as *anni* 'coldness', *amai* 'hunger' and *maroro* 'thirst'. These nouns only represent transient sensations, however, and contrast with those nouns in sections 3.1.1 and 3.1.2, which represent permanent aspects of our being. Thus, it is not surprising that they should be expressed by free form nouns.

3.1.4. Unsuffixed derived abstract nouns

There are a great many other abstract nouns in Paamese that also express feelings and sensations which never receive direct pronominal suffixes. Rather than trying to find semantic explanations for the behaviour of such nouns, it is possible in the case of these nouns to invoke a formal explanation. In Section 2.3 above, I described nouns referring to body parts which are compounded on the basis of an indirectly suffixed noun. In a similar way, there is also a regular nominalising suffix *-en* in Paamese which itself never accepts direct pronominal suffixes. This suffix derives abstract nouns from verbs, as exemplified in Table 19.

Table 19. Derived abstract nouns in *-en*

Verb		Derived noun	
sou	'sing'	*souen*	'a song'
selūs	'talk'	*selūsien*	'a speech'
lohloh	'run'	*lohlohien*	'a race'
tūn	'chat'	*tūnien*	'a story'
an	'eat'	*anien*	'a meal'

This process is quite productive in Paamese, and it can be used to derive nouns referring to abstract aspects, as well as products of our bodies and other nouns with more concrete reference. All nouns carrying the nominalising suffix *-en* behave as free form nouns, regardless of their semantics. Table 20 lists derived nouns which have reference relating to the body.

Table 20. Derived nouns in *-en* which express body-related concepts

Verb		Derived noun	
redem	'think'	*redemien*	'thought/idea'
mesai	'sick'	*mesaien*	'disease'
vurvur	'rough'	*vurvuroen*	'leprosy'
māh	'painful'	*māhien*	'pain/suffering'
matil vōvoi	'dream'	*matil vōvoien*	'dream'

3.2. Orientation nouns: suffixed and unsuffixed

The following four nouns indicate position with respect to something else in which the possessor represents the point of reference for the position: *ē-n* 'its interior', *ne-n* 'among it', *hehe-n* 'under it' and *tou-n* 'behind it'. Nouns expressing orientation and which carry pronominal suffixes in this way are widespread among Micronesian languages. Ewe (Ameka 1995) also treats orientation as "inalienable". In Paamese, these four items are the only words expressing orientation that behave in this way. Of these only the first is at all widely attested, and then it is mostly found occurring in a series of idiomatic constructions as the subject of a verb of feeling or emotion, as in:

(32) *Ēk* *vobongoni.*
interior:1SG 3SG:REAL:forget:TR:3SG
'I forgot it.' (Literally: 'My interior forgot it.')

(33) *Ēk* *vati.*
interior:1SG 3SG:REAL:want:3
'I want it.' (Literally: 'My interior wants it.')

Orientation in Paamese is ordinarily indicated by a range of quite different constructions. In many cases, position is indicated by means of prepositions (Crowley 1982: 195-210), or by a separate class of un-inflected location nouns followed by a preposition that relates it to a following unsuffixed noun (Crowley 1982: 161-162). Table 21 lists forms

Table 21. Unsuffixed orientation nouns and their accompanying prepositions

nesa en X	'above X'	*tav en X*	'across X'
netan en X	'below X'	*hae ran X*	'outside X'
neim en X	'inside X'	*soutin ran X*	'far from X'
luhi en X	'between X'	*vesesal ven X*	'near X'

of the latter type, in which the preposition *en* normally expresses the locative, *ran* the ablative and *ven* the directional (i.e. 'towards').

3.3. Impressions and imprints of body parts

A further subcategory of directly suffixed nouns in Paamese are those that express the physical impression or imprint made by a body part on something else. This relationship can also be considered to be "inalienable" in that the physical imprint of the body part could not exist without the body part that made it in the first place. The entire corpus includes only four examples of this kind of noun, i.e. *ulunge-n* 'his/her end of the bed where the pillow goes', *vē-n* 'his/her footprints', *vuli-n* 'his/her role in society/regular sleeping place' and *sue-n* 'its disturbed ground (dug up by snout of pig)'.

A further possible noun of this type might be the noun *tavunge-n* 'its fallen fruit (when a rat, flying fox or bird has chewed a fruit, causing it to fall to the ground and become inedible)'. The possessor in the case of this noun is the animal that has chewed the fruit to cause it to fall. This example is problematical in that it does not involve an imprint in the same way, though normally the teeth marks of the animal are visible on the fallen fruit, indicating that it should not be eaten.

Similarly, *uve-n* 'sleeplessness caused by someone else staying up all night' may be considered as belonging in this set.[12] Although this does not represent a physical impression, there is nevertheless an impact on the individual of someone else's behaviour. This noun occurs in the following idiom involving the verb of impact *muas* 'hit':

(34) *Uven* *mē-Vouleli* *keil* *muas*
sleeplessness:3SG people-Vouleli PL 3SG:REAL:hit
īr.
1PL:INCL
'We are tired from the Vouleli people being up all night.'

While a number of nouns expressing imprints and impressions are expressed as directly suffixed nouns in Paamese, there is a similar number which are expressed as linked nouns, e.g. *vuli-* 'hole left by something', *valenge-* 'hollowed out part', *tahe-* 'wash (of vessel)', *vone-* 'mark'. The main difference between suffixed and linked nouns of this semantic type is that suffixed nouns refer to imprints left by the referents of animate nouns, whereas linked nouns refer to imprints left by the referents of inanimate nouns.

We also find the unsuffixed noun *vaila* 'footprints', which is synonymous with one of the nouns listed at the beginning of this section. Evans (1995) notes that footprints set in concrete are considered to be alienable in Mayali. However, I have not been able to establish any semantic correlate for these two grammatically different words for 'footprint' in Paamese.

3.4. Suffixed parts of inanimate wholes

There are three directly suffixed nouns in Paamese which allow inanimate "possessors" and which indicate a part of a whole, that is, *ue-n* 'its handle/string', *kahi-n* 'its dry bark (found at joints of cane plants)' and *ingi-n* 'its edge' . The possessor in these cases is the whole of which the referent of the suffixed noun is a part, e.g.

(35) *ue-n* *atuvoi*
handle-3SG basket
'handle of basket'

(36) *kahi-n* *eau*
dry:bark-3SG bamboo
'dry bark at joints of bamboo'

Part-whole relationships involving inanimate nouns in Paamese are also normally expressed by the linked noun construction, so the appearance of these two nouns in this morphological category is anomalous. In fact, there are three apparently synonymous words in Paamese for 'handle', falling into three separate grammatical classes: the suffixed noun *ue-n*, the free form noun *apak* and the linked noun *aroe-*.

3.5. Best food

There is a single noun referring to food that takes direct pronominal suffixes, i.e. *vuse-n* 'meat part of his/her meal (as against staple, that is, taro, yam, rice etc)'. Ordinary edible possession is expressed by means of the edible possessive constituent *ān*, e.g.

(37) *vuas ā-n*
pig POSS:ED-3SG
'his pork'

This directly suffixed noun, however, indicates the meat part of the meal when one has a special craving for it. (In fact, the word *vuse-n* can also be used in a secondary sense to refer to the craving itself that one has for meat.)

As mentioned in 2.2.1, the possessive constituent *ān* can be used to indicate something that is specially intended for an individual, a semantic relationship which could conceivably be held to include also the best part of a meal. Given that in the case of foods, *ān* already indicates ordinary possession, another construction would be needed to express the favourite part of the meal. It is not unreasonable to suggest that the inalienable possessive construction has been called in for this purpose.

There is a single noun indicating an item of traditional clothing which is one that would not normally have been removed from one's body (at least in public), i.e. *tinivuse-n* 'his penis sheath'. All other items of clothing that are less intimately worn, however, are expressed by unsuffixed nouns, and possession is indicated by attaching the pronominal suffixes to a possessive constituent. In the case of possession involving items of clothing, the domestic possessive constituent *emon* can be used in alternation with the neutral possessive constituent *onen*. Thus:

(38) *eisin emo-n/one-n*
shirt POSS:WEAR-3SG/POSS-3SG
'his/her shirt'

This noun *tinivuse-n* 'his penis sheath' could also be included in the category of inalienably possessed nouns for primarily formal reasons, rather than for semantic reasons. The noun *tinivuse-n* is possibly morphologically complex. The initial element *tini-* is a linked noun meaning 'used leaves for wrapping food for cooking'. Thus, the word *tinivuse-n*, resegmented as *tini-vuse-n*, could literally mean 'used cooking leaves that one wraps around one's favourite meat', in which one's penis is presumably the meat.[13]

3.6. Kin terms

Sections 3.1 to 3.5 have for the most part dealt with very small classes of nouns not referring to body parts which accept direct pronominal suffixes like many body part nouns. There is one remaining semantic grouping of nouns entering into the same grammatical relationship which is somewhat larger, however, and this is the category of kin terms. Blood kin relationships are more likely to enter into the directly suffixed construction, while kin by marriage are more likely not to accept direct pronominal suffixes. It will be seen in the discussion which follows, however, that there is a considerable degree of unpredictability about the behaviour of many nouns in this semantic field.

3.6.1. Directly suffixed kin terms

Table 22 sets out ten forms referring to kinship relations which express possession by direct pronominal suffixation. These are all blood relations, with the exception of *asō-n* 'his/her spouse', which is a relationship derived by marriage.

Table 22. Directly suffixed kin terms

tevi-n	'his/her grandparent'[a]	*lati-n*	'his/her mother'
hāvi-n	'his/her grandchild'	*tame-n*	'his/her father'
nati-n	'his/her child'	*matue-n*	'his/her maternal uncle'
tue-n	'his/her same sex sibling'	*asō-n*	'his/her spouse'
mone-n	'her brother'[b]	*tītā-n*	his/her child; its offspring

a. Note that only the primary meanings of these kinship terms are given here. Many of these terms refer by extension to other kinship relations as well.

b. The first five nouns in this table differ structurally from the following five in that they have the option of distinguishing the gender of the referent when the possessor is singular by means of special forms of the pronominal suffixes, i.e.

	Unspecified gender	Male kin	Female kin
1SG	*-k*	*-kuli*	*-kahin*
2SG	*-m*	*-mali*	*-mahin*
3SG	*-n*	*-nali*	*-nahin*

3.6.2. Indirectly suffixed kin terms

In addition to the suffixed kin terms in Table 22, there are some kinship terms which can be expressed as free form nouns in Paamese. Some of the kinship terms listed in Table 22 have referentially identical free form equivalents that are totally different in form. Table 23 lists such semantically equivalent forms which belong to the two distinct grammatical categories.

Table 23. Suffixed and unsuffixed kin terms expressing identical relationships

Unsuffixed	Suffixed	
itau one-n	*lati-n*	'his/her mother'
nāna one-n	*lati-n*	'his/her mother'
māma one-n	*lati-n*	'his/her mother'
itet one-n	*tame-n*	'his/her father'
tāta one-n[a]	*tame-n*	'his/her father'
avu one-n	*tevi-n*	'his/her grandparent'
tomali one-n	*tevi-nali*	'his/her grandfather'
tomahin one-n[b]	*tevi-nahin*	'his/her grandmother'
avov one-n	*matue-n*	'his/her maternal uncle'
tuak one-n[c]	*tue-n*	'his/her brother'
tovuli one-n	*asō-n*	'his wife'

a. The forms *māma* 'mother' and *tāta* 'father' are more common today, while the other forms are much less frequently used.

b. The forms *tomali* and *tomahin* are only rarely used in Paamese. Somewhat more common are the compounded forms *avu tomali* 'grandfather' and *avu tomahin* 'grandmother'. The form *avu* is the most frequently used word, and it is used equally for grandparents of either sex.

c. This form no longer means 'his/her same sex sibling' when it is used as a free form noun, but it refers to the brother of either a male or a female.

Of the synonymous pairs listed, it is only the free forms (without the possessive constituent) which are used as address terms. This is not to say however, that the indirectly suffixed forms in Table 23 are used only as address terms. These forms do allow possessors, with the suffixes attached to a possessive constituent as shown.

The only kin terms in Table 22 that do not have free form alternatives are *hāvi-n* 'his/her grandchild' and *nati-n* 'his/her child'. In Paamese culture, people with whom one is in either of these relationships are normally only addressed directly by personal name, so address forms are not needed for this relationship. It is culturally unacceptable to address by name any of those people with whom one is in the relationships included in Table 23. Such people are normally addressed only by the kin term itself. The existence of the free forms *tuak* 'brother' and *tovuli* 'wife' are exceptions in that the Paamese normally address their own brothers, and wives directly by name rather than by a kin term. However, the term *tuak* 'brother' is frequently used by males as an address term for one's friends, or to somebody who one wants to be friendly with but whose name one does not actually know (in much the same way as 'mate' is used in New Zealand and Australian English). The existence of the free form *tovuli* for the same meaning can be explained as it is a relationship derived by marriage. Thus, *tovuli* 'wife' behaves regularly, while the synonymous *asō-n* 'his/her spouse' from the previous set is anomalous in accepting direct suffixation.

This brings us to those kin relationships that are only ever expressed in Paamese by means of free form nouns. Table 24 sets out the forms of this type in my corpus.

Table 24. Kin relationships never expressed by means of directly suffixed nouns

māse one-n	'his/her sibling-in-law'
uan one-n	'his/her sibling-in-law'
toulet one-n	'his/her sibling-in law'
ahineli one-n	'his/her daughter-in-law'
ahoi one-n	'his/her fiancé(e)'
meteilau one-n	'his/her nephew/niece'

One could argue that one's in-laws, as they are only 'acquired' through marriage, are considered to be 'separable' kinship relationships, in contrast with the blood relationships included in Table 22. The difference between *asō-n* 'his/her spouse' from Table 22 and *ahoi* 'fiancé(e)' from Table 24 could be explained in the same way. One is permitted to change one's intended spouse in Paamese culture before marriage, but once married, one is more or less stuck for life. Divorce is almost never practised. The terms *avu*, *tomali* and *tomahin* in Table 23 do not just refer to grandparents, but are also used to refer to parents-in-law, and could therefore be considered to be separable in the same way for this reason.

If this explanation for the distribution of kinship terms between the sets indicated in Tables 22-24 is in fact correct, we would still be left with the problem of why *matue-n* 'his/her maternal uncle' is directly suffixed, while reciprocal term *meteilau* 'nephew/niece' is not. The form *tevi-n* 'his/her grandparent' in Table 22 can also be used to refer to one's parents-in-law in the same way as *avu*, *tomali* and *tomahin* from Table 23, yet it accepts direct pronominal suffixes and the synonymous forms do not.

4. Inalienable-alienable interchange

It is often possible to refer to the referents of inalienably possessed nouns in Paamese either in relation to a particular individual, or as being detached from an individual, either literally or metaphorically. In some Oceanic languages, it is possible for any directly suffixed noun to appear without a possessive suffix when the referent is considered as alienable rather than inalienable. For instance, in Fijian, Lynch (1973: 74) reports a contrast between the following:

(39) *na* *yaca-qu*
ART name-1SG
'my name'

(40) *na* *no-qu* *yaca*
ART POSS-1SG name
'my namesake'

Although there is a variety of structural strategies in Paamese which allow this semantic detachment to be expressed, there is much less regularity than in languages such as Fijian, with only some nouns being able to occur without suffixes, and then being used only rarely. These strategies for the alienable expression of inalienables will be discussed in the following sections.

4.1. Forms reflecting **a*

There is a small number of directly suffixed nouns listed in the various sets of nouns above which exist in formally related pairs, in which there is a derived unsuffixed form with an initial syllable which takes the shapes *o*-before *u*-, *e*- before *i*-, and *a*- elsewhere. This syllable derives historically from the proto-Oceanic common noun phrase marker **a*, which has become morphologically reanalysed as part of the root with quite a number of nouns in Paamese, chiefly with historically disyllabic roots that would have become monosyllabic in modern Paamese by regular phonological change (Crowley 1985).[14] When this syllable is added to a directly suffixed noun, there is no longer any possibility of that noun accepting a pronominal suffix. Of the total corpus of just over eighty directly suffixed nouns in Paamese, only about a quarter of these have

Table 25. Suffixed nouns with formally related free forms

Suffixed form		Unsuffixed form
vati-n	'his/her head'	*a-vat*
hili-n	'his/her hair'	*a-hil*
mete-n	'his/her eye'	*a-met*
ralinge-n	'his/her ear'	*a-raling*
nā-n	'his/her face'	*a-na*
loho-n	'his/her tooth'	*a-loh*
mē-n	'his/her tongue'	*a-me*
sī-n	'his/her bone'	*a-si*
hē-n	'his/her limb'	*a-he*
vango-n	'his/her belly'	*a-vang*
vite-n	'his/her navel'	*a-vit*
vote-n	'his/her buttocks'	*a-vot*
vile-n	'her vagina'	*a-vil*
rā-n	'his/her blood'	*a-ra*
tā-n	'his/her excrement'	*a-ta*
ise-n	'his/her name'	*e-is*
uve-n	'his/her staying up all night'	*o-uv*

derived free forms of this type. Table 25 represents an exhaustive listing of such forms in the corpus.

To the forms which exist in regularly derived pairs listed in Table 25, we can also add the suffixed forms *ongo-n* 'his/her mouth' and *ō-n* 'his penis'. These also have unsuffixed equivalents, though they are irregular in their derivation. The unsuffixed form of *ongo-n* 'his/her mouth' is *ango*, while the unsuffixed form of *ō-n* 'his penis' is *oa*.

The pairs of suffixed and free form nouns in the lists in Table 25 have identical reference. There are, however, two words from the set of nouns described in section 3.3 as referring to imprints of body parts which enter into the same structural relationship. However, the meaning of the unsuffixed noun is not identical with that of the suffixed noun, though the meanings are clearly related. Thus, corresponding to *ulunge-n* 'end of bed where his/her pillow goes' we find *oulung* meaning 'pillow'. The suffixed noun *vuli-n* 'his/her place' corresponds to the unsuffixed noun *avul* 'hole'.

Although the two sets of nouns in Table 25 have identical reference, there is still a very clear semantic distinction between the two. The forms with pronominal suffixes are seen very clearly as inalienably possessed body parts, or as some inherent and inseparable part of a particular individual or individuals. On the other hand, when the free form of these nouns is used, the referent of the noun is being spoken of without reference to any particular individual. Thus compare the following examples:

(41) *Hili-n houlu.*
hair-3SG much
'He/she has a lot of (i.e. long) hair.'

(42) *Ahil houlu mul en atan.*
hair much 3SG:REAL:exist LOC ground
'There is a lot of hair on the ground.'

(43) *Ise-n mē-Voum keil houlu vi Maki.*
name-3SG Paamese PL many 3SG:REAL:be Maki
'The name of many Paamese is Maki.'

(44) *Maki vi eis tā tenout Voum.*
Maki 3SG:REAL:be name one P/Wh:PLC Paama
'Maki is a Paamese name.'

These free forms can be used when the speaker for some reason wants to avoid associating a body part with a particular individual (for instance, because it might perhaps be incriminating or embarrassing). The first statement below is therefore less embarrassing than the second one because it enables the speaker to avoid associating the referent of the body product noun with any particular individual:

(45) *Naveas ata tā.*
1SG:REAL:step:in excrement one
'I have stepped in (a piece of) excrement.

(46) *Naveas tām.*
1SG:REAL:step:in excrement:2SG
'I have stepped in your excrement.'

The non-suffixed forms of these nouns are also used when the noun is not being used to refer to part of the body of any particular individual, but of a generic class of individuals. The two nouns which participate in this kind of construction are linked by the preposition *ten* which expresses purpose. Thus:

(47) *sōv ten ahil*
soap PURP hair
'shampoo'

(48) *tokita ten amet*
doctor PURP eye
'optician'

(49) *tokita ten aloh*
doctor PURP tooth
'dentist'

This construction is also used in the case of idioms such as the following, in which there is no reference to the body part of any particular individual:

(50) *asi ten āv*
bone PURP firewood
'shin' (Literally: 'firewood bone')

(51) *muas avot*
hit buttocks
'practise sodomy' (Literally: 'hit buttocks')[15]

These unsuffixed forms of inalienably possessed nouns are also sometimes used as insults or as joking epithets to people, to indicate that the person is characterised in some unusual way by that particular part of their body. Thus:

(52) *Ametemau, avu!*
eye:extent grandmother
'What big eyes you have grandmother!'

(53) *Asa, oa?!*
what penis
'What is it, you prick?!'

That the use of these free form nouns has this particular effect perhaps derives from the fact that the body part is being metaphorically detached from the body and held up for scrutiny and ridicule.

4.2. Forms without reflexes of **a*

The majority of suffixed body part nouns in Paamese only ever occur with a suffixed prominal possessive form. Such nouns have no equivalent free form that can be used in the kinds of alienable constructions that were just discussed in section 4.1. There is no satisfactory conditioning factor which determines whether a suffixed noun in Paamese will have an unsuffixed equivalent or not. Most of those forms which do have unsuffixed equivalents have monosyllabic roots in their surface forms, but there are many forms in Table 4 above with monosyllabic roots which do not enter into this alternation. Similarly, there is no discernible semantic conditioning factor involved here.

If the speaker wants to use any of these other suffixed nouns in sentences in which the body part is semantically dissociated from its inalienable possessor, then the noun must be used in its third person singular possessive form. Thus, the example below would be used to refer to deodorant for armpits generally, as well as deodorant for some particular person's armpit, and it is impossible to distinguish between these two interpretations formally in Paamese.

(54)	*seda*	*ten*	*hingen*
	deodorant	P/Wh	armpit:3SG

'underarm deodorant/deodorant for his/her armpit'

Similarly, if such body part nouns are used jokingly or insultingly as epithets, then this must also be based on the third person singular possessed form:

(55)	*Vulingasinemau!*
	nose:3SG:extent

'What a nose!'

Suffixed nouns which refer to aspects of a person other than body parts can also be used in a way that is semantically alienable, but they too must carry the semantically empty third person singular suffix. Thus:

(56)	*Ialu*	*lotās*	*en*	*ninin.*
	1DL:INCL	1DL:INCL:REAL:one:only	in	spirit:3SG

'You and I are together in spirit.'

In addition to referring in this way to disembodied body parts, the special free forms referred to in section 4.1 can also be used with extended meanings that are not directly related to our bodies at all. We therefore find the following examples, which have the extended meanings as indicated:

vati-n	'his/her head'	*avat*	1.	a head (detached from a living being)
			2.	determination
			3.	leader
rā-n	'his/her blood'	*ara*	1.	some blood
			2.	red wine
ise-n	'his/her name'	*eis*	1.	a name
			2.	word
			3.	writing

Although these free form nouns can be considered to be formally like alienable nouns in that they do not have pronominal suffixes, and also in that their referents are semantically dissociated from the body part of any particular individual, they still cannot take part in possessive constructions in the way that we would expect for any ordinary alienable noun. While it is possible to conceive of a wide variety of situations in which body parts and products could be alienably possessed, it is only possible to use these particular nouns with an alienable possessor when the noun is used in reference to anything *but* a body part. Thus:

(57) *ara* *ema-k*
blood/wine POSS:POT-1SG

can only mean 'my red wine' and never 'my blood' (even if one had a bottle of blood which, for whatever perverse reason, one wanted to drink). Similarly:

(58) *avat* *ona-k*
head/determination/leader POSS-1SG

can only mean 'my determination' or 'my leader' and could not refer to a guillotined head that one might have claimed as a personal souvenir. Finally:

(59) *eis* *ona-k*
name POSS-1SG

can be used to mean 'my writing' but never 'my name' (even if it is not the name I use for myself, but merely a name that I happen to have chosen for someone else). In fact, the only specially derived free form nouns of this type that occur freely with alienable possessors are the two semantically unpredictable forms *avul* 'hole' and *oulung* 'pillow' mentioned in section 4.1.

In order to express alienable possession over body parts and products which have unsuffixed equivalents, it is grammatically impossible to avoid specifying some kind of an inalienable possessor. It is necessary therefore, to express both an inalienable possessor and an alienable possessor within a single possessive phrase, with the latter of course being expressed by means of a suffix attached to the appropriate possessive constituent. This can be expressed by means of the formula:

[POSSESSED NOUN-SUFFIX + POSSESSOR NOUN] +
[POSSESSIVE CONSTITUENT-SUFFIX (POSSESSOR NOUN)]

with the following variant when there is a pronominal possessor:

[POSSESSIVE CONSTITUENT-SUFFIX] + [POSSESSED NOUN-SUFFIX + POSSESSOR NOUN]

Thus:

(60) *eaten* *vuas* *āk*
liver:3SG pig POSS:ED:1SG
'my pork liver' (Literally: 'my pig's liver')

(61) *lohon* *vuas* *onen* *asuv*
tooth:3SG pig POSS:3SG chief
'the chief's tusk' (Literally: 'the chief's pig's tooth')

(62) *āk* *tin* *buluk*
POSS:ED:1SG intenstines:3SG cow
'my offal' (Literally: 'my cow's intestines')

In situations where the speaker is genuinely unable to specify the actual nominal possessor, it is possible just to use the third person singular suffixed form of the noun as a "dummy" possessor, with an associated possessive constituent still expressing the alienable possessor. For instance, if one wanted simply to say 'my liver (to eat)' without specifying that it is pork liver, chicken liver, fish liver or the liver of a cow, it would be possible to say the following:

(63) *āk* *eaten*
POSS:ED:1SG liver:3SG
'my liver' (Literally: 'my its liver')

4.3. Inalienability and phrasal nouns

A closely related construction to the one just described that is found in only the two following examples:

(64) *sī-n* *ten* *eisilu-m*
bone-3SG P/Wh back-2SG
'your backbone' (Literally: 'your its bone of the back')

(65) *sī-n* *ten* *vati-n* *mesau*
bone-3SG P/Wh head-3SG fish
'the fish's procumbent spine' (Literally: 'fish's its bone of the head')

This construction involves a part-whole relationship, with the preposition *ten* (which also marks the purposive) linking the part to the whole. Although there is no formally expressed alienable possessor as in the preceding sets of examples, the part-whole relationship behaves in the same way as a relationship of alienable possession. The part, in this case *sī-n* 'its bone', is expressed with an obligatory third person singular pronominal suffix, and the noun does not occur in its free form of *asi*.[16]

There are some noun phrases in Paamese which structurally have the form of an inalienably possessed noun phrase, but which are used idiomatically. Although the possessor in such constructions is not interpreted as applying to any particular suffixed noun but represents a metaphorical extension of the meaning from possession of a generic class, we nevertheless find that such constructions are expressed formally as directly suffixed constructions. The corpus includes the following examples of this type:

(66) *tā-n* *aman*
excrement-3SG bird
'kind of breadfruit'

(67) *tā-n* *hōs*
excrement-3SG horse
'kind of yam'

(68) *tā-n* *vuas*
excrement-3SG pig
'tobacco stick'

(69) *tā-n* *ahu*
excrement-3SG turtle
'pumice'

(70) *tā-n* *vares*
excrement-3SG flying fox
'kind of wild yam'

(71) *vati-n* *vares*
head-3SG flying fox
'kind of wild yam'

(72) *mete-n* *āi*
eye-3SG wood
'prow of canoe'

(73) *ō-n* *buluk*
penis-3SG bull
'whip'

(74) *luo-n* *temat*
vomit-3SG devil
'kind of insect'

(75) *uti-vile-n* *Makerit*
seed-vagina-3SG Margaret
'kind of breadfruit'

There are also two examples of this kind of construction in Paamese in which the 'possessor' noun itself carries a pronominal suffix for possession:

(76) *vati-n* *ō-m*[17]
head-3SG penis-2SG
'your glans penis' (Literally: 'your penis' head')

(77) *mete-n* *ō-m*
eye-3SG penis-2SG
'your (male) urethral opening' (Literally: 'your penis' eye')

Examples such as those listed above are so removed semantically from their forms as possessive constructions that it is even possible for them to be compounded in the same way as ordinary monomorphemic nouns, e.g.:

(78) *tin-mē-n* *buluk*
island:cabbage-tongue-3SG cow
'kind of island cabbage' (Literally: 'cow's tongue island cabbage')[18]

4.4. Compounds involving inalienable nouns

There is one other kind of construction involving suffixed nouns in Paamese in which the referent of the noun is considered to be separate from the 'possessor', and is thus semantically alienable. This is the compound construction in which an ordinarily suffixed noun carries no suffix, but is instead morphologically bound to a following stative root.[19] Such compound nouns can carry any of the following three meanings:

(i) a general class of people characterised by having a body part in that particular state;
(ii) being in that state itself;
(iii) an actual body part that is characterised by being in that state, without stating whose body part it is.

Thus, from *vati-n* 'his/her head' and the stative root *to* 'bald', it is possible to derive the compound *vatito*, which means any of the following:

1. bald person
2. baldness
3. a bald head

Not all suffixed nouns are attested as occurring in this kind of compound construction. None of the suffixed nouns with kinship reference are able to occur in compound constructions of this kind. Of those suffixed nouns which can form compounds in this way, we can include those nouns that have separate unsuffixed forms (as listed in Table 25), as well as the nouns listed in Table 26.

Table 26. Suffixed nouns that can be compounded with statives

ave-n	'his/her body'	*ase-n*	'his external genitalia'
hirē-n	'his/her neck'	*ē-n*	'his interior'
eisili-n	'his/her back'		

Sometimes, compound nouns that are derived in this kind of construction can also have extended meanings that only indirectly relate to the meanings of the nouns that they are derived from, as illustrated by the following:[20]

(79) *av-keih*
body-strong
'pimple without a head'

(80) *av-tin*
body-hot
'perspiration'

Another construction in Paamese in which unsuffixed forms of inalienably possessed nouns can take part is a compound construction involving verbs. This construction is one in which a noun that is ordinarily suffixed for possession is morphologically bound to a preceding verbal form, with the noun taking no possessive suffixes. This construction is not productive, and is only attested with a handful of verbs. Thus, for example, *hili-n* 'his/her hair/its feathers' appears in *rātā-hil* (cut-feather) 'preen feathers' and *vati-n* 'his/her head' appears in *kalialī-vat* (shake-head) 'shake one's head in disagreement'.

4.5. Inalienable-alienable interchange with kin terms

With regard to kin terms, it is more difficult to conceive of these nouns as being semantically alienable. It is possible, however, to find sentences in English such as the following in which no specific possessor is expressed: *I want a son*. It is clear that if there were to be a possessor in such a sentence, then it would be 'I'. Explicitly including this possessor is the only structural option in Paamese for expressing such a sentence:

(81)

Inau	*ēk*	*vat*	*natukuli*	*tāi.*
1SG	interior:1SG	3SG:REAL:want	child:1SG:MASC	one

'I want a son.' (Literally: 'I my interior wants a my son.')

There are equational constructions also such as the following in English: *Those two are brothers*, and *Those two are father and son*. In such constructions, the kin term is being used simply to express an abstract kinship relationship, without applying it to any particular individual. In

Paamese, it is possible for unsuffixed kin terms to occur in a parallel kind of construction without the need to express any possessor, as in:

(82) *Keilu luvi tuak.*
3DL 3DL:REAL:be brother
'Those two are brothers.'

At the same time, it is possible to express this kind of sentence with one of those directly suffixed kin terms by adding a possessive suffix that agrees with the person and number of the subject. Thus:

(83) *Īr vasī rovi tuer*
1PL:INCL all 1PL:INCL:REAL:be brother:1PL:INCL
en meten Ahi.
LOC eye:3SG God
'We are all brothers in the eyes of God.' (Literally: 'We are all our brothers in the eyes of God.')

In the case of suffixed kin terms for which there is no free form equivalent, such constructions can only ever be expressed with a pronominal possessive constituent. Thus:

(84) *Keilu luvi tāta keilu min natnali.*
3DL 3DL:REAL:be father 3DL DAT child:3SG:MASC
'Those two are father and son.' (Literally: 'Those two are father with his son.')

(85) *Oreliato āk vī tītān.*
egg:chicken POSS:ED:1sg 3SG:REAL:have offspring:3SG
tai.
PERF
'My egg has a chicken in it (and I was going to eat it!)' (Literally: 'My chicken egg for eating has its offspring.')

5. Conclusions

The introductory section of this paper included a brief reference to the question of whether it is grammar or semantics that is prior in establishing the categories of a language. My own position on this, at least with respect to the question of inalienability in Paamese, falls somewhere between these two extremes. Despite my best efforts in this paper, I have not been able to relate fully the distribution of suffixed and unsuffixed nouns in the Paamese lexicon to the difference between inalienable and alienable possession. Why should *ale-n* 'his/her lower leg' accept direct pronominal suffixes, while *haroma* 'his/her shin' does not? Similarly, if central aspects of our "selves" such as *nini-n* 'his/her soul/spirit' accept direct suffixation, why is the same not true for *alual* 'life force'? These are questions that I confess to having no answer for, and, unless there are semantic explanations for these facts hidden deep within Paamese cosmology, I am prepared to say that there is not a total match-up between formal and semantic categories in this language.

However, the grammatical construction of direct pronominal suffixation of possessors as against the attachment of suffixes to a nearby, but grammatically separate, possessive constituent is a fairly iconic way of expressing semantic inalienability, given that inalienability also involves the concept of direct attachment in some way. If the construction is genuinely iconic, then we should perhaps expect it to be maximally utilised in the grammar, rather than just being applied haphazardly. In treating the subject of inalienability as a discrete topic as I have in this paper, we are forced to look much more closely at the relationship between semantic inalienability and the structural category of directly suffixed nouns. Many semantic explanations for facts have become obvious to me which before had previously appeared exceptional. Thus, while I do not claim that all of these structural facts can be explained from a semantic basis, the vast majority certainly can.

A prototypical directly suffixed noun in Paamese is therefore one that is considered to be in an inalienable semantic relationship with another noun having an animate referent. An inalienable relationship holds between two things if, under normal circumstances, the referent of the "possessed" noun does not exist independently of the referent of the "possessor" noun. Thus, the following generally accept direct pronominal suffixation in Paamese, for purely semantic reasons:

(i) permanent external body parts rather than transient external growths on the body;
(ii) body products exuded under normal rather than transient body functions;
(iii) internal body parts perceived to be central to emotions and to life itself rather than internal body parts that are removed while

butchering an animal, reserved for eating, or impediments which are removed while eating;

(iv) blood kin rather than kin acquired by marriage.

In addition to these open-ended classes of inalienably possessed nouns, we also find the following much smaller sets of nouns which fall under the same semantic generalisation, and therefore accept direct pronominal suffixation:

(v) most nouns expressing aspects of one's individuality;
(vi) some nouns expressing relative position;
(vii) most nouns expressing imprints and impressions made by something;
(viii) a few parts of wholes;
(ix) some nouns expressing the best part of a meal.

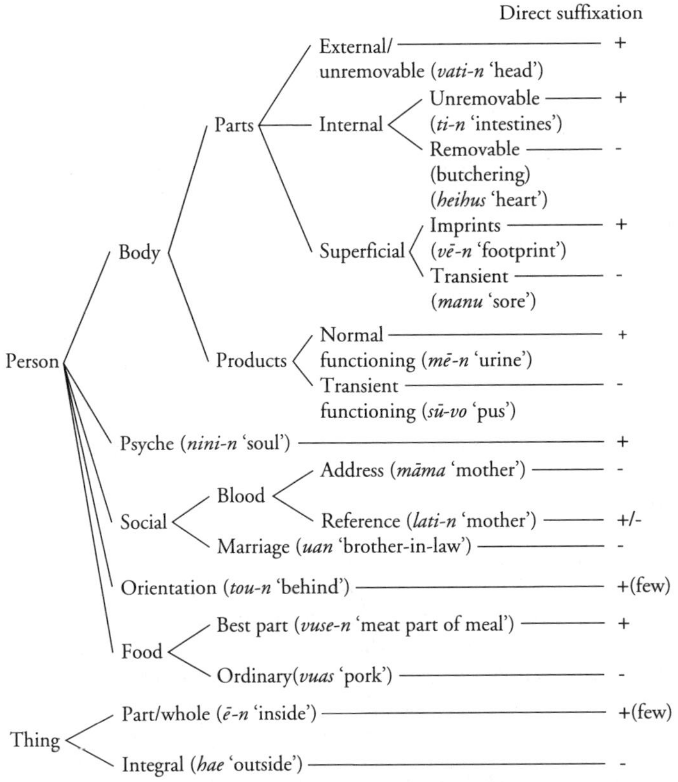

Figure 3. Overall statement of the distribution of directly suffixed nouns in Paamese

A more detailed summary of the semantic criteria in the distribution of suffixed and unsuffixed nouns in Paamese is presented in Figure 3. This summary ignores those nouns which are suffixed or unsuffixed for purely formal rather than semantic reasons.

Notes

1. This paper is based on research carried out during the years of 1976-78 and 1983-90, during which I was resident in Vanuatu. I would like to thank Elaine Good, John Lynch, Hilary Chappell and Bill McGregor for helpful comments on earlier drafts. All conclusions, of course, remain my own.
2. Underlying forms in Paamese generally differ from surface forms. For a discussion of Paamese underlying phonology and the rules that connect this level of analysis to surface forms, see Crowley (1982: 12-43). For a discussion of how the Paamese orthography relates to the phonological system of the language, refer to Crowley (1982: 12-18).
3. The following abbreviations are used in this discussion: 1 – first person; 2 – second person; 3 – third person; SG – singular; DL – dual; PCL – paucal; PL – plural; POSS – possessive constituent; ED – edible possession; POT – potable possession; REAL – realis; DIS – distant; PERF – perfective; TR – transitive; LOC – locative; PURP – purposive; COM – comitative; P/Wh – part/whole; MASC – masculine; RED – reduplication; WEAR – wearable; PLC – place; ART – article.
4. These possessive suffixes are partly irregular in their morphology. The roots undergo some unpredictable changes with the addition of certain pronominal suffixes. Thus, for instance, this form in the second person singular is *ono-m*, while in the third person singular it is *one-n*. For details of this morphology, see Crowley (1982: 111-112). See also Osumi (1995).
5. Most other nouns are not semantically compatible with all four possessive constituents as we find with this noun. However, many nouns can occur with two different possessive constituents with predictably different meanings.
6. The alternations between front and back vowels in this example and earlier examples of the same noun with different pronominal suffixes is predictable. See Crowley (1982: 106-107) for details of the morphophonemic alternations.
7. The appearance of the vowel in the suffix *-ne* here is predictable according to the phonological rules of the language referred to above.
8. In Crowley (1982: 104-106), this fact prompted the present writer to equate *-n* in these kinds of contructions with the "construct suffix" of Micronesian languages. However, it is also possible to analyse this as a particular use of the third person singular possessive suffix. There are some kinship terms (which are referred to below) which have special forms of the singular suffixes to indicate gender, and the *-n* in such constructions behaves identically with these nouns.
9. Paamese males are circumcised just before puberty. Despite the fact that there is clearly pain involved, at least traditionally when the operation was carried out without the benefit of local anaesthetic, foreskins should be considered as being separable body parts in Paamese culture.

10. Sensations and feelings in Paamese are also often expressed by means of the suffixed nouns *uli-n* 'his/her seat of emotions' and *ēn-* 'his/her interior', which will be discussed below. These forms do not refer to specific internal organs however.
11. Pork fat is relished by Paamese speakers and so is regarded as a potentially edible body part.
12. The Paamese people believe that other people staying up all night can cause one to sleep badly, even if they are in another village and their noise cannot be heard.
13. Although this would suggest either jocularity or euphemism, the context in which it was recorded suggested it had no unusual connotations. This item of clothing has not been regularly worn on Paama for seventy-five years or so, and the term itself is now extremely rare in any case.
14. The southern dialect of Paamese reflects this original common noun phrase marker as invariant *a-* with all nouns.
15. The form *kiskisvot* 'index finger', involving a compound of *kiskis* 'poke' and the unsuffixed form of *vote-n* 'his/her buttocks', could perhaps be explained in the same way. Nouns retaining the historical common noun phrase marker **a* generally lose this in compounds (Crowley 1985: 142-144).
16. At the same time, however, these examples should be compared with example (48) in which *asi ten āv* (rather than *sīn ten āv*) means 'shin' (literally: 'firewood bone').
17. These two examples are presented with second person singular possessors in order to make it clear which possessive suffix expresses the possessor of the body part that is referred to by this phrasal construction.
18. There is, however, another compound noun in Paamese that is formed without the possessor being expressed:

    tōhilsesāh
    chicken:hair:hairy:caterpillar

    and not as:

    **tōhilin* *sesāh*
    chicken:hair:3SG hairy:caterpillar

    A possible explanation could be the higher degree of animacy in the case of the *buluk* 'cow' as against the *sesāh* 'hairy caterpillar', with higher animates requiring the expression of a possessor and lower animates not.
19. Adjectives constitute a very small word class in Paamese. Most statives are expressed formally as verbs.
20. Sometimes, such compounds also express meanings that are totally removed from the semantic field of body parts, e.g. *met-hatetamen* (eye-big-PL) 'kind of fish with large eyes', *hirē-vu* (neck-bulge) 'heron'.

References

Ameka, Felix
1995 "Body parts in Ewe grammar", [this volume.]
Clark, Ross
1976 *Aspects of proto-Polynesian syntax.* (*Te Reo* Monographs.) Auckland: Linguistic Society of New Zealand.
nd List of proto-North Central Vanuatu reconstructions. [Unpublished MS.]
Crowley, Terry
1982 *The Paamese language of Vanuatu.* (Pacific Linguistics Series B Number 87.) Canberra: Pacific Linguistics.
1985 "Common noun phrase marking in proto-Oceanic," *Oceanic Linguistics* 24 (1/2): 135-193.
1991 "Linked nouns in Paamese: grammatical iconicity in part-whole constructions," *Language and linguistics in Melanesia* 22: 21-49.
1992 *A dictionary of Paamese.* (Pacific Linguistics Series C Number 121.) Canberra: Pacific Linguistics.
Evans, Nick
1995 "The syntax and semantics of body part incorporation in Mayali," [this volume.]
Halim, Amran – Lois Carrington – Stephen A. Wurm (eds.)
1982 *Papers from the third International Conference on Austronesian Linguistics, Volume 1: Currents in Oceanic.* (Pacific Linguistics Series C Number 74.) Canberra: Pacific Linguistics.
Lichtenberk, Frantisek
1985 "Possessive constructions in Oceanic languages and in proto-Oceanic". In: Andrew Pawley – Lois Carrington (eds.), 93-140.
Lynch, John
1973 "Verbal aspects of possession in Melanesian languages." *Oceanic Linguistics* 12 (1/2): 69-102.
1982 "Towards a theory of the origin of the Oceanic possessive constructions." In Amran Halim – Lois Carrington – Stephen A. Wurm (eds.), 243-268.
Milner, George B.
1972 *Fijian grammar.* Suva: Government Press.
Osumi, Midori
1995 "Body parts in Tinrin," [this volume.]
Pawley, Andrew – Lois Carrington (eds.)
1985 *Austronesian linguistics at the 15th Pacific Science Congress.* (Pacific Linguistics Series C Number 88.) Canberra: Pacific Linguistics.
Schütz, Albert J.
1985 *The Fijian language.* Honolulu: University of Hawaii Press.
Wilson, William H.
1982 *Proto-Polynesian possessive marking.* (Pacific Linguistics Series B Number 85.) Canberra: Pacific Linguistics.

Body parts in Tinrin

Midori Osumi

1. Introduction[1]

This paper examines body parts in the morpho-syntactic structure of Tinrin. Tinrin is an Austronesian language spoken by about 400 people in the southern part of New Caledonia, mainly in the reserves of Petit Couli and Grand Couli; the speakers of Tinrin are dispersed and mixed with speakers of Xârâcùù in Lafoa, Katrikoin, Pocquereux and Sarramea. There is also a related language Ha Mea in the Kouaoua area (Méchin, Méa Mébara and Fachin), with about 300 speakers (Rivierre 1981).

Typologically Tinrin is characterised by VOS word order[2], prepositions and right-branching. It makes extensive use of tense-aspect markers and verbal modifiers which can be combined to express subtle differences in meaning. Singular, dual, and plural personal pronouns are distinguished in first, second and third persons.[3] They also contrast between inclusive and exclusive in the first person non-singular. There are also two impersonal pronouns.

In the following, we first examine the possessive constructions of Tinrin. Tinrin exhibits multiple types of possessive constructions, as do most Oceanic languages (Lichtenberk 1985). Then, we will have a closer look at the nouns referring to body parts, especially how they fit into these possessive constructions. Thirdly, the use of some body part nouns as prepositions indicating a relative location is discussed.

2. Possessive constructions

Archetypical possessive constructions are grammatical constructions which involve two nominals and indicate that the referent of one of these nominals is possessed by the other. In this discussion, one of the nominals is referred to as the noun possessed and the other as the possessor, although the relationship expressed is not necessarily that of true possession, but can be of many other types.

In many languages, when there is an inherent relationship between two referents such as the kinship or body part relationship, it is realised in constructions encoding inalienability, while non-inherent relationships such as temporary ownership or voluntary association occur in constructions encoding alienability (Chappell – McGregor 1989; Mosel 1984). Tinrin also makes a distinction between alienable and inalienable possession. In fact, it shows a variety of possessive construction types, which will be discussed in the following sections.

2.1. Alienable possession

Alienable possession in Tinrin can be realised in various ways, which on the basis of structural criteria can be divided into two types: Nouns which do not obligatorily occur in inalienable possessive relationships (which will be described later) can occur in constructions where the noun possessed is followed by a possessive preposition which in turn is followed by the pronominal or nominal possessor. Most of these nouns can also enter into another construction, in which the noun possessed occurs in apposition to a possessive classifier.

2.1.1. Constructions with a possessive preposition

There are three possessive prepositions which can occur in this construction: *nrâ*[4] 'of (general, temporary possession)', *rre/rrê* 'of (personal living need)', and *ò* 'of (fire)'. They occur between two nominals (the possessor NP[5] can be pronominal) as in the following formula.

$$\begin{matrix} \text{NP} \\ \text{poss'd} \end{matrix} \quad \left\{ \begin{matrix} \textit{nrâ} \\ \textit{rre / rrê} \\ \textit{ò} \end{matrix} \right\} \quad \begin{matrix} \text{NP} \\ \text{poss'r} \end{matrix}$$

The choice of the preposition is determined by the noun possessed. The majority of nouns can occur in possessive constructions with *nrâ*. It normally suggests a temporary ownership or voluntary/controlling relationship as in the following example:

(1) *wake nrâ nrü*
job of 2sg
'your job'

Nrâ also suggests the entity's involvement in an event. In the following example, *nrâ* is used to indicate the agent and the object of nominalised active verbs (when both the subject and object are present, *nrâ* before the object is occasionally dropped):

(2) *bee ta (nrâ) pù nrâ ri*
CMPZ kill (O) flying:fox S 1pl:inc
'our killing (of) flying foxes'

The use of *nrâ* to indicate the agent of verbal nouns as above suggests the active voluntary or controlling relationship, as does the Tolai (Austronesian) *ka*-possession, which Mosel (1984: 37-47) considers as the most "established" (alienable) possession as opposed to inherent (inalienable) possession.

Rre and *rrê* are variants of the same morpheme (*rre* occurs with *ò*, *rrê* with *mwâ* and *wâ*, conditioned by the nasality of the vowel in the preceding syllable), indicating personal possession, which is related to some essential means of livelihood. The possessed NPs are restricted to *mwâ* 'hut', *wâ* 'boat' and *ò* 'pot'. When *mwâ* 'hut' is reduplicated as *mwâmwâ* meaning 'assembly house', it no longer occurs with *rrê* but occurs with the general preposition *nrâ*, because *mwâmwâ* is not possessed personally for living in, but is given a public use (for a meeting or a party). Boats are essential for catching fish, or carrying important items of subsistence such as food from other places. Pots are also important, in preparing food.

(3) *mwâ rrê kevi*
hut of 1pl:exc
'our (not including you) hut'

(4) *wâ rrê mê-pwogarra*
boat of pl-white:people
'white people's boat'

(5) *ò rre toni*
pot of Tony
'Tony's pot'

The possessor marked by *rre*/*rrê* can be inanimate, which is an essential constituent of the entity possessed as in the following examples where *mwâ* in (6), *wâ* in (7), *ò* in (8) refer to things which contain something, and *arròò* 'water', *farrawa* 'flour', *u* 'yam' in each phrase refer to constituents in them.

(6) *mwâ rrê arròò*
container of water
'bottle of water'

(7) *wâ rrê farrawa*
boat of flour
'boat carrying flour'

(8) *ò rre u*
pot of yam
'yam dish cooked in the pot'

The possessed NPs which can occur with the preposition *ò*, are restricted to *nre* 'fire, firewood' or compounds with *nre*. The possessive relationship expressed by *ò* is thus related to fire, which is possessed for the purpose of warming oneself or cooking something. The possessor expressed by this construction can also be something that is cooked or burnt with fire.

(9) *nre ò ru*
fire of 1dl:inc
'fire of us two'

(10) *nre ò ò*
fire of pot
'the fire with which the pot is cooked'

These possessive prepositions are summarised as follows:

nrâ	temporary ownership or voluntary relationship
rre/rrê	personal living needs; house, boat, pot
ò	the use of fire for warming or cooking purpose

Although the above three prepositions are all used to express alienable possession, (that is, the possessed noun which occurs in the possessive constructions with these prepositions does not obligatorily enter into these constructions, but can stand by itself), the use of *rre/rrê* and *ò* differ from that of *nrâ*. Compared to *nrâ* which can occur with almost any alienable noun, *rre/rrê* and *ò* can occur only with a few nouns as discussed above. In fact, the nouns referring to 'house', 'pot' or 'fire' often occur as inalienable nouns in other languages as essential or valued possessions (Mosel 1984: 45). Therefore, it can be presumed from the semantic point of view that *nrâ* expresses the most alienable possession among the three prepositions.

nrâ 'general'	*rre/rrê* 'house, pot'
	ò 'fire'
more alienable ⟵	⟶ less alienable

2.1.2. *Constructions with a possessive classifier*

Other types of alienability are expressed in constructions where the possessed noun occurs in apposition with a possessive classifier. Possessive classifiers are a group of nominals whose referents are themselves inalienably possessed by a pronominal or nominal possessor in the same way as bound nouns which we will discuss later. That is, they are either suffixed by a pronominal possessor or bound to a nominal possessor directly following them (cf. next section).

When possessive classifiers occur in apposition with another nominal phrase head, they determine the nature of the possessive relationship between the noun possessed (which is a specifier of the generic possessive classifier) and the possessor. These constructions are seen commonly among Melanesian languages, and are generally used to express alienable possession. In many languages, certain classifiers extend their primary meanings, such as the alimentary classifier also being used to express subordinate possession (Lichtenberk 1985; Mosel 1984; Pawley 1973: 163).

In these constructions, the possessive classifier with its pronominal suffix often precedes the specifier.[6] In Tinrin, it is placed variably before or after the specifier (the noun possessed) as shown in (a) and (b) below. Also, the nominal possessor is not cross-referenced elsewhere in the construction as in Hawaiian (Lichtenberk 1985).

(a)	NP poss'd	POSS.CLFR-NP poss'r
(b)	POSS.CLFR-NP poss'r	NP poss'd

These constructions are illustrated by following examples:

(11) *wî* *ere-rò*
banana fruit-1sg
'my banana to eat'

(12) *ere-rò* *wî*
fruit-1sg banana
'my banana to eat'

Tinrin has seven possessive classifiers, each of which expresses a particular type of socially determined relationship with respect to an entity:

e-	generally starches, to be eaten[7]
ere-	generally fruit, to be eaten
hwee-	generally meat or egg, to be eaten
odho-	things to be drunk
hwiie-	things to be chewed
êê-	plants growing on his/her land, or to be planted
hêê-	belongings

The same object can occur in several possessive constructions depending on the nature of the relationship between it and the possessor.[8] The possessive classifier *hêê-* seems to be a general classifier used when none of the other, more specific constructions is appropriate. The following examples illustrate that *nawa* 'coconut' can occur with four different possessive classifiers:

(13) *nawa* *odho-toni*
coconut drink-Tony
'Tony's coconut to drink'

(14) *nawa* *hwee-toni*
coconut meat-Tony
'Tony's coconut to eat'

(15) *nawa* *êê-toni*
coconut plant-Tony
'Tony's coconut tree (on his land or to be planted).'

(16) a. *nawa* *hêê-toni*
coconut belong-Tony
'coconut belonging to Tony'

Nawa can also occur in the construction using the possessive preposition *nrâ* which was discussed earlier, when *nawa* is temporarily possessed or related to the person in some loose way:

(17) a. *nawa nrâ toni*
coconut of Tony
'Tony's coconut (temporarily possessed by, or related to Tony)'

The difference in meaning between the noun phrases using *hêê* (16a) and *nrâ* (17a) is not very clear, but the construction using the possessive preposition (17a) seems to suggest a looser association of the thing possessed (coconut) and its possessor (Tony). That is, in the noun phrase (17a), the coconut may not belong to Tony, but may be used by Tony for some temporary purpose. Formally, the constructions (13)-(16) are distinguished clearly from that of (17), as those of (13)-(16) allow the inversion of word order (i.e. the classifier can be preposed in front of the possessed noun), and the latter cannot (i.e. the preposition can only occur between the possessed noun and the possessor). Thus, (16b) in the following is grammatical, while (17b) is not.

(16) b. *hêê-toni nawa*
belong-Tony coconut
'coconut belonging to Tony'

(17) b. **nrâ toni nawa*
of Tony coconut
'Tony's coconut'

2.2 Inalienable possession

Nouns that occur in inalienable possessive constructions are generally those whose referents have no independent existence of their own, but are always related to something else. While alienable possession is expressed in various ways, inalienable possession in Tinrin is always expressed in the juxtaposition of the two nouns (the possessor can be pronominal) that are bound to each other in the following way, except for the nouns that occur in the link structure which will be discussed later in this section:

NP – NP
poss'd poss'r

In this structure, the noun possessed always precedes the possessor; it is either suffixed by a possessor pronoun or bound to a nominal possessor directly attached to it. Note that nouns which occur in this structure may also occur in a compounding construction, being bound to another root, either noun or verb, which will not be discussed in this paper.

There are three types of nominals that occur in this construction: possessive classifiers, location nouns and the others which are called bound nouns in this paper.

2.2.1. Possessive classifiers

Possessive classifiers have already been discussed in the previous section. They serve to express alienable possession when they occur with another noun in apposition. They themselves, however, are always bound to their possessor nouns, in which sense they are inalienably possessed. They are semantically distinguished from other inalienably possessed nouns, however, in that while other nouns are possessed in some inevitable and inextricable way by the possessors, the referents of possessive classifiers are things such as food, drink, plants, whose relationship to the possessor is not something inherent, and whose possessor may change from time to time. The reason why these classifiers are bound to their possessor nouns is that eating, drinking and planting crops are essential for existence and are therefore marked as inalienable.

As the meanings of each possessive classifier are stated in the previous section, I will only give some examples in the following:

(18) *u nrorri e-nrü*
1sg give food-2sg
'I gave you food.' (Literally: 'I gave your food.')

(19) *nrâ bwê nrâ hwee-ri ânrâha*
3sg lot sm meat-1pl:inc here
'We have lots of meat here.' (Literally: 'There are lots of our meat here.')

(20) *nrâ fi nrî tòwò ere-tito*
3sg go PURP gather fruit-Tito
'He went to pick up fruits for Tito.' (Literally: 'He went to pick up Tito's fruits.')

Possessive classifiers are also formally distinguished from other bound nouns by the fact that they can derive collective nouns by being prefixed with *nrî-* as follows:

nrî-e-nrî	(COL-food-3sg)	'his food (starches)'
nrî-ere-rò	(COL-food-1sg)	'my food (fruit)'
nrî-hwee-nrî	(COL-meat-3sg)	'his food (meat)'
nrî-odho-ri	(COL-drink-1pl:inc)	'our drink'
nrî-hwiie-komu	(COL-chew-1dl:exc)	'our (we two excluding you) chews (sugarcane)'
nrî-êê-rri	(COL-crop-3pl)	'their crops'
nrî-hêê-nrü	(COL-belonging-2sg)	'your belongings'

2.2.2. *Location nouns*

Location nouns are those which refer to places and which can mark locative case without prepositions. Some location nouns can also occur with prepositions. They can also be subjects or objects like other nouns. Some of these nouns are morphologically bound to another nominal, and express location only with respect to something else, which is expressed in the second NP slot of the structure. They are the following words:

nrùwù-	'inside'
padrêrrê-	'side'
wò-	'back, under, side'
nrînawâ-	'between'
nraamwârrâ-	'front'
drô-	'on, surface'
tê-	'at, home of'
nrîtruwirri-	'middle, half way'
nrîrri-	'edge, at the end'
nrîpoo-	'hollow part'
hodri-	'top edge, tip'
mee-	'end, edge'
êrrê-	'content, inside'
nrime-	'face'
tròò-	'back'

The following examples will illustrate the use of these location nouns:

locative:

(21)	*toni*	*nrâ*	*tuo*	*nrî*	*padrêrrê-tave*
	Tony	3sg	put	3sg	side-bed

'Tony put it beside the bed.'

(22)	*treanrü*	*rri*	*hara*	*nrùwù-mwâ*
	people	3pl	eat	inside-house

'People ate inside the house.'

subject:

(23)	*nrâ*	*drorri*	*nrâ*	*padrêrrê-tave*
	3sg	straight	sm	side-bed

'The side of the bed is straight.'

object:

(24)	*u*	*jorri*	*nrùwù-mwâ*	*a*	*harru*
	1sg	see	inside-house	AT	good

'I saw the nice inside of the house.'

2.2.3. Bound nouns

The rest of the nouns that occur in this construction are the nouns which we call bound nouns. This subclass of nouns consists of various inalienable nouns such as nouns referring to kinship (e.g. *audrê-* 'father', *aunê-* 'mother'), body part (e.g. *ô-* 'head', *nrime-* 'face', *nrîfò-* 'mouth'), a secretion, or product of the body or a plant (e.g. *tufârrî-* 'saliva', *fwîî-* 'excrement'), personal attributes or properties (e.g. *farrî-* 'name', *gwâ-* 'occupation'), parts of a plant or some inanimate things (e.g. *dri-* 'leaf', *kùrrù-* 'stem'), effect/cause of human affairs (e.g. *fwîrrî-* 'effect of speech, magic', *ûû-* 'cause, reason') and a collectivity/genus of something (e.g. *pwò-* 'package', *nrûû-* 'field', *fomee-* 'genre, sort') (Osumi 1995).

The bound noun *fwîî-* 'excrement' has a different form, *fù*, when used as a free noun, that is, when the referent is not attributable to a particular human being or animal.

There are nouns which have referents similar to the categories as those in the above, and yet do not belong here, but belong to the other category of nouns which will be discussed in the next section. The same thing can be said about body part nouns, which will be discussed in detail later in the paper.

Some of the bound nouns are illustrated in the following examples:

(25)	*toni*	*nrâ*	*sùveharru*	*afiraa-nrî*
	Tony	3sg	love	wife-3sg
	'Tony loves his wife.'			

(26)	*ne*	*farrî-treanrü*
	what	name-person
	'What is the name of the person?'	

(27)	*nrâ*	*vajù*	*nrâ*	*hi-rò*
	3sg	sick	sm	leg-1sg
	'My leg is hurting.'			

2.2.4. *Link noun structure*

Other nouns which occur in inalienable possessive constructions are those which we call link nouns in this paper. They behave differently from the other nominals discussed previously. They are obligatorily linked to a pronoun or the head of another nominal phrase with which they are in a possessive relationship by means of the link morpheme *-nrâ-*.

NP	– *nrâ*	– NP
poss'd	LINK	poss'r

When the possessor is the the first person singular pronoun, the link morpheme *-nrâ-* is often suppressed, in which case they appear exactly like bound nouns as illustrated in *panrea-rò* (aunt-1sg) 'my aunt'.

-Nrâ- is identical in form to one of the general possessive prepositions. Therefore, link nouns are formally identical to free common nouns followed by a possessive preposition *nrâ*, except with the first person singular possessor. It is difficult to differentiate these two constructions, but I consider the link noun construction as encoding inalienable possession, as the link nouns are bound to the possessor nouns by the *-nrâ-*, forming a phonological unit with the primary accent on the first syllable of the link noun (and not on that of the possessor noun). Also they cannot be separated from their possessors. Semantically, the link noun constructions also differ from those using the preposition *nrâ*, because the link nouns refer to things that overlap with the same categories as the bound

nouns, while the preposition *nrâ* serves to indicate the typical alienable and loose relationship between two entities as described in 2.1.1.

A few nouns occur both as free nouns and link nouns. *moo* refers to 'man' (and never 'husband') when it is used as a free noun, but as a link noun, it refers to 'husband', in which case *moo-* cannot occur independently without being related to another nominal phrase head, to which it is phonologically attached. The following examples illustrate this:

(28) *wa moo mwâ*
DET man M:DIST
'that man'

(29) *moo-nrâ-sonya*
husband-LINK-Sonya
'Sonya's husband'

The link noun constructions may be considered as being on the boundary between bound noun constructions and alienable possessive constructions, and the fact that the link morpheme *nrâ* can be suppressed with the first person pronominal possessor suggests that the relationship of the noun possessed to its first person singular possessor is semantically closer than to other person possessors, perhaps from the speaker's egocentric perspective.[9] (Cf. Ameka 1995.)

The three possessive constructions of bound, link nouns and free nouns with a preposition are summarised along the scale of inalienable and alienable possession as shown in Figure 1.

bound	link	free	
NP-NP	(1sg) NP-(*nrâ*)-NP (others) NP-*nrâ*-NP	NP *rre/rrê* NP NP *ò* NP	NP *nrâ* NP

inalienable ⟵ ⟶ alienable

Figure 1. Scale of alienability of possessive constructions

As mentioned before, referents of link nouns often belong to similar categories of things as those of bound nouns. They include: kin (e.g. *moo-* 'husband', *hùwù-* 'child'), body parts (e.g. *mwâ-* 'brain', *ao-* 'cheek'), secretions or products of the body (e.g. *kòfio-* 'perspiration', *nruâ-* 'dirt (of body)'), personal attributes or properties (e.g. *mòrrò-* 'life', *fôdrò-* 'idea, thought'), part of plants or inanimate things (e.g. *wii-* 'fibres', *wù-* 'shoot'), effects or results of human affairs (*ahwai-* 'effect, result') and products of some inanimate things (*kû-* 'smoke'). We will discuss the link nouns referring to body parts and some related things in detail in the next section.

Figure 2 summarises the morphemes which appear between the noun possessed and its possessor when a noun forms a possessive relationship with another NP.

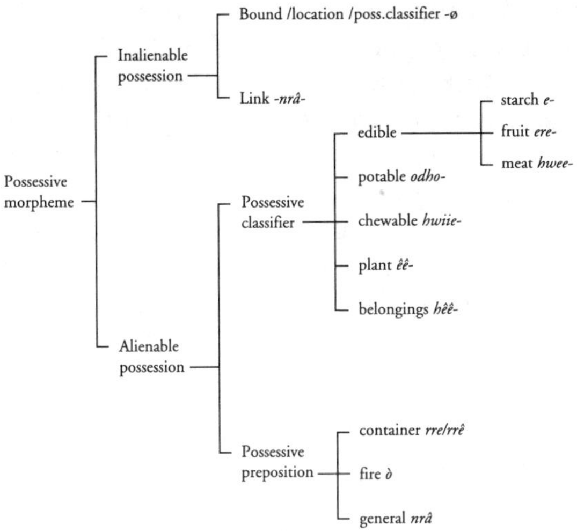

Figure 2. Possessive morphemes in Tinrin

3. Body part nouns

In this section, we examine body part nouns in Tinrin. They generally occur in constructions encoding inalienable possessions. However, possessive relationships are not expressed uniquely with all the body parts. As we saw in the previous section, they are expressed either by bound nouns or link nouns. We will examine in the following what are the members of each category of nouns, and discuss whether the structural distinction between them correlates with semantic or formal factors. In the following discussion, the term "body part nouns" is used in the broader sense, applied also to body products, parts of a plant or objects, properties and concepts related to human affairs.

3.1. Body part nouns occurring as bound or link nouns

Body part nouns occur as either bound nouns or link nouns, and they never enter into other possessive constructions except when the body part is regarded as an object to eat or drink, in which case they occur with a possessive classifier.[10]

(A)-(F) below present exhaustive lists of both bound and link nouns referring to body parts, parts of plants or objects, secretions or body products, and personal attributes or properties. Bound and link nouns are presented next to each other so that they can be readily compared.

(A) External body part nouns. Table 1 lists the morphologically simple bound and link nouns referring to external parts of the body.

The following bound nouns referring to external body parts are morphologically complex. They involve two bound nouns, a bound noun preceded by a location noun, or a bound noun preceded by a bound noun and location noun in the last example.

bound + bound:

nrîfò-pe- 'anus' (< *nrîfò-* 'mouth' + *pe-* 'bum')
virri-hi- 'fingernail' (< *virri-* 'shell' + *hi-* 'leg')
virri-ô- 'skull' (< *virri-* 'shell' + *ô-* 'head')
wâ-ô- 'hair on the head' (< *wâ-* 'hair' + *ô-* 'head')
wâ-bê- 'hair on the arm' (< *wâ-* 'hair' + *bê-* 'arm')
wâ-hi- 'hair on the leg' (< *wâ-* 'hair' + *hi-* 'leg')
wave-bê- 'elbow, wrist' (< *wave-* 'joint' + *bê-* 'arm')
wave-hi- 'ankle' (< *wave-* 'joint' + *hi-* 'leg')
ûû-bê- 'shoulder' (< *ûû-* 'base, origin' + *bê-* 'arm')
ûû-mââ- 'upper part of thigh' (< *ûû-* 'base, origin' + *mââ-* 'thigh')

location + bound:

padrêrrê-wêjô- 'the wings of the nose' (< *padrêrrê-* 'side' + *wêjô-* 'nose')
nrùwù-wêjô- 'nostrils' (< *nrùwù-* 'inside' + *wêjô-* 'nose')
nrùwù-hi- 'sole of foot' (< *nrùwù-* 'inside' + *hi-* 'leg')
mee-hi- 'finger' (*mee-* 'edge, end' + *hi-* 'leg')

bound + location + bound:

wâ-mee-fwîî- 'tail' (< *wâ-* 'hair' + *mee-* 'edge' + *fwîî-* 'excrement')

Table 1. External body part nouns

bound	link
ô- 'head'	*ao*- 'cheek'
nrime- 'face'	*ôfaò*- 'horn (of cow)'
nrîfò- 'mouth'	*jawe*- 'tail'
wâ- 'hair'	*nrî*- 'shell or skin (of locust)'
ôrrôme- 'eye'	*mirrô*- 'crest'
bê- 'arm, fin'	
hi- 'leg'	
î- 'body'	
pe- 'bum'	
mââ- 'thigh'	
arrohi- 'calf'	
gùwù- 'body, stem'	
bwadri- 'little finger'	
bwara- 'forehead'	
warrabê- 'tongue'	
tròò- 'back'	
wòrò- 'penis'	
gurru- 'testicles'	
wòùrrù- 'nape of neck'	
puu- 'buttock'	
fònôô- 'breast'	
tiifò- 'lip'	
wâdrewò- 'neck'	
wêjô- 'nose'	
wù- 'belly'	
fòdrêrrê- 'ear'	
meetô 'breast'	
vewò- 'teeth'	
sùrrùfò- 'jaw'	
ôgi- 'skin'	
êrrê- 'body in the shell, tuber'	
virri- 'hell'	
nôbi- 'fin, tail'	

The following two words are morphologically complex link nouns. They are made up of a bound noun or a location noun as the first element and a link noun as the second:

bound + link:

nrîfò-pi- 'the opening of vagina' (< *nrîfò*- 'mouth' + *pi*- 'vagina')

location + link:

mee-pi- 'clitoris' (< *mee*- 'edge, end' + *pi*- 'vagina')

Table 2. Internal body part nouns

bound	link
veha- 'heart'	*mwâ-* 'brain'
pwari- 'stomach'	*wòù-* 'bladder, kidney'
dru- 'liver'	*mwâdrù-* 'rib'
nô- 'bone'	*wii-* 'tendon'
wave- 'joint'	*harra-* 'string (of meat)'
wînrâ- 'bowels'	*pi-* 'vagina'
verrewâwarra- 'groin'	

(B) Internal body parts. Table 2 shows the morphologically simple bound and link nouns referring to internal parts of the body.

The following internal body part nouns are morphologically complex bound nouns. They involve two (or three) bound nouns as follows:

bound + (bound) + bound

nô-nrîfò- 'jawbone' (<*nô-* 'bone' + *nrîfò-* 'mouth')
nô-bê- 'bone of the arm' (<*nô-* 'bone' + *bê-* 'arm')
nô-hi- 'bone of the leg' (<*nô-* 'bone' + *hi-* 'leg')
nô-ûû-bê- 'shoulder plate' (<*nô-* 'bone' + *ûû-* 'base' + *bê-* 'arm')
nô-tròò- 'spine' (<*nô-* 'bone' + *tròò-* 'back')

The following word is a morphologically complex link noun. It involves a bound noun as the first and a link noun as the second element:

bound + link:

wââ-tra- 'vein' (< *wââ-* 'root'+ *tra-* 'blood')

(C) Parts of a plant or an object. Table 3 shows the known morphologically simple bound and link nouns referring to parts of plants or inanimate objects.

In addition to the list above, there is the following bound noun which involves a free noun as the first element and a bound noun as the second.

free + bound:

fònribê- 'fork of tree' (< *fònri* 'river' + *bê-* 'arm')

(D) Secretions or body products. The known monomorphemic bound and link nouns denoting secretions and bodily products are listed in Table 4.

Table 3. Parts of plants and objects

bound	link
wo- 'fruit'	*wii-* 'fibres'
wage- 'flower'	*hùwù-* 'bud, sprout'
êrrê - 'tuber'	
dri- 'leaf'	
wââ- 'root'	
noo- 'spine'	
kùrrù- 'stem'	
ôgi- 'bark'	
wave- 'joint, knot'	
ô- 'top of tree, or tuber'	
bê- 'twig'	
hêrrê- 'content'	
bwîî- 'rest'	
nrii- 'the remains, rest'	

Table 4. Secretions and bodily products

bound	link
tufârrî- 'saliva'-	*kòfio-* 'perspiration'
fwîî- 'excrement'	*nruâ-* 'dirt (of body)'
koo- 'vomitus'	*tra-* 'blood'
nô- 'mucus, juice'	*tòghòrrò-* 'boil'
wiafò- 'saliva, foam'	*hadhu-* 'wart'
huuvi- 'foam'	
nrirre- 'snivel'	
puu- 'smell'	

The following bound noun involves two bound nouns:

bound + bound:

nô-wòrò- 'sperm' (< *nô-* 'juice' + *wòrò-* 'penis')

There are also three morphologically complex link nouns which refer to a secretion or body products. *Nô-pi-* is made up of the bound noun as the first element and the link noun as the second.

bound + link:

nô-pi- 'liquid from vagina' (< *nô-* 'juice' + *pi-* 'vagina')

The following two words have the bound noun as the first element but the verb as the second.

bound + verb:

nô-mâ- 'urine' (< *nô-* 'juice' + *mâ* 'to urinate')
nô-ti- 'milk' (< *nô-* 'juice' + *ti* 'to suck')

(E) Personal attributes or properties. Bound and link nouns designating personal attributes or properties are listed in Table 5.

Table 5. Personal attributes and properties

bound	link
wii- 'soul, shade'	*mòrrò-* 'life'
gwâ- 'occupation'	*fôdrò-* 'idea, thought'
farrî- 'name'	*îgâ-* 'dream'
terre- 'way, manner'	*vajù-* 'sickness, death'
hôdri- 'fault'	
gee- 'lie'	
wò- 'living place, trace'	
erre- 'habitation'	
dòwò- 'clothes'	

(F) Effects, results and origins of human affairs. Table 6 lists the monomorphemic bound and link nouns from this semantic domain.

Table 6. Effects, results and origins of human affairs

bound	link
fwîrrî- 'effect of speech'	*ahwai-* 'result'
ûû- 'cause, reason'	
êrrê- 'origin, essence'	
pue- 'the ruins, site'	
sii- 'price'	

3.2 Semantic and formal criteria for bound and link nouns

We have listed the body part and related nouns exhaustively in the previous section. In this section, the semantic and formal criteria to categorise nouns into bound and link nouns will be examined.

Concerning the external body part nouns ((A) in the previous section) and parts of a plant or an object (C), it is obvious that most of these nouns are expressed in the form of bound nouns. In (A), 33 bound nouns are counted while there are only five listed as link nouns. Among link nouns, only *ao*- refers to a human body part 'cheek', while others refer to body parts of animals, birds and insects. This suggests that human external body parts are expressed in bound nouns, except 'cheek'. The reason why 'cheek' behaves differently from other human body part nouns may be that 'cheek' is just a surface area on the face, and may not be considered as a specific body part in the language.

The only words among bound nouns which refer exclusively to animal body parts are *nôbi*- 'fin, tail', *virri*- 'shell' and *êrrê*- 'body in the shell, tuber'. *Nôbi*- and *virri*- share the similar meanings of *jawe*- 'tail' and *nrî*- 'shell or skin' of link nouns. The comparison of these words gives a clear semantic distinction for the two noun categories. *Nôbi*- refers to 'fin, tail' which is actually an inseparable part integrated into the body shape as for animals such as snakes, lizards and fish, while *jawe*- refers to 'tails of quadrupeds', which are visually distinct and cuttable (as with the dog). *Virri*- and *nrî*- both refer to 'shell', but the referent of *virri*- is the hard one like a 'sea shell'. When it is compounded with *ô*- 'head' (*virri-ô*-), it refers to 'skull'. The entity referred to by *virri*- is the part of the body which is essential, forming the main frame of the body or body part totally inseparable from the body. On the other hand, *nrî*- is the one which is rather soft like the shell of a locust which such insects shed from time to time. Another animal body part which is expressed in bound noun form is *êrrê*-. *Êrrê*-refers to the body in a shell or tuber like a potato, and again, it is the most important and essential body part for the animal or the plant.

Thus, the animal body part nouns which occur as bound nouns all refer to the essential and totally inseparable body parts. This fact confirms that the bound noun constructions express the most inalienable possessive relationship. List (C) also shows that most of the plant/inanimate parts occur as bound nouns. The only nouns which are link nouns are 'fibre' and 'bud, sprout'. It is understandable that 'fibre' is less inalienable as it is often detached from plants, and used as cord. It is more difficult to explain why *hùwù*- 'bud' is a link noun. *Hùwù*- 'bud' has other meanings: 'son' as a kinship term, in which case it still is a link noun, and 'a small thing' when used as a free noun. This word and *moo*- which was referred to earlier, occurring as a free noun meaning 'man', and as a link noun meaning 'husband' share the same behaviour, both with the meaning as a free form being the primary one, that has extended later on to indicate the kinship relation and the body part.

In contrast to the external body parts, relatively many nouns referring to internal organs (under (B)) occur as link nouns. These nouns often occur in alienable possessive constructions in other languages (cf. McGregor 1995). This is because these parts are normally invisible and are observed only when their possessor is dead and has been opened up, in which case they are no longer inalienably possessed by the possessor (Crowley 1995). Those internal body parts which are found in the list of bound nouns (heart, stomach, liver, bone, joint, bowels) are the organs that are of primary importance and essentially related to bodily condition, emotion and feeling. In many Oceanic languages, the liver is regarded as the seat of the emotions (Crowley 1995); heart, stomach, and bowels are also used to express human moods, in various languages (Koh – Wilkins 1989). Bones and joints form the basic structure of the body.

As for body (or plant) products (discussed under (D)), those designating 'saliva', 'excrement', 'vomitus', 'foam' (of a fit), 'snivel', 'mucus/juice' and 'smell' belong to bound nouns, while those referring to 'perspiration', 'dirt', 'boil', 'wart' and 'blood' occur as link nouns. Those which occur as bound nouns may be considered to be inevitably related or attributable to their possessor, while perspiration, dirt, boil, wart are the bodily products which may not be so "personal" and are just temporary manifestations of the body (Crowley 1995). As mentioned earlier, *fwĩĩ-* 'excrement' has a free noun equivalent *fù*, which form is used when it is "separated" from its possessor, or is no longer attributable to a person or animal. Apart from this attributability, there is no difference in meaning between these two words.

The noun for 'blood' also occurs in an alienable possessive construction in other Austronesian languages, as blood is generally separated from the body when one sees it. *Pwêvò* 'egg' is not in the list as it normally occurs as a free noun, as it often does in other languages (Mosel 1984: 42). This is because the egg is usually separated from the possessor when one sees it, and it is normally considered as 'food' which means it will take an alimentary possessive classifier *hwee-* (cf. previous section). When one wants to specify the kind of egg, one uses the link noun construction as in *pwêvò-nrâ-meò* (egg-LINK-chicken) 'chicken's egg'.

Concerning the nouns referring to personal attributes or properties (listed under (E)), the criteria are not so clear. The possible explanation may be that nouns which occur as link nouns (life, idea, dream and sickness) are something transient and often vague or abstract, while those which occur as bound nouns (shade/soul, name, manner, fault, lie, occupation, living place, habitation, clothes) may be regarded as inherent or closely associated with a human being. One's soul (*wii-* also means 'picture') is considered to be the same as the person themself; occupation is often pre-determined by the society (by totem or clan); so are names and manners, which are non-transferrable properties (Mosel 1984: 44). Fault and lie may also be regarded as human propensities. Living place, habitation and clothes are very personal/intimate properties of the possessor.

Link nouns on the other hand refer to things that are general phenomena which are transient. All of them can occur as verbs as well, without any morphological derivation, while those of bound nouns cannot.

Those nouns under (F) refer to an effect, result or origin of some human affairs. The relationship of cause-effect is generally inseparable. *Ahwai-* occurring as a link noun is the derived noun meaning 'that which follows' from the verb *hwai* 'to follow'. Nouns denoting 'impression or mark' made by someone (or by some body part) or by some activity are inalienably possessed in some languages (Crowley 1995); but in Tinrin, 'impression, mark' is indicated by a productive prefix *fò-* which derives nouns from either nouns or verbs, or other bound morphemes. *Fò-* has several related meanings 'mark, impression, trace, opening' made by the object or action expressed by the root of the word: e.g. *fò-sarra* (mark-knife) 'knife cut', *fò-seghe* (mark-to stab) 'scar by stabbing', *fò-fwòtra* (trace-year) 'age', *fò-arròò* (opening-water) 'bay, spring', *fò-drââ* (opening-day) 'dawn'. As this is a prefix, it is never suffixed by a pronominal possessor.

There are many bound nouns that can occur with both animate and plant/inanimate possessors. They are listed in Table 7 with the two different meanings corresponding to the possessor.

Table 7. Bound nouns with animate and inanimate possessors

	animate possessor	plant/inanimate possessor
ô-	'head'	'top part, front'
nrîfò -	'mouth'	'mouth (of a river)'
êrrê-	'body in the shell'	'tuber, essence, origin'
bê-	'arm, fin'	'twig, border, edge'
tròò-	'back'	'backside'
hi-	'leg'	'leg (of table)'
nô-	'bone, bodily liquid'	'juice, liquid'
gùwù	'body part'	'body of something'
î-	'body'	'body (of axe)'
wave-	'joint'	'knot, pause'

The words listed in Table 7 remain bound nouns even when the possessor is not animate. We can explain this by assuming that the primary sense of those nouns is the one on the left, and that the use and meaning with inanimate possessors are the extended, or anaphorical ones. These extended uses of body part terms are also reported in other Oceanic languages (cf. Crowley 1995), and in African languages such as Ewe (see Ameka 1995).

There is a word *wii-*, which occurs exceptionally in both bound and link noun constructions. When used as a bound noun, it refers to 'soul, shade, picture' (and the possessor is normally animate but can be inanimate as in 'the shade of the tree'), but when used as a link noun, it means 'fibre, vine, tendon' (the possessor can be either animate or inanimate) as illustrated in the following examples.

(30) *u jorri wii-nrî ânrâmwâ*
1sg see soul-3sg there
'I saw his soul (shade) there.'

(31) *uei verru wii-nrâ-nrî varaha*
1sg:FUT plait fibre-LINK-3sg like:this
'I will plait its fibres like this.'

The above examples show that the bound noun construction is used when the referent is something inherent/inseparable or attributable to the possessor, whereas the link noun construction is used when the referent is something already separated from the possessor and not attributable any more as, for example, fibres which are already removed from the possessor, or when the speaker is not interested in its former possessor at all. Furthermore, whether the possessor is animate or not does not influence the construction, but what matters is the relationship to the possessor.

Table 8 is the list of a few link nouns that occur with animate and plant/inanimate possessors.

Table 8. Link nouns with animate and plant/inanimate possessors

	animate possessor	plant/inanimate possessor
wii-	'tendon'	'cord, vine,fibre'
hùwù-	'child'	'bud, small thing'
nruâ	'dirt of body'	'dust'

We have so far discussed the semantic criteria in distinguishing bound and link nouns. As for the compounded body part nouns, however, it is the formal criteria rather than semantics that determine the nominal category. In the compounds, it is the last element which determines whether the noun is bound or link: that is, if the compound is made up of two elements, and the second element is a bound noun, then, the whole word behaves as a bound noun. When the second element is a link noun as in *nrîfò-pi-* (mouth-vagina, 'the opening of the vagina') or a verb as in *nô-mâ-* (liquid, to urinate, 'urine'), the word is then regarded as a link noun. The first element is always a bound noun in words of both categories, whether it is a location noun or a bound noun (except with *fònrî-bê-* (river-arm 'fork of tree'). In fact, bound nouns are often compounded with other words as the first element. Thus, words like *nô-mâ-* 'urine' which could be a bound noun from the semantic point of view, are link nouns, since the formal criteria take precedence over semantics in determining the category of the word.

Another point to mention about the form of these nouns is that in Tinrin there is no morphological device to derive free form nouns from bound or link nouns which exist in some other Oceanic languages such as Tolai and Paamese (Mosel 1984; Crowley 1982: 110).

3.3. Body part nouns used as prepositions

In Tinrin location is indicated by either locative prepositions or location nouns. As discussed in an earlier section, location nouns are characterised by the fact that they are nouns but often mark the locative case without being preceded by a preposition.

Some body part nouns can also indicate location as in some other New Caledonian languages such as Iaai (Ozanne-Rivierre – Moyse-Faurie 1983). They function like a preposition, and the following noun bound to them behaves like the object of the preposition. In the following examples, *nrime-* 'face' and *tròò-* 'back' are used like prepositions, meaning respectively 'in front of' and 'at the back of':

(32) *nrâ fwi nrâ ù nrime-nrî*
3sg exist sm tree face-3sg
'There is a tree in front of it.'

(33) *maamùrrù rri nrâ mwage tròò-mwâ*
children 3pl PROG play back-house
'Children are playing at the back of the house.'

Tròò- also occurs in a compound *tròò-hara* (back-eat) 'afternoon (after eating)' in which *tròò-* has a temporal meaning.

Êrrê-, meaning 'body in the shell, content, tuber, essence' as a bound noun, marks locative case in a rather limited way: It occurs only with certain verbs in specific contexts - often with adversative effect, as we see in the following examples:

(34) *toni nrâ sê aròò êrrê-ri*
Tony 3sg throw water content-1pl inc
'Tony threw water on us (and we became wet).'

(35) *toni nrâ mâ êrrê-wa ù*
Tony 3sg urinate content-DET tree
'Tony pissed on that tree.'

(36) *nrâ tufârrî êrrê-treanrü*
3sg spit content-people
'He spat on the people.'

(37) *nrâ voi êrrê-sôbwi*
3sg defecate content-mat
'He defecated on the mat.'

Verbs which occur with *êrrê-* are limited to meanings such as 'to spit, urinate, throw water, pour water, defecate', which are the actions that bring about certain physical effects at the location expressed by the bound noun and its adjunct. In the above examples, the locations referred to either got all wet or spoiled. When the speaker is not concerned whether the place is affected or not, other locative expressions such as *ru* 'in, at', or *drô-* 'surface, on' may be used to indicate a location.

Verbs such as *tuo* 'to put' which do not drastically change conditions in a place, do not occur with *êrrê-*.

(38) **u tuo tâ êrrê-borrù*
1sg put taro content-bowl
'I put taros in a bowl.'

The verb *truu* 'to stay, rest' can occur with *êrrê-*, only if the following nominal phrase refers to some kind of container that is hollow inside. In this case, the referent of the nominal phrase may not be so much affected physically, but it certainly "holds" something in it, which differs from other prepositional phrases which do no more than indicate the location. Compare the following sentences:

(39) *nrâ truu êrrê-borrù*
3sg stay content-bowl
'It is in the bowl.' (Literally: 'The bowl contains it.')

(40) **nrâ truu êrrê-sôbwi*
3sg stay content-mat
'It is in the mat' (Literally: *'The mat contains it'.)

This peculiar use of *êrrê-* we have just discussed would definitely be the reflection of the inalienable meaning of *êrrê-* as a bound noun, as its prepositional use is the secondary function. As a bound noun, the relationship between the noun possessed and the possessor is inherent or inseparable, whereas in the above constructions, the objects of the verb (e.g.'water' of 'throw water', or 'urine' of 'urinate' (this object is implicit in the verb) become in contact with another object (locative) which is marked by *êrrê-* – for example, 'a mat', and these two objects become inseparable (for example, the place gets soiled/stained by something).

A possessive nominal *hêê-* 'belonging' is used to indicate the beneficiary in some contexts as follows:

(41) *nra fi ve hêê-ri nra*
3sg go take belong-1pl:inc PST
'She went away for our benefit.' (Literally: 'She went to take our belongings.')

The above usage of *hêê-* is similar to that of the preposition *tε-n* in Cèmuhî (Central New Caledonia), which is used to mark the addressee or the beneficiary; this comes from an inalienable noun *dε̄ -n* 'his property, his goods' (Ozanne-Rivierre – Moyse-Faurie 1983: 119). However, *hêê-* retains its nominal meaning and cannot occur in phrases such as **hidrò hêê-nrî* (say belonging-3sg) 'say to/for the benefit of him.' It is also reported that in Nemi (North New Caledonia) *hi-n* '(his) hand' is used to indicate the beneficiary: i.e. give his hand = 'give to him' (Ozanne-Rivierre – Moyse-Faurie 1983). In Tinrin, there is no similar use of 'hand';

instead, *nrorri hire bê-audrê-nrü* (give, branch, hand-mother-2sg) 'give your mother's hand' means 'help your mother' (cf. English 'give your mother a hand'). This is the same notion used in an expression *nrorri e-nrü wî* (give, your food, banana) 'give your food banana' to indicate 'give you a banana.'

4. Conclusions

In this paper, we have examined the possessive constructions of Tinrin, and the behaviour of body part nouns in detail.

We have dealt with several possessive constructions which can be grouped largely into alienable and inalienable possessive types. However, the difference between them is not so clear cut. The evidence points to a continuum with two poles of alienability/inalienability. We have observed a close correlation between these construction types and the semantics of the nouns involved.

Possessive prepositions express alienable possession, which, however, shows a hierarchical order as discussed in section 2.1.1. There are also constructions using various possessive classifiers. They may be placed between constructions with prepositions and inalienable constructions on this continuum, though they are included under alienable possession. As we saw in the examples (13) - (16), the same entity can occur in several constructions depending on the nature of the possessive relationship.

Similarly, there are two types of possessive constructions expressing inalienable possession. In the first type, two nouns (the possessed and the possessor) are juxtaposed. In the second, they are linked by the link morpheme *-nrâ-*. As discussed in 2.2.4., link nouns are bound to possessors, but have structural as well as semantic features of alienable possession (for example, 'the shell of a locust' which is shed or removed from the body).

The structure of these constructions shows the high degree of iconicity of the relationship between the possessed and the possessor: the linguistic distance between the possessed and the possessor is greater in the alienable construction (Haiman 1985: 130).

A close look at the semantics of body part nouns which overlap in both bound and link noun categories reveals that the bound noun constructions express the most inalienable and intimate relationship, whereas link noun constructions do the opposite. Although on the surface, simi-

Table 9. Possessive constructions and nouns possessed

alienable	structure	references of the noun possessed
↑	NP PREP NP	
	nrâ	general, temporarily possessed
	rre/rrê	house, boat, pot
	ò	fire
	NP POSS:CLFR- NP	
	hêê-	belongings
	e-	starches in general to be eaten
	ere-	fruit in general to be eaten
	hwee-	meat, egg in general to be eaten
	odho-	things to be drunk
	hwiie-	things to be chewed
	êê-	plants growing or to be planted
	NP - LINK- NP	kinship – husband, cousin, aunt
	-nra-	external body part of animals
	(*-nra-*)	separable, not integrated
	1sg poss'r	plant part detachable
		internal organs
		body products – temporary or non-attributable
		personal attributes or properties transient
		result of human activity – only derived noun
	NP - NP	
	bound nouns	
		kinship – parent, siblings, wife
		external human body part
		non-human body part integrated
		plant part
		some internal organs related to emotions, bodily condition
		body products- essential
		personal attributes, properties – permanent, personal
		effect/cause of human activity
		collectivity/genus
	possessive classifiers	belongings, food, drink, chew, plant
↓	location nouns	relative position
inalienable		

lar sets of nouns occur in both constructions, the ones that occur in link noun constructions have 'separable', 'transient', 'non-personal' elements in their meanings compared to the 'inherent', 'intimate' or 'personal' character of bound nouns.

The degree of the inalienability is often related to the person/human/animacy of the possessor, in addition to the semantics of the noun possessed. In the case of link noun constructions, the choice of first person singular possessor can suppress the link morpheme, thus creating the same structure as bound nouns (more inalienable). The animacy of the possessor often determines the meaning of the possessed (Tables 7 and 8), and in some words determines its form as well. The majority of external body part nouns, which generally have 'human' possessors, behave as bound nouns, while those which have 'animal' possessors behave as link nouns.

Table 9 summarises the possessive constructions in terms of structural and semantic correlations.

Notes

1. This paper is based on research undertaken in the course of my Ph.D. studies at Australian National University. The data was collected on three field-trips between 1983 to 1987.
2. SVO word order also has developed from the topicalisation of the subject (Osumi 1995).
3. There are three distinct forms (free/focal, subject, object/possessive) for the first person singular pronoun, and one or two forms for the other personal pronouns. Impersonal pronouns have only subject forms:

	free/focal	subject	object/possessive
1sg	*nro*	*u*	*rò*
2sg	*nrü*	*ke*	*nrü*
3sg	*nrî*	*nrâ*	*nrî*
1dl(inc)	*haru*	*ru*	*ru*
1dl(exc)	*komu*	*komu*	*komu*
2dl	*kou*	*kou*	*kou*
3dl	*nrorru*	*rru*	*rru*
1pl(inc)	*hari*	*ri*	*ri*
1pl(exc)	*kevi*	*kevi*	*kevi*
2pl	*wiri*	*wiri*	*wiri*
3pl	*nrorri*	*rri*	*rri*
impersonal	*hêrrê* 'someone, they' *va* 'it, there'		

4. The following orthographic devices are used to represent Tinrin sounds in the paper: pw [pʷ], bw [ᵐbʷ], fw [fʷ], mw [mʷ], kw [kʷ], gw [ᵑgʷ], hw [hʷ], dh [ð], tr [ʈ], dr [ɳɖ], nr [ɳ], ny [Ê], rr [ɽ], s [ʃ], gh [ɣ], è [ɛ], ù [ɯ], ò [ɔ], î [ĩ], ê [ɛ̃], â [ã], ô [ɔ̃], û [ũ], ü [ɯ̃]. Long vowels are shown by duplication of vowels.
5. The following abbreviations are used in the paper: AT – attribute marker; CMPZ – complementizer; COL – collective; DET – determiner; dl – dual; exc – exclusive; FUT – future; inc – inclusive; LINK – link morpheme; M.DIST – mid-distant deixis; NP – nominal phrase; O – object; pl – plural; POSS:CLFR – possessive classifier; poss'd – possessed; poss'r – possessor; PROG – progressive; PST – past; S – subject; sg – singular; sm – subject marker; 1 – first person; 2 – second person; 3 – third person.
6. Ozanne-Rivierre – Moyse-Faurie (1983: 119) illustrates this with New Caledonian languages. Pawley (1972: 34) also states that the possessive plus pronoun sequence always occurs preposed to the specifier in EO languages of his data. In Lenakel (Southern Vanuatu) it is placed variably before/after the possessed noun (Lynch 1978).
7. Edible and potable possessive classifiers have the corresponding verbal forms depending on the kind of food or drink (Osumi 1995).
8. The same situation is noted by Lynch (1973: 86-88) with many other Melanesian languages such as Aroma, Lenakel, Fijian and Paamese.
9. Seiler (1983: 17) states that the most prototypical structure of juxtaposition is that with a personal or possessive pronoun, in the first person, thus coreferent with EGO, as the most intimate, most immediate (possessive) relation.
10. There is no extensive use of a classifier when used with body part nouns to indicate bodily afflictions, as is reported for Paamese (Crowley 1995).

References

Ameka, Felix
1995 "Body parts in Ewe grammar", [this volume.]

Chappell, Hilary – William McGregor
1989 "Alienability, inalienability and nominal classification", *Berkeley Linguistic Society* 15: 24-36.

Crowley, Terry
1982 *The Paamese language of Vanuatu.* (Pacific Linguistics Series B Number 87.) Canberra: Pacific Linguistics.
1995 "Inalienable possession in Paamese grammar", [this volume.]

Green, Roger C. – M. Kelly (eds.)
1972 *Studies in Oceanic culture history* (Volume 3, Number 13.) Honolulu: Pacific Anthropological Records.

Haiman, John
1985 *Natural syntax*. Cambridge: Cambridge University Press.

Koh, Ann – David Wilkins
1989 Body parts in Malay discourse – a preliminary investigation. [Unpublished MS.]

Lichtenberk, Frantisek
1985 "Possessive constructions in Oceanic languages and in proto-Oceanic". In: Andrew Pawley – Lois Carrington (eds.), 93-140.

Lynch, John
1973 "Verbal aspects of possession in Melanesian languages", *Oceanic Linguistics* 7 (1): 69-102.
1978 *A grammar of Lenakel.* (Pacific Linguistics Series B Number 55.) Canberra: Pacific Linguistics.
McGregor, William
1995 "The grammar of nominal prefixing in Nyulnyul", [this volume.]
Mosel, Ulrike
1984 *Tolai syntax and its historical development.* (Pacific Linguistics Series B Number 92.) Canberra: Pacific Linguistics.
Osumi, Midori
1995 *Tinrin grammar.* (Oceanic Linguistics Special Publication No. 25) Hawaii: University of Hawaii Press.
Ozanne-Rivierre, Françoise – Claire Moyse-Faurie
1983 "Subject case markers and word order in New Caledonia and Loyalty Islands languages". In: *Papers from the Third International Conference on Austronesian Linguistics.* (Pacific Linguistics Series C Number 77.) Canberra: Pacific Linguistics. 113-151.
Pawley, Andrew
1972 "On the internal relationship of Eastern Oceanic languages". In: Roger C. Green – M. Kelly (eds.), 1-142.
1973 "Some problems in proto-Oceanic grammar", *Oceanic Linguistics* 12: 103-181.
Pawley, Andrew – Lois Carrington (eds.)
1985 *Austronesian linguistics at the 15th Pacific Science Congress.* (Pacific Linguistics Series C Number 88.) Canberra: Pacific Linguistics.
Rivierre, Jean-Claude
1981 "Linguistique". In: *Atlas de la Nouvelle-Calédonie et Dépendances.* Paris: Office de la Recherche Scientifique at Technique d'Outre Mer. Planche 19.
Seiler, Hansjakob
1983 *Possession as an Operational Dimension of Language.* Tübingen: Narr.

Part IV

Languages of Asia

Inalienability and the personal domain in Mandarin Chinese discourse

Hilary Chappell

1. Introduction

The main goal of this article is to examine the double subject construction in Mandarin Chinese discourse, a structure typologically characteristic of many languages in the East Asian region,[1] and argue that it expresses an inalienable relation in terms of the personal domain (cf. Bally 1926 [1995]). It is proposed that there are two main subtypes of the double subject construction whose form and discourse features are the main topic of analysis, particularly with regard to intonational features. In addition to these two structures, it is also shown that inalienable possession can be expressed by the use of certain discourse strategies related to a reduced or truncated form of the double subject construction. For the purposes of the discourse analysis, data belonging to the spoken narrative genre from the Chinese Pear/Guava Stories (see Appendix B) as well as conversational and written texts will be adduced as evidence.[2] Although this paper is not the first to suggest that the double subject construction expresses inalienability, it is nonetheless the first to attempt a systematic account of this phenomenon in terms of discourse, semantic and intonational properties.

The double subject construction belongs to a spectrum of syntactic structures found in Mandarin Chinese and many other languages of the Asian region which have become generally labelled as the 'topic-comment' construction (cf. Chao 1968; Li – Thompson 1976, 1981). In fact, Li – Thompson (1976: 480) and Lin (1981: 30) suggest that double subject sentences are prototypical topic-comment sentences. Although the grammatical criteria proposed in many analyses are insufficient to uniquely define the topic-comment, it can be described in a general manner as a loosely formed structure of NP + S, characterised by the fact that there is no necessity for an argument relation to hold between the topic NP and the main verb of the comment (Chao 1968: 95-104; Li – Thompson 1976). Some analyses have also described the main function of the topic as providing the frame of reference for the comment (cf.

Barry 1975, Gundel 1987). In the first example which follows, the topic noun phrase, *zhèi-ge fàang'ān* 'this proposal', cannot be considered to be a core argument of the verb phrase in the predicate *méi yǒu yìjiàn* 'to not have objections':

(1) Topic-comment construction

N		//	S(ENTENCE)			
#*Zhèige*	*fāng'àn*	//	*wǒ*	*hé*	*tā*	*dōu*	*méi*
this:CL	proposal	//	1sg	and	3sg	all	NEG

yǒu yìjiàn
have objection

'As for this proposal, neither s/he nor I have any objections.'[3]

Double subject sentences can also be superficially analysed as formed by a loose conjunction of NP + S. On closer examination, however, it is clear that they differ from the general form of the topic-comment presented in (1) in that the sentence acts as predicate to the initial NP as in (2) below rather than as a totally disjunct comment sentence. In other words, double subject sentences are characterised by a semantically related sentential predicate. (This is elaborated upon in section 4 below.)

In Chinese linguistics, this type of topic-comment construction has been labelled *zhǔ-wèi wèiyǔ jù* 'sentences with a subject -predicate predicate' ; "double nominative" (Mullie 1932, 1933; Teng 1974; Tsao 1978, 1982) and "double subject" (Hashimoto 1971; Li – Thompson 1976, 1981: 92-95; Modini 1981), due to the feature of two NPs juxtaposed in sentence-initial position, where both have a subject-like role with respect to the verb in terms of semantics. For example, in (2) below, both *tā* '3sg' and *yǎnjing* 'eye' could be regarded as the subjects of the verb *jìnshi* 'short-sighted'. There is also general agreement among analyses that the predicate is typically stative or intransitive as in (2).[4]

(2) Double subject construction

NP_1	NP_2	VP
WHOLE	PART	STATIVE PREDICATE
#*Tā*	*yǎnjing*	*jìnshi*
3sg	eye	short-sighted

'S/he is short-sighted.'

The semantic relationship between the two sentence-initial NPs has been described as one of inalienable possession with the first NP representing the whole and the second, the part.[5] There is, however, no overt marking of this semantic relation nor of any dependency relation between the two noun phrases, which conforms to the typological features of Chinese (see section 2 below) in contrast to Japanese double subject sentences (see Hosokawa 1995; Tsunoda 1995). The semantic function of this type of topic-comment construction is to characterize a person or the "whole" by means of a predication referring to the part (q.v. Chao 1968: 96; Teng 1974; Modini 1981: 13). With reference to this structure, Chao states that "the commonest subtype of an S-P predicate is one which describes the state or characteristic, less commonly an event, about the main subject. A personal subject may have an S-P predicate in which the subject represents a part of the body and the 'small' predicate some physiological or psychological condition or property." Other analyses such as Tsao (1979, 1982) and Tsunoda (1995) have considered the contrastive and relating functions of the topic-comment structure for Mandarin and the double subject construction for Japanese respectively.

This paper is organised in the following way: A brief typological description of Mandarin Chinese is given in section 2, following on from the introduction in section 1. Section 3 provides an overview on the topic of inalienability in Mandarin Chinese grammar, contrasting nominal and clausal syntax in section 3.1 and discussing the syntactic and semantic characteristics of the double subject construction in section 3.2. In section 4, the main discourse analysis of inalienable possession in colloquial narratives is presented. Section 5 considers some further data in the form of conversational texts and written narratives, followed by the conclusion in section 6 which summarises the findings for all the corpora.

In this paper, I adopt the generally held view in Chinese Linguistics that the possessor [or NP_1] can be pronominal as well as nominal for this construction type. Such a standpoint can be traced back as far as the earliest analyses treating the double subject construction such as Mullie (1932, 1933) or seen in more recent treatments such as Chao (1968), Teng (1974) and Li – Thompson (1976). As a consequence of this, in the main analysis in section 4, two main hypotheses are stated regarding the form of coding for inalienable possession in Mandarin: It is argued that double subject constructions of the kind exemplified by (3) are rare in colloquial narratives where the clause contains two fully lexical NPs designating the whole and the part in the same intonation unit (Chafe 1987).[6]

(3) *Xiàng bízi cháng*
elephant nose long
'Elephants have long noses.'
OR: 'As for elephants – their trunks are long.'[7]

The Pear/Guava Stories data for Mandarin Chinese show that fully lexical possessor NPs, that is, NPs designating the whole, do not turn up in clause-initial slot directly preceding the possessed NP *within the same intonation unit*. In other words, in Mandarin Chinese, it is typical for the possessor not to be overtly lexically encoded in a pair of juxtaposed NPs when the possessed noun is a body part or related term but rather to be encoded as a pronominal form within the same intonation unit. The first hypothesis will be stated in the following terms:[8]

(i) *The subtype of the Mandarin double subject construction with a pronominal possessor preceding a lexical possessum coincides with a single intonation unit and grammatical clause. It has the basic structure*:

$$\text{PRONOUN}_1 \text{ – NOUN PHRASE}_2 \text{ – VERB PHRASE}_{\text{(Stative)}}.$$

There is a corollary to this first hypothesis:

(ii) *The preferred strategy for coding an inalienable relation in a Mandarin double subject construction is not to mention the possessor in the same intonation unit unless in pronominal form.*

In contrast to this, the discourse data show that for double subject constructions with nominal possessors, there is an apparently disparate group of examples containing two lexical NPs which are sometimes separated by an intonation unit break, (indicated by //) sometimes by a rhetorical particle and at other times by a lexical filler such as an adverb or verb taking a sentential complement (represented by X in the hypothesis stated below). Nonetheless, the majority of examples belonging to this second subtype of the double subject construction share the feature of the lexical possessor NP falling in the preceding intonation unit, regardless of its grammatical role. This leads to postulation of the second main hypothesis:

(iii) *The subtype of the double subject construction with a nominal possessor typically contains two lexical NPs in the relation of inalienable possession occurring in consecutive intonation units that belong to the one clause. It has the basic structure*:

NOUN $PHRASE_1$ (X) // NOUN $PHRASE_2$ – VERB $PHRASE_{[Stative]}$

Note that as an alternative strategy to both the above structures, the possessor need not be encoded within the same clause as the possessum when it has already been established as a discourse entity. In this case, it may be viewed as omission of verbalisation (Chafe 1987), that is, as undergoing the process known as zero anaphora. The data also show, significantly in this respect, that the body part term is always linked to a preceding "possessor" NP not more than two clauses distant, patterning then in the same way as the double subject construction with a nominal possessor which spans two intonation units. Hence, we can consider this particular instance of zero marking to be anaphoric. Examples of this type have been classified and labelled as a "reduced form of the double subject construction" in this analysis since their discourse and syntactic properties are remarkably similar to the subtype of the double subject construction with a nominal possessor and thus, we believe, worthy of inclusion for this reason.

The third claim argued for in section 4 concerns the characterisation of the semantic domain encoded by the two NPs in the double subject construction in terms of the discourse data. The relation of inalienable possession between the two sentence-initial NPs is confirmed in this section by the data taken from colloquial narratives, where examples of the double subject construction prove to be restricted to those where N1 refers to a person and N2 to the semantic category of parts of the body and its physical condition as well as psychological states.[9] There was also one example referring to one's homeland *Zhōngguó* 'China', which fits in with the semantic domain for inalienability as defined by Bally (1926) and thus, does not constitute an exception.

This contrasts to the treatment of inanimate nouns representing ordinary posssessions and concrete objects, considered in section 4.2. We compare the coding for these in the Pear/Guava Stories with relational terms to show that Mandarin Chinese excludes emotions, clothing and other simple possessions from the domain of inalienability for animate possessors in the double subject constructions.

2. Typological characteristics of Mandarin Chinese

Mandarin belongs to the Sinitic sub-family of languages of the Sino-Tibetan language family. It is a tonal language of the analytic or isolating type with a basic S-V-O word order, which some analyses claim is drifting to S-O-V (cf. *pro* Li – Thompson 1974, 1975; Huang 1978 and *contra* Sun – Givon 1985; Light 1979).

The domain of Mandarin Chinese stretches territorially from the northeast of the People's Republic of China in Manchuria, southwards through Beijing and Nanjing, then southwest to Sichuan and Yunnan provinces. It is also designated as the official language in Taiwan, Republic of China while it is one of four official languages in Singapore. In contrast to this, the non-Mandarin Sinitic languages are largely concentrated in the east and southeast of mainland China: Wu (for example Shanghainese) in Zhejiang and Jiangsu provinces; Min (for example, Amoy Hokkien) in Fujian province and Taiwan; Yue (for example, Cantonese) in Guangdong and Guangxi provinces and Hakka which is found scattered throughout the Yue and Min areas. Particularly Cantonese and Hokkien predominate in many overseas Chinese communities in Southeast Asia, Australia and North America.

In general, the structural principle of Chinese grammar is for the modifier to precede the modified. Hence, attributive adjectives and relative clauses precede their head noun while not only adverbs but also locative, benefactive and instrumental phrases precede the main verb.[10]

Mandarin possesses little inflectional or derivational morphology. Verbs are not marked for tense, number or person; nouns are not marked for number (singular, dual or plural) and neither nouns nor pronouns are marked for case. Most of the few affixes and grammatical morphemes in Mandarin are atonal suffixes, one of the few productive ones being the particle *-men*. It is used to pluralise singular pronouns, shown under (a) of Table 1. This function is spreading however to the animate category of nouns indicating professions and occupations as in (b) of Table 1. The other main suffixes in Mandarin are used to modify verbs aspectually (for a detailed discussion, see Chappell 1992c) or to nominalise nouns and form their diminutives.

Although there may be a paucity of affixes in Mandarin, we do find on the other hand highly productive word compounding processes. The latter is typical of loan words, particularly calques and neologisms. Consider example (4) with the morpheme *diàn* 'electricity' and (5) with the suffix *-zhǔyì* 'ideology':

Table 1. Mandarin pronoun paradigm

(a) Mandarin pronoun paradigm

	Singular	Plural	
i.	*wǒ*	*wǒ*men	'1 pl. exclusive'
	ø	*zán*men	'1 pl. inclusive'[a]
iia.	*nǐ*	*nǐ*men	
Polite forms for 2nd person:			
iib.	*nín*	*nín*men	
iii.	*tā*	*tā*men[b]	

(b) Nouns referring to occupation or profession:

xuésheng	*xuésheng*men
'student'	'students'
lǎoshī	*lǎoshī*men
'teacher'	'teachers'
tóngzhì	*tóngzhì*men
'comrade'	'comrades'[c]

a) Note that the unsuffixed form *zán* can also be used to mean '1 pl. inclusive'. In some Mandarin (non-standard) dialects, the inclusive forms may be used to refer to 1sg.

b) Of rare occurrence are the polite 3rd person forms *tān* and *tānmen* (see Chao 1968: 641). Note also that *nínmen* is not typical of the spoken language).

c) Note that this suffix is not used for nouns designating inanimates such as *zhuōzi* 'table' (**zhuōzimen*).

(4) diàn-*chuán* electric-transmit 'telex'; diàn-*nǎo* electric-brain 'computer'; diàn-*yǐng* electric-shadow 'film'

diàn*yǐng*-*yuàn* electric:shadow-court 'cinema'

diàn-*chuán*	diàn-*nǎo*	diàn-*yǐng*
electric-transmit	electric-brain	electric-shadow
'telex'	'computer'	'film'
diàn*yǐng-yuàn*		
electric:shadow-court		
'cinema'		

(5)

mínzhǔ-zhǔyì	*wúzhèngfǔ*-zhǔyì	*zìyóu*-zhǔyì
people:master-ideology	no:government-ideology	free-ideology
'democracy'	'anarchism'	'liberalism'

A special feature of Chinese and of most East and Southeast Asian languages is the use of the measure word or classifier when enumerating a noun. The classifier *tiáo* in (6) is used for nouns designating long and thin flexible objects such as ties, ribbons, rivers, ropes and snakes. The classifier construction in Mandarin has the form Numeral – Classifier – (Adjective) – Noun:

(6) *liăng tiáo shé* *yì tiáo jiù kùzi*
two CL snake one CL old trousers
'two snakes' 'one old pair of trousers'

Complex verb phrases are exemplified in (7) by *jì bú tài qīngchu* 'unable to recall clearly' and verb complements of extent with *de* 'to V to the extent that'. Resultative verb compounds in (8) indicating action and result also characterise the Mandarin predicate. Formation of these two kinds of predicates is, similarly to word compounding exemplified above, a highly productive process.

(7) Complex verb phrase

Wŏ zìjĭ yĕ jì bú tài qīngchu
I self also recall NEG too clear
'I myself can't recall too clearly.' (Pear II.3: 17)

nòng de húli-hútu; qì de yào-sĭ
make EXT confused angry EXT want-die
'to make someone confused'; 'extremely angry'

(8) Resultative verb compounds: V_1(Action)-V_2(State)

chuān-pò
wear-tattered
'to wear something until it is in tatters'

zhuāng-măn
pack-full
'pack something until it is full'

cāi-duì
guess-right
'to guess something correctly'

kàn-huā
look-blur
'to look until the eyes become blurred'

Strictly speaking, there is no adjective class in Mandarin. Adjectives or property designators are a subclass of the verb. This can be tested by the fact that when used predicatively, they can be modified by both adverbs (degree and other categories) and aspectual marking. The predicative adjective *pàng* 'fat' in (9) is modified by the inceptive aspect marker *le* and the adverb *yòu* 'again':

(9) *#Nĭ de haízi yòu pàng le!*
2sg DE child again fat INC
'Your child has got fat again!' OR: 'Your child has got even fatter!'

Hence, despite the absence of complex inflectional morphology which led many 19th century European linguists such as Schleicher (1863) to view Chinese as having no grammar, it can be seen that Mandarin not only possesses complex predicates and productive word compounding processes but also many special features such as aspect marking of the verb as opposed to tense inflections and complex noun phrases with classifier modification.

It was pointed out above that Mandarin has the basic word order of S-V-O. Relevant for the present analysis is its description as a topic-prominent language by Li – Thompson (1976) due to the high frequency of the topic-comment construction, a structure which can be considered as basic to the language as the S-V-O construction. To complicate matters further, constructions with a preposed object NP, also known as "left dislocation" and "pseudo-passive" with the form Object NP – [Subject NP] – Verb are equally common as are sentences where the object is preposed before the verb (but not into sentence-initial position) for contrastive effect: S-O-V. These are exemplified by (10) and (11) respectively with the O-NP in plain font:

(10) Left dislocation or pseudo-passive

Kǎoshìjuàn	*wǒ*	*dōu*	*kàn*	*-le*	*yíbiàn*
exam:paper	I	all	read	PFV	one-time

'The exam papers I've gone over once.' (Notes: June 1990)

(11) Contrastive S-O-V

Wǒ	yìdiǎnr	kùnyì	*dōu*	*méi*	*yǒu*
1sg	little	sleepiness	all	NEG	have

'I wasn't one bit sleepy.' (Bing 18: 267)
[literally: 'I any sleepiness did not have']

In addition to this, another construction known as the disposal or *bǎ* construction exhibits S-O-V word order with the marker *bǎ* introducing a typically referential and definite object NP and the predicate expressing the result state affecting this object NP (S-BA-O-V) (see Thompson 1973; Chappell 1992a).

(12) *Bǎ* construction: S- BA-O-V

Tā	*bǎ*	*yì-lóu*	*quánbù*	*dài-zǒu-le*
3sg	BA	one-basket	all	carry-leave-INC

'He carried off one whole basket.' (Pear I.11: 66)

When we include the double subject construction, the topic of this paper, it can be seen that nearly all types of word order are possible for simple clauses in Mandarin: S-V-O; S-O-V; S-BA-O-V; O-(S)-V; topic-comment and S-S-V for the double subject.

The use of passive constructions in Mandarin differs from that in English in two main ways, particularly with respect to the spoken language: First, passives are subject to the special semantic restriction of expressing misfortune in most contexts of spoken genres (see Chappell 1986a). Note, however, that this is an areal feature of the Asian region, so-called 'adversity' passives also being found in Vietnamese, Thai, Lao, Cambodian and Japanese (see Clark 1974a, 1974b, Wierzbicka 1979b). Second, object-preposing or left-dislocation constructions which are semantically neutral provide an alternative strategy to the syntactic passive. These structures function in a similar mode to the passives of many European languages in foregrounding the object or patient NP into sentence-initial position, as in (10) above. This also serves to restrict the range of usage of the syntactic passives formed with the exponents *bèi*, *ràng* and *jiào* (see Chappell to appear).

On the level of discourse, Mandarin shares with other languages of the Asian area a high frequency use of zero anaphora in preference to pronominal anaphora once a discourse entity has been established in a narrative or conversational discourse (see Chen 1986 for a detailed analysis of referent tracking in Mandarin). This point is taken up again in section 4 below. The example in (13) shows how once *tā* 'he' has been overtly expressed in the first clause in Mandarin, it need no longer be mentioned in the following four clauses, contrasting with the English translation and its use of *five* further instances of pronominal anaphora, including the possessive pronoun *his*, which have been italicised in the translation. Furthermore, the English only uses zero anaphora twice in the coordinate clauses joined by *and*:

(13) *Ránhòu* tā *jiù* *shì* ... //
after 3sg then be

diē-dǎo *zuò-zài* *dì-shang* *jiù* *bǎ* *nèige*
fall-topple sit-at ground-on then BA that:CL

kùguǎn *juǎn* //
trouser:leg roll

dàgài *jiǎo* *shāng-le,* //
probably foot wounded-INC

kùguǎn	*juǎn-qilai*	*mō*	*-le*	*yíxià*
trouser:leg	roll-up	rub	PFV	once

'After he'd fallen over and ø was sitting on the ground, *he* rolled *his* trouser leg .... Probably *his* foot was hurt, *he* rolled up *his* trouser leg and ø rubbed it.' (Pear I.6: 88-92)

As a consequence of this, verbs and predicates which share the same subject NP can be chained together in Mandarin to form complex serial verb constructions. There is one brief example of this in (14) with the simple juxtaposition of two verbs sharing the same subject argument: *tā* ... *diē-dǎo zuò-zài dì-shang* 'he'd fallen over and was sitting on the ground'.

The typological sketch has presented some salient grammatical features of Mandarin Chinese with respect to morphology, nominal and verbal syntax; major grammatical constructions and discourse properties which diverge markedly from English and other European languages. More detailed descriptions of the linguistic characteristics of Mandarin are to be found in Chao (1947, 1968) and Li – Thompson (1978, 1981). A discussion and overview of different syntactic means for expressing inalienability in Chinese follows, including genitive noun phrases, noun incorporation into predicate complements and the double subject construction.

3. Inalienability and the personal domain in Mandarin Chinese

There is typically no obligatory overt morphological marking of alienable versus inalienable possession at the level of nominal syntax in most of the languages of the East Asian and Southeast Asian region (see also Clark 1995), although some analyses have made this claim for Mandarin with respect to the absence or presence of the genitive marker (discussed below). In contrast to this, alienable versus inalienable possession is frequently found coded as a primary classification in many language families, for example, Australian Pama-Nyungan languages; Austronesian languages such as Paamese and Tinrin (see, respectively, Crowley 1995 and Osumi 1995) and African languages such as Ewe and Acholi (see respectively, Ameka 1995 and Bavin 1995).

At the level of syntax, there is, however, in addition to the double subject construction discussed in sections 3.2 and 4 below; a spectrum of sentence constructions in Mandarin for which the claim of expressing the part-whole relation or inalienable possession has been made. These include the following:

(a) retained object constructions (see Lü 1965; Chu 1976) including one type of the *bǎ* disposal construction with a retained object (Thompson 1973), exemplified by (14);
(b) the passive construction with a retained object or "passive of bodily effect" , exemplified by (15), where the retained object acts like an incorporated noun not permitting attributive modification (Chappell 1986b);
(c) discontinuous double objects in intransitive predicates discussed by Mullie (1933), Teng (1974) and Modini (1981), exemplified by (16).

(14) *Bǎ* construction with a retained object:

Liǎng rén jiù dòng-shǒu máng-qilai, bǎ yú
two people then start:work busy-INC BA fish

guā-le lín
scrape-PFV scale

'The two of them then got busily to work, scaling the fish.' (Gui 1984: 143) [literally: 'took the fish and scraped its scales off']

(15) Passive of bodily effect with a retained object:

Zhǔguǎn de Déguórén ... *dào Nèi-Ménggǔ kāi*
be:in:charge L German to Inner-Mongolia fly

fēiji hōng yáng wán, ràng lǎoxiāng yòng bùqiāng
plane shoo sheep play PASS villager use rifle

dǎ-diào-le yìtiáo tuǐ
hit-off-PFV one:CL leg

'The German who was in charge ... went to Inner Mongolia, flying in a plane to shoo the sheep away for fun, and had one of his legs shot off by gun-wielding villagers.' (Beijing 1986: 24)

(16) Discontinuous double objects:

Fùqin ... zhǐ shì bái tóufa *shìhu yòu*
father only be white hair seem again

duō-le jǐ- gēn
many-PFV few- strand

'As for father ... he just had a few more white hairs.'
(literally: 'Father – it was just that his white hair seemed to have increased by a few more strands.' (Sophie 21: 17)

Note that in all these constructions, including the double subject, the NP designating the "whole" precedes the NP designating the "part" without exception which seems to conform to a principle of general preceding specific in Mandarin, elsewhere described in terms of universe-scope relations (see Barry 1975). It is interesting that the same case applies for a non-related Australian language, Djaru (Tsunoda, pers. comm.).

To survey some of these constructions in Mandarin Chinese, we begin by examining both nominal and clausal syntax in this section. First we look at the use and function of the Mandarin genitive marker *de* in nominal constructions in section 3.1 in order to determine whether its presence or absence can create two classes of nouns in Mandarin Chinese resulting in just such a grammaticalised distinction between alienable and inalienable. In section 3.2, we consider this possibility for clause-level syntax and present a detailed discussion of the argumentation concerning the syntax and semantics of the double subject construction with reference to previous studies which provides the background for section 4, where the main analysis of the double subject construction is carried out within a discourse framework.

3.1. Nominal syntax: Genitive and zeromarked complex noun phrases

According to Nichols (1986), there are two main types of morphological marking found in the languages of the world which she labels "head" and "dependent" marking. This can be exemplified by means of the genitive constructions in Hungarian and English with Hungarian representing head marking through the use of a pronominal suffix *-a* on the head noun 'house' in contrast to English which represents dependent marking through use of the suffix *-s* on the dependent possessor noun 'man':

(17) *az* *ember* *hàz*-a
the man house-3sg
'the man's house'[11]

Although Nichols does not include "isolating" languages such as Chinese in her sample, Mandarin can nevertheless be classified as possessing a dependent-marked genitive construction using the suffix-like particle *de* to mark the dependent noun:

(18) Genitive construction: NP_1 *de* NP_2

nèige	*xiānsheng*	*de*	*fángzi*
that:CL	gentleman	GEN	house

'that gentleman's house'

This construction in (18) is used to express ordinary possession and ownership in Mandarin.

The first NP slot can be filled by a pronoun. Possessive pronouns in Chinese are typically formed by suffixing *de* to the basic pronoun:

(19) Pro_1 *de* NP_2

tāmen	*de*	*cháhú*;	*wǒ*	*de*	*jìsuànjī*
they	GEN	teapot	I	GEN	computer

'their teapot' 'my computer'

The fact that the genitive particle *de* clearly patterns in with the preceding dependent noun and not its following head can be tested with ellipsis: 'the gentleman's' or 'his' in (20) are both possible in Mandarin but not **de fángzi* [*GEN-house] (see also Dragunov 1960).

(20) *nèige*	*xiānsheng*	*de* ;	*tā*	*de*
that:CL	gentleman	GEN	3sg	GEN

'the man's' 'his'

From a purely syntactic point of view, in neither of the two main realisations of the genitive construction (i) with a possessive pronoun or (ii) with a lexical noun as the dependent possessor noun, need the marker *de* be invariably present. In fact, in many cases where we would expect to find the genitive marker, it proves to be absent. This phenomenon has been described in some analyses, for example, Hashimoto (1971), as the optional ellipsis of the genitive marker, within a generative framework. In other grammars and analyses of Mandarin, it has also been suggested or claimed that the absence of the genitive marker *de* signals the relation of inalienable possession (Dragunov 1960, Rygaloff 1971, Egerod 1985, Tiee 1986 *inter alia*). In the following discussion, it will be shown that this view oversimplifies the issue to a certain degree. As a starting point, however, let us first outline the descriptions of the use of *de* given in some of these analyses:

Tiee (1986: 127) notes that "when personal pronouns are used as the equivalents of English possessives before the nouns for personal relations or before nouns of place, the marker *de* may be used but is usually omitted" but he does not attribute any meaning contrast to this phenomenon.[12] Chao (1968: 289) and Li – Thompson (1981: 115) take a similar standpoint on *de,* while Dragunov (1960: 52-54) points out that it is particularly kinship terms, but also body parts and parts of a whole which do not require marking by the possessive suffix *de* . Under kinship terminology, he subsumes neighbours, friends, and village members. For Dragunov, the absence of genitive *de* in a PRONOUN + NOUN construction expresses 'identity' (*Identität*) rather than attribution (1960: 54). He contrasts the following sets of examples in (21) and (22). According to Dragunov, nouns such as 'father' and 'neighbour' in (21) do not require the presence of *de* whereas nouns such as 'house', 'book' and 'horse' in (22) must take what he calls an obligatory possessive suffix *de* after the relevant pronoun. In other words, kin terms are inalienably coded through zeromarking and material possessions are alienably coded through marking by *de*:

(21) Dragunov: Examples where *de* is not required:

wǒ fùqin; *nǐ línshèjiā*
I father you neighbour
'my father' 'your neighbour'

(22) Dragunov: Examples where *de* is obligatory

tā de fángzi; *wǒ de shū*,
3sg GEN house I GEN book
'his/her house' 'my book'

nǐ de mǎ
you GEN horse
'your horse'

He also points out what seem to be a few exceptions although they are equally "inseparable from the person", as he remarks. These are nouns which like those in (21) do not require *de*:

(23) *nǐ liángxīn*, *tā hún*
you conscience, s/he soul
'your conscience' 'his/her soul'

Terms such as 'conscience', 'soul' and 'mind' correspond however to core categories in many languages that are treated as inalienable by the morphology or the syntax of the given language (see prolegomena to this volume; Bally 1926 [1995] and Chappell – McGregor 1989).

This notwithstanding, Dragunov (1960: 52, and footnote 2) weakens his line of argumentation by proceeding to suggest that either euphony or emphasis is responsible for determining whether or not the possessive suffix *de* will be used in the case of kinship terms and related semantic categories.

In contrast to both Tiee and Dragunov, Rygaloff (1971: 145) attempts a uniform transformational account of nominals formed by *de*. Nonetheless, he too points out that *de* is an optional marker of determination in the case of "inalienable possession" where the head noun is a body part or kin term. For some reason (not discussed), he restricts the modifying constituent to a singular pronoun.

Nevertheless, on examining textual material, it can be seen that examples of zeromarked genitive expressions containing NPs designating a material possession are not rare: Examples (24) and (25) are taken from the Chinese Pear/Guava Stories (see also section 4 to follow):

(24) *tāmen bǎ* tā chēzi *fú-qilai*
they BA 3sg bike support-up
'They stood up his bike.' (Pear: I.0: 132)

(25) *Bù xiǎoxīn*, tā màozi *bèi chuī-diào -le*
NEG careful 3sg hat PASS blow-away-PFV
'When he wasn't paying attention, his hat was blown away.' (Pear I.12: 50)

From the two examples above, it can be seen that neither of the NPs – *tā chēzi* 'his bike' nor *tā màozi* 'his hat' – uses the genitive marker *de*. This refutes, in an effective way, any hypothesis stating that *de* invariably codes alienable possession while zeromarking codes inalienability (see also discussion below).

Chappell – Thompson (1992) carried out a discourse study of genitive expressions in Mandarin using a large database containing 440 complex relational noun phrases from informal spoken and written texts. Using a variable rules analysis, they found that semantic and pragmatic factors influencing the choice of *de* are not independent of each other. Out of 23 different variables hypothesised as potentially determinative in the use of *de*, seven were shown to be statistically significant. Three of these factors were structural, pertaining to the possessed or head noun (NP_2), specifically, the addition of extra linguistic material in the form of an attributive modifier, locative postposition or a second genitive expression in tandem: These three factors all strongly disfavoured the use of *de* by speakers. This was accounted for in terms of the structural heaviness of the expression leading to omission of *de*. A fourth factor was pragmatic and pertained to givenness or recency of mention (Chafe 1987) of the entire genitive expression. This similarly disfavoured the use of *de* in the case of given referents: The rationale for this appears to be that once the relation or association is established, it need not be signalled again by use of *de*.

The three remaining factors all involved semantic properties of the possessed noun, NP_2. All possessed nouns were classified into the categories of kin, body parts, spatial orientiation, location, abstract or concrete terms. Only spatial orientation, kin and location terms were shown to be statistically significant, with spatial orientation least likely to co-occur with *de* at the top of the continuum. For the semantic properties coded for the possessor noun, NP_1, there was only a partial indication that first and second person possessors tend to favour absence of *de*, while third person and other human possessors favour its presence.

Chappell – Thompson (1992) interpret this data in terms of a high correlation between the iconicity or conceptual closeness of possessor and possessed, economic motivation (see Haiman 1983, 1985) and the absence of explicit marking of the possessive relationship by *de*. They show that although the presence or absence of the genitive marker *de* cannot be defined in terms of a categorical rule, the probability of its use or non-use by speakers can be ranged along a continuum with respect to the interaction of the inherent semantics of both possessor and possessed nouns. At one end of the continuum, the combination of first person pronominal possessors with a term for spatial orientation as the possessed item is most likely to be coded *without* the genitive marker, with kin terms as the possessed noun, second highest on this scale, followed closely by location terms. At the other end of the continuum, the combination of concrete and abstract nouns as either possessor or possessed almost invariably leads to use of the genitive marker *de*. In the middle, it is surprising to find that terms for parts of the body as possessed items strongly favour co-occurrence with *de*, that is, the 'alienable' marking strategy. Contra Dragunov (1960) and Rygaloff (1971), this appears to provide evidence that this construction cannot be analysed in terms of inalienability, since it is specifically terms for body parts which are considered to be prototypically inalienable (cf. Nichols 1988, Chappell – McGregor 1989 but *contra* see Chappell – McGregor 1995).

In Ewe, a Kwa language of West Africa, a similar continuum holds with body part terms typically coded in the alienable possessive construction while kin and spatial orientation terms are treated as inalienables (see Ameka 1995). In fact, Ameka views terms for spatial orientation as the most inalienable relation which correlates well with the Mandarin analysis where such referents are viewed as the conceptually closest.

In summary, there is a convergence of pragmatic and cognitive factors at work in determining whether or not *de* will be used; not only the inherent semantics of the nouns but also the givenness of the entire noun phrase; the presence of modifiers; the presence of a second genitive noun or a postposition all influence its use.

To conclude, the expression of inalienable posssession cannot be defined at the level of nominal syntax for Mandarin in terms of genitive NPs versus zeromarked NPs, even though a tendency has been observed for inherently relational nouns such as those referring to spatial orientation and kin not to require overt genitive marking when combined with possessor nouns treated as conceptually close, such as first person pronouns. A range of factors interacts in a complex manner to determine the use or omission of *de*.

Although we have countered the inalienability hypothesis at the level of the noun phrase by establishing that the presence or absence of the Mandarin genitive marker *de* does not signal the existence of two distinct morphological classes of nouns in Chinese, the study of reference to parts of the body and other relational terms at the level of clausal syntax in Chinese does, however, turn up interesting correlations between coding and semantics. This is the topic of the next section.

3.2. Clause-level syntax: Genitive subject sentences versus double subject constructions

Teng (1974), Barry (1975), Li – Thompson (1976, 1981) and Modini (1981) have all suggested that the double subject construction in Mandarin expresses inalienable possession as the particular semantic relation holding between the first NP in sentence-initial position representing the "whole" and the second NP representing the "part". Of these, Teng (1974), Barry (1975) and Modini (1981) go further to attempt a classification of topic-comment constructions into the double subject construction; the partitive construction (see section 4.3) and the general or "topicalised" construction where the topic is completely disjunct from the comment, providing its general frame of reference.[13] This last type corresponds to sentences beginning with *regarding*, *concerning* or *as for* in English and is exemplified by both sentence (1) in the introduction and (28) below. An example of each of the other two types is also presented below.

(26) Double subject construction:

Yīnwèi,	*wǒ*	nèige	kùzi,	kǒudài	*hěn*	*duàn*	*ne*
because	I	that:CL	trousers	pocket	very	short	RP

'Because those trousers of mine, they had very shallow pockets.' (HK: 9)

(27) Partitive construction:

Wán	*le,*	sānge	jǐngchá,	yíge	*kāi*	*chē*	*de* ...
finish	ASP	three:CL	police	one:CL	drive	car	L

zhuī	*chē*	*qù*	*le*
chase	car	go	ASP

'After that, three police, one drove the car ... to chase after the bus.' (HK: 11)

(28) Generalised construction:

Wǒ	*shuō*	guónèi	*ne,*	*dàochù*	*xūyào*	*de*	*shíhou,*
I	say	in:country	RP	everywhere	need	L	time

zěnme	*yě*	*kàn-bù-dào*	*jǐngchá*
how	also	see-NEG-achieve	police

'I mean, back home, whenever they're needed, how come you never see the police?' (HK:6)

Most of these analyses agree implicitly or explicitly that the predicate of topic-comment sentences is sentential (NP-V) while, more specifically, the main verb of the double subject construction has to be stative or intransitive. Note that the same constraint applies to the double nominative construction in Japanese (see Tsunoda 1995 and Kuno 1970). For Modini (1981: 13), the double subject construction expresses attribution of some state or property of the second NP to the first-mentioned NP, while Teng (1974: 463), similarly to Chao (1968: 96) quoted above, describes its use as referring to temperament and physical condition.[14]

In addition to this, some analyses have argued that the 'double subject' construction NP_1 NP_2 VP is a reduced form of a simple S-V construction [NP_1 DE NP_2] – VP where the subject slot is filled by a genitive noun phrase, overtly coded as such by the genitive marker *de* – for example, Hashimoto (1971) and Tang (1972) for Chinese and Kuno (1970) for analogous constructions in Japanese. The opposing view that these represent two separate constructions has been put forward in analyses such as Teng (1974) and Li – Thompson (1976).

First we briefly recapitulate the arguments used in favour of the non-derivability of double subject sentences and review some of the relevant literature which enables a more detailed description of the semantic, syntactic and discourse features of this construction to be presented. Simultaneously, the form and function of simple S-V constructions with genitive subjects are also outlined in this section due to their significance for the double subject construction in determining the syntactic status of its two sentence initial NPs. The discussion of genitive NPs given in the previous section is used as a basis for this. Consider first the following additional example of a double subject construction:

(29) #*Tā bízi tiē-zài chuānghu – shang*
3sg nose press-at window-on
'Her nose was pressed against the window.'

The problem for the present analysis is how to analyze the relationship between the two sentence-initial NPs: *tā* '3sg' and *bízi* 'nose'. Analyses which argue for the double subject construction as representing one kind of topic-comment relation would consider *tā* '3sg' as topic, and *bízi*' 'nose' as subject of the comment (cf. Chao 1968; Teng 1974; Li – Thompson 1976; Modini 1981). In fact, some might say that the English rendering of this relationship might better be captured by the translation: 'As for her, her nose was pressed against the window'.

In the opposing view, these two sentence-initial NPs are treated as a reduced genitive construction, resulting from the deletion of the genitive marker *de* (cf. Hashimoto 1971), that is, *tā bízi* 'her nose' is generated from *tā de bízi* 'her GEN nose'. This is akin to the argumentation for "possessor ascension" where the possessor is raised out of a genitive NP and thereby promoted into an argument role (see Chappell – McGregor 1995).

Teng (1974: 456-459) argues that this sentence type which he names 'double nominative' cannot be derived from possessive sentences, as claimed by Hashimoto (1971) and Tang (1972). Using the following two related sentences, Teng postulates two different underlying structures. The following two figures are reproduced from Teng (1974: 461):

(30) Double subject construction

NP	NP	VP
Tā	*dùzi*	*è*
3sg	stomach	hungry

'He is hungry'

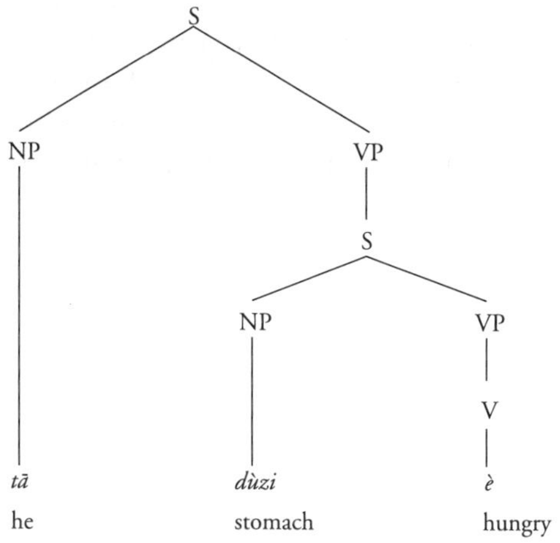

Figure 1. Double subject construction: NP NP VP

(31) S-V construction with genitive subject

NP	DE	N	VP
Tā	*de*	*dùzi*	*yuán*
3sg	GEN	stomach	round

'His stomach is round'

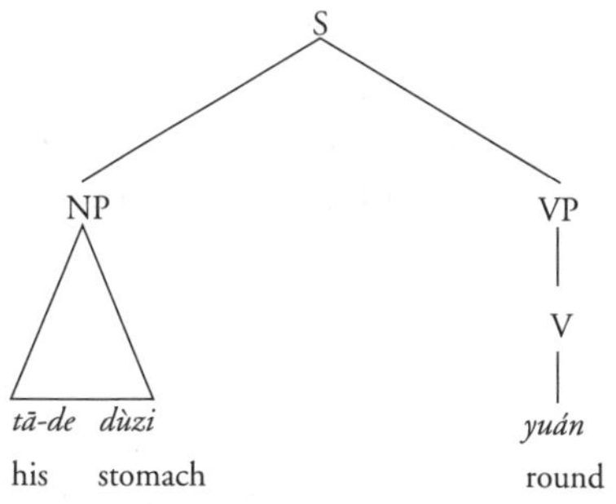

Figure 2. S-V construction with genitive subject: NP DE NP VP

Sentence (30) is analyzed as having a sentence acting as its predicate, rather than a noun, verb or adjective (that is, stative verb) while (31) contains a genitive NP acting as its subject and a stative verb as its predicate. In Teng's analysis, the structure underlying (30) cannot be derived from that of (31), since the relevant NPs are dominated by different nodes.

Several further arguments have been proposed to account for these different underlying structures that are worth briefly summarising here to show that double subject constructions are not a case of sentences with a reduced genitive NP as subject. Furthermore, these criteria will provide a useful diagnostic for determining which sentences are examples of the double subject construction in the Pear/Guava Stories data examined in section 4 which presents the main analysis.

(i) Teng (1974: 468), Li – Thompson (1976: 481) and Tsao (1979: 34, 126-129) all point out that many double subject sentences have no genitive counterparts whatsoever. Consider Teng's examples (39a) and (40a) reproduced here as (32) and (33):

(32) Double subject construction:

Tāmen	*shéi*	*dōu*	*méi*	*lái*
they	whichever:one	all	NEG	come

'None of them came.'[15]

(33) S-V sentence with genitive subject:

**Tāmen*	*de*	*shéi*	*dōu*	*méi*	*lái*
they	GEN	whichever	all	NEG	come

Similarly, the following two elicited examples given in (34) and (35) sound semantically odd with a genitive NP as subject, as the possessed head noun, a body part, is interpreted as personified and thus as potentially belonging to a context where it is detached from the "owner". Note that example (30) above containing *tā dùzi è* [3sg-stomach-hungry] is the common and usual way of expressing 'S/he is hungry' in Mandarin.

(34)

#?*Tā*	*de*	*dùzi*	*è*[16]
3sg	GEN	stomach	hungry

'His/her stomach is hungry.' (This is awkward since it is viewed as a separate event from that of the person being hungry)

(35) #?*Tā de bízi tiē-zài chuānghu – shang*
3sg GEN nose stick-at window – on
'Her (detached) nose was stuck to the window.'

Example (35) is the corresponding genitive of the double subject example in (29) above.

(ii) Mullie (1933), Teng (1974: 458) and Modini (1981: 10) show that only the double subject construction allows adverbial modification to split the two inalienably related nouns.[17] This is not possible when the clause-initial NP contains a genitive, as in (37):

(36) Double subject construction:
Tā yòu // tóu téng le
3sg again // head ache PFV
'He has a headache again.' (Teng's (10a))

(37) S-V sentence with a genitive subject:
**Tā de yòu tóu téng le*

(38) *Tā de tóu // yòu téng le*
3sg GEN head // again ache PFV
'His head is aching again.'

As outlined in the typological sketch of Mandarin in section 2, adverbs typically occur directly preceding the predicates they modify. In the genitive sentence (38), the adverb *yòu* 'again' modifies the verb *téng* 'ache' acting as the predicate. This rule can apply, however, even where the predicate is sentential, that is, contains an idiom in the form of an S-V sentence, as in example (36) with the adverb *yòu* 'again' directly preceding *tóu téng* 'head-ache'. For sentences with genitive subjects, direct adverbial modification of the head noun is not permitted, shown by (37). Chao (1968: 98) also points out that the genitive NP cannot be expanded by insertion of adverbs. Only the genitive sentence in (38) is grammatically acceptable, neatly showing the distinct syntactic behaviour of double subject sentences versus sentences with genitive NPs at their head.

(iii) In other analyses, such as Chao (1968), Li – Thompson (1976), Jeng (1978: 328) and Tsao (1978: 185; 1979: 87), it has been similarly observed that there can be an intonation break between the "topic" noun and the "subject" noun of the predicate for double subject sentences. This can be verified on insertion of a rhetorical particle such as *a (~ya*), *ba*, *ne* or *me* (see Chappell 1992b). These two last criteria are also important for the argument to be developed below, as many of the examples from the Pear/Guava Stories data were found to contain adverbs and rhetorical particles (RP) which separate off the topic by intervening between the two nouns of the double subject, as *a* in (39) which follows:

(39)	*#Lǐ*	*Yífēng*	*a,*	*fùmǔ*	*dōu*	*bú*	*zài*	*le*
	(name)		RP	parents	all	NEG	be:at	INC

'As for Li Yifeng, her parents have both passed away.'

Needless to say, the insertion of a rhetorical particle into a genitive NP is not a possible strategy in Mandarin, shown by the unacceptability of (40):[18]

(40)	**Lǐ*	*Yífēng*	*de*	*a*	*fùmǔ*	*dōu*	*bú*	*zài*	*le*
	(name)		GEN	RP	parents	all	NEG	be:at	INC

Hence, these three criteria clearly distinguish double subject sentences from simple S-V sentences with genitive subjects in terms of syntactic behaviour.[19] Moreover, the semantic and discourse relationship between the two nouns representing whole and part for the two types of construction can consequently be expected to be quite different, belying the use of different discourse strategies by speakers.[20]

Finally, it needs to be noted that the notion of topic has been considerably revised in more recent literature to be treated purely as a discourse-level notion (cf. Givón 1983; Barry 1975 and Tsao 1979 for Chinese). This point of view suggests that defining the first constituent of a particular sentence type as the topic is not relevant at the sentence-level for this can only be ascertained within the larger boundaries of a stretch of discourse by examining, for example, how long certain NPs "persist" in a text, how many clauses separate each "mention" (see particularly Givón 1983) and the form each "mention" takes – zero, pronominal or nominal (see Tomlin 1987). This article, however, restricts itself to defining the discourse, semantic and intonational characteristics of double subject constructions.

4. The analysis of double subject constructions in the Chinese spoken narrative genre

4.1. Double subjects in the Pear/Guava Stories

For the purposes of this analsyis, the Chinese Pear/Guava Stories were used as the main database. These are a set of 21 elicited colloquial narratives collected by Mary Erbaugh at National Taiwan University, Taipei, in 1976 (see section 10). The interviews were carried out following the protocol given in Chafe (1980).[21] In addition, two short conversational texts were used to compare results (see section 5).

In the transcription of the Pear/Guava narratives, equivalent to 2½ hours of taping, 66 clauses with noun phrases containing complex relational expressions as head nouns were separated out of the data, regardless of the presence or absence of the genitive marker *de*.[22] That is, the 66 noun phrases contained both genitive NPs marked by *de* and juxtaposed NPs unmarked by *de*.

Of the total number of relational expressions, it was found that possession was typically not marked at all by the genitive particle *de* for 48/66 terms, that is, 72.7% of terms referring to parts of the body and related categories were not linked to the term for the overt possessor by the genitive particle *de*. In other words, the preferred syntactic structures for complex relational NPs in this domain are $NP_1 NP_2$ or $PRO_1 NP_2$ but not the structures NP_1 *de* NP_2 or PRO_1 *de* NP_2. Prima facie this would seem to lend some support for the traditional rules of usage for Mandarin *de* based on elicited data but only in terms of a tendency *not as a categorical rule* (see Chappell – Thompson 1992), the tendency being the absence of *de* with relational terms.[23]

Of the 66 terms, 42/66 were in clause-initial position which earmarked them for checking as potential double subject sentences while 24/66 were in object or oblique grammatical roles.[24] Amongst the 42 clause initial NPs with relational head nouns, 36 were zeromarked with the form $NP_1 - NP_2$. Hence, there were only 6 genitive NPs in clause-initial position marked by *de*, out of a total of 18 genitive structures in all possible syntactic positions (see section 6 below for a discussion of these).

Of most significance for this study, 22 of the 36 juxtaposed NPs formed part of a double subject or topic-comment construction. This constitutes 1.2% of the 1804 intonation units forming the 21 narratives in the Pear/Guava Stories, a small percentage when calculated in this manner. It is more significant, however, to view the total number of double subject sentences as a proportion of intonation units containing relational NPs as these are the places in the discourse where the speaker has the option of choosing this construction to express inalienable possession. The percentage using this method as a basis is 33.3% (22/66). The double subject sentences were identified as such on the basis of the criteria discussed in section 3.2 above. The remaining 14/36 examples with juxtaposed NPs in clause-initial position were all S-V-O and thus, were not further considered.

Of the 22 examples of double subject sentences, seven examples had a pronominal possessor; four had two fully lexical nominals juxtaposed in sentence-initial position while eight examples, although containing two lexical NPs, did not have these contiguously placed. Three remaining examples had the two relational terms in different clauses, in what were identified as 'reduced double subject' sentences.[25] The three main kinds of syntactic structures with examples from the corpus are given in (41), (42) and (43):

(41) $PRONOUN_1$ NP_2 VP [7 examples]

Yīnwèi	tā	gèzi	*yòu*	*xiǎo*,	//
because	3sg	build	again	small	

'Because he was smaller in build, as well.' (Pear I. 6: 63)

(42) NP_1 NP_2 VP [12 examples]

Nèige	nǚ	háizi,	//	gèzi	*xiǎnde*	*bǐjiào*	*dà*
that:CL	girl	child		build	appear	rather	big

'The girl, she appeared to be quite big in build.' (Pear I.0: 200)

(43) NP_1/PRO_1 and NP_2-VP in different clauses [3 examples]

Nèi sānge	xiǎo	háir	*yě*	*zǒu*	*le*,	//
that three:CL	small	child	also	leave	PFV	
kàn-qilai (...)	shēntǐ	*hěn*	*jiànkāng*	*de*	*yàngzi*	
look-INCH	body	very	healthy	DE	appearance	

'Those three children left too, they looked very healthy.' (Pear I.0: 135-137)

This information is summarised in Table 2 which follows.

On the basis of the data analysis, a refinement of the syntactic configuration for the double subject construction is proposed: For the examples with a pronominal NP_1, the double subject sentences showed one major difference from the other group with a lexical NP_1 and NP_2: The syntactic form $PRONOUN_1 - NP_2 - VP$ consistently fitted into one intonation

Table 2. Distribution of 66 complex relational NPs

66 complex relational NPs					
Oblique/object NPs:		24			
Clause-initial NPs:		42			
		66			
42 Clause-initial NPs					
1. 6	GENITIVE: NP_1 *de* NP_2				
2. 36	JUXTAPOSED: NP_1-NP_2				
42					
36 Clause-initial juxtaposed NPs: NP_1-NP_2					
(i)	SVO	14			
(ii)	Double Subject	22	a.	PRO_1 -NP_2-VP	7
			b.	NP_1-NP_2-VP	12
			c.	PRO_1/NP_1-X-NP_2-VP[a]	3
Totals:		36			22

a. The symbol 'X' is used to indicate any kind of lexical filler, including a clause, separating the two NPs.

unit (Chafe 1987), whereas examples with the form NP_1- NP_2 – VP or NP_1 – X- NP_2 – VP typically had NP_1 and NP_2 in different intonation units.

There were only two examples given as (84) and (85) in Appendix A, section 8 below, which did not conform to this rule of one intonation unit. I do not consider these to be exceptions, however, for two reasons: (a) lexical material clearly separated NP_1 from NP_2 (b) the possessor NP in each case was given, being mentioned explicitly in one of the preceding intonation units, with the effect that the double subject construction had a contrastive function. Jeng (1978: 328) has described the prosodic feature of an intonation break separating the topic from its following microsentence (or sentence-predicate) as a kind of demarcation line reinforcing the syntactic and semantic independence of topic from comment. The spoken data from the Pear/Guava Stories upholds his description exemplified with elicited data.

The data containing two lexical NPs: NP_1 – NP_2 – VP can be viewed as a spectrum, at one of whose ends is found the smallest gap between NP_1 and NP_2 of a mere intonation break[26] while at the other end, non-prosodic segmental material such as rhetorical particles, adverbs and verbs that take sentential complements separate NP_1 and NP_2.[27]

The possibility of an adverb or rhetorical particle preceding possessum NP_2 and thereby separating it from its possessor, NP_1 (see section 3.2 above) is exemplified by (44) where a rhetorical particle *ne* separates the term *nán háizi* 'boy' from *liǎn* 'face' and by (45) where an adverb *yídìng* 'certainly' separates *nóngcháng zhǔrén* 'farmer' from *xīn-li* 'heart':[28]

(44) Nèige xiǎo nán háizi *ne,* //
that:CL little male child RP
liǎn-shang *méi yǒu biǎoqíng*
face-on NEG have expression
'The little boy, well, there was no expression on his face whatsoever.' (Pear II.7/8: 77-78)

(45) Nèige nóngcháng zhǔrén *yídìng ...* //
that:CL farm owner certainly
xīn-li *juéde zhèige,* //
heart-in feel this:CL
wū! zhèi sānge zhēn shì dǎn-dà-bāo-tiān
RP this three:CL really be gall-big-cover-heaven
'The owner of the farm certainly ... must have felt that this – well! – these three were really incredibly brazen.' (Pear I.7: 107-108)

It is proposed that examples with two lexical NPs, regardless of whether the material intervening between the two NPs is prosodic or segmental, should be collapsed as $NP_1 - X - NP_2 - VP$ and treated as one and the same phenomenon spanning at least two intonation units. Jeng (1978: 327) similarly points out that the intonation break may coincide with the boundary of an information unit. The definition of intonation unit used here follows Chafe (1987: 22) who defines it in terms of "a sequence of words combined under a single, coherent intonation contour, usually preceded by a pause" with each intonation unit containing one piece of active information. In this analysis, X can be realised by any of the following items:

(46) X = intonation break
rhetorical particle
adverb
verb taking a sentential complement

In fact, this analysis extends to include three examples where an entire intonation unit separates the inalienably possessed term NP_2 from the possessor NP_1. An example of this is given in (47):

(47) Nèi sānge xiǎo háir *yě zǒu le* //
that three:CL small child also leave PFV
kàn-qilai hěn qīngsōng yúkuài //
look-INCH very relaxed happy
uh shēntǐ *hěn jiànkāng de yàngzi*
uh body very healthy GEN appearance
'Those three children left too, they looked very carefree, happy, and uh, healthy.' (Pear I.0: 135-137)

The significance of the inclusion of such examples is that it clearly points to the special syntactic treatment of certain categories of nouns in Mandarin Chinese as inalienable possessa in contrast to other categories of nouns such as ordinary material possessions (see also sections 4.2 and 6 below). If the two lexical NPs representing whole and part are viewed as belonging to two different intonation units, then in all these cases, there is in reality zero anaphora of the possessor NP within the intonation unit and clause containing the possessum NP. This method of analysis is supported by claims made by Chafe (1987) with respect to the English spoken narrative genre. It tallies perfectly with the "one new concept at a time" constraint Chafe proposes (1987: 32) whereby a maximum of one new piece of information is expressed in each intonation unit. That is, the full lexical forms for both possessor and possessed entities are unlikely to occur in the same intonation unit, since lexical forms are typically used to identify new referents or concepts ("previously inactive" in Chafe's terminology). According to Chafe, "old" or "given" referents, for example, those already established in the discourse, tend to be "attenuated", that is, either pronominalised or not overt.

This leads in turn to the question of whether the structure $NP_1 - X - NP_2 - VP$ is fundamentally the same structure as that of $PRONOUN_1 - NP_2 - VP$, contained in a single intonation unit. If we momentarily regard both as possible syntactic configurations of the double subject construction, then on the basis of the Pear/Guava Story narratives, the preference is for omission, that is, zero anaphoric reference of the possessor NP. Although the corpus of examples is small, the preferred strategy in just over two-thirds of the examples of double subject sentences (15/22 or 68.2%) in the Chinese Pear/Guava Stories is *not to mention the possessor in the same intonation unit unless in pronominal form* as was the case for the remaining 7/22 examples. This was also confirmed by data from two further conversational texts (see section 5.1 below).

To back up this argument, note that Teng (1974: 458, footnote 3) has pointed out that for double subject sentences with pronominal possessors, a pause usually may not fall between the pronoun possessor and the possessed noun as in (48). It must be made after the juxtaposed noun phrase *tā fùqin* 'his father'. Teng's examples (a) and (b) are reproduced below as (48) and (49):[29]

(48) **Tā // fùqin zài Rìběn zuò shì*
3sg // father at Japan do business
'His/her father is working in Japan.'

(49) *Tā fùqin // zài Rìběn zuò shì*
3sg father // at Japan do business
'His/her father is working in Japan.'

I conclude that there are two subtypes of the double subject construction in narrative discourse semantically related by virtue of expressing inalienability at this level of syntax and structurally related in sharing many syntactic features and constraints such as a stative or intransitive predicate.

The structure PRONOUN$_1$ – NP$_2$ – VP which fits into a single intonation unit can be identified as the first subtype of the double subject construction.[30] The hypothesis made on the basis of the Chinese Pear/Guava Stories with regard to this structure is formulated in the following manner:

(i) *The subtype of the Mandarin double subject construction with a pronominal possessor preceding a lexical possessum coincides with a single intonation unit. It has the basic structure*: PRONOUN$_1$ – NOUN PHRASE$_2$ – VERB PHRASE$_{(\text{Stative})}$.

Thus, in the case of single intonation unit utterances, cognitively corresponding to the expression of one new piece of information, it is typical for the possessor not to be lexically coded in a pair of juxtaposed NPs heading a double subject construction but rather to be coded in pronominal form. The corollary to this hypothesis is next stated:

(ii) *The preferred strategy for coding an inalienable relation in a Mandarin double subject construction is not to mention the possessor in the same intonation unit unless in pronominal form.*

The configuration NP_1 – X // NP_2 – VP, where the possessor is not made explicit within the same intonation unit as the possessed term, is treated here as a reduced form of the double subject construction since it shares the features of a stative predicate and coding of an inalienable relation. The omission of the possessor NP can be explained at the level of discourse as a feature of discourse organisation and 'information flow'. In the reduced form of the double subject construction, the overt possessor NP_1 is however not more than two clauses distant. Example (13), reproduced here as (50) for the reader's convenience, shows the greatest distance – that of two clauses – between possessor and possessed NPs in the data examined, that is, between *tā* '3sg' and *jiǎo* 'foot':

(50) *Ránhòu* tā *jiù* *shì* ... //
after 3sg then be
diē-dǎo *zuò-zài* *dì-shang* *jiù* *bǎ* *nèige*
fall-topple sit-at ground-on then BA that:CL
kùguǎn *juǎn* //
trouser:leg roll
dàgài jiǎo *shāng* *le*,//
probably foot wounded INC
kùguǎn *juǎn-qilai* *mō* *-le* *yíxià*
trouser:leg roll-up rub PFV once
'After he'd fallen over and was sitting on the ground, he rolled his trouser leg ... . Probably his foot was hurt, he rolled up his trouser leg and rubbed it.' (Pear I.6: 88-92)

The rationale behind this discourse strategy can be explained in the following way: When the referent of a possessor term has already been established in the immediate narrative as a discourse entity, it is not necessary for speakers to specify it again through lexical coding within the same intonation unit as its possessum NP. The inherently relational nature of inalienably possessed items means that, in Chinese discourse, the existence of a possessor is inferrable from the basic semantic properties of concepts such as parts of the body and all those belonging to the personal domain. In this way, relational terms are accorded special treatment in discourse: there is no need to continually overtly specify who the possessor is, as long as the concept remains "activated" in the minds of speaker and hearer (Chafe 1980, 1987) nor do they need to keep track of the possessor through use of pronominal anaphora or other lexical means within a given segment of discourse. This leads to postulation of the second main hypothesis:

(iii) *The subtype of the double subject construction with a nominal possessor typically contains two lexical NPs in the relation of inalienable possession occurring in consecutive intonation units that belong to the one clause. It has the basic structure*: NOUN PHRASE$_1$ (X) // NOUN PHRASE$_2$ – VERB PHRASE$_{[\text{Stative}]}$.

A parallel can be made here with the use of the definite article in preference to the possessive pronoun to mark the personal domain and inalienable possessa in the dative constructions of many Indo-European languages (cf. Bally 1926 [1995]). Consider the German dative construction in (51) where the owner of the body part *Zähne* 'teeth' need not be coded by the possessive pronoun *deine* 'you', although this is a possible alternative. Neumann (1995) makes a stronger case in pointing out, that "in German, the possessive relation need not be made explicit: there is no possessive pronoun when the possessor is mentioned elsewhere in the clause". Typically for these dative constructions, the definite article directly precedes the relational term with the "owner", in this case, of the teeth being directly inferrable from the subject pronoun *du* '2sg:NOM':

(51) *Du* *sollst* *dir* *die* (/*deine*) *Zähne* *putzen*
2sg should 2sg:DAT the:Pl (/2sg:POSS) teeth clean
'You should clean your (*the) teeth.' [Literally: 'You should to-yourself clean the teeth']

As Chafe points out (1987: 26), the two typical treatments of given concepts, that is, concepts which have already been activated in the minds of speaker and hearer and established in the prior discourse is (1) to pronominalise or (2) to omit from verbalisation, with the latter representing the 'maximum degree of attenuation'. This is clearly the option preferred in Chinese (see also section 2 above).

4.2. The semantic category of the possessors and possessa

4.2.1. The extent of the personal domain

The possessor NPs in juxtaposed noun phrases of the form NP$_1$- NP$_2$ referred exclusively to the main persons in the story – the little boy, the three other boys, the girl and the farmer.[31] In addition to this, there was one NP which referred to 'adults', *dàrén*, and another NP referring to generic 'you', *rén*. No possessor NP referred to an inanimate or animate non-human entity as the topic NP_1 in a double subject construction. This in itself is interesting since there are several props in the story which could be encoded as topic of a double subject sentence, for example, the bicycle, the hat, the goat and the stolen basket of pears.

Concerning possessa and the characterisation of the personal domain in Chinese, most terms referred to either parts of the body or closely associated features such as a person's build, age or health: six examples contained a term referring to the heart, *xīn*, in the metaphorical sense of general psychological state of that person, five examples contained a term for the build or shape of the body, for example, *gèzi*, *gèr* and four had 'foot', *jiǎo* . There were also two examples each with *shēntǐ* , 'body, health', and *niánjì*, 'age', in addition to those above. These accounted for 20/22 examples. Finally, there was one example with *dòngzuò* meaning 'movement' and one with *Zhōngguó* 'China'. Again, it is worth noting that the six examples in the Pear/Guava Stories of terms for emotions and mental activities were *never* coded as double subjects, not even in juxtaposed NPs, but were found in genitive noun phrases with *de*. This contrast was borne out in the examination of written narratives, particularly the literary narrative discussed in section 5.2.1 below .

It has been pointed out in section 3.1 that terms for material possessions as NP_2 are found in discourse coded both as possessa in genitive constructions with the genitive marker *de* or without the presence of *de* in juxtaposed NPs. This contrasts with elicited data which typically yields only the genitive construction for material possessions and concrete objects as possessa.

In the 21 narratives of the Pear/Guava Stories collection, there were 18 complex NPs containing *màozi*, 'hat'. This included three, however, where it was, contrary to the traditional rules of usage for *de,* part of juxtaposed PRO_1 – NP_2 noun phrase, *tā màozi* , 'his hat', with zero marking for the genitive in contrast to 15 genitively-marked NPs: NP_1 *de* NP_2. However, none of these three examples of *tā màozi*, 'his hat', formed double subject sentences.[32]

Confining the analysis to this data, the claim can be made that clothing, concrete objects and emotions as opposed to personality traits and physical condition are excluded from the double subject construction expressing inalienable possession of the personal domain in Mandarin Chinese.

4.2.2. Inanimate part-whole expressions

There was, in addition, a total of 35 genitive and juxtaposed NPs where inanimate nouns filled both positions of NP_1 and NP_2. These consisted of 18 genitive NPs and 17 juxtaposed NPs.[33] Not surprisingly, the grammatical role of object or oblique was the function in which the majority of these inanimate NPs were found, a reflection of the animacy hierarchy.[34] For the juxtaposed NPs, 16/17 were grammatical objects including 3 obliques and only one had the role of grammatical subject, being part of a conjoined subject, presented in (52).

(52) *Nèige* *qiú* *hé* nèige pāizi
that:CL ball and that:CL bat
shàngmian *lián-zhe* *yìtiáo* *xiàn*
top:side connect-CONT one:CL string
'The ball was connected to the upper side of the bat by a thread.' (Pear I.0: 121)

This example was not classified as a double subject construction, although similar in the stativity of the verb and the non-agent role of the subject for the following reason: Of the total number of juxtaposed complex inanimate NPs, 15/17 had a locative expression representing the head noun, such as *shàngmian*, 'upper side, top' in (52) or *lǐtou*, 'inside'. Only two contained terms for a part of a bicycle, both referring to its saddle, *zuò* (or front rack). An example of this type of juxtaposed NP in the role of grammatical object is given in (53) below:

(53) *Fàngzài* *tā* *nèige* jiǎotàchē qiánzuò *shang*
place:on 3sg that:CL bicycle front:seat on
'(He) put it on the front rack of his bicycle.' (Pear I.11: 68)

In addition to these 35 examples of inanimate NPs, there were 3 complex partitive NPs also with inanimate head nouns. Two of these could be classified as subjects of topic-comment sentences and are thus discussed in section 4.3 which follows.

4.3. Partitive constructions

In section 3.2 above, the topic-comment construction was classified into three types. In addition to double subject sentences, the main focus of this paper, two other types of structures are generally recognised. To briefly recapitulate, these are partitive sentences which also have a juxtaposed NP sentence-initially, and the general category of topicalised sentences.

In the data, 12 examples with zeromarked partitive noun phrases with a person as the head noun were noted, that is sentences where NP_1 represents a collective and NP_2, individual items or members of this collective.[35] Teng (1974: 368) and Barry (1975: 4-5) both describe this type of double subject sentence as representing a 'subset relation'.

Most of the examples in the Pear/Guava Stories data that have such a partitive subject are however S-V-O in form or contain transitive verbs in their predicate (9/12). Apart from one S-O-V example, the remaining two could be classified as a type of double subject sentence. They conform to the analysis given in section 4.1 by virtue of the fact that the predicate is stative, being filled by a nominal (in both examples, a numeral and classifier) and by the presence of an intonation unit break between the topic NP_1, *sānge nánhái* 'three boys', and the second NP representing the subset, *yíge* 'one' or 'each'. All twelve examples refer exclusively to the scene of the three boys each eating or holding a pear – their reward for helping the hero of the story after his bike crash. The two partitive double subject sentences are reproduced below:

(54) *Tāmen jiù...* // sānge rén
3pl then ... // three:CL person
yì rén *yíge*
one person one:CL
'Then they ... the three of them each had one.' (Pear II.5: 70-71)

(55) *Ránhòu ne, yì tái yǎn kànjiàn*
after RP one raise eye see
sānge nánhái yì rén *yíge* //
three:CL boy one person one:CL
'After that, well, as soon as (he) looked up, he saw three boys, each had one (of his pears).' (Pear II.6: 54-55)

Finally, note that we could collapse this partitive category with the double subject sentences representing a part-whole relation discussed above and increase our total number to 24 examples for this corpus of data. However, the standpoint adopted in this paper is that the partitive relation, though related, is not semantically identical to the inalienable part-whole relation. The main point in this section is to show that sentences expressing a partitive relation also conform to the characteristic features of the double subject construction, particularly to the feature of an intonation break between the NP designating the collective and the NP designating the subset.

5. Double subject constructions in Chinese conversational and literary genres

In addition to the Chinese Pear/Guava Stories, another two short conversational texts were examined for the treatment of relational terms in the personal domain in order to check the first set of findings based on the spoken narrative genre. The first transcription, 'Bing', concerns a daughter recounting a three-day camping trip to the countryside in Australia to her father. The second, 'Education', is a conversation between three adults about the difficulties of enrolment in China's senior secondary education system and the hardships of studying abroad when separated from one's family. They are respectively 20 minutes and 13 minutes in length.

Eight additional double subject sentences were found: There were two double subject sentences with lexical head NPs in different intonation units, conforming to the first claim; two double subject sentences with pronominal possessors and four reduced double subject sentences.[36] These four all contained relational terms as the subject NP with the possessor being overtly mentioned in a preceding clause not more than two clauses distant. Each conversation is briefly analysed separately below.

5.1. Conversational texts

5.1.1. Conversation 1: 'Bing Bing goes camping'

Conversation 1, 'Bing', had only four double subject sentences out of a total of 523 intonation units (0.765 %). This is quite high however when we consider that altogether there were only 12 intonation units with relational NPs having the form NP_1 (*de*) NP_2 out of which double subject sentences could potentially be formed (33.3% of total). In fact, this is the same proportion as for double subject sentences in the Pear/Guava Stories (see section 4.1 above). Ten of these relational NPs were in juxtaposition with their possessor NPs (that is, zeromarked for the genitive) and two were marked by genitive *de*.

The relational NPs could be classified in the following manner: seven with kin or social relations as head noun; two with body part terms as head noun and three with inanimate head nouns. The example of the double subject sentence with two lexical NPs in different intonation units has one of the two body part terms as NP_2. It is found in (56) below:

(56) *Nèige* *gǒu* *ba,//* *hái* shēn *huái* *juéjì*
that:CL dog RP even body possess unique:skill
'As for that dog, it was very talented.' (Bing 29: 420)

In addition to this, there were eight clauses containing relational terms where the possessor was not explicit (zero anaphora). Three of this type were able to be classified as reduced double subject sentences in that they all contained a stative predicate. Two examples had an inanimate "whole" as the subject, reproduced in (57) while the other one had a body part term, reproduced in (58):

(57) Inanimate part-wholes in reduced double subject clauses:
(On eating dessert of icecream and fruit:)

Ránhòu *gěi* *nǐ* *gēzài* bīngqílín *pánggbiānr* //
after for 2sg put:at icecream side
... wèidao *mǎn* *bù* *cuò* *de*
... taste full NEG bad DE
wèidao *hǎo-jí* *le*!
taste good-extreme INC
'Then they put it (the fruit) on the side of the icecream for you ... the taste is really not bad, the taste is great!.' (Bing 17: 245-249)

(58) Body part term in a reduced double subject clause

ěrduo *yě* *shì* *dāla* *-zhe*
ear also be droop CONT
'Its ears were droopy too.' (referring to a little black dog) (Bing 25: 370)

5.1.2. *Conversation 2: 'China's Education System'*

Conversation 2, 'Education', also contained four double subject sentences out of a total of 474 intonation units (0.84%). However, the percentage of double subject sentences in relation to the total number of intonation units with relational NPs is again relatively high at 12.9% (4/31). The 'Education' transcription contrasted with both the Pear/Guava Stories and 'Bing' in containing a large proportion of kin terms in relational NPs. In fact, out of a total of 31 relational NPs with the structure NP_1(*de*) NP_2, 23 referred to kin and social network , including place of work (21 without *de*, 2 with *de*); seven referred to inanimate part-whole relations and one to a person's studying abilities. The two examples of double subject sentences with kin terms both had pronominal possessors. These are given in (59) and (60):

(59) Nǐ nǚ'er *gòu* *lucky* *de,* *wǒ* *shuō,* //
2sg daughter enough lucky DE 1sg say
'Your daugher is really lucky, I mean.' (Ed. 18: 270)

The second example had inverted word order – a type of 'afterthought' construction, which repeats the topic of the given passage (which is also the subject of the clause) after the predicate. This provides backup evidence for the PRO_1 NP_2 structure normally falling within the one intonation unit and, hence, Chafe's 'one new concept at a time' constraint, since even when a speaker inverts the word order as in (61), the pronominal form and head noun are kept together:

(60) *Gòu* *yìsī* *a,*// nǐ nǚ'er
quite terrific RP 2sg daughter
'Really terrific, your daughter.' (Ed. 1: 2-3)

The third example had two lexical NPs in different intonation units and so, conformed to specifications of the first hypothesis. It characterised a young high school boy in terms of studying abilities:

(61) Nà háizi, // xuéxi *tè* *bàng*
that child study extremely good
'That child was extremely good at his studies.' (Ed. 17: 252-3)

Finally, out of two cases of zero anaphora of the possessor, one clause could be considered to be a reduced double subject, given in (62).[37] This counts as the fourth example:

(62) Reduced double subject
Wǒ *shuō* tā *juéde* *nèijiù* //
1sg say 3sg feel guilty

Běnlái xīnqíng *jiù* *bù* *hǎo*
original feelings then NEG good

'I mean he had a guilty conscience, and he wasn't in good spirits in the first place.' (Ed.28: 435-436)

5.2. Written texts

5.2.1. Shāfēi Nǚshì de Rìjì [Diary of Miss Sophie]

In addition to these conversational texts, two further texts of written narrative were briefly consulted in order to find out if double subject sentences might have a higher frequency in this genre, as opposed to unplanned spoken narrative and conversation. The converse proved to be true in the case of the first text chosen, a 42-page short story written in the vernacular in the late 1920s by one of China's most famous modern writers, Ding Ling, *nom de plume* of Jiang Bingzhi. This is her well-known story *Shāfēi Nǚshì de Rìjì* [Diary of Miss Sophie]. There were only nine double subject sentences in this text, six of these being reduced double subject forms with the possessor NP in the preceding clause or sentence and three with pronominal possessors in the same clause. Hence, the double subject construction as a representative type of topic-comment structures would appear to be very much a feature of spoken language as opposed to written. The small number of examples in this literary narrative also conformed to our predictions stated as hypotheses above.

The majority of complex relational NPs were genitive structures, comprizing a total of 178 out of 229 relational NPs, that is, 77.7 %, particularly for body part terms and emotions. This is not surprising given that the story is written in first person and describes in detail the psychological state of the heroine as well as the physical appearance and personality traits of her friends in the form of diary entries. In other words, the genitive is the preferred structure when particular emotions, personal characteristics and physical attributes are being focussed on for the purposes of description. This is backed up by the findings of a discourse study of the genitive in Mandarin (see Chappell – Thompson 1992) where terms for abstract concepts, including emotions, perceptions and thoughts were almost categorically coded with the genitive marker *de*. Recall that the number of genitive versus zeromarked juxtaposed structures in the Pear/Guava Stories were in inverse proportion to this with

48/66 relational NPs *not* being marked by *de*, that is, only 27% were genitives. (See Section 6 below on the function of the genitive in the Pear/Guava Stories.)

5.2.2. Běijīng Rén [Chinese Lives]

Secondly, one personal narrative was selected at random from the collection *Běijīng Rén: Yìbǎi ge Pǔtōngrén de Zìshù* [Chinese Lives: One hundred personal narratives from ordinary people]. This was *Yǐngzi* 'Shadow/Son of film' (1986: 388-392). Although this collection of narratives has been edited out of tape recordings made by the two authors, it remains nonetheless close to its original colloquial form. This may account for the fact that the selected piece patterned more like the spoken narratives and conversational texts than the short story *Diary of Miss Sophie* which can be classified as a narrative in literary style. The short narrative, five pages in length, that was selected from this anthology revealed five double subject clauses out of 101 full sentences. This is quite a high proportion in comparison with the short story by Ding Ling which contained only nine examples of this construction type spread over 42 pages of text. When we consider this in terms of the total number of relational NPs which was 17, the proportion becomes significantly higher (29.4%). Two of these had pronominal possessors and three were reduced double subject sentences all with the possessor NP in either the preceding clause or the preceding sentence.

Two conclusions can be made on the basis of this comparison of written and spoken materials:

(a) In general, the frequency of double subject sentences is higher in the colloquial narratives and conversational discourse examined than in the literary genre, namely the written narrative by Ding Ling. This generalisation includes the edited spoken narratives.
(b) The personal domain in Mandarin does not include emotions and feelings in either genre of the colloquial or the literary narrative: Where a term for an emotion or feeling was the head noun, the genitive marker *de* was categorically used in this corpus of data.

6. Semantic function of alienable possessa in genitive structures

When body part and relational terms turn up as possessa in genitive constructions in the Chinese Pear/Guava stories, they are given a different semantic treatment from when they occur as the inalienably possessed item in a double subject construction. This corresponds to the use of the genitive in many other languages (cf. Bally 1926 [1995]) and thus to the general semantic concept of alienability (see Chappell – McGregor 1989): I argue the case for Mandarin in this section.

In NP_1 *de* NP_2 genitive constructions in Mandarin, body parts and other relational terms are treated as individuated entities rather than as implicitly related to the person who constitutes the "whole". As such, they are found to be adjectivally modified in, for example, descriptive passages, something that does not occur for body part terms in juxtaposed NPs occurring clause-initially in double subject constructions. Teng (1974: 463) also points out that the double subject construction "rarely refers to physical description which is usually rendered by possessive patient", a finding which tallies well with the analysis of the Pear/Guava Stories data given here. By "possessive patient", Teng means nouns occurring as the head of genitive object noun phrases.

In the corpus of colloquial narratives examined, there were only 18 genitive NPs containing body part and relational terms out of the total of 66 NPs. This gives a percentage of 27.3%. Significantly, eleven of these were in object or oblique grammatical roles and only seven in subject position where they were not semantically agents: The cross-linguistic picture shows that inalienable possessa such as body parts are dispreferred in agent roles (see Allen 1964). In the following two examples, the semantic interpretation of a body part treated as an individuated entity is clearly evident. Not surprisingly, both belong to passages describing the appearance of the fruit picker in (63) and that of the girl on the bike in (64).

(63) Nèige rén de yǎnjing *bú xiàng shì*
that:CL person L eye NEG resemble be
gōngzuò
work
'That person didn't look like he wanted to work.' (One could tell from looking at the expression in his eyes.) (Literally: 'That person's eyes didn't look like they wanted to work.') (Pear I.0: 34)

(64)

Yīnwèi	tā	de	tóufa	*shì*	*hēi*	*yánsè*	*de*
because	3sg	GEN	hair	be	black	colour	DE

'Because her hair was black.' (Pear I.0: 96)

In (63), the body part term in the genitive subject is used metaphorically for the fruit picker's attitude to work while (64) occurs in a context where the narrator is providing a detailed description of the appearance of the little girl who comes riding along on her bike and consequently focuses on the girl's features and clothing. From Bally's analysis of the personal domain (1926), it is predictable that the genitive construction would also be used in such contexts in Mandarin Chinese when particular parts of the body are being viewed as individuated entities and are not being related to the whole, that is, the person.[38] Hence, speakers have recourse to two strategies for body part terms used in clause-initial position: In the case of detailed description of a person's appearance or in the use of metaphor, syntactic forms with a genitive subject are chosen. However, in cases where the narrators want to say something about the condition of a person in terms of a part of the body, then the body part term unmarked by genitive *de* will be chosen.

Similarly, in (65), the little boy is described as focusing his gaze on a particular part of the body to examine his injuries after the bike tumble. The NP containing the body part term is in the grammatical role of object and again, the genitive construction is used:

(65)

Tā	*shuāi-dǎo*	*dì-shang*	*le*	*Zìjǐ*	*kàn*	zìjǐ	de	jiǎo
3sg	fall-over	ground-on	INC	self	look	self	L	foot

'He had fallen over onto the ground and was gazing at his foot.' (Pear 8: 49-51)

7. Conclusion

This analysis of the double subject construction has presented two new findings which I believe are significant with respect to its syntactic, semantic and intonational properties as shaped by the requirements of discourse. This has lead to a new characterisation of the double subject construction in terms of its realisation in both single intonation unit utterances and larger segments of discourse such as complex clauses. In addition to this, its semantic property of expressing inalienable possession, noted by many previous analyses, has been confirmed for the first time on the basis of discourse data. The use of zeromarked juxtaposed noun phrases in double subject clauses was also semantically contrasted with that of the genitive in Mandarin.

In the analysis of the three main texts comprising the 21 spoken narratives of the Pear/Guava Stories and two short conversational texts, 109 relational head NPs were found with the form NP_1-NP_2. Of these, 30 (27.5%), formed double subject sentences, nine having a pronominal possessor in the same intonation unit as the possessed item; and fourteen with a lexical possessor in a different intonation unit from the head noun. If, as argued in the analysis above, we include seven reduced double subject sentences, that is, stative intransitive clauses with the possessed head noun as subject but no overt possessor ([ø POSSESSOR NP_1] POSSESSED $NP_{2(Relational)}$ – $VP_{(Stative)}$), we then arrive at the total number of 30.[39] We present this in tabular form below for all three texts:

Table 3. Classification of double subject sentences in three sets of spoken texts

	TYPE I[a]	TYPE II	TYPE III	
Text:	PRO_1-NP_2	NP_1-NP_2	ø – NP_2-VP	Total/Text
PEAR	7	12	3	22
BING	0	1	3	4
EDUCATION	2	1	1	4
Total/Type:	9	14	7	30

a. Type I = Single intonation unit with pronominal possessor NP_1; Type II = Possessor NP_1 in preceding intonation unit separated by intonation break only, rhetorical particle, adverb or verb taking a sentential complement; Type III = Reduced double subject sentence with possessor NP_1 in a preceding clause, not more than 2 clauses distant.

The double subject construction in Mandarin functions to express an inalienable relation in the personal domain, typically describing a person's physical or psychological condition through a related part of the body or aspect of the personality, but not emotions. It can also refer to kin relations. The first subtype of the double subject construction fits into a single intonation unit having a pronominal possessor and a stative predicate. The second subtype with two lexical noun phrases typically contains possessor and possessed in two different but consecutive intonation units. This is in accordance with Chafe's "one new concept at at time" constraint with respect to the introduction of new referents and the manner of their coding within the basic linguistic unit Chafe has labelled "the intonation unit": In the first subtype of the double subject construction, the possessor term is given and only the possessed noun represents a new referent, hence both can be coded into the same intonation unit, while in the second subtype where both NPs can be lexical, the intonation unit break separates the two nominals, avoiding information overload.

Discourse organisation also plays an important role, allowing the term for the possessor to be omitted (also described as undergoing zero anaphora), once this entity has been established in the discourse. Examples of this phenomenon were labeled as reduced double subject constructions. The effect of this patterning is that, in general, two fully lexical NPs designating the possessor and possessum in double subject structures rarely occur in the same intonation unit in our spoken data.

On the basis of the data from two main genres – spoken narrative and conversation, the treatment of relational terms has also been shown to differ significantly from that of other categories of nouns. They are coded with a higher frequency as inalienable than other terms such as material possessions and inanimate entities. Terms for body parts, attributes and kin were shown to be accessible to immediate treatment as inalienable through coding in either juxtaposed NPs with the form NP_1-NP_2 at the head of a double subject clause or as sentence-initial nouns in stative clauses with the possessor omitted, that is, in reduced double subject sentences. This latter possibility was explained in terms of their discourse status as "inferrable" entities.

Body part and relational terms in genitive NPs, NP_1-*de*-NP_2, were accounted for in terms of the genitive construction serving to individuate parts of the body for the purposes of physical description or focus. These possibilities of expression including both double subject and genitive structures can be viewed as major discourse strategies available to speakers of Mandarin Chinese for coding inalienability in the personal domain as well as possession, the choice depending on communicative intent. The findings presented in this study concerning the patterning of Mandarin with respect to the coding of alienable and inalienable possession are a result of approaching the problem from a new and different angle that has examined the conjunction of nominal and clausal syntax at the levels of semantics, discourse and prosody.

8. Appendix A: Total of 30 examples of double subject sentences from the three texts

I. PRONOUN$_1$ NOUN PHRASE$_{2(\text{Relational})}$ VP$_{(\text{Stative})}$ [9 examples]

(66) *Yīnwèi* tā gèzi *yòu* *xiǎo*, //
because 3sg build again small
'Because he was smaller in build, as well, ...' (Pear I. 6: 63)

(67) *Wǒ xiǎng*, tā xīn-li *jiù* *huì*
I think 3sg heart-in then be:likely
juéde *hěn* *qíguài*, //
feel very strange
'I think that he would have felt very strange.' [Literally: 'In his heart, he felt very strange.'] (Pear I.4: 94)

(68) Tā rén *tài* *xiǎo* //
3sg person too small
'He was too small in build.' (Pear I.7: 40)

(69) *Wǒ xiǎng*, tā xīn-li *zài* *dānxīn*, *ba!* //
I think 3sg heart-in PROG worry RP
'You know, I think he was worrying. (Pear I.7: 61)

(70) *Yīnwèi* tā jiǎo *yě* *diē-shāng* *le*, //
because 3sg foot also fall-wound INC
'Because his foot had also been injured in falling over.' (Pear I.7: 72)

(71) *Búguò* tā dòngzuò *bǐjiào* *bènzhuō* *yìdiǎn*
however 3sg movement rather clumsy a:little
'However his movements were rather clumsy.' (Pear II.1: 9)

(72) *Nàzhǒng* *shuǐguǒ* *hǎoxiàng* *shì* wǒmen
that:CL fruit seem be 1pl
Zhōngguó méi yǒu //
China NEG have
'It seems that that type of fruit is something (our) China does not have.' (Pear I.9: 7)

(73) Nĭ nǚ'er *gòu* *lucky* *de*, *wǒ* *shuō* //
2sg daughter enough lucky DE 1sg say
'Your daugher is really lucky, I'd say.' (Ed. 18: 270)

(74) *Goù* *yìsī* *a*, // nĭ nǚ'er //
quite terrific RP 2sg daughter
'Really terrific, your daughter.' (Ed. 1: 2-3)

II. NOUN PHRASE$_1$ NOUN PHRASE$_{2(Relational)}$ VP$_{(Stative)}$ [14 examples]

a. Intonation break separates NP$_1$ from NP$_2$: (5/14 examples)

(75) *uh* ... nèige nü-háizi, // gèzi *xiănde* *bĭjiào* *dà*
uh that:CL girl-child build appear rather big
'Uh ... the girl, she appeared to be quite big in build.' (Pear I.0: 200)

(76) Nèige xiăo-hái ... // gèr *hĕn* *xiăo*
that:CL small-child build very small
'That little child ... he was very small in build.' (Pear II.6: 17-18)

(77) *Bāng* *tā* *máng* *de* *shì* sānge nán-háizi, //
help 3sg work NOM be three:CL male-child
niánjì *xiănde* *bĭjiào* *dà*
age appear rather big
'There were three boys helping him, they seemed to be quite a bit older.' (Pear I.0: 198-199)

(78) *Suŏyi*, *tā* *qí* *de* *shíhòu* *bù* *néng*
so 3sg ride L time NEG can
xiàng yíbàn dàrén nèige yàngzi, //
resemble ordinary adult that:CL way
liăng jiăo *nénggòu* *tā*,
two feet can pedal
'So when he rode, it wasn't like the way adults normally do it, with both feet pedalling.' (Pear I.6: 32-34)

(79) Nà háizi, // xuéxi *tè* *bàng*
that child study extremely good
'That child was extremely good at his studies.' (Ed. 17: 252-3)

b. Rhetorical particle separates NP1 from NP_2: (2/14 examples)

(80) Nèige xiăo nán-háizi *ne*, //
that:CL little boy-child RP
liăn-shang *méi* *yŏu* *biăoqíng*
face-on NEG have expression
'The little boy, well, there was no expression on his face whatsoever.' (Pear II.7/8: 77-78)

(81) *Nèige* *gŏu* *ba*, // *hái* shēn *huái* *juéjì*
that:CL dog RP even body possess unique:skill
'That dog, it was really talented.' (Bing 29: 420)

c. Adverb separates NP_1 from NP_2: (3/14 examples)

(82) Nèige nóngchăng zhŭrén *yídìng* ... //
that:CL farm owner certainly
xīn-li *juéde* *zhèige*, //
heart-in feel this:CL
wū! *zhèi* *sānge* *zhēn* *shì* *dăn-dà-bāo-tiān*
RP this three:CL really be gall-big-cover-heaven
'The owner of the farm certainly ... must have felt that this – well! – these three were really incredibly brazen.' (Pear I.7: 107-108)

(83) Nèige nán-háizi *jiù* ... //
that:CL male-child then
yīnwèi, xīn ... // *huāng* *me*
because heart confused RP
'The boy then ... because he was confused you see.' (Pear I.13: 46-48)

(84) *Wŏ* *xiăng* zhèige nóngrén *kĕnéng*
1sg think this:CL farmer perhaps

xīn-li *juéde ei,* // *shăo-le* *yìlŏu* *guŏzi*
heart-in feel hey, be:short-PFV one:CL fruit
'I think that farmer perhaps had the feeling, hey!, he was missing a basket of fruit.' (Pear I.4: 89-90)

d. Verb taking sentential complement separates NP_1 from NP_2: (4/14 examples)

(85) Nèige nán-háizi *hăoxiàng* gèzi *hái bĭjiào xiăo,* //
that:CL male-child seem build still rather small
'The boy seemed to be rather small in build. ('Pear I.6: 31)

(86) *Ràng* rén *juéde* ... // *uh* xīn-lĭtou
make people feel uh heart-inside
kànjiàn zhèige fēngjĭng fēicháng píngjìng
see this:CL scenery extremely calm
'It makes you feel, um, having seen the scenery, extremely calm.' (Pear I.0: 18)

(87) ... *nèige dài shuĭguŏ de xiăo* háizi, //
... that:CL carry fruit REL small child
kànqi niánjì *xiănde zuì xiăo*
seem age appear most small
'As for that small child carrying the fruit, he appeared to be the youngest in age.' (Pear I.0: 195-196)

(88) *Ránhòu* tā *en* ... //
after 3sg um
hăoxiàng jiăo *yĕ nòng* ... //
seem foot also do
nòng-shāng le, // *hái nòng-téng le* //
do-wound INC also do-hurt INC
'Afterwards he, um, it seemed that his foot was ... was also injured as well as painful.' (Pear I.2: 41-44)

III. Reduced double subject sentences
[ø POSSESSOR NP_1] // NOUN PHRASE$_{2(\text{Relational})}$ VP$_{(\text{Stative})}$
[7 examples with possessor noun in a preceding clause]

(89) *Ránhòu*, tā *jiù shì* ... // *diē-dǎo zuò-zài*
after 3sg then be fall-topple sit-at
dì-shang jiù bǎ nèige kùguǎn juǎn ... //
ground-on then BA that:CL trouser:leg roll
dàgài jiǎo *shāng le*
probably foot wounded INC
'After he'd ... fallen over and was sitting on the ground, he then rolled his trouser leg – ... probably he'd hurt his foot.' (Pear I.6: 89-91)

(90) Tā *zài huí shù-shang* // *Huí shù-shang*
3sg again return tree-on return tree-on
de shíhou,
L time
yīnwèi shēntǐ *hěn dà* ... .
because body very big
'He went back up the tree again. When he went back up the tree, because his body was very big ...' (Pear II.7/8: 17-18)

(91) Nèi sānge xiǎo-háir *yě zǒu le*, //
that three:CL small-child also leave PFV
kàn-qilai hěn qīngsōng yúkuài, //
look-INCH very relaxed happy
uh shēntǐ *hěn jiànkāng de yàngzi*
uh body very healthy GEN appearance
'Those three children left too, they looked very carefree, happy and uh, healthy.' (Pear I.0: 135-137)

(92) (On eating dessert of icecream and fruit:)
Ránhòu gěi nǐ gēzài bīngqílín *pánggbiānr* //
after for 2sg put:at icecream side
... wèidao *mǎn bù cuò de*
... taste full NEG bad L
'Then they put it (the fruit) on the side of the icecream for you ... the taste is really not bad.' (Bing 17: 245, 248)

(93) wèidao *hǎo-jí* *le*!
taste good-extreme INC
'The taste is great!' (Bing 17: 249) [directly follows example (92) in the text, as in (57) above]

(94) ... *bù* *zhīdào* *nǎr* *lái-le* yìtiáo hēi gǒu (...)//
NEG know where come-PFV one:CL black dog
ěrduo *yě* *shì* *dāla* *-zhe*
ear also be droop CONT
'... a little black dog turned up out of nowhere. (...) Its ears were droopy too.' (Bing 25: 370)

(95) *Wǒ* *shuō* tā *juéde* *nèijiù* //
1sg say 3sg feel guilty
Běnlái xīnqíng *jiù* *bù* *hǎo*
original feelings then NEG good
'I mean he had a guilty conscience, and he wasn't in good spirits in the first place.' (Ed.28: 435-436)

9. Appendix B

The outline of the Pear Story is reproduced here from Chafe (1980: xiii-xiv) for the convenience of the reader.

The film begins with a man picking pears on a ladder in a tree. He descends the ladder, kneels, and dumps the pears from the pocket of an apron he is wearing into one of three baskets below the tree. He removes a bandana from around his neck and wipes off one of the pears. Then he returns to the ladder and climbs back into the tree.

Toward the end of this sequence we hear the sound of a goat, and when the picker is back in the tree, a man approaches with a goat on a leash. As they pass by the baskets of pears, the goat strains toward them, but is pulled past by the man and the two of them disappear in the distance.

We see another closeup of the picker at this work, and then we see a boy approaching on a bicycle. He coasts in toward the baskets, stops, gets off his bike, looks up at the picker, puts down his bike, walks toward the baskets, again looking at the picker, picks up a pear, puts it back down, looks once more at the picker, and lifts up a basket full of pears. He puts the basket down near his bike, lifts up the bike and straddles it, picks up the basket and places it on the rack in front of his handle bars, and rides off. We again see the man continuing to pick pears.

The boy is now riding down the road, and we see a pear fall from the basket on his bike . Then we see a girl on a bicycle approaching from the other direction. As they pass, the boy turns to look at the girl, his hat flies off, and the front wheel of his bike hits a rock. The bike falls over, the basket falls off, and the pears spill out onto the ground. The boy extricates himself from under the bike, and brushes off his leg.

In the meantime we hear what turns out to be the sound of a paddleball, and then we see three boys standing there, looking at the bike boy on the ground.

The three pick up the scattered pears and put them back in the basket. The bike boy sets his bike upright, and two of the other boys lift the basket of pears back onto it. The bike boy begins walking his bike in the direction he was going, while the three other boys begin walking off in the other direction.

As they walk by the bike boy's hat on the road, the boy with the paddle ball sees it, picks it up, turns around, and we hear a loud whistle as he signals to the bike boy. The bike boy stops, takes three pears out of the basket, and holds them out as the other boy approaches with the hat. They exchange the pears and the hat, and bike boy keeps going while the boy with the paddleball runs back to his two companions, to each of whom he hands a pear. They continue on, eating their pears.

The scene now changes back to the tree, where we see the picker again descending the ladder. He looks at the two baskets, where earlier there were three, points at them, backs up against the ladder, shakes his head, and tips up his hat. The three boys are now seen approaching, eating their pears. The picker watches them pass by, and they walk off into the distance.'

10. Primary sources of data

I. Main texts:

(a) Chinese Pear/Guava stories (ca. 2 ½ hrs)

The Chinese Pear/Guava Stories is a transcription in Mandarin Chinese based on two and a half hours of taping (unpublished transcription ms.) made by Francisco Y-w Hwang on behalf of Mary Erbaugh, University of Oregon, referred to by the abbreviation 'Pear' after language examples in the analysis above. This set of 21 narratives was collected by Mary Erbaugh in March 1976 at National Taiwan University (NTU) in Taipei, ROC.

The 20 participants in the interviews comprised 15 women college students or graduates at NTU and also 5 women teachers. Following the protocol set out in Chafe (1980), all the young women were shown the film and interviewed within an hour of seeing it. They were requested not to discuss the film in the intervening period.

Two of the narratives are made by the interviewer herself, before and after seeing the film. All but one of the interviewers are native speakers of Mandarin (known as *guóyŭ* in Taiwan, ROC), being raised in monolingual Mandarin families. The "exception" was a fully bilingual Taiwanese-born speaker, whose proficiency enabled her to become a teacher of Mandarin to foreigners at the Stanford Center, NTU (Mary Erbaugh, pers. comm.).

I would like to thank Mary Erbaugh for making these transcriptions available.

(b) Bing Bing goes camping (20 mins)

This conversation was recorded and transcribed by Liu Mingchen in April 1988 in Melbourne, following the guidelines of Du Bois et al (1992) at the request of the author. A twelve-year old girl, Liu Bing Bing is the main speaker in this conversation. At the time of the conversation, Liu Bing Bing has just arrived in Australia from Beijing where she grew up. She recounts her experiences during a three-day camping trip with her English class and teachers to the Victorian countryside (Southeastern Australia) to her father. Abbreviated as 'Bing'.

(c) China's education system (13 mins)

A second conversation between three adult native speakers of Mandarin was recorded and transcribed in September, 1988, in Melbourne, following the guidelines of Du Bois et al (1992), at the request of the author. The conversation takes place at the home of a married couple from Beijing. A third person, a woman friend, also from North China (Hebei province) drops in one Saturday evening to see them and relate the latest news from China. All three are in the 40 -50 years age range. They have come to Australia for study or work purposes. The visitor has, however, come on her own. The conversation mainly concerns the difficulty her daughter has experienced, like so many other students in China, in gaining a place at a polytechnic senior college (*Zhong Zhuan*) in China. In addition, other current problems in the education system in China are discussed as well as the increasingly popular trend to study abroad. Abbreviated as 'Education'/ 'Ed.'.

II. Other texts consulted

(d) Written narrative: A short story

Ding Ling. 1928/1956. *Shāfēi Nǚshì de Rìjì* [The Diary of Miss Sophie] pp.2-44 in *Dīng Líng Xuǎnjí* [Selected Stories of Ding Ling]. Hong Kong: Wenxue Chubanshe. Abbreviated as 'Sophie' .

(e) Written narrative: A novel

Chen Ruoxi. 1984. *Guī* [Return].Taipei: Lianhe Bao Congshu.

(f) Written narratives based on edited spoken texts

Zhang Xinxin – Sang Ye. 1986. *Běijīng Rén: Yìbǎige pǔtōng rén de zìshù* [Chinese Lives: One hundred personal narratives from ordinary people] Shanghai: Shanghai Wenyi Chubanshe

(g) Notes from a Chinese colleague

A collection of short notes in Mandarin Chinese, gathered during the period March – July 1990 at La Trobe University, Melbourne by the author, addressed to her by a colleague from Beijing. Abbreviated as 'Notes'.

(h) *A conversation entitled "Hong Kong police"*

Recorded in Hong Kong with native speakers of Mandarin in July, 1990, and transcribed by Hongyin Tao, University of California at Santa Barbara. Abbreviated as "HK".

Notes

1. See Modini (1981) for an overview of this construction in East Asian languages. For analyses of double subject constructions in unrelated languages, see Burridge (1995) on Middle Dutch, Hosokawa (1995) on Yawuru and Tsunoda (1995) on Japanese.
2. I thank Wallace Chafe, Jack Du Bois, Bill McGregor, Hongyin Tao, Sandra Thompson and Tasaku Tsunoda for their comments and discussion at various stages of writing this paper. As author, I remain solely responsible for the interpretation of the data. A shorter version of this paper was presented at the 23rd International Conference on Sino-Tibetan Languages and Linguistics, held at the University of Texas at Arlington in October, 1990.
3. The transliteration system used in this paper for the Mandarin language examples is the *pīnyīn* system of romanisation. This was officially adopted in People's Republic of China in 1958 for this purpose and is the system most commonly used in Chinese linguistics. The following *pīnyīn* symbols on the left hand side correspond to the given phonetic values in the International Phonetic Alphabet on the right hand side:

   [b] = [p]; [p] = [p']; [d] = [t]; [t] = [t']; [g] = [k]; [k] = [k']
   [z] = [ts]; [c] = [ts']; [j] = [tɕ]; [q] = [tɕ']; [zh] = [tʂ]; [ch] = [tʂ']
   [f] = [f]; [s] = [s]; [x] = [ɕ]; [r] = [ʐ]; [sh] = [ʂ]; [h] = [χ]
   [m] = [m]; [n] = [n]; [ng] = [ŋ]; [l] = [l]
   [w] = [w]; [y] = [j]; [-r] = [ɚ].
   [i] = [i] except after initials *zh*, *ch* and *sh*; *z*, *c* and *s* where [i] = [ɿ]
   [e] = [ɛ]; [a] = [a]
   [u] = [u] except after initials *j*, *q* and *x* where [u] = [y]; [o] = [ɔ].

   Tone is phonemic in Mandarin. The four tones are marked by the following diacritics on the nucleus vowel of the syllable rhyme:

	Description		*Y.R. Chao relative tone pitch*
Tone 1:	High level	ā	55
Tone 2:	High rising	á	24
Tone 3:	Low falling-rising	ǎ	214
Tone 4:	High falling	à	51

   The absence of any tone diacritic indicates the neutral tone (q.v. Chao 1968), for example, an unstressed syllable which does not receive its full tonal value. In addition to this, many grammatical and discourse particles are atonal. Tone sandhi has been indicated for the combinations of syllables which have citation tones in isolation as follows: Tone 3 + Tone 3 → Tone 2 + Tone 3 [ǎ + ǎ → á + ǎ] ; and for individual lexical items such as *yì* 'one, a' which changes to Tone 2 before another Tone 4 [*yì* + à → yí + à].

Any example preceded by the symbol # represents elicited data. The abbreviations used in the interlinear glossing are as follows: BA = exponent of the pretransitive or 'disposal' construction in which grammatical objects are positioned preverbally; BEI = exponent of the passive construction; CL = classifier; CONT = continuous aspect marker, *-zhe*; EXT = marker of the extent construction 'so much that'; INC = marker of inceptive aspect (inception of a new state of affairs); INCH = marker of the inchoative aspect, *-qilai* ; L = linking particle *de* which enables attributive modification, also functioning as a subordinator and as a genitive marker; N = noun; NEG = negative marker; NOM = nominaliser, *de*; NP = noun phrase; O = object; PASS = colloquial passive markers other than *bèi*; PFV = perfective aspect marker; PROG = progressive aspect marker: *zài*; Q = quantifier; P = discourse particle; S = subject noun; V = verb; 1sg = first person singular pronoun; 2sg = second person singular pronoun; 3sg = third person singular pronoun. Note that the symbol "//" is used to denote an intonation unit break.

4. For details of defining criteria of the double subject, see discussion in section 3.2 below.
5. Modini (1981) examines inalienable possession and double subject constructions in several East Asian languages including Mandarin. In outlining the syntactic characteristics of the double subject construction, Teng (1974) also points to the possibility of such a semantic relationship but ultimately rejects it.
6. Extrapolation to other genres is attempted in section 5 of this analysis. As the database proves to be quite small, however, for conversational texts, we confine most of our comments and conclusions purely to the spoken narrative genre.
7. Example (3) is example (68) from Li – Thompson (1976). See Li – Thompson (1976) and Gundel (1987) for a discussion of topic-comment constructions within a cross-linguistic perspective.
8. With respect to the analysis above, if there was no genitive marker *de* present, I treated the combination of *Pronoun + Noun + Stative Verb* as a double subject construction.
9. Note however that it can also be used for both kin and inanimate part-whole relations, for example, *Shù yèzi dà* 'As for the tree, its leaves are big.' (This is a reproduction of example (23) from Li – Thompson 1976). However, none of this type of double subject construction was found in the Pear/Guava Stories data. Inanimate part-whole expressions were found typically in object and oblique positions (see sections 4.2 and 4.3 above for further discussion of inanimate appositional NPs). The reason why no kin terms were found in the Pear/Guava Stories is that such a possible relationship between any of the characters was not part of the story or at least relevant in a silent film. Double subject constructions with kin terms or inanimates as the head noun are, however, found in the conversational texts discussed in section 5.
10. A more detailed typological description would however need to discuss locatives of destination goal which iconically follow the main verb (see Tai 1975); similarly resultative and extent constructions which are also postverbal express the end state of affairs.
11. This is example (2) in Nichols (1986: 57).
12. Acceptability of the N-N structure is not only determined by factors such as nominal versus pronominal but also by the person of the pronoun (1,2,3) and number (singular versus plural) in conjunction with the semantic category of the noun. See Chappell – Thompson (1992).

13. Chao (1968) also carries out a semantically-based classification of the different kinds of S-P predicates (his label for topic-comment sentences). See also Shen (1987) who applies the framework of Dik's functional grammar in the analysis of double subject sentences and Chappell (1994: section 5.7) for an analysis of double subject constructions as a syntactic means of expressing the language universal of (inalienable) possession in Mandarin.
14. Teng (1974: 463) is the most explicit of these analyses in specifically stating the following: "Only state and process verbs ... are observed to occur in the sentence predicate construction. This kind of predicate only specifies an event which affects a patient, or a patient experiencing a certain state."
15. Teng (1974: 468) describes this example as having a "subset relation" (q.v. section 4.3 below).
16. Modini (1981: 9) also labels the similar example of *?tā de tóu téng* 'His/her head aches' as marginal in acceptability.
17. Note, however, that Li – Thompson (1976: 487-88; foonote 9) object to adverb placement as a test of VP-hood.
18. Note that such a sentence might occur where the speaker hesitates momentarily and uses [*a*] or some other similar morpheme as a pause particle. In this case, it would not have the same rhetorical or semantic value as [*a*] in (39) and does not, therefore, vitiate our argument. Our point of view here largely coincides with Tsao (1978: 184-185) who uses the possibility of insertion of a pause particle as a means of identifying topic in Chinese. See his examples (45) and (46).
19. Li – Thompson (1976: 481) adduce what would be a fourth criterion, namely that of coreferential deletion. In topic-comment sentences, coreferential deletion is controlled by the topic whereas in genitive sentences it is controlled by the head noun of the genitive subject NP (erstwhile subject of the predicate sentence in an analogous topic-comment construction). For reasons of space, we will not include it here.
20. See section 6 below on the discourse properties of sentences with genitive subjects. Chao (1968: 95-96) remarks on the difference in meaning between double subject sentences and sentences with genitive NPs at their head that the "transformation" from one to the other "may either change the style or even the sentence value. Thus, while *Ta ell.dou roan* always means 'He is gullible'. *Ta de elldou roan* may mean either (1) 'He is gullible.' or (2) 'His ear is (physically) soft.' "
21. A resumé of the story from Chafe (1980) is reproduced in Appendix B of section 9 above. Note also that for the purposes of the data analysis, the preface at the beginning of each narrative has been discounted. This is where the interviewer explains the task to each of the 19 women students from National Taiwan University. As the explanation is merely repeated, more or less verbatim, before each narration, its exclusion is essential to avoid skewing the results. The Chinese Pear/Guava Stories are so named due to the fact that the majority of storytellers identify the unripe green fruit in the silent film as guavas instead of pears (see also Erbaugh 1990).
22. However, as indicated, the sixty-six terms sifted out from the Pear/Guava Stories data included not only terms for body parts and emotions but also those for related categories in the personal domain such as the build or shape of a person's body, a person's age, clothing, actions and friends. The rationale for initially including all these categories was to allow for comparison of syntactic behaviour across categories, noting that all were possible candidates for treatment as inalienables according to Bally (1926

[1995]). This was also crucial for independently obtaining a definition of inalienability for Chinese. Note that all complex relational NPs, regardless of grammatical role were included.

It is also important to note that excluded from consideration were compound verbs, adjectives and adverbs as well as four-character idioms containing body part terms from the final corpus. Examples of these categories are: *huí -tóu* [return-head] 'to turn around'; *wān-shēn* [bend-body] 'to bend over'; *zhuān-xīn* [exclusive-heart] 'with concentration'; *shāng nǎojīn* [wound-brain] 'to rack one's brains'; *cá jiān ér guò* [rub-shoulder-and-pass] 'to brush past someone'.

23. This is discussed in section 3.1 above.
24. The descriptive term "clause-initial' is deliberately vague as to whether the body part head noun acts as subject of an S-V-O sentence or as subject of a sentential predicate in a topic-comment construction. In many examples, the body part term was the subject of an embedded sentence or the pivot noun in a pivot construction, that is, simultaneously the object of the matrix clause and subject of the following embedded clause. That topic-comment sentences are not restricted to basic declarative sentences is pointed out by Li – Thompson (1976: 483) who give an example of embedding. This evidence is also used to argue for the basicness of topic-comment constructions in their analysis.
25. All examples in the corpus of double subject sentences from the spoken texts are presented in Appendix A of section 8.
26. This was generally indicated in the transcription by double oblique lines or by a new line or comma, following in the latter case, Du Bois et al's guidelines (1993) for transcription work.
27. The possibility of such configurations is foreshadowed in Chao (1968), Teng (1974), Li – Thompson (1976) and Modini (1981). The 12 relevant examples from the Pear/Guava Stories are listed under Group II in Appendix A of section 8.
28. Both body part terms in these two examples occur with a locative postposition *liǎn-shang* 'face-on' and *xīn-li* 'heart-in' which is common in the data. We consider these nonetheless to form double subject constructions while acknowledging this feature requires further investigation.
29. Teng (1974: 458, footnote 3) considers such examples as (49) to have had the genitive marker deleted in the surface.
30. As pointed out in note 8 above, this view conforms to standard practice in Chinese Linguistics, where both pronominal and nominal possessors have generally been subsumed under the one construction type of the double subject, for example, in Mullie (1932, 1933), Chao (1968) and Li – Thompson (1976, 1981).
31. A resumé of the story is given in Appendix B of section 9 above.
32. In two of these cases, it was the preposed direct object of a *bǎ* disposal construction (S-BA-O -V) (see Chappell 1992a). In the other, it was the subject of a *bèi* passive but not of a double subject construction. One of the *bǎ* constructions is exemplified below:

Qí-guo	*qù*	*bǎ*	tā	màozi	*zhuàng-diào*	*le*
ride-past	go	BA	3sg	hat	knock-down	PFV

'Riding past, (she) knocked his hat off.' (Pear I.7: 67)

33. This excludes three appositional NPs contained in attributive modifiers which grammatically require *de* to be omitted.
34. With regard to the group of 18 genitive NPs, 14 were objects and only 4 subjects. As they do not fit the criteria for a double subject construction, we will not discuss them further.
35. Teng (1974) and Lü (1965) both distinguish the partitive relation from the part-whole relation in the manner described above in the text. I follow suit in the belief that the two relations, though similar, are not semantically equivalent.
36. With such a small number of double subject sentences in each transcription, it is not possible to absolutely confirm the findings presented above. Nonetheless, no counterexamples to the hypotheses were found.
37. 'Bing' also contained 4 partitive sentences while 'Education' had 8 partitive constructions.
38. See also Malczewska – McGregor (1990) for a detailed analysis of the personal domain and individuation in Polish dative constructions.
39. The total number of double subject constructions constitutes 1.1% of the combined total of intonation units of the three texts (30/2801).

References

Allen, W. Sidney

1964 "Transitivity and possession", *Language* 40: 337-343.

Ameka, Felix

1995 "Body parts in Ewe grammar", [this volume.]

Anderson, John M. – Charles Jones (eds.)

1974 *Historical linguistics.* Amsterdam: North Holland.

Bally, Charles

1926 "L'expression des idées de sphère personnelle et de solidarité dans les langues indo-européennes" , in: Franz Fankhauser – Jakob Jud (eds.), 68-78.

[1995] "The expression of concepts of the personal domain and indivisibility in Indo-European languages", translated by Christine Béal – Hilary Chappell, [this volume.]

Barry, Roberta

1975 "Topic in Chinese: An overlap of meaning, grammar and discourse function", *Papers from the Parasession on Functionalism.* Chicago: Chicago Linguistic Society, 1-9.

Bavin, Edith

1995 "Body parts in Acholi: alienable and inalienable distinctions, and extended uses", [this volume.]

Burridge, Kate

1995 "Degenerate cases of body parts in Middle Dutch", [this volume.]

Chafe, Wallace

1987 "Cognitive constraints on information flow", in: Russell S. Tomlin (ed.), 21-51.

Chafe, Wallace (ed.)
1980 *The Pear Stories: cognitive, cultural and linguistic aspects of narrative production.* Norwood, NJ: Ablex.

Chao, Yuen Ren
1947 *A Mandarin primer*. Cambridge, Mass.: Harvard University Press.
1968 *A grammar of spoken Chinese*. Berkeley: University of California Press.

Chappell, Hilary
1986a "Formal and colloquial adversity passives in standard Chinese", *Linguistics* 24.6: 1025-1052.
1986b "The passive of bodily effect in standard Chinese", *Studies in Language* 10.2: 271-296.
1992a "Causativity and the *ba* construction in Chinese", in: Hansjakob Seiler – Waldfried Premper (eds.), 509-530.
1992b "Strategies for the assertion of obviousness and disagreement in Mandarin Chinese discourse: A semantic study of modal particle *me*", *Australian Journal of Linguistics* 11.1: 39-65.
1992c "Towards a typology of aspect in Sinitic languages", *Zhongguo Jingnei Yuyan ji Yuyanxue* [Journal of Chinese Languages and Linguistics] 1: 67-106 (Taipei: Academia Sinica).
1994 "Mandarin semantic primitives", in: Cliff Goddard – Anna Wierzbicka (eds.), 109-147.
To appear *Analytic syntax in standard Chinese. A semantic analysis of passive, causative, dative and benefactive constructions.*

Chappell, Hilary – William McGregor
1989 "Alienability, inalienability and nominal classification", *Proceedings of the Berkeley Linguistics Society* 15: 24-36.
1995 "Prolegomena to a theory of inalienability.", [this volume.]

Chappell, Hilary – Sandra A. Thompson
1992 "Semantics and pragmatics of associative *de* in Mandarin discourse" *Cahiers de Linguistique Asie Orientale* 21: 199-229.

Chen, Ping
1986 *Referent tracking in Chinese*. [Unpublished doctoral dissertation, University of California at Los Angeles.]

Chen, Ruoxi
1984 *Guī.* [Return]. Taipei: Lianhe Bao Congshu.

Cheng, Robert L. – Ying-che Li – Ting-chi Tang (eds.)
1978 *Proceedings of symposium on Chinese linguistics.* (2nd edition) Taipei: Student Book Company.

Chu, Chauncey
1976 "Conceptual wholeness and the retained object", *Journal of Chinese Linguistics* 4.1: 14-23.

Clark, Marybeth
1974a " Passive and ergative in Vietnamese", in: Nguyen Dang Liem (ed.), 75-88.
1974b "Submissive verbs as adversatives in some Asian languages", in: Nguyen Dang Liem (ed.), 89 – 110.
1995 "Where do you feel? Stative verbs and body part terms in mainland Southeast Asia", [this volume.]

Crowley, Terry
1995 "Inalienable possession in Paamese grammar", [this volume.]
Ding Ling
1928 *Shāfēi Nǚshì de Rìjì* [Diary of Miss Sophie], reprinted in: Ding Ling (1956), 1-42.
1956 *Dīng Líng xuǎnjì* [Selected Stories of Ding Ling]. Hong Kong: Wenxue Chubanshe.
Dragunov, Aleksandr. A.
1961 *Untersuchungen zur Grammatik der modernen chinesischen Sprache* [Investigations into the grammar of the modern Chinese language]. Translated by Wolfgang Lippert. Berlin: Akademie Verlag. [Originally published in 1952 in Russian by the USSR Academy of Sciences, Moscow.]
Du Bois, John – Stephan Schuetze-Coburn– Danae Paolino – Susanna Cumming
1993 "Outline of discourse transcription", in: Jane A. Edwards – Martin D. Lampert (eds.).
Edwards, Jane A. – Martin D. Lampert (eds.)
1993 *Talking data: Transcription and coding methods for language research.* Hillsdale, N.J.: Lawrence Erlbaum .
Egerod, Søren
1985 "To have and have not – The category of possession in East Asian languages and the role it plays in the syntactic core." Paper presented at the 18th International Conference on Sino-Tibetan Languages and Linguistics, Bangkok.
Erbaugh, Mary
1990 "Mandarin oral narrative style compared with English and Greek", *Journal of the Chinese Language Teachers Association* 25.2: 21-42.
Fankhauser, Franz – Jakob Jud (eds.)
1926 *Festschrift Louis Gauchat.* Aarau: Sauerländer.
Givón, Talmy (ed.)
1983 *Topic continuity in discourse: A quantitative cross-linguistic study.* Amsterdam: John Benjamins.
Goddard, Cliff – Anna Wierzbicka (eds.)
1994 *Semantic and lexical universals.* Amsterdam: John Benjamins.
Gundel, Jeannette
1987 "Universals of topic and comment structure", *Another Indiana University Linguistics Club Twentieth Anniversary Volume.* Bloomington, Indiana: Indiana University Linguistics Club.
Haiman, John
1983 "Iconic and economic motivation" *Language* 59: 781-819.
1985 *Natural syntax.* Cambridge: Cambridge University Press.
Hansson, Inga-Lill (ed.)
1971 *A symposium on Chinese grammar.* (Scandinavian Institute of Asian Studies Monograph Series 6) Copenhagen: Studentlitteratur.
Hashimoto, Anne Yue
1971 "Mandarin syntactic structures", *Unicorn* 8: 1-154. Princeton: Chinese Linguistic Project.
Huang, Shuan-fan
1978 "Historical change of prepositions and the emergence of SOV word order", *Journal of Chinese Linguistics* 6: 212-242.

Jeng, Heng-hsiung
1978 "Topic and subject in Chinese, English and Bunun", in: Robert L. Cheng – Ying-che Li – Ting-chi Tang (eds.), 317-348.

Kuno, Susumo
1970 "The position of locatives in existential sentences", *Linguistic Inquiry* 2.3: 333-378.

Lehmann, Winfred (ed.)
1978 *Syntactic typology: Studies in the phenomenology of language*. Austin: University of Texas Press.

Charles N. Li (ed.)
1975 *Word order and word order change*. Austin: University of Texas Press.
1976 *Subject and topic*. New York: Academic Press.

Li, Charles N. – Sandra A. Thompson
1974 "Historical change of word order: A case study in Chinese and its implications", in: John M. Anderson – Charles Jones (eds.), 199-217.
1975 "The semantic function of word order: A case study in Mandarin", in: Charles N. Li (ed.), 163-195.
1976 "Subject and topic. A new typology of language", in: Charles N. Li (ed.), 457-489.
1978 "An exploration of Mandarin Chinese", in: Winfred P. Lehmann (ed.), 223-266.
1981 *Mandarin Chinese A functional reference grammar*. Berkeley: University of California Press.

Light, Timothy
1979 "Word order and word order change in Mandarin Chinese", *Journal of Chinese Linguistics* 7: 149-180.

Lin, Helen
1981 *Essentials of modern Chinese grammar*. Boston: Cheng and Tsui.

Lü, Shuxiang
1965 "Yǔwén zhājí [Notes on Language]", *Zhongguo Yuwen* 4.137: 287-292.

Malczewska, Beata – William McGregor
1990 Body parts and the notion of the personal domain in Polish. [Unpublished MS]

Modini, Paul
1981 "Inalienable possession and the 'double subject' constructions in East Asian", *Cahiers de Linguistique Asie Orientale* 9: 5-15.

Mullie, Joseph M.
1933 "Le double nominatif en chinois [The double nominative in Chinese]", *T'oung Pao* 30: 231-236.
1932 *The structural principles of the Chinese language*. Peiping: The Bureau of Engraving and Printing.

Nichols, Johanna
1986 "Head-marking and dependent-marking grammar", *Language* 62.1: 56-119.

Neumann, Dorothea
1995 "The dative and the grammar of body parts in German", [this volume.]

Nguyen Dang Liem (ed.)
1974 *South-East Asian Linguistic Studies Volume 1: Pacific Linguistics Series C-31*. Canberra: Department of Linguistics, RSPacS, ANU.

Osumi, Midori
1995 "Body parts in Tinrin", [this volume.]

Rygaloff, Alexis
1971 "Le syntagme nominal complexe, la fonction de *de*, et les valeurs de *you*", in: Inga-Lill Hansson (ed.), 145-156.

Shen, Jiaxuan
1987 "Subject function and double subject construction in Mandarin Chinese", *Cahiers de Linguistique Asie Orientale* 16.2: 195-211.

Seiler, Hansjakob – Waldfried Premper (eds.)
1992 *Partizipation: Das sprachliche Erfassung von Sachverhalten* [Participation: The linguistic comprehension of events]. (Language Universals Series 6) Tübingen: Gunter Narr Verlag.

Sun, Chao-fen – Talmy Givon
1985 "On the so-called SOV word order in Mandarin Chinese: A quantified text study and its implications", *Language* 61.2: 329-351.

Schleicher, August
1863 *Die Darwinische Theorie und die Sprachwissenschaft* [Darwinian theory and linguistics]. Weimar: Hermann Böhlau.

Tang, Ting-chi
1972 *A case grammar of spoken Chinese*. Taipei: Haiguo Book Co.

Tai, James H-Y
1975 "On the two functions of place adverbials in Mandarin Chinese," *Journal of Chinese Linguistics* 3.2: 154-179.

Teng Shou-hsin
1974 "Double nominatives in Chinese", *Language* 50.3: 455-473.

Tiee, Henry
1986 *A reference grammar of Chinese sentences*. Tucson: University of Arizona.

Thompson, Sandra A.
1973 "Transitivity and some problems with the *ba* construction", *Journal of Chinese Linguistics* 1.2: 208-221.

Tomlin, Russell
1987 "Linguistic reflections of cognitive events", in: Russell S. Tomlin (ed.), 455-479.

Tomlin, Russell (ed.)
1987 *Coherence and grounding in discourse*. Amsterdam: John Benjamins.

Tsao Feng-fu
1978 "Subject and topic in Chinese", in: Robert L. Cheng – Ying-che Li – Ting-chi Tang (eds.), 165-196.
1979 *A functional study of topic in Chinese. A first step towards discourse analysis*. Taipei: Student Book Company.
1982 "The double nominative construction in Mandarin Chinese", *Tsing Hua Journal of Chinese Studies* 14.1/2: 275-297.

Tsunoda, Tasaku
1995 "The possession cline in Japanese and other languages", [this volume.]

Wierzbicka, Anna
1979a "Ethnosyntax and the philosophy of grammar", *Studies in Language* 3.3: 313-383.
1979b "Are grammatical categories vague or polysemous? (The Japanese 'adversative' passive in a typological context.)", *Papers in Linguistics* 12: 111-162.
Zhang Xinxin – Sang Ye
1986 *Běijīng Rén: Yìbǎige pǔtōng rén de zìshù* [Chinese Lives: One hundred personal narratives from ordinary people] Shanghai: Shanghai Wenyi Chubanshe.

Where do you feel? – Stative verbs and body-part terms in Mainland Southeast Asia

Marybeth Clark

1. Introduction

When describing physical and emotional feelings, the languages of the Mainland Southeast Asia linguistic area commonly use a distinctive construction.[1] This construction consists of a stative/adjectival verb which occurs in close association with a body-part term which in turn represents a part of an animate whole which is the subject of the stative verb, i.e.: Animate Subject ^ Stative Verb ^ Body-Part Noun (*Nam ^be-sick ^head* = 'Nam has a headache'), where the verb describes a state which is located in the body part but where the subject whole is expressed as undergoing that state. This construction is used to describe both physical and metaphorical feelings, as in these sentences: *Lan be-hungry stomach* = 'Lan is hungry in the stomach,' *Youa be-lonely liver* = 'Youa is lonely in the liver: homesick,' *Arunee be-good heart* = 'Arunee is happy (in the heart).'[2]

Such a construction raises the conceptual questions of part-whole relationships and inalienability as well as the possibility of body-part-noun incorporation in the stative verb. This particular construction, although making an explicit expression of inalienability by means of possessor-whole promotion to subject, has as its intention to emphasise the subjectivity of inalienability rather than to make a distinction between alienability and inalienability. Further, it appears that, via lexical juxtaposition and compounding, incorporation of the possessee body part in the stative verb does take place, although different expressions are in different stages of lexical binding.

After describing some general typology and the use of stative verbs in the languages under discussion (section 1), I will explain the stative-verb-body-part construction and discuss the questions of inalienability and noun incorporation (section 2). In succeeding sections I will elaborate on the expressions, first for physical states then for metaphorical states.

1.1. The languages

The languages of Mainland Southeast Asia through long contact with each other form a definitive linguistic area. Besides some phonological characteristics such as back unrounded vowels and a strong tendency to develop tone distinctions,[3] there are many shared grammatical features. Some of the features most relevant to this paper are a basic Subject-Verb-Object sentence order, general monosyllabicity with compounding and loss of affixation, nominal classifiers, and stative verbs: descriptive words/adjectives which function as main verbs, needing no other verb to form a grammatical sentence. It is this last characteristic which is the focus of this paper: animate states of being or feeling which are associated with particular parts of the body and occur in the pattern described above.

The languages which I have used to illustrate this phenomenon include the following, with particular emphasis on Vietnamese.

Austroasiatic languages:
  Viet-Muong: Vietnamese [Viet];
  Mon-Khmer: Khmer [Khm]; Mountain Mon-Khmer of Vietnam: Chrau and Rengao [Reng];
Tai languages:
  Central (Bangkok) Thai and Nung (NE Vietnam);
Hmong [Hmg] (Hmong-Mien/Miao-Yao language family).

I have looked only at languages within the clearly defined linguistic area, which does not include the western Tai languages nor the western or far southern Austroasiatic languages. I have not looked at the fringe area Sino-Tibetan languages, nor have I looked at the small group of Mainland Austronesian languages although their grammar tends to be more characteristic of Mainland typology than it does of Pacific Austronesian typology.

1.2. Word order and noun phrase typology

Verbal affixation (causative, resultative, reciprocal, "passive" (which forms statives), nominalising) in Mon-Khmer languages has been common throughout the family, but there is a tendency especially in Khmer (Cambodian) and southern Mountain Mon-Khmer for affixes to become less productive or even frozen (see especially Huffman 1970: 311 and D.M. Thomas 1969). In the other – tonal – languages there is no affixation or inflection. In fact none of the languages have any inflection whatsoever for case or for person, tense or aspect. In other words these are isolating languages with heavy dependence on word order and context and a high degree of zero anaphora with little redundancy compared with many agglutinating languages.

As indicated above, primary word order in the area is Subject-Verb-Object, more explicitly diagrammed here with Indirect Object, Locative and Means (Starosta 1988: 126) and illustrated by the Vietnamese sentences in (1)-(5).[4] (Phrase indicators are under the phrase head or under the preposition in the exocentric Prepositional Phrase.)

S ˆ V ˆ O (ˆ IO and/or Loc) (ˆ Mns) or sometimes:
S ˆ V ˆ IO ˆ O

(1)	*Nàng*	*giặt*	*quần-áo*	*rồi*	
Viet	she	launder	clothes	already	
	S	V	O		

'She washed the clothes already.'

(2)	*Nàng*	*gửi*	*quần-áo*	*cho*	*tôi*
	she	send	clothes	to	I
	S	V	O	IO(PP)	

'She sent some clothes to me.'

(3)	*Nàng*	*để*	*quần-áo*	*trong*	*họp*
	she	put	clothes	inside	box
	S	V	O	L	

'She put the clothes in a box.'

(4)	*Nàng*	*đưa*	*quần-áo*	*cho*	*tôi*	*trong*	*một*	*cái*	*họp*
	she	hand	clothes	to	I	inside	one	thing	box
	S	V	O	IO		Mns			

'She handed the clothes to me in a box.'

(5)	*Nàng*	*đưa*	*(cho)*	*tôi*	*mấy*	*cái*	*áo*	*mới*
	she	hand	to	I	some	thing	shirt	new
	S	V	IO	O				

'She handed me some new blouses.'

In the languages under discussion descriptive attributes follow the head noun. Such attributes include other nouns, stative verbs (which comprise the set of adjectives), other verbs, prepositional phrases. Restrictive attributes, such as numerals and quantifiers, fall into a subareal northeast-southwest pattern (see for example Jones 1970: 2-3). In Vietnamese, Hmong and the northeast Tai and Mon-Khmer languages, restrictive attributes occur before the head noun, which is usually a classifier noun. Thai and Khmer, on the other hand, have a different order: Head Noun–Numeral–Classifier Noun, with the classifier noun frequently absent in Khmer, (9) and (11) below. Compare (6) and (7) with (8) and (9).[5]

Except for Hmong, possessor noun phrases are seen as descriptive modifiers and often are headed by a noun of possession. In Hmong, animate possessors are apparently seen as restrictive modifiers, occurring before the possessed noun with nearly always a classifier noun intervening. Compare (10) with the other phrases. (See Clark 1989: 184-187.)

(6)	*ba*	*con*	*chó*	*đen*	*của*	*tôi*
Viet	three	CLF:anm.	dog	black	belonging	I
	'my three black dogs'					

(7)	*slám*	*tú*	*má*	*đáhm*	*?*	cáu
Nung	three	CLF:anm.	dog	black		I
	'my three black dogs'					

(8)	*măa*	*dam*	*săam*	*tua*	*khɔ̆ɔŋ*	*phŏm*
Thai	dog	black	three	CLF:anm.	thing(of)	I:MASC
	'my three black dogs'					

(9)	*ckae*	*kmaw*	*bəy*	*Ø*	*rəbah*	*kñom*
Khm	dog	black	three		thing(of)	I
	'my three black dogs'					

(10)	*kuv*	*(peb)*	*tug/tus*	*dev*	*dub*
Hmg	I	three	CLF:anm.	dog	black
	'my (three) black dog(s)'				

(11)	*ceik*	*pii*	*ceik*	*pii*	*snət*
Khm	banana	two	banana	two	clf:hand
	'two bananas'		'two hands of bananas'		

(12)	a. *xe*	*tôi*	*chiếc*	*xe*	*?(của)*	*tôi*
Viet	vehicle	I	CLF:vehicle	vehicle	belonging	I
	'my car'		'the car belonging to me', 'my car'			

b.	*chén*	*tôi*	*cái*	*chén*	*?(của)*	*tôi*
	ricebowl	I	CLF:thing	bowl	belonging	I
	'my ricebowl'		'the ricebowl belonging to me', 'my ricebowl'			

One further remark regarding classifier nouns might be relevant. Besides the role of classifying other nouns they play a major role in specifying definiteness, as can be seen in the phrases above. In Vietnamese at least, they also play a role in a stylistic preference for disyllabic symmetry and in this role often occur or do not occur according to the surrounding syllables rather than to grammatical functions, as seen in (12). When the classifier noun occurs, as in the second phrases of (12), the omission of the possessive noun is unusual. Stylistic factors also play a role in the ordering of object and dative noun phrases.

These factors of word order and absence of morphological complexity must be taken into account in any analysis of the stative-verb-body-part construction.

1.3. Stative verbs

Although stative verbs in the languages of Mainland Southeast Asia may occur in cut-and-dried intransitive sentences such as the Vietnamese sentence in (13), in which they describe the state of the Patient/Theme subject, they more often occur with some kind of modification or intensification, as in the sentences following.[6]

(13) *Tôi buồn*
Viet I be-sad
'I'm sad.'

(14) *Mấy hôm này tôi mệt quá*
Viet several day this I be-tired very-very
'These last few days I've been very tired.'

(15) *Tiếng Việt giàu đẹp là một*
Viet language Vietnamese be-rich beautiful be one
điều rõ-ràng
matter clear
'It is clear that Vietnamese is a rich and beautiful language.'
('That Vietnamese language is rich and beautiful is a clear thing.')

(16) *Tus ncej tsev lwj tag*
Hmg stick post house be-rotten finished
'The house posts are all rotten.'

(17) *Nqaij npuas qab dua nqaij nyug*
Hmg flesh pig be:delicious surpassing flesh cow
'Pork tastes better than beef.'

(18) *Krouc nih thom haəy lqαα phααŋ*
Khm orange this be:large already be-pretty also
'These oranges are large, and pretty too.'

(19) *Ramo bOq pataw ramaq*
Reng cow grandfather rich:man be-fat
'The rich man's cows are fat.'

(20) *Áhn thủ mưhn đồ dả*
Nung clf:thing head he be:bald already
'His head is already bald.'

Stative verbs in Hmong can be intensified by reduplication, in which the duplicated syllable is accompanied by an extremely high sustained tone.

(21) *Tus poj Hmoob Dawb tus ntiag tsho*
Hmg anm. female Hmong white strip front jacket
xiav-xī:ā li
be:blue:blue so
'That White Hmong woman's jacket trim is very very blue.'

Simple reduplication of stative verbs sometimes accompanied by tone change also occurs in Vietnamese, but here such reduplication lessens rather than strengthens the force of the described state (Nguyen K.T. 1975: 189-190):

(22) *đỏ* *đo đỏ*
Viet be red be reddish

mặn *măn mặn*
be salty be slightly salty

gầy *gầy gầy*
be thin be somewhat thin (of body)

As mentioned above in 1.2, when stative verbs modify nouns they occur after the noun, as shown by 'black dog' in (6)-(10) and 'a clear thing' in (15) above and in these phrases:

(23)

Viet:	*người*	*cao*		*người cao*	
	person	be:tall		'a tall person'	
	răng	*trắng*		*răng trắng*	
	tooth	be:white		'white teeth'	
	cổ	*dài*		*cổ dài*	
	neck	be:long		'a long neck'	
Hmg:	*taws*			*loj*	*taws loj*
	foot			be:big	'big feet'
	peb	*tus*	*niam*	*lauj*	*peb tus niam lauj*
	three	anm.	mother	be:old	three old women
	siab			*zoo*	*siab zoo*
	liver			be:good	'a kind heart'
Thai:	*taa*	*cèp*		*taa cèp*	
	eye	be:sore		'sore eyes'	
	cay	*dii*		*cay dii*	
	heart	be:good		'a heart in good condition'	

2. Stative verbs and body-part terms

2.1. The stative-verb-body-part construction

The construction discussed in this paper has the reverse order to that of Noun-Adjective:

(Subject +) *Stative Verb + Body Part Noun*

and can be understood as 'Subject has condition described by verb located in that body part (of Subject)'. Some of these expressions are illustrated here:

(24)

Viet	*đau*	*bụng*	*đau bụng*
	be:sick, to hurt	stomach	'be sick in the stomach'
	đẹp	*người*	*đẹp người*
	be:beautiful	person	'be beautiful in body'
	nặng	*mặt*	*nặng mặt*
	be:heavy	face	'be heavy in the face: be offended'
	trắng	*răng*	*trắng răng*
	be:white	tooth	'be white in the teeth: be young'

Hmg	*mob*	*plab*	*mob plab*
	be:sick	stomach	'be sick in the stomach'
	mob	*siab*	*mob siab*
	be:sick	liver	'be sick in the liver: hurt, sorry'
	puv	*siab*	*puv siab*
	be:full	liver	'be full in the liver: be satisfied'
	zoo	*siab*	*zoo siab*
	be:good	liver	'be good in the heart: be happy'

Note that Vietnamese 'white teeth' and Hmong 'a kind heart' in (23) become respectively 'white in the teeth: young' and 'good in the heart: happy' when the order of stative verb and body part are reversed.

How this stative-verb-body-part construction occurs in a simple sentence is illustrated in (25)-(28), the (a) sentences with a body-part term and the (b) sentences without. The meaning of a physical state, (25), and a few abstract states as in (26), is much the same with or without the body part; the presence of the body part tends to focus, emphasise, personalise the state described.

(25) Viet

a. *Tôi đói bụng*
I hungry stomach
'I'm (really) hungry.'

b. *Tôi đói*
I hungry
'I'm hungry.'

(26) a. *Tôi buồn lòng*
Viet I sad intestines, abdomen
'I'm (really) unhappy.'

b. *Tôi buồn*
I sad
'I'm sad.'

On the other hand, there is a distinct difference with many abstract expressions, as shown clearly in (27) and (28). The body part is seen as the locus/seat of feeling and plays the core role in forming idiomatic expressions. Although it is possible for the (a) sentences to imply a physical state, its use almost always refers to the metaphorical state.

(27) a. *Tôi đau lòng*
Viet I sick intestines
'I'm broken-hearted.' (or 'I'm sick in the belly.')

b. *Tôi đau*
I sick
'I'm sick.'

(28) a. *Nws mob siab*
Hmg 3PER sick liver
'She's distressed' or 'Her liver is diseased (she's sick in the liver).'

b. *Nws mob*
3PER sick
'She's sick.'

This construction raises two questions of theory regarding grammatical relationships. One is the question of the part-whole relationship between the body part and the understood subject, i.e. the question of inalienability. The other question is one of the relationship between the stative verb and the body-part term and the possibility of noun incorporation in the verb. Although the two questions are closely linked in this construction, I will address them separately and then illustrate the construction in terms of what appears to be the major distinction in the classification of these expressions: actual physical states versus abstract metaphorical states.

2.2. The question of inalienability

A stative verb followed by a body-part term, as in the a. sentences in (25)-(28), describes the state of that body part with respect to its whole which is the subject of the stative verb. The sentences in (29) are all grammatical and have much the same meaning but the one in (29a) – the stative-verb-body-part construction – is the much preferred construction, seemingly a strategy to focus on the subject whole as undergoing the state described by the verb although the verb presumably describes the state of the particular body part not the whole person. The same applies to the sentences in (30).

(29) Viet
a. *Tôi (bị) đau bụng*
I suffer sick stomach
'I ache/am sick in the stomach.'

b. *Bụng tôi bị đau*
stomach I suffer sick
'My stomach aches.'

c. *Tôi bị đau ở bụng*
I suffer sick in stomach
'I'm sick in my stomach/abdomen.'

(30) Thai
a. *Khǎw cèp taa*
3PER sore eye
'She is sore in the eyes' or 'She has sore eyes.'

b. *Taa khǎw cèp*
eye 3PER sore
'Her eyes are sore.'

The metaphorical expressions are for the most part idiomatic and usually cannot take the form of possessed body part plus possessor with the same meaning, as in (29b) and (30b) above, nor an explicit marking of location, as in (29c), as may many of the physical-state expressions. This is shown here in (31)-(32). There are exceptions as shown in (33b) and (34b), although note the meaning change in the Hmong sentences in (34).

(31) Viet
a. *Nàng đau lòng lắm*
she sick intestines very
'She's heart-broken.'

b. **Lòng nàng (bị) đau*
intestines she suffer sick
'Her heart is broken.'

(32) a. *Phǒm sâw cay mâak*
Thai I sad heart very
'I'm very sad.'

b. **Cay phǒm sâw mâak*
heart I sad very
'My heart is very sad.'

(33) a. *Phǒm sǐa cay*
Thai I spoiled,broken heart
'I feel heavy hearted.'

b. *Cay phǒm sǐa*
heart I spoiled
'My heart is heavy.'

(34) a. *Nws zoo siab kawg*
Hmg 3PER good liver very:much
'She's very good in the liver: She's very happy.'

b. *Nws (lub) siab zoo kawg*
3PER thing liver good very
'Her liver is very good: She has a kind heart.'

The subject focus on the possessor whole in all these sentences is the result of what is commonly known as possessor raising (see Mithun 1984, 1995: section 3), that is, the promotion of the possessor from its position as dependent modifier of the possessee, as *tôi* in (29b), to grammatical subject of the sentence, as in (29a). Such a process involves the relation of inalienability between the subject whole and the body part and raises the question of the status of inalienability in this construction.

These languages have no regular grammatical devices for distinguishing between alienability and inalienability. Such grammatical criteria as morphological marking or affixation for inalienability are non-existent. Nominal juxtaposition is more a function of modification than inalienability, as 'my car/bowl' in (12) above. (See Chappell – McGregor 1989: 26 regarding inalienable possession by juxtaposition.) On the whole the concept of alienability versus inalienability seems to be irrelevant. The focus on the subject whole provides an explicit expression of inalienability; however, I believe the presence of inalienability in these sentences is an incidental consequence of possessor promotion for the purpose of emphasising the subjective value of the state described by the verb as affecting the *whole person* (as opposed to the whole body). The sentence in (29a) could well be translated, 'I am sick and the focal point of my sickness is my stomach.' Likewise for the abstract state in (35):

(35)	*Tôi*	*rất*	*buồn*	*lòng*
Viet	I	very	sad	intestines

'I'm very sad and where I feel my sadness is in my belly.'

There is just no evidence in these languages to support the notion of obligatory and overt coding of the alienable-inalienable distinction in terms of syntax.

Before going on to the discussion of noun incorporation, I would like to make clear that, although all the expressions classified in this paper are of the inalienable type described above, a stative-verb-body-part construction in itself does not necessarily demonstrate inalienability. It is possible to have a stative verb with a body-part term which has no reference to a specific whole and has only an indirect attributive reference to the subject of the stative verb, as in (36). As shown in (37), the 'external' or alienable subject of such Vietnamese expressions cannot be animate. The Thai example is from Nacaskul (1976: 888).

(36)

Viet	*sướng*	*tai*	*sướng tai*
	be:happy	ear	'(something external) be pleasant to the ear'
	sướng	*mắt*	*sướng mắt*
	be:happy	eye	'(something external) be pleasing to the eye'
	đẹp		*đẹp mắt*
	be:pretty		'(something external) be pleasing to the eye'
Thai	*khay*	*taa*	*khay taa*
	be:far	eye	'(something external) be out of sight'

(37)	**Cô*	*ấy*	*đẹp*	*mắt*
Viet	Miss	that	pretty	eye

('She's pleasing to look at (pleasing to the eye).')

2.3. The question of noun incorporation

Because of the close semantic and grammatical association between the stative verb and the body part in this construction there is the consideration of possible verb incorporation of the body part noun. By incorporation is meant that the incorporated noun and the verb which incorporates it act as a single grammatical unit, that is, as a qualified stative verb.

There are several reasons for hesitation in considering this construction to have incorporated nouns. For one thing there are instances where the probability of incorporation seems marginal; for example, sentences in which classifier nouns can intervene (see (53) below), although this may be true only for physical states.

In Hmong when the animate entity to which the body part belongs is not the subject of the stative verb it may occur as the dependent possessor of the body part, as in (38) and (39), that is, the possessor has not been promoted. (Example (39) is from PXNNH: 35.)

(38)	*Kuv*	*paub*	*hais*	*tias*	*nyuab-nyuab*	*koj*	*siab*	*kawg* ...
Hmg	I	know	say	that	troubled-troubled	you	liver	much

'I know it is a great worry to you.'

(39)	*Poj-niam*	*me-nyuam*	*khuam*	*yus*	*siab*	*hwv*
Hmg	wife	child	caught	I(humble)	liver	very:much

'I miss my wife and children very much.'

Although these sentences are not parallel to the construction under discussion since they lack possessor promotion to subject, the question of noun incorporation in the verb still arises. In the sentences here the whole possessive phrase of possessor and possessed body part appears to play the same role as the body-part term alone in the sentences under discussion. This and the fact that the whole stative-verb-plus-possessed-body-part phrase is modified by the sentence-final adverb argue for incorporation in the verb of the entire possessive phrase. This argument is strengthened by the fact that the possessive phrase does not have the usual form of a location noun phrase, the only grammatical role the phrase could have with the stative verb.

Another cause for hesitation in the consideration of noun incorporation is that the semantic feeling of these sentences is more akin to the translation of the Hmong sentence in (34a) above, repeated here in (40), in which the implication is that the body part is the head of a separate noun phrase representing the location of the condition described by the stative verb. For example, in the Vietnamese sentence in (41) the sense 'please be satisfied in your heart to ...' is closer to the meaning of this construction in these languages than the sense 'please be heart-happy to ...', although the two words act as a grammatical and semantic unit, and again, the body part is lacking the usual form of a location case relation.

(40)	*Nws*	*zoo*	*siab*	*kawg*
Hmg	3per	good	liver	very-much
	'She's very good in the liver: She's very happy.'			

(41)	*Xin*	*cô*	*vui-lòng*	*trả-lời*	*thư*	*này*	*trước*	*ngày*	*15*
Viet	request	Miss	pleased	answer	letter	this	before	day	15
	'Pleased be happy/willing to answer this letter before the 15th.'								

Finally, the typological nature of the languages themselves argues against any certainty of incorporation since they have no morphological indicators for incorporation.

However, compounding is a very productive process in the area. Aside from inflectional affixing in Mon-Khmer, there are two sorts of compounding, both of them involving lexical derivation not inflection. One is the compounding of two words which are independent words but may act as a single lexical item as in (42), where the nominalising noun part of the *noun+verb* compound occurs again as the classifier noun when specificity is required, and in (43), which is simply a *noun+noun* compound to form another noun.

(42)	*rooŋ - rien*	*rooŋrien*	*sɔ̌ɔŋ*	*rooŋ*	N+V = N
Thai	building - study	school	two	building	
	'school'	'two schools'			

(43)	*paj - ntaub*	*peb*	*daim*	*pajntaub*	N+N = N
Hmg	flower - cloth	three	sheet	needlework	
	'needle-worked cloth'	'three pieces of needlework'			

The other kind of compound is where only one or neither of the two morphemes of the compound can stand alone. This kind is particularly productive for Chinese loans in Vietnamese.[7] No morpheme in (44) can stand alone. (Bound morphemes are indicated by square brackets.) In the first compound in (45), the first morpheme is bound and in the second compound the second morpheme is bound.

(44)	*giáo-viên*	*bãi-thị*	[V]+[N] = N, [V]+[N] = V
Viet	teach - acting:person	stop - market	
	'teacher'	'to go on strike (market vendors, general)'	

(45)	*đại- sứ*	*bệnh -viện*	[V]+N = N, V+[N] = N
Viet	be:great - envoy	be:ill - institute	
	'ambassador'	'hospital'	

It appears that compounding is involved in the idiomatic metaphorical expressions and some of the physical expressions in the stative-verb-body-part construction. It is compounding of the first sort mentioned above where two independent words combine to form a single expression. What is crucial here is the fact that two grammatically different lexical items – a stative verb and a noun signifying location – combine to act as a single grammatical unit whose function is to describe and locate a particular state. It is this fact which makes it very likely that these expressions are indeed noun-incorporated verbs.

In fact Mithun, in her 1984 discussion of noun incorporation in many languages, gives four types of noun incorporation, the first of which is lexical compounding in which the noun "loses its syntactic status as an argument of the sentence" (1984: 849). (See also Evans 1995.) Mithun's second type, involving the "manipulation of case", shares with the first type the essential compounding nature and loss of argument but further affects the clause by replacing the lost argument with another (oblique) argument from the clause (1984: 856), that is, the incorporated noun. The most common type of incorporated noun is the direct object, as 'taro' in 'She sat (and) taro-ground' = 'She was sitting grinding taro.' In the stative-verb-body-part construction however the incorporated noun is the locus of the stative verb in much the same way as that described by Mithun (1995: section 3.1) for Mohawk, a highly inflected North American Iroquoian language. In the Mainland Southeast Asian languages, as in the typologically different Mohawk, noun incorporation is achieved by possessor promotion. That is to say, the possessor of the body-part reference of the stative verb becomes the subject of the stative verb, as discussed above. In discussion of Blackfoot, a North American Algonquian language which promotes possessors of body parts, she states that noun incorporation of body parts "allows affected persons to assume a primary role" (1984: 858). This is precisely what is happening in the Southeast Asian languages: possessors assume a primary role of affected persons.

To illustrate again possessor promotion in the construction, the two sentences from (29) are repeated here in (46). In (46a), the possessed body part with its modifying possessor constitute the subject of the stative verb in a straightforward statement regarding the possessor's stomach. In (46b), the possessor has been promoted to the subject argument, replacing the body part which appears to have compounded with the verb. As stated earlier, the preferred sentence is the one in (46b) with its focus on the whole self as being the undergoer of the state described in the now location-qualified verb. It is easy to see in (46) and similarly in (50) below that possessor promotion in the stative-verb-body-part construction encodes far more subjectivity than reference simply to the possessed body part itself.

(46) a. *Bụng tôi (bị) đau*
Viet stomach I suffer sick
'My stomach aches.'

b. *Tôi (bị) đau bụng*
I suffer sick stomach
'I ache/am sick in the stomach.' (< 'I stomach-ache.')

Therefore it seems that these expressions fit the specifications of Mithun's Type II noun incorporation (1984: 856, 859). However, these expressions seem not to be so tightly bound in some cases and, contrary to Mithun's claim that such incorporated nouns are unmarked for definiteness (1984: 856, 859), sometimes have intervening classifiers, as in (53) below, and at least one case where the body part is modified by another stative verb: (77). This is not surprising, however, if Mithun is correct in her hypothesis that different types of noun incorporation represent historical development of incorporation. With such a process one would naturally expect to see compounding and noun incorporation in different stages of binding, with Type I loose-bound simple juxtaposition preceding Type II incorporation. This is certainly true of these stative-verb-body-part expressions, some of which are felt by speakers to be a single word and others of which are more freely manipulated.

Stative-verb-body-part expressions are divided here first into types having to do with physical states described by the expression and then into metaphorical states described by the expression, with metaphorical states subdivided into non-inherent states versus characteristic states.

3. Physical states and body parts

Physical states are as a rule actually located in the particular part of the body mentioned, as shown in the following examples. For this reason the physical-state-body-part expressions tend to be productive and are not so likely to be idiomatic. Nonetheless the construction itself, with potential body-part-noun incorporation and promotion to subject of the body part possessor, lends itself to idiomaticity. Following are some examples of physical-state-body-part expressions. The Khmer data is from Huffman (1970: 295).

(47)

Viet	*đói* hungry	*bụng* stomach	*đói bụng* 'hungry'
	đau sick, to hurt	*bụng* stomach	*đau bụng* 'sick in the stomach'
	đau sick	*chân* leg	*đau chân* '(be) hurt in the leg, footsore'
	đau sick	*cổ* neck	*đau cổ* 'sore throat'
	đau sick	*đầ u* head	*đau đau* 'head ache, hurt on the head'
	nhức be:aching	*đầu* head	*nhức đầu* 'have a headache'
	nóng hot	*đầu* head	*nóng đầu* 'feverish'
	cay spicy, peppery	*mắt* eye	*cay mắt* 'stinging, burning in the eyes'
Khm	*chɨɨ* ill, hurt	*kbaal* head	*chɨɨ kbaal* 'have a headache'
	chɨɨ ill, hurt	*pŭəh* stomach	*chɨɨ pŭəh* 'have an upset stomach'
	baq broken	*cəəŋ* foot, leg	*baq cəəŋ* 'have a broken leg

	kdaw	*kluən*	*kdaw kluən*
	hot	body	'feel hot, feverish'
	sruəl	*kluən*	*sruəl kluən*
	comfortable	body	'be well'
Thai	*pùat*	*hǔa*	*pùat hǔa*
	ache	head	'have a headache'
	cèp	*taa*	*cèp taa*
	sore	eye	'have sore eye(s)'
Hmg:	*mob*	*taub-hau*	*mob taub-hau*
	sick	head	'ache in the head, feverish'
	mob	*plab*	*mob plab*
	sick	stomach	'have a stomach ache'

Following are some sentence examples. Where there are both (a) and (b) sentences, the (a) sentence is with a stative verb and body part term, the (b) sentence is with a stative verb only as predicate.[8]

(48) Viet
a. *Mấy giờ chúng-ta đi ăn?*
how-much hour we:INCL go eat

Tô rất đói bụng
I very hungry stomach

'What time are we going to eat? I'm really hungry.'

b. *Tôi đói lắm*
I hungry very

'I'm very hungry.'

(49) Viet
Con tôi đau bụng có một tuần rồi
offspring I sick stomach have one week already

'My child has been sick for one week already.'

(50) Viet
a. *Mẹ nhức đầu lắm*
mother ache head very

'Mother has a bad headache.'

b. *Đầu mẹ nhức lắm*
head mother ache very

'Mother's head aches badly.'

(51)	*Anh*	*ấy*	*bị*	*đau*	*răng*	*hai*	*ngày*	*rồi*
Viet	brother	that	suffer	hurt	tooth	two	day	already

‘He’s had a toothache for two days.’

(52)	a. *Mûawaanníi*	*phǒm*	*tháŋ*	*pùat*	*hǔa*	*tháŋ*	*cèp*	*taa*
Thai	yesterday	I	both	hurt	head	both	sore	eye

‘Yesterday I had both a headache and sore eyes.’ (‘I ached in the head and was sore in the eyes.’)

b. *Taa*	*khǎw*	*cèp*
eye	3PER	sore

‘Her eyes are sore.’

(53)	a. *Tú*	*slứ*	*chéhp*	*áhn*	*tõng*	*lái*
Nung	CLF:anm.	tiger	sore	CLF:thing	stomach	very

‘The tiger was very sore in his stomach.’

b. *Áhn*	*tõng*	*tú*	*slứ*	*chéhp*	*lái*
CLF:thing	stomach	CLF:anm.	tiger	sore	very

‘The tiger’s stomach was very sore.’

(54)	a. *JAh*	*nu*	*jIq*	*badUk,*	*nu*	*thay*	*’way*	*paq*	*hyE*
Reng	if	2PER	sick	stomach	2PER	should	stay	at	home

‘If you are sick in your stomach, you should stay home.’

b. *Hyit*	*jIq*
Hyit	sick

‘Hyit is sick.’

(55)	a. …	*saq*	*sa*	*něh*	*cồ*	*ji*	*candúl*
Chrau		go	eat	3PER	who	sick	stomach

‘… and goes to eat people who have stomach aches.’

b. …	*an*	*vu*	*vâng*	*ji*	*vầy*
	make	they	don’t	sick	so

‘… make them not sick.’

(56)	*Kñom*	*kdaw*	*kluən*	*haəy-nɨŋ*	*chɨɨ*	*kbaal*
Khm	I	hot	body	and	ill	head

‘I feel hot and have a headache.’

(57)	*Koj*	*mob*	*plab*	*los*	*hov*	*ntev*	*lawm?*
Hmg	you	sick	stomach	so-then	how-much	long	already

'So how long have you been sick?'

(58)	*Kuv*	*tus*	*me-nyuam*	*tseem*	*mob-mob*	*taub-hau*
Hmg	I	anm.	child	still	sick-sick	head

'My child is still very sick.'

Another characteristic of the area is the use of patterned four-syllable idiomatic expressions. Sometimes these expressions combine stative verbs and body-part terms, as shown in the following Vietnamese example and an example from one of the Hmong languages of China, eastern Guizhou Miao.[9]

(59)

Viet	*đỏ*	*da*	*thắm*	*thịt*
	red	skin	bright-red	flesh

'glowing with health'

Miao	*faŋ*^{55}	*nze*^{55}	*faŋ*^{55}	*mɛ*^{13}
	yellow	ear	yellow	face

'be pale and thin'

There are a few physical-state-body-part expressions which describe the inherent state of a body or a part of the body or which describe features which are perceived to be particularly characteristic of an individual. It would seem that the possibilities for such expressions are large, but even the small number given here are hardly in common usage. (The Chrau example is from D.D. Thomas 1971: 125.)

(60)

Viet	*bẹp*	*tai*	*bẹp tai*
	flat	ear	'have flattened ears ' (cf. abstract meaning)
	mỏng		*mỏng tai*
	thin		'have fine ears'
	điếc		*điếc tai*
	deaf		'deaf (also: deafening)'

	ngắn short	*cổ* neck	*ngắn cổ* 'be short in the neck'
	lớn big	*người* person	*lớn người* 'tall, large (of person)'
Hmg	*qeg* short	*taub* head	*qeg taub* 'short in stature'
	lag deaf	*ntseg* ear	*lag ntseg* 'deaf'
Chrau:	*jong* long	*ti* arm	*jong ti* 'long-armed'

4. Metaphorical states and body parts

4.1. Non-inherent metaphorical states

The metaphoric use of stative verbs with body parts is interesting not only for its particular grammatical structure in these languages but for its cultural relatedness to many languages of the world. It is common to perceive feelings to reside in particular parts of the body, characteristically internal organs. Most languages employ a number of different body parts for particular expressions, although tending to have a favourite location for the seat of feeling. In English, for example, the favourite organ for feelings is the heart, but the skin, gut, bones and liver also occur in expressions of abstract states. Some such metaphorical expressions in English are shown in (61) and (62); the sentence in (62) is doubly metaphorical with the abstract use of *cut*.

(61) *heartbroken* = extremely disappointed
heartless = without compassion or pity
sweetheart = one loved by another
thin-skinned = oversensitive
have guts = have courage

(62) *She cut me to the bone when she said that.*
'What she said hurt my feelings very deeply.'

In Mainland Southeast Asia the relationship between feelings and a particular part of the body is an especially close one, as illustrated in the proliferation and frequent use of stative-verb-body-part compounds in the expression of states which apply to animate beings. In contrast to physical states, which actually reside in the body part named, states which might be more properly characterised as being in the mind are metaphorically located in an organ in the trunk of the body.

There is a great deal of variation in these abstract expressions, some occurring rarely, some being more tightly compounded than others, some with particular body parts being more productive, and one expects that new idioms will arise. Phuong Tonthat (pers. comm.) found more than forty compounds with just the one Vietnamese term for 'intestines' given here. In Vietnamese the seat of feeling is usually the intestines although the stomach may occur in more colloquial use. In Thai it is usually the heart and in Hmong the seat of feeling and thinking is the liver. Matisoff (1978: 210-213) states that in Tibeto-Burman (the western edge of the core linguistic area) the heart, liver and gall bladder are asociated with emotions but the intestines apparently not. Following are a few examples from some of the core languages:

(63)

Viet	*vui* happy	*lòng* intestines, gut	*vui lòng* 'pleased (to)'
	thỏa satisfied		*thỏa lòng* 'satisfied, content'
	sẵn prepared		*sẵn lòng* 'ready, willing'
	khó difficult		*khó lòng* 'difficult, feel reluctant'
	phiền vexed, bothered		*phiền lòng* 'feel troubled'
	đau sick		*đau lòng* 'broken-hearted'
	bận busy		*bận lòng* 'worried, preoccupied'
	hết finished		*hết lòng* 'devoted'

	bằng equal	*lòng* intestine	*bằng lòng* 'satisfied, in agreement'
	thật true	*bụng* stomach, abdomen	*thật lòng/bụng* 'sincere'
	nóng hot, impatient	*ruột* intestines	*nóng lòng/ruột* 'impatient'
	nóng hot	*mắt* eye	*nóng mắt* 'angry, furious'
Hmg	*kho* lonely	*siab* liver	*kho siab* 'lonely, homesick'
	chim angry		*chim siab* 'angry'
	npau angry (to boil)		*npau siab* 'angry'
	ntxhov restless, disturbed		*ntxhov siab* 'disturbed'
	nyuaj/nyuab difficult		*nyuaj siab* 'worried, troubled'
	zoo good		*zoo siab* 'happy'
	kaj bright, light		*kaj siab* 'satisfied, pleased, refreshed'
	tas finished		*tas siab* 'satisfied'
	mob sick, ache		*mob siab* 'hurt, sorry, distressed'
	khuam caught, hung up		*khuam siab* 'lonely, missing others'
	chob pierced		*chob siab* 'inwardly offended, grieved'
Khm	*sapbaay* be:happy	*cət* heart	*sapbaay cət* 'happy, content, glad'
	pibaaq be:difficult		*pibaaq cət* 'be unhappy'

Actually, in Thai the term used for the seat of feeling is only part of a compound which designates the heart organ. Here are three examples of a stative-verb-plus-'heart' compound and an expression with 'teeth':

(64)

Thai	*sâw*	*cay*	*sâw cay*
	sad	heart	'sad'
	dii		*dii cay*
	good		'happy'
	thùuk		*thùuk cay*
	correct		'satisfied'
	khèt	*fan*	*khèt fan*
	repulsed	teeth	'be repulsed or put off by something' (physically: teeth set on edge as with lemon)

Sometimes seats of feeling can be distinguished from actual body organs by classifiers (Huffman – Tran 1980: 389 and Bertrais 1979: 391):

(65)	*trái/quả*	*tim*	*con*	*tim*
Viet	fruit:shaped	the:heart	animate	seat:of:feeling
(66)	*nplooj*	*siab*	*lub*	*siab*
Hmg	leaf:shaped	the:liver	GENERAL:CLF	seat:of:feeling

Following are sentence examples utilising some of the metaphorical expressions. (The sentence in (69) is from Nguyen V.K. 1984: 904 and in (70) from Huffman 1970: 296.) Again, (b) sentences are with only the stative verb as predicate.

(67) Viet

a. *Cô cũng đi thì bà rất vui lòng*
Miss also go then Mrs. very happy intestines
'If you go too she'll be very glad.'

b. *Bà đó rất vui*
Mrs. that very happy
'She's very happy.' or 'She's fun to be with.'

(68) Viet

Khi nào nghĩ đến bài-thi tôi bị đau lòng
time which think to examination I suffer sick intestines
'Whenever I think of the examinations my heart sinks.'

(69)	*Tôi*	*rất*	*sẵn*	*lòng*	*tha-thứ*			
Viet	I	very	ready	intestines	forgive			
	'I'm fully prepared to forgive.'							
(70)	*Tŋay*	*nih*	*kñom*	*sapbaay*	*cət*	*nah*		
Khm	day	this	I	happy	heart	very		
	'Today I'm very happy.'							
(71)	*Kuv*	*zoo*	*siab*	*kawg*				
Hmg	I	good	liver	very:much				
	'I'm very happy.'							
(72)	a. *Kuv*	*nyuab*	*siab*	*rau*	*kuv*	*tus*	*muam*	*rauqhov …*
Hmg	I	difficult	liver	to	I	anm.	sister	because
	'I'm worried about my sister because …'							
	b. *Ntawv*	*Askiv*	*nyuaj*	*heev*				
	(written)paper	English	difficult	very				
	'English is very difficult.'							
(73)	*Mûuawaanníi*	*khăw*	*sâw*	*(cay)*	*mâak*			
Thai	yesterday	3PER	sad	heart	very			
	'She was very sad yesterday.'							

Following are some four-syllable idiomatic expressions. Lauj Pov Vaj gave Jaisser (1990) the second meaning for the second Hmong expression.

(74)				
Viet	*buồn*	*chân*	*buồn*	*tay*
	sad	leg/foot	sad	hand
	'not to know what to do with one's hands'			
Hmg	*lwj*	*siab*	*lwj*	*plaw*
	decayed	liver	decayed	heart
	'be wholly upset by grief, emotion'			
	tas	*siab*	*tas*	*ntsws*
	completed	liver	completed	lungs
	'be patient; wholehearted, with all the heart'			

4.2. Characteristic metaphorical states

Although most stative-verb-body-part expressions refer to characteristics which are not an inherent feature of the animate being who is the subject of the stative verb, there are expressions which refer to features which are inherent or which are perceived to be particularly characteristic of that individual. The greater number of these characteristic states are abstract or metaphorical rather than physical or actual. I have classified as abstract those physical features which are not predictable in the combined meanings of the state and the body part. Thus I have classed 'deaf in the ears' as a physical state but 'heavy in the ears', also meaning 'deaf', as an abstract state, and 'white in the teeth' meaning 'young' as abstract. Given here are some examples.

(75)

Viet	*cao* high, tall	*tay* hand	*cao tay* 'skilful, capable'
	già old		*già tay* 'skilful, capable'
	mát cool		*mát tay* 'skilful'
	mau/nhanh fast		*mau/nhanh tay* 'quick of hand, quick'
	to big	*đầu* head	*to đầu* 'be a big shot'
	đẹp beautiful	*mặt* face	*đẹp mặt* 'glorious, honoured'
	mỏng thin	*môi* lip	*mỏng môi* 'talkative, gossipy'
	trang white	*răng* tooth	*trang răng* 'still young'
	bẹp flat	*tai* ear	*bẹp tai* 'opium smoker'
	nặng heavy		*nặng tai* 'hard of hearing'
	dài long	*lưng* back	*dài lưng* 'lazy, idle'

	thật	*lòng/bụng*	*thật lòng/bụng*
	true	intestines/stomach	'honest, sincere person'
	sáng	*dạ*	*sáng dạ*
	bright	abdomen	'intelligent'
	tối		*tối dạ*
	dark		'slow-witted'
	to	*gan*	*to gan*
	big	liver	'bold, daring'
	cả		*cả gan*
	oldest, biggest		'bold, daring'
	non		*non gan*
	young,tender		'chicken-hearted'
Hmg	*kub*	*siab*	*kub siab*
	hot	liver	'earnest, zealous'
	qaug		*qaug siab*
	weak		'easily deceived by others'
	ruam		*ruam siab*
	dumb, mute		'stupid, dull-witted'
	sov		*sov siab*
	warm		'sympathetic person'

Following are some Vietnamese four-syllable idiomatic expressions, the first from Nguyen D.H. (1966: 83) and the last two from Liem (1970: 43, 44).

(76) *tre người non dạ*
Viet young person young stomach
'be young and immature'

thành tâm thiệt ý
pious heart true idea
'be honest, pious'

trắng da dài tóc
white skin long hair
'be beautiful'

A few sentence examples follow. The body part in (77) (from Khai 1972: 196) is itself modified, which weakens the argument for the body part being incorporated in the verb. Throughout this discussion there has been the implication of degrees of incorporation and this example reinforces this notion, a notion which is consistent with a theory of different stages in historical development of noun incorporation, as well argued by Mithun (1984: 872-877). (The example in (80) is from Bertrais 1979: 372.)

(77)	*Bà*	*ta*	*giàu*	*óc*	*tưởng-tượng*	
Viet	Mrs.	our	rich	brain	imaginative	
	'She is imaginative (rich in imaginative brain).'					
(78)	*Cô*	*đó*	*mỏng*	*môi*	*quá*	*trời*
Viet	Miss	that	thin	lip	very:very	sky (ultimate)
	'That girl babbles incessantly.'					

(79)	*Nws*	*kub*	*siab*	*lug*	*kawm*	*ntawv*	
Hmg	3PER	hot	liver	INTENSIVE	study	book	
	'She's completely earnest about studying.'						
(80)	*Tus*	*ntawd*	*ruam*	*ncauj*	*tsis*	*ruam*	*siab*
Hmg	anm.	that:one	dumb	mouth	not	dumb	liver
	'That person is mute not stupid.'						

5. Some expressions with nonstative verbs

Sometimes nonstative verbs are combined with body part terms to form stative and other predicates, or stative verbs are combined with body part terms as above but form nonstative predicates. These two types will not be discussed here but a few examples will illustrate. (The Khmer examples are from Nacaskul (1976: 879, 886); the Thai sentence in (82) was given by Anthony Diller (personal communication).)

(81)

Viet	*nhát*	*gan*	*nhát gan*
	cut, stab	liver	'timid, shy'
	mất	*lòng*	*mất lòng*
	lose	intestines	'hurt, offended'

	có have		*có lòng* 'generous'
	thuộc belong:to		*thuộc lòng* 'know by heart'
	vững firm		*vững lòng* 'persevere'
Hmg	*khis* notch, chip	*siab* liver	*khis siab* 'pained, grieved'
	rhiab tickle		*rhiab siab* 'apprehensive, to shudder'
	zais hide		*zais siab* 'hide one's feelings'
	rau put:into		*rau siab* 'try'
	xu miss:the:mark		*xu siab* 'not like something'
	nkag crawl:(to)		*nkag siab* 'understand'
	to pierce	*taub* head	*to taub* 'understand'
Thai	*sǒn* focus:on	*cay* heart	*sǒn cay* 'interested'
	tòk drop		*tòk cay* 'frightened, startled'
	khâw enter		*khâw cay* 'understand'
	màn persevere	*sây* intestines	*màn sây* 'have aversion to, thoroughly disgusted with'
	thùuk correct	*khɔɔ* neck	*thùuk khɔɔ (kan)* 'share interests (together)' (requires Plural subject)

Khm	*phlec*	*trəciək*	*phlec trəciək*
	forget	ear	'forget'
	pruəy	*cɣt*	*pruəy cɣt*
	worry	heart	'worry'

(82) *Mɯ̂a dâyyin khráŋ rɛ̂ɛk sadùt hǔu ...*
Thai when hear time first trip:over ear

'On the first hearing, it (the expression) sounds odd (literally: it "stumbles" in the ear),' or 'The first time I heard (the expression) it seemed odd to me [but later ...].'

Two four-syllable idiomatic expressions follow, the Vietnamese example from Nguyen D.H. (1966: 155) and the Chrau example from D.M. Thomas (1978: 282).

(83)

Viet	*ghi*	*lòng*	*tạc*	*dạ*
	engrave	intestines	engrave	stomach

'remember (favour) forever'

Chrau	*chĕq*	*nuih*	*chĕq*	*tôr*
	set	heart	set	ear

'get angry'

6. Conclusion

The languages of Mainland Southeast Asia use stative/adjectival verbs to make statements describing the state of Patient/Theme subjects, that is, in their simplest form: *Arunee be-sick* = 'Arunee is sick'. There is a distinctive construction commonly used in the area which consists of the same type of stative-verb sentence plus a body-part term which locates the state described by the verb, that is, *Arunee be-sick stomach* = 'Arunee is sick in the stomach.' The body-part term represents a part of the subject animate entity, thus: Animate-Whole Subject Noun ^ Stative Verb ^ Body-Part Noun. Such a construction raises the conceptual questions of the relationship between the subject whole and the body part and the status in this construction of inalienability as well as the possibility of body-part-noun incorporation in the stative verb.

In these languages there is no regular linguistic basis for a distinction between alienability and inalienability. Furthermore, in this construction probably such a distinction is irrelevant. Through the process of the raising of the possessor whole to subject this construction is marked as one of inalienability. Nonetheless the intention of this process is not to make an alienable-inalienable distinction but to focus on the subject whole as undergoing the state described by the verb, the encoding of subjectivity as the basis for inalienability. The stative verb, although in fact describing the state of the body part, semantically and grammatically describes the state of the whole being who is the possessor of the body part. In the sentence *Arunee be-sick stomach*, it is Arunee who undergoes the state of sickness and the description of sickness is qualified by the location in her body of the sickness: 'Arunee is stomach-sick.' The factor of subjectivity is particularly relevant in metaphorical expresssions, for example, *Kim be-sick intestines* 'Kim is broken hearted', which comprise the greater number and variety of these stative-verb-body-part expressions.

Many of the stative-verb-body-part expressions are clearly compounds, those which are borderline compounds are at least idiomatic while some 'compounds' are distinguishable enough in their two parts to be productive in forming other compounds. In a few cases classifiers can occur between the verb and the body-part noun or the body-part noun can be the head of a noun phrase. Such variation as this is consistent with the view expounded by Mithun (1984: 872-877) that noun incorporation develops historically along a certain path. In the case of languages such as these which do not have morphological marking for incorporation, the historical path can be seen as one moving from expressions whose morphemes are relatively free to those whose morphemes are closely bonded. Thus an examination of this particular stative-verb-body-part construction gives insights on problems of inalienability, compounding, and noun incorporation, and historical factors involved in these processes, as well as a window on the attitudes towards bodily feelings in these societies.

Notes

1. I want to thank Hilary Chappell and Bill McGregor for helpful comments on earlier versions of this paper. Also helpful were seminar discussants in the Research School of Pacific Studies at the Australian National University, where one version of the paper was presented. The fallible interpretation of comments and of data is my own. I am indebted to Preecha Juntanamalaga and Anthony Diller for the Bangkok Thai data as well as ideas, and I am grateful to Phuong Tonthat, Lo Giang Tran and Long Nguyen

for helpful advice on some of the Vietnamese data. The White Hmong of Laos is the language used here and many persons over time have helped me, but especially my teacher Vangkoua Cheurtong, Neng Chue Yang, Tong Vang, Youa Vang, Sai Xiong, Youa Yang and more currently Chu Lee. Besides sources individually cited, data was culled from Dao 1964, Heimbach 1979, D.D. Thomas 1971, and Thompson 1965.

2. There seems to exist very little literature on this particular phenomenon of Mainland Southeast Asian languages. What there is consists of scattered small bits in grammars and textbooks and miscellaneous examples not remarked on. This paper attempts to introduce and analyse one stative-verb-body-part phenomenon and omits a number of interesting verb-body-part phenomena. For example, both Hmong and Thai have productive metaphorical expressions of the reverse order – Body Part + Stative Verb (Jaisser 1990 and Anthony Diller, personal communication) which are not touched on here. Chappell (1995) discusses the double subject construction in Mandarin Chinese, a topic-comment construction suggestive of the construction discussed in this paper: Possessor Whole ^ Body Part ^ Stative Verb, wherein the verb refers to the part in characterising the whole. Matisoff (1985: 427-428) posits in some Sino-Tibetan languages the historical semantic development of stative verbs from body-part terms, e.g. 'armhole' → 'tickle' → 'ticklish' → 'bashful'.
3. Most Mon-Khmer languages have developed phonemic register, according to some scholars, and, in some, affixes have become frozen and reduced, but the languages have not yet developed full tone distinctions. The other languages used here – Vietnamese, Thai, Nung and Hmong – have fully developed tone systems.
4. I have used the standard orthography for Vietnamese, the romanised orthography developed in Laos for Hmong (see Smalley 1976: 87-101), the romanised orthography used in Huffman 1970 for Khmer and one romanised version of Thai. The orthographies for Chrau, Rengao and Nung are those used in the references cited. In general, diacritics represent tones in Vietnamese, Nung and Thai, roughly as follows:

á	high/high rising	à	low
Ø	mid/mid high, level	â	falling (Thai)
ả	mid rising	ạ	low (with glottal)
ã	mid rising (with glottal)		
ă, ĕ, â, ê, ô	vowels (Vtn, Chrau, Nung)		
à	falling intonation (Chrau)		

The Hmong Romanised Popular Alphabet uses VV for Vng (the only final consonant in Hmong) and consonant symbols word finally for tones:

-b	high level	-g	low falling breathy
-j	high falling	-s	low
-v	mid rising	-m	low, final glottal
Ø	mid level	-d	low rising

5. Abbreviations used in the glosses are: anm. = animal, animate; CLF = Classifier noun; INCL = Inclusive, as in first person inclusive; MASC = Masculine; 2PER = 2nd person singular; 3PER = 3rd person singular.
The Nung and Khmer phrases were culled from Saul – Wilson 1980 and Huffman 1970 and 1973, respectively. The Khmer phrases in (11) are from Huffman 1973: 492.
6. I have adapted the citation glosses for classifiers to indicate their general meaning. The sentence in (15) is from Nguyen K.T. 1975: 205, in (16) from PXNNH: 33, in (17)

from PXNNH: 40, in (18) from Huffman 1970: 44, in (19) from Gregerson 1979: 91, and in (20) from Saul – Wilson 1980: 70.

7. These Sino-Vietnamese compounds take the Chinese word order of descriptive modification preceding the head noun, as opposed to the Vietnamese order of postposed descriptive modification.
8. The sentence in (50b) is from Nguyen K.T. 1975: 196; the sentences in (53) from Saul – Wilson 1980: 71, in (54) from Gregerson 1979: 17, 90, in (55) from D.M. Thomas 1978: 273, 270 and in (56) from Huffman 1970: 290.
9. The Vietnamese expression is from Nguyen V.K. 1984: 355. The Miao four-syllable expression is from P'an – Ts'ao 1972: 214; superscript numbers represent tones: 55 = high level tone, 13 = low rising tone.

References

Bertrais, Yves

1979 *Dictionnaire Hmong – Français*. Bangkok: Imprimerie de l'Assomption.

Bradley, David (ed.)

1989 *Papers in South-East Asian linguistics* No. 11: *South-East Asian syntax*. (Pacific Linguistics A77) Canberra: Australian National University.

Chappell, Hilary

1995 "Inalienability and the personal domain in Mandarin Chinese discourse", [this volume].

Chappell, Hilary – William McGregor

1989 "Alienability, inalienability and nominal classification", *Proceedings of the Annual Meeting of the Berkeley Linguistics Society* 15: 24-36.

Clark, Marybeth

1989 "Hmong and areal South-East Asia", in: David Bradley (ed), 175-230.

Dao Dang Vy

1964 *Viet-Phap tu dien. Dictionnaire Vietnamien-Francais*. Saigon: Nha Sach Khai Tri.

Evans, Nick

1995 "The syntax and semantics of body part incorporation in Mayali", [this volume].

Gregerson, Kenneth

1979 *Predicate and argument in Rengao grammar*. (Publications in Linguistics 6.1) Dallas: Summer Institute of Linguistics and the University of Texas at Arlington.

Heimbach, Ernest E

1979 *White Hmong – English dictionary*. (Southeast Asia Program Data Paper 75). Ithaca: Cornell University Department of Asian Studies.

Huffman, Franklin E

1970 *Modern spoken Cambodian*. New Haven: Yale University Press.

1973 "Thai and Cambodian – A case of syntactic borrowing?", *Journal of the American Oriental Society* 93.4: 488-509.

Huffman, Franklin E. – Tran Trong Hai

1980 *Intermediate spoken Vietnamese*. Ithaca: Cornell University Southeast Asia Program.

Jaisser, Annie C.
1990 "De*liver*ing an introduction to psycho-collocations with *siab* in Hmong", *Linguistics of the Tibeto-Burman Area* 13 (1): 159-178.
Jenner, Philip N. – Laurence C. Thompson – Stanley Starosta (eds)
1976 *Austroasiatic studies.* (Oceanic Linguistics Special Publication No.13) Honolulu: University Press of Hawaii.
Jones, R.B. Jr.
1970 "Classifier constructions in Southeast Asia", *Journal of the American Oriental Society* 90.1: 1-13.
Khai, Buu
1972 *A formalized syntax of Vietnamese.* (Georgetown University Ph.D dissertation.) Ann Arbor: University Microfilms International 73-4190.
Liem, Nguyen Dang
1970 *Four-syllable idiomatic expressions in Vietnamese.* (Pacific Linguistics D6). Canberra: Australian National University.
Matisoff, James A.
1978 *Variational semantics in Tibeto-Burman. The "organic" approach to linguistic comparison.* (Occasional Papers of the Wolfenden Society on Tibeto-Burman Linguistics, VI) Philadelphia: Institute for the Study of Human Issues.
1985 "Out on a limb: *arm*, *hand*, and *wing* in Sino-Tibetan", in: Graham Thurgood – James A. Matisoff – David Bradley (eds), 421-450.
Mithun, Marianne
1984 "The evolution of noun incorporation", *Language* 60.4: 847-894.
1995 "Multiple reflections of inalienability in Mohawk", [this volume.]
Nacaskul, Karnchana
1976 "Types of elaboration in some Southeast Asian languages", in: Philip N. Jenner – Laurence C. Thompson – Stanley Starosta (eds), 873-889.
Nguyen Dinh Hoa
1966 *Vietnamese – English dictionary.* Tokyo: Tuttle.
Nguyen Kim Than
1975 "An outline of Vietnamese grammar", *Vietnamese Studies* 40: Linguistic essays. (Hanoi: Foreign Languages Publishing House), 148-217.
Nguyen Van Khon
1984 *Viet-Anh tu-dien Vietnamese-English dictionary.* Hong Kong: Knowledge Publishing Co.
P'an Yuan-en – Ts'ao Ts'ui-yun
1972 "Four-syllable coordinate constructions in the Miao language of eastern Kweichow" [tr. from Chinese], in: Herbert C. Purnell (ed.), 211-234.
Purnell, Herbert C (ed.)
1972 *Miao-Yao linguistic studies.* (Southeast Asia Program Data Paper 88) Ithaca: Cornell University Department of Asian Studies.
[PXNNH] Rooj Ntawv Hmoob (Hmong Academy)
1977 *Phau xyaum nyeem ntawv Hmoob* [Text for learning to read Hmong]. Bangkok: Assumption Printing Press.
Saul, Janice E. – Nancy Freiberger Wilson
1980 *Nung grammar.* Dallas, Texas: Summer Institute of Linguistics.

Smalley, William A.
1976 *Phonemes and orthography: Language planning in ten minority languages of Thailand.* (Pacific Linguistics C-43). Canberra: Australian National University.

Starosta, Stanley
1988 *The case for case. An outline of lexicase grammatical theory.* London and New York: Pinter.

Thomas, David D.
1971 *Chrau grammar.* (Oceanic Linguistics Special Publication No.7.) Honolulu: University Press of Hawaii.

Thomas, Dorothy M.
1969 "Chrau affixes", *Mon-Khmer Studies* III: 90-107.
1978 "The discourse level in Chrau", *Mon-Khmer Studies* VII: 233-295.

Thompson, Laurence C.
1965 *A Vietnamese grammar.* Seattle: University of Washington Press.

Thurgood, Graham – James A. Matisoff – David Bradley (eds)
1985 *Linguistics of the Sino-Tibetan area: The state of the art.* (Pacific Linguistics C-87). Canberra: Australian National University.

The possession cline in Japanese and other languages

Tasaku Tsunoda

1. Aim of the paper

The aim of this paper is to propose a cline which represents the degree of closeness/attachedness – physical and/or psychological – between the possessor and the possessee (Ultan 1978) , ranging from inalienable possession to alienable possession.[1] This cline, which I shall term "Possession Cline", indicates that the distinction between inalienable and alienable possession is not clear-cut but is a matter of degree. This paper exemplifies this cline with examples from Japanese, regarding honorifics and expressions of possession. It cites additional examples from English and two Australian Aboriginal languages: Warrungu (also spelt Warungu) of north Queensland and Djaru of Western Australia.

The format of this paper is as follows: Section 2 provides an outline of Japanese morphosyntax, paying special attention to honorifics and expressions of possession. section 3 introduces the Possession Cline, followed by section 4, which illustrates its morphosyntactic correlates with examples from the four above-mentioned languages. Section 5 summarises the main points of this paper.

When I refer to works on Japanese, I shall mainly draw on those written in English. This is largely in consideration of non-Japanese readers.

2. Outline of Japanese morphosyntax

This section provides a brief account of Japanese morphosyntax in general in 2.1, of expressions of possession in 2.2, and of honorifics in 2.3.

Japanese is mainly spoken in Japan. It is often said to be related to the Altaic languages, but its genetic affiliation is still unknown (cf. Shibatani 1987: 856).

2.1. Japanese morphosyntax in general

Japanese grammar is sketched, for example, in Kuno (1973) and Shibatani (1987), while it is described in detail in Martin (1975) and Hinds (1986). In the following, I shall summarise mainly those points of Japanese grammar which are relevant to the ensuing discussions.

Japanese is largely suffixing, with some prefixation. It is postpositional, and SOV is the most usual order of transitive sentences. Verbs and most (though not all) adjectives inflect for tense and aspect but nouns generally lack a distinction regarding number.

Cases are indicated by case postpositions, and the case-marking system is nominative-accusative. Certain case postpositions – *ga* NOM and *o* ACC, in particular – are often ellipsed in the spoken language.[2] Furthermore, case postpositions are obliterated (though not obligatorily in every instance) when they are followed by a modal-thematic postposition, such as *wa* 'topic' or *mo* 'too, also'.[3] For example, when *wa* is added, case postpositions behave as follows:

(1)	(i)	nominative	*ga* + *wa* → **ga wa*, *wa*
		accusative	*o* + *wa* → **o wa*, *wa*
	(ii)	genitive	*no* + *wa* → *no wa* (as an independent NP)
			wa (as a genitive phrase)
	(iii)	other case postpositions	
		dative	*ni* + *wa*→ *ni wa*, *wa*
		instrumental	*de* + *wa* → *de wa*, *wa*
			...

That is, *ga* NOM and *o* ACC obligatorily disappear when *wa* TOP is added. Thus:

(2) *Taroo ga* (or: *Taroo-wa*, but * *Taroo-ga-wa*)
Taroo NOM (Taroo-TOP Taroo-NOM-TOP)
warat-ta[4]
laugh-PAST
'Taroo laughed.'

When *wa* is attached, the genitive *no* remains if the possessee noun is deleted. The erstwhile genitive noun phrase now behaves as an independent NP. Thus, compare (3) and (4):

(3)

Zoo	*no*	*hana*	*ga*	*naga-i*
elephant	GEN	nose	NOM	long-PRES/FUT

'The elephant's nose is long.' (adapted from Mikami 1960: 12)[5]

(4)

Zoo	*no*	*wa*	*naga-i*
elephant	GEN	TOP	long-PRES/FUT

'The elephant's is long.'

No GEN disappears if the possessee noun remains.[6] Thus, compare (3) with (5):

(5)

Zoo	*wa*	*hana*	*ga*	*naga-i*
elephant	TOP	nose	NOM	long-PRES/FUT

'The elephant's nose is long.' (Mikami 1960: 9)[7]

Other case postpositions can either remain or disappear when *wa* is attached. For example, see (12) and (13), which involve *ni* DAT.

The issues surrounding *wa* TOP are highly controversial. Nonetheless, one thing is certain: many sentences sound unnatural unless they contain *wa*. Its use/non-use is conditioned by numerous factors. One such factor is the sentence type; a sentence expressing judgement requires *wa*, while one of description does not (cf. Shibatani 1987: 873). Another factor involved is information structure: Kuno (1973 :37-61) proposes that, roughly speaking, *wa* TOP indicates "topic/theme" or "contrast", and that *ga* NOM expresses "exhaustive listing" ('It is (only) ... that ...') or "neutral description", which lacks all these pragmatic nuances. Kuno's account of *wa* and *ga* can be refined by means of what is essentially identical with Silverstein's (1976) noun phrase hierarchy (cf. Shibatani – Tsunoda 1982: 106, Tsunoda 1983: 76). That is, (with other things being equal) *wa* marks "topic/theme" at the high end of the hierarchy, and "contrast" at the low end, while *ga* indicates "exhaustive listing" at the high end, and "neutral description" at the low end. (The distinctions are not clear in the middle of the hierarchy.) Thus:

(6)

	pronoun	noun
	1st 2nd 3rd	kinship human animate inanimate proper
wa	topic/theme	contrast
ga	exhaustive listing	neutral description

I have included the preceding paragraph in order (i) to warn the reader about what might appear to be an inconsistent appearance of *wa*, (ii) to hint at the reasons why certain examples are translated 'It is ... (only) ... that ...', and why certain examples are unnatural as they stand (for example, (3)), and (iii) to present my view of *wa* and *-ga*. In the following, I shall not provide literal (and perhaps more accurate) translations regarding *wa* or *ga* unless there is a reason to do so.

To conclude this subsection, I should note that Japanese discourse is highly elliptical. This phenomenon is briefly sketched in Kuno (1973: 16-18), and is discussed extensively in Hinds (1982).

2.2. Expressions of possession

Expressions of possession in Japanese are discussed in Suzuki (1968), Kuno (1973:73-95), Martin (1975: 193-198), Takahashi (1975), Takahashi – Yaku (1984) and Hinds (1986: 134-142). Hinds (1986:134) states as follows: "There are two primary ways to express possession in Japanese. The first involves a construction in which the possessor is linked to a following possessed item with the [genitive postposition – TT] *no*. The second involves a verbal construction in which the possessor and the possessed become arguments of a small set of verbs which may be used to indicate possession."

No GEN can be used for both alienable and inalienable possession. In this respect, Japanese lacks an alienable-vs.-inalienable distinction. Examples of *no* include 'elephant's nose' in (3), and:

(7) *Taroo no hon*
Taroo GEN book
'Taroo's book'

(8) *Taroo no me*
Taroo GEN eye
'Taroo's eyes'

The verbs that can express possession include *mot-* 'have', *syoyuusu-* 'possess', *i-* and *ar-* 'be, exist' (Hinds 1986: 183), and also *su-* 'do'. *Mot-* 'have' and *syoyuusu-* 'possess' take the NOM+ACC case frame. (Recall, however, that an NP may be elliptical, and that a case postposition may be obliterated by *wa* or the like.) When used finitely, they generally occur in the progressive form, which consists of a participle and the verb *i-* 'be', used as an auxiliary verb in examples (9) and (10):

(9)	*Watasi*	*wa*	*kuruma*	*o*	*mot-te*	*i-ru*
	I	TOP	car	ACC	have-PTCPL	be-PRES/FUT

'I have a car.' (Hinds 1986: 138)

(10)	*Watasi*	*wa*	*kabu*	*o*	*syoyuusi-te*	*i-ru*
	I	TOP	stock	ACC	possess-PTCPL	be-PRES/FUT

'I possess stock.' (Hinds 1986: 138)

Here, as in some other examples, my glosses differ somewhat from those of the source.

I- and *ar-* 'be, exist' may be used to express possession (as well as existence). The possessee is marked by *ga* NOM. The possessor is generally indicated by *ni* DAT, but it can be expressed by *ga* NOM. These two verbs are differentiated in terms of the animacy of the possessee. Thus:

(11)	Possessor	Possessee	Verb
	DAT (or, NOM)	NOM (human, animate, excluding inanimate)	*i-*
	DAT (or, NOM)	NOM (human, inanimate, excluding nonhuman animate)	*ar-*

As can be seen from (11), possession in Japanese can be expressed by the DAT+NOM case frame, as in many other languages (cf. Tsunoda 1981b, 1985).

Martin (1975: 195) states that "The possessor [is] marked optionally by *ni* or *-ga*". However, my own judgement on this matter is slightly different. That is, the possessor is generally indicated by the dative, rather than by the nominative. For example:

(12)	*Taroo*	*ni*	(or:	*Taroo-ni-wa*;	or:	*Taroo-wa*)
	Taroo	DAT	(	Taroo-DAT-TOP		Taroo-TOP)

kodomo *ga* *i-ru*
child NOM be-PRES/FUT

'Taroo has a child.'

(13)	*Taroo*	*ni*	(or:	*Taroo-ni-wa*,	or:	*Taroo-wa*)
	Taroo	DAT	(	Taroo- DAT-TOP		Taroo-TOP)

kodomo ga ar-u
child NOM be-PRES/FUT
(As above.)

In my judgement, in contrast with Martin's view, the possessor may be indicated by the nominative rather than by the dative only under limited circumstances, for instance, when the possessor is the interrogative word *dare* 'who'.

(14) *Dare ni* (or: *dare-ga*) *kodomo ga*
who DAT (who-NOM) child NOM
ar-imas-u ka?
be-POL-PRES/FUT Q
'Who has a child?' (Martin 1975:195)

The verb *syoyuusu-* 'possess' will be further discussed in 4.4.1, *mot-* 'have' in 4.4.3, and *i-* and *ar-* 'be, exist' in 4.4.4. As noted above, *su-*, which literally means 'do', can express possession in a rather special use; see 4.4.2.

Hinds (1986: 178) states that "There is no syntactic distinction between alienable and inalienable possession in Japanese". However, in my view, this distinction does exist in Japanese, although it is not always clear-cut. Manifestations of this distinction are discussed in section 4. See 4.4.2, 4.4.4, 4.5, 4.7 and Figure 3, in particular.

2.3. Honorifics

Japanese honorifics are discussed in Kuno (1973: 18-22), Martin (1975: 331-354), Harada (1976), Ooishi (1983), and Minami (1987). Following Harada (1976: 502), the types of Japanese honorifics can be shown as follows. (Harada's terms are indicated in parentheses.)

(15) Honorifics
 (i) Polite honorifics (Performative honorifics)
 (ii) Respect honorifics (Propositional honorifics)
  (ii-a) Subject respect honorifics (Subject honorifics)
  (ii-b) Object respect honorifics (Object honorifics)[8]

Categories (i) and (ii) are not mutually exclusive, while (ii-a) and (ii-b) are. The honorific prefix *o-/go-* is used in all of the categories listed above. It can be attached to both nouns and verbs. Its use with nouns is frequent, but is not obligatory. Certain nouns do not take this prefix, for example *ame* 'rain' in (17). The predicate generally contains certain honorific elements, often in addition to the honorific prefix attached to nouns.

There are levels of politeness such as informal or unmarked, polite and superpolite (Kuno 1973: 19). The polite level involves, for instance, the verbal suffix -(*i*)*mas-* if the predicate is a verb, as in example (17). The informal level involves no such suffix as in example (16).

(16) *Ame ga hut-ta*
rain NOM fall-PAST
'It rained.' (Harada 1976: 502)

(17) *Ame ga hur-imas-ita*
rain NOM fall-POL-PAST
'It rained.' (Harada 1976: 502)

A more literal, and no doubt unnatural, translation of (17) would be something like: 'I politely state (to you) that it rained'. As this literal translation indicates, polite honorifics are used to make the sentence sound polite. Also, as Harada's term "performative honorifics" implies, these expressions do not express respect for any argument of the sentence, although it is possible to say that they express respect for the addressee.

In contrast, respect honorifics denote respect for an argument of the sentence, as Harada's term "propositional honorifics" indicates. I shall use the term "respectee" to refer to the person to whom respect is directed. In translating respect honorifics, I shall avoid clumsy English sentences and simply indicate the respectee with capital letters.

Subject respect honorifics contain elements such as the following in the predicate (Harada 1976: 504):

(i) *o*-ROOT-INFINITIVE *ni nar-*, or, ROOT-(*r*)*are-*, if the predicate is a verb, and
(ii) *o-*/*go*-ROOT if the predicate is an adjective.

Ni is the dative postposition, and the verb *nar-* literally means 'become'; see (29).

Compare (18) and (19), both adapted from Kuno (1973: 20). The latter contains the elements for subject respect honorifics.

(18) *Tanaka-san ga Takesita-san o mat-u*
Tanaka-Mr. NOM Takeshita-Mr ACC wait-PRES/FUT
'Mr. Tanaka waits for Mr. Takeshita.'

(19) *Tanaka-san ga Takesita-san o o-mat-i*
Tanaka-Mr. NOM Takeshita-Mr. ACC HON-wait-INF
ni nar-u (or: *mat-are-ru*)
DAT RESP-PRES/FUT (wait-RESP-PRES/FUT)
'MR. TANAKA waits for Mr. Takeshita.'

Object respect honorifics contain the element *o*-ROOT-INFINITIVE *su-* in the predicate (Harada 1976: 504). (The verb *su-* literally means 'do'; see (104) and (105).) Thus, compare (18) with (20), adapted from Kuno (1973: 20):

(20) *Tanaka-san ga Takesita-san o o-mat-i*
Tanaka-Mr. NOM Takeshita-Mr. ACC HON-wait-INF
su-ru
RESP-PRES/FUT
'Mr. Tanaka waits for MR. TAKESHITA.'

In (20), the respectee is the DO, but object respect honorifics can also be used to show respect to an IO (Harada 1976: 530):

(21) *Tanaka-san ga Takesita-san ni hon o*
Tanaka-Mr. NOM Takeshita-Mr. DAT book ACC
o-okur-i su-ru
HON-send-INF RESP-PRES/FUT
'Mr. Tanaka sends a book to MR. TAKESHITA.'

Furthermore, respect honorifics can be used to show respect to a possessor that is marked by *no* GEN (Harada 1976: 526, 529), for example:

(22) *Tanaka-san ga Takesita-san no o-nimotu*
Tanaka-Mr. NOM Takeshita-Mr. GEN HON-baggage
o o-mot-i su-ru
ACC HON-carry-INF RESP-PRES/FUT
'Mr. Tanaka carries MR. TAKESHITA's baggage.' (adapted from Harada 1976: 502)

In examples such as (22), it might appear on the basis of form as though Mr. Takeshita's baggage (that is, his possessee) were respected. However, it is really Mr. Takeshita (that is, the possessor in the genitive) whom respect is directed at. That is, the respect is expressed, so to speak, indirectly via the possessee. (Takesi Sibata (pers. comm.) suggests that the degree of respect conveyed by such expressions is higher than otherwise because of their indirectness.)

I shall use the term "Possessor Respect" to refer to those respect honorifics in which respect is directed at the possessor in the genitive. Naturally, the genitive postposition may be obliterated, say, by *wa*. Indeed, the entire genitive phrase may be ellipsed, being contextually given or previously indicated in discourse.

Possessor Respect is common and perfectly natural in Japanese, although I do not know whether or not such a phenomenon is common crosslinguistically.

I shall use the term "Direct Respect" to refer to those respect honorific expressions which do not involve Possessor Respect and in which respect is directed at a subject or an object, for example, (19) to (21).

In the following, I shall be concerned with Possessor Respect, rather than with Direct Respect.

Harada (1976: 529) states to the effect that Possessor Respect is confined to the object, and is inapplicable to the subject. However, it is in fact applicable to the subject as well. This will be exemplified in 4.1 and further discussed in 4.2.

Respect honorifics can be used in reference, as the examples given above demonstrate. They can also be used in address, for example:

(23) *O-nimotu* *o* *o-mot-i*
HON-baggage ACC HON-carry-INF
si-mas-u
RESP-POL-PRES/FUT
'(I) will carry (YOUR) baggage.' (Harada 1976:529)

Generally, the use of the honorific prefix *o-/go-* on noun(s) alone does not complete a respect honorific sentence. The predicate of such a sentence must contain respect honorific element(s), as in the examples given above. That is, honorification is most highly concentrated in the predicate (Minami 1987: 142).

Use or non-use of, and degree of politeness in, honorifics depend on numerous – mostly sociolinguistic – factors, for example, channel, situation, speaker or writer's age, sex, education, social background, respectee's person, social status, and so on (Minami 1987). For example, honorifics tend to be used more by older people than by younger people; more by women than by men; more when the respectee is the second person (that is, addressee) than when the respectee is the third person; and so on. Reference to the imperial family, who hold the highest social status in the Japanese society, tends to involve honorifics more than reference to other people.

Honorifics referring to the imperial family used to employ special expressions that were virtually unintelligible to ordinary people. However, after World War II, ordinary honorific expressions are used for the imperial family as well.

3. Possession Cline

3.1. Introduction

In the media reports on the last days of the previous emperor, from the latter half of 1988 to the beginning of 1989, I often observed examples of Possessor Respect such as the following. (I shall cite these examples exactly as observed in the media reports.)

(24) *Heika* *no* *go-taion* *wa* *motono* *zyootai*
emperor GEN HON-temp TOP previous state
ni *modor-are-mas-ita.*
DAT return-RESP-POL-PAST
'The EMPEROR's temperature has returned to its previous state.' (taken from TV news)

(25) *Zentaitosite* *tennooheika* *no* *go-yoosu* *wa*
on:the:whole emperor GEN HON-state TOP
otitu-ite *irassyar-u*
stable-PTCPL be:RESP-PRES/FUT
'On the whole the EMPEROR's condition is stable.' (taken from TV news) (*irassyar-* 'be:RESP' is the suppletive form of *i-* 'be' (Harada 1976: 506).)

(26) *O-karada* *ga* *yowat-te* *or-are-ru*
HON-body NOM weaken-PTCPL be-RESP-PRES/FUT
'The (EMPEROR's) body has become weak.' (taken from TV news)

(27) *Naizoo* *ga* *taihen* *o-tuyo-i*
internal:organ NOM very HON-strong-PRES/FUT

yoo-da
seem-be:PRES/FUT
'The (EMPEROR's) internal organs seem to be very strong.' (taken from a magazine article)

(28) *Ketuatu wa agar-are-mas-ita*
blood:pressure TOP rise-RESP-POL-PAST
'The (EMPEROR's) blood pressure has risen.' (taken from TV news)

(29) *Nyoo no de ga kiwamete waru-ku*
urine GEN flow NOM extremely bad:ADV
nat-te or-are-ru
become-PTCPL be-RESP-PRES/FUT
'The (EMPEROR's) urine flow has become extremely bad.' (taken from a newspaper article)

(In (29), the verb *nar-* is used in its literal sense 'become'. Similarly for (31).)

Looking at these and many other, similar examples, I realised that Possessor Respect was not always possible. For example, in my judgement, the following two examples, which I composed, sound unnatural.

(30) **Tennoohei no go-yootei ga zisin de*
emperor GEN HON:villa NOM earthquake INST
o-tubure ni nar-imas-ita.
HON-collapse:INF DAT RESP-POL-PAST
'The EMPEROR's villa collapsed in an earthquake.'

(31) **Heika no syoyuuti ga taihuu de mizubitasi*
emperor GEN property NOM typhoon INST flooding
ni nar-are-mas-ita
DAT become-RESP-POL-PAST
'The EMPEROR's land was flooded in a typhoon.'

3.2. Possession Cline

While examining examples such as (24) to (31) and analogous examples of Possessor Respect, I realised that their acceptability depended on the type of possessee involved. I then realised that different types of possessee could be arranged on a cline. Specifically, my own intuition indicated that acceptabilty of Possessor Respect correlated with a cline which can be shown as follows:

(32) Possession Cline:

body part > inherent attribute > clothing > (kin) >
pet animal > product > other possessee

In other words, Possessor Respect is more acceptable when it involves a possessee higher on the cline than when it involves a possessee lower on the cline.

I then discussed this cline with more than a dozen native speakers of Japanese (including the people mentioned in Note 1). I also presented an early version of this paper to a monthly meeting of Nagoya Kotoba no Tsudoi [Nagoya Language Circle], on the 6th May 1989. The audience of the meeting included about 40 native speakers of Japanese. Those fifty-odd native speakers of Japanese virtually unanimously agreed with me regarding:

(i) the correlation between this cline and the degree of acceptability of Possessor Respect, and
(ii) the relative order of possessee types on the cline.

Furthermore, I conducted a questionnaire survey regarding this correlation. I obtained 73 responses. Despite the problems mentioned in the fourth paragraph of 4.1, the result strongly suggests that Japanese people on the whole share my intuitions, including those of the fifty-odd people consulted.

This cline is a refinement of the well-known inalienable/alienable distinction. More specifically, it represents the degree of closeness/attachedness – physical and/or psychological – between the possessor and the possessee. It should be noted that this cline indicates that the inalienable/alienable distinction is not clear-cut, but is a matter of degree.

As mentioned above, this distinction is widely known. I shall cite two fairly non-controversial characterisations of this distinction.

(i) Employing the concept of "conceptual distance", Haiman (1985: 130) states:

> It seems intuitively clear that conceptual distance is greater when possession is *alienable* than when it is not: possessor and possessum are not indissolubly bound together where possession is alienable, either in fact or in the perception of the speaker. Typically, inalienable possession is indicated when the possessum is a body part, a kinsman, or personal at-

tribute: all of these denotata are viewed as permanently associated with the possessor.

(ii) Nichols (1988:568) states as follows:– "inalienable possession is inborn, inherent, not conferred by purchase; alienable possession is, roughly, ownership, socially and economically conferred."[9]

Another characterisation – one by Dixon (1980) – is cited in 4.3. The possession cline presented in this paper has been arrived at independently.[10]

A cline similar to the one I propose is the one advanced by Haiman (1985: 135), who states as follows:- "Intuitively, the hierarchy of 'inalienable' is: body parts less alienable than kinsmen; kinsmen less alienable than artefacts".[11] However, on the basis of some crosslinguistic morphosyntactic data, Haiman (1985: 136) modifies his hierarchy and presents it as follows:

$$\left\{ \begin{array}{c} \text{kinsmen} \\ \text{body parts} \end{array} \right\} > \text{artefacts}$$

Largely on the basis of data from the languages of native North America, Nichols (1988: 572) puts forward the following hierarchy:

1. Kin terms and/or body parts
2. Part-whole and/or spatial relations
3. Culturally basic possessed items (for example arrows, domestic animals)

In the following, I shall comment on the individual possessee types, some of which are discussed in Chappell – McGregor (1989), among others.

The body part type naturally includes nouns denoting body parts. I tentatively assign excreta, for example, sweat, tears, faeces and semen to this type; I do not have any evidence to separate them as an independent type. This paper mainly looks at human possessors only, but the preceding remarks apply to animate and non-human possessors (for example, dogs) as well.[12]

The inherent attribute type (hereafter, simply "attribute") includes inherent personal attributes such as propensity, nature, weight, height, size, shape, state of health, temperature, blood pressure, energy, bodily functions (for example, movement, reaction, urination), consciousness and so on. In certain languages, for example, Djaru (Tsunoda 1981a: 194), this type will also include name, spirit, shadow, footprint and dream. (See Chappell – McGregor (1989).)

"Body part" and "attribute" are both in the nature of inalienable possession. Nevertheless, according to my intuition, "body part" is (perceived to be) closer or more attached to the possessor than is "attribute". My intuition is shared by virtually all of the fifty-odd people mentioned above.

"Clothing" refers to clothes, spectacles, ties, hats, shoes and earrings worn and attached to the body. According to the usual classification, they are alienable possessees. Nonetheless, they are physically (and probably psychologically as well) very close to the possessor, and they are almost body parts – since they are worn and attached to one's body.[13] However, they are of the "other possessee" type when they are not worn but are kept, say, in a wardrobe.

For a reason which will be explained in the last paragraph of 4.1, I did not include "kin" in my study of Possessor Respect. It is because of this that I parenthesised "kin" on the Possession Cline as presented in (32). In many languages, kin is grammatically treated as inalienable (Nichols 1988: 573), and, as seen above, both Haiman and Nichols place kin, together with body parts, at the highest position on their respective hierarchies. Recall, however, that Haiman suggests that body parts are less alienable than kin, as do Hyman (1977) and Durie (1987); cf. Note 11. There are also many languages in which kin is grammatically treated as alienable such as in a large number of Australian Aboriginal languages (Dixon 1980: 293), for example, Warrungu (Tsunoda 1974: 96) and Djaru (Tsunoda 1981a: 194), not to mention Japanese. Dixon (1980: 293) is cited in 4.3.

My own intuition indicates that "kin" should be placed between "clothing" and "pet animal". This echoes Haiman (1985), Hyman (1977) and Durie (1987) in that "kin" is more alienable than "body part". My intuition, which is shared by several native speakers of Japanese I have discussed the issue with, is reflected in the phenomena dealt with in section 4. See Figure 3, in particular.

Consanguineous kinsfolk, for example, parents, grandparents and siblings have a Janus-like nature. They resemble typical inalienable possessees such as body parts and inherent attributes in that they are inborn and inherent. (Naturally, this does not apply to spouses, in-laws or adopted children; cf. Chappell – McGregor (1989).) However, they also resemble the usual alienable possessees in that (physically at least) they are not attached to the possessor's body. In this respect, "kin" is less close or less attached to the possessor than is "clothing".

The "pet animal" type includes pet animals, pet birds and domestic animals (Nichols 1988: 565) as well. They seem to be fairly close to the possessor; he/she usually feels attached to them (Kimiaki Takahashi, pers. comm.). Similarly for "product". For instance, I feel more attached to the book I wrote than to the books I merely possess. Similarly, craftspeople will be more proud of their works than of the goods they bought at a shop. All other possessees belong to the last type, "other possessee", to which the possessor is least close or least attached.

This cline is, needless to say, not definitive. Thus, it will be possible to further divide the types listed above, or to set up additional types. Different cultures may require different classifications. Nonetheless, for the purpose of the present paper, this classification is adequate as it stands.

4. Morphosyntactic correlates of Possession Cline

4.1. Possessor Respect in Japanese (1)

As noted in section 2.3, the use of Japanese honorifics and the degree of politeness depend mainly on sociolinguistic factors. Consequently, acceptability judgements for a given example exhibit considerable variation. Despite all this, the fifty-odd people mentioned in 3.2 unanimously agree with me that the degree of acceptability of Possessor Respect correlates with the Possession Cline.

Additional support is provided by the result of the questionnaire survey I conducted on Possessor Respect. The questionnaire comprises 30 examples: 6 sentences actually used in the media reports on the last emperor and 24 sentences I composed. (I obtained responses from 73 people.)[14] In order to keep constant the variables in the questionnaire, I used examples in which (i) the possessor is the emperor, (ii) the possessor is overtly expressed, (iii) the possessor is marked by *no* GEN, and (iv) the predicate contains respect honorific element(s). (But a possessee noun is not attached with the honorific prefix *o-/go-* in every example. As noted in 2.3, the use of this prefix with nouns is common, but is not obligatory.)

In the questionnaire, I asked to have each example judged as "Natural", "Unnatural", or "Hard to decide". In order to estimate the degree of acceptability of the examples, I assigned 2 points to a judgement "Natural", 1 point to a judgement "Hard to decide", and nil to a judgement "Unnatural", and calculated the average figure for each example. If a given example is judged as natural by all the respondents, then its figure will be 2. If a given example is judged as unnatural by all the respondents, then its figure will be zero.

In compiling the questionnaire, I naturally aimed to examine the relevance of the Possession Cline to the degree of acceptability of Possessor Respect. However, subsequent discussions with some of the respondents have revealed that, in many instances, a sentence which I had expected to be judged as natural, was judged to be unnatural for a reason that was irrelevant to the cline itself. I shall list some of such reasons: (i) Example (33) is unnatural as it stands, but will be natural if *ga* NOM is replaced by *wa* TOP. (ii) Example (33) will be natural if the noun *kami* 'hair' takes the honorific prefix *o*-. (iii) Example (38) is unnatural, but will be natural if *no* GEN is replaced by *ni* DAT. (Recall that the verb *ar*- 'be, exist' generally takes the DAT+NOM case frame when it expresses possession; cf. (11).) (iv) Example (41) is unnatural, but it will be natural if the noun *aiken* 'pet dog' takes the honorific prefix *go*-. (v) One respondent (a speaker of the Tokyo dialect) considers all the examples involving -(*r*)*are*- 'RESP' as unnatural on the ground that this suffix is not used in his dialect.

As noted above, the people I had a subsequent discussion with agreed that, despite all these problems, the degree of acceptability of Possessor Respect on the whole correlated with the Possession Cline. Thus, it has become apparent that, in view of these problems, not too much importance should be placed on the statistical result obtained in this survey. (Takashi Murakami, pers. comm.)

Figure 1 presents the average figure for each of all the relevant examples in the questionnaire. It is significant that these figures on the whole correlate with the cline – despite (i) individual variations (largely due to sociolinguistic factors), and (ii) the problems mentioned above.

Examples of, and comments on, each possessee type follow.

[1] *Body part*:
The degree of acceptability of Possessor Respect is highest with body part nouns, for example:

Q(33)	*Saikin*	*tennooheika*	*no*	*kami*	*ga*	*sukkari*
	recently	emperor	GEN	hair	NOM	completely

	body part	attribute	clothing	pet animal	other possessee
2.0					
1.9					
1.8					
1.7					
1.6					
1.5		x			
1.4	x				
1.3	x				
1.2		x			
1.1					
1.0			x		
0.9		x			
0.8					
0.7					
0.6			x		
0.5		x		x	
0.4				x	x x
0.3					x
0.2					
0.1					
0.0					

Figure 1. Possessor respect (1)

siro-ku *nar-are-mas-ita*
white-ADV become-RESP-POL-PAST (1.4/2) (TT)
'Recently the EMPEROR's hair has become completely white.'

("Q" indicates that this example is taken from the questionnaire. The figure "1.4/2" means that the average figure for this example is 1.4, given that the full mark is 2. "TT" indicates that I composed this example.)

Q(34) *Tennooheika no asi wa mekkiri yowa-ku*
emperor GEN leg TOP considerably weak-ADV
nar-are-mas-ita
become-RESP-POL-PAST
'The EMPEROR's legs have become very weak.' (1.3/2) (TT)

Other examples include (26) ('body') and (27) ('internal organs').

[2] *Attribute*:
Possessor Respect is somewhat less acceptable with this type than with "body part".

Q(35)=(24) (1.2/2) (taken from a TV news)

Q(36)=(25) (1.5/2) (taken from a TV news)

Q(37) *Heika no go-taityoo wa o-yowar-i*
emperor GEN HON-state TOP HON-weaken-INF
ni nat-te i-ru
DAT become-PTCPL be-PRES/FUT (0.5/2)
(taken from a TV news)
'The EMPEROR's condition has weakened.'

Q(38) *Heika no isiki ga o-ar-i*
emperor GEN consciousness NOM HON-exist-INF
ni nar-u
DAT RESP-PRES/FUT
'The EMPEROR is conscious.' (literally: 'The EMPEROR's consciousness exists.') (0.9/2) (taken from a newspaper article)

Other examples include (28) ('blood pressure') and (29) ('urine flow').

[3] *Clothing*:
Possessor Respect is less acceptable with "clothing" than with "attribute". In my own judgement, the examples below border on the category of "unnatural", and I will mark them with a question mark.

Q(39) ?*Tennooheika no o-boosi ga sukosi huru-ku*
emperor GEN HON-hat NOM a:little old-ADV
nar-are-mas-ita
become-RESP-POL-PAST
'The EMPEROR's hat has become a little old.' (0.6/2) (TT)

Q(40) ?*Heika no megane ga totemo miryokuteki-de*
emperor GEN spectacles NOM very charming-PTCPL
irassya-imas-u
be:RESP-POL-PRES/FUT
'The EMPEROR's glasses are charming.' (1.0/2) (TT)

As alluded to in 3.2, in situations where the clothing, etc. referred to is not worn, but is placed, say, in a wardrobe, that is, when it is of the "other possessee" type, these sentences are less acceptable than when they are of the "clothing" type.

[4] *Pet animal*:
The acceptability is decidedly very low, and most of the respondents judged the examples as unnatural. I shall mark them with an asterisk.

Q(41) **Tennooheika no aiken ga byooki ni*
emperor GEN pet:dog NOM ill DAT
nar-are-ta
become-RESP-PAST
'The EMPEROR's pet dog has become ill.' (0.4/2) (TT)

Q(42) **Heika no uma ga yuusyoo-nasa-imas-ita*
emperor GEN horse NOM victory-do:RESP-POL-PAST
'The EMPEROR's horse won a race.' (0.5/2) (TT)

[5] *Product*:
I realised the significance of this type after reading Takahashi (1975) too late to include it in the questionnaire. My own intuition, which is shared by the fifty-odd people mentioned in 3.2, indicates that Possessor Respect is even less acceptable with "product" than with "pet animal".

(43) **Tennooheika no go-tyosyo ga tentoo ni*
emperor GEN HON-writing NOM shop DAT
de-te irassya-imas-u
appear-PTCPL be:RESP-POL-PRES/FUT
'The EMPEROR's book has appeared in shops.' (TT)

(It is well-known that the previous emperor had publications.)

[6] *Other possessee*:
The acceptability is definitely lowest. Most of the respondents judged the examples as unnatural.

Q(44)=*(30) (0.4/2) (TT)

Q(45)=*(31) (0.3/2) (TT)

Q(46) **Tennooheika no o-kuruma ga*
emperor GEN HON-car NOM
kosyoo-nasa-imas-ita
breakdown-do:RESP-POL-PAST
'The EMPEROR's car broke down.' (0.4/2) (TT)

I have thus far illustrated the correlation between the Possession Cline and the degree of acceptability of Possessor Respect.

As noted in 3.2, this cline represents the degree of closeness or attachedness – physical and psychological – between the possessor and the possessee. To employ terms used in related works, this cline represents "conceptual distance" (Haiman 1985:130, see 3.2 above), "physical closeness" (Wierzbicka 1979: 330), and "contiguity" and "attached[ness]" (Fox 1981: 326) between the possessor and the possessee.

What lies behind this correlation between the Possession Cline and the degree of acceptability of Possessor Respect seems to involve a psychological motivation: The closer or more attached the possessee is to the possessor, the more we tend to identify it with him or her. Furthermore, the more we tend to identify the possessee with the possessor, the more we want to, or have to, show our respect to it, in accordance with our respect for the possessor.

I parenthesised "kin" on the cline as presented in (32) and excluded this type from the discussion of Possessor Respect. I now state the reason for this exclusion. Consider:

(47)	*Yamada-si*	*no*	*reisoku*	*ga*	*daigaku*	*ni*
	Yamada-Mr.	GEN	son:RESP	NOM	university	DAT
	hair-are-ta.					
	enter-RESP-PAST					

'Mr. Yamada's son entered a university.' (Adapted from Ooishi 1983:127)

(In this particular translation, the respectee is not indicated by capital letters.)

As Ooishi points out, in instances such as (47), in which not only the possessor but also the possessee are humans, potentially the respectee can be either the possessor or the possessee (or possibly both). I agree with Ooishi's view that generally the respectee is the possessor (that is, Mr. Yamada in (47)), rather than the possessee (that is, his son in (47)). However, the other alternative can never be ruled out, and such sentences are not unequivocal instances of Possessor Respect. It is in view of this that I excluded "kin" from the discussion of Possessor Respect, and that I consequently parenthesised it on the cline in (32).

4.2. Possessor Respect in Japanese (2)

All the examples of Possessor Respect in 4.1 are intransitive sentences and the possessee is an intransitive subject ("Si"). However, Possessor Respect is not restricted to Si, nor is Direct Respect. For example, (22) is an instance in which the possessee is a DO.

Interestingly enough, acceptability of Possessor Respect varies depending on the syntactic status of the possessee. I shall look at a transitive subject ("St"), DO and IO, in addition to Si.

[1] *The possessee as an Si*:
As we saw in 4.1, Possessor Respect is natural at the high end of the cline, but becomes barely acceptable with "clothing", and is unnatural below this type.

[2] *The possessee as a DO*:
Possessor Respect is possible. Indeed, it is natural throughout the cline, from "body part", as in example (48), to "other possessee", as in examples (49) to (51).

Q(48) *Zizyo ga tennooheika no te o*
lady:in:waiting NOM emperor GEN hand ACC
o-tor-i si-ta
HON-take-INF RESP-PAST (1.5/2) (TT)
'A lady in waiting took the EMPEROR's hand (say, to assist him).'

Q(49) *Tanaka-san wa tennooheika no go-yootei o*
Tanaka-Mr. TOP emperor GEN HON-villa ACC
o-mikake si-ta
HON-glance:INF RESP-PAST
'Mr. Tanaka glanced at the EMPEROR's villa.' (0.8/2) (TT)

The score for (49), that is, 0.8/2, is not really high. However, this is no doubt due to a reason which is irrelevant to the Possession Cline itself. For example, Taroo Kageyama (pers. comm.) and a few other people suggest that the verb *mikake-* 'glance' is not suitable for this sentence. (The reason for this is not clear to me. I find (49) perfectly acceptable. He states that the following example, which contains a different verb, is perfectly natural, and I share his judgement of this sentence:

(50) *Tanaka-san wa tennooheika no go-yootei o*
Tanaka-Mr. TOP emperor GEN HON-villa ACC
o-mamor-i si-ta
HON-guard-INF RESP-PAST
'Mr. Tanaka guarded the EMPEROR's villa.'

The following example, adapted from Harada (1976: 526), is also perfectly natural.

(51) *Yamada-san wa heika no o-nimotu o*
Yamada-Mr. TOP emperor GEN HON-baggage ACC
o-mot-i si-ta
HON-carry-INF RESP-PAST
'Mr. Yamada carried the EMPEROR's bag.'

[3] *The possessee as an IO*:
Possessor Respect involving an IO is natural throughout the cline, as is the case with the DO. I have not been able to find any difference between them.

Q(52) *Zizyo ga tennooheika no te ni*
lady:in:waiting NOM emperor GEN hand DAT
o-boosio o o-watas-i si-ta
HON-hat ACC HON-hand-INF RESP-PAST (1.4/2) (TT)
'A lady in waiting handed a hat to the EMPEROR's hand.'

Q(53) *Tanaka-san wa tennooheika no go-yootei ni sono*
Tanaka-Mr. TOP emperor GEN HON-villa DAT that
nimotu o o-okur-i si-ta
baggage ACC HON-send-INF RESP-PAST (1.5/2) (TT)
'Mr. Tanaka sent that baggage to the EMPEROR's villa.'

[4] *The possessee as a St*:
Possessor Respect involving a St (hereafter, "St Possessor Respect") is at best barely acceptable, in sharp contrast with the Si, DO and IO. I have been unable to compose any example which sounded natural.

Q(54) ?*Tennooheika no odayakana me wa itumo*
emperor GEN gentle eye TOP always

kokumin no kokoro o yasuraka-ni si-te
people GEN mind ACC peaceful-ADV make-PTCPL
irassya-imas-ita (or: *or-are-mas-ita*)
be:RESP-POL-PAST (be-RESP-POL-PAST) (0.8/2) (TT)
'The EMPEROR's gentle eyes always made the people's mind peaceful.'

Q(55) ?*Heika no go-byooki ga kokumin ni*
emperor GEN HON-illnes NOM people DAT
ookina syoogeki o o-atae ni nat-ta
large shock ACC HON-give:INF DAT RESP-PAST
(or: *atae-rare-ta*)
(give-RESP-PAST)
'The EMPEROR's illness gave the people a big shock.' (0.6/2) (TT)

Example (54) involves "body part", and (55) "attribute", the two highest types on the cline. Despite this, they are barely acceptable in my judgement, hence the question mark. Needless to say, St Possessor Respect below this type is unnatural. The following example, which involves "other possessee", at the low end on the cline, is unnatural in my judgement, hence the asterisk.

Q(56) **Tennooheika no syoyuuti ga sono tiiki*
emperor GEN property NOM that area
no kaihatu o huseg-are-ta
GEN development ACC prevent-RESP-PAST
(or: *o-huseg-i ni nat-ta*)
(HON-prevent-INF DAT RESP-PAST)
'The EMPEROR's land prevented the development of that area.' (0.4/2) (TT)

As has just been seen, St Possessor Respect is unnatural. Needless to say, this phenomenon is uncommon. None of the previous works I consulted provides an example. Also, in the examples collected from the media reports, there was no example of St Possessor Respect which could be considered a "true" or prototypical transitive sentence. (For prototypical transitive sentences, see Hopper – Thompson (1980), and Tsunoda (1985, 1994).)

The degrees of acceptability of Possessor Respect involving the four grammatical relations respectively, are presented in Figure 2. (A solid line means "natural", a broken line "barely acceptable", and no line "unacceptable".) These results can also be shown as follows:

(57) DO, IO > Si > St

All of the fifty-odd people mentioned in 3.2 unanimously agree with me on these results as well.

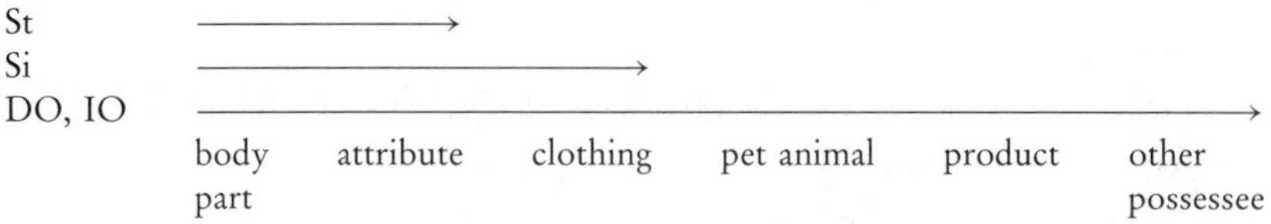

Figure 2. Possessor respect (2)

The very low acceptability of St Possessor Respect is striking. First, recall that reference to the imperial family tends to involve honorifics more than that to ordinary people (2.3). And yet, St Possessor Respect is at best barely acceptable even when the respectee is the emperor. Second, as seen above, when the possessee is a body part, Possessor Respect is natural with an Si, DO, and IO. And yet, it is at best barely acceptable with an St.

The following two facts are worth mentioning in this connection.

(i) Transitive sentences such as (54) to (56) become natural once the honorific elements coding respect are removed from the verb. This indicates that it is the presence of these elements that makes these sentences unnatural. In other respects, these sentences are normal. Thus, compare (54) with (58); (55) with (59); and (56) with (60).

Q(58) *Tennooheika no odayakana me wa itumo*
emperor GEN gentle eye TOP always
kokumin no kokoro o yasuraka-ni
people GEN mind ACC peaceful-ADV
si-te i-mas-ita
make-PTCPL be-POL-PAST (1.3/2) (TT)
'The emperor's gentle eyes always made the people's mind peaceful.'

Q(59) *Heika no go-byooki ga kokumin ni ookina*
emperor GEN HON-illness NOM people DAT large

syoogeki	*o*	*atae-ta*
shock	ACC	give-PAST

'The emperor's illness gave the people a big shock.' (1.7/2) (TT)

Q(60)

Tennooheika	*no*	*syoyuuti*	*ga*	*sono*	*tiiki*	*no*
emperor	GEN	property	NOM	that	area	GEN
kaihatu	*o*	*huse-ida*				
development	ACC	prevent-PAST	(1.4/2)			

'The emperor's land prevented the development of that area.'

(ii) Even if there is no honorific element coding respect on the verb, certain body part nouns cannot occur in a genitive phrase modifying a St. Thus, in my judgement, the following sentences are unnatural.

(61)

?*Tennooheika*	*no*	*te*	*ga*	*hana*	*o*	*tun-da*
emperor	GEN	hand	NOM	flower	ACC	pick-PAST

'The emperor's hand picked flowers.' (TT)

(62)

?*Tennooheika*	*no*	*asi*	*ga*	*booru*	*o*	*ket-ta*
emperor	GEN	foot	NOM	ball	ACC	kick-PAST

'The emperor's foot kicked a ball.' (TT)

Instead, versions such as the following would be preferred, in which the possessor occurs as the St and a body part in an adverb phrase:

(63)

Tennooheika	*ga*	*te*	*de*	*hana*	*o*	*tun-da*
emperor	NOM	hand	INST	flower	ACC	pick-PAST

'The emperor picked flowers with his hand.' (TT)

(64)

Tennooheika	*ga*	*asi*	*de*	*booru*	*o*	*ket-ta*
emperor	NOM	foot	INST	ball	ACC	kick-PAST

'The emperor kicked a ball with his foot.' (TT)

In my view, instances such as (61) and (62) are not nearly as acceptable as those such as (58) and (59). It is difficult to pinpoint the difference between them. When a body part noun or related category functions as an instrument, it generally occurs in an adverb phrase, as in (63) and (64), rather than as a St, as in (61) and (62), and, only when a body part noun does not function as an instrument, can it occur as a St, as in (58) and (59). (I suspect that this observation applies to many other languages.) This uncommon occurrence of body parts as a St is no doubt relevant to the uncommonness of St Possessor Respect.

It seems that no previous study of Japanese honorifics has pointed out the very low acceptability of St Possessor Respect. In fact, most studies of Japanese honorifics mainly look at their pragmatic or sociolinguistic aspects, and/or their morphological aspect, and they do not pay much attention to their syntactic side, with exceptions such as Harada (1976) and Shibatani (1978).

Harada (1976) is, to the best of my knowledge, the only previous work that pays attention to syntactic differences between Possessor Respect and Direct Respect (although he does not use such terms). He states (1976: 529) to the effect that Possessor Respect is possible with an object but not with a subject. While his observation is indeed insightful, it is incomplete, for the following two reasons:

(i) As seen above, the distinction between a Si and a St is important in this respect, but Harada does not make this distinction. The two examples that he gives in order to substantiate his claim are both intransitive sentences. He does not consider transitive sentences in this connection. (As noted above, St Possessor Respect is uncommon. It is, no doubt, due to this that it has escaped Harada's attention.)

(ii) As we have seen above, Possessor Respect with an Si is possible and, in fact, natural with the types high on the cline. However, the above-mentioned examples Harada provides concern the two lowest types: one is someone's writing (that is, "product") and the other is baggage ("other possessee"). He does not look at Possessor Respect with an Si which involves the types high on the cline.

4.3. Possessor ascension

There is a possible link between the findings presented in 4.2 and a phenomenon known as possessor ascension; see, for example, Fox (1981), Blake (1984), Durie (1987) and Nichols (1988: 578-579). This phenomenon is also called possessor promotion (Hyman 1977) and possessor raising (Munro – Gordon 1982: 95).

Possessor ascension is illustrated by the following pair of examples from English (adapted from Wierzbicka 1979: 350):

(65) *John kissed Mary's lips.*

(66) *John kissed Mary on the lips.*

The possessor (*Mary*) is in a genitive phrase in (65), but it has "ascended" or been "promoted" or "raised" to the (direct) object status in (66). (Naturally, such pairs of sentences exhibit pragmatic differences; see Wierzbicka's discussion.)

It has been suggested or demonstrated that, in certain languages, the acceptability of possessor ascension correlates with the following hierarchy of grammatical relations (see also Fox (1981) and Durie (1987: 375, 389) in this connection):

(67) DO > Si > St

A cursory survey of the literature indicates that the hierarchy in (67) is manifested at least in two different ways:[15]

(68) DO > Si, St

(69) DO, Si > St

The Blackfoot of North America, Gumbaingar of Australia (Fox 1981: 324-325), Chamorro of South Pacific (Durie 1987: 389), and English seem to allow possessor ascension for the DO only, that is, (68). In languages such as Haya of east Africa (Hyman 1977: 109), Lardil of Australia (Fox 1981: 327-328) and Acehnese of Indonesia (Durie 1987: 374, 389) among others, possessor ascension seems to be possible with the DO and the Si, but impossible with the St, that is, (69). For such languages, there is no evidence to place the DO higher than the Si on the hierarchy.

It seems that a different hierarchy of grammatical relations is operative behind possessor ascension in certain other languages. Thus, Japanese exhibits a phenomenon which may be termed possessor ascension. This process is acceptable with the Si but unacceptable with the St, DO, and also IO:[16]

(70) Si > DO, St

These grammatical relations are not differentiated in an analogous phenomenon in Djaru and Warrungu of Australia:

(71) Si, DO, St

It seems to me that, whichever hierarchy is operative, the acceptability of possessor ascension or an analogous phenomenon in a given language (in terms of a given grammatical relation) correlates with the Possession Cline, among other factors.[17] In fact, such a view is implicit in Hyman (1977: 104-105, 107, 112, 113), Fox (1981: 326, 328), Blake (1984), and

Durie (1987). Thus, Hyman (1977: 107) on possessor ascension in Haya states as follows: "possessor promotion can take place only if the possessor is affected by the action of the verb". He notes that this process is acceptable with body parts and clothing attached to the body (1977: 104-105), but impossible with, for example, a stick (1977: 101, 106), that is, "other possessee", the type lowest on the Possession Cline. (Another version of Hyman's view is cited in Note 11.) Similarly, employing the concept of "contiguity", Fox (1981: 326) provides the following explanation:

> The mechanism for Possessor Ascension, according to the hypothesis based on contiguity, can be seen as follows. Body parts, and often pieces of clothing, are physically attached to and contiguous with their possessors, and thus when a body part or an article of clothing ... is affected by an action, its possessor is necessarily affected by that action as well. Being thus affected by the action, the possessor can be viewed as a full participant in the action, like a direct object ...

Thus, compare the following pairs of examples, in each of which the possessor has "ascended" from the genitive phrase status (*his*) to the direct object status:

(72) *I hit his leg.*

(73) *I hit him on the leg.* (Fox 1981: 326)

(74) *I hit his wine bottle.*

(75) **I hit him on the wine bottle.* (Fox 1981: 326)

Durie (1987: 388) puts forward a very similar view: "Presumably [possessor ascension – TT] requires that the affectedness of the [possessee – TT] be attributable to the possessor. Because my head hurts I hurt too. One cannot so easily say that because my child [that is, "kin" – TT] hurts or is sick that I also hurt or am sick". See also Blake (1984: 451).

Having suggested that the acceptability of possessor ascension correlates with the Possession Cline, I shall now substantiate this suggestion, with examples from English, Japanese and Djaru.

In English, possessor ascension seems to be possible with the DO only, as noted above. In terms of the Possession Cline, this phenomenon is attested in the three highest types on the cline, that is, "body part", for example (73); "attribute", for example (76); and "clothing", for example (77) (although it may not be possible with all of the nouns in these types):

(76) *I know him by name.* (Cf. *I know his name.*)

(77) *I grabbed him by the sleeve.*

As noted above, Japanese exhibits a phenomenon which may be termed possessor ascension (although no such term has been used in the literature on Japanese). Thus, compare the following pair of examples, taken from Kuno (1973: 69). (The translations are Kuno's.)

(78) *John no o-toosan ga sin-da*
John GEN HON-father NOM die-PAST
'John's father died' or 'It is John's father who died.'

(79) *John ga o-toosan ga sin-da*
John NOM HON-father NOM die-PAST
'It is John whose father died.'

Another pair of examples which I composed:

(80) *Taroo no hyoozyoo ga akaru-i*
Taroo GEN expression NOM bright-PRES/FUT
'Taroo's expression is bright' or 'It is Taroo's expression that is bright.'

(81) *Taroo ga hyoozyoo ga akaru-i*
Taroo NOM expression NOM bright-PRES/FUT
'It is Taroo whose expression is bright.'

As a final pair of examples:

(82) *Zoo no hana ga naga-i*
elephant GEN nose NOM long-PRES/FUT
(Mikami 1960: 12)
'The elephant's nose is long' or 'It is the elephant's nose that is long.'

(83) *Zoo ga hana ga naga-i*
elephant NOM nose NOM long-PRES/FUT
'It is the elephant whose nose is long.'

Note that, in (79), (81), and (83), both of the possessor and the possessee are in the nominative. Kuno terms such constructions "multiple subject constructions". (See Kuno (1973: 62-78) for a discussion of this and related phenomena.) It may thus look as if these sentences are instances of possessor ascension. (Indeed, Kuno (1973: 71) considers sentences such as (79), (81), (83) as derived from those such as (78), (80), (82).) However, it is not really certain that they can be appropriately analysed as involving possessor ascension. There are two reasons for this. (i) Despite the nominative marking of the possessor and also despite Kuno's term "multiple subject construction", the syntactic status of the two nominative nouns, that is, possessor and possesee, is not clear. For example, it is not clear whether or not the nominative possessor has attained subjecthood. (Masayoshi Shibatani, pers. comm.) (ii) Alternative analyses are possible. (Taroo Kageyama, pers. comm.) For instance, it is possible to say that the nominative marking on the possessor has not replaced the genitive marking but that it is there from the beginning. Such an analysis is supported by many specialists in the field (see Note 3.).

In the following, I shall use the term "possessor ascension" for this Japanese phenomenon only tentatively.

The acceptability of possessor ascension in Japanese correlates with the Possession Cline. (Masayoshi Shibatani, pers. comm.). Possessor ascension is possible throughout the cline. It sounds natural towards the high end of the cline, for example, (83) ("body part"), (81) ("attribute"), and (79) ("kin"), but it is less natural, though not entirely unacceptable, towards the lower end. A pair of examples follows which involves "other possessee", the type at the lower end:

(84) *Taroo no bessoo ga tubure-ta*
Taroo GEN resort:house NOM collapse-PAST
'Taroo's resort house collapsed.'

(85) ?*Taroo ga bessoo ga tubure-ta*
Taroo NOM resort:house NOM collapse-PAST
Intended meaning: 'It is Taroo whose resort house collapsed.'

As seen in 4.1 and 4.2, the acceptability of Possessor Respect, too, correlates with the cline. There is, however, a difference between this phenomenon and possessor ascension. Namely, the cline is, so to speak, "steep" in the case of Possessor Respect, but it is almost "flat" in the case of possessor ascension. That is, the difference in the degree of acceptability between the high end and the low end of the cline is considerable in the case of Possessor Respect (cf. Figure 1), but it is very small in the case of possessor ascension.

Now, as stated above, possessor ascension in Japanese is conditioned by the hierarchy (70). That is, although Kuno (1973) does not discuss this, in my view possessor ascension in Japanese is possible with the Si only, as exemplified above, and it is not possible with the St, DO or IO. In each of the following pairs of examples, the version in which possessor ascension has applied, is untranslatable in the intended sense. Thus, examples involving an St:

(86) *Taroo no me ga Hanako no sugata o*
Taroo GEN eye NOM Hanako GEN sight ACC
torae-ta
catch-PAST
'Taroo's eyes caught sight of Hanako.'

(87) **Taroo ga me ga Hanako no sugata o*
Taroo NOM eye NOM Hanako GEN sight ACC
torae-ta
catch-PAST

Involving a DO:

(88) *Taroo ga Hanako no te o but-ta*
Taroo NOM Hanako GEN hand ACC hit-PAST
'Taroo hit Hanako's hand.'

(89) **Taroo ga Hanako o te o but-ta*
Taroo NOM Hanako ACC hand ACC hit-PAST

Involving an IO:

(90) *Taroo ga Hanano no te ni hon o*
Taroo NOM Hanako GEN hand DAT book ACC
watas-ita
hand-PAST
'Taroo placed a book in Hanako's hand.'

(91) **Taroo ga Hanano ni te ni hon o*
Taroo NOM Hanako DAT hand DAT book ACC
watas-ita
hand-PAST

Chinese, too, has double subject constructions; see Chappell (1995) and the references cited therein. According to Chappell's description, the situation seems to be as follows:

(i) Possession can be expressed either by the pattern *Possessor GENITIVE Possessee* or by the apposition of the possessor and the possessee.
(ii) The genitive marking is acceptable throughout the Possession Cline. The apposition is attested only in the types high on the the cline. That is, the acceptability of the apposition correlates with the cline.
(iii) The apposition is more likely to occur with Si than with DO, St, and oblique functions, while the genitive marking is more likely to occur with the latter than with Si. The resultant pattern conforms with the hierarchy of (70).
(iv) What are termed double subject constructions are those instances in which the possessor and the possessee occur in apposition and also in Si function.

I now turn my attention to a parallel phenomenon in Australian Aboriginal languages. Dixon (1980: 293) states as follows:

> Almost all Australian languages have two distinct ways of indicating possession. A genitive suffix can be added to the possessor noun (and also to any adjective modifying it) to indicate "alienable possession" – something that is not an intrinsic part of the possessor – or a kinship relation. ... Inalienable possession – something being an inseparable part of something else – is shown by just apposing possessor and possessed nouns, without any special suffix such as genitive.

See also Blake (1977: 40-41).

Note that, unlike "body part", "kin" is treated as alienable in Australian languages. This conforms with Haiman (1985), Hyman (1977) and Durie (1987); see 3.1.

Examples from Djaru of Western Australia follow. For detailed discussions, see Tsunoda (1981a: 196-203). (This language has enclitic pronouns as well as free pronouns. The enclitic pronouns crossreference free NPs. They are generally attached to the carrier morpheme *nga*, whose sole function is to carry them. The case-marking system is, very roughly speaking, nominative-accusative for enclitic pronouns, and ergative-absolutive for free pronouns and nouns – except that the demonstratives *nyawa* 'this' and *nyila* 'that' lack case-marking. The third person singular enclitic pronoun for the nominative and that for the accusative are phonologically zero and will be ignored in the following examples.)

In Djaru, apposition to express possession is limited, in terms of the Possesion Cline, to "body part" and "attribute", that is, it is limited to typical inalienable possession. "Clothing" and other types lower on the cline, including "kin", cannot utilise apposition; the possessor must be attached with a case suffix – generally the dative(/genitive) for free pronouns and the dative(/genitive)-1 for nouns. (The latter have two datives: dative-1 and dative-2.) An example of the dative(/genitive) involving "kin":

(92) *nganingu-lu* *ngawiyi-lu* *nga=lu* *pali winy-a* *watyapli-ø*
1SG:DAT-ERG father-ERG C-3PL:NOM find-PAST white:man-ABS
'My ancestors (lit. fathers) found the white man.'

(Dative pronouns, though not dative nouns, take an additional case suffix – in (92), the ergative suffix – showing case agreement with the possessee noun. *Pali winy-* is a compound verb. An equal sign ("=") indicates that the morpheme that follows is an enclitic. A hyphen is used for other, usual morpheme boundaries.)

The apposition of the possessor and the possessee is attested with all of the cases of the free NPs (that is, absolutive, ergative, dative, etc.) and also with all of the grammatical relations, that is, Si, St, DO, IO, and adverb phrases. In other words, there is no restriction on the use of apposition in terms of cases and grammatical relations. Examples of Si, which involves the absolutive case:

(93) *ngatyu-ø* *nga=rna* *milngali-ø* *yan-an*
1SG-ABS C-1SG:NOM tear-ABS go-PRES
'My tears are flowing', that is, 'I am weeping.'

(94) *ngatyu-ø* *nga=rna* *yiri-ø* *yakayarri–ø*
1SG-ABS C-1SG:NOM name-ABS Yagayari-ABS
'My name is Yagayari.'

Examples of the DO, which again involves the absolutive case:

(95) *ngatyu-ngku* *nga=rna* *mawun–ø* *langka–ø* *pung-an*
1SG-ERG C-1SG:NOM man-ABS head-ABS hit-PRES
'I hit the man's head.'

(96) *ngatyu-ngku* *nga=rna* *nyila* *mawun–ø* *yiri–ø*
1SG-ERG C-1SG:NOM that man-ABS name-ABS

ngarra man-an
know-PRES
'I know that man's name.'
(*ngarra man-* 'know' is a compound verb)

An example of the St, which involves the ergative case:[18]

(97) *mawun-tu kunyarr–ø pung-an marla-ngku*
man-ERG dog-ABS hit-PRES hand-ERG
'A man hits a dog with (his) hand.'

Examples of adverb phrases follow. (98) involves the locative case and (99) the ablative case.

(98) *ngatyu-ngku nga=rna=nyanta makarta–ø yaan-an*
1SG-ERG C-1SG:NOM-3SG:OBL hat-ABS put-PRES
mawun-ta langka-ka
man-LOC head-LOC
'I put a hat on the man's head.'

(99) *kungulu–ø yan-an nyanunginy-ngu langka-ngu*
blood-ABS go-PRES 3SG-ABL2 head-ABL
'Blood is flowing from his head', that is, 'He is bleeding in the head.'

(Free pronouns have two ablative cases (ablative-1 and ablative-2), while nouns have just one ablative case.)

As noted above, with "body part" and "attribute", apposition is employed to indicate possession. However, a dative suffix is occasionally used as in example (100):

(100) *nga=lu nyanunga=lu karrun-an-i*
C-3PL:NOM 3PL:DAT-CLITIC have-CONT-PAST
milyilyi–ø
brain-ABS
'They (Aboriginal people) had brains.'

(Clitics such as *=lu* provide a modal qualification. Their meanings are not necessarily understood well, and, in this paper, they are simply glossed 'CLITIC'. See Tsunoda (1981a: 204-211).)

(101) *nyununga yiri–ø, timngayarri–ø*
2SG:DAT name-ABS Dimngayari-ABS
'Dimngayari is your name!'

The conditioning factor regarding such a use of the dative as against apposition is not known. (See Dixon's (1977: 362) explanation for an analogous phenomenon in Yidiny of north Queensland.)

What has been stated above about Djaru – regarding (i) the distribution of apposition and the genitive(/dative)-marking and (ii) the occasional use of the genitive(/dative)-marking for inalienable possession – applies to Warrungu (Tsunoda 1974: 576-599) and presumably many other Australian languages.

Like many other Australian Aboriginal languages, Kalkatungu and Pitta-Pitta of western Queensland display examples parallel to those illustrated above for Djaru. Blake (1984) regards them as instances of possessor ascension. However, I am not certain whether such an analysis is adequate for Djaru or Warrungu.

To sum up the discussion of this subsection, possessor ascension and analogous phenomena are conditioned by a certain hierarchy of grammatical relations. But the actual form of the hierarchy varies from language to language; the DO occupies the highest position in English; the DO and the Si in Haya; and the Si in Japanese; but these grammatical relations are not distinguished in Djaru or Warrungu. Whichever hierarchy is operative in a given language, such a phenomenon correlates with the Possession Cline. Namely, if the genitive (or dative) marking on the possessor is to be suspended at all, then it is more likely to be so at the high end of the cline. It is more likely to be retained at the low end of the cline. The results of this subsection are partially shown in Figure 3.

4.4. Possession verbs in Japanese

Have in English, for instance, can be used for all of the possessee types on the Possession Cline. That is, its use does not correlate with the cline. As noted in 2.2, Japanese has at least five verbs that can express possession. Although this does not seem to be noted in previous studies, the use of these verbs is conditioned by the Possession Cline.

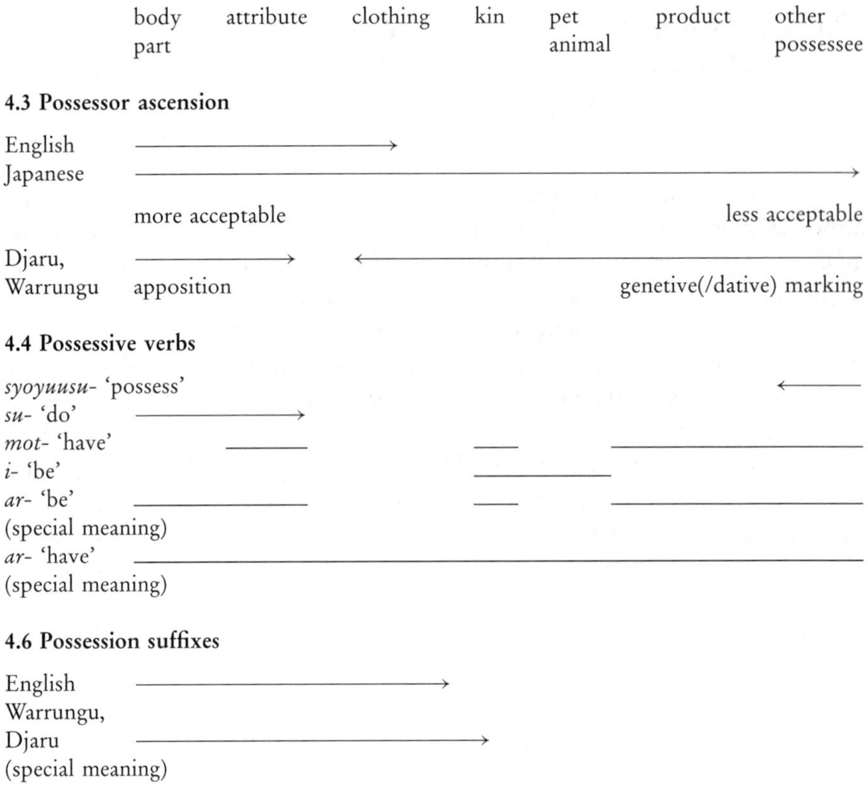

Figure 3. Other correlates of the Possession Cline

4.4.1. Syoyuusu- *'possess'*

Syoyuusu- 'possess' is formal and bookish. As noted in 2.2, it takes the NOM+ACC case frame (if no NP is elliptical and if no case postposition is obliterated). When used finitely, it usually occurs in the progressive form (that is, a participle plus *i-*, 'be'). (When used nonfinitely, for example, in a relative clause, it can occur in a non-progressive form.)

In view, *syoyuusu-* 'possess' is limited to "other possesee", that is, the type at the low end of the cline, for example (10) ('stock'). The use of this verb is unacceptable when used with "product", the type immediately above "other possessee":

(102)	?*Yamada-sensei*	*wa*	*ookuno*	*tyosyo*	*o*
	Yamada-professor	TOP	many	writing	ACC

syoyuusi-te *i-ru*
possess-PTCPL be-PRES/FUT

Example (102) is unnatural in the "product" sense as in 'Professor Yamada has many publications (of his own)', although it may be marginally acceptable in the "other possessee" sense as in 'Professor Yamada possesses many books written by other people'. Similarly, this verb cannot be used with the "pet animal" category as in example (103), or with other types higher on the cline.

(103) **Yamada-san* *wa* *inu* *o* *syoyuusi-te*
Yamada-Mr. TOP dog ACC possess-PTCPL
i-ru
be-PRES/FUT
Intended meaning: 'Mr. Yamada has/keeps dogs.'

In order to express the intended meaning of (103), the verb *kaw-* 'keep (animals)' has to be used.

The use of *syoyuusu-* 'possess' may be said to correlate with the Possession Cline in that it is confined to one end (that is, the low end) of the cline.

4.4.2. Su- *'do'*

Although this is not mentioned in Hinds' (1986: 138) discussion of possession in Japanese, the verb *su-*, which literally means 'do', can express possession. It takes the NOM+ACC case frame (if no NP is elliptical and if no case postposition is obliterated). This is irrespective of whether it has its literal meaning or describes possession. Examples of the literal meaning 'do' include:

(104) *Kodomo-tati* *wa* *yakyuu* *o* *su-ru*
child-PL TOP baseball ACC do-PRES/FUT
'The children play baseball.'

(105) *Kodomo-tati* *wa* *yakyuu* *o* *si-te*
child-PL TOP baseball ACC do-PTCPL
i-ru
be-PRES/FUT
'The children are playing baseball.'

Examples of *su-* expressing possession include:

(106) *Naomi wa ao-i me o si-te*
Naomi TOP blue-PRES/FUT eye ACC do-PTCPL
i-ru
be-PRES/FUT
'Naomi has blue eyes.' (Kageyama 1980b: 12-13)

(107) *Taroo wa naga-i kami o si-te*
Taroo TOP long-PRES/FUT hair ACC do-PTCPL
i-ru
be-PRES/FUT
'Taroo has long hair.' (adapted from Kageyama 1980b: 12)

(108) *Titi wa kura-i hyoozyoo o si-te*
father TOP dark-PRES/FUT expression ACC do-PTCPL
i-ta
be-PAST
'Father had a grim expression on his face.' (adapted from Kageyama 1980b: 12)

(109) *Hanako wa yo-i seikaku o si-te*
Hanako TOP good-PRES/FUT nature ACC do-PTCPL
i-ru
be-PRES/FUT
'Hanako has a good nature.'

The use of *su-* 'do' to express possession is rather peculiar. There are three reasons for this: (i) Its literal meaning 'do' does not imply anything like possession. (ii) There is an aspectual restriction on its use to code possession. That is, when used finitely, it must occur in the progressive form, as in examples (106) to (109). (But when used nonfinitely, this restriction does not apply.) In contrast, this restriction does not apply when *su-* is used in its literal sense. Thus, *su-* in (104) is not in the progressive form. (I should note that there is no tense restriction irrespective of whether *su-* has its literal meaning 'do' or expresses possession.) (iii) As Kageyama (1980b: 12) points out, the possessee noun must be accompanied by an adjective or some other qualifier, without which these sentences are ungrammatical. Thus, compare (106) with (110). The latter is nonsensical.

(110) **Naomi wa me o si-te i-ru*
Naomi TOP eye ACC do-PTCPL be-PRES/FUT

Now, it has been suggested, though not always explicitly, that this special use of *su*- 'do' to express possession is limited to inalienable possession (Takahashi 1975: 11, Kageyama 1980b: 12-13). Indeed, it is confined to typical inalienable possessees: "body part" in examples (106) and (107), and "attribute" in examples (108) and (109).

The situation with "clothing", the type immediately below "attribute", is rather complicated (Taroo Kageyama, pers. comm.) and is described by McCawley (1978: 250) in the following way : "Japanese has a large number of words that can be rendered in English as 'put on', differing from each other with regard to the place where the article is worn and the manner in which it is put on". (For further details, see Kageyama (1980a: 78-102).) Examples are *kabur*- 'wear on the head' in (102), *ki*- 'wear on the body, involving arms and hands' in (103) and *hak*- 'wear on the feet or legs'.

(111) *Taroo wa boosi o kabut-ta*
Taroo TOP hat ACC wear-PAST
'Taroo put a hat on.'

(112) *Taroo wa syatu o ki-ta*
Taroo TOP shirt ACC wear-PAST
'Taroo put a shirt on.'

Now, the verb *su*- 'do' can be used to mean 'wear, put on, have on' but this is only under certain conditions, that is, when none of the 'wear' verbs can be used (Kageyama 1980a: 92). Consider examples (113) and (114):

(113) *Taroo wa nekutai o si-ta*
Taroo TOP tie ACC do-PAST
'Taroo put a tie on.'

(114) *Hanako wa tebukuro o si-ta*
Hanako TOP gloves ACC do-PAST
'Hanako put gloves on.'

Otherwise, *su*- cannot mean 'wear'. This is shown by the following sentences which are nonsensical.

(115) **Taroo wa boosi o si-ta*
Taroo TOP hat ACC do-PAST

(116) **Taroo wa syatu o si-ta*
Taroo TOP shirt ACC do-PAST

Compare (115) with (111); and (116) with (112).

Similarly, sentences involving *su-* are nonsensical with 'kin' and other types lower on the cline.

Now, sentences such as (113) and (114) – and, naturally, those such as (111) and (112) as well – can occur in the progressive form. Thus, compare (113) with (117); and (114) with (118):

(117) *Taroo wa nekutai o si-te i-ru*
Taroo TOP tie ACC do-PTCPL be-PRES/FUT
(or: *i-ta*)
(be-PAST)
'Taroo has (or, had) a tie on.'

(118) *Hanako wa tebuku roo si-te i-ru*
Hanako TOP gloves ACC do-PTCPL be-PRES/FUT
(or: *i-ta*)
(be-PAST)
'Hanako has (or, had) gloves on.'

On the basis of examples such as (111), (112), (117) and (118) it may appear as though the special use of *su-* 'do' to express possession extends to "clothing". However, in my view, *su-* for 'wear' should be distinguished from *su-* for possession. There are three reasons for this:

(i) As seen above, the use of *su-* for 'wear' is limited.
(ii) *Su-* for possession must occur in the progressive form when used finitely, as in examples (106) to (109). In contrast, *su-* for 'wear' does not have to do so as in examples (113) and (114), although it can, for example, in (117) and (118).
(iii) With the *su-* for possession, the possessee must be accompanied by a qualifier, compare (106) with (110). On the other hand, with the *su-* for 'wear', the possessee does not have to be accompanied by a qualifier, as in examples (113), (114), (117) and (118), although it may be.

To conclude, the special use of *su-* 'do' to express possession is limited to "body part" and "attribute", the typical inalienable possessees. These two types occupy the high end of the Possession Cline. In this sense, its use may be said to correlate with the cline.

Note also that the use of *su-* 'do' for possession contrasts with *syoyuusu-* 'possess', which is confined to the lowest type on the cline.

4.4.3. Mot- *'have'*

Unlike *syoyuusu-* 'possess', which is formal and bookish (4.4.1), *mot-* 'have' is style-neutral. However, like the former, *mot-* 'have' takes the NOM+ACC case frame (if no NP is elliptical and if no case postposition is obliterated). When used finitely, it generally occurs in the progressive form. (When used nonfinitely, it can occur in a non-progressive form.)

Unlike *syoyuusu-* 'possess' and *su-* 'do' for possession, the use of *mot-* 'have' does not really correlate with the Possession Cline. Nonetheless, its use/non-use is conditioned by the possessee types on the cline.

First, normally *mot-* 'have' cannot be used with the "body part" category:

(119) **Naomi wa ao-i me o mot-te*
Naomi TOP blue-PRES/FUT eye ACC have-PTCPL
i-ru
be-PRES/FUT
Intended meaning: 'Naomi has blue eyes'.

In order to express the intended meaning, *su-* 'do' for possession must be used. See (106). The verb *mot-* can, however, be used with the "attribute" category:

(120) *Taroo wa marena sainoo o mot-te*
Taroo TOP rare ability ACC have-PTCPL
i-ru
be-PRES/FUT
'Taroo has a rare talent.'

But this verb cannot be used with the "clothing" category:

(121) **Taroo wa boosi o mot-te i-ru*
Taroo TOP hat ACC have-PTCPL be-PRES/FUT
Intended meaning: 'Taroo has a hat on.'

For the verbs that can be used with "clothing", see 4.4.2.

Example (121) can mean 'Taroo owns a hat' (he may not necessarily have it on at the time of speaking; it may be kept in a wardrobe). In this case, the hat belongs to the "other possessee" type.

Mot- can be used with "kin":

(122) *Taroo wa rippana oya o mot-te*
Taroo TOP respectable paren ACC have-PTCPL
i-ru
be-PRES/FUT
'Taroo has respectable parents.'

However, this verb cannot be used with the "pet animal" category:

(123) **Taroo wa inu o mot-te i-ru*
Taroo TOP dog ACC have-PTCPL be-PRES/FUT
Intended meaning: 'Taroo has/keeps a dog.'

In order to express the intended meaning, the verb *kaw-* 'keep (animals)' must be used instead.

Mot- 'have' can be used with the "product" category:

(124) *Yamada-sense iwa ookuno tyosyo o*
Yamada-professor TOP many writing ACC
mot-te i-ru
have-PTCPL be-PRES/FUT
'Professor Yamada has many publications (of his own).'

It can also be used with "other possessee"; see example (9) with 'car'.

To sum up, the use of *mot-* 'have' does not correlate with the Possession Cline, but its use/non-use is conditioned by the possessee types on the cline.

4.4.4. I- *and* ar- *'be, exist'*

As noted in 2.2, with these verbs, the possessor takes the dative (or the nominative under limited circumstances) and the possessee the nominative. (Recall that a case postposition may be obliterated by *wa* TOP.)

These verbs may express existence or possession. When they express possession, *ar-* can be used for human possessees and for inanimate possessees, but not for nonhuman animate possessees; and *i-* can be used for human possessees and nonhuman animate possessees, but not for inanimate possessees. That is, in terms of the Possession Cline, *i-* can be used for the categories of kin (see example (12) with 'child') and pet animal only, since all other types are inanimate. In the following, I shall concentrate on *ar-*.

As alluded to in the preceding paragraph, the use of *ar-* 'be' is conditioned by the possessee types on the Possession Cline, although this is not noted in previous works such as Takahashi – Yaku (1984) (Taroo Takahashi, pers. comm.).

First, *ar-* 'be' (when expressing possession) can be used with "body part", as in example (125), and with "attribute", as in example (126):

(125) *Taroo wa siraga ga ar-u*
Taroo TOP grey:hair NOM be-PRES/FUT
'Taroo has grey hair.'

(126) *Taroo wa sainoo ga ar-u*
Taroo TOP ability NOM be-PRES/FUT
'Taroo is talented.'

However, it cannot be used with "clothing". Consider (127):

(127) *Taroo wa syatyu ga ar-u.*
Taroo TOP shirt NOM be-PRES/FUT

This sentence cannot have the "clothing" reading: 'Taroo has a shirt on'. For the verbs that can be used with "clothing", see 4.4.2. This sentence can mean 'Taroo owns a shirt' (he may not necessarily have it on at the time of speaking; it may be kept in a wardrobe). In this case, the shirt belongs to the "other possessee" type.

Ar- 'be' can be used with "kin" as in examples (13) and (14) with 'a child' but it cannot be used with "pet animal":

(128) **Taroo wa inu ga ar-u*
Taroo TOP dog NOM be-PRES/FUT
Intended meaning: 'Taroo has/keeps a dog.'

In order to express the intended meaning, the verb *kaw-* 'keep' must be used.

Ar- 'be' can be used with "products":

(129) *Yamada-sensei wa ookuno tyosyo ga*
Yamada-professor TOP many writing NOM
ar-u
be-PRES/FUT
'Professor Yamada has many publications (of his own).'

It can be used with "other possessees" as well as in (127).

Regarding the use of *ar-* (when expressing possession), the following facts are worth noting:

(i) This verb can be used with "kin", but not with "pet animal", despite the fact that both are animate.
(ii) It can be used with "body part", but not with "clothing", despite the fact that "clothing" is attached to the body, almost like "body part".

When *ar-* is used with certain body part nouns, the sentences can have a special meaning, for example:

(130)	*Hukumoto-sensyu*	*wa*	*asi*	*ga*	*ar-imas-u ...*
	Hukumoto-athlete	TOP	foot/leg	NOM	be-POL-PRES/FUT

'Hukumoto is a fast runner (he runs faster than the average).'
(I recorded (130) from a radio broadcast of a baseball match.)

(131)	*Taroo*	*wa*	*atama*	*ga*	*ar-u*
	Taroo	TOP	head	NOM	be-PRES/FUT

'Taroo is brainy (that is, brainier than the average).'

(132)	*Taroo*	*wa*	*me*	*ga*	*ar-u*
	Taroo	TOP	eye	NOM	be-PRES/FUT

'Taroo makes good judgements (that is, better judgements than the average).'

Such examples all describe a function or ability of the body part concerned. Furthermore, they all mean 'more ... than the average'.

Parallel examples are found with certain "attribute" nouns as well:

(133)	*Taroo*	*wa*	*sintyoo*	*ga*	*ar-u*
	Taroo	TOP	height	NOM	be-PRES/FUT

'Taroo is tall (that is, taller than the average).'

(134)	*Konisiki*	*wa*	*taizyuu*	*ga*	*ar-imas-u*
	Konishiki	TOP	weight	NOM	be-POL-PRES/FUT

'Konishiki is heavy (that is, heavier than the average).'
(*Konishiki* is the ring name of a sumo wrestler who weighs as much as 230kg.)

A similar situation obtains with the English verb *have*. When used with certain "body part" and "attribute" nouns, the sentences can have a special meaning. Examples of "body part":

(135) *He has an eye for paintings.*

(136) *He has an ear for music.*

Like the analogous Japanese examples, these English examples describe a function or ability of the body part concerned, and, a function or ability better than the average. (It is probably because these sentences describe a function or ability of the body part, rather than the body part itself, that these body part nouns occur in the singular – despite the fact that people normally have two eyes and two ears.) An example of "attribute":

(137) *He has a temperature.* (That is, higher than the norm.)

Note that this special meaning of *temperature* contrasts with its literal meaning as exemplified in instances such as:

(138) *The nurse checked my temperature.*

In Japanese and apparently in English as well, these special meanings are restricted to "body part" and "attribute", that is, they are limited to typical inalienable possessees. I shall return to these special meanings in 4.7.

What has been discussed in 4.4.1 to 4.4.4 is summarised in Figure 3.

4.5. *Noun-1 GENITIVE Noun-2* of Japanese

As seen in 2.2, the postposition *no* GEN expresses possession in the following pattern:

(139) Possessor *no* Possessee.

Examples include (3), (7) and (8).

There is a pattern which superficially looks similar to that of (139). Namely:

(140) Noun-1 *no* Noun-2

This pattern can express various meanings but, on the whole, it can be said that the Noun-1 describes a characteristic of Noun-2 (Suzuki (1968: 28-29), Takahashi (1975: 11), Kageyama (1980b: 13), Ikudoo Tajima (pers. comm.), and Chappell and McGregor (pers. comm.).)

It is not exactly justifiable to say that the degree of acceptability of the pattern (140) correlates with the Possession Cline. Nonetheless, its use or non-use and also the meaning it expresses vary depending on the possessee type on the cline. There are two interesting facts about this pattern:

(i) Mainly when used with "body part" and "attribute", that is, typical inalienable possessees, this pattern expresses possession. In this case, the possessee precedes the possessor, in contrast with (139), the usual genitive construction, in which the possessee follows the possessor.
(ii) This pattern is either disallowed or has a special meaning when used with certain "body part" and "attribute" nouns.

Examples of each possessee type and comments on them follow.

[1] *Body part*

(141) *hige no otoko*
beard GEN man
'a bearded man'

(142) *nikibi no syoonen*
pimple GEN boy
'a pimpled boy'

(143) *ao-i me no syoonen*
blue-PRES/FUT eye GEN boy
'a blue-eyed boy'

(144) *naga-i kami no syoozyo*
long-PRES/FUT hair GEN girl
'a long-haired girl'

(145) *aka-hana no tonakai*
red-nose GEN reindeer
'a red-nosed reindeer'[19]

Note that the possessee precedes the possessor. Thus, compare (141) and (144) with (146) and (147), respectively. The latter are instances of the usual genitive construction, that is, (139), in which the possessor precedes the possessee.

(146) *otoko no hige*
man GEN beard
'a man's beard'

(147) *syoozyo no naga-i kami*
girl GEN long-PRES/FUT hair
'a girl's long hair'

[2] *Attribute*

(148) *hage no otoko*
baldness GEN man
'a bald man'

(149) *yo-i seisitu no otoko*
good-PRES/FUT nature GEN man
'a good-natured man'

(150) *kura-i hyoozyoo no otoko*
dark-PRES/FUT expresssion GEN man
'a grim-faced man'

[3] *Clothing*

(151) *megane no otoko*
spectacles GEN man
'a spectacled man'

(152) *naga-i sukaato no syoozyo*
long-PRES/FUT skirt GEN girl
'a long-skirted girl' (Takahashi 1975: 11)

(153) *aka-i kaban no syoozyo*
red-PRES/FUT bag GEN girl
'a girl who has, or, is carrying a red bag' (Takahashi 1975: 11)

When used with "clothing", this pattern does not necessarily indicate possession. Thus, (153) can refer to a girl who is carrying someone else's bag. I should also note that examples such as (151) to (153) cannot be used to describe "other possessee", whereby, for instance, the girl does not have the skirt on, but has left it in a wardrobe.

[4] *Kin*
The pattern of (140) is, on the whole, less acceptable with "kin" than with the types discussed above.

(154) *?ooki-i kazoku no otoko*
large-PRES/FUT family GEN man
Intended meaning: 'a man who has a large family'

In my judgement, (154) is somewhat less acceptable than the examples involving a possessee type higher on the cline. Next, consider (155):

(155) *rippa-na titioya no musuko*
great-ADN father GEN son

Example (155) cannot mean 'a son who has a great father'. That is, it is unacceptable as an instance of pattern (140). Its meaning is 'a great father's son'. That is, this example has to be considered an instance of (139), the usual genitive construction.

[5] *Pet animal*
Pattern (140) is possible, though not highly acceptable. Furthermore, it is possible only when the possessor is physically close to, or, contiguous with, the pet animal.

(156) *?siro-i inu no otoko*
white-PRES/FUT dog GEN man
'a man who has a white dog with him'

This example is unacceptable if the possessor is not accompanied by the dog.

[6] *Product*
Pattern (140) is unnatural.

(157) *?takusanno tyosyo no gakusya*
many writing GEN scholar
Intended meaning: 'a scholar who has many publications'

(158) *?rippa-na sakuhin no syokunin*
great-ADN work GEN craftsman
Intended meaning: 'a craftsman who has produced splendid works'

These examples are not natural in the intended meanings. But they may be marginally acceptable when they refer to someone who is carrying his/her works/products.

[7] *Other possessee*
When used with this type, the resultant examples appear to be generally considered as instances of (139), the usual genitive construction. Consider example (159):

(159) *ooki-i* *koozyoo* *no* *otoko*
large-PRES/FUT factory GEN man

The usual interpretation of (159) is one based on (139): 'a man who belongs to a large factory' (for example, he might be an employee there), rather than one based on (140): 'a man who has a large factory'. However, with "other possessee", a reading based on (140) is also possible, presumably depending on the relationship between Noun-1 and Noun-2. Consider (160):

(160) *ooki-i* *ie* *no* *otoko*
large-PRES/FUT house GEN man

This example can have a reading based on (140): 'a man who owns a large house'. In addition, it can have a reading based on (139), the usual genitive construction: 'a man who belongs to, who lives in a large house'. In this reading, he does not necessarily own the house. He is merely associated with the house in one way or another.

To sum up, (i) the acceptability of the patterns of (139) and (140) and the meaning they express vary depending on the possessee type, and (ii) the pattern of (140) describes possession in the main with "body part" and "attribute", that is, with the typical inalienable possessee. In this case, the possessee precedes the possessor, in contrast with (139), the usual genitive construction.

It is interesting to note that, in contrast with certain examples given above, instances involving certain "body part" or "attribute" nouns are unacceptable. Thus, compare (143) with (161); (144) with (162); and (149) with (163). (The same applies to their English translations; see 4.6.1.)

(161) **me* *no* *syoonen*
eye GEN boy
'*an eyed boy'

(162) **kami* *no* *syoozyo*
hair GEN girl
'*a haired girl'

(163) **seisitu no otoko*
nature GEN man
'*a natured man'

The unacceptability of these examples will be dealt with in 4.7.

4.6. Possession suffixes

4.6.1. -ed *adjectives of English*

As can be seen from the English translations of some of the Japanese examples given in 4.5, the pattern *Possessee-ed Possessor* expresses possession. Numerous examples are given in Hirtle (1970) and several examples in Hudson (1975). Many of the relevant Japanese examples given above are in fact a Japanese translation of an example provided by Hirtle and/or by Hudson. I shall cite additional examples, and also a couple of unacceptable examples.

[1] *Body part*

(164) *a bearded man* (Hudson 1975: 70)

(165) *a moustached general*

(166) *a blue-eyed boy* (Hirtle 1970: 22)

(167) *a red-headed boy* (Hirtle 1970: 27)

(168) *a rough-faced youth* (Hirtle 1970: 28)

[2] *Attribute*

(169) *a talented girl*

(170) *a good-natured man* (Hirtle 1970: 21)

(171) *an absent-minded professor*

(172) *a hard-featured girl* (Hirtle 1970: 28)

[3] *Clothing*

(173) *a uniformed commissionaire* (Hudson 1975: 71)

(174) *a helmeted motorcyclist* (Hudson 1975: 71)

(175) *a white-coated attendant* (Hudson 1975: 71)

(176) *a white-hatted cabman* (adapted from Hirtle 1970: 30)

[4] *Kin*

(177) **a wifed man* (Hudson 1975: 71)

No acceptable example available.

[5] *Pet animal*
No example available.

[6] *Product*
No example available.

[7] *Other possessee*

(178) *moneyed men* (Hirtle 1970: 21)

(179) *a landed proprietor* (Hirtle 1970: 29)

(180) *an estated country gentleman* (Hirtle 1970: 29)

(181) *a many-carred family* (Hirtle 1970: 31)

(182) **a carred man* (Hudson 1975:71)

Hudson (1975: 71) notes as follows: "the "possession" which is always shown by *ed*-adjectives seems to be INALIENABLE possession". However, he also notes that this pattern is possible with clothing, and lists examples (173) to (175). That is, he admits that this pattern is not confined to inalienable possession.

On the basis of the examples available and also Hudson's statement, it seems that the use of *-ed* adjectives is confined to the three highest types on the Possession Cline, in other words, "body part", "attribute", and "clothing" – except for the four exceptions attested with "other possessee", that is, (178) to (181).

Hudson (1975: 69) states that "no general restriction can be stated to distinguish the nouns which allow [this pattern – TT] from those that do not". Obviously, his statement needs to be re-examined in the light of the Possession Cline.

I tentatively present the acceptability of this pattern as in Figure 3.

As seen in 4.5, the pattern *Noun-1 GENITIVE Noun-2* of Japanese is not allowed with a certain "body part" or "attribute" nouns. The same applies to the parallel examples of the English *-ed* suffix; see (161) to (163) and their English translations. Consider also:

(183) **an eyed boy* (Hirtle 1970: 29)

(184) **a headed boy* (Hirtle 1970: 27)

(185) **a featured girl* (Hirtle 1970: 28)

I shall return to this issue in 4.7.

4.6.2. 'Having' suffix of Warrungu and Djaru

Many Australian Aboriginal languages have an adjective/noun-stem-forming suffix 'having, with'. (See Tsunoda (1976) on Warrungu and also other papers for Topic A in Dixon (ed.) (1976).) In Warrungu (Tsunoda 1974: 142-152, 162-170; 1976) and in Djaru (Tsunoda 1981a: 196, 227-229) and apparently in many other Australian languages, the use of this suffix correlates with the Possession Cline. Thus, in Warrungu and in Djaru, this suffix has been attested to extend from "body part" to "kin". With other types, the use of this suffix is not possible unless the possessee is attached to, close to, or, in the vicinity of, the possessor. This situation is presented in Figure 3.

Examples of the Warrungu 'having' suffix (*-tyi* after a consonant, *-yi* after a vowel) follow, involving a body part ('grey hair'), an attribute ('illness'), clothing ('clothes'), and kin ('spouse'), respectively. (The case-marking system of Warrungu is nominative-accusative for pronouns, and absolutive-ergative for nouns and demonstratives.)

(186)

pirngka-yi-tu	*kamu-ø*	*pitya-lku*	*pama-ngku*
grey:hair-HAVING-ERG	water-ABS	drink-PURP	man-ERG

'The grey-haired men (that is, old men) want to drink water.'
(Adjectives generally take a case suffix – the ergative in (186) – showing case agreement with the noun they qualify.)

(187)

murran-tyi-ø	*nyula*	*wun-an*
illness-HAVING-ABS	3SG:NOM	lie-PAST/PRES

'She is lying ill.'

(188)

yarru-ø=kul	*kampi-yi-ø*	*ngaya*
this-ABS-only	clothes-HAVING-ABS	1SG:NOM

nyina-n
sit-PAST/PRES
'I am sitting with only these clothes on.'
(The 'having' suffix would be expected to be attached to *yarru* as well; see (195).)

(189) *nyula* *tyakuli-n* *tyulpun-tyi-ku*
3SG:NOM be sorry-PAST/PRES spouse-HAVING-DAT
'He felt sorry for the married person (literally: one with a spouse).'

Examples of the Djaru 'having' suffix (*-tyaru* after a consonant, *-yaru* after a vowel) follow.

(190) *nyila* *ngaringka-ø* *yamparra-yaru-ø*
that woman-ABS long:hair-HAVING-ABS
'That woman is long-haired.'

(191) *ngatyu-ø* *nga=rna* *ngaringka-yaru-ø*
1SG-ABS C-1SG:NOM woman-HAVING-ABS
'I am married.'

(In Djaru, sentences such as (191) can be used even when the posessor is not accompanied by the kinsman/kinswoman concerned. It seems that the same applies to instances such as (189) of Warrungu, although this is not definitely certain.)

In Warrungu and in Djaru, certain "body part" and "attribute" nouns attached with the 'having' suffix may have a special meaning; see 4.7.

4.7. Notes on "body part" and "attribute"

As noted in 4.5 regarding the *Noun-1 GENITIVE Noun-2* pattern of Japanese, and in 4.6.1 concerning *-ed* adjectives of English, these patterns are unacceptable when used with certain nouns of 'body part' or 'attribute' type. Thus, in English, *a blue-eyed boy*, *a red-headed boy*, and *a hard-featured girl* are acceptable. However, they become unacceptable once the qualifier is removed (Hirtle 1970: 22, Kageyama 1980b: 12-13), for example **an eyed boy*, **a headed boy*, and **a featured girl*. It might appear as if it is the presence/absence of a qualifier, for example *blue*, that is relevant to the acceptability/unacceptability of these examples. However, this view is not adequate, for there are expressions such as the following, which lack a qualifier and yet are acceptable (Hirtle 1970: 22): *a bearded man*, *a pimpled boy*, *a whiskered man*, and *a talented girl*. Note also that the expressions of the latter type can contain a qualifier, for example *a brown-whiskered man* (Hirtle 1970 :30, footnote 12) and *a grey-bearded man*. Needless to say, they are acceptable without a qualifier. These remarks all apply to the parallel Japanese examples (cf. Suzuki 1968: 29, Takahashi 1975: 11, Kageyama 1980b: 12-13.)

Regarding the acceptability/unacceptability of these expressions, explanations such as the following have been proposed.

Hirtle (1970: 27-30) points out that we do not say in English **an eyed boy*, since *eye* is inherent in *boy*, that is, a boy normally has eyes. Hudson (1975) puts forward a similar view.

Regarding Japanese, Takahashi (1975: 11) elaborates on Suzuki (1968: 29) and maintains that those examples without a qualifier are possible when the presence of, or a characteristic of, the possessee is a characteristic of the possessor.

Kageyama (1980b: 12-13), on Japanese and English, observes that information contained in expressions such as **a haired girl*, **an eyed boy*, and **a headed boy* convey is redundant, since people inherently possess hair, eyes and a head. With the addition of a qualifier, however, they become meaningful, since they describe states which deviate from the typical situation. Referring to Grice's (1975) conversational implicatures, Kageyama concludes that linguistic descriptions of typical states of affairs are disallowed due to their redundant nature.

These explanations account for the acceptability or unacceptability of the examples concerned. Furthermore, they account for the acceptability of examples such as the following, that is, those without a qualifier: *a bearded man*, *a whiskered man*, *a pimpled boy*, and *a talented girl*. Not all men have a beard or whiskers, for example.

To sum up, the acceptability/unacceptability of the examples in question depends on whether or not the possession of the possessee concerned is usual, normal or typical for a given possessor in a given situation, or, perhaps alternatively, whether or not the presence of that possessee can be considered a characteristic of his or hers.[20]

As is clear by now, the presence/absence of a qualifier is not really crucial. This is demonstrated by crosslinguistic evidence as well. Thus, compare the Warrungu example *pirngka-yi* 'grey hair-having' (cf. (186)) and the English *grey-haired*; and the Djaru *yamparra-yaru* 'long hair-having' (cf. (190)) and the English *long-haired*. They are all perfectly acceptable. While the Warrungu and the Djaru examples do not contain a qualifier, their English equivalents do.

In the following discussion, it will be useful to distinguish two kinds of inalienable possession. Thus:

(i) 'not everyone'-type possession:
 (i-a) possession of blue eyes, long hair, long legs, a good nature;
 (i-b) possession of a beard, whiskers, pimples, illness and;
(ii) 'everyone'-type possession: possession of just eyes, hair, legs, nature.

It is needless to say that the distinction between these two types may not always be clear-cut. Note also that addition of a qualifier turns an 'everyone'-type possessee into a 'not everyone'-type possessee.

The works cited above in effect maintain that expressions of 'everyone'-type possession are redundant and pointless, and consequently disallowed. Indeed, their explanations are adequate as far as the examples they look at are concerned. There is, however, a related issue that they do not deal with. Namely, contrary to what might be expected on the basis of their explanations, there are expressions which, superficially at least, describe 'everyone'-type possession and yet are not pointless. There are two circumstances under which such expressions are used:

Circumstance (i): 'Everyone'-type possession is expressed without a qualifier. Such expressions describe something special/unusual/marked about that 'everyone'-type item. That is, they have a special meaning.

Circumstance (ii): A normal kind of 'everyone'-type possession is expressed. There is nothing special/unusual/marked about the possessee, and yet, when an adequate context has been established, such an expression is not pointless, but is relevant and meaningful. The acceptability of such expressions is highly context-dependent.

I have made these observations in Tsunoda (1974, 1976) on Warrungu, and subsequently in Tsunoda (1981a) on Djaru. I shall first cite examples from these languages, before turning to Japanese and English.

[1] *Warrungu*

The 'having' suffix (cf. 4.6.2 above) can be used with 'not everyone'-type possession, for example *pirngka-yi* 'grey hair-having', for example (186), *murran-tyi* 'illness-having', for example (187), *tyityi-yi* 'sore-having', and *kuympirra-yi* 'cicatrice-having' (*kuympirra* is a cicatrice on the chest which only initiated men have) (Tsunoda 1974: 142-152, 162-170, Tsunoda 1976). When this suffix is used with 'everyone'-type possessees, it is generally used in the circumstance (i). There is one example of circumstance (ii).

Circumstance (i):

This suffix is used to indicate that something is unusual, abnormal, or wrong with the 'everyone'-type possessee concerned. Examples include *tyina-yi* (lit. 'foot-having') 'sore-footed', *mampu-yi* ('back-having') 'sore-backed', *pampa-yi* ('stomach-having') 'having diarrhoea, wanting to go to the toilet', *pulu-yi* ('belly-having') 'satisfied with food, having diarrhoea, wanting to go to the toilet, pregnant', *tyingku-yi* ('body hair-having') 'hairy (having more hair than usual)', *walngka-yi* ('breath-having') 'short-winded, tired', and *kuna-yi* 'faeces-having') 'having diarrhoea'. Sentential examples follow:

(192) *ngaya* *nyina-n* *pulu-yi-ø*
1SG:NOM sit-PAST/PRES belly-HAVING-ABS
'I am sitting, satisfied with food.'

(193) *ngaya* *walngka-yi-ø*
1SG:NOM breath-HAVING-ABS
'I am tired.'

(194) *nguna-ø* *pama-ø* *kuna-yi-ø*
that-ABS man-ABS faeces-HAVING-ABS
'That man has diarrhoea.'

Circumstance (ii):
The only example available is:

(195) [If I had two penises, I would look for two women, but:]
ngaya *nyunkul-tyi-ø* *tyumpi-yi-ø*
1SG:NOM one-HAVING-ABS penis-HAVING-ABS
'I have (only) one penis.'

This sentence expresses a usual kind of 'everyone'-type possession. It describes the normal number of penises a man possesses; it does not say anything unusual about it. And yet, in this context, it is meaningful to specify its number.

[2] *Djaru*
What has been stated about the Warrungu 'having' suffix applies to the Djaru equivalent, except that I have not found any really good examples of circumstance (ii) (Tsunoda 1981a: 196, 227-229).

Examples of the circumstance (i), that is, 'everyone'-type with a special meaning, include *punpulu-yaru* ('body hair-having') 'hairy', *munta-yaru* ('belly-having') 'pregnant', *mankirr-tyaru* ('ear-having') 'wise, sensible'. (The ear is regarded as the seat of intelligence (Tsunoda 1981a: 229).)

[3] *Japanese*
We saw in 4.4.4 that, when the verb *ar-* 'be, exist' is used with certain 'body part' and 'attribute' nouns, the sentences have a special meaning: 'more ... than the average', for example (130) ('foot'), (131) ('head'), (132) ('eye'), (133) ('height'), and (134) ('weight'). As is obvious now, these nouns are all of the 'everyone'-type.

We saw in 4.5 that expressions such as the following are unacceptable:

(196) = (161) **me no syoonen*
eye GEN boy
'*an eyed boy'

(197) = (162) **atama no otoko*
head GEN man
'*a headed man'

They all concern 'everyone'-type possessees. Their unacceptability is predicted by the explanations cited above.

However, there are examples such as the following:

(198) *hana no boozu*
nose GEN monk
'the big-nosed monk' (Hujio Minami, pers. comm.)
(This expression refers to a character in a certain novel.)

(199) *medama no Mattyan*
eye GEN Mattyan
'Big-eyed Mattyan'
(*Mattyan* is an actor's nickname.)

Such instances refer to an 'everyone'-type body part. They do not contain any qualifier. And yet, unlike instances such as (196) and (197), they are acceptable. Note that they describe something unusual about the body part concerned. That is, they are instances of the circumstance (i).[21] These examples all concern body parts. I have not found any examples involving attributes.

Examples of the circumstance (ii), that is, those parallel to 'I have (only) one penis', occur in Japanese as well (Taroo Takahashi, pers. comm.). They describe a normal kind of 'everyone'-type possession, and yet they are meaningful in an appropriate context. An example I composed is given in (200):

(200) [Talking about a person who is totally inebriated]

Taroo wa asi ga ar-u.
Taroo TOP foot NOM be-PRES/FUT
Jibun de aruk-ase-ro!
oneself INST walk-CAUS-IMPERA

'Taroo has feet. Make him walk by himself!'

[4] *English*
As we saw in 4.4.4, when the verb *have* is used with certain "body part" and "attribute" nouns (without a qualifier), the sentences have a special meaning; see (135) to (137).

I have not found parallel examples involving an *-ed* adjective. The following examples, which were provided by Jodi Dalton, employ the preposition *with*:

(201) *the girl with the eyes*
'the girl who has the most beautiful eyes in the group'

(202) *the man with the belly*
'the man who has the biggest belly in the group'

(203) *the man with the foot*
'the fastest man in the group'

Dalton notes that the article has to be the definite article, rather than the indefinite article, and that these examples all mean 'the person who has the X-*est* in the group'. Chappell and McGregor (pers. comm.) suggest that a more likely interpretation of (203) will be: 'the man with something unusual or wrong about his foot (perhaps its large size)'. (This difference in the interpretations may reflect regional differences; Dalton is a native speaker of American English, while Chappell and McGregor are native speakers of Australian English.)[22]

Examples analogous to 'I have (only) one penis' are possible in English as well. Thus, this sentence itself and examples analogous to the English translation of (200) are possible as long as the context is adequate.

I shall now recapitulate the main points of this subsection. Expressions of possession of body parts and attributes (typical inalienable possessees) which people normally have – that is, 'everyone'-type possessees – are usually unacceptable. This is due to the redundant nature of the information they convey. There are, however, two circumstances under which what superficially appear to be expressions of 'everyone'-type possession are acceptable and not pointless. In one circumstance, such expressions indicate that there is something unusual about the possessee concerned. Because of this unusualness, they are no longer pointless. In a sense, they have gone through a shift of meaning (Hirtle 1970: 30), and have joined the 'not everyone'-type.[23] In the other circumstance, such expressions do not describe anything unusual about the possessee concerned. And yet, they are meaningful and not pointless when an adequate context has been built up.

5. Summary and conclusion

This paper has proposed a cline that represents the closeness/attachedness – physical and/or psychological – between the possessor and the possessee, and has illustrated its morphosyntactic manifestations.

The acceptability of Possessor Respect in Japanese, the Japanese verb *su-* 'do' expressing possession, the *-ed* adjectives of English and the 'having' suffixes of Warrungu and Djaru correlates with the cline. They extend from the high end of the cline towards the low end; the cut-off point varies. The same seems to apply to possessor ascension and analogous phenomena. In contrast to this, the Japanese verb *syoyuusu-* 'possess' is confined to the low end of the cline.

The verbs *mot-* 'have', *i-* 'be', *ar-* 'be' and the pattern *Noun-1 GENITIVE NOUN-1* of Japanese do not really correlate with the cline, but the use of each of them is conditioned by the possessee types on the cline.

Expressions of possession of a body part or attribute – typical inalienable possessions which people normally have – are usually unacceptable, due to the redundancy of information. Nonetheless, these expressions may become acceptable if they express that there is something unusual about the inalienable possession or if an adequate context has been built up.

It seems likely the Possession Cline represents one of the ways humans perceive the real world, that is, it represents the way they perceive the distance between the possessor and the possessee. This paper has suggested how such a perception may be reflected in human languages. (See also Fox (1981: 340) and Haiman (1985: 130) for further relevant discussons.)

Postscript

In [4] *English* of 4.7, I stated that I had found no example in which the suffix *-ed* was added to an 'everyone-type' possessee and had a special meaning. I have since discovered that the suffix *-y* can be used in this way. I shall list some of the examples I have collected.

(i) *Body parts:*
brainy, bosomy, chesty, hairy, hearty, throaty, handy, leggy, meaty, bloody, sweaty
(ii) *Attributes:*
weighty (Bernard Comrie, pers. comm.), *healthy*

Note that these expressions all have a special meaning. Thus, *brainy* means 'intelligent, clever'; compare it with the Japanese example (131) ('Taroo is brainy'). *Weighty* means 'heavy'; compare it with the Japanese example (134) ('Konishiki is heavy'). *Healthy* means 'having, conducive to, good health'.

A cursory inspection of the examples available suggests that the patterns *Noun-y* and *Noun-ed* exhibit a complementary distribution in terms of their derivational possibilities and meanings. Thus, consider the table.

	'everyone'-type possessee	'not everyone'-type possessee
noun-*ed*	no example (?)	*bearded, moustached,* etc. *blue-eyed, read-headed, grey-haired,* etc. no special meaning
noun-*y*	*brainy, bosomy, chesty, hairy, leggy,* etc. special meaning	no example (?)

That is, *-ed* can be attached to 'not-everyone' type possessees, and the resultant adjectives do not have a special meaning. But *-ed* does not seem to be suffixed to 'everyone'-type possessees. At least, those putative adjectives such as *brained, haired, legged,* etc. which would be considered to consist of a noun (not a verb) and *-ed,* are not listed in the dictionaries consulted. In contrast, *-y* can be added to 'everyone'-type possessees and the resultant adjectives always (?) have a special meaning. But *-y* does not seem to be affixed to 'not everyone'-type possesseees. At least, those putative adjectives such as *beardy, moustachy,* etc. and *read-heady, grey-*

hairy, etc. which would be cosidered to consist of a noun and *-y* are not listed in the dictionaries consulted.

Notes

1. I am grateful to Setsuko Andoo, Greville G. Corbett, Jun-ya Ishihara, Hujio Minami, Tsunao Ogino, Keiichiroo Okutsu, Masayoshi Shibatani, Tooru Shionoya, Takesi Sibata, Hiroshi Sugita, Seiju Sugito, Taroo Takahashi, Mie Tsunoda, Toshihiro Yamada, and an anonymous referee for their advice and/or their comments on earlier versions of this paper. In particular, I wish to thank Taroo Kageyama, and the editors of the volume, Hilary Chappell and Bill McGregor, for providing detailed and valuable comments on an early version of the paper and also for providing reference to some relevant works. I benefited from a discussion with Takashi Murakami (Faculty of Education, University of Nagoya) on the validity of the statistical data used in this paper; and from a discussion with Keiichiroo Tsuji (Faculty of Letters, University of Nagoya) regarding psychological aspects of the Possession Cline. In addition, I wish to thank the people who responded to my questionnaire (cf. 4.1), and also the audience of the meeting mentioned in 3.2. People and institutions that provided assistance for my work on Warrungu and that on Djaru are gratefully acknowledged in Tsunoda (1974) and in Tsunoda (1981a), respectively.
2. Abbreviations employed include the following: ABL = ablative; ABS = absolutive; ACC = accusative; ADV = adverbal as against adnominal; C = carrier; CAUS = causative; CON = continuative; DAT = dative; DO = direct object; ERG = ergative; GEN = genitive; HON = honorific; IMPERA = imperative; INF = infinitive; INST = instrumental; IO = indirect object; LOC = locative; NOM = nominative; OBL = oblique; PAST/PRES = past/present; PL = plural; POL = polite; PRES, present; PRES/FUT = present/future; PTCPL = participle; PURP = purposive; Q = question; RESP = respect; SG = singular; Si = intransitive subject; St = transitive subject; TOP = topic; 1 = first person; 2 = second person; 3 = third person.
3. This account of the behaviour of case postpositions and thematic/modal postpositions is transformational. Naturally, a non-transformational account, too, is possible (and is supported by many specialists of the field). In such a view, for example, *wa* is not attached to any case postpositions. Nor does it obliterate any. It is there from the beginning.
4. In the transliteration of Japanese, the orthography I employ for citing examples and that I use for names of people, places and publishers are slightly different from each other.
5. As it stands, example (3) sounds unnatural. It becomes natural if *wa* TOP is added (with the subsequent deletion of *ga* NOM). Similarly for some other examples. However, this sentence is natural in the following reading: 'It is (only) the elephant's nose that is long'. I shall return to this issue shortly.
6. This view, which considers the topic phrase in (5) as the topicalised version of the genitive phrase in (3), is due to Mikami (1960: 12). However, there is an alternative view, that is, the topic phrase in (5) is the topicalised version of a nominative phrase. That is, (5) corresponds to a sentence with the NOM+NOM case frame rather than one with the GEN+NOM frame – see (78) to (83). A third view is also possible, that

is, *wa* in (5) has not obliterated the case postposition *no* GEN or *ga* NOM. No case postposition was there; *wa* is there from the beginning; see Note 3. As is clear by now, analysis of sentences such as (5) is extremely controversial.

7. A more literal and more accurate translation of (5) will be: 'As for the elephant, its nose is long.'
8. In this paper, I use the concept of grammatical relations in a loose sense. An explicit approach to this issue is discussed in Comrie (1981: 59-64) in general terms, and in Shibatani (1978) regarding Japanese.
9. I am grateful to Chappell and McGregor (pers.comm.) for drawing Haiman (1985) and Nichols (1988) to my attention.
10. I should note here that I was aware of typological studies of possession such as Clark (1978) and Ultan (1978).
11. Similarly, Hyman (1977: 114) states as follows: "body parts must be higher [than kin – TT] in the ... hierarchy, since doing something to somebody's body will affect that person more directly (i. e., more physically, if you will) than doing something to that person's kin". Likewise, Durie (1987: 388) in effect considers body parts less alienable than kinsmen. His view is cited in section 4.3.
12. Although this paper does not deal with this, the whole-part relationship of inanimate entities (for example, a table and its legs, and a house and its roof) is very similar to the relationship between a possessor and his/her body parts, semantically at least. However, Wierzbicka (1979: 319) demonstrates that body part syntax and whole-part syntax exhibit a difference between them in French, Italian, etc.
13. For relevant discussions, see Wierzbicka (1979) and Fox (1981). The latter's view is cited in 4.3.
14. Since this survey is not a sociolinguistic one, I did not make deliberate efforts to exhaust all the sociolinguistic variables. Nonetheless, the respondents to the questionnaire cover a fairly wide range, including both sexes, from 20 to 70 years of age. Most of them had a secondary education, and many of them a tertiary education. In terms of the social background, they include university students, academics, university administrative officers, housewives, company employees, retired people, and so on.

    I cite, in this paper, all of the relevant examples – 24 in all – from the questionnaire. The remaining six examples have turned out not to be directly relevant to the issues examined in the paper.
15. The hierarchy of (67) is crosslinguistically operative in another area of grammar, that is, incorporation or compounding of a noun and a verb (Comrie 1978: 337). This applies to Japanese as well (Tsunoda 1984: 79).
16. Durie (1987: 389) states that, in Malagasy of Madagascar and Cebuano of the Philippines, possessor ascension is allowed with the subject only. However, he does not make clear whether this applies to the Si only, to the St only, or to both of them.
17. Hyman (1977: 104, 109) notes that other relevant factors include the nature of the verb and the nature of the possessee.
18. What is termed the ergative case of Djaru can be divided into the ergative and the instrumental. If this analysis is adopted, (97) is not an instance of apposition; *mawun-tu* is ergative, while *marla-ngku* is instrumental. In this case, the hierarchy of grammatical relations involved is that of (69) rather than that of (71).

In Tsunoda (1981a: 57, 105) on Djaru, I distinguish between the ergative and the instrumental (although they have the same allomorphy). The reasons for this are as follows:

(i) The ergative marks a transitive subject (semantically, typically an agent), while the instrumental refers to instruments (semantically, always inanimates, including body parts).

(ii) The instrumental can never be cross-referenced by an enclitic pronoun, while the ergative can – even when its referent is inanimate.

(iii) The ergative can only occur in transitive sentences, while the instrumental can occur in both transitive and intransitive sentences.

19. As is obvious from the English translations, this Japanese pattern parallels the English pattern involving an *-ed* adjective. See 4.6.1.

20. The preceding discussions have only looked at "body part" and "attribute" but they are relevant to "clothing" as well. Thus, the expression *a coated cabman* would be unnatural in a society where cabmen normally wore a coat, but it would be natural in one where they did not (Hirtle 1970:30, Note 12). Furthermore, it may even be possible that there are languages in which these observations are relevant to other possessee types as well, although I am not aware of any. For example, it is conceivable that there is a language in which **a wifed man* is (grammatically, not socially) unacceptable (since adult men usually have a wife), while *a wifed boy* is (grammatically) acceptable (since usually boys do not have a wife).

21. Taroo Kageyama (pers. comm.) points out a difference between instances such as (198) and those such as (196) and (197). Namely, (198) is something like a nickname. (Note that (199) is really a nickname.) It cannot have a nonspecific reading, that is, 'not a particular monk with a big-nose, but any monk with a big nose'. In the nonspecific reading, (198) is unacceptable, like (196) and (197).

22. It seems to me that English expressions such as (201) to (203), and (135) to (137), and the Japanese expressions such as (130) to (134) (and perhaps (198) and (199) as well), that is, those instances which have a special meaning, tend (though not necessarily consistently) to imply a positive evaluation, while the opposite is generally the case with the Warrungu and the Djaru equivalents, for example, 'diarrhoea', 'sore feet', 'sore back'. Wierzbicka (1979: 355-58) suggests that languages tend to have fixed expressions for a negative/bad perspective rather than a positive/good perspective. This is on the whole true with the Warrungu and the Djaru examples, but this does not apply to the Japanese examples or the English examples.

23. What lies behind this semantic shift is probably a Gricean-like principle such as the following: Do not say anything obvious. If you are going to talk about an obvious thing, add a special meaning to it so that it will be worth talking about.

References

Asher, R. E. et al. (eds.)
1994 *Encyclopedia of Language and Linguistics.* 10 vols. Oxford: Pergamon Press.
Blake, Barry J.
1977 *Case marking in Australian languages.* Canberra: Australian Institute of Aboriginal Studies.
1984 "Problems for possessor ascension: some Australian examples", *Linguistics* 22: 437-453.
Chappell, Hilary
1995 "Inalienability and the personal domain in Mandarin Chinese discourse", [this volume.]
Chappell, Hilary – William McGregor
1989 "Alienability, inalienability and nominal classification", *Proceedings of the Fifteenth Annual Meeting of the Berkeley Linguistics Society*, 24-36.
Clark, Eve V.
1978 "Locationals: existential, locative, and possessive constructions", in: Joseph H. Greenberg et al. (eds.), 85-126.
Cole, Peter (ed.)
1978 *Pragmatics.* (Syntax and Semantics 9) New York: Academic Press.
Cole, Peter – Jerry Morgan (eds.)
1975 *Speech acts.* (Syntax and Semantics 3) New York: Academic Press.
Comrie, Bernard
1978 "Ergativity", in: Winfred P. Lehmann (ed.), 329-394.
1981 *Language typology and linguistic universals.* Oxford: Blackwell.
Comrie, Bernard (ed.)
1987 *The world's major languages.* London: Croom Helm.
Dixon, R.M.W.
1977 *A grammar of Yidiny.* Cambridge: Cambridge University Press.
1980 *The languages of Australia.* Cambridge: Cambridge University Press.
Dixon, R.M.W. (ed.)
1976 *Grammatical categories in Australian languages.* Canberra: Australian Institute of Aboriginal Studies, and New Jersey: Humanities Press.
Durie, Mark
1987 "Grammatical relations in Acehnese", *Studies in Language* 11: 365-399.
Fox, Barbara
1981 "Body part syntax: towards a universal characterization" *Studies in Language* 5: 323-342.
Greenberg, Joseph H. et al. (eds.)
1978 *Syntax.* (Universals of Human Language 4) Stanford: Stanford University Press.
Grice, H.P.
1975 "Logic and conversation", in: Peter Cole – Jerry Morgan (eds.), 41-58.
Haiman, John
1985 *Natural syntax.* Cambridge: Cambridge University Press.
Harada, S.I.
1976 "Honorifics", in: Masayoshi Shibatani (ed.), 499-561.

Hinds, John
1982 *Ellipsis in Japanese*. (Current Inquiry into Language, Linguistics and Human Communication 43) Edmonton: Linguistic Research.
1986 *Japanese*. (Croom Helm Descriptive Series) London: Croom Helm.
Hirtle, W.H.
1970 "-Ed adjectives like 'verandahed' and 'blue-eyed' ", *Journal of Linguistics* 6: 19-36.
Hopper, Paul J. – Sandra A. Thompson
1980 "Transitivity in grammar and discourse", *Language* 56: 251-299.
Hudson, R.A.
1975 "Problems in the analysis of ed-adjectives", *Journal of Linguistics* 11: 69-72.
Hyman, Larry H.
1977 "The syntax of body parts", in: *Haya grammatical structure*, (Southern California Occasional Papers in Linguistics 6), 99-117. Los Angeles: University of Southern California.
1995 "The syntax of body parts in Haya", [this volume.]
Kageyama, Taroo
1980a *Goi no koozoo*. [Structure of lexicon.] Tookyoo: Shoohakusha.
1980b "Goiimiron no mondai – goibunkai to tenkei". [Issues in lexical semantics – lexical decomposition and prototype.] *Gengo Bunka Kenkyuu* 6:1-20. Oosaka: University of Oosaka.
Kuno, Susumu
1973 *The structure of the Japanese language*. Cambridge, Mass.: MIT Press.
Lehmann, Windfred P.
1978 *Syntactic typology*. Austin: University of Texas Press.
Martin, Samuel E.
1975 *A reference grammar of Japanese*. (Yale Linguistic Series.) New Haven: Yale University Press.
McCawley, James D.
1978 "Conversational implicature and the lexicon", in: Peter Cole (ed.), 245-259.
Mikami, Akira
1960 *Zoo wa hana ga nagai* [The elephant has a long nose]. Tookyoo: Kuroshio.
Minami, Hujio
1987 *Keigo* [Honorifics]. Tookyoo: Iwanami.
Munro, Pamela – Lynn Gordon
1982 "Syntactic relations in Western Muskogean: a typological perspective", *Language* 58: 81-115.
Nichols, Johanna
1988 "On alienable and inalienable possession", in: William Shipley (ed.), 557-609.
Ooishi, Hatsutaroo
1983 *Gendai keigo kenkyuu* [Studies in modern honorifics]. Tookyoo: Chikuma.
Shibatani, Masayoshi
1978 *Nihongo no bunseki* [Analysis of Japanese]. Tookyoo: Taishuukan.
1987 "Japanese", in: Bernard Comrie (ed.), 855-880.
Shibatani, Masayoshi (ed.)
1976 *Japanese generative grammar*. (Syntax and Semantics 5). New York: Academic Press.

Shibatani, Masayoshi – Tasaku Tsunoda
1982 "Gengo ruikeiron no saikin no dookoo" [Recent trends in language typology], *Gengo*, 11 (4): 100-108. Tookyoo: Taishuukan.
Shipley, William (ed.)
1988 *In honor of Mary Haas: From the Haas Festival Conference on Native American Linguistics.* Berlin: Mouton de Gruyter.
Silverstein, Michael
1976 "Hierarchy of features and ergativity", in: R.M.W. Dixon (ed.), 112-171.
Suzuki, Yasuyuki
1968 "Meishi to meishi no kumiawase" [Combinations of a noun and a noun], *Bulletin of Tokyo Seitoku College* 2: 25-40. Tookyoo: Tookyoo Seitoku Tanki Daigaku.
Takahashi, Taroo
1975 "Bunchuu ni arawareru shozokukankei no shushusoo" [Aspects of possession relationship expressed in sentences], *Kokugogaku* 103: 1-17.
Takahashi, Taroo – Shigeko Yaku
1984 "'... ga aru' no yoohoo" [The use of ... '*ga aru*'], *Kenkyuu Hookokushuu* [Research Reports] 5: 1-42. Tookyo: National Language Research Institute.
Tsunoda, Tasaku
1974 *A grammar of the Warungu language, North Queensland.* M.A. thesis. Melbourne: Monash University.
1976 "The derivational affix 'having' in Warungu", in: R.M.W. Dixon (ed.), 214-225.
1981a *The Djaru language of Kimberley, Western Australia.* Canberra: Pacific Linguistics.
1981b "Split case-marking patterns in verb-types and tense/aspect/mood", *Linguistics* 19: 389-438.
1983 "Gengo ruikeiron" [Language typology], *Gengo Seikatsu* 383: 64-79. Tookyoo: Chikuma.
1984 "Nookaku to taikaku"[Ergative and accusative], *Gengo* 13 (3): 71-80. Tookyoo: Taishuukan.
1985 "Remarks on transitivity", *Journal of Linguistics* 21: 385-396.
1994 "Transitivity", in: R.E. Asher et al. (eds.), vol 9, 4670-4677.
Ultan, Russell
1978 "Toward a typology of substantival possession", in: Joseph H. Greenberg et al. (eds.), 11-49.
Wierzbicka, Anna
1979 "Ethno-syntax and the philosophy of grammar", *Studies in Language* 3: 313-383.

Part V

Languages of North America

Multiple reflections of inalienability in Mohawk*

Marianne Mithun

1. Introduction

Discussions of the concept of inalienability arise most often in two contexts: descriptions of nominal possessive constructions, and discussions of grammatical patterns sometimes termed "possessor raising" or "promotion". Although the semantic distinctions reflected by these constructions may be closely related, they are not necessarily equivalent.

Both kinds of constructions appear in Mohawk, an Iroquoian language of northeastern North America. Mohawk is a polysynthetic language with elaborate prefixation, suffixation, and noun incorporation. All verbs contain pronominal prefixes referring to their core arguments. In addition to person, gender, and number, the prefixes distinguish participants according to their roles as agents or patients. Compare the first person pronominal prefixes in the verbs in (1)-(3):

(1) k*ateweyʌ̨́hstha?*
k-*ate-weyʌ̨́-hst-ha?*
1:SG:AGT-know:how-CAUSATIVE-HABITUAL
'*I* am studying, learning.'

(2) wak*atshenų́:ni*
wak-*at-shenų́:ni*
1:SG:PAT-SEMI:REFLEXIVE-happy:STATIVE
'*I* am happy.'

(3) *te*khey*atų?nektʌ̨́:nis*
te-khey-*atų?nektʌ̨ni-s*
DUALIC-*1:SG:AGT/F:SG:PAT*-frighten-HABITUAL
'*I* scare *her*.'

The Mohawk case system is based on the semantic roles of participants in events and states, but it is fully grammaticised. In any particular context, only one case or case combination is grammatical.

Nouns contain a formally similar set of pronominal prefixes. Some nouns appear with agent prefixes and some with patient prefixes, but the motivation behind the case distribution is different from that in verbs and less transparent.

(4) ra*ksà:ʔa*
ra-*ksaʔ-a*
M:SG:AGT-child-DIMINUTIVE
'boy'

(5) e*ksà:ʔa*
e-*ksaʔ-a*
F:SG:AGT-child-DIMINUTIVE
'girl'

(6) ká*khwaʔ*
ka-khw-aʔ
N:SG:AGT-food-NS
'food, meal'

(7) ro*kstʌ́:ha*
ro-*kstʌ-ha*
M:SG:PAT-old-DIMINUTIVE
'old man'

(8) ako*kstʌ́:ha*
ako-*kstʌ-ha*
F:SG:PAT-old-DIMINUTIVE
'old woman'

(9) o*ʔnyų́kseri*
o-*ʔnyųkseri*
N:SG:PAT-onion
'onion'

2. Nominal possession

Possessive constructions appear less often in Mohawk than in many languages because of the existence of alternative structures that accomplish some of the same functions. Kinship, for example, is generally expressed by means of stative verbs that relate participants. In most kin terms, both kin are represented within the pronominal prefix, whether they are first, second, or third person masculine or feminine. Where there is a culturally significant age difference between the two kin in question, a transitive prefix is used, the older kin encoded as agent, the younger as patient.

(10) rak*hsótha*
rak-*hsot-ha*
M:SG:AGT/1:SG:PAT-grandparent-DIMINUTIVE
'*my* grandfather'

(11) rake*ʔníha*
rake-*ʔní-ha*
M:SG:AGT/1:SG:PAT-father-DIMINUTIVE
'*my* father'

(12) rak*htsì:ʔa*
rak-*htsi-ʔa*
M:SG:AGT/1:SG:PAT-older:sibling-DIMINUTIVE
'*my* older brother'

(13) ri*ʔkʌ́:ʔa*
ri-*ʔkʌ-ʔa*
1:SG:AGT/M:SG:PAT-younger:sibling-DIMINUTIVE
'*my* younger brother'

(14) ri*yʌ́:ʔa*
ri-*yʌ-ʔa*
1:SG:AGT/M:SG:PAT-son-DIMINUTIVE
'*my* son'

(15) ri*yaterè:ʔa*
ri-*yatere-ʔa*
1:SG:AGT/M:SG:PAT-grandchild-DIMINUTIVE
'*my* grandson'

When the two kin are of culturally equivalent ages, the relationship is reciprocal, and the kin are jointly represented by a dual or plural pronominal prefix followed by a reflexive/reciprocal marker.

(16) akw*atateʔkʌ̨ʔokų́:ʔa*
akw-*atate-ʔkʌ̨-ʔokų-ʔa*
1:EXCL:PL:AGT-RECIPROCAL-sibling-PLURAL-DIMINUTIVE
'*my* brothers and sisters' (Literally: 'We are siblings to each other.')

(17) ų̨ty*araʔsè:ʔa*
ų̨ky-*ar-aʔse-ʔa*
1:EXCL:DU:PAT-SEMI:REFLEXIVE-cousin-DIMINUTIVE
'*my* cousin' (Literally: 'We two are cousins to each other.')

(18) ų̨ty*atyóha*
ų̨ky-*at-yo-ha*
1:EXCL:DU:PAT-SEMI.REFLEXIVE-brother:in:law-DIMINUTIVE
'*my* brother-in-law' (Literally: 'We two are brothers-in-law to each other.' (male ego))

A second reason for the comparative rarity of nominal possessive constructions in Mohawk is the fact that many entities are identified by means of morphological verbs rather than nouns.

(19) *tsi thonų̨hwétshaʔ*
tsi t-hon-ų̨hwets-haʔ
at CISLOCATIVE-M:PL:AGT-spend:the:night-HABITUAL
'*their* bedroom' (Literally: 'where they spend the night')

Part/whole relationships are also often represented by morphological verbs rather than by pairs of nouns.

(20) *tsi yotekhwahraʔtsherá:teʔ*
tsi yo-te-khw-hra-ʔtsher-at-eʔ
at N:SG:PAT-SEMI:REFLEXIVE-food-set-NOM-be:edged:STATIVE
'(the) table's edge' (Literally: 'where (the) table edges')

(21) *watenhohaniyų̨táhkhwaʔ*
w-ate-nhoh-niyų̨t-ahkw-haʔ
N:SG:AGT-door-hang-INSTRUMENTAL-HABITUAL
'door hinge' (Literally: 'One door-hangs with it.')

Possessive constructions consisting of pairs of nominals do of course exist. As in many polysynthetic languages, the possessor is specified by a pronominal prefix on the possessed noun, a pattern sometimes referred to as head-marking (Nichols 1988). The nominal identifying the possessor is unmarked for the relationship.

(22) *ohkwá:ri* ra*néhų*
o-hkwari ra-*nehų*
3:N:SG:PAT-bear *3:M:SG:AGT*-hide
'bear's skin', 'bearskin' (Literally: 'bear his-skin')

(23) *tsyò:kwaris* raotí*khwaʔ*
tsyoʔkwaris raoti-*khw-aʔ*
blackbird *M:PL:PAT*-food-NS
'blackbirds' food' (Literally: 'blackbird their-food')

The prefix indicating the possessor appears sometimes in the agent case as in (22), sometimes in the patient case as in (23). The choice of case for the possessor reflects a distinction much like inalienability.

2.1. Agent possessors

Agent pronouns occur with most nouns referring to body parts, including: *-yaʔt-* 'body', *-nųʔtsi* 'head', *-nųʔwar-* 'hairline', *-kʌ̨ʔkwar/-kʌ̨ʔtstar-* 'forehead', *-ahrahųhs-* 'temple', *-kųhs-* 'face', *-kahr-* 'eye', *-ʔnyųhs-* 'nose', *-tsiʔts-/-riʔts-* 'lip', *-hsųhkar-* 'upper lip (area below nose)', *-ʌ̨ʔnahs-* 'tongue', *-nawir-/-noʔts-* 'tooth', *-yʌ̨ht-* 'gum', *-nhoʔkw-/-nhoskw-* 'cheek', *-nyaʔkw-* 'jaw', *-ʔrhyotsh-* 'chin', *-ahųht-* 'ear', *-hnyaht-* 'neck (whole)', *-nyar-/-nyahkarokų* 'nape', *-hnyaʔs-/-nyaʔt-* 'throat, inside', *-nyaʔkwa-* 'throat, outside', 'Adam's apple', *-nyahkar-* 'throat, middle', *-hnʌ̨hs-* 'shoulder', *-nerahr-* 'shoulder blade', *-nʌ̨tsh-* 'arm', *-hnųh-* 'armpit', *-hyohs-* 'elbow', *-nʌ̨tshawiʔt-* 'wrist', *-ahtsy(ʌ̨ʔn)-* 'hand', *-hsy-* 'palm', *-tsihkw-* 'fist', *-nʌ̨hst-* 'knuckle', *-hsnųhs-* 'finger', *-tsiweyųhkar-/-hyųhkar-* 'thumb', *-ʌ̨tskwʔn-* 'chest', *-kųhkwar-* 'nipple', *-nʌ̨ʔaht-* 'ribs', *-hsihųt-/-hsiahųt-* 'midriff', *-hsyaʔkar-* 'waist', *-ryahs-* 'torso', *-nekwʌ̨ʔt-* 'belly', *-neriʔtst-* 'navel', *-ʔkats-* 'body from waist down', *-hsų-* 'back', *-rųhkweʔn-* 'lower back', *-hoʔkw-/-hnats-* 'buttocks', *-neskwahr-* 'hip', *-hsin-* 'leg', *-hnitsh-* 'thigh', *-kwitsh-* 'knee', *-rehets-* 'back of knee', *-nyʌ̨t-* 'shin', *-senųn-* 'calf', *-hsineko/t-/-hsinekwʌ̨ʔt-* 'ankle', *-tsirahoʔkw-* 'ankle, back', *-hsiʔt-* 'foot', *-aʔkotar-*

'heel', *-rat-* 'heel (fleshy part)', *-yaʔkwatáhkwʌ̨* 'arch', *-askweʔy-* 'ball of foot', *-rakwaht-/-nųny-* 'sole', and *-hyakwir-* 'toe'. Interestingly, all of these nouns consistently appear with a locative suffix when possessed, even in those contexts where no locative would be used in English.

(24) k*kahrà:ke*
k-*kahr-aʔ-ke*
1:SG:AGT-eye-NS-LOCATIVE
'(on) *my* eye'

(25) k*eneriʔstà:ke*
ke-*neriʔst-aʔ-ke*
1:SG:AGT-navel-NS-LOCATIVE
'(on) *my* navel'

(26) k*aʔkotarà:ke*
k-*aʔkotar-aʔ-ke*
1:SG:AGT-heel-NS-LOCATIVE
'(on) my heel'

Nouns referring to animal body parts such as *-ʔnyųkwarist-* 'snout', *-tsinareʔt-* 'hoof', *-itahs-* 'tail' and *-aʔkohs-* 'tail feathers' similarly appear with agent pronouns and locatives.

2.2. Patient possessors

Patient prefixes appear with all other kinds of possessions. (The form of the prefix changes with the phonological context.)

(27) ak*hnà:taʔ*
ak-*hnaʔt-aʔ*
1:SG:PAT-bag-NS
'*my* purse'

(28) akè*:sere*
ake-*ʔsere*
1:SG:PAT-car
'*my* car'

(29) akw*athǻhsterʌ̨*
akw-*athǻhsterʌ̨*
1:SG:PAT-pants
'*my* pants'

Locative suffixes can appear with such nouns if a location is specifically designated, but they are not used otherwise.

(30) *akhnaʔtà:*ke
ak-hnaʔt-aʔ-ke
1:SG:PAT-bag-NS-*LOCATIVE*
'*on* my purse'

2.3. The distinction

A difference between the use of agent and patient prefixes can be seen in several pairs of terms based on the same root.

(31) k*hnyaʔsà:ke*
k-*hnyaʔs-aʔ-ke*
1:SG:AGT-throat-NS-LOCATIVE
'*my* throat'

(32) ak*hnyà:saʔ*
ak-*hnyaʔs-aʔ*
1:SG:PAT-throat-NS
'*my* collar'

(33) ke*ʔnyų̨kserà:ke*
ke-*ʔnyų̨kser-aʔ-ke*
1:SG:AGT-onion-NS-LOCATIVE
'*my* big toe' (onionlike in appearance, according to speakers)

(34) ake*ʔnyuų̨́kseri*
ake-*ʔnyų̨kseri*
1:SG:PAT-onion
'*my* onion'

As is typical of languages with an alienable/inalienable distinction, disconnected body parts are generally classified as alienable. With the noun for 'fingernail', for example, either an agent or a patient possessor can be used, but the agent form is generally used when the fingernail is still attached, while the patient form is more often used after it has been cut off.

(35) k*tsiʔerà:ke*
k-*tsiʔer-aʔ-ke*
1:SG:AGT-nail-NS-LOCATIVE
'(on) *my* nail(s)'

(36) ak*tsiʔerà:ke*
ak-*tsiʔer-aʔ-ke*
1:SG:PAT-nail-NS-LOCATIVE
'(on) my nail(s)' (especially after cutting)

In a legend about a headless horseman, the noun referring to the horseman's head contained a patient possessor after decapitation.

Not all attached body parts appear with agent prefixes, however. The noun for 'hair', for example, appears with patient possessors.

(37) ake*nuų́hkwis*
ake-*nųhkwis*
1:SG:PAT-hair
'*my* hair'

There is some variation with nouns for 'eyebrows' and 'eyelashes', perhaps because these, like nails, are encountered both attached and unattached.

(38) ke*naʔwasà:ke* or ake*naʔwá:saʔ*
ke-*naʔwas-aʔ-ke* ake-*naʔwas-aʔ*
1:SG:AGT-eyebrow-NS-LOC *1:SG:PAT*-eyebrow-NS
'(on) *my* eyebrow(s)' '*my* eyebrow(s)'

(39) k*kahrehtà:ke* or ak*kahrehtà:ke*
k-*kahreht-aʔ-ke* ak-*kahreht-aʔ-ke*
1:SG:AGT-eyelash-NS-LOC *1:SG:PAT*-eyelash-NS-LOC
'(on) *my* eyelash(es)' '*my* eyelashes'

Yet attachment is not the only factor determining the case of the possessor. A number of nouns referring to major internal organs usually appear with patient possessors.

(40) awé:*ri*
aw-*eri*
1:SG:PAT-heart
'my heart'

(41) akw*athwʌ̨́hsaʔ*
akw-*athwʌ̨hs-aʔ*
1:SG:PAT-liver-NS
'*my* liver'

(42) akw*atstyéhseri*
akw-*atstyéhseri*
1:SG:PAT-kidney
'*my* kidney(s)'

(43) ak*tsinà:kwaʔ*
ak-*tsinaʔkw-aʔ*
1:SG:PAT-lung-NS
'*my* lung(s)'

(44) ake*nekwʌ̨́hsaʔ*
ake-*nekwʌ̨hs-aʔ*
1.SG.PAT-blood-NS
'*my* blood'

(45) ák*hstyʌ̨*
ak-*hstyʌ̨*
1:SG:PAT-bone
'*my* bone(s)'

A noun for 'skin' also appears with patient possessors.

(46) ak*ihnà:ke*
ak-*ihn-aʔ-ke*
1:SG:PAT-skin-NS-LOCATIVE
'on *my* skin'

The distribution of agent and patient possessors is clearly not random. If a new term for a body part were discovered, the appropriate forms of possessive pronouns would be easy to predict with relatively little chance of error. Yet a single feature distinguishing the two categories is not immediately obvious. Body parts that appear with agent prefixes are permanently attached and have visible surfaces. Speakers are seldom conscious initially of the formal distinction between the two kinds of possession, but when it is brought to their attention, they sometimes suggest that the crucial factor may be one of control. They note that one does not control one's hair or internal organs in the same way one can control arms and legs (see also Ameka (1995) on this property for Ewe body part terms). They then point out, however, that many body parts requiring agent possessors are not fully controlled, either. One moves one's navel or one's temple only when moving part or all of the rest of the body.

In fact, the factor underlying the choice of agent pronouns for some possessors seems to be based on a conceptual identity between animate beings and their parts. When speakers refer to their arms, they refer to themselves at the arm, a fact reflected by the appearance of locatives with body part nouns. Arms may be seen as integral, palpable, permanent extensions of a being in a way that hair is not. Internal organs are not only rarely under conscious voluntary control, they are also seldom encountered as parts of living bodies, a fact that may explain their exclusion from this category. (See also Crowley (1995) for a similar phenomenon in Paamese.)

3. "Possessor promotion" by noun incorporation

For a number of languages, an inalienability distinction has been identified in a second kind of construction, often termed "possessor raising" or "promotion". In such constructions, the possessor of an entity seems to assume the grammatical role of that entity. Instead of 'That dog bit *my leg*', for example, in which the direct object is the leg, English speakers can say 'That dog bit *me* in the leg', where the direct object is the possessor of the leg. Such constructions are appropriate only with certain kinds of possessions. English speakers do not say, for example, 'He bit me in the dog' for 'He bit my dog'.

3.1. The Mohawk construction

Mohawk contains a construction, termed noun incorporation, in which a noun stem is compounded with a verb root to yield a larger derived verb stem, like *-khw+ųni* 'meal+make' > *-khųni* 'cook'. Often events that seem to affect possessions, especially body parts, are expressed by verbs with incorporated nouns. The incorporated noun refers to the possession, but the grammatical patient of the whole verb is the possessor. In the Mohawk examples in (47)-(49) the patient of each intransitive verb is the owner of the body part, not the body part itself.

(47) *te*wak*ahsyó:tanos*
te-wak-*ahsyot-anos*
DUALIC-*1:SG:PAT*-hand-cold:STATIVE
'My hands are cold.' (Literally: '*I* am coldhanded.')

(48) wak*hsuʔnųhtanų́:waks*
wak-*hsuʔnųht-nųhwak-s*
1:SG:PAT-back-ache-HABITUAL
'My back aches.' (Literally: '*I* backache.')

(49) *te*yako*ʔnyų́:kweks*
te-yako-*ʔnyų-kwek-s*
DUALIC-*3.F:SG:PAT*-nose-blocked-HABITUAL
'Her nose is stuffed up' (Literally: '*She* is noseblocked.')

In the transitive verbs in (50)-(52), the grammatical patient is the possessor of the body part, rather than the body part itself, even though the body part is the object most immediately affected.

(50) *waʔk*he*kʌ̨ʔstaráyaʔke/*
*waʔ-k*he-*kʌ̨ʔstar-ayaʔk-e/*
PAST-1:SG:AGT/*3.F:SG:PAT*-forehead-hit-PUNCTUAL
'I slapped *her* face' (Literally: 'I faceslapped *her.*')

(51) *tá:*k*hswaket*
tak-hsw-ket
2:SG:AGT/*1:SG:PAT*-back-scratch
'Scratch *my* back!' (Literally: 'Backscratch *me.*')

(52) *wa*hi*kųhsohareʔ*
wa-hi-*kųhs-ohare-ʔ*
PAST-1:SG:AGT/*3.M:SG:PAT*-face-wash-PUNCTUAL
'I washed *his* face.' (Literally: 'I facewashed *him.*')

Noun incorporation of this type appears at first to reflect the same distinction of inalienability as nominal possession. In many languages, the set of nouns incorporated in this way consists chiefly of those referring to body parts. Mohawk nouns for the hand, various parts of the back, the nose, the forehead, and the face all require agent possessive prefixes when they appear alone.

Noun roots requiring agent possessors in nominal constructions are not the only ones that can be incorporated into verbs in Mohawk, however. A wide variety of noun roots requiring patient possessors can be incorporated into the same verbs that appear in the above examples.

(53) *waʔkeʔ*serehtó*hareʔ*
*waʔ-ke-*ʔsereht-*ohare-ʔ*
PAST-1:SG:AGT-*car*-wash-PUNCTUAL
'I washed the *car*.' (Literally: 'I *car*washed.')

(54) *waʔke*nų̨hsó*hareʔ*
*waʔ-ke-*nų̨hs-*ohare-ʔ*
PAST-1:SG:AGT-*house*-wash-PUNCTUAL
'I cleaned the *house*.' (Literally: 'I *house*cleaned.')

Possessors of the car and the house cannot appear as core arguments of the derived verb, however.

(55) **wa*hi*ʔserehtóhareʔ*
*wa-*hi*-ʔsereht-ohare-ʔ*
PAST-1:SG:AGT/*3.M:SG:PAT*-car-wash-PUNCTUAL
*'I carwashed *him*.' (for 'I washed his car.')

(56) **waʔk*he*nų̨hsóhareʔ*
*waʔ-k*he*-nų̨hs-ohare-ʔ*
PAST-1:SG:AGT/*3.F:SG:PAT*-house-wash-PUNCTUAL
*'I housecleaned *her*.' (for 'I cleaned her house.')

Instead, constructions like *raò:sere waʔkenóhareʔ* 'his-car I-washed' or *raò:sere waʔkeʔserehtóhareʔ* 'his-car I-car-washed', and *akonų́hsoteʔ waʔkenóhareʔ* 'her-house-standing I-washed' or *akonų́hsoteʔ waʔkenų̨hsóhareʔ* 'her-house-standing I-house-washed' would be used. Although a vast number of nouns are incorporated in Mohawk, it appears that only possessors of inalienably possessed entities may assume the roles of core arguments of the resulting verbs.

A closer look reveals that these two sets of nouns roots, those that require agent possessors, and those whose possessors serve as core arguments when they are incorporated, are not the same. Recall that the noun roots for 'hair' and 'skin' do not appear with agent possessors. Yet these roots are often incorporated into verbs whose agents or patients are their possessors.

(56) *te*yako*nuų́hkeri*
te-yako-*nųhkw-ri*
DUALIC-*F:SG:PAT*-hair-curly:STATIVE
'*She* has curly hair.' (Literally: '*She* is curlyhaired.')

(57) *waʔk*hen*ų́hkwahreʔ*
*waʔ-k*he-*nųhkw-hre-ʔ*
PAST-1:SG:AGT/*3.F.PAT*-hair-cut-PUNCTUAL
'I cut *her* hair.' (Literally: 'I haircut *her*.')

(58) ten*ihnará:kʌ̨*
te-n-*ihn-rak-ʌ̨*
1:INCL-DU-skin-light-STATIVE
'*Our* skin is fair.' (Literally: '*We* are skinfair.')

(59) *waʔtkųhnó:rarakeʔ*
waʔ-t-kų-ihn-orarak-ʔ
PAST-DUALIC-1:SG:AGT/*2:SG:PAT*-skin-pinch-PUNCTUAL
'I pinched *your* skin.' > 'I pinched *you*.' (Literally: 'I skinpinched *you*.')

Internal organs are incorporated in the same way. The noun root for 'blood' normally requires a patient possessor, but the root appears in verbs whose grammatical patient is the owner of the blood.

(60) *te*yako*nekwʌ̨hsatsikhè:tareʔ*
*te*yako-*nekwʌ̨hs-tsikheʔt-ar-eʔ*
DUALIC-*3:F:SG:PAT*-blood-sugar-in-STATIVE
'*Her* blood is sweet.' > '*She* is diabetic.' (Literally: 'She is blood-sweet.')

Noun roots for 'voice' and 'kiss' appear incorporated into verbs whose core arguments are their owners, although these noun roots never appear with agent possessors in nominal constructions.

(61) *riwʌ̨nayʌ̨té:ri*
ri-*wʌ̨n-yʌ̨teri*
1:SG:AGT/*M:SG:PAT*-voice-know:STATIVE
'I know *him* (by his voice).' (Literally: 'I voice-know *him*.')

(62) *wa*hi*ahsohkwanų̨hsko?*
wa-hi-*ahsohkw-nų̨hsko-?*
PAST-M:SG:AGT/*1:SG:PAT*-kiss-steal-PUNCTUAL
'He kissed *me*.' (Literally: 'He kiss-stole *me*.')

3.2. Alternative motivations

Mohawk thus contains two constructions that mirror distinctions one might term inalienability: nominal possession and noun incorporation. The distinction reflected in the two is not the same, however. The discrepancy stems from the fact that the two structures have different motivations. In possessive constructions, noun roots requiring agent possessors refer to body parts that are identified with their owners, while those requiring patient possessors are classified as discrete possessions. In incorporation, noun roots narrow the scope of their associated verbs. As was seen in the verbs to 'car-wash' and to 'house-clean', incorporated nouns need not name body parts. Carwashing and housecleaning are special kinds of cleaning: recurring, conceptually unitary activities worthy of conventional labels. Incorporated nouns need not represent possessions at all:

(63) *wʌ̨hniseráksʌ̨*
w-ʌ̨hniser-aksʌ̨
N:SG:AGT-day-bad:STATIVE
'The weather is bad.'

Possessors appear as grammatical patients of verbs containing incoporated nouns only when the event as a whole affects that person significantly. Mohawk speakers can say 'I facewashed him' but not 'I carwashed him'. They can say 'I carstole him' because the effect of the theft on the owner is usually of more interest than its effect on the car.

(64) *waʔhiʔserehtanʌ́hskoʔ*
waʔ-hi-ʔsereht-nʌhsko-ʔ-
PAST-1:SG:AGT/3.M:SG:PAT-car-steal-PUNCTUAL
'I stole his car' (Literally: 'I carstole him.')

The argument structure of the derived verb depends upon the meaning of the verb stem as a whole, not on the category of the incorporated noun.

3.3. Exploitation of the device

Incorporation of certain body parts has resulted in a partial semantic partitioning within the lexicon. A large number of verb stems contain the incorporated noun root *-ʔnikųhr-* 'mind', for example. These are predicates whose effect on animate participants is specifically mental.

(65) *raʔ*nikųhr*í:yo*
*ra-*ʔnikųhr-*iyo*
M:SG:AGT-*mind*-good:STATIVE
'He is patient.'

(66) *yakoʔ*nikųhr*hʌ̀:ʔʌ*
*yako-*ʔnikųhr-*hʌʔ-ʌ*
F:SG:PAT-*mind*-down-STATIVE
'She has forgotten, is depressed.'

(67) *wakeʔ*nikųhr*ákšʌʔs*
*wake-*ʔnikųhr-*aksʌ-ʔs*
1:SG:PAT-*mind*-bad-DAT
'I am grieving.'

Another set of verbs contain the incorporated noun root *-yaʔt-* 'body'. These denote events or states with a physical effect on animate beings.

(68) *shako*yaʔt*ʌ́hawe*ʔ
*shako-*yaʔt-*ʌhaw-e*ʔ
M:SG:AGT/F:SG:PAT-*body*-hold-STATIVE
'He is holding her.'

(69) *wahųwaya*ʔt*isákha*ʔ
*wa-hųwa-*yaʔt*-isak-ha*ʔ
PAST-M:PL:AGT/M:SG:PAT-*body*-seek-ANDATIVE-PUNCTUAL
'They went to look for him.'

(70) *sha*ʔ*ky*aʔt*yé:n*ʌʔ*ne*ʔ
*sha*ʔ*-k-*yaʔtʔ*yen*ʌʔ*-ne*ʔ
COINCIDENT-1:SG:AGT-*body* lie-INCHOATIVE-PUNCTUAL
'when I fell'

4. Conclusion

Distinctions identified as "inalienability" are most often reported in two kinds of constructions in languages: nominal possession and structures often termed "possessor raising". Constructions of both kinds exist in Mohawk, and both seem at first to reflect such a distinction. A closer look reveals that the distinctions they encode are subtly different.

Nominal possession, shown in Mohawk by pronominal prefixes on possessed nouns, can mark a special relationship of identity between the possessed and the possessor. Agent case prefixes are used for body parts that are considered palpable extensions of their owners. They are typically visible, attached, and under the control of their possessors. Patient case prefixes are used with other kinds of possessions, including not only separate objects, but also internal (invisible, uncontrollable) body parts, such as lungs and kidneys, and detachable body parts, such as hair and nails.

Noun stems referring to body parts may be incorporated into Mohawk verbs whose grammatical agents or patients are the owners of the body parts. These owners do not owe their status as core arguments to their identity with the incorporated body parts, however. The owners assume the role of grammatical agents or patients when they are the most salient agents or patients of the event or state as a whole.

What at first may appear to be a simple distinction of inalienability thus actually reflects two subtly different notions: identity on the one hand, and salient involvement on the other.

Notes

*I am especially grateful to the following Mohawk speakers who have shared their expertise on their language: Leatrice Beauvais, Geraldine Cross, Mary Cross, Annie Deer, Glenda Deer, Beverly Delormier, Myrtle Diabo, Margaret Edwards, Shelley Goodleaf, Margaret Hill, Josephine Horne, Annette Jacobs, Carolee Jacobs, Connie Jacobs, Frank Jacobs Jr., Georgina Jacobs, Verna Jacobs, Hazel Lazare, Dorothy Lazore, Leonora Montour, Madeleine Montour, Mae Montour, Frank Natawe, Rita Phillips, Muriel Rice, Evelyn Sawyer, and Geraldine Standup, of Caughnawaga, Quebec; Judy Caldwell, Hilda Gabriel, Elizabeth Nelson, Hattie Nelson, Minnie Nelson, and Mary Nicholas, of Oka, Quebec; Louise Cook, Veronica Cook, Helen Edwards, Josephine Gabriel, Christina Jock, Bessie Lazore, Margaret Lazore, Margaret MacDonald, Mary MacDonald, Joyce Sharrow, Marie Sunday, Mary Thompson, Mildred White, and Rosemarie White, of Ahkwesahsne, Quebec/Ontario; and Ruth Isaacs of Six Nations, Ontario.

The symbols used here generally represent their normal IPA values. Obstruents are automatically voiced before other voiced sounds. The symbol *y* represents a glide. The colon [:] represents vowel length, an acute accent [´] stress and high or rising tone, and a grave [`], stress and falling tone. The Polish hook [̨] represents nasalisation. The following abbreviations appear: AGT = grammatical agent; DU = dual number; EXCL = exclusive; person; F = feminine gender; INCL = inclusive; LOC = locative; M = masculine gender; N = neuter/zoic gender; NOM = nominaliser; NS = nominal suffix; PAT= grammatical patient; PL = plural number; SG = singular number; 1= first person; 2 = second person; 3 = third person.

References

Ameka, Felix
1995 "Body parts in Ewe grammar", [this volume.]
Crowley, Terry
1995 "Inalienable possession in Paamese grammar", [this volume.]
Nichols, Johanna
1988 "On alienable and inalienable possession", in: William Shipley (ed.), 557-609.
Shipley, William (ed.)
1988 *In honor of Mary Haas: from the Haas Festival Conference on Native American Linguistics*. Berlin: Mouton de Gruyter.

On the grammar of body parts in Koyukon Athabaskan

Chad Thompson

1. Introduction

1.1. Scope of paper

In the following paper, I look at alienably and inalienably possessed nouns in Koyukon, an Athabaskan language spoken in the interior of Alaska. Alienable and inalienable possession are semantic and pragmatic notions tied to a generalised degree of association between the possessor and possessed, and to the degree of salience or individuation of the possessed apart from the possessed. The notions of association and salience are culturally defined, although as the papers in this volume show, they are consistent in many ways from language to language. For example, hands and feet are typically inalienably possessed, while urine and warts are typically not (see Chappell – McGregor 1989).

I use the requirement of an overt possessor as the principle diagnostic for inalienable possession. Below is a summary of the terminology I use as it relates to Koyukon:

(a) *Inalienably possessed* nouns require an overt possessor in all cases, marked by pronominal prefixing and/or the presence of a nominal possessor. Kinship terms and body part terms as bound words are typical of this class.
(b) *Alienably possessed* nouns may or may not have an overt possessor. Most other nouns belong to this class.
(c) *Unpossessible nouns* cannot have an overt possessor without some morphological modifications. Terms for animals, people, and loan words are typical of this class.

I will discuss below the morphological means by which inalienable nouns may be "alienated", and unpossessible nouns may be "possessed".

Although the necessity of an overt possessor is the chief diagnostic for inalienable possession, there are a few additional structural manifestations which I will also discuss below. These include the presence or absence of the possessive suffix, *-e'*, the means of expressing possession with a full proposition (for example, "I have a brother"), and morphological changes in the verb resulting from incorporation.

Like most semantic and pragmatic notions, the notion of alienability is not categorial; while some nouns are clearly inalienable, and others clearly alienable, some nouns are ambiguous in terms of their alienability. The structural manifestations of alienability, however, are categorial. Borderline cases, therefore, will be treated differently from one language to the next. For example, internal organs are treated in some languages as alienable, presumably because they are not usually seen attached to a living body (see Crowley 1995 and McGregor 1995 for more detailed discussions). However, in other languages, including Koyukon, internal organs are inalienably possessed, presumably because they are integral parts of a whole and cannot exist as functioning units without some possessor.

1.2. Background information

Koyukon belongs to the Athabaskan language family, which along with Eyak and Tlingit comprises the NaDene stock. The Athabaskan language family covers most of interior Alaska, a large portion of the interior of northwestern Canada, parts of the Southwest United States, and part of the Pacific Coast in northern California and southern Oregon. Koyukon is spoken by about seven hundred people in the interior of Alaska along parts of the Yukon and Koyukuk Rivers. There are three dialects of Koyukon: Lower, Central and Upper; all the data below are from Central Koyukon, but the analysis is the same for all three dialects. In almost all of the Athabaskan languages, most body parts and all kinship terms are inalienably possessed.

The Koyukon data contained in the following are from several sources. Eliza Jones, a Koyukon linguist and speaker, provided me with much of the data, for which I am most grateful. She is not to be blamed for any mistakes which I may have made, however. Some of the following data are from Jones 1983, Attla 1983, Thompson – Axelrod – Jones 1983, and Axelrod 1988. I will be using the Koyukon practical orthography for all of the data in this paper.

Koyukon is a polysynthetic language with complex verb morphology and simple noun morphology. Nouns are, in fact, only marked for possession. Koyukon does not mark case on subjects nor on direct object nouns and independent pronouns, as is shown in (1) through (4). The same set of bound pronouns is used for direct objects, oblique objects, and possessors (with the exclusion of the reflexive affix). Such identity of pronouns has been observed in other languages, although sometimes possessive prefixes are identical with agent pronominal affixes (as in Allen 1964 and Mithun 1995). Consider (1) through (5) below:

(1) *Denaa sołt'aanh nee-ł 'aanh*[1]
man woman THM-CL see
'The man is looking at the woman.'

(2) *Sołt'aanh denaa nee-ł- 'aanh*
woman man THM-CL see
'The woman is looking at the man.'

(3) *See yaan' John ne- tl- 'aanh*
1s only John THM 1s:CL see
'I'm the only one looking at John.'

(4) *John see yaan' nee-ł- 'aanh*
John 1s only THM-CL see
'John is looking at me only.'

(5) *Se-leeg-e' ehonh*
1s-dog:POS is:eating
'My dog is eating.'

(6) *Se-nee-ł- 'aanh*
1s-THM-CL see
'S/he is watching me.'

(7) *Se-ts'e he- naayh*
1s:AL THM-talking
'S/he is talking to me.'

Table 1 is a list of the Koyukon pronominal prefixes.

Table 1. Koyukon pronominal prefixes

	Subject	Object	Indirect Object	Possessor
1s	*s-/ge-/tl-*	*se-*	*se-*	*se-*
2s	*ne-/ee-*	*ne-*	*ne-*	*ne-*
3s:DIRECT[a]	ø	*ye-*[b]	*ye-*[b]	*ye-*[b]
		ø[c]	ø[c]	ø[c]
		heye-[d]	*heye-*[d]	*heye-*[d]
3s:INVERSE[a]	*ye-*	*be-*	*be-*	*be-*
3:INDEFINITE	*k'e-*	*k'e-*	*k'e-*	*k'e-*
1p	*ts'e-*	*denaa-*	*denaa-*	*denaa-*
2p	*ʉh-*	*yʉh-*	*yʉh-*	*yʉh-*
3p:DIRECT	*he-*	*heye-*[b]	*heye-*[b]	*heye-*[b]
		hebe-[c]	*hebe-*[c]	*hebe-*[c]
		hʉ-[d]	*hʉ-*[d]	*hʉ-*[d]
3p:INVERSE	*heye-*	*hebe-*	*hebe-*	*hebe-*
REFLEXIVE	–	*hode-*	*'ede-*	*de-*[b]
RECIPROCAL	–	*neeɫ-*	*neeɫ-*	*neeɫ-*

a. The choice between direct and inverse prefixes is not relevant to the present discussion. It is discussed in some detail in Thompson 1989a.
b. Used only with a third person subject.
c. Used only with a first or second person subject.
d. Used only with a third person plural subject.

2. Alienable and inalienable possession

2.1. Alienable possession in Koyukon

Possessors–both nouns and pronominal prefixes–precede possessed nouns in Koyukon. Koyukon nouns are also typically marked as alienably possessed by the addition of the suffix *-e'* to the noun. With the addition of this suffix, a stem final fricative is voiced, and a stem final stop consonant is unaspirated.[2] Compare the unpossessed nouns in (8) and (10) with the possessed ones in (9) and (11). A stem ending in a vowel takes only the glottal stop *-'* as a possessive suffix, as is seen in (13).

(8) *teɫ* 'socks'

(9) *se-tel-e'*
1s-socks-POS
'my socks'

(10) *gets* 'mittens'

(11) *ne-gedz-e'*
2s-mittens-POS
'your mittens'

(12) *saaye* 'knife'

(13) *be-zaaye-'*
3s:knife-POS
'his/her knife'

A stem initial fricative is also voiced when preceded by a possessor, as is seen by comparing (15) with (17). Both this voicing, and the voicing caused by the addition of the suffix *-e'*, is because of the intervocalic position of the consonant. A stem initial fricative is voiced, however, even when the possessor is a noun ending in a voiceless segment. This voicing seen in (18) and (19). A stem initial stop consonant does not change when the noun is possessed as is seen in (9) and (11) above.

(14) *ƚeek* 'dog'

(15) *se-leeg-e'*
1s:dog-POS
'my dog'

(16) *hʉtl* 'sled'

(17) *be-ghʉdl-e'*
3s-sled-POS
'his sled'

(18) *Dick ghʉdl-e'*
Dick sled-POS
'Dick's sled'

(19) *Dick leeg-e'*
Dick dog-POS
'Dick's dog'

2.2. Inalienable possession in Koyukon

In Koyukon, most nouns for body parts and kinship terms must be grammatically possessed–the exceptions are listed in section 2.3. For example, the stem for 'head' is *-tlee'*. This stem is bound, however, and one cannot say **tlee'* in isolation. If one is speaking of a detached animal's foot or sees no need to specify the possessor, one uses the bound pronoun *k'e-* 'something' as a possessive prefix. The prefix *k'e-* in Koyukon is generally an indefinite or nonreferential pronoun.[3]

(20) *k'e- tlee'*
K'E head
'a/the head'

When referring to human body parts without specifying the possessor, one uses the prefix *denaa-*, which can function as an indefinite prefix or a first person plural prefix. This prefix is used exclusively for human referents, while *k'e-* is used only for non-human possessors. See (21):

(21) *denaa- tlee'*
IMPR head
'our heads', 'someone's head', 'a (human) head'

(22) **tlee'* 'head'

The body parts of animals are also structurally treated as inalienably possessed, as is seen in (23) through (25).

(23) *k'e-ggoołge-'*
K'E-scale-POS
'(fish) scales'

(24) *k'e-daa -'*
K'E-antler-POS
'an antler, antlers'

(25) *k'e-k'ʉh*
K'E-fat
'(animal) fat'

Certain plant parts are also inalienably possessed. The complete list of inalienable plant parts (as far as I know) is given in (26) - (36) below.

(26) *k'e-ken-e'*
K'E-stump-POS
'(tree) stump' (cf. *kenh* 'base')

(27) *k'e-tloo' okk'aat*
K'E-branch on
'a knot hole'

(28) *k'e-lot'oodze'*
K'E-bark
'bark'

(29) *k'e-looken-e'*
K'E-branch-POS
'a branch (on trunk)'

(30) *k'e-tloo-'ʉ*
K'E-branch-POS
'a branch'

(31) *k'e-tlaadl -e'*
K'E-wood:chip-POS
'wood chips'

(32) *k'e-tsaan'*
K'E-blade/grass
'grass'

(33) *k'e-t'on'*
K'E-leaf
'leaf'

(34) *k'e-ggedl -e'*
K'E-big:root-POS
'big roots' (cf. *hʉyh* 'little roots')

(35) *k'e-deneeg-e'*
K'E-seeds-POS
'seeds'

(36) *k'e-dekene-'*
K'E-stem-POS
'stem' (cf. *dekenh* 'stick')

It is not just attached parts that are considered inalienably possessed. For example, the stems in (37) through (39) below are inalienably possessed:

(37) *k'e-ghaaze'*
K'E-egg
'egg'

(38) *k'e-t'on'*
K'E-leaf
'leaf, leaves'

(39) *be-tleet'eez*
3s-dandruff
'his/her dandruff'

Furthermore, some things which were never a part of a person or animal but which are closely associated with it are inalienably possessed. 'Track', for example, is inalienably possessed while 'trail' is not. Animals' houses are inalienably possessed while human houses (*yeh*) are not. The term for a person's voice is also inalienably possessed. Examples (40) and (42) through (46) are the non-body part/non-kinship terms which are inalienably possessed in Koyukon. As far as I know, these are the only inalienably possessed nouns which are not body parts, plant parts, or kinship terms.

(40) *k'e-kk'e*
K'E-track
'a track, tracks'

(41) *tene* 'trail'

(42) *be-laagge'*
3s- voice
'his/her voice'

(43) *k'e-'onh*
K'E-den
'a den'

(44) *k'e-kaanh*
K'E-den/lodge
'a (for example, beaver's) lodge'

(45) *k'e-t'иh*
K'E-nest
'a nest'

(46) *k'e-k'el -e'*
K'E-clothes-POS
'clothes'

Souls and spirits are also inalienably possessed:

(47) *be-yeege'*
3s-spirit
'his/her spirit, picture'

(48) *be-yeets*
3s-breath/breath:of:life
'his/her breath of life'

The other class of inalienably possessed nouns is that of kinship terms. Kinship terms behave differently from body part nouns, but like nouns denoting body parts, kinship terms need a possessor.[4]

(49) *be-to'*
3s-father
'his/her father'

(50) **to'* 'father'

One significant way in which the possession of body parts differs from the possession of kin is that kinship terms require a possessive prefix whether or not there is a full noun as possessor. In (51) and (52), the third person *be-* is obligatorily prefixed to the noun. Where the possessed noun is a body part term, no prefix is allowed if the full noun occurs as possessor – see (54) and (55). It is open to question as to whether nouns possessed in such a way are phonologically bound to their possessor, but if they are not, then the only inalienably possessed nouns to occur phonologically unbound are body part terms when they are preceded by a nominal possessor.

(51) *denaa be-to'*
man 3-father
'the man's father'

(52) *sołt'aanh be-kkun'*
woman 3-husband
'the woman's husband'

(53) *be-lo'*
3s-hand
'his/her hand'

(54) *John lo'*
John hand
'John's hand'

(55) *gguh kkaa'*
rabbit foot
'a/the rabbit's foot'

2.3. Exceptions

In (37) through (39) above, I demonstrated that some nouns denoting separable parts may be inalienably possessed. Not all things which are outside the body although originating in it are inalienably possessed, however. See the terms for body products (56) through (64) below which are not treated as inalienable possessions:

(56) *lekkone* 'blood'

(57) *deł* 'blood'

(58) *tson'* 'excrement'

(59) *doots* 'round hard excrement'

(60) *tseetseł* 'soft excrement'

(61) *łets* 'urine'

(62) *tsehtoo'* 'tears'

(63) *saakk* 'saliva'

(64) *tl'oo* 'milk'

Nouns denoting certain temporary or abnormal parts of a body are not inalienably possessed. It may be that they are alienably possessed because they are not part of the normative associations with a human body. See also Crowley (1995) for more discussion on such parts.

(65) *saał* 'wart'

(66) *ts'ets* 'boil'

(67) * łoot* 'scab'

(68) *dezet* 'scar'

(69) *heł* 'pus'

(70) *ełget* 'open sore'

Words for 'hair' can also occur unpossessed.

(71) *tlooh* 'hair'

(72) *deegoyh* 'grey hair'

The words for 'foot' and 'skin' may occur as separate stems with an alternation in the meaning of the stem. For example, the word for a sole to be sewn onto a boot is an unpossessed form of the stem for 'foot'. The sole of a foot, however, is inalienably possessed.

(73) *k'e-kkaa'*
K'E-foot
'(a) foot'

(74) *kkaa* 'sole (for a boot)'
(cf. *be-kkaa-tl'ogha*
3s-foot-under
'his/her sole (of foot)')

(75) *k'e-leł*
K'E-skin
'(a) skin, hide, pelt'

(76) *łeł* 'belt'

The stem *tl'enh* can mean either 'bone' or 'leg'. As 'leg' the noun is inalienably possessed. As 'bone', however, it is not.

(77) *tl'enh* 'bone (not attached to any being)'

(78) *be-tl'en -e'*
3s-bone/leg-POS
'his/her leg; his/her bone (attached)'

Finally, although the word for 'hand' *-lo'* is inalienably possessed, the word for 'fist' *dzestl* is not. This may be the case because fists are salient entities in themselves, while hands are not. The term 'fist' may be alienably possessed because it is a temporary manifestation. As suggested above, such an explanation can also be used for the alienability of examples (65) through (70).

In most instances, body parts and products are closely associated with their possessors, therefore one would expect to see them inalienably possessed, while expecting those with a lesser degree of association to be alienably possessed. For example, feet are likely strongly associated with their possessors, while boils and other abnormal growths are not. Close association with the possessor also explains the inalienability of the nouns given in (40) and (42) through (48) above.

Some body parts tend to have a special salience or individuation of their own apart from the body of the possessor, and these are more likely to be alienably possessed. Included in this class would be hair, bones, fists, and bodily products.

2.4. The use of the possessive suffix

The possessive suffix *-e'* does not always occur with an alienably possessed noun. I claim in this section that semantic/pragmatic factors enter into the choice of whether or not to use this suffix, but also that in many cases the absence of the suffix is due to as yet unknown historical factors. Below are two examples of stems lacking the possessive suffix:

(79) *be-konh*
3s-stomach
'his/her stomach'

(80) *n-onh*
2s-mother
'your mother'

Many body parts end in *-e'* (61 out of 106), but since most body parts are inalienably possessed, it is not readily apparent whether or not they have a possessive suffix. It can be surmised that they take this suffix, however, because in compound body parts, only the final stem ends in *-e'*. For example, in (81) the word for 'eye' ends in *-e'*, and in (83) the noun for 'head' ends in a glottal stop. In the compound forms (82) and (84), however, these stems appear without *-e'* and *-'*.

(81) *be-nogh-e'*
3s-eye-POS
'his/her eye(s)'

(82) *be-noh zen -e'*
3s-eye black-POS
'his/her pupil(s)'

(83) *k'e-tlee-'*
K'E-head-POS
'a head'

(84) *k'e-tlee-ghaa'*
K'E-head-grease
'head oil (of fish)'

We have already seen that the word for 'belt' is *łeł*, and that when this word is inalienably possessed as *-leł*, it means 'skin'. When the possessed form means 'belt', however, it takes the alienable possessive suffix. See (85) and (86):

(85) *be-leł*
3s-skin
'his/her/its skin, hide, pelt'

(86) *be-lel -e'*
3s-skin-POS
'his/her belt'

In general, loan words and words for animals do not take the alienable possessive suffix, as shown in (88) below. I explain the presence of the prefix *k'e-* in section 2.5.

(87) *sookaanee* 'bread'

(88) *se-k'e-sookaanee*
1s-K'E-bread
'my bread'

Animals and people (excluding kin) are not generally thought of as possessible, as they are considered to have a spirit of some type. When they are marked for possession, no possessive suffix is used. See (89) and (90) below. Again the use of the prefix *k'e-* will be explained in section 2.5. Without the possessive suffix and the prefix *k'e-*, it sounds more as if the possessed animal were a pet. This type of possession is seen in (91), which is questionable for cultural and not linguistic reasons. I believe that loan words and names for wild animals fall into the same class because they are both, in a word, alien. When possessed, both loan words and words for wild animals lack the possessive suffix and, as we will see in section 2.5, require the indefinite prefix *k'e-*.

(89) *ggʉh* 'rabbit'

(90) *se-k'e-ggʉh*
1s-K'E-rabbit
'my rabbit [which I caught]'

(91) *?se-ggʉgh -e'*
1s-rabbit-POS
'my [pet] rabbit'

Dogs, however, are considered quite possessible, so the word for 'dog' takes the suffix when possessed.

(92) *łeek* 'dog'

(93) *se-leeg-e'*
1s-dog-POS
'my dog'

As I stated, I cannot find any principle or principles which would account for every presence or absence of the possessive suffix. In general, however, it can be said that nouns on either end of the alienability continuum, inalienably possessed and unpossessible nouns, are less likely to take the suffix than those in the middle. The fact that both ends of this continuum are marked the same is not surprising, however, since what is inalienably possessed by one is unpossessible by another.

Table 2 shows the percentage of inalienable body parts, kinship terms, animal body parts and plant parts which take the possessive suffix. It can be said that kinship and body part terms as a group do have statistically significantly fewer cases of possession without a suffix and are more like alienably possessed nouns in this regard.

I interpret these figures as indicating that generic (or cultural) salience plays some role in the choice of whether to have a possessive suffix on inalienably possessed nouns–the farther away from humans, the more likely there is to be a suffix. If, however, one is far enough away to be

Table 2. The possessive suffix on inalienably possessed nouns

	Suffix	No Suffix	Total
Kinship Terms	9 30%	21 70%	30 100%
Human Body Parts	61 57.5%	45 42.55%	106 100%
Animal Body Parts	22 84.6%	4 15.4%	26 100%
Plant Parts	10 90.9%	1 9.1%	11 100%
Alienably Possessed Nouns[a]	49 98%	1 2%	50 100%

(Chi-Square = 27.416 $p \leq .001$)

a. The alienably possessed nouns counted include fifty randomly selected nouns for tools, clothing, food, and structures. No loan words or nominalisations are included. The one word in this group not taking a possessive suffix is *yeh* 'house', which is discussed in this section.

alien, then the likelihood of a suffix again decreases. This is the reason that "unpossessible" nouns and loan words take no possessive suffix. As mentioned above, terms for wild animals take no possessive suffix, while those for pets do.

One further factor for the presence or absence of a possessive suffix is not directly a semantic one. Words which are historically derived from locationals or postpositions of location, or which are compounds ending in such a morpheme, often do not take a possessive suffix (see also Ameka 1995 on terms for spatial orientation in Ewe). For example, *-dzeyet* 'inner ear' is probably a compound derived from *dzeh* 'ear' plus *yeet* 'in'. Two further examples are given in (94) and (95) below, including the one plant part I am aware of which does not take a possessive suffix, found in (94):

(94) *k'e-tloo' okk'aat*
K'E-branch on
'knothole'

(95) *be-tl'o-yet*
3s-buttocks-in
'his/her lap'

The same explanation applies to the word for 'house' *yeh* which is alienably possessed, but which takes no possessive suffix. The word is derived from the word for 'under, down'; traditional Koyukon winter houses were semi-subterranean. See (96) below.

(96) *be-yeh*
3s-house
'his/her house'

Not all locative-related nouns however, lack a possessive suffix. For example, compare the postposition in (97) with the noun in (98).

(97) *be-kk'el*
3s-beside
'beside it/him/her'

(98) *be-kk'el-e'*
3s-side-POS
'his/her side'

In short, the choice of whether or not to use a possessive suffix depends in part on the degree of alienation from the possessor, the degree of generic salience on the part of the possessed, the locational nature of the possessed, and historical factors which are not at present clear. None of these factors, however, appears to be primary.

2.5. The alienation of inalienable nouns

In Koyukon the possession of body parts which are not part of the possessor is indicated by the use of the prefix *k'e-*. The word *nelaane* 'meat, flesh' in (99) may or may not appear possessed. In its unpossessed form, it refers to flesh which is no longer part of a living being. In (101), the fact that the flesh is not the possessor's own flesh is indicated by the prefix *k'e-*. In (100), on the other hand, no *k'e-* is used because the flesh is part of the possessor's body.

(99) *nelaane* 'meat, flesh'

(100) *be-nelaane'*
3s-meat/flesh
'his/her [own] flesh'

(101) *se-k'e-nelaane'*
1s-K'E-meat/flesh
'my [animal's] meat'

As stated above, it is considered unusual to possess animals other than dogs, so one also uses the double possessive form exemplified in (101) when talking about the possession of some game animal which one has killed or trapped. Notice below that (102) does not use *k'e-* while (103) does. Remember also that terms for wild animals do not take the possessive suffix, as is seen in (103).

(102) *se-leeg-e'*
1s-dog-POS
'my dog'

(103) *se-k'e-deneege*
1s-K'E-moose
'my moose'

When speaking of possessing animal body parts other than *nelaane* 'meat, flesh', two *k'e-*'s are used – one to show that the term for an animal is alienable, and one to show that the body part term is alienable. In examples (106) and (109) below, two *k'e*'s are used, while only one *k'e-* is used in (105) and (108) because the animal is designated by an overt noun–in such cases the *k'e-* marks the noun for the animal as alienable (see section 2.3 above). In (104) and (107) no *k'e-* is used, because no change to alienable possession is involved.

(104) *se-tlee'*
1s-head
'my [own] head'

(105) *se-k'e-dineega tlee'*
1s-K'E-moose head
'my moose head'

(106) *se-k'e-k'e-tlee'*
1s-K'E-K'E-head
'my [animal's] head'

(107) *ne-kkaa'*
2s-foot
'your [own] foot'

(108) *ne-k'e-gguh kkaa'*
2s-K'E-rabbit foot
'your rabbit foot'

(109) *ne-k'e-k'e-kkaa'*
2s-K'E-K'E-foot
'your [animal's] foot'

An interesting use of the alienating function of the possessive *k'e-* is seen in the possession of loan words. Words which are not part of the native vocabulary take a *k'e-* prefix when possessed. Examples (110) and (111) below contain alienably possessed nouns which are native to Koyukon, *oyh* 'snowshoes' and *hutl* 'sled'. Examples (112) through (114) contain borrowed nouns – (112) is from English while (113) and (114) are borrowed from Russian.[5]

(110) *se-'oye'*
1s:snowshoes
'my snowshoes'

(111) *be-ghuudl-e'*
3s-sled-POS
'his/her sled'

(112) *se-k'e-flashlight*
1s-K'E-flashlight
'my flashlight'

(113) *be-k'e-kiloosge*
3s-K'E-cup
'his/her cup'

(114) *ne-k'e-saasee*
2s-K'E-clock/watch
'his/her clock/watch'

2.6. Predicative possession

Koyukon has a verb indicating possession or ownership. This verb is used quite frequently with alienably possessed nouns. See (115) and (116) below:

(115) *łeek es-ø-t'aanh*
dog 1s:SUBJ-CL-have
'I have (own) a dog.'

(116) *hʉtl e-ø-t'aanh*
sled 3-CL-have
'S/he has (owns) a sled.'

This verb is not, however, used with inalienably possessed nouns. With kinship terms, possession is indicated with the "areal" form of the verb 'to be', which translates as 'exists'. (For a discussion of the areal prefix, see Thompson 1989b.)

(117) *S-ode hoo-ø-laanh.*
1s-older:sister area-CL-be
'I have an older sister.' (but **soda ast'aanh*)

(118) *Be'ot hoo-ø-laanh.*
3-wife area-CL-be
'He has a wife.'

It is a little more difficult to have a possessive predicate with body parts, since only their absence is remarkable. (Absence is generally marked for both alienably and inalienably possessed nouns with the uninflected predicate *kkelaa* 'gone, missing'.) The few cases I have found of body parts with a possessive predicate use the verb 'to be' like kinship terms, instead of the common verb of 'to have'. For example, consider (119) below which has been taken from Attla (1989:152).

(119) *Dotson' be-kkaalghʉl tokk'ee yaan' nelaanh*
raven 3s-toes three only be
'Ravens only have three toes.'

3. Incorporation

3.1. The incorporation of nouns into the verb

Incorporation is not as productive a process as it is in many other Amerindian languages. The number of nouns which may be incorporated is limited as is the number of verbs which allow incorporation.

As far as I know, one has to memorise which incorporations are allowable and which are not. This area is in need of further study as, in fact, only one paper has been written on Koyukon incorporation (Axelrod 1988). The examples in this section are all from Axelrod (1988), although I deviate from her analysis somewhat.

Koyukon allows incorporation to indicate manner (120), location/direction (121), patientive object (122), and patientive subject (123). Givón (1984: 129) notes that patients, manner, instrument, and location/direction are commonly incorporated into a verb. Example (121) is also remarkable for having incorporated the stem for an animal's young, *-ggoyh* along with the indefinite possessive prefix *k'e-*. Axelrod (1988) notes that this word is the only one she is aware of with an animate noun incorporated. As we will see further below, terms for body parts, although inalienably possessed do not take any prefix when incorporated.

(120) *Ko# ne-de-taanh*
belly THM-CL-lie
'S/he is staying humbly (Literally: on her belly).'

(121) *To# k'e-ggoyh#hu- de-ghee-ø-lo*
water K'E-young:area-THM:PRF CL-put:plural:objects
'It (a duck) put its young into the water.'

(122) *To- ts'eeyh#ghee-ø-tonh*
water boat-PRF-CL-put:long:object
'He launched a boat (put it into the water).'

(123) *Ho- to# le-ø-yo*
up:bank water PRF-CL-sg:go
'The water rose over the bank.'

Inanimate agents (such as the wind) may also be incorporated, as is seen in (124) below:

(124) *No- 'ełts'eeyh#ye- ghee-ł- gheł*
back wind-3s:DO-PRF-CL-move:long:object:abruptly
'The wind knocked it down.'

3.2. The Koyukon classifiers and the voice system

Koyukon verbs need to be specified in the lexicon as taking one of four "classifiers": ø, *de-*, *ł*, or *le-*. The term "classifier" is the conventional name for these prefixes, but they do not classify anything (for a discussion of these morphemes in Athabaskan, see Krauss 1969). As part of the specification of the verb in the lexicon, they are essentially meaningless. Under certain conditions related to voice and valence, however, the classifiers change.

The type of classifier change I will discuss here is when a ø changes to *de-*, or a *ł* a *le-*.[6] This happens in middle voice derivations – agentless passives, reflexives, reciprocals, self benefactives – as well as in intransitive distributives or "perambulatives," and intransitive iteratives. A more developed explanation of these various functions can be found in Thompson (1989b). Examples (125) and (126) below are normal active, transitive verbs. Example (125) has a ø-Classifier. In the passive (126), the ø has changed to *de-*. The same change is seen in the reflexive (127) and the reciprocal (128).

(125) *Y-ee-to-ø-ts'eyh* [Normal Active]
3s:DO-once-FUT-CL-pinch [ø classifier]
'S/he will pinch him/her once.'

(126) *Ee-to-de-ts'eyh* [Agentless Passive]
once-FUT-CL-pinch [*de-* classifier]
'S/he will be pinched once.'

(127) *Ho-d-ee-to-de-ts'eyh* [Reflexive]
RFL-RFL-once-FUT-CL-pinch [*de-* classifier]
'S/he will pinch him/herself once.'

(128) *neeł-h-ee-to-de-ts'iyh* [Reciprocal]
RECPR-3p:SUBJ-once-FUT-CL-pinch [*de-* classifier]
'They will pinch each other once.'

Self-benefactive verbs trigger this classifier change. Self-benefactives are obviously related to reflexives. In reflexives, however, the agent and patient are coreferential, while in self-benefactives, the patientive object is not coreferential with the agent. Instead, the agent is somehow understood to benefit from his or her own action. Compare (129) below with the self-benefactive in (130).

(129) *Le-tl- baats*
TNS-1s:CL-boil [*ł* classifier]
'I boiled it.'

(130) *Daa-l- ge-baats* [Self-benefactive]
THM-TNS-1s:CL-boil [*le-* classifier]
'I boiled it for myself.'

Axelrod (1988) reports that certain noun incorporations involve this same classifier change to *de-*. When body parts are incorporated into a transitive verb, indicating "the movement of a body part in a typical manner", there is a classifier change provided that the possessor is the subject. The unincorporated form (131) has a ø classifier, while the incorporated form (132) has a *de-* classifier.

(131) *De-lo' nee-nee-ø-'onh*
3r:hand AL-TNS-CL-put [ø classifier]
'S/he put her/his [own] hand there, at rest (in response to a request).'

(132) *Nee-lo'-ee- t'onh*
AL-hand-TNS-CL:put (*de-* classifier)
'S/he put her/his [own] hand there, at rest.'

Other incorporated patients do not normally trigger a classifier change. For example, see (133) where there is no classifier change for the incorporated form:

(133) *Nee-haał-nee-ø-'onh*
AL-trap-TNS-CL-put
'S/he set a trap.'

The classifier change cannot occur when the possessor of the incorporated body part is not the subject. In (134) below, there has been no classifier change for this reason:

(134) *Y-e-tl'o-'etl- hʉl*
3s-on-buttocks-TNS:CL-hit [*ł-* classifier]
'S/he hit him/her on the buttocks.'

There is also no classifier change if the incorporated body part is an instrument rather than a patient. As far as I know, instrument incorporation in Koyukon is fairly rare. Consider (135):

(135) *Ggeydekk e- kkaa-'oo-ł- kkoyh*
ball to-foot-THM-CL-kick
'S/he is kicking the ball.'

There are a few incorporated body parts, however, which do not trigger a classifier change even when they are patients and belong to the agent/subject of the clause. I suggest that there is no classifier change in these because they code states and not actions; reflexives and reflexive-like constructions usually require some change of state or place. Consider (136) and (137) where no classifier change has occurred:

(136) *H-e-ts'uts-dee-ł-'o'*
area-on-elbow -THM-CL-put
'S/he has his/her elbow propped on a surface.'

(137) *H-e-ggu-dee-ł-'o'*
area-on-knee-THM-CL-put
'S/he is kneeling.'

In summary, I suspect that incorporated patientive body parts occur with a reflexive classifier change because a change of state of a part is also a change of state of the whole. Even though noun stems denoting body parts are not incorporated with a possessive prefix, their inalienably possessed nature triggers the classifier change.

As suggested above, this classifier change is associated with derivations typical of the middle voice, an association noted by Bally (1926 [1995]) for Indo-Europoean languages. Bally's explanation for the association is the same as mine–in his words "a person involved in an action may be viewed as a whole regardless of the constituent parts".

4. Conclusion: the nature of inalienability

I have claimed that inalienability is a semantic/pragmatic notion with several structural consequences. The requirement of an overt possessor marked by pronominal prefixing is the major structural consequence which I use as a diagnostic for testing inalienability. The other structural consequences I have discussed include the presence or absence of the alienable possessive suffix *-e'*, the use of the indefinite prefix *k'e-* with inalienable possessions, the use of predicative possession, and the triggering of a classifier change upon incorporation of patientive body part terms.

Although the allowability of an unpossessed noun stem for some categories discussed in section 2 and the triggering of a classifier change discussed in section 3 are two different processes, they both are tied to the notion that body parts necessarily have a possessor.

The question of whether an item is considered inalienable, I believe, is tied to three factors: (1) the degree of its connectedness to the possessor (or the degree of its separability), (2) the degree of its cultural association with the possessor, and, (3) the degree of its salience apart from the possessor.

All three of these factors are important. Items which are not detachable from a body, such as hands and feet, are inalienable, while those which can exist apart from the body, such as saliva or blood, are not inalienably possessed. Items which are closely associated with a possessor, such as antlers, dens or dandruff, also tend to be inalienably possessed, even if they are detachable or separate. Finally, items which are salient by themselves, such as bones or belts made of skin, tend not to be inalienably possessed.

Notes

1. Below is a list of the abbreviations used in the examples:
   # = Boundary between disjunct and conjunct verb prefixes; AL= Allative; CL = Classifier; DIR = Direct; DO = Direct Object; FUT = Future; IMPR = Impersonal; INV = Inverse; POS = Possessive Suffix; THM = Thematic (Derivational) Prefix; TNS = Tense/Mode/Aspect; PRF = Perfective; RCPR = Reciprocal; RFL = Reflexive; SUBJ = Subject; 1s = First Person Singular; 2s = Second Person Singular; 3s = Third Person Singular; 3p = Third Person Plural.
2. In truth, all word final stops in Koyukon are unaspirated. They are marked in the Koyukon orthography as being aspirated.
3. The full range of functions of the prefix *k'e-* is a fascinating area in itself. It can be used as an indefinite non-human pronoun. It can code arguments which are generically (rarely anaphorically) predictable but which are left unspecified because they are unimportant or because one wishes to avoid saying them. It can be used to indicate a great distance or a stretch of time. Arguments coded by this prefix tend to have a low degree of individuation; it generally codes only non-humans but can be used to code a mass of people (such as a war party). One can sum up many of these functions by calling *k'e-* an alienating prefix. I have treated this prefix in more detail in Thompson 1991.
4. There are two kinship terms which occur without a possessor, *eetaa'* 'father' and *eenaa'* 'mother'. Although neither of these words has a possessor, the possessor is understood to be the speaker.

5. *Keloosge* is borrowed from the Russian *kruzka*, and *saasee* is borrowed from the Russian *casy*.
6. In general, there is an association between valence and the choice of classifier independent of any of the derivations described in this section. The *ł* classifier tends to be used with transitive verbs, the *le-* and *de-* classifiers with intransitive verbs, while the ø classifier is unmarked in this regard. While such a correlation exists in a general way, the choice of classifier, independent of the derivations described, is unpredictable.

References

Allen, W. Sidney
1964 "Transitivity and possession", *Language* 40: 337-343.
Ameka, Felix
1995 "Body parts in Ewe grammar", [this volume.]
Attla, Catherine
1983 *Sistiy yʉgh noholnik ts'in': as my grandfather told it.* Fairbanks: Alaska Native Language Center and the Yukon Koyukuk School District.
1989 *Bakk'aatʉgh ts'ʉghʉniy: stories we live by.* Fairbanks: Alaska Native Language Center and the Yukon Koyukuk School District.
Axelrod, Melissa
1990 "Incorporation in Koyukon Athabaskan", *International Journal of American Linguistics* 56: 179-195.
Bally, Charles
1926 "L'expression des idées de sphère personnelle et de solidarité dans les langues indo-européennes", in: Franz Fankhauser – Jakob Jud (eds.), 68-78.
[1995] [Reprinted in translation, this volume].
Chappell, Hilary – William McGregor
1989 "Alienability, inalienability and nominal classification", *Proceedings of the 15th annual meeting of the Berkeley Linguistics Society*, 24-36.
Crowley, Terry
1995 "Inalienable possession in Paamese grammar", [this volume].
Fankhauser, Franz – Jakob Jud (eds.)
1926 *Festschrift Louis Gauchat.* Aarau: H. R. Sauerländer.
Givón, Talmy
1984 *Syntax: a functional-typological introduction.* Amsterdam/Philadelphia: John Benjamins.
Jones, Eliza
1983 *Junior dictionary for Central Koyukon Athabaskan: Dinaakkanaaga ts'inh hʉyoza.* Anchorage: National Bilingual Materials Development Center.
Krauss, Michael E.
1969 "On the classification of the Athapaskan, Eyak, and the Tlingit verb", *Indiana University Publications in Anthropology and Linguistics* Memoir 24 (Supplement to *International Journal of American Linguistics* 35(4): 49-83).
McGregor, William B.
1995 "The grammar of nominal prefixing in Nyulnyul", [this volume.]

Mithun, Marianne
1995 "Multiple reflections of inalienability in Mohawk", [this volume.]
Thompson, Chad L.
1989a "Pronouns and voice in Koyukon Athabaskan: a text-based study", *International Journal of American Linguistics* 55: 1-23.
1989b Voice and obviation in Athabaskan and other languages. PhD thesis, University of Oregon.
1991 "The low topicality prefix *k'i-* in Koyukon", *Studies in Language* 15 (1): 59-84.
Thompson, Chad L. – Melissa Axelrod – Eliza Jones
1983 *Koyukon Language Curriculum: Scope and Sequence.* Nenana: Yukon-Koyukuk School District.

Part VI

Languages of Europe

Degenerate cases of body parts in Middle Dutch

Kate Burridge

1. Introduction[1]

Medieval medical texts provide a treasure trove for those interested in the syntactic behaviour of body parts and are ideally suited for a study such as this. The texts examined here consist of Middle Dutch medical treatises from the 14th and 15th centuries (see Appendix 2 for details of these texts). Middle Dutch belongs to the West Germanic family of languages. In reality, it represents a compound of different dialects both northern and southern spoken in the Netherlands during the Middle Ages. This present study focusses on two of these dialects in particular – Brabantish, the language of the southern medieval duchy of Brabant, and Hollandish, the language of the former province of Holland (now divided into North and South Holland).

While it is true that some medical writers of the time discuss body parts as separate entities – as if they are somehow detached from their owners – most use the familiar construction of inalienability which codes the possessor as a separate dative argument and expresses the body part with the definite article (see Appendix 1 for a description of the Dutch case system at this time; the reader might also like to consult Neuman 1995 and Manoliu-Manea 1995 for details of this same construction in related Indo-European languages).

But the texts here also contain a surprising array of variant body part constructions. Some involve identical case marking of body part and possessor; that is, where *both* part and whole are expressed in either the dative or the nominative case. Yet other examples express the body part in the dative and the possessor in the nominative. It is perhaps tempting to dismiss these sorts of constructions as simply performance errors – evidence that speakers at this time were losing grip of case functions and endings. Until now, this is precisely how they have been viewed – nothing more than the consequence of a crumbling case system. It is argued here however, that to plead simply error of case assignment is to ignore what is in fact an interesting feature of the morpho-syntax of Dutch at this time.

This paper goes on to link these constructions with a number of other types; for example, accusative subjects (which if discussed in the literature at all are also typically dismissed as scribal errors), impersonal verbs, ethic datives, and the so-called datives of self-interest / concern / (dis)advantage / misfortune etc. It is argued that all these, together with the construction of inalienability, are manifestations of one and the same syntactic phenomenon. They all illustrate a means of encoding entities not actively involved in the situation or event described, but which are nonetheless affected by it in some way. The various body part constructions found here, far from being scribal slips, encode fine semantic distinctions which have to do with prominence and different degrees of involvement. The eventual breakdown of cases has now meant a loss of these semantically expedient devices.

In sum, the paper concludes that it is misleading to view the Middle Dutch body part constructions purely in terms of inalienability. We are not simply dealing here with a relationship of identity – a small group of entities which can be thought of as being inalienably possessed. The constructions have much wider theoretical implications for the overall organisation of the grammar of the language at this time.

2. The morpho-syntax of body part constructions

The texts here are devoted almost exclusively to the sorts of things that can happen to the human frame, through sickness, disease and other 'accidents' of the body. This section examines, firstly from the perspective of the body part, some of the different types of grammatical constructions which medieval Dutch writers used to describe these unhappy events.

2.1. The 'part'

Situations and events which involve body parts do so in a number of different ways (see Wierzbicka 1979 and Neuman 1987 for a far more comprehensive account of types of bodily actions and their expression in modern European languages). Depending on the degree of involvement/affectedness, the body part receives a different grammatical role; for example, it may be represented as a core argument (a subject or object) or as a peripheral argument (a locative object). The following are the typical constructions found in the texts here (the body part has in each instance been highlighted):[2]

a) *Body part as central undergoer – affected from without:* A body part can be outwardly affected; that is, affected in a literal sense. For example, something outside the body (usually expressed in the relation of subject) acts on or causes an effect on the body part, to the extent that it may even undergo some sort of change. The body part is then the central undergoer and appears as the direct object of the sentence, coded in the accusative case. Consider the plight of the body part in the following two examples:

(1) *Doen querteleerde ic hem ende ondecte*
then trepanned I him:ACC and uncovered
hem dat hersenbecken
him:DAT the cranium:ACC
'Then I trepanned him and uncovered his cranium.'

(2) *Si namen* die hersenen *ende leid*se *in*
they took the brains:ACC and laid:them:ACC in
een silveren scale
a silver dish
'They took the brains and laid them in a silver dish.'

b) *Body part as peripheral undergoer – affected from without:* A body part can also be acted upon so that it is affected in a more indirect way. It is then coded not as a central participant but as a location relative to where the activity is taking place; that is, as the object of a preposition.

(3) *Hem viel een balck* op siin hooft
him:DAT fell a beam on his head
'A beam fell on his head.'

(4) *Teghen dronckenscap cout water ghegoten* op sijn
against drunkenness cold water poured on one's
ghemacht *is oec goet*
groin is also good
'Cold water poured on one's groin is also good for drunkenness.'

c) *Body part affected from within:* The events described above all affect the body in an exterior sense; that is, something from outside is acting on the body. A body part can also be inwardly affected, such as in a physical or emotional process/state. These sorts of involuntary events are caused by something within the body itself, even if perhaps triggered by some external force (for example, food eaten or a remedy applied) – according to medieval humoralism, they are usually ascribed to the quality and varying mixtures of the patient's four humours. The body part is then typically expressed in the nominative case, although as we shall later see there are some interesting variants of this particular construction.

(5) *Is den zieken* dat lichaem *bestopt*
is the patient:DAT the body:NOM blocked
'If the patient's bowel is blocked.'[3]

(6) *Den sieken verandert* siin varuwe
the patient:DAT changes his colour:NOM
'The patient's colour changes.'

2.2. The 'whole'

Whenever something happens to a body part, it must necessarily also involve the body part's owner. This is the special relation which exists between body and person and which is the subject of study in this present volume of papers. As the different examples above have already shown, sometimes this fact is left unstated and there is no mention of the person, while at other times it is made explicit. Consider now the different body part expressions, this time from the perspective of the person. (Note, that in each instance the type of event and the degree to which the body part is affected will determine whether it is coded either as a core argument or as a locative object; that is, according to (a-c) described above). In each of the examples below, the body part owner has been highlighted.

(a) *No mention of person:* If the involvement of the body part is central, the person need not be mentioned at all. The body part is then treated as any other detached entity and is expressed either with a definite article (example (7)) or else without (example (8)).

(7) *Doen moeste ic dat ghesonde hersenbecken drielen*
then must I the healthy cranium:ACC drill
'Then I must drill the healthy cranium.'

(8) *Men salre oec mede bestriken slaep*
one should:there also with smear temple:ACC

ende *kele*
and throat:ACC

'One should also smear the temple and the throat with this.'

This sort of impersonal account occurs in the texts, but it is much more usual to find mention of the person (see section 6 on the topicality of the patient in medical texts).

(b) *Person encoded as location:* If the involvement of the person is viewed as indirect, this is captured by encoding him/her as a location relative to the event (that is, as the object of a preposition). The body part, still as an independent entity, is typically expressed with a definite article. Again this is an infrequent construction in the texts:

(9) *Dat* *reynicht* in vrouwen *die* *moder*
that cleans in women the mother:ACC

'That cleans the womb in women.'

(c) *Person appears as a possessor:* The activity may still be viewed primarily as involving the body part, but the person can appear as possessor; that is, as either a possessive pronoun modifier (example (10)) or else as a full noun phrase in a genitive construction (example (11)):

(10) *Men* *sal* *nemen* *een* *veder* *ende* *stekense* *dan*
one should take a feather and stick:it then
in sine *nose*
in his nose

'One should take a feather and then stick it in his nose.'

(11) *Dat* [...] *conforteert* des menschen *zenen*
that [...] comforts the person:GEN sinews:ACC
bouen alle *dinghen*
above all things

'That soothes the person's sinews above all things.'

(d) *Person encoded as dative argument and possessor*: The involvement can be made more explicit by assigning the person a distinct participant role. The possessor is then doubly coded as a dative object and as a possessive pronoun modifier of the body part:

(12) Sijn *camerganc* *sel* hem *reyn* *worden*
his chamber:going:NOM shall him:DAT pure become

'His (bowel) movement will become pure.'

(13) *Als ghi dat werct soe seldi* den
when you that work so shall:you the
sieken *siin oren stoppen*
patient:DAT his ears:ACC stop
met catoene
with cotton
'When you operate that (that is, the drill), then you shall stuff the patient's ears with cotton wool.'

(e) *Person encoded as dative argument only:* The involvement of the person can be viewed as primary. As in (d) above, the person is marked with the dative and is thereby assigned a distinct participant role. The body part, however, is no longer seen as an independent entity, but rather as an aspect of the person. It is expressed accordingly with a definite article.

(14) *Dat sterket* hem *die leden*
that strengthens him:DAT the limbs:ACC
'That strengthens his limbs.'

(15) *Doen nam ic een driele ende drielde* hem
then took I a drill and drilled him:DAT
een gat int ghesonde been
a hole:ACC in:the healthy bone
'Then I took a drill and drilled a hole in his healthy bone.'

To get a better picture of the differences between (a)-(e), compare the following examples. As we move down these sentences, so the writer's description focusses increasingly on what is happening to the person.

(16) *Dat suuert harde dat hoeft*
that cleans well the head:ACC
'That cleans the head well.'
(no mention of person, body part expressed as accusative object with definite article)

(17) *Dat suuert siin hoeft*
that cleans his head:ACC
'That cleans his head.'
(person expressed as possessive pronoun modifier of accusative object body part)

(18) *Hi suuert hem siin hoeft*
it cleans him:DAT his head:ACC
'It cleans his head.'
(person doubly coded as dative object and possessive pronoun modifier of accusative object body part)

(19) *Dat suuert hem die hersen*
that cleans him:DAT the brain:ACC
'That cleans his brain.' (that is, what it does to the person is clean the brain)
(person coded as dative object, body part expressed as accusative object with definite article)

The last of these different construction types has been described in the literature as a construction encoding inalienability; that is, the special part-whole relationship which exists between body part and owner (see Fox 1981 and Chappell – McGregor 1989, 1995). The person is defined as a central participant and the inseparable status of the body part is expressed by the preceding definite article and its inability to be modified (I should point out here that there was no evidence of any modification in the texts examined here, although of course in the case of 'dead' languages negative occurrence is never conclusive). In many languages, the construction of inalienability can, in addition to body parts, also involve other types of bodily 'accessories'. In Middle Dutch, the entities which enter into the construction seem to include anything which has any close association with the body; for example, speech, mind, hair, nails, effluvia, sores, wounds, sickness, fever etc. Examples like the following were recorded here:

(20) *Als enen die naghele afvallen*
when one:DAT the nails:NOM fall-off
'When one's nails fall off.'

(21) *Dat doet hem dat haer dicke worden*
that makes him:DAT the hair:ACC thick grow
'That makes his hair grow thickly.'

(22) *Dat heelt hem dat seer*
that heals him:DAT the sore:ACC
'That heals his sore.'

(23) *Daer hem dat euel wast*
where him:DAT the evil:NOM grows
'Where his sickness grows.'

(24) *Dat verteert hem die hetten*
'that soothes him:DAT the heats:ACC
'That relieves his temperature.'

(25) *Hi scarpt hem den sin*
he sharpens him:DAT the mind:ACC
'It sharpens his mind.'

For reasons which will become apparent, however, I feel the concept of inalienability is inadequate to account for these particular constructions. For one, the construction is *not* confined to entities which have this kind of identity or part-whole relationship.

3. Exceptional body part syntax

In addition to the constructions we have just seen, the texts present a confusing array of other possible body part constructions. First impression suggests that they are probably little more than the chaotic product of a crumbling case system. Closer inspection, however, reveals that they are much more interesting than straightforward errors.

3.1. Dative body part NPs

There are a number of examples of body parts which themselves appear in the dative case. Consider the following. (Note, examples (26)–(28) do not mention any body part possessor, while (29)–(33) express the body part possessor in the nominative – in each instance, the dative body part has been placed in roman font).

(26) *Als* den lichaem *heet is*
when the body:DAT hot is
'When the bowel is hot.'

(27) Den lichaem *is swaer*
the body:DAT is heavy
'The bowel is heavy.'

(28) Sinen pols *sloech sterkeliic*
his pulse:DAT beat strongly
'His pulse beat strongly.'

(29) *Die* den mont *stincket* ...
who:NOM the mouth:DAT stinks
'He whose mouth stinks ...'

(30) *Die* den maghe *vercout is* ...
who:NOM the stomach:DAT cold is
'He whose stomach is cold ...'

(31) *Die* den kele *siec is* ...
who:NOM the throat:DAT sick is
'He whose throat is sick ...'

(32) *Wie* den arm *sweert* ...
who:NOM the arm:DAT hurts
'He whose arm hurts ...'

(33) *Die* den buuc *is gheheven* ...
who:NOM the stomach:DAT is swollen
'He whose stomach is swollen ...'

Because the dative and accusative distinction is collapsing for the singular at this time (see Appendix 1), the forms above could conceivably represent either case.[4] Examples like (34) below, however, involve plural body parts which unambiguously appear in the dative case.

(34) *Sie heelt wel denghenen die* den voeten
it heals well those who the feet:DAT
dicwil swellen
often swell
'It heals well those people whose feet often swell.'

It is also clear from sentences like (35) below that we are dealing here with the dative case. This example is taken from another (probably eastern) dialect which retains the dative/accusative distinction.

(35) *Wan* deme maghen *vorkoldet* *is*
when the stomach:DAT cold is

'When the stomach has become cold.'

But, when you compare this example with (36) from the same text, the whole picture begins to look very confusing.

(36) *De minsche* deme sin maghe *is vorkoldet*
the person who:DAT his stomach:NOM is cold

'The person whose stomach is cold.'

Example (36) is once more the familiar construction of inalienability with the body part possessor appearing in the dative case, and the body part in the expected nominative case. Why do we find this switching of cases? Are examples like (35) simply errors of case marking, as has traditionally been assumed, or is there some underlying system to all of this? If they are errors, then there are some striking coincidences to be explained when we come to consider the sentences in the next section.

3.2. Accusative subjects

Grammarians have long noted the appearance of what they describe as "accusative subjects" in Middle Dutch. Van der Horst (1981: 21), for example, writes "occasionally the subject of a sentence has the accusative form" [my translation]. Like the dative body parts above, they are commonly attributed to the decaying condition of the case system at this time. Van der Horst claims that they typically appear post-verbally and in subordinate clauses, and he appeals to something like Sapir's (1921) original notion of "drift" to account for them; namely, the increasing importance of position in the sentence and the diminishing feeling for cases as an indication of syntactic function. (Note, the same argument of drift has been used to explain changes in English like *It is I* → *It is me* and *Whom did you see* → *Who did you see*). For the three reasons given below, however, I think this account is inadequate to explain these accusative subjects in early Dutch:

(i) Examples of accusative subjects are far too frequent in the body of texts here to dismiss as simply errors of case assignment.
(ii) I could find no such correlation between position and clause type to support Van der Horst's proposal.

(iii) If these forms were simply the result of a declining case system, then their appearance should be chaotic and unsystematic. As I will show below, this is, however, not the case.

In Middle Dutch, only the masculine nouns preserve a distinction between the nominative and accusative (see Appendix 1). However, with the ambiguity of some case forms and also the confusion of genders (see note 4), it is not really clear whether we are dealing with the accusative or dative case a lot of the time. (For instance, note 4 showed an example of a feminine 'subject' noun *hersen* with the definite article *den* – is this accusative or dative? It is impossible to know). Probably we should think of the forms in roman font in the following as simply being in the oblique.

(37)

Dien	doec	*sal*	*bernen*	*sonder*	*te*	*bederven.*
the	cloth:ACC?	shall	burn	without	to	spoil

'The cloth should burn without spoiling.'

(38)

Als	den	pot	*beginnt*	*te*	*sieden.*
when	the	pot:ACC?	begins	to	boil

'When the pot begins to boil.'

It is apparent from the examples in the texts here, that these oblique forms express pragmatic considerations of topicality and focus (see Burridge (1992a) for more details). Consider the sentences in (39) below. This is an extract from a discussion on wine-making. The topic *wijn* here is highlighted by both the use of the oblique case and a left dislocation structure. (Note, in the left-dislocation construction, the coreferential pronoun *die* is not in an oblique case):

(39)

Dien	wijn	*soude*	*so*	*sterc*	*worden*	[...]	*ende*
the	wine:ACC?	should	so	strong	become	[...]	and

dien	wijn	*die*	*van*	*desen*	*wijngaert*	*comen*
the	wine:ACC?	which	from	this	vineyard	come

sal	die	*sal*	*smaken*	*al*	*waert*	*goeden*	*clareyt*
shall	that-one	shall	taste	as:if	were:it	good	claret

'The wine should become so strong ... and the wine which comes from this vineyard, that will taste as if it were good claret.'

We will be returning to the question of topicality in sections 6 and 7. Meanwhile, consider what these constructions have in common with the earlier oblique body part constructions in examples (26)–(35). In the texts here, examples like (37)–(39) always involve sentences of low transitivity (for example, stative predicates, intransitives, reflexives and passives), where there is no argument filling the role of agent or actor. Without exception, all of the so-called 'accusative' subjects occurring in the texts here are themselves acted upon in some way, and this is what triggers the oblique case rather than the nominative. The appearance of the oblique is not random, but is the direct consequence of the semantics of the sentence verb – it signals the non-active involvement of an entity.[5]

This is also true of the dative body part constructions given earlier. Sentences (26)–(35) all involve something happening to the body. The cause may be felt to lie within body itself (the combination of the patient's humours, for example), but the body is still a passive undergoer. It is perhaps useful to digress a little here and consider something of the current medical opinion which also strongly emphasised the vulnerability of the body to external forces. Illness was usually linked to the supernatural – either demonological influences or the wrath of celestial powers (where diseases were believed to be sent as retribution for sins and indiscretions). It is not surprising that people arrived at such fantastic conclusions about the workings of the body and disease etiology. What is a disease after all? There are symptoms and there are sick patients, but there is often nothing tangible in disease itself. It seems to arrive out of the blue and then just as mysteriously transmits itself from person to person, affecting some while leaving others curiously untouched. The mystery would have been even greater during the middle ages when physicians had such little knowledge of physiology and few sophisticated instruments to guide them; recall that at this time it was illegal to even open up a human body. These facts are nicely reflected in the syntax. The absence of an expected nominative subject and the use instead of an oblique case captures the passive role of body and person in processes and states believed to be controlled by outside forces (see also Wierzbicka 1979: 369-377 on the portrayal of 'the unknown' in syntax).

When the affected person is also made explicit, it is possible to have structures where *both* body part and possessor appear in the dative. Consider the following two sentences. Here the effect on the body part and the entire person has been highlighted by the double dative.

(40) Hem *wart* den lichaem *weec*
him:DAT becomes the body:DAT soft
'His stool becomes soft.'

(41) Mi *is* den buuc *so gheladen*
me:DAT is the stomach:DAT so full
'My stomach is so full.'

As a last example, consider (42). Again we have a sentence with a non-action verb and no grammatical subject. (Although not physically contiguous with the body, it is not hard in this case to see 'work' as another 'bodily accessory').

(42) *Want* den arbeit *sal* mi *wesen*
for the work:ACC/DAT? shall me:DAT be
groot ende *swaer*
great and arduous
'For my work will be great and arduous', or: ' For the work will great and arduous for me.'

The varying use of cases in all these examples encodes different emphases and promotes different perspectives of a situation, all to do with affectedness and involvement. To dismiss it as simply a performance error, is to miss an important feature of Dutch morpho-syntax at this time.

4. Impersonal verb constructions

Middle Dutch, like all the Germanic languages of the time, possesses a construction known as the impersonal verb construction. Traditionally, this has been described as involving a group of exceptional verbs which lack arguments with any of the usual subject properties; that is, nominative case marking and verb agreement. All arguments are oblique and the verb itself stands in the third person singular, regardless of the number or person of its arguments. This is the characteristic which earns it the description impersonal (cf. the *me thinks* construction in early English). The following are some examples:[6]

(43) Den godleken *twivelt ende* den anderen *wondert*
the godly:DAT doubts and the others:DAT wonders
'The godly ones doubt and the others wonder.'

(44) Hem *dunct datter engheen ghewin aen en*
him:DAT thinks that-there no gain on not
leghet
lies
'He thinks that there is no gain in it.'

The similarity between impersonal constructions like the above and the sentences involving 'errors' of case assignment is striking. Compare (43) and (44) with two of the oblique subject constructions given earlier.

(26) *Als* den lichaem *heet is*
when the body:DAT hot is
'When the bowel is hot.'

(37) Dien doec *sal bernen sonder te bederven*
the cloth:ACC? shall burn without to spoil
'The cloth should burn without spoiling.'

The similarity is no coincidence – the use of the dative/accusative for arguments of impersonal verbs has long been pointed out as a device for signalling entities 'unvolitionally/unself-controllably involved in the situation' (see McCawley 1976: 194 for Old English). It makes no sense then to distinguish between (43) and (44) as impersonal constructions, and (22) and (37) as scribal slips. What this does, is throw into question the validity of assuming, as has traditionally been the case, that the impersonal verb construction involves only a very small handful of syntactically aberrant verbs. In fact, I am not the only one to question the traditional impersonal-personal distinction. Some years back, Van der Horst (1985) also proposed that the impersonal form was potentially a feature of all verbs, and arose as a natural consequence of their lexical meaning. There is substantial support for this in the texts here. Consider, for example, the two sentences below (both appeared in the texts):

(45) a. Hem *walght*
he:DAT vomits
'He vomits.'

b. Hij *walght*
he:NOM vomits
'He vomits.'

The verb 'to vomit' is not considered among the list of impersonal verbs of Dutch at this time. And yet a sentence like *hem walght* is indistinguishable from a recognised impersonal construction like *hem dunct* given in (44). The activity suggested by *walghen* is, like any bodily function, not something over which we have a lot of control and the use of the dative here reflects this. Sentence (45a) is simply semantically more marked than (45b) by capturing this fact.

There is another piece of evidence which suggests that the impersonal-personal distinction should be abandoned – the existence of impersonal verbs which *do* have verb agreement. For example, early English *me think* or Middle Dutch *mij dunke* both show the first person singular verb ending and not the third person singular expected of an impersonal construction (although in this particular example it is conceivable that the Middle Dutch involves a third person subjunctive form). In the literature on Germanic impersonal verbs, these sorts of examples are often cited as evidence for the disappearance of the impersonal construction via the reanalysis of impersonal objects as subjects (se, for example, Lightfoot 1979). But, as far as Dutch goes, what these examples do is only bring impersonals closer together in structure to the supposed errors cited earlier here. This suggests still more strongly that all these oblique subject constructions should be subsumed under the one syntactic umbrella – the notion of impersonal verbs as a distinct class on its own is not a valid one, at least for Dutch (and I suspect probably also for the other Germanic languages).

In sum, I suggest that at one stage in early Dutch, there was a productive system for signalling the involuntary involvement of the subject by placing it in an oblique case. This system gave rise to 'impersonal' expressions which sometimes showed person/number agreement on the verb (as in *den voeten swellen* 'the feet:DAT swell' where the verb is plural to agree with 'feet') or, as a more marked variant, showed no agreement and inflected for third person singular. This gave the classical *me thinks* type 'impersonal' construction (effectively, there was so much levelling of verbal inflections in Dutch at this time, that this distinction is anyway not always apparent) – of course this will mean that those verbs whose meanings typically express non-volitional activities (like 'to be hungry/thirsty') will more usually appear in this construction.

All this changes in the 17th century, however, with the eventual breakdown of the case system. This means then the loss of this semantically expedient system, and its replacement with a much less transparent one (see also McCawley 1976: 212 on the loss of Old English impersonals).

With only remnants of cases left, the language has to find other ways of handling these sorts of distinctions.

5. The construction of inalienability

Underlying all the different constructions we have seen so far is a single common denominator; namely, the use of the dative case (or simply oblique, since in many places it is not clear which case is involved) to promote entities that are indirectly affected in some way by a situation, event or state of affairs. The fact that we so frequently find the dative marking the body part possessor is a direct consequence of the fact that, as many have pointed out before, body parts are physically contiguous with their owners and therefore what affects the part will necessarily affect the whole. The dative case with its original meaning of 'direction towards' is a natural candidate to express this. Not suprisingly then, the construction involving dative persons is by far the favourite body part expression in the texts here. Persons in these medical texts are after all quite literally patients and the whole discourse generally revolves around their suffering! But the dative is not used simply to express the relationship of identity which exists between body part and owner. This is only part of the story – its function is much wider. The concept of inalienability does not make clear what is really going on here. For example, it will not explain the presence of dative persons in constructions like the following:

(46) *Ende siin meester hadde* hem *den scacht*
and his physician had him:DAT the shaft
uut ghetoghen ende tyser was hem *in*
out drawn and the:iron was him:DAT in
siin hersen bleven
his brain stuck
'And his physician had drawn the shaft out of him, but the iron still remained in his brain.'

Both clauses involve dative person pronoun NPs (highlighted), but the dative is not simply coding a relation of identity here. (It is anyway doubtful whether the relationship between the man and the arrow shaft could even be viewed in this way). What is crucial here and what motivates the appearance of the dative forms in both clauses, is the fact that we are dealing with an event which is affecting the entire person. It is the plight of the patient which is the main focus in the description here and the fact that the events are happening beyond his control – the patient is quite incapable of acting in this situation. It seems then that these sorts of constructions have not so much to do with inalienable possession; that is, whether or not something belongs to a group of entities which could be thought of as inalienably possessed, but rather whether or not a speaker/writer views body part owners as involved participants in the discourse and chooses to promote them as such. Compare the following two extracts from the one treatise:

(47) a. *Nem die wortel ende hanct* den zieken aen
take the root and hang the patient:DAT on

den hals. *Hi sal ghenesen.* Galienus *orcont*
the neck he shall recover Galienus says

van enen dien *hijt* aen den hals *hinc.*
of one who:DAT he:it on the neck hung

Ende als hijt afnam had hijt onghemac
and when he-it removed had he:the disease

ende als hijt hem anden hals
and when he:it on the neck

hinc en had hijt niet.
hung not had he:it not

'Take the root and hang it around the patient's neck. He shall recover. Galienus reports of someone whose neck he hung it around. And when he removed it, he had the illness, and when he hung it around his neck, he didn't have it.'

b. *Daer ene vrouwe in kindelbedde leghet* [...]
where a woman in childbed lies [...]

hangt den wortel an horen
hang the root on her

hals *ieghen bose ghespaente*
neck against evil spirits

Hi iaghet den duuel vanden mensche.
he chases the devil from:that person

'When a woman lies in labour ... hang the root around her neck against evil spirits. It chases the devil from that person.'

Why should the patient be coded by the dative in (47a) but not in (47b)? Consider the situation being described in (a). The physician is not simply hanging a herbal root around the patient's neck, but is doing something to the patient that affects him significantly. The root cures him of the sickness – if removed, the sickness returns. So the patient is central here and the writer is able to focus on this fact by expressing him in each instance as a dative argument. The event described in (b), however, has nothing like the same force. The root here is a protective rather than a curative agent as in (a) and it does not directly affect the patient in the same way. Appropriately, she is not expressed as a participant in the situation but is only coded as the possessive adjective modifier of *hals* 'neck'. The presence/absence of the dative case is nicely able to capture the different type of involvement by the patient in each instance.

6. The topicality of body part owners

Body part owners are highly topical. As human beings they are anyway expected to appear high on the topicality hierarchy (see Givon 1976), but in these medical texts of course, they have a particularly high profile – usually the whole discourse centres around what is happening to these people. Grammatical evidence which indicates their topicality is the fact that they (a) are typically pronominalised (b) appear early in the sentence (that is, in topic position) and (c) typically control coreferential deletion.[7] Consider, for example, the extract below (the body part constructions have been highlighted):

(48) *Wanneer een mensche in arbeyt is, soe*
whenever a person at work is so

is hem sijn herte *moere,*
is him:DAT his heart:NOM tired

ende – wille gherne drinken, soe sijn hem
and – wants much to:drink so are him:DAT

sijn aderen *open.*
his veins:NOM open

Drinct hi dan inder hitten, soe gaet
drinks he then in:the heat so goes

hem *die dranc* tusschen vel ende vleesch.
him:DAT the drink between skin and flesh

Soe *bestaet* hem *daer* *na* een hitten.
then exists him:DAT there after a heat:NOM
'So whenever a person works, then his heart is tired and (he) very much wants to drink, then his veins are open: if he drinks then in the heat, then the drink will go through his skin and flesh. Then he will experience a temperature.'

Normally grammatical subjects control coreferential deletion (as is strictly the case in both modern Dutch and modern English). In this example, however, the dative person *hem*, *not* the grammatical subject *sijn herte* (that is, the whole, not the part) is coreferential to the missing subject anaphor in the conjoined clause, indicated here by the dash (see Burridge 1986: 63-66 for other examples of topic control of deletion).

The topicality of body part owners is evident in another aspect of their grammatical behaviour. Consider examples like the following:

(49) *Die kinder sal men binden die arme*
the children:NOM shall one:NOM bind the arms:ACC
dat si niet en cnauwen
that they not not scratch
'One shall bind the children's arms so that they can't scratch.' [Literally: 'Children – one should bind the arms ...']

(50) *Als een vrouwe een kint in haren*
when a women:NOM a child:NOM in her
lichaem doot is ...
body dead is ...
'When a woman's child is dead in her body...' [Literally: 'When a woman – a child is dead in her body']

(51) a. *Die vrouwen ioket hi dat melc*
the women:NOM increases it:NOM the milk:ACC
'It increases women's milk.' [Literally: 'Women – it increases the milk']

On the surface at least, these constructions with their double nominatives are remarkably reminiscent of Chinese-style double subjects (see Chappell 1995). They differ from the earlier body part constructions in that the body part owners are in the nominative. Sometimes, though, other versions of the same text will show the more familiar dative possessor, as in (51b) below:

b. *Den vrouwen oket he de melic*
the women:DAT increases it:NOM the milk

'It increases women's milk.'

Constructions like the above place the body part owner in initial topic position and are clearly topic-comment oriented in organisation. They differ from usual left-dislocation constructions, however, where a focussed item is moved into initial position leaving some sort of pronoun copy in its place (see example (39) above). These constructions lack any following coreferential proform. The initial topics are syntactically unintegrated into the sentence and resemble something like a 'hanging topic'.

Although examples like these are frequent in the present texts, it is not clear whether they represent a productive device in the language at this time for signalling topic or whether they represent what traditionally is referred to as 'anacoluthon' – where a speaker/writer begins a sentence with an NP which s/he nominally places in the nominative case (the least marked case) but continues with a different subject (something like Chafe's 1976: 51-52 notion of 'premature subject'). Either way, the topicality of these initial NPs is clear – they represent something foremost in the speaker's/writer's mind (context could not be supplied here because of space).[8]

In these medical texts, concessive relative clauses involving body and person also frequently contained an initial nominative-marked topic (in each instance the body part owners; cf. examples (29)–(33) given earlier). For example:

(52) *Wat vrouwen die melc in den*
what women:NOM the milk:NOM in the
borsten verhardet is ....
breasts hardened is

'Whichever women's milk has hardened in the breasts ...'

(53) *Wie die oghen seer sien ...*
who:NOM the eyes:NOM sore are

'Whose eyes are sore ...'

Compare this last example with the following sentence taken from the same text. It shows instead an initial dative pronoun:

(54) *Wien die oghen seer sien ...*
who:DAT the eyes:NOM sore are

'Whose eyes are sore ...'

Like the double-subject type constructions mentioned earlier, these concessive relatives with initial nominatives also display an obvious topic-comment framework, with the nominative case marking the initial non-subject topic (the body part owner).

7. Concluding remarks

Traditionally, a number of different (so-called sentence) dative uses have been recognised. Curme (1904 [1970]) for example, distinguishes in modern German what he calls (1) the dative of reference (denoting persons to whom the statement holds true); (2) the dative of interest (denoting persons who directly benefit or are somehow disadvantaged by the situation, sometimes termed "the dative of advantage / disadvantage"); and (3) the ethic dative (denoting persons who have an interest in the situation but whose involvement is more detached than in 1 and 2). All these traditional types of dative share, together with the inalienable constructions and the impersonal and oblique subject constructions described earlier, the same intrinsic meaning; namely, the non-active involvement of a person in an event. All show the dative as an essentially pragmatically controlled device to bring to prominence the idea of a person (or personified entity) that is in some way involved (typicaly indirectly) in an activity. The dative case with its original meaning of 'direction towards' is ideally suited to express this idea – as all the studies on grammaticalisation are now showing, the full lexical meaning of a item is rarely totally bleached out of the new grammaticalised form (see, for example, Hopper – Traugott 1993). As is the case here, the original semantics of the dative still shows through.

In sum, body part constructions show the following hierarchy of possibilities in the texts here. As we move down the list, greater prominence is given to the involvement of the person (cf. English: *he looked into my eyes* versus *he looked me in the eyes*):

(a) Mention of the body part without possessor
(b) Possessor expressed in a locative phrase and independent of the body part constituent.
(c) Possessor coded as a possessive pronoun or genitive modifier of the body part
(d) Possessor doubly coded as a dative person and as a possessive pronoun modifier of the body part

(e) Possessor coded as a dative person independent of the body part constituent

In these constructions, the body part is usually expressed either as an accusative object (if it is the direct undergoer) or as a locative object (if the effect is indirect). Where bodily processes or states are involved, the body part appears in the nominative case, although a more marked variant of this has a dative marked body part. This last dative construction presents a diferent view of the situation – it reflects the fact that the body is still a natural undergoer in these events. This means that it is possible to have constructions with both body part and person expressed in the dative.

The findings here suggest that there are at least three things which seem to motivate the appearance of a dative person in construction types like (d) and (e) above and the appearance generally of oblique 'subjects':

(i) the affectedness/involvement of an entity in an activity
(ii) the lack of control an entity has in an activity
(iii) the topicworthiness of an entity

All these three will interact to bring about the appearance of a dative possessor in body part constructions. It stands to reason that the special relationship which exists between body and person would make the dative construction a favourite one – particularly in medical texts like the ones here, where the plight of the person, literally as medical 'patient', is in focus. The following extract makes very clear the force of the dative in these texts:

(55)

Heeft	*die*	*sieke*	*hooft*	*sweer*	*ende*
has	the	patient:NOM	head	pain:ACC	and

swimelinghe	*ende*	*siin*	*oghen*	*worden*
dizziness:ACC	and	his	eyes:NOM	become

root	*ende*	*siin*	*aensicht*	*al*	*ontsteken*
red	and	his	face:NOM	all	inflamed

ende	*hi*	*verliest*	*alle*	*siin*	*verstandenisse*
and	he:NOM	loses	all	his	understanding:ACC

ende	*keert*	*ter*	*kelen*	*bitter*	*coleren*	*ende*
and	vomits	thru:the	throat	bitter	bile:ACC	and

bi	*aventueren*	*vloyt*	hem	tbloet	ter
by	adventure	flows	him:DAT	the:blood:NOM	thru:the

nosen	ende	ten	oren	*uut*	*ende*	hem
nose	and	thru:the	ears	out	and	him:DAT

ontvalt siin sprake *ende* -
disappears his speech:NOM and -

verliest siin stemme *ende* hem siin
loses his voice:ACC and him:DAT his
aensicht *root is ende*
face:NOM red is and
hem die puysten *uut broddelen*
him:DAT the pustules:NOM out break

ende hi sinen appotiit verliest ...
and he his appetite loses ...

'If the patient has head pain and dizziness and his eyes become red and his face all inflamed and he loses all his understanding and vomits bitter bile from the throat and it happens that his blood flows out through his nose and his ears and his speech disappears and (he) loses his voice and his face is red and pustules break out on him and he loses his appetite ...'

The patient is quite obviously being affected here! The writer is not simply describing each of the appalling symptoms, but he is describing and focussing on what is happening to the person. The phrase *bi aventueren* shows that it is all happening quite beyond the person's control and it is interesting to see that from this phrase on, the patient is encoded as a separate dative argument. (Note, also that it is the dative person which controls the coreferential deletion here). It certainly loses both semantic and stylistic content to translate each of these unpleasant bodily events with simply a possessive construction, as has been necessary in the loose English translation underneath. It is anyway not possible with the phrase concerning *puysten* 'pustules'. This can only translate as something like 'pustules break out on him'. The concept of inalienability, although relevant, is not adequate to account for these constructions: it captures nothing of the rhetorical force which this dative construction clearly had in Middle Dutch. Only relics of this construction still survive in the language today.

It stands to reason that a writer would choose this construction above a simple possessive construction like the genitive to emphasise the personal involvement of a possessor. But it is not difficult to see how a general possessive might then develop out of this. We finish by looking at examples where precisely this has occurred – the dative has actually grammaticalised into the usual marker of possession in the language.

8. Dative of possession

In a number of modern German dialects (for example, Swiss German, Alsatian, Pennsylvania German and in colloquial Standard German), the genitive survives only in relic forms. In these languages, possession is indicated by a possessive dative construction, formed with the possessor in the dative case in association with a possessive adjective preceding the entity possessed. Consider, for example, the following from Pennsylvania German:

(56) *Des is em Dadi sein Aarmschtuul*
this is the Grandpa:DAT his armchair
'This is Grandpa's armchair.'

(57) *Des is em Gaul sein Schwans*
this is the horse:DAT his tail
'This is the horse's tail.'

(58) *Es waar dem sein Hund*
it was this:one:DAT his dog
'It was this one's dog.'

and colloquial standard German:

(59) *Das ist dem Mann sein Fahrrad*
that is the man:DAT his bicycle
'That is the man's bicycle.'

The construction is also found in colloquial modern Dutch although with the breakdown of the case system the possessor is no longer dative. (Note, the possessive adjectives *z'n* and *d'r* are reduced forms of the full pronoun forms *zijn* 'his' and *haar/hun* 'her'/'their' respectively).

(60) *Dat is mijn broer z'n vrouw*
that is my brother his wife
'That is my brother's wife.'

(61) *Dat is die vrouw d'r kind*
that is the woman her child
'That is the woman's child.'

In the above examples (56)–(61), the dative NP (the possessor) is no longer a clausal argument, that is, a complement, but a nominal modifier or attributive. It is easy to see how the possessive dative would have grammaticalised out of the sort of dative of involvement expressions we have been looking at. Take, for example, a construction like modern German *Dem Mann ist sein Haus abgebrannt* (literally: 'the-man:DAT-is-his house-burned-down'; best rendered in English as something like 'the man had his house burn down on him'). Here the loss and injury to the man is emphasised by placing him in the dative as a separate clausal argument. This is what Curme (1904 [1970]: 501-502) would have called the 'dative of interest', where the dative emphasises the person to whose advantage or disadvantage the action or event results. This construction then becomes *Dem Mann sein Haus ist abgebrannt* (literally: 'the-man:DAT-his-house-is-burned-down'; that is, 'the man's house burned down') where 'the man' is understood now as a modifier of 'the house'. The closer relationship between the man and his house is now shown by the contiguous word order. In very early German texts it is sometimes unclear what meaning is intended, as both complement and attributive readings are possible. Examples like the following (from Lockwood 1968: 21) would have provided the way in for the dative to take over from the already declining genitive as a new marker of possession.

(62)	*Thaz*	*ih*	*druhtine*	*sinan*	*sun*	*souge*	(Otfrid)
	that	I	Lord:DAT	his	son:ACC	may:suckle	

'That I may suckle the Lord's son', or 'That I may suckle for the Lord his son'

In this example, it is unclear whether the dative NP is a possessive (that is, nominal attribute) modifying 'son', or a separate argument of the verb. That is, do we translate the phrase as – 'the Lord's son' (a simple possessive) or 'for the Lord' (where the Lord's personal involvement is made more prominent)? It is very probable that this sort of ambiguity exists in Middle Dutch also, although word order often makes it clear which interpretation is meant.

In both modern colloquial German and Dutch, the colloquial dative possessive is not possible unless the possessor is animate. For example, the following two sentences are ungrammatical:

(63)	**Das*	*ist*	*dem*	*Haus*	*sein*	*Dach*
	that	is	the	house:DAT	its	roof

(64)	**Dat*	*is*	*dat*	*huis*	*z'n*	*dak*
	that	is	the	house	its	roof

['That is the roof of the house.']

The restriction is directly explainable by its origins. As we have seen, this possessive construction has evolved out of the original personal dative, a construction which principally involved animate (or at least personified) entities; that is, entities capable of showing personal interest/involvement in an event (cf. in English, the awkwardness (ungrammaticality?) of something like ?**I kicked the chair in the leg*). Its associations with the dative are still felt; for example, it still preserves something of its original force and is perceived by speakers to give more prominence to the possessor than any of the alternative possessive constructions. As the construction develops further, however, and as it loses its link with the original dative construction, so we would expect it to include inanimate entities in its range. This has already happened in a languages like Afrikaans – grammaticalisation is now complete and this has become the *only* possessive construction available. For example, the following are both perfectly grammatical in Afrikaans:

(65) *Die huis se dak*
the house its roof
'The roof of the house'

(66) *Drie uur se werk*
three hour its work
'Three hour's work'

Though the meaning of this construction is first and foremost that of 'possession / ownership', as with possessive constructions generally, it can express a variety of other relations between nouns. This is clear in example (66) and also in examples like (67) below where, as in English, the expression can be interpreted in a number of different ways; for example, 'the photo of the boy'; 'the photo the boy possesses' or 'the photo the boy took'.

(67) *Die seun se foto*
the boy his photo
'The boy's photo'

Increasing generalisation of this sort indicates further grammaticalisation of this dative construction. Still another indication is the reanalysis of *se*: Originally a third person singular pronoun, it has now become a fixed item in this construction, regardless of gender and number of the possessor (cf. Burridge 1992a for a discussion of a similar development in Pennsylvania German).

Appendix 1: The case system of Middle Dutch

1. The articles

1.1. The definite article

The definite article in Middle Dutch has its origin in the set of demonstrative pronouns. The paradigm given below, therefore, contains forms for both the definite article and demonstrative pronouns. Even in Middle Dutch, though, the whole system was showing signs of breaking down. Case syncretism, collapse of gender and number distinctions were already apparent in the documents of the time.

Table 1.

		Masc.	Fem.	Neut.
Singular	N	*die (de)*	*die (de)*	*dat ('t)*
	A	*dien (den)*	*die (de)*	*dat ('t)*
	G	*dies (des/'s)*	*dier(e) (der)*	*dies (des)*
	D	*dien (den)*	*dier(e) (der)*	*dien (den)*
Plural (all genders)	N	*die (de)*		
	A	*die (de)*		
	G	*dier(e) (der)*		
	D	*dien (den)*		

(a) The three genders are neutralised in the plural.

(b) The reduced forms which are provided in the brackets may only appear as the definite article. The full *ie-* forms may be used as both the definite article and the demonstrative pronoun.

1.2. The indefinite article

Table 2.

		Masc.	Fem.	Neut.
Singular	N	*een (ene)*	*een/ene*	*een (ene)*
	A	*enen*	*een/ene*	*een (ene)*
	G	*eens*	*eenre/ere*	*eens*
	D	*enen*	*eenre/ere*	*enen*

(a) The indefinite article in Middle Dutch has its origin in the numeral *een* 'one'.

(b) The forms given in the brackets for both masculine and neuter are rare. Alternative forms given for the feminine appear with equal frequency.

2. Pronouns

2.1. The interrogative pronouns

Table 3.

		Masc.	Fem.	Neut.
Singular	N	*wie*	*wie*	*wat*
	A	*wien*	*wien/wie*	*wat*
	G	*wies/wes (wiens)*	*wies/wes/wier(e)*	*wies/wes (wiens)*
	D	*wien*	*wien/wier(e)*	*wien*
Plural (all genders)	N	*wie*		
	A	*wien/wie*		
	G	*wies/wes/wiere/wier*		
	D	*wien*		

(a) Originally, there were no morphologically distinct feminine and plural forms in the language. These were identical with the masculine forms. Under analogy with the demonstrative pronoun *die*, however, the alternative forms given in Table 3 gradually developed.

2.2. The demonstrative pronouns

The simple demonstrative pronouns are identical to the stressed full forms of the definite article. They are, therefore, listed in Table 1 above. In addition to these, Middle Dutch also has compound demonstratives *dese* 'this (one)'/'these (ones)' and *ghene* 'that (one)'/'those (ones)'.

Table 4.

		Masc.	Fem.	Neut.
Singular	N	*dese*	*dese*	*ditte/dit*
	A	*desen*	*dese*	*ditte/dit*
	G	*des*	*deser(e)/derre*	*des*
	D	*desen*	*deser(e)/derre*	*desen*
Plural (all genders)	N	*dese*		
	A	*dese*		
	G	*desere/deser/derre*		
	D	*desen*		

(a) The form *derre* is the most usual form for the Sing. Fem. genitive and dative cases, and the Plur. genitive case.

(b) Occasionally *dese* is found for the Sing. Masc. accusative and dative cases.

2.3 Personal pronouns

Table 5.

		1st person	2nd person	
Sing.	N	*ic/icke*	*du*	
	A	*mi*	*di*	
	G	*mijns*	*dijns*	
	D	*mi*	*di*	
Plur.	N	*wi(e)*	*ghi (-i)*	
	A	*ons*	*u/ju/jou*	
	G	*onser/onse/ons*	*uwer/uwes*	
	D	*ons*	*u/ju/jou*	
		3rd person singular		
		Masc.	Fem.	Neut.
	N	*hi (-i)*	*soe/si (-se)*	*het/hit (-(e)t)*
	A	*heme/hem (en(e))*	*hare/haer (-se)*	*het/hit (-(e)t)*
	G	*syns (-s)*	*hare/haer (-er(e))*	*syns (-s)*
	D	*heme/hem (-em)*	*hare/haer (-er(e))*	*heme/hem (-em)*
		3rd person plural (all genders)		
	N	*si (-se)*		
	A	*hem/hen (-se)*		
	G	*haers/hare/haer (-er(e)/-re)*		
	D	*hem/hen/him (-en)*		

(a) Alongside the set of full pronouns, Middle Dutch also has a set of corresponding reduced enclitic forms. These are given in brackets beside their corresponding full forms.

(b) The second person plural form *ghi* was used also as the polite form of address. By the early 17th century, however, *du* had virtually dropped out of the language in favour of *ghi* (which was then in turn replaced by *jij (je)* and a new second person plural form *jullie*). A new form *U* emerged during the 17th century as well.

(c) In subject-verb inversion, the initial *d* of enclitic pronouns *du* and *di* assimilates totally to the preceding *t* of the verbal ending as in *doestu (doest du).*

(d) When *ghi* is used enclitically, the final *t* of the preceding verb and the initial voiced velar fricative of *ghi* assimilate totally to give *d* (except when they are preceded by another *t*) – *segdi, moochdi* but *laetti.*

Appendix 2: Middle Dutch sources

Braekman, W.L

1987 "Een merkwaardige collectie secreten uit de vijftiende eeuw [A curious collection of secrets from the fifteenth century]", *Verslagen en Mededelingen van de Koninklijke Academie voor Nederlandse Taal- en Letterkunde* 2: 270-287.

Braekman, W.L (ed.)

1970 *Middelnederlandse Geneeskundige Recepten: Een bijdrage tot de geschiedenis van de vakliteratuur in de Nederlanden* [Middle Dutch medical recipes: a contribution to the history of technical literature in the Netherlands]. Gent: Koninklijke Vlaamse Academie.

1975 *Medische en technische middelnederlandse recepten: Een tweede bijdrage tot de geschiedenis van de vakliteratuur in de Nederlanden* [Medical and technical Middle Dutch recipes: a second contribution to the history of technical literature in the Netherlands] Gent: Koninklijke Academie voor Taal en Letterkunde.

(N.B. Both the Braekman volumes contain a compilation of texts from the 14th–16th century)

Daems, W.F. (ed.)

1967 *Boec van medicinen in Dietsche* [Book of medicine in Dutch]. (Hollandish ca. 1300) Een Middelnederlandse compilatie van medisch-farmaceutische literatuur [A Middle Dutch compilation of medical-pharmaceutical literature]. [Thesis, Leiden.]

Frencken, H.G. (ed.)

1943 *T Bouck van Wondre* [The book of wonder]. (Brabantish ca 1513) [Thesis, Rijksuniversiteit Leiden.]

Leersum, E.D. (ed.)

1928 *Het 'Boec van Surgien' van Meester Thomaes van Scellinck* [The 'Book of Surgery' by Master Thomaes of Scellinck]. (Brabantish 1343) Amsterdam: Opuscula Selecta Neerlandicorum de Arte Medica.

Vanderwiele, L.J. (ed.)

1970 *Middelnederlandse Versie van de 'Circa Instans' van Platearius* [(A) Middle Dutch version of the 'Circa Instans'] (Hollandish 1387) Oudenaarde: Sanderus.

Notes

1. I am grateful to the editors of this present volume, Hilary Chappell and Bill McGregor, whose critical comments and advice helped to clarify a number of the issues presented here. I also wish to express my thanks to Bruce Donaldson and Wim Hüskin for their comments and suggestions.
2. I will only be indicating the grammatical relations for body part and possessor, using the following abbreviations: (NOM)inative; (ACC)usative; (DAT)ive; (GEN)itive.
3. In early Dutch both words *lichaem* and *lijf* had the general meaning of 'body', as they still do in Modern Dutch. However, they could also be used – presumably euphemistically – to mean 'stomach' and more specifically 'bowel'. In this way, they could also refer to actual bowel movements, and this is the meaning in a number of examples here.

4. The distinction had collapsed for the masculine but remained distinct for feminine and neuter nouns (see Appendix 1). However, at this time the traditional masculine/feminine/neuter gender distinction was also collapsing into a two-way system. Many nouns belonged to more than one gender which meant that gender was not always a reliable guide here. In the following, for instance, *hersen* (usually a feminine noun) receives a masculine/neuter ending on its modifying article – *denghenen die den hersen verwout is* 'those:DAT who the brain:DAT is frenzied'.
5. From the late 16th century on things change – the appearance of these accusative subjects becomes more frequent and indeed far less systematic. For example, you start to find accusative subjects of transitive verbs with overt objects. Dialect differences also begin to emerge. In the north, with the levelling of the nominative, constructions like these eventually disappear. In Flemish, however, the accusative eventually grammaticalises as the masculine marker for all cases.
6. Note, this does not include the weather constructions (*it rains* etc.) which have also been described as impersonal.
7. Elsewhere (Burridge 1986) I have argued for the prominence of the notion topic in the organisation of Middle Dutch syntax.
8. Like the body part constructions (49)–(51), most of the double-subjects involve a part-whole relationship between the initial NP and some NP later in the sentence (usually the grammatical subject), but not all. For example, *Avicenna seit dat die wonden van den pesen gheen dinck quader en is noch meer deeret dan coude dinghen* 'Avicenna says that wounds of the tendons (topic) – nothing is worse, nor is more harmful than cold things'.

References

Burridge, Kate
1986 "Topic-Prominence in Middle Dutch", *Australian Journal of Linguistics* 6: 57-75.
1992a "Creating grammar: examples from Pennsylvania German, Ontario", in: Kate Burridge – Werner Enniger (eds.), 199-241.
1992b *Some aspects of syntactic change in Germanic: with particular reference to Dutch.* Amsterdam: John Benjamins.

Burridge, Kate – Werner Enninger (eds.)
1992 *Diachronic studies on the languages of the Anabaptists.* Bochum: Brockmeyer.

Chafe, Wallace L. (ed.)
1976 "Givenness, contrastiveness, definiteness, subjects, topics, and point of view", in: Charles Li (ed.), 25-56.

Chappell, Hilary
1995 "Inalienability and the personal domain in Mandarin Chinese discourse", [this volume.]

Chappell, Hilary – William McGregor
1989 "Alienability, inalienability and nominal classification", *Proceedings of the Berkeley Linguistic Society* 15:24-36.
1995 "Prolegomena to a theory of inalienability", [this volume.]

Curme, George O.
1904 *A Grammar of the German Language*. New York: Frederick Ungar.
[1970] [Reprinted New York, Fredrick Ungar.]
Fox, Barbara
1981 "Body part syntax: towards a universal characterization", *Studies in Language* 5: 323-342.
Givon, Talmy
1976 "Topic, Pronoun, and Grammatical Agreement", in: Charles Li (ed.), 149-188.
Horst, Johannes Martinus van der
1981 *Kleine Middelnederlandse syntaxis*. Amsterdam: Huis aan de drie grachten.
1985 "Verkenning van de onpersoonlijke constructies", *Tijdschrift voor Nederlandse taal- en letterkunde* 101: 34-63; 81-92
Hopper, Paul – Elizabeth Traugott
1993 *Grammaticalisation.* Cambridge: Cambridge University Press.
Li, Charles (ed.)
1976 *Subject and Topic: a new typology of language*. New York: Academic Press.
Lightfoot, David
1979 *Principles of Diachronic Syntax.* Cambridge: CUP.
Lockwood, William B.
1968 *Historical German Syntax.* Oxford: Clarendon Press.
Manoliu-Manea, Maria
1995 "Inalienability and topicality in Romanian. Pragma-semantics of syntax", [this volume.]
McCawley, Noriko A.
1976 "From OE/ME 'impersonal' to 'personal' constructions: what is a subject-less S", in: Sanford B. Steever – Carol A. Walker – Salikoko S. Mufwene (eds.), 192-204.
Neumann, Dorothea
1987 *Objects and spaces: A study in the syntax and semantics of the German case system*. (Studien zur deutschen Grammatik 32) Tübingen: Gunter Narr.
1995 "The dative and the grammar of body parts in German", [this volume.]
Sapir, Edward
1921 *Language: An introduction to the study of speech*. New York: Harcourt Brace.
Steever, Sanford B. – Carol A. Walker – Salikoko S. Mufwene (eds.)
1976 *Papers from the Parasession on Diachronic Syntax*. Chicago: Chicago Linguistics Society.
Wierzbicka, Anna
1979 "Ethnosyntax and the philosophy of grammar", *Studies in Language* 3.3: 313-383.

Inalienability and topicality in Romanian: Pragma-semantics of syntax

Maria Manoliu-Manea

As has been already pointed out in crosslinguistic descriptions, the semantico-syntactic feature 'inalienable possession' is not a simple reflection of a 'state of affairs' dealing with inseparable terms, but rather an expression of a certain interpretation of the world we talk about in which the part and the whole are presented as being linked by an intrinsic relation of solidarity rather than possession.[1] Consequently everything which affects the part is viewed as acting simultaneously upon the whole.[2] Although limited to Romanian data, the present paper is aimed at bringing supporting evidence in favour of the hypothesis that coding of whole-part relations is governed not only by morphosyntactic and semantic patterns, but also by discourse and cognitive conditions such as (a) topicality and (b) centrality. Topicality will be used in relation to topic, roughly defined as the entity 'speakers want to talk about'; the topic is identifiable across sentence-boundaries because it is usually shared by several sentences.[3] Centrality defines the figure in the mental and, implicitly, linguistic reconstruction of the event. In Langacker's terms, the figure is the structure perceived as "standing out" from the remainder (the ground) and accorded special prominence as the pivotal entity around which the scene is organised (1987: 120). Centrality does not necessarily coincide with topicality. The figure may correspond to the rhematic part, to the focalised constituent or to a newly introduced topic (subtopic).[4]

As I hope to demonstrate in this paper, the lowest degree of part autonomy is encoded by constructions in which the NP_{whole} is envisaged as being both *topical* and *central*.

As soon as the NP_{whole} is no longer topical and central, the part may be presented as having various degrees of autonomy as shown by the fact that in part-centered constructions the NP_{part} may occur with various determiners (epithets, demonstratives or possessives) or it may be subjectivised.

After a brief introduction dealing with those morphosyntactic features which are important in encoding whole:part relations in Romanian (namely case markers and their pragma-semantic structure), and a brief description of semantic features of predicate classes, the evidence supporting this hypothesis will focus on the following phenomena:

(a) The choice between	S	V	DAT_{whole}	ACC_{part}
and	S	V	ACC_{whole}	PP_{part};
(b) The choice between	S_{part}	V	DAT_{whole}	
and	S_{whole}	V	PP_{part}.	

1. A typological characterisation of Romanian

1.1. Word order and topics

It is claimed that Romanian shares with other Romance languages an important set of typological features, among them, a predominant S V O order. However, recent studies have pointed out that contemporary Romanian is characterised by a strong tendency toward verb-initial structures (see Dobrovie-Sorin 1989),which may account for the fact that postverbal subjects have a slightly higher frequency than preverbal subjects (see Manoliu 1987). Romanian has a higher flexibility in word ordering than any other Romance language, which is probably due to two main factors: (i) typologically, it belongs to the central and eastern European languages which have better preserved nominal inflection (see below, 1.1.1.) and (ii) it has shown an increasing sensitivity to topicality markers. Pronominal clitics occur in preverbal positions in conjunction with finite forms (indicative, subjunctive, conditional) and, consequently, they may refer to topical constituents even when they are not subjectivised. Moreover, clitic copies of NPs have become the most frequent markers of discourse prominent entities such as topicalised or focalised NPs.

Since in the following sections, case selection is one of the most important aspects characterising the grammar of 'inalienable possession', it is necessary to describe briefly the structure of the Romanian noun declension.

1.1.1. Cases and roles

1.1.1.1. Case markers

Unlike its Romance cognates, Romanian still has a three-case noun declension. The paradigm of the feminine distributional class of nouns has two forms, namely, one for the nominative-accusative singular and another for the genitive-dative singular, which is also the only form for the plural (see Table 1).

Table 1. Case forms of nouns

1st declension (feminine nouns): e.g. *fată* 'girl'		
SG:	NOM/ACC	*fată*
	GEN/DAT	*fete*
	VOC	*fato!* (spoken Rom.)/*fată*!
PL:	NOM/ACC/GEN/DAT	*fete*
	VOC	*fetelor!* (with a definite article)
2nd declension (masculine nouns): e.g. *copil* 'child'		
SG:	NOM/ACC/GEN/DAT	*copil*
	VOC	*copile!*
PL:	NOM/ACC/GEN/DAT/VOC	*copii*

The definite article may also be the sole bearer of the NP case markers, either as a postnominal bound morpheme (see Table 2), or as a proclitic free morpheme, before proper nouns and kinship terms (see Table 3).[5]

Table 2. Noun declension with an enclitic definite article

masculine nouns: e.g. *copil*UL (Literally: child-the 'the child')		
	Singular	Plural
NOM/ACC	*copilUL*	*copiiI*
GEN/DAT	*copilULUI*	*copiiLOR*
VOC	*copilULE!*	*copiiLOR!*
feminine nouns: *mama* (Literally: mother-the 'the (my) mother')		
	Singular	Plural
NOM/ACC	*mamA*	*mameLE*
GEN/DAT	*mameI*	*mameLOR*

The preposition *pe* 'on' has become a compulsory marker for human direct objects before proper nouns, kinship terms, personal pronouns (emphatic forms) and other pronouns when referring to persons. It may

Table 3. Proper-noun declension

	MASC	
	e.g. *Ion* 'John'	*Andrei* 'Andrews'
NOM/ACC	*Ion*	*Andrei*
GEN/DAT	*LUI Ion*	*LUI Andrei*
VOC	*Ioane!*	*Andrei!*
	FEM	
	e.g. *Maria*	*Carmen*
NOM/ACC/VOC	*Maria*	*Carmen*
GEN/DAT	*MariEI*	*LUI Carmen*
VOC	*Mario!* (substandard)	

also introduce common nouns, if left-dislocated or in postverbal positions, when they belong to foregrounded events (see Manoliu-Manea, 1988b).

(1) *l-* *am* *văzut* *PE* *Ion/copil*
him:ACC have:I seen ON John/child
'I saw John/ the child.'

The clitic copy of the prepositional direct object is also compulsory before proper nouns, kinship terms, and common nouns if specific but not followed by a definite description (see *l-:* 'him:ACC' in (1)).

Personal pronouns have a three-case system (as shown in Table 4, in which the paradigm of the 3rd person is given).

Table 4. Personal pronouns (third person)

	Singular			
	Conjoint		Disjoint	
MASC	MASC	FEM	MASC	FEM
NOM	*ø*		*el*	*ea*
DAT	*(î)i*		*lui*	*ei*
ACC	*(î)l*	*o*	*(pe) el*	*ea*
	Plural			
	Conjoint		Disjoint	
	MASC	FEM	MASC	FEM
NOM	*ø*		*ei*	*ele*
DAT	*le*		*lor*	
ACC	*(î)i*	*le*	*(pe) ei*	*ele*

1.1.1.2. Case and topic hierarchy.

Case-assignment in the area of whole-part relations follows closely the topic hierarchy of pragmatic roles as defined by Givon (1984: 139). Topic hierarchy is a ranking order of the various semantic roles according to their probability of becoming the more continuous topic in discourse, and it may be formulated as follows:

Agent > Dative > Patient > Locative > Instrument > Manner.

The Patient (also called Undergoer) is the affected participant, 'being in a state' or 'undergoing a change in state'. The Dative is an Experiencer (see, for example, the subject of verbs of knowledge or feelings) or a Beneficiary (when involved in receiving-giving activities). In Romance languages, the Agent corresponds most frequently to the subject (of an active construction); the Experiencer to the subject or to the indirect object (bearing a dative case or a prepositional marker); the Patient is the prototypical role of the direct object in the accusative, and the Locative prefers prepositional constructions.

In Romanian, with two-place predicates expressing activities which may be envisaged as affecting both the part and the whole, the case markers are assigned according to the topic hierarchy. For example, when the topical whole selects the accusative (Patient), then the part takes a prepositional locative marker (Locative), as in (2). When the topical whole selects a dative marker (Experiencer), then the part takes an accusative marker (Patient), as in (3)

	V	ACC_{whole}	PP_{part}					
(2)	*spală*	*-te*	*pe*	*mâini*	*şi*	*vino*	*la*	*masă*!
	wash	you:REFL:ACC	on	hands	and	come	to	table!

'Wash your hands and come to the table!'

	DAT_{whole}	V	ACC_{part}	
(3)	*şi*	*a*	*vopsit*	*păr*ul
	3:REFL:DAT	has:she	dyed	hair:the:ACC
	şi	*arată*	*mai-bine*	
	and	looks:she	better	

'She dyed her hair and she looks better.'

With verbs expressing a change in state (inchoatives), the topic hierarchy seems to be violated, since the topical whole may take a dative marker and then the part corresponding to the 'focus' is subjectivised (see (4)).

	DAT_{whole}	V:REFL			NOM_{part}:
(4)	*mi*	*s-*	*a*	*î nroşit*	*faţa*
	to:DAT	REFL:3SG	has	got:red	face:the

am stat prea mult la soare
have:I stayed too much in sun

'My face became red (sun-burnt). I was in the sun too long.'

This 'violation' may account for the postverbal position of the NP-subject, which is the preferred order for direct objects (for details see below 3.).

1.2. Predicate features

If the semantic features of the NPs belonging to the area of whole-part relations (parts of the body, kinship terms) have been given serious consideration in the Romance domain, little attention has been paid to the constraints imposed upon the choice of role markers by the inherent features of the verb, although in more recent approaches, roles have been assigned to the verb (since they are determined by the interpretation of the entire event) and copied onto the noun (see however, Cristea 1974). In an ethno-syntactic framework, Wierzbicka (1988: 169) also emphasises the importance of the 'event category' in accounting for the constructions under discussion. The following categories of predicates – which have been cross-linguistically identified as being grammatically relevant – have been taken as a primary filter in accounting for the relation between syntactic subcategories and semantic differences:[6]

(1) Dynamic: stative verbs such as *a fi* 'to be' are typical instances of non-dynamic predicates, while predicates expressing activities are characterised by the feature "dynamic": *a merge* 'to go', *a lovi* 'to hit'.
(2) Causative: characterises predicates referring to activities which cause an object to undergo a certain change; their "normal primary term" is an agent.
(3) Change in position (movement) characterises verbs such as *a ridica* 'to raise', *a coborî* 'to go down', 'to lower'.

(4) Inchoativity refers to the fact that the activity has a certain point which coincides with the beginning of a new state: e.g. *a slăbi* 'to slim', *a se îngrăşa* 'to get fat'.

Further refinements have been introduced in order to account for specific constraints on case-assignment to NPs referring to "inalienable possession". For example, separation characterises predicates that refer to activities by which the part is separated from the whole. For example, *mi-a tăiat degetul* 'to me (it) (has) cut the finger'.

The analysis of predicate features will focus on the constraints imposed on encoding the lowest degree of autonomy, that is on the way in which the match between topicality and centrality of the whole is realised.

2. Central whole versus central part

2.1. S V O: ACC_{whole} PP_{part}

Unlike in other Romance languages, in Romanian the topical whole may be expressed by a direct object (in the accusative). The accusative construction is, in fact, the regular choice with two-place dynamic predicates, when the whole represents the figure, that is, it is central and the part is viewed as an 'aspect' of the whole.

	V	ACC_{whole}	PP_{part}	
(5)	*spală-*	*te*	*pe*	*mâini*
	wash:IMPER	you:ACC	on	hands

şi vino la masă!
and come to table!

'Wash your hands and come to the table!'

In (5), the addressee (the whole) corresponds both to the topic and to the figure. The NP 'you' plays the role of the subject in both sentences (sign of its topicality), but in the first sentence it also functions as a direct object, which means that 'you' is treated as the central entity affected by the activity in question (one has to have clean hands in order to avoid catching any disease). The part is expressed by a locative prepositional construction (external to the core of the sentence) and is presented as a restrictive specification of the affected whole, an interpretation which is responsible for its incompatibility with any further specification (that is with a definite article and/or a modifier):

(6) **spală- te pe mâinile murdare/tale*!
wash you:ACC on hands:the dirty/yours
'Wash your dirty hands!'

In an ethno-syntactic framework (a model which describes semantic areas by a set of primitives, formulated in terms of simple sentences), the meaning conveyed by the accusative-locative construction may be roughly formulated as follows:

(a) something happened to Z (in which Z represents the whole)
because X (or Z) did something to Y (the part)
and I think of it (what happened to Y) as something that happened to Z and
not to something other than Z.

(See Wierzbicka's analysis of English and German accusative of the whole (1988: 172-173, 198-204)).

In other Romance languages, such as Spanish, the whole takes a dative marker even when the part is introduced by a preposition:

For example:

(7) *le golpeé en la nariz* (Spanish)
him:DAT hit:I in the nose
'I hit him in the nose.'

2.2. S V DATIVE$_{whole}$ DIRECT OBJECT$_{part}$

The extended use of dative case markers for the topical whole is one of the most striking characteristics of Romance languages.[7]

(8) *lave- toi les mains*
wash:IMP you:DAT the hands
'Wash your hands!' (French)

(9) *lava- te las manos!*
wash:IMP you:DAT the hands
'Wash your hands!' (Spanish)

In (8) or (9), the dative construction encodes the interpretation according to which the whole is marginal but topical. Consider (10), in which 'you' is the topic:

(10) *lave- toi les mains*
wash:IMPER you:DAT the hands
et viens déjeuner!
and come:IMPER have:lunch:INF
'Wash your hands and come to have lunch!' (French)

The noun referring to the whole is assigned the role of an Experiencer/Benefactive, while the part corresponds to the direct object – the prototypical expression of a Patient. According to Wierzbicka (1988: 171-174), the dative construction may express two interpetations of the event. When the possessive adjective is absent and the NPpart takes a definite article, the dative construction is associated with the idea that the part of the body is viewed as an aspect of the person (as in (8)). When the possessive is co-occurrent, the part of the body is viewed as an object separate from, but related to, the whole (see (11)):

(11) *Pierre s' est lavé sa tête (sale)*
Peter himself:DAT AUX washed his head (dirty)
'Peter washed his (dirty) head.'

According to Wierzbicka (1988: 172), in (8) the dative of the whole is related to the following interpretation of the world:

(b) something happened to part Y of Z's body
because X did something to it
I am thinking of it (what happened to Y) as something that happened to Z
not to something other than Z

while in (11), the interpretation of the event corresponds to the following formula:

(c) something happened to part Y of Z's body
because X did something to it
it (what happened to Y) can be thought as something that happened to Z

In Romanian, the dative construction of the whole is also possible, but only if the part is central and topical, as in (12):

(12) *spală-* *ţi* *mâinile*
wash you:DAT hands:the:ACC
cu *sănt* *murdare* *rău*!
because are dirty badly!
'Wash your hands, because they are very dirty!'

The difference between formula (a) on the one hand, and formulae (b) and (c) on the other, can account for the semantic distinction between the accusative and the dative of the whole by postulating a change in the point of departure when reconstructing mentally and linguistically the event. In other words, the Romanian accusative-locative construction is whole-centered, while the dative-accusative construction is part-centered.

Since in the dative-accusative construction, the part is topical and central (and has thus a higher degree of discourse saliency than the whole), it may be assigned a higher degree of autonomy, which can account for the fact that it always takes a definite article (see (12)) and can take a determiner and a modifier (see (13)).

(13) *spală-* *ţi* *mâinile* *astea* *murdare*!
wash yourself:DAT hands these dirty
'Wash these dirty hands of yours!'

3. Central part versus marginal part: Accusative versus prepositional phrase

3.1. Causatives

With two-place causative predicates, the match between topicality and centrality of the whole is encoded in the accusative-locative construction, that is:

$$\text{NP:ACC}_{\text{whole}} \;\; \text{V} \;\; \text{PREPOSITIONAL PHRASE}_{\text{part}}$$

3.1.1. The semantic area

The accusative construction (see 1.1.), which is the first choice when the whole is topical, may be used not only with NPs referring to body parts, but also with NPs denoting clothing when in contact with the human body. Consider example (14):

(14) *mă* *închei* *la* *cămaşă*
me:ACC button:I at shirt
'I button (up) my shirt.'

which presupposes that the subject 'I' is wearing the shirt.

See also the French construction

S ACC_{person} V $PP_{part/clothing}$

as represented by (15), which is acceptable only when the person is wearing the piece of clothing in question:

(15) *il* *la* *tire* *par* *la* *manche*
he her:ACC pulls by the sleeve
'He pulls her by the sleeve.' (Prévost, in Grevisse 1986: 950)

But unlike in Romanian, in (15) the PP is treated as an Instrumental, as indicated by the preposition *par* 'by', rather than as a Locative.

The accusative construction may be also used for expressing the lowest degree of part autonomy in reference to other categories of objects, such as clothing, a few domestic objects or vehicles driven by the possessor, such as cars (*maşină*), chariots (*car*) and bicycles.

(16) *scurteaz-* *o* *la* *mâneci*!
shorten it:FEM:SG:ACC at sleeves!
'Shorten its sleeves!' (talking about a dress)

(17) *ţine-* *o* *de* *toartă*!
hold it:FEM:SG:ACC by handle!
'Hold it by the handle!' (talking about a cup)

(18) *a* *lovit* *maşina* *în* *dreapta*
has:he hit car:the:ACC in right:the
'He hit the car on the right side.'

The dative construction (see section 2.2) is also acceptable with causative predicates, if the part is topical:

(19) *îmi inchei cămaşa*
me:DAT button:I shirt:ACC:the
'I am buttoning my shirt.'

Compare (19) with the Spanish example in (20):

(20) *me he desabrochado la camisa*
me:DAT have:I unbuttoned the shirt
'I unbuttoned my shirt.'

where the whole is topical.

In Romanian, the dative construction is not always associated with the part-whole relation. It is also possible when referring to kinship, or simple possession, with the condition that the 'possessor' is topical, but not central.[8] For example, when somebody is helping me to pack, I can ask him to button my shirts by saying:

(21) *te rog încheie- mi cămăşile,*
you:ACC ask:I button:you:IMPER me:DAT shirts:the:ACC
că eu am o groază
because I have a terrible:amount
de lucruri de făcut
of things to do:SUPINE
'Please button my shirts!, because I have to do a lot of (other) things.'

(22) *îmi ud grădina*
1st:REFL:DAT water garden:the:ACC
'I am watering my garden.'

A similar situation may be found in Italian or Portuguese, where datives are assigned to all kinds of possessions:

(23) *il professore ci guardò il compito*
the professor us:DAT looked the paper:ACC
'The professor looked at our papers.' (Italian)

(24) *o profesor me corrigué as redações*
the professor me:DAT corrects the papers:ACC
'The professor is correcting my papers.' (Portuguese)

3.1.2. Case assignment and metaphorical meanings

The difference in topicality and, implicitly, in case-assignment, may reflect a wider semantic distinction at the level of the entire sentence. The metaphorical meaning is limited to only one of the constructions, usually to the accusative-locative combination.[9]

Compare (25) with (26):

(25)

mă	*frec*	*la*	*ochi,*	*ca*	*săvăd*	*mai*	*limpede*
me:ACC	rub:I	at	eyes	to	see:SUB	more	clearly

'I am rubbing my eyes to see more clearly.' (usually because I do not believe that what I see is true)

(26)

nu	*ţi*	*mai*	*freca*	*ochii,*
not	you:DAT	continuously	rub	eyes,

că	*ai*	*să-*	*i*	*infectezi*
because	are:you:going	to	them	infect:you:SUBJ

'Do not go on rubbing your eyes, because you're going to infect them.'

The accusative of the whole is also limited to the figurative meaning if the 'possession' refers to a piece of clothing in contact with the human body. Example (27) is a common idiomatic expression and does not presuppose that the hat is actually on the possessor.

(27)

m-	*ai*	*lovit*	*în*	*pălărie*
me:ACC	have:you	hit	in	hat

Literally: 'You hit me in the hat.'

This means 'what you are saying (or doing) is so stupid that you fail to hit the target', since you hit me on the hat and not on the head. The dative construction has only the strict physical meaning in which 'the hat' is not viewed as intrinsically related to the person but as the sole Patient:

(28)

mi-	*ai*	*lovit*	*pălăria*
me:DAT	have:you	hit	hat:the:ACC

'You hit my hat.'

3.1.3. Reflexive causatives

With reflexives of transitive verbs expressing the fact that the whole suffers or is hurt, the accusative construction is more frequently the expression of "a non-agentive subject": the whole did not act deliberately in order to affect the patients as in (29):

(29)	*m-*	*am*	*udat*	*la*	*picioare*
	1sg:REFL:ACC	got:I	wet	at	feet

'My feet got wet.' (usually because I stepped into a puddle)

An agentive subject is more likely to choose the dative of the whole. Compare (30) and (31):

	ACC$_{whole}$	V		PP$_{part}$		PP$_{instrument}$		
(30)	*m-*	*am*	*lovit*	*la*	*picior*	*de*	*o*	*piatră*
	me:ACC	have	hit	at	leg	of	a	stone

fiindcă	*era*	*întuneric*
because	it:was	dark

'I hit myself on the leg by bumping into a stone because it was dark.'

	DAT$_{whole}$	V		ACC$_{part}$	PP$_{instrument}$		
(31)	*mi-*	*am*	*lovit*	*piciorul*	*de*	*o*	*piatră*
	me:DAT	have:I	hit	leg:the:ACC	in	a	stone

?fiindcă	*era*	*întuneric*
because	it:was	dark

Literally: 'I had my leg hit a stone (?because it was dark).'

The event in example (31) cannot be interpreted as the effect of walking down a dark road, but of a voluntary activity (with various purposes). A construction in which the whole is presented as an Agent assigns a certain discourse prominence to the whole, so it is less appropriate for encoding a relation of 'solidarity' between the whole and the part envisaged as mere 'affected participants'.

3.2. [+Separation]

Verbs such as *a tăia* 'to cut' may express the separation of the part from the whole if paired with the dative of the possessor:

(32) *i- au tăiat un deget*
him:DAT have:they cut a finger:ACC
'They cut off one of his fingers.'

When such verbs express an activity which does not separate the part, but affects the whole in a limited way, as indicated by the locative construction of the possessum, the 'owner' takes the accusative marker as in (33):

(33) *m- am tăiat la deget*
me:ACC have:I cut on finger
'I cut myself on the finger.'

In this case, the domain of the part is strictly limited to the body, and does not include clothing, because such activities cannot be envisaged as presenting the piece of clothing as a restrictive location on the human body. This is shown by (34):

(34) **m- ai tăiat la cămaşă*
me:ACC have:you cut on shirt
('You cut me on (my) shirt.')

3.3. Transitive verbs of movement

Verbs which express a movement of the part (such as *a ridica* 'to raise', *a coborî* 'to lower', *a închide* 'to close', *a deschide* 'to open') may also occur as two place-predicates in a structure such as:

$S_{Agent-whole}$ V O_{part}

If the Agent has the same referent as the whole to which the part belongs, the NP_{whole} (either adnominal or dative) is present only when it conveys a higher degree of part-autonomy. Compare (35), (36) and (37):

(35) *Petru ridică mâna*
Peter raises hand:the
'Peter is raising his hand(arm).'

(36) *Pierre lève la main* (French)
Peter raises the hand
'Peter is raising the hand.'

(37) *Pierre lève sa main*
Peter raises his hand
'Peter is raising his hand.' (French)

In example (36), there is no possessive adjective referring to the whole, while in example (37), the possessive (*sa* 'his:FEM') has the same referent as the subject, Peter.

In Wierzbicka's words (1988: 178), constructions such as (35) and (36) express an 'indirect action in which the part of the body is active and viewed as an aspect of the person'. The occurrence of the possessive adjective has been connected with an insistence upon the 'voluntary character of the activity' (for details see Hatcher 1944, Kayne 1975: 167-169), and, consequently, with a higher degree of part autonomy (see also Wierzbicka 1988: 177). This definition of the semantic information brought into the picture by the co-presence of the possessive can hardly account for utterances such as (38), however:

(38) *Il avait, sans le savoir, réuni ses pieds, croisé ses mains sur sa poitrine.*
(Malraux, in Grevisse 1986: 950).
'He had put his feet together, without being aware of it, crossed his hands on his chest'.

In order to account for utterances such as (37) and (38), it seems more appropriate to consider that the occurrence of the possessive brings in the idea that the relation between the part and the whole is presented as being more relaxed and, consequently, the activity may be envisaged either as being imposed on the part by an agentive whole or as an activity of the part which is not controlled by the whole (as in (38)).

In Romanian an increase in the degree of agentivity is marked by the co-presence of a reflexive dative for the whole. Compare ((39) and (40):

(39) *a întors capul*
has:he turned head:the:ACC
'He turned his head.'

(40) *şi- a întors capul*
3sg:REFL:DAT has:he turned head:the:ACC
ca să nu- l mai vadă
to SUBJ not him:ACC no:more sees:SUBJ
'He turned his head not to see him any more.'

In (39), the possessive adjective is less likely to occur, as is shown by (41a):

(41) a. **?a* *întors* *capul* *lui/său*
has:he turned head:the:ACC his
'He turned his head.'

In (40), since the presence of a reflexive dative for the whole is a marker of a higher degree of intentionality, the possessive is more acceptable when the whole is under contrastive focus with the intentional stress on *său* 'his'.

b. *?şi:-* *a* *întors* *capul* *său*
3sg:REFL:DAT has:he turned head:the:ACC his:REFL
'He turned his head.'

3.4. Verbs of perception

Stative or causative predicates represented by verbs of perception usually co-occur with the dative of the whole.

(42) *i-* *am* *văzut* *profilul*
her:DAT have:I seen profile:the:ACC
'I saw her profile.'

(43) **am* *văzut* *-o* *la* *profil*
have:I seen her:ACC at profile
('I saw her profile.')

The NP_{part} may be determined by an adjective:

(44) *i-* *am* *privit* *minile* *(albe)*
her:DAT have:I looked hands:the:ACC (white)
'I looked at her white hands.'

The constraint in question is due to the fact that 'perception' cannot be conceived in Romanian as an activity that simultaneously focusses on the whole and is confined to one of its parts.

4. Subjectivation of the part: Nominative vs. PP

4.1. Causatives

S_{part} DAT_{whole} $V_{causative}$

The high degree of part autonomy encoded in constructions in which the whole is assigned a dative marker is compatible with the subjectivisation of the part, either with active or passive forms of the verb.

4.1.1. Active predicates

With two-place predicates (causatives such as *a lovi* 'to hit' or verbs of directional movement such as *a coborî* 'to lower'), the subjectivised part is presented as a specification of the topical whole which belongs to the newly introduced information, although it usually occurs in preverbal position. For example, in (45) the topic is the person, 'me', which is encoded by the subject in the first and third sentence:

(45)

mergeam	*prin*	*întuneric*		
walked:I:IMPF	through	dark		
mâna	*mi*	*s-*	*a*	*lovit*
hand:the:NOM	me:DAT	3sg:REFL	has	hit
de	*ceva*	*tare,*	*necunoscut*	
of	something	hard,	unknown	
şi	*m-*	*am*	*oprit*	*speriat*
and	me:ACC	have:I	stopped	frightened

'I was walking through the dark .... My hand hit something hard, unknown, and I stopped, frightened.'

See also (46), in which the directional verb *a coborî* 'to go down' preserves its active status and does not co-occur with a reflexive of the part:

(46)

mâna	*i-*	*a*	*coborât*	*pe*	*genunchii*
hand	him:DAT	has	gone:down	on	knees:the
Anei					
Ann:GEN:the:GEN					

'His hand went down on Ann's knees.'

Such constructions are the mark of a more relaxed relation between the part and the whole, that is they refer to an unintentional activity, in which the part is envisaged as acting by itself.

4.1.2. Passivisation

With the dative of the whole, the topical part may be subjectivised in a passive construction:

(47)	*piciorul*	*mi-*	*a*	*fost*	*rănit*	*de*	*o*	*piatră*
	leg:the:NOM	me:DAT	has	been	hurt	by	a	stone

'My leg has been hurt by a stone.'

Given the fact that the difference between active and passive rests upon the reversal of the figure-ground alignment (see Langacker 1987: 141, 351-352), such a construction seems less acceptable than its active counterpart (30), since it is in contradiction with the low rank occupied by an inanimate Patient on the topic hierarchy and the high rank of an animate Experiencer. Its thematic structure may be roughly represented as follows:

S ('leg' = NOM:part:Undergoer – secondary topic or contrastive focus)
DAT('I' = whole:Benefactive – Topic) V:Passive by PP ('a stone' : Cause)

However, (47) is not impossible because it contains a rhematic external cause. It may be perfectly acceptable in a situation such as the one described by (48), in which the part is a newly introduced secondary topic:

(48)	*piciorul*	*mi-*	*a*	*fost*	*rănit*	*de*	*o*	*piatră*
	leg:the:NOM	me:DAT	has	been	hurt	by	a	stone

atât	*de-tare*	*încât*	*a*	*trebuit*
so	strongly	that	have:I	had

să	*mă*	*opresc*	*să-*	*l*	*bandajez*
SUBJ	me:ACC	stop:SUB	SUBJ	it:ACC	bandage:SUBJ

'My leg was so badly hurt by a stone, that I had to stop to bandage it.'

The same applies to (49), in which the part is focalised:

(49) *piciorul mi- a fost rănit*
leg:the:NOM me:DAT has been hurt
de o piatră, nu capul
by a stone, not head:the
'My leg was hurt by a stone, not my head.'

4.2. Statives of 'suffering' (V_s)

Stative predicates such as *a durea* 'to ache', 'to be in pain', *a mânca* 'to itch', which express the fact that 'someone feels something bad in his body because something bad is happening in a certain part of his/her body', are whole-centered and, therefore, they are compatible only with the accusative of the whole.[10] The part may be conceived as the Source or Cause of the negative sensation in question and is then subjectivised. Consider the structure ACC_{whole} V_s S_{part} as illustrated by (50), in which the causal relation is given priority:

(50) *mă doare sufletul*
me:ACC aches soul:the:NOM
'My soul aches, I am suffering.'

The subject referring to the part has a special discourse status: it does not represent the topical constituent, it occurs in postverbal positions and, in spoken Romanian, it does not always control the number of the verb, as in (51):

(51) *mă doare picioarele*
me:ACC aches legs:the:PL
'My legs ache.'

The pronominal clitic referring to the whole has a higher degree of discourse prominence. It refers to the topic and occupies the sentence-initial position. The NP_{part} may be fronted only if focalised:

(52) *capul mă doare, nu piciorul*
head:the:NOM me:ACC hurts, not leg:the:NOM
'It is the head which aches (me) not the leg.'

or presented as a specification of the whole, belonging to the rhematic part of the discourse:

(53) *capul* *mă* *durea,*
head:the:NOM me:ACC ached:IMPF
nasul *îmi* *curgea,*
nose:the:NOM me:DAT ran:IMPF
ochii *mi* *se* *înroşiseră*
eyes:the:NOM me:DAT 3sg:REFL got:red:IMPF
într- *un* *cuvânt,* *eram* *într-* *o* *stare* *jalnică*
in a word was:I:IMPF in a state pitiful
'My head was aching, my nose was running, my eyes got red, ... in a word, I was in a pitiful state ...'

The part may be also envisaged as a Locative ('where a bad thing is happening') and, consequently, it is expressed in a prepositional phrase as in (54).

ACC_{whole} V_s PP_{part}
(54) *mă* *doare* *în* *suflet,*
me:ACC aches in soul,
când *îl* *văd* *aşa-de* *palid*
when him:ACC see:I so pale
'It is painful for me to see him so pale.'

By losing its subject characteristics, the NP_{part} acquires a lower degree of discourse prominence, which may be interpreted as a sign of a lower degree of autonomy. It is also a way of eliminating the contradiction between a structure with an inanimate subject and an animate Patient in accusative, on the one hand, and the topic hierarchy, on the other.

The higher degree of part autonomy in constructions with a subjectivised part can account for the fact that the NP_{part} may take determiners:

(55) *de* *ieri* *gâtul* *ăsta* *idiot*
since yesterday neck:the this stupid
mă *doare* *fără* *întrerupere*
me:ACC aches without interruption
'Since yesterday my stupid neck has been aching continuously.'

The difference in case assignment may be paired with a lexical distinction. Compare (55) with (56):

(56) *mă* *doare* *în* *gât*
me:ACC aches in throat
'My throat aches.'

A prepositional construction such as (56) refers only to the 'throat', while (55) may very well mean either 'my neck aches' or 'I have a sore throat', because in Romanian the same word, *gât*, is used in both cases.

In expressions such as (57), the subjectless sentence has only a figurative meaning:

(57) *mă* *doare* *în* *cot*
me:ACC aches in elbow
'I do not care at all.'

4.3. "Activity" predicates

Verbs such as *a curge* 'to run', *a merge* 'to go, to work', which refer to a movement without a natural end-point, may also co-occur only with subjectivised parts of the body for the whole cannot be viewed as acting upon the part:

(58) *îmi* *curge* *nasul*
me:DAT runs nose:the:NOM
'I have a runny nose.'

(59) *îi* *merge* *mintea*
him:DAT goes mind:the:NOM
'(S)he is very smart.'

4.4. Directional movements

Unlike *a coborî* 'to lower' ,'to go down' or *a ridica* 'to raise', verbs expressing directional activities such as *a cădea* 'to fall', *a creşte* 'to grow' and *a ieşi* 'to come out' accept only a subject which refers to the part because they cannot occur in causative constructions in which the whole can be viewed as 'doing something to the part':

(60) *i-* *a* *căzut* *părul*
him:DAT has fallen:out hair:the:NOM
'His hair fell out.'

(61) *i-* *a* *crescut* *părul*
him:DAT has grown hair:the:NOM
'His hair has grown.'

(62) *i-* *a* *ieşit* *primul* *dinte*
him:DAT has come first:the tooth:NOM
'He got his first tooth' (when talking about a baby).'

Consider also the figurative expression:

(63) *i-* *a* *ieşit* *sufletul*
him:DAT has come:out soul:NOM:the
'He is exhausted.'

which is also confined to the dative of the whole.

The one-place predicates of 'movement' are therefore co-occurrent with postverbal subjectivised parts, except when the part is focalised, in which case, it may precede the verb as in (64):

(64) *părul* *i-* *a* *căzut*
hair:the:NOM him:DAT has fallen:out
nu *dinţii*
not teeth:the:NOM
'His hair fell out, not his teeth.'

4.5. Inchoatives of natural changes

4.5.1. Active forms of the verb

4.5.1.1. DAT_{whole} Vi S_{part}

Verbs expressing a natural change in colour or physical condition are usually part-centered[11] and, consequently, the topical whole takes a dative marker:

(65) *i-* *a* *albit* *părul*
him:DAT has become:white hair:the:NOM

şi	*arată*	*mai*	*bine*
and	looks:he	more	well

'His hair became white and he looks better.'

As shown by (65), 'he' – the subject of the second sentence – is the topic of the discourse, although it corresponds to the indirect object in the first sentence.

The part-centered construction may accept a determiner, although not very frequently:

(66)

i-	*a*	*albit*	*părul*	*mătăsos,*	
her:DAT	has	whitened	hair:the	silky	
i-	*a*	*slăbit*	*faţa*	*ei*	*frumoasă*
her:DAT	has	slimmed	face:the	her	pretty

'Her silky hair whitened, her pretty face slimmed.'

(67)

i-	*a*	*albit*	*până*	*şi*	*puţinul*	*păr*
him:DAT	has	whitened	even	and	thin:the	hair
care	*i-*	*a*	*mai*	*rămas*		
which	him:DAT	has	still	left		

'Even his thin hair, which was still left, whitened.'

4.5.1.2. S_{whole} Vi PP_{part}

The interpretation according to which the activity in question is conceived as the result of a deeper change affecting the whole, such as 'getting older', 'being or becoming sick', 'being or becoming ashamed' is encoded by a construction in which the 'centrality' of the whole is marked by subjectivisation:

(68)

a	*slăbit*	*la*	*faţă*	*(*faţa*	*albă)*
has:he	slimmed	at	face	(*face:the	white)

'His (*white) face slimmed.'

Therefore, with inchoatives, the difference between the constructions

(i) S_{whole} V PP_{part}

and

(ii) S_{part} V DAT_{whole}

rests on a change in the term chosen as the figure of the sentence and, consequently, in the degree of autonomy assigned to the part. In (68) the whole is central and excludes any determination of the part, while in (65) and (66) the whole is marginal.

4.5.2. Reflexive inchoatives

Reflexive constructions are the favourite choice for inchoatives when the change is viewed as an activity which presupposes an intense participation of the referent expressed by the subject (be it whole or part). Verbs which refer to the loss of a normal colour or weight, to a movement in agreement with gravity, to a change in age do not require a reflexive copy of the subject, for example, *a albi* 'to become white' referring to the loss of hair colour; *a îngălbeni* 'to become yellow'; *a păli* 'to become pale' referring to loss of the normal face colour; *a slăbi* 'to become slim' – loss of normal weight'; *a coborî* 'to go down'; *a îmbătrîni* 'to get old' and its antonym *a întineri* 'to become younger'.

On the contrary, verbs which are viewed as referring to activities with a surplus of intensity – such as *a se înroşi* 'to become red' – because of too much exposure to the sun or because of shame; *a se îngrăşa* 'to put on weight'; *a se umfla* 'to swell'; *a se ridica* 'to go up' – occur only in reflexive constructions, when no external agent is envisaged.

4.5.2.1. Degrees of dynamism

The choice of an active or reflexive construction with the subjectivised part is governed by the behaviour of the verb when the subject refers to the whole, that is by the degree of intensity assigned to the involvement of the subjectivised participant.[12]

Compare (69) with (70):

(69) *Ion a slăbit*
John has lost:weight
'John lost weight.'

(70) *i- a slăbit vederea*
him:DAT has weakened sight:the
'His sight weakened.'

Both (69) and (70) are active constructions, because in the Romanian ethno-cognitive pattern, 'to lose weight' has been considered a rather passive 'suffering' of the whole.

With a verb such as *a se înroşi* 'to become red', the reflexive may refer to either the subjectivised whole as in (71) or the subjectivised part as in (72):

(71)

Ion	*s-*	*a*	*înroşit*
John:NOM	3sg:REFL	has	blushed

când	*a*	*văzut*	*-o*
when	has:he	seen	her:ACC

'John blushed when he saw her.'

(72)

mâinile	*i*	*s-*	*au*	*înroşit*
hands:the	him:ACC	3sg:REFL	have	turned:red

'His hands turned red.'

4.5.2.2. Reflexive and the Locative of the Part

When the subject refers to the whole, the inchoatives may be divided into two syntactic subcategories, according to their compatibility with the Locative of the part:

(a) Verbs which are compatible with the Locative of the part:

(73)

s-	*a*	*înroşit*	*la*	*faţă*
3sg:REFL	has	become:red	at	face

'He has become red in the face.'

(74)

m-	*am*	*îngrăşat*	*la*	*corp*
me:ACC	have:I	got:fat	at	body

'I fattened at my body.'

(b) Verbs which are incompatible with the Locative of the part:

(75)

**s-*	*a*	*îngroşat*	*la*	*încheieturi*
3sg:REFL	has	thickened	at	joints

('His joints thickened.')

(76)

**s-*	*a*	*umflat*	*la*	*picioare*
3sg:REFL	has	swollen	at	legs

('His legs have swollen.')

In the first group of predicates, the locative is commonly presupposed by the parallel construction in which the restrictive specification is lacking:

(77) *s-* *înroşit* *când* *l-* *a* *văzut*
3sg:REFL become:red when him:ACC has seen
'(S)he turned red (blushed) when ((s)he) saw him.'

(78) *s-* *a* *îngrăşat*
3sg:REFL has got:fat
'He put on weight.'

When a non-presupposed specification is needed, the dative construction is selected, exactly as in the case of the second group of predicates:

(79) *i* *s-* *au* *înroşit* *mâinile*
him:DAT 3sg:REFL have become:red hands:the:NOM
'His hands turned red.'

As in the case of non-reflexive inchoatives, the difference in the term chosen as subject rests on a change in the figure of the event:
Let us compare utterances (80) and (81):

(80) a. *mi* *s-* *a* *înroşit* *faţa*
me:DAT 3sg:REFL has got:red face:the

b. *am* *stat* *prea* *mult* *la* *soare*
have:I stayed too much in sun
'My face became red (sunburnt). I was in the sun too long.'

(81) a. *m-* *am* *înroşit* *la* *faţă*
me:ACC have:I become:red at face

b. *când* *l-* *am* *văzut* *intrând*
when him:ACC have:I seen coming:in
'I blushed when I saw him coming in.'

In (80) the topic corresponds to the person 'I' (the whole), represented both by the dative in sentence (a) and by the subject in sentence (b), but the figure is 'my face', which is expressed by the subject of (a) – (80) may be uttered by somebody looking at his/her own face in a mirror.

In (81), the person 'I' corresponds both to the topic and to the figure. The NP 'I' plays the role of the subject in both (a) and (b) (sign of its topicality), but in (a) it also functions as a direct object, which means that 'I' is treated as the central entity affected by the activity in question, in spite of the fact that the effects of the sun are visible only on a limited part of the person. The NP_{part} – treated as marginal – belongs to a prepositional phrase (PP).

5. Conclusions

(1) In order to describe the structures encoding 'inalienability', various parameters have to be considered – (a) syntactic constructions (specifically case assignment); (b) semantic features of predicate classes; (c) semantic features of NPs referring to the whole; (d) semantic features of the NP_{part} and their compatibility with predicate-classes and (e) discourse features such as 'topicality' and (f) cognitive features, such as 'centrality'.

(2) In Romanian, 'inalienability' (rather the lowest degree of part autonomy in the whole-part relation) can be expressed only by constructions in which the whole is topical and central. The topical whole may correspond to the direct object (in accusative) and/or to the subject (with two-place predicates), to the indirect object (dative) or to the subject (with one-place predicates).

(3) The accusative of the whole always refers to a topical constituent, while the dative may refer either to a topical or a non-topical whole. The main semantic difference between the two constructions consists in the fact that the former is whole-centered, while the latter is part-centered.

(4) If the construction S V O_{whole} (PP_{part}) is confined to 'inalienability', a part-centered construction can be associated with various degrees of 'autonomy'. When the dative of the whole is not topical, and the part is determined by a definite description, the construction S V $DAT_{whole}O_{part}$ presupposes a higher degree of 'autonomy'. It can express a relation between the whole and the part in which the part of the body is viewed as an object separated from, but related to, the person. The dative construction is not limited to whole-part relations. It may also refer to kinship, or simple possession, when the 'possessor' is topical.

(5) The locative of the part (PP) is co-occurrent only with a topical whole, which may be expressed either by a direct object or a subject.

(6) The choice of case markers is also governed by the predicate class.

(a) The accusative of the whole co-occurs with dynamic two-place verbs such as *a lovi* 'to hit' or *a spăla* 'to wash' and with stative predicates expressing intense negative sensations. As predicted by their whole-centered meaning, metaphorical expressions comprising such verbs are confined to the accusative-locative combination.

(b) With one-place predicates such as 'inchoatives' referring to 'natural changes in size or colour', 'centrality' is marked by subjectivisation, namely, when the discourse is part-centered, then the topical whole becomes an indirect object, while the part corresponds to the subject. If the discourse is whole-centered, then the topical whole corresponds to the subject, while the part is expressed by a locative prepositional phrase.

(c) Predicates expressing unintentional, natural, directional movements of the part, such as *a ieşi* 'to come out' or *a creşte* 'to grow, in which the whole cannot be envisaged as acting upon the part are always part-centered and, consequently, they are incompatible with a subjectivised whole.

(d) Predicates referring to movements for which the whole can be held responsible (such as *a ridica* 'to raise' or *a coborî* 'to lower') may be associated with a higher degree of autonomy in two ways:

(i) when the whole is subjectivised, they may co-occur with a dative reflexive of the whole in order to express an insistence upon the agentive character of the whole or,

(ii) when the part is subjectivised (and fronted), the construction "insists" upon the non-agentive character of the whole.

(7) The subjectivisation of the part does not affect the topical character of the whole. The subject referring to the part has a special discourse and syntactic status – it does not represent the topical constituent; with most of the inchoatives or with statives (such as *a durea* 'to ache', *a mânca* 'to itch'), the subjectivised part occurs in postverbal positions, as a rhematic constituent, and does not always control the number of the verb. The pronominal clitic referring to the whole has a higher degree of discourse prominence, refers to the topic and occupies the sentence-initial position.

(8) The choice of an active or reflexive construction for 'inchoatives' is governed by a semantic factor, namely by the degree of intensity assigned to the involvement of the subjectivised participant in the event.

Notes

1. For the definition of 'inalienability' as 'solidarity' rather than 'possession' or 'belonging', see Riegel (1984).
   I am deeply grateful to my friends and colleagues, Glanville Price and David Wilkins, as well as to the editors of the present volume, Hilary Chappell and Bill McGregor, who read earlier versions of this paper and made many valuable suggestions.
2. See, for example, Wierzbicka (1988) and Chappell – McGregor (1995).
3. Marandin (1988: 68- 69) gives a quite complete list of the concepts involved in the definitions of the discourse topic: (i) la notion (primitive) "être à propos de": "tel discours ou fragment de discours est à propos de x"; (ii) l'idée qu'une communication est communiquée à propos de x; (iii) l'idée de cohérence, de continuité: "tel discours est cohérent parce qu'il porte sur le même objet, qu'il développe le même thème"; (iv) l'idée d'importance, de saillance d'un objet ou d'un sujet dans la conscience d'un

locuteur ou dans son discours; (v) l'idée de perspective; (vi) l'idée (plus récemment introduite) de limitation d'un domaine de discours (ou du champ de conscience d'un locuteur) et de pertinence ("relevance"). [Translation: (i) the (primitive concept) "to be about x": a certain discourse or a segment of a discourse "is about x"; (ii) the idea that a certain communication is communicated in relation to x; (iii) the idea of (text-) coherence: "a certain discourse is coherent because it says something about the same object, it develops the same theme"; (iv) the idea of importance or saliency of an object or a subject in the speaker's consciousness or discourse; (v) the idea of perspective; (vi) the idea (more recently advanced) of delimiting a domain in the discourse or in the speaker's consciousness, and, finally, the idea of relevance.

A good discussion of the concept of 'discourse topic' may be found in Brown – Yule (1984), chapters 3 and 4. For the concept of 'sentence topic', see Lambrecht (1986).

4. "Rhematic" refers to rheme, that is the constituent which represents "what is asserted about the topic". In other models, the rheme corresponds to the focus (see Ebneter 1976; Lambrecht 1986). In the present article, the term "focalisation" refers to the procedure by which "contrastive focus" is expressed; in other words, a focalised constituent is the constituent which refers to a candidate other than the one expected as being appropriate for the predication in question (see Lakoff 1971).
5. For details concerning the conditions governing the choice of proclitic markers, see Manoliu-Manea 1985.
6. These semantic features have been identified on the basis of the following works: Lakoff (1972), Manoliu-Manea (1977), Dowty (1979), and Wierzbicka (1988).
7. The bibliography on the matter is overwhelmingly extensive: Bally (1926), Frei (1939), Hatcher (1944a and b), Langacker (1968), Ivic (1970), Cartagena (1972), Diffloth (1974), Leclère (1978), Seiler (1983), Kliffer (1984), Riegel (1984), Guéron (1985), Herslund (1988), Hanon (1988) – to mention only the most referred to studies. For the dative of the whole in German and Middle Dutch, see Neumann (1995) and Burridge (1995) respectively.
8. See more recently Van Peteghem (1988: 45-46) which analyses the Romanian dative constructions.
9. An important characteristic of the Romanian adnominal construction is the replacement of the possessive adjective in the third person (*său*) with the dative of the personal pronoun (*lui* 3sg:DAT), when the possessor-whole is focalised and/or does not belong to the domain of discourse (is less known or was not mentioned previously. See Popescu – Tasmowski De Ryck (1988).
10. An identical construction may be found in Polish. As Wierzbicka (1988: 419- 421) has pointed out: "intense sensations such as pain or itchiness require the accusative.' In other words, with verbs expressing such intense sensations, the accusative construction is the expression of the following semantic structure: 'Z feels something bad in his/her body (or: in part Y of his/her body) because of something bad that is happening in some parts of his/her body – not because someone is doing something to it' (1988: 421).
11. For a similar interpretation of inchoatives as being part-centered in German, see Neumann (1995).
12. For a pragmatic description of Romanian reflexive constructions, see more recently Manoliu-Manea (1988a).

References

Bally, Charles

1926 "L'expression des idées de sphère personnelle et de solidarité dans les langues indo-européennes", in: Franz Fankhauser – Jakob Jud (eds.), 68- 78. Translation by Christine Béal – Hilary Chappell [this volume.]

Brown, Gillian – George Yule

1984 *Discourse analysis.* Cambridge, London, New York: Cambridge University Press.

Burridge, Kate

1995 "Degenerate cases of body parts in Middle Dutch" [this volume.]

Cartagena, Nelson

1972 *Sentido y estructura de las construcciones pronominales en español.* Chile: Universidad de Concepción.

Cazelles, Brigitte – René Girard (eds.)

1987 *Alphonse Juilland. D'une passion l'autre.* Stanford: Anma Libri.

Colón, G. – R. Kopp (eds.)

1976 *Mélanges de langues et de littératures romanes offerts à Carl Th. Gossen.* Bern-Liège: Franke-Marche Romane.

Cristea, Theodora

1974 "Remarques sur le datif possessif en roumain et en français", *Bulletin de la Société Roumaine de Linguistique Romane* 10: 5-14.

Deanović, Mirko – Rudolf Filipović – Petar Guberina – August Kovačec – Valentin Putanec – Petar Šimunović – Vojmir Vinja (eds.)

1985 *Mélanges de linguistique dédiés à la mémoire de Petar Skok (1881-1956).* Zagreb: Académie Yougoslave des Sciences et des Arts.

Diffloth, Gérard

1974 "Body moves in Semai and French", in: Chicago Linguistics Society 10: 128-138.

Dobrovie-Sorin, Carmen

1989 "Le système auxiliaire du roumain: de la morphologie à la syntaxe" (Paper presented at the 19th International Congress of Romance Linguistics and Philology, Santiago de Compostela, 4 – 9 Sept, 1989).

Dowty, David

1979 *Word meaning and Montague grammar.* Dordrecht: Reidel.

Ebneter, Theodor

1976 "Thema und Fokus im Französichen", in: G. Colón – R. Kopp (eds.), 195-214.

Fankhauser, Franz – Jakob Jud (eds.)

1926 *Festschrift Louis Gauchat.* Aarau: H.R. Sauerländer.

Frei, Henri

1939 "Sylvie est jolie des yeux", in: Albert Séchéhaye et al (eds.), 185-192.

Grevisse, Maurice

1986 *Le bon usage. Grammaire française avec des remarques sur la langue française d'aujourd'hui.* (12th revised edition). Gembloux (Belgium): Duculot.

Guéron, J.

1985 "Inalienable possession, pro-inclusion and lexical chains", in: H.-G. Obenauer – J. Y. Pollock (eds.), 43-86.

Hanon, Suzanne
1988 "Qui à quoi? Réflexions sur la possession inaliénable et le verbe *avoir* en français", *Revue Romane* 23.2: 161-177.
Hatcher, Anna Granville
1944a "'Il me prend le bras' vs. 'il prend mon bras'", *The Romanic Review* 35: 156-164.
1944b "'Il tend les mains' vs. 'Il tend ses mains'", *Studies in Philology* (Chapel Hill) 41: 457-481.
Herslund, Michael
1988 *Le datif en français*. Louvain, Paris: Peeters.
Ivič, Milka
1970 "On the part-whole relation and its linguistic consequences", in: Roman Jakobson – Shigeo Kawamato (eds.), 281- 286.
Jakobson, Roman – Shigeo Kawamato (eds.)
1970 *Studies in general and oriental linguistics in honor of Shiro Hattori*. Tokyo: TEC Co.
Kayne, Richard
1975 *French syntax. The transformational cycle*. Cambridge, Mass. and London, England: M.I.T. Press.
Kliffer, Michael D.
1983 "Beyond syntax: Spanish inalienable possession" *Linguistics* 21: 759-794.
1984 "Interpenetration of linguistic levels: French inalienable possession", *Lingua* 62: 187-208.
Lakoff, George
1971 "On generative semantics", in: D.D. Steinberg – L.A. Jakobovits (eds.), 232–296.
1972 *Irregularity in Syntax*. New York: Holt.
Lambrecht, Knud
1986 *Topic, Focus and the Grammar of Spoken French*. [Unpublished Ph.D. dissertation, University of California, Berkeley.]
1987 "Sentence focus, information structure, and the thetic-categorial distinction", *Proceedings of the Berkeley Linguistic Society* 13: 366-382.
Langacker, Ronald W.
1968 "Observations on French possessives", *Language* 44: 51-75.
1987 *Foundations of cognitive grammar 1. Theoretical prerequisites*. Stanford, CA: Stanford University Press.
Leclère, Christian
1978 "Sur une classe de verbes datifs", *Langue française* 39: 66-75.
Manoliu-Manea, Maria
1977 *Tipologie şi istorie. Elemente de sintaxă comparată romanică*. Bucharest: Universitatea Bucureşti (see also the Spanish version, *Tipología e historia: Elementos de sintaxis comparada románica*. Madrid: Gredos, 1985).
1982 "Un cas désespéré de la syntaxe romane: les possessifs. Syntaxe et pragmatique du langage", in: Aina Moll – Jaume Vicens (eds.), 259-268.
1985 "Genetic type versus areal coherence: Rumanian case markers and the definite article", in: Mirko Deanović et al (eds.), 301-308.
1987 "The myth of the agent: roles and communicative dynamism in Romance", in: Brigitte Cazelles – René Girard (eds.), 261- 275.

1988a "Pragmatique et sémantique du passif: L'agent et le réfléchi roman", *Revue Romane* 23. 2: 198-210.
1988b "Verbal complementation and discourse strategies. Direct-object constructions in Romanian", *International Journal of Rumanian Studies*, 6.2: 53-68.

Moll, Aina – Jaume Vicens (eds.)
1982 Actes, XVI Congrés Internacional de Linguística i Filologia Romàniques, Palma de Mallorca, 7-12 d'abril de 1980, 1. Sessions plenaries i taules rodones. Palma de Mallorca: Moll.

Neumann, Dorothea
1991 "The dative and the grammar of body parts in German", [this volume.]

Obenauer, H.-G. – J. Y. Pollock (eds.)
1985 *Grammatical representation.* Dordrecht: Reidel.

Peteghem, Marleene van
1988 "Attributs nominaux à complémentation verbale en roumain", *International Journal of Rumanian Studies*, 6.2: 33-52.

Popescu-Ramirez, Liliana – Liliane Tasmowski De Ryck
1988 "Thématicité et possessivité en Roumain" (Paper presented at the M.L.A. Annual Convention, New Orleans, December 27-30, 1988).

Riegel, M.
1984 "Pour une redéfinition linguistique des relations dites de 'possession' et d' 'appartenance'", *L'information grammaticale* 23. 3-7.

Seiler, Hansjakob
1983 *Possession as an operational dimension of language.* Tubingen: Narr.

Séchéhaye, Albert et al (eds.)
1939 *Mélanges de linguistiques offerts à Charles Bally.* Geneva: Georg et Cie. sa.
[1972] [Reprinted Geneva: Slatkine].

Silverstein, Michael
1981 "Case marking and the nature of language", *Australian Journal of Linguistics* 1: 227-246.

Steinberg, D.D. – L.A. Jakobovits (eds.)
1971 *Semantics. An interdisciplinary reader in philosophy, linguistics, and psychology.* London and New York: Cambridge University Press.

Wierzbicka, Anna
1988 *The semantics of grammar.* Amsterdam and Philadelphia: Benjamins.

The dative and the grammar of body parts in German

Dorothea Neumann

1. Introduction

This paper investigates the syntactic constructions which are used in German to describe body parts and their possessors. The grammar of German is of interest in the discussion of alienable and inalienable part-whole relations because German has a construction in which the possessor of a body part is marked by the dative case (see (1) below), whereas an inanimate possessor is marked by the genitive.[1] There is therefore the possibility that this syntactic difference is based on a distinction between alienable and inalienable possession. Inalienability is here understood as the property of a noun denoting an object which in a particular language is treated as low in autonomy (cf. Manoliu-Manea 1995). For any given language, the set of such inalienably possessed nouns must be lexically defined, although it has a certain amount of semantic coherence (Nichols 1988: 568).

The aim of this study is to show that the constructions used in the grammar of body parts in German – of which the dative of the possessor is only one – are not triggered by the occurrence of a particular set of lexical items, however defined, but reflect a number of different semantic relations which can hold within a part-whole relation when the possessor or whole is animate, but not when it is an inanimate object.

Following this introduction, section 2 places the dative of the possessor within the German case system in general, explaining the topic function of the dative in subsection 2.3. Section 3 describes the role of inference in the interpretation of syntactic constructions and shows how it applies to German transitive constructions in 3.1, locative constructions in 3.2, and the dative of the possessor in 3.3.

The main part of this paper is section 4, which discusses the grammar of body parts in detail. It begins by showing, in 4.1, the parallels between animate and inanimate part-whole constructions in German. The constructions which are peculiar to possessor-body part relations are investigated in 4.2, according to the roles in which the possessed parts can occur: as moving object in 4.2.1, as object undergoing a change in 4.2.2, as location in 4.2.3.

A moving/changing object would normally be expected to be the subject (marked by the nominative), or the direct object (marked by the accusative) if there is an agent causing the motion/change; the role of location is indicated by a preposition (there is no locative case in German); the unmarked case for the possessor is the genitive. These case markings can, of course, be used with body parts and their possessors when a speaker chooses to ignore the special relation between an animate possessor and its part, but this option is not of interest here.

When a body part noun occurs in these roles, there are several different constructions to encode the part-whole relation with each of them. The body part may be demoted to a prepositional phrase, with the possessor becoming the subject. When the body part remains the subject, the possessor may have to be in the dative, or may have to be in the genitive, similarly when the body part is part of a prepositional phrase; when the body part is the direct object, the possessor usually has to be in the dative, but when the possessor is also the subject, this dative may or may not be necessary. The body part is never, as happens in Middle Dutch (see Burridge (1995)) marked by the dative when it is the affected object. The semantic conditions for these constructions, which are not interchangeable, will be detailed in section 4 below.

Section 5 shows a use of the dative which may seem to be an extension of the "possessive" dative, but has nothing to do with possession, and section 6 places the so-called *dativus commodi*[2] in the same context.

2. The dative in German

2.1. The free dative

The dative in German has been called the "personal" case (cf. Brinkmann 1962: 440). It is restricted to animate noun phrases in all its free uses, i.e. when it marks noun phrases whose case marking is not triggered by another element in the clause. There are several types of such datives (for a detailed discussion, see Neumann (1987)). An example of the dative of the possessor is (1), discussed as (5a) below:

(1) *Der Stein fällt dem Mann auf den Kopf*
The brick$_{\mathrm{NOM}}$ falls the man$_{\mathrm{DAT}}$ on the head$_{\mathrm{ACC}}$
'The brick falls on the man's head.'

This dative is also called "dative of pertinence".[3] Other free datives are, for example, the *dativus commodi/incommodi*, which marks the person for whom an event is "good" or "bad", and the "ethic dative" (Abraham (1973); Heidolph et al (1980: 387)) as in:

(2) *Das sind mir Gauner!*
that are me$_{\mathrm{DAT}}$ rascals$_{\mathrm{NOM}}$
'They are rascals, I must say.'

2.2. The obligatory dative

In its obligatory uses, the dative is not restricted to animate noun phrases. There are certain predicates, such as the verbs of giving and telling, which require an animate recipient, and this is marked by the dative; but there are also verbal and adjectival predicates requiring the dative which are neutral with regard to animacy, for example predicates expressing symmetry, such as *gleichen* 'be alike':

(3) *Die Absturzstelle glich einem Trümmerfeld*
the crash:site$_{\mathrm{NOM}}$ was:equal a rubble:field$_{\mathrm{DAT}}$
'The crash site was like a field of rubble.'

There are also predicates with the dative that do not even allow animacy, such as in (4):

(4) *Dem Gas war ein Reizstoff beigemischt*
the gas$_{\mathrm{DAT}}$ was an irritant$_{\mathrm{NOM}}$ admixed
'There was an irritant admixed to the gas.'

There are also a number of prepositions which obligatorily take the dative, such as instrumental *mit* 'with' and locational prepositions indicating position. The function of the dative to mark animacy can therefore at most be a partial function.

2.3. The topic function of the dative

The dative, especially the dative of the possessor, has also been described as a *Gegensubjekt* [alternative subject] (von Weiss (1962)). This characterisation was intuitive, not based on a definition or analysis of "subject", and apparently was not intended to be applied to other uses of the dative. But it is valid also for the obligatory dative if it is interpreted as a kind of topic.

The term *topic* is here used as a semantic concept: whatever the semantic relation between a topic and another noun phrase in a clause, it is always a relation in which the topic is dominant in the sense that what is being said about the other noun phrase is part of what is being said, or sayable, about the topic. With verbs of giving and telling there are two noun phrases that can, in a way, be called topics: while the subject (X), in the nominative, is said to do something to an object (Y): causes Y to be haveable/ knowable, and so is dominant over Y, the recipient (Z), marked by the dative, comes to be able to have or know the same thing and so is also dominant in relation to Y.

The subject X does not cause Z to have/know Y,[4] so the fact that Z comes to have/know Y is not part of what is sayable about X: X is not dominant over Z. The subject is the actual topic, the recipient is a potential topic, and their spheres intersect: part of what is being said about the subject is also part of what is sayable about the recipient.[5] An object that is semantically dominated by both subject and dative referent need not always be specified: with 2-place verbs such as *gleichen* 'be alike' and *ähneln* 'be similar', something that is not explicitly said about an object X_{NOM} but sayable (a feature of X) is compared to a feature of object Y_{DAT}; in *helfen* 'help', which in German also takes a dative, the agent does something unspecified but specifiable which the dative referent may want to have done.

2.4. The dative of the possessor

As what is said about a body part is often, if not always, part of what is sayable about the possessor, the dative of the possessor fits naturally into this pattern. Compare the dative construction (1), repeated here as (5a), with (5b) and (5c):

(5) a. *Der Stein fällt dem Mann auf den Kopf*
the brick$_{NOM}$ falls the man$_{DAT}$ on DEF[6] head$_{ACC}$
'The brick falls on the man's head.'

b. *Der Stein fällt auf das Dach des Autos*
the brick$_{NOM}$ falls on the roof$_{ACC}$ the car$_{GEN}$
'The brick falls on the roof of the car.'

c. *Der Stein fällt auf den Kopf des Mannes*
the brick$_{NOM}$ falls on the head$_{ACC}$ the man$_{GEN}$
'The brick falls on the man's head.'

The inanimate possessor in (5b) must be in the genitive. With animates the genitive construction can be used as well as the dative construction, but there is a difference in meaning: in (5c) the body part seems to be seen in isolation and might even be detached from the body, whereas in (5a) the possessor is felt to be more affected, more involved in the event: something (affectedness) is sayable about the possessor when something else (contact) is being said about the part. In contrast to the genitive, the so-called possessive dative is thus able to indicate not only that there is a link between animate whole and body part, but that in this relation what is being said about the part is to be understood as a comment on the whole.

Syntactically, the dative of the possessor marks the animate possessor of a "body part or other object" (Heidolph et al 1980: 368); the possessed item may be the subject or direct object of a clause, or part of an adverbial phrase (this means that practically only a possessed item in the genitive or as another dative noun phrase is excluded).

2.5. Investigation procedure

The distribution of this possessive dative – one of the few instances where two cases can alternate, as in (5a) and (5c) above – has always invited the comparison with the genitive (for example, von Weiss 1962), so that the, or a, meaning of the dative is derived from the perceived semantic difference between the two cases. The underlying assumption is that both cases indicate a having-relation (Brinkmann 1962: 403 – 404), although in different ways.

The approach chosen here is different, starting not from the syntactic environment of the dative/genitive alternation alone, but by investigating body part nouns systematically in the semantic roles of moving object, patient (object caused to move or caused to undergo a change) and location, and examining how the part-whole relation is coded syntactically within these roles. The aim is to explain the distribution of the possessive dative not only by comparison with the possessive genitive, but by determining its place in the grammar of body parts in general.

Throughout the discussion a semantic metalanguage similar to that developed by Anna Wierzbicka from *Semantic primitives* (1972) onwards is used in preference to a more abstract terminology to explain the central ideas, but no attempt has been made to translate all the meanings discussed into this language. The main difference between the semantic analysis offered here and that of Anna Wierzbicka's lies in the differentiation between explicit and inferred meanings.

3. Inferences

The idea of inference is an important concept used in this paper. It is adopted from Diver (1964). Diver reduces the function of the three cases nominative, accusative and dative (in Latin) to distinguishing the most active participant (nominative), the least active (accusative), and the one that is neither (dative). All other case roles such as recipient or affected object are derived by inference from the combination of case markings and lexical items in a clause. As an explanation of the use of cases, Diver's theory is unsatisfactory; but the idea of stripping a case, and in fact any syntactic element or lexical item, of everything that is not essential to its meaning because it can be inferred from other sources has proved very fruitful.

If we want to understand how language manages to convey all possible meanings with a finite number of linguistic signs, inferences must be considered as an important part of linguistic communication. Inferences are based on the information transmitted by linguistic signals on the one hand and on the hearer's knowledge of what the signals refer to in the real world on the other. When, for example, contact between two objects is described, the hearer infers from the nature of the objects whether there is an effect through contact, even if only the spatial relation is explicitly stated. Inferences of this kind depend on the individual hearer's real-world knowledge and are not always predictable.

3.1. Inferences from syntactic constructions: from transitive constructions

There are also inferences which the hearer is invited to make through certain syntactic constructions. A transitive construction, for example, involves a relation between subject and object (in German marked by nominative and accusative respectively). The verb may or may not make this relation explicit.

When the verb specifies a relation, as, for example, in English *cover*, *touch*, or German *treffen* 'hit (a target)', it is not necessary to infer either agency of the subject or an effect on the object.

When the verb only specifies an effect on the object, for example, *melt* in:

(6) *X melts butter (Y)*,

we infer that the subject X causes the effect on Y, although the verb does not specify what the subject does to cause it: if X is animate, we infer that it acts in a specifiable way so that it causes the butter to melt; if X is a force of nature, such as heat or sun, we infer that it has a specifiable quality that can cause the effect. This may override our real-world knowledge: whatever X the speaker chooses (for example, *X-rays* in (6)), we infer that X causes the effect, or rather we assume that that is what the speaker wants us to think, because Y is marked as the direct object.

When the verb specifies only a motion or action of the subject (X), we are left to infer the effect on the object (Y), for example, whip (cream), kick (the dog). Verbs of this type (known as *impact* verbs) describe motion of a part (body part or instrument) of the agent without displacement of the whole. If this motion establishes a relation to another object, this will be a relation of contact, not relative position. Although the verb specifies the action of the subject only, an effect on the object can be inferred from the accusative marking on the object. We infer that by establishing contact with the object the subject causes something to happen to the object. In English, impact verbs can take a direct object also when the only result is contact, as in *kick the wall*; but in German there must be some further effect (contact *per se* is described by a locative construction corresponding to *kick against the wall*). The following discussion of German impact verbs, which is intended to show that inferences are governed by statable rules, is therefore not automatically transferrable to English.

The nominative-accusative construction can semantically be interpreted as indicating a relation between two entities in which the one marked by the nominative is dominant. The subject is said to move, extend (in both of these the space which we think of when we think of the subject X may come to include the space of another object), or cause a change to happen, whereas the accusative object is the entity to which the subject moves or extends, or that undergoes a change: what can be said about the object is part of what is being said about the subject. When an object is in the dative, however, what can be said about the dative noun phrase is related to, but not part of what is being said about the subject.

When an impact verb has an inanimate direct object, this is understood to undergo a change of state (as, for example, *cream*), or sometimes of place; or it may be caused to produce another effect, as in *beat a drum*; if the effect on the object is not predictable, it has to be specified in addition to the verb, for example, *kick across the room*, *beat to a pulp*. The predictable effect of contact on an animate object is that the animate being comes to feel something.

If the direct object is inanimate and an effect through contact is neither predictable nor explicitly stated, as in

(7) ? *Der Mann tritt den Stuhl*
the man_{NOM} kicks the $\text{chair}_{\text{ACC}}$
'The man kicks the chair.'

the transitive construction is normally not acceptable in German. If a hearer tries to assign an interpretation to (7), it can only be in terms of animacy; interestingly the most likely way of doing this is not by assuming that the chair in (7) must somehow be personified and come to feel something, but by assuming that the agent treats the chair as if it were animate, that is, the inference is something that is sayable about the subject, not about the direct object.

3.2. Inferences from spatial relations

Spatial relations between objects are usually described by prepositional phrases. German spatial prepositions take a noun phrase (Y) in the dative or in the accusative case. For an analysis of these prepositions, see Neumann (1987). Examples:

(8) *Die Leute stehen auf der Strasse (Y)*
the people$_{NOM}$ are:standing PREP the road$_{DAT}$
'The people are standing on the road.'

(9) a. *Die Leute (X) gehen auf die Strasse (Y)*
the people$_{NOM}$ are:walking PREP the road$_{ACC}$
'The people are walking onto the road.'

b. *Die Leute (X) stehen bis auf die Strasse(Y)*
the people$_{NOM}$ are:standing as:far:as PREP the road$_{ACC}$
'The people are queuing as far as the road.'

Example (8) describes position, (9a) and (9b) motion and extension respectively.

Motion and extension are usually described by the same devices, not only lexically (many motion verbs are used to describe extension: *berühren* 'touch', *laufen* 'run', *steigen* 'rise'), but also syntactically; so the accusative after the preposition in (9a) and (9b) – motion/extension – contrasts with the dative in (8) (position).

When an object is said to move or extend, the space we think of when we think of the object extends across some distance. The accusative in the prepositional phrase, like the accusative marking on the direct object, indicates that what is being said about the object Y (here: the spatial relation to Y) is part of what is being said about the subject X. The space we think of as related to Y comes to be part of the space we think of as X, because X is said to move or extend through several points in space, only one of which is related to Y (the same is, of course, true if X is not the subject, but a (direct) object). Because of this, the directional prepositional phrase often has the force of ascribing the meaning of directional motion to a verb, even if, as in (11) below, the verb does not make motion specific.

When the position of an object X is asserted, as in (8), the relation to another object Y is something that is being said about X, but it is not part of what is being said, because nothing further is said about X. The space that we think of as X (the space dominated by X) is said to be PREP Y, i.e. part (in (8) the part called "on") of the space dominated by Y, and it continues to be thought of as the space dominated by Y. In this situation, Y is marked by the dative.

The prepositional phrase does not distinguish between contact and displacement: this must be inferred from the context.

(10) *Der Mann (X) tritt gegen den Stuhl*
the man$_{\mathrm{NOM}}$ kicks PREP the chair$_{\mathrm{ACC}}$
'The man kicks the chair.'
(contact but no displacement of X)

(11) *Das Auto (X) knallt an einen Baum*
the car$_{\mathrm{NOM}}$ bangs PREP a tree$_{\mathrm{ACC}}$
'The car crashes into a tree.'
(displacement of X and contact)

When there is contact, an effect can often be inferred; but the construction does not specify which of the objects involved is affected, and it does not present the effect as caused by the moving object. The effect of sensation, which is essential in making some of the direct objects of impact verbs acceptable, can here be inferred about an animate subject, for example, *the man* in (10), but it is not usually inferred about an animate location, as in (12b):

(12) a. *Der Stein (X) fällt auf das Auto (Y)*
the brick$_{\mathrm{NOM}}$ falls PREP the car$_{\mathrm{ACC}}$
'The brick falls on the car.'

b. ? *Der Stein fällt auf den Mann*
the brick$_{\mathrm{NOM}}$ falls PREP the man$_{\mathrm{ACC}}$
'The brick falls on the man.'

The construction in (12b) does not exactly exclude the inference of sensation from the animacy of the noun phrase: this inference is always possible when an animate being is involved; but it does not invite it. When an animate being, here *the man*, is presented as a location, it is an object in space like any other object. A sentence like (12b) would be natural as a statement about the movement of a brick; but as a statement about an animate being it is odd, even in English, where the idea of (12b) would normally be expressed using the verb *hit*, for example, 'The (falling) brick hits the man', so that the animate being becomes the direct object. German would instead use a transitive verb which specifies the effect, for example, *erschlagen* 'kill', *verletzen* 'injure', rather than *treffen* 'hit (a target)', because *treffen* is not an impact verb and does not suggest the same force as the English verb does.

3.3. Inferences from the dative construction

Usually a location in relation to an animate being, as the one intended in (12b), will be defined by a body part, with the possessor in the dative, as in

(13) a. *Der Stein fällt dem Mann auf den Kopf*
the $brick_{NOM}$ falls the man_{DAT} on DEF $head_{ACC}$
'The brick falls on the man's head.' = (5a)

When the noun phrase in the prepositional phrase denotes a part and the possessor or whole is marked by the genitive, there is again no difference between animate and inanimate nouns in the type of inferences invited:

b. *Der Stein fällt auf das Dach des Autos*
the $brick_{NOM}$ falls on the $roof_{ACC}$ the car_{GEN}
'The brick falls on the roof of the car.' = (5b)

c. *Der Stein fällt auf den Kopf des Mannes*
the $brick_{NOM}$ falls on the $head_{ACC}$ the man_{GEN}
'The brick falls on the man's head.' = (5c)

In (13b) and (13c) there is presumably an effect on the part, and it is possible to make an inference about the whole, but this inference can only be a re-interpretation of the effect on the part (damage) as an effect on the whole. There is no suggestion in the construction that the speaker wants to say that the animate possessor in (13c) feels something or is aware of what is happening, that is, the speaker does not make allowance for the animacy of the possessor; as this is unusual when speaking of an animate being, the genitive construction can convey the impression that the body part is somehow detached from the whole.

When the alternative strategy of marking the possessor by the dative is used instead, as in (13a), we understand that the body part is in its proper place, not detached, and that the possessor is likely to feel something and consequently becomes aware of what is happening. As the comparison with the genitive construction in (13c) shows, these inferences are not simply derived from the fact that the possessor is animate. They are inferences which the hearer is invited to make by the construction.

The dative cannot be used with inanimate possessors:

(14) **Der Stein fällt dem Auto aufs Dach*
the $brick_{NOM}$ falls the car_{DAT} PREP:DEF $roof_{ACC}$
'The brick falls on the roof of the car.'

More precisely: if it is used with an inanimate noun phrase, it has the force of ascribing some kind of animacy to the inanimate possessor (in (14), *the car*), just as the transitive construction in (6) has the force of ascribing some kind of agency to a subject.

It is this force of the dative that requires explanation. To say that the dative is always used with animates and that therefore any noun phrase with which it occurs will be considered animate is begging the question, as we want to know why it is that the dative marks the animate noun phrase in the first place. Even the fact that the relation between an animate being and part of its body differs from the relation between an inanimate object and its parts is not sufficient: what we want to know is what aspect of the difference it is that is reflected in the syntax.

4. The part-whole relation in the syntax

When we think of a concrete object, we not only think of its identity, but also, for example, of the space where it is, of its potential to move, to affect other objects, in short, of the possible relations to other objects into which it might enter. The mention of an object evokes an expectation of what is potentially sayable about it. It is because much more is sayable about animate beings than about inanimate objects that animate noun phrases tend to be the subjects and topics of clauses.

A part of an object Z can be defined semantically as part of what we think of when we think of Z.

4.1. General features of the part-whole relation

Stylistically, a part may stand as representative of the whole (*pars pro toto*, for example, the roof of a house, the engine of a car, the head or face of a person), so that a statement about the part is identical in meaning to a statement about the whole, for example, *counting people/heads*, *the car/engine breaks down*, and either way is acceptable; only a limited number of body part nouns can be used in this way. A change in such a representative part is preferably predicated of the whole:

WHOLE $(Z)_{NOM}$ VERB

rather than

PART $(Y)_{NOM}$ WHOLE $(Z)_{GEN}$ VERB.

(15) *Die Bäume färben sich*
the $trees_{NOM}$ are:colouring $REFL_{ACC}$
'The trees are changing colour (leaves).'

(16) *Sie errötete*
she_{NOM} became:red
'She blushed (face).'

(17) *Er war über Nacht ergraut*
he_{NOM} AUX over night become:grey
'He had turned grey overnight (hair.)'

The part may be commented on independently, as in (18a) and (19a) below; but even when there is no identification between whole and part, something that would be sayable about the part is very often presented as something that is being said about the whole, because the whole is more salient in the speaker/hearer's mind and he/she perceives it as dominant over the part. The comment about the part is then subordinated to the whole, so that the whole becomes the subject, and the part with its comment becomes the complement of a verb such as *have*. This strategy is acceptable (but not obligatory) with animate and inanimate possessors alike, depending on what the speaker wants to focus on:

(a) PART $(Y)_{NOM}$ WHOLE $(Z)_{GEN}$ VERB or:
(b) WHOLE $(Z)_{NOM}$ have ADJ PART $(Y)_{ACC}$.

(18) a. *Die Augen der Kinder glänzen*
the $eyes_{NOM}$ the $children_{GEN}$ are:shining
'The children's eyes are shining.'

(19) a. *Der Henkel der Tasse ist blau*
the $handle_{NOM}$ the cup_{GEN} is blue
'The handle of the cup is blue.'

(18) b. *Die Kinder haben glänzende Augen*
the $children_{NOM}$ have shining $eyes_{ACC}$
'The children have shining eyes.'

(19) b.

Die	*Tasse*	*hat*	*einen*	*blauen*	*Henkel*
the	cup$_{\text{NOM}}$	has	a	blue	handle$_{\text{ACC}}$

'The cup has a blue handle.'

This construction is not restricted to any particular parts, and with animate possessors it occurs with a large range of possessed as well as other items, compare *Mr Miller has a nice wife* and *I hope we'll have nice weather.*[7]

When there is a change in the part, again the part itself may be commented on:

(20)

Das	*Rückfenster*	*des*	*Autos*	*wird*	*schmutzig*
the	rear:window$_{\text{NOM}}$	the	car$_{\text{GEN}}$	is:getting	dirty

'The rear window of the car is getting dirty.'

(21)

Ihre	*Hände*	*werden*	*rot*
3SGF:POSS	hands$_{\text{NOM}}$	become	red

'Her hands become red.'

(22)

Das	*Haar*	*des*	*Vaters*	*war*	*grau*	*geworden*
the	hair$_{\text{NOM}}$	the	father$_{\text{GEN}}$	AUX	grey	become

'The father's hair had turned grey.'

Alternatively, the comment on the part can be subordinated to the whole with a verb such as *bekommen* 'come to have', as in (23):

(23)

Sie	*bekommt*	*rote*	*Hände*
she$_{\text{NOM}}$	comes:to:have	red	hands$_{\text{ACC}}$

'She comes to have red hands.'

Even with animate possessors this is less common than the use of *haben* to describe a state.

Sometimes the comment is equally applicable to the part and the whole, for example when there is contact, as contact involves part of an object only, or when the verb describes a posture (or stance) such as *stand* or *lie*, and this posture can be both that of the whole and that of the part; when there is a choice between possessor (examples (a)) and part (examples (b)) for the role of subject/direct object, the possessor is usually preferred.

Intransitive:
(a) WHOLE (Z)$_{\text{NOM}}$ VERB PREP PART (Y)
(b) PART (Y)$_{\text{NOM}}$ WHOLE (Z)$_{\text{GEN}}$ VERB

Transitive:
(a) AGENT (X)$_{NOM}$ VERB WHOLE (Z)$_{ACC}$ PREP PART (Y)
(b) AGENT (X)$_{NOM}$ VERB PART (Y)$_{ACC}$ WHOLE (Z)$_{GEN}$

Note that the definite article (DEF) is sufficient and the possessive relation is not explicitly stated:

(24) a. *Die Negative bitte nur am Rande anfassen!*
the negatives$_{ACC}$ please only PREP:DEF edge$_{DAT}$ handle
'Please handle negatives by the edges only.'

b. ??*Bitte nur den Rand der Negative anfassen*
please only the edge$_{ACC}$ the negatives$_{GEN}$ handle

(25) a. *Das Auto steht mit den*
the car$_{NOM}$ is:standing PREP(INSTR) DEF
Vorderrädern im Wasser
front:wheels$_{DAT}$ in:the water$_{DAT}$
'The car is standing with its front wheels in the water.'

b. ?*Die Vorderräder des Autos stehen im Wasser*
the front:wheels$_{NOM}$ the car$_{GEN}$ are:standing in:the water$_{DAT}$
'The front wheels of the car are standing in the water.'

(26) a. *Sie legten den Verletzten*
they placed the injured:man$_{ACC}$
mit dem Kopf auf einen Stein
PREP(INSTR) DEF head$_{DAT}$ on a rock$_{ACC}$
'They placed the injured man with his head on a rock.'

b. ??*Sie legten den Kopf des Verletzten*
they placed the head$_{ACC}$ the injured:man$_{GEN}$
auf einen Stein
on a rock$_{ACC}$
'They placed the injured man's head on a rock.'

(27) a. *Er stösst mit dem Fuss*
he hits PREP(INSTR) DEF foot$_{DAT}$
an eine Kiste
PREP a box$_{ACC}$
'He hits his foot against a box.'

b. ??*Sein* *Fuss* *stösst* *an* *eine* *Kiste*
3SGM:POSS $foot_{NOM}$ hits PREP a box_{ACC}
'His foot hits a box.'

Here there is some difference in acceptability between (25b) on the one hand and (26b) or (27b) on the other; it is, however, not a difference between animate and inanimate possessors, but between parts that can be seen as individual objects and relational terms (see Nichols 1988: 572-573): *edge* in (24), like *top, middle, inside*, specifies a subspace of the whole. Such subspaces are often described by adverbs,[8] for example, German: *unten* 'at the bottom', *oben* 'on top', *links* 'on the left', English: *underneath*. An individually definable part such as *wheels*, by contrast, can be seen, and commented on, as an object in its own right, and so the noun can more easily occur as a topic, that is, be a grammatical subject or direct object. Body part nouns may either be used as a spatial subspecification of the whole, or as denoting individual objects.

Items worn on the body are treated like parts of the body, if the action described by the verb can apply to either the part or the whole, as in (28a):

(28) a. *Er* *packt* *den* *Jungen* *beim* *Kragen*
he_{NOM} seizes the boy_{ACC} PREP:DEF $collar_{DAT}$
'He seizes the boy by the collar.'

instead of

b. *Er* *packt* *den* *Kragen* *des* *Jungen*
he_{NOM} seizes the $collar_{ACC}$ the boy_{GEN}
'He seizes the boy's collar.'

When the part, but not the whole is thought of as moving, the motion is predicated of the part:

PART $(Y)_{NOM}$ WHOLE $(Z)_{GEN}$ VERB,

as in:

(29) *Die* *Gardine* *des* *Schlafzimmers* *flattert* *im* *Wind*
the $curtain_{NOM}$ the $bedroom_{GEN}$ is:fluttering in:the wind
'The curtain of the bedroom is fluttering in the wind.'

(30) *Das Dach des Hauses ist eingefallen*
the $roof_{NOM}$ the $house_{GEN}$ AUX fallen:in
'The roof of the house has collapsed.'

(31) *Der Kopf der schlafenden Frau sinkt.*
the $head_{NOM}$ the sleeping $woman_{GEN}$ sinks
auf die Stuhllehne
onto the chair:back
'The sleeping woman's head sinks onto the back of the chair.'

(32) *Sein linkes Augenlid zuckt*
3SGM:POSS left $eyelid_{NOM}$ is:twitching
'His left eyelid is twitching.'

In (29)-(32) neither the whole nor the moving part is active: the motion is seen as uncontrolled and uncontrollable and has therefore no relation to the whole. A body part that moves in this way is described in the same way as the part of an inanimate object: it becomes the subject of the clause, as in (31), and the possessor is in the genitive.

When a part of an inanimate object is said to move, the possessor is marked by the genitive, regardless of whether the motion is spontaneous (when the part is the subject, as in (29)-(30)), or caused by an agent (when the part is the direct object, as in (33)),

AGENT $(X)_{NOM}$ VERB PART $(Y)_{ACC}$ WHOLE $(Z)_{GEN}$:

(33) *Der Wind dreht die Flügel der Mühle*
the $wind_{NOM}$ turns the $wings_{ACC}$ the $windmill_{GEN}$
'The wind turns the sails of the windmill.'

With animates, however, there is a difference, and this will be discussed in the next subsection.

4.2. Features of the possessor-body part relation

4.2.1. Motion of body parts

The motion of a body part is rarely perceived as uncontrollable, like inanimate motion, and presented as unrelated to the possessor, as in (31)-(32):

PART $(Y)_{NOM}$ WHOLE $(Z)_{GEN}$ VERB.

Usually it is seen as active motion and so something that is sayable about the animate whole: motion in part of what we think of when we think of Z (the whole) is something that Z does:

POSSESSOR $(Z)_{NOM}$ VERB BODY PART $(Y)_{ACC}$.

(34) *Der Hund hebt das Bein*
the dog_{NOM} lifts DEF leg_{ACC}
'The dog lifts its leg.'

(35) *Der Mann senkt unwillkürlich den Kopf*
the man_{NOM} lowers involuntarily DEF $head_{ACC}$
'The man automatically lowers his head.'

(36) *Das Kind faltet die Hände*
the $child_{NOM}$ folds DEF $hands_{ACC}$
'The child folds his/her hands.'

In (34)-(36) the motion results in a change of position or posture; syntactically the body part noun becomes the direct object, the possessor the subject (in German the possessive relation need not be made explicit: there is no possessive pronoun when the possessor is mentioned elsewhere in the clause). The possessor is thought of as initiating the change and as controlling the movement of the part, even if, as in (35), this is not conscious control. This subject-object construction is not fully transitive because the patient is part of the agent. It cannot (normally) be passivised.

When there is no change, the animate possessor in German is not directly said to move the part; but the motion, which may or may not be controlled, is still described as action of the whole:

POSSESSOR $(Z)_{NOM}$ VERB PREP(INSTR) BODY PART $(Y)_{DAT}$.

(37) *Die alte Frau wackelt mit dem Kopf*
the old $woman_{NOM}$ is:wobbling PREP(INSTR) DEF $head_{DAT}$
'The old woman's head is wobbling.'

(38) *Die Kinder klappern mit den Zähnen*
the $children_{NOM}$ are:chattering PREP(INSTR) DEF $teeth_{DAT}$
'The children's teeth are chattering.'

(39) *Die Jungen baumeln mit den Beinen*
the boys$_{NOM}$ are:dangling PREP(INSTR) DEF legs$_{DAT}$
'The boys are dangling their legs.'

(40) *Das Mädchen zuckt mit den Schultern*
the girl$_{NOM}$ shrugs PREP(INSTR) DEF shoulders$_{DAT}$
'The girl shrugs her shoulders.'

In English, controllable motion ((39)-(40)) is treated like controlled motion ((34)-(36)), using a transitive construction, but uncontrolled motion, as in (37)-(38), cannot be attributed to the possessor as it is in German. In both English and German, the possessor becomes the subject if the motion is thought of as active.

This active construction is not possible with inanimate wholes, even when the inanimate object is able to move: a moving part may be treated as an autonomous object, as in (41):

(41) a. *Die Räder des Autos drehen sich*
the wheels$_{NOM}$ the car$_{GEN}$ are:turning REFL$_{ACC}$
'The wheels of the car are turning.'

but as the motion of the part is not dependent on the whole, it cannot be presented as action of the whole:

b. **Das Auto dreht die Räder*
the car$_{NOM}$ turns DEF wheels$_{ACC}$
'The car turns its wheels.'

Motion of a body part which is caused or controlled but in which the animate being is not active is not usual, but it occurs in real life, as described in (42a), for example:

AGENT (X)$_{NOM}$ VERB POSSESSOR (Z)$_{DAT}$ BODY PART (Y)$_{ACC}$.

(42) a. *Der Mann(X) schüttelt dem Jungen(Z) die*
the man$_{NOM}$ shakes the boy$_{DAT}$ DEF
Hand(Y)
hand$_{ACC}$
'The man shakes the boy's hand.'

This motion of the body part Y cannot be attributed to the possessor Z, because it is not initiated by Z, but by X; all the same it is not unrelated to the possessor: it is something that is sayable about Z because part of the relation between Z and Y (while Z is alive) is that Z is able to move Y, and so Z is dominant in this relation; and although Z is not actually moving Y, this potential is indicated by the choice of the dative case for the possessor. The situation is different when the possessor, though animate, is lifeless, as in (42b):

b. **Der Mann schüttelt dem Toten die Hand*
the man_{NOM} shakes the $dead{:}man_{DAT}$ DEF $hand_{ACC}$
'The man shakes the dead man's hand.'

A dead body has no potential for action. The situation intended in (42b) can only be described by the genitive construction:

AGENT $(X)_{NOM}$ VERB PART $(Y)_{ACC}$ WHOLE $(Z)_{GEN}$:

c. *Der Mann schüttelt die Hand des Toten/Schlafenden*
the man_{NOM} shakes the $hand_{ACC}$ the $dead/sleeping{:}man_{GEN}$
'The man shakes the dead/sleeping man's hand.'

When the verb does not specify motion, but the result of motion, i.e. a change in posture, the potential for action is not relevant:

(43) a. *Die Mutter(X) faltet dem Kind(Z) die Hände (Y).*
the $mother_{NOM}$ folds the $child_{DAT}$ DEF $hands_{ACC}$
'The mother folds the child's hands.'

b. *Die Mutter faltet dem Toten die Hände*
the $mother_{NOM}$ folds the $dead{:}man_{DAT}$ DEF $hands_{ACC}$
'The mother folds the dead man's hands.'

In (43a) and (43b) the dative is acceptable also when the animate possessor is dead. Posture, that is, the position of parts relative to each other, is part of what we think of when we think of an intact animate being, whether it is alive or dead. A change in posture (and so a change in outward appearance) is something that is sayable about an animate being even when it could not cause such a change itself and when it is not aware of it, because the postures which are possible are still determined by the way the parts can move as parts of the whole. In (43b) the change of posture in Y is part of what is sayable about Z, whereas the motion of Y in (42b) is not.

The genitive construction would be acceptable in (43) as well as in (42c), because the dominance of a lifeless whole over a body part is only spatial and can more easily be ignored than with a living being:

c. *Die Mutter faltet die Hände des Toten*
the $mother_{NOM}$ folds the $hands_{ACC}$ the $dead{:}man_{GEN}$
'The mother folds the dead man's hands.'

When the possessor is fully alive, this genitive construction is not acceptable, because it only focusses on the agent X, without indicating that something different is also sayable about the possessor Z, namely that part of Z undergoes a change; note that this is independent of any experience the possessor Z may have. An exception to this unacceptability of the genitive may be if the speaker chooses to focus on the action as an activity, not a change in the object, for example, to indicate that the agent (who is not the possessor) is oblivious of his/her surroundings; however, the verb in (45) = (43d) could only with difficulty be interpreted as activity.

(44) *?Der Mann schüttelt die Hand des Jungen*
the man_{NOM} shakes the $hand_{ACC}$ the boy_{GEN}
'The man keeps shaking the boy's hand.'

(45) ?**Die Mutter faltet die Hände des*
the $mother_{NOM}$ folds the $hands_{ACC}$ the
Kindes
$child_{GEN}$
'The mother folds the child's hands.'

The dative also occurs when there is movement of a body part in which the possessor is not active, which is not controlled by an outside agent either, but which is thought of as caused by something that is sayable about the possessor. The construction then is:

POSSESSOR $(Z)_{DAT}$ VERB BODY PART $(Y)_{NOM}$ (PREP CAUSE (W)).

Examples:

(46) *Mir hüpft das Herz vor Freude*
me_{DAT} is:leaping DEF $heart_{NOM}$ PREP joy
'My heart is leaping with joy.'

(47) *Der Frau zittern die Hände*
the $woman_{DAT}$ are:trembling DEF $hands_{NOM}$
'The woman's hands are trembling (with fatigue, fear).'

(48) *Dem Kind fallen die Augen zu*
the child$_{DAT}$ are:falling DEF eyes$_{NOM}$ shut
'The child's eyes are falling shut (with tiredness).'

To sum up: with motion, the dative of the possessor occurs when the motion of a body part is something that is sayable about the possessor, but is initiated or caused by something other than the possessor, that is, when part of the possessor undergoes a motion or a change of posture: this is something that is sayable about the possessor because of the relation between the whole and its parts, and it is not simply a reversal of what is being said about the subject/causer. The dative is not used when the motion of the part is initiated by the possessor, whether consciously or not (then the possessor is active and appears as the subject instead), and it is not used when there is a change of place resulting from motion which is neither initiated by the animate being (active) nor caused (passive), but spontaneous, as in (31).

4.2.2. Change of state

We think of an animate being as actively moving, but change is not seen as active, as something that the animate being does. This seems to be true of all languages. A change of state in a body part is not a change of state of the possessor in the same way as motion of the part is motion of the whole. When there is a change in the part, it is therefore not surprising that the whole does not appear as the subject in the nominative (unless the part is representative of the whole, as in (15)-(17) above).

BODY PART (Y)$_{NOM}$ POSSESSOR (Z)$_{GEN}$ VERB (CHANGE OF STATE):

(49) a. *Der Finger des Jungen schwillt an*
the finger$_{NOM}$ the boy$_{GEN}$ swells up
'The boy's finger swells.'

b. **Der Junge schwillt mit*
the boy$_{NOM}$ swells PREP(INSTR)
dem /am Finger an
DEF/PREP(LOC) finger$_{DAT}$ up

(50) a. *Die Hände der Frau werden rot*
the hands$_{NOM}$ the woman$_{GEN}$ are:getting red
'The woman's hands are getting red.'

b. **Die* *Frau* *wird* *rot* **an*/**in*/**mit*
the $woman_{NOM}$ is:getting red PREP(LOC/INSTR)
den *Händen*
DEF $hands_{DAT}$

If the change in the part is caused by some inner state of the possessor, this is something which, like the motion described in (46)-(48), is part of what can be said about the possessor, although it has to be predicated of the part; so the part is the subject, but the possessor is in the dative:

POSSESSOR $(Z)_{DAT}$ VERB BODY PART $(Y)_{NOM}$.

(51) *Dem Jungen knurrt der Magen*
the boy_{DAT} is:grumbling DEF $stomach_{NOM}$
'The boy's stomach is grumbling (with hunger.)'

(52) *Die Augenlider werden ihm schwer*
DEF $eyelids_{NOM}$ are:getting him_{DAT} heavy
'His eyelids are getting heavy (he is getting very tired).'

(53) *Die Ohren müssen ihm klingen*
DEF $ears_{NOM}$ must him_{DAT} be:ringing
'His ears must be burning.'

The situation of (49) can be described with a dative construction as well ((54) below), but the process of (50/55), which is due to external causes, cannot:

(54) Dem Jungen schwillt der Finger an
the boy_{DAT} swells DEF $finger_{NOM}$ up
'(It happens to the boy that) the boy's finger swells.'

(55) **Der Frau werden die Hände rot*
the $woman_{DAT}$ are:getting DEF $hands_{NOM}$ red

A change of state in the part caused by an agent is, like caused motion, expressed by a transitive construction. In German, the possessive relation does not have to be specified when the motion is initiated by the possessor:

(56) *Das Kind faltet die Hände*
the $child_{NOM}$ folds DEF $hands_{ACC}$
'The child folds his/her hands.'

but when the causer is distinct from the possessor, the possessor occurs in the dative:

(57) *Die Mutter faltet dem Kind die Hände*
the mother$_{NOM}$ folds the child$_{DAT}$ DEF hands$_{ACC}$
'The mother folds the child's hands.'

With a change of state, the dative of the possessor is necessary not only when the agent and the possessor are distinct, but even when they are identical:

AGENT (X)/POSSESSOR (Z)$_{NOM}$ VERB (CHANGE OF STATE) POSSESSOR (Z)$_{DAT}$ BODY PART (Y)$_{ACC}$.

(58) a. ??*Die Mutter wäscht das Gesicht des Kindes*
the mother$_{NOM}$ washes the face$_{ACC}$ the child$_{GEN}$

b. *Die Mutter wäscht dem Kind das Gesicht*
the mother$_{NOM}$ washes the child$_{DAT}$ DEF face$_{ACC}$
'The mother washes the child's face.'

(59) a. **Der Junge hat ein/sein Bein gebrochen*
the boy$_{NOM}$ AUX INDEF/3SGM:POSS leg$_{ACC}$ broken
'The boy has broken a leg/ his leg.'

b. *Der Junge hat sich das Bein gebrochen*
the boy$_{NOM}$ AUX REFL$_{DAT}$ DEF leg$_{ACC}$ broken
'The boy broke his leg.'

(60) a. **Du hast die/deine Haare gefärbt*
you$_{NOM}$ AUX DEF/2SG:POSS hair$_{ACC}$ dyed

b. *Du hast dir die Haare gefärbt*
you$_{NOM}$ AUX REFL$_{DAT}$ DEF hair$_{ACC}$ dyed
'You have dyed your hair.'

As this is not necessary with motion and change of posture, the function of the dative in (59b) and (60b) cannot simply be to indicate the possessive relation.

The transitive construction asserts that what is said about the subject includes something that is sayable about the object. When the subject is the possessor and the object is a body part, the possessor is not only an agent, but also the possessor of the patient, and this must be taken into account in German, otherwise the relation between animate being and body part is not complete. The nominative marks the possessor as the cause of the change; the dative reflexive indicates that not only the cause of the change, but also the effect on the part is something that is sayable about the whole. This is not the same as the whole undergoing the change: but it is a change in what we think of when we think of the whole.

In German it is not possible, as it is in Romanian (see Manoliu-Manea 1995), to make such a change of state in the part a predicate of the whole, relegating the affected part to a prepositional phrase. The body part noun may be demoted to become the object of a preposition when it is moving ((37)-(40)), or when it is seen as a location, i.e. part of the space of the whole (in the same way as the part of an inanimate whole (24)-(28)), but not when it is undergoing a change.

4.2.3. *Body part as location*

There are many ways in which an animate being can be dominant in a relation. Being able to feel something and being able to perceive something are only two of these relations. In the examples discussed so far sensation and awareness, even though they can never be excluded when a living being is involved, did not play a role in determining the acceptability of a construction. In the initial dative example (1) (repeated as examples (5a) and (13a)), this role was assumed to be relevant; but in fact other constructions of the same pattern, involving directional locational phrases and body part nouns, can be fully acceptable with the dative of the possessor even when there is no contact between objects and so no sensation, or no awareness on the part of the possessor:

(i) Intransitive:
MOVING OBJECT $(X)_{NOM}$ VERB POSSESSOR $(Z)_{DAT}$ PREP BODY PART (Y).

(ii) Transitive:
AGENT $(W)_{NOM}$ VERB POSSESSOR $(Z)_{DAT}$ MOVING OBJECT $(X)_{ACC}$ PREP BODY PART (Y).

In this construction the preposition must specify a relation to an object as extended in space[9] (such as *auf* 'on', *unter* 'under'), not a relation to its position (*zu* 'to', *bei* 'near'), and the statement of the spatial relation cannot be the main purpose of the utterance (compare ((68a) and 68b)).[10]

(61) *Die Kinder setzen dem Schneemann einen Hut*
the children$_{\text{NOM}}$ put the snowman$_{\text{DAT}}$ a hat$_{\text{ACC}}$
auf den Kopf
PREP the head$_{\text{ACC}}$
'The children put a hat on the snowman's head.'

(62) *Der Ball rollt dem Jungen unter den Beinen*
the ball$_{\text{NOM}}$ rolls DEF boy$_{\text{DAT}}$ PREP DEF legs$_{\text{DAT}}$
durch
through
'The ball rolls through (under) the boy's legs.'

(63) *Der Junge sieht dem Vater (heimlich)*
the boy$_{\text{NOM}}$ looks the father$_{\text{DAT}}$ (secretly)
über die Schulter
PREP DEF shoulder$_{\text{ACC}}$
'The boy (secretly) looks over his father's shoulder.'

On the other hand, the dative of the possessor is not acceptable in the following examples:

(64) a. **Das Kind sitzt der Mutter*
the child$_{\text{NOM}}$ is:sitting the mother$_{\text{DAT}}$
auf dem Schoss
PREP DEF lap$_{\text{DAT}}$

Instead:

b. *Das Kind sitzt auf dem Schoss*
the child$_{\text{NOM}}$ is:sitting PREP DEF lap$_{\text{DAT}}$
der Mutter
the mother$_{\text{GEN}}$
'The child is sitting on the mother's lap.'

(65) **Der Junge steht dem Vater*
the boy$_{\text{NOM}}$ is:standing the father$_{\text{DAT}}$
hinter der Schulter
PREP DEF shoulder$_{\text{DAT}}$
('The boy is standing behind his father's shoulder.')

(66) **Der Hut sitzt dem Mann auf dem Kopf.*
the hat$_{NOM}$ is:sitting the man$_{DAT}$ PREP DEF head$_{DAT}$
('The hat is sitting on the man's head.')

Examples (64)-(66) are intended to describe position, whereas (61)-(63) describe a direction. German has a set of locational prepositions which take the dative when indicating position, the accusative when expressing direction. But the difference in case-marking is not one between position and motion, or between an unchanged position and a change. As indicated above (examples (8) and (9a/b)), extension patterns with motion, not with unchanged position: when the subject X is said to extend or move through space, the space related to an object Y (PREP Y) is part of what is being said about X, so Y is in the accusative. This is also relevant for the acceptability of the possessive dative:

(67) *Das Wasser reicht ihm bis ans Knie*
the water$_{NOM}$ reaches him$_{DAT}$ as:far:as PREP:DEF knee$_{ACC}$
'The water reaches up to his knee.'

The dative of the possessor is not formally excluded by the presence of another dative (as might seem to be the case in (64)-(66)): in (62), *Der Ball rollt dem Jungen*$_{DAT}$ *unter den Beinen*$_{DAT}$ *durch* 'the ball rolls through (under) the boy's legs', the body part noun following the preposition, *den Beinen*, is in the dative, but the whole phrase with *durch* indicates direction, and the possessive dative is acceptable. It is also acceptable when the position of an object relative to a body part is not asserted,but unlike the unacceptable examples (64)-(66) and (68a), something else is being said about the object while it is in that position, as in (68b):

(68) a. *Ein Blatt liegt ihm auf der Wange*
a leaf$_{NOM}$ is:lying him$_{DAT}$ PREP DEF cheek$_{DAT}$
('A leaf is lying on his cheek.')

but:

b. *Ein Blatt klebt ihm auf der Wange*
a leaf$_{NOM}$ is:sticking him$_{DAT}$ PREP DEF cheek$_{DAT}$
'A leaf is sticking to his cheek.'

In (62) and (68b), the space related to the body part is not directly subordinated to what is being said about the subject and the noun *Wange* is not in the accusative; but the verb describes more than position; therefore the position, and so the space related to *Wange* in (68b) and the space through which the subject moves in (62) is subordinated to what is being said about the subject. The function of the dative of the possessor is to assert that in this situation the body part, although it is part of what is being said about the subject, is also part of what can be said about its animate possessor, because there is a relation of spatial dominance of the whole over the part. This spatial dominance is a very complex relation which we take for granted when we talk of an animate being but which we never really define: it is not only that the location of the body part depends on the location of the whole and changes when the whole moves, and that the possessor is able to move a body part at will; the way a body part can move or be made to move and its potential postures are determined by its relation to the whole as well. An inanimate whole does not dominate its parts in this way.

The genitive (as in (5b) above) only indicates that there is a relation between two entities, it does not necessarily present one as dominant over the other.

When an object X is said to move or extend into a space that is thought of as the space of an animate being Z, X itself (and what is being said about it) may be said to become part of the space or sphere of Z. Sometimes this relation is purely spatial, as, for example, in (61), where *the hat* becomes part of what we see when we see *the snowman*; in (62) and (63) the direction of a motion or action is defined in relation to the space of Z. But often other relations can be inferred from such purely spatial relations, so that in a wider sense the moving object X becomes part of that part of the world with which the animate being Z is able to interact.

5. Inferences from the dative construction

When a purely spatial interpretation is possible, as in (61)-(63), the dative of the possessor is acceptable whether the possessor is dead or alive. But when the possessor is alive and able to act and interact with its environment, many more inferences can be made from this spatial relation. Inferences that can be made about the animate possessor are: coming to feel, coming to perceive, coming to know, and coming to (be able to) have or do, and several of these may be possible at the same time. In all these relations the animate being is dominant, that is, something becomes sayable about the animate being which involves the (moving) object, not the other way round. Consider example (69):

(69) *Der Apfel fällt dem Mädchen*
the $apple_{NOM}$ falls the $girl_{DAT}$
in den Schoss
PREP DEF lap_{ACC}
'The apple falls into the girl's lap.'

In (69) the girl comes to feel something, comes to see the apple, perhaps comes to be aware of the person who threw it, comes to have the apple, comes to be able to do something to or with the apple (eat it, throw it, hit somebody). This is often described as the possessor being "affected"; when affectedness is not further specified, this makes it appear as if the possessor were somehow passive, that something happens to the possessor. Other researchers, such as von Weiss (1962: 260), on the other hand, claim that the possessor is potentially active, again without specifying. As the list of possible inferences above shows, both may be correct, even in relation to the same event. The dative does not specify what it is that is sayable about the animate possessor. It only indicates that because there is a specifiable relation between the referent of the dative noun phrase and a body part (or other object), when something is said in relation to this object, something unspecified but specifiable can be inferred (that is, comes to be sayable) about the dative referent.

When these inferences are predictable, the range of nouns that can occur in the prepositional phrase is much wider than the set of body part nouns. For a purely spatial interpretation, all nouns referring to objects that have a place on the body and move with the possessor are acceptable, such as clothing, ornaments, harness (in poetic language this may be extended to objects which are thought of as the place where the possessor is, for example, 'house', 'cradle'). Different types of inference allow different types of extensions: sensation requires contact with the body, visual perception depends on position of an object in front of the animate being, and so on; but if it is possible to infer what the relation between the dative referent and the subject/object is, the spatial relation need not be defined in terms of a body part:

(70) *Der Brief flatterte mir auf den Schreibtisch*
the $letter_{NOM}$ fluttered me_{DAT} PREP DEF $desk_{ACC}$
'The letter fluttered onto (arrived on) my desk (I came to see it).'

(71) *Der Fisch geht dem Fischer ins Netz*
the fish$_{\text{NOM}}$ goes the fisherman$_{\text{DAT}}$ PREP:DEF net$_{\text{ACC}}$
'The fish comes to be in the fisherman's net (he comes to have it).'

The dative here has the same semantic function as it has in its obligatory uses with verbs of giving (coming to have), telling (coming to know), showing (coming to perceive); there seems to be no class of verbs of enabling as distinct from giving, but German has adjective-derived verbs meaning 'make easier/harder/possible', which are three-place verbs with the dative (the adjectives themselves may also be construed with a dative); the purely locational meaning is lexicalised in German in three-place verbs that describe the putting on (and taking off) of clothes and other objects which have a given place in relation to the body, such as *anziehen* 'put on', *einflössen* 'make flow in', that is, 'make drink'. There is no class of three-place verbs describing sensation: sensation can be directly caused by contact and is therefore described by transitive verbs.

The negative meanings (take away, steal from, withhold, put a hand over a person's eyes, place obstacles in a person's way) are in German also expressed using the dative of the recipient (coming not to have, be able to do, perceive and so on).

6. Dativus commodi/incommodi

The *dativus commodi/incommodi*, the 'dative of advantage/disadvantage' – the term comes from the grammarians of Roman antiquity – marks the person for whom an event is good or bad (see also note 2).

When a body part moves or undergoes a change of state or changes posture, this is something that is sayable about the possessor, whether it is good or bad for the possessor, initiated by the possessor or caused by something else. The ideas of "good" and "bad" are possible inferences; they can make a dative construction acceptable – and so be invited inferences – when the only relation between the referent of the dative noun phrase and an event that can be inferred is that the person would have, or would not have, wanted it to happen. Only an event, that is, a change, can be said to be good or bad for a person in this way, not motion or the unchanged state of an object.

The concepts of "good for" and "bad for" an individual do not refer to absolute values nor to the judgment of the speaker, but have to do with the individual referent's attitude towards individual objects and events, and his/her expectations of control over his/her part of the world. What happens against this expectation of control is "bad" for the person; if however the change is something the possessor would have wanted to happen, it is "good". The change in the position of a part, which can otherwise be presented as an independent motion, as in example (72):

(72) *Der Kopf der schlafenden Frau*$_{\text{GEN}}$
the head$_{\text{NOM}}$ the sleeping woman$_{\text{GEN}}$
sinkt auf die Stuhllehne
sinks onto the chair:back
'The sleeping woman's head sinks onto the back of the chair.'

becomes something sayable about the possessor when it is thought of as happening in spite of the possessor's attempt at control:

(73) *Der Kopf sank ihr immer wieder*
DEF head$_{\text{NOM}}$ sank her$_{\text{DAT}}$ always again
auf die Stuhllehne
PREP the chair:back$_{\text{ACC}}$
'Her head kept sinking onto the back of the chair (although she tried to prevent it).'

Insofar as objects that are not body parts can be expected to be under the control of a person, something that happens involving these objects can also be said to be good or bad. For example:

(74) *Der Frau ist eine Fliege in die*
the woman$_{\text{DAT}}$ AUX a fly$_{\text{NOM}}$ PREP DEF
Suppe gefallen
soup$_{\text{ACC}}$ fallen
'It happened to the woman that a fly fell into the soup.' (She would not have wanted it to happen to the location: bad.)

(75) *Der Frau ist die schöne Vase auf den*
the woman$_{\text{DAT}}$ AUX DEF lovely vase$_{\text{NOM}}$ PREP the
Boden gefallen
floor$_{\text{ACC}}$ fallen
'It happened to the woman that the lovely vase fell on the floor.' (She would not have wanted it to happen to the subject: bad.)

(76) *Der Junge putzt der Mutter die Schuhe*
the boy$_{\text{NOM}}$ cleans the mother$_{\text{DAT}}$ DEF shoes$_{\text{ACC}}$
'The boy cleans the shoes for his mother.' (She would have wanted it to happen to the object: good.)

When there is no change, even though there may be an effect which is good or bad for an affected object, this dative construction cannot be used:

(77) **Der Mann hat der Frau das Kind*
the man$_{\text{NOM}}$ AUX the woman$_{\text{DAT}}$ DEF child$_{\text{ACC}}$
geschlagen
beaten
'The man has beaten the woman's child.'

Coming to have and coming to be able to do are usually good for a person, the opposite is bad. For example:

(78) *Den Dieben fielen 2.000 Mark in die*
the thieves$_{\text{DAT}}$ fell 2,000$_{\text{NOM}}$ DM PREP DEF
Hände
hands$_{\text{ACC}}$
'DM 2,000 fell into the thieves' hands.' (They netted DM 2,000, which was good for the thieves, though bad for the owner.)

(79) *Diese Bemerkung nahm dem Redner den Wind*
this remark$_{\text{NOM}}$ took the speaker$_{\text{DAT}}$ the wind$_{\text{ACC}}$
aus den Segeln
PREP DEF sails$_{\text{DAT}}$
'This remark took the wind out of the speaker's sails.' (bad)

The good/bad-interpretation is often possible with dative constructions that would be acceptable for other reasons, because there may be several relations holding at the same time in which the animate being is dominant over an object. As examples (78) and (79) and also (70) and

(71) above show, the dative construction is productive, and it allows a rich and colourful variety of expressions as an alternative to the lexical verbs which are obligatorily construed with the dative.

7. Conclusion

An animate possessor is in more ways than one semantically dominant in the relation to its body part. Syntactically this is indicated by marking the possessor by the nominative when it dominates what is being said in relation to the part, by the dative when the body part it dominates is part of what is being said about something else, or when what is happening to the body part is caused by something else, even if this is not explicitly mentioned. The dative then indicates that although something else is the topic, the possessor is also in a topic-relation to the part. When the possessor is a direct object and so is itself part of what is being said about something else, it is, of course, marked by the accusative.

The cases do not have to encode any specific meanings such as "animacy" or "affectedness", it is sufficient to assume that they indicate the dynamics of the relations between noun phrases. The details are fleshed out by invited inferences. The functions of the cases are clearly distinct: the nominative noun phrase is dominant in its relation to what is being said about it (including the accusative object, but not the indirect object); the accusative marks a noun phrase as part of what is being said about something else; the dative noun phrase is dominant in its relation to at least part of what is being said about the subject, that is, something specifiable (but not specified) is sayable about the dative noun phrase because of what is explicitly being said about the subject. The meanings the cases can convey through these dynamics are surprisingly precise.

An animate noun phrase occurs more often as a subject than other noun phrases not only because an animate being is more active – that is only a part of its special character – but because more is sayable about it. It occurs much more often in the dative than other noun phrases, because there are many different relations possible between it and other objects in all of which the animate being is dominant, and these relations may come into play even when they are not made explicit and when the animate being is not directly involved in an event described. Within the possessive relation between the animate being and its body parts several such relations usually hold at the same time.

Notes

1. Contrasting the dative with the genitive in German is a simplification, as the genitive is rarely used in everyday speech; with inanimates, compounding is very common instead: *das Autodach* (5b), *die Schlafzimmergardine* (29), but this does not affect the argument.
2. Technical term from the grammar of Latin; this use of the dative is widespread in Indo-European languages; see Isačenko (1965: 20); Heidolph et al (1980: 368-369) distinguishes between *dativus commodi* (transitive construction with an agent) and "dative of responsibility" (intransitive); Abraham (1973: 8) describes only Heidolph's "dative of responsibility" as *dativus commodi.*
3. See, for example, von Polenz (1969), Isačenko (1965); this kind of dative is also split into a "dative of pertinence" and a "sympathetic dative" in Abraham (1973: 6f).
4. This distinction is not always considered relevant, see Wierzbicka (1988: 427-429), where the semantic explications of the dative construction are similar to the interpretation proposed here for German, but the labels CAUSING TO ... suggest direct causation.
5. This explication of the dative was developed in the author's Ph.D. thesis, published as Neumann (1987).
6. The definite article is identified as a morpheme by DEF only where it needs to be distinguished from the possessive.
7. Brinkmann (1962); Bally (1926 [1995]), examples (41) onwards, does not state whether in French the construction is restricted to animates.
8. For an analysis of these adverbs in German see Neumann (1987: 85-88).
9. For a detailed analysis of German spatial prepositions, see Neumann (1987: Chapter 4).
10. There is clearly some variation in acceptability: Examples (64a) and (66), but not (65), sound acceptable to some speakers; for me, they are acceptable only if qualified in some way. Compare (68a) and (68b) below.

References

Abraham, Werner
1973 "The ethic dative in German", in: Ferenc Kiefer – Nicholas Ruwet (eds.), 1-19.

Bally, Charles
1926 "L'expression des idées de sphère personnelle et de solidarité dans les langues indo-européennes", in: Franz Fankhauser – Jakob Jud (eds.), 68-78.
[1995] "The expression of concepts of the personal domain and indivisibility in Indo-European languages", translated by Christine Béal – Hilary Chappell [this volume.]

Brinkmann, Hennig
1962 *Die deutsche Sprache: Gestalt und Leistung.* [*The German language: Its form and function*]. Düsseldorf: Schwann.

Burridge, Kate
1995 "Degenerate cases of body parts in Middle Dutch", [this volume.]

Diver, William
1964 "The system of agency of the Latin noun", *Word* 20, 178-196.
Engel, Ulrich – Paul Grebe – Heinz Rupp (eds.)
1969 *Festschrift für Hugo Moser*. Düsseldorf: Schwann.
Fankhauser, Franz – Jakob Jud (eds.)
1926 *Festschrift Louis Gauchat*. Aarau: H. R. Sauerländer
Heidolph, Karl Erich – Walter Flämig – Wolfgang Motsch
1980 *Grundzüge einer deutschen Grammatik* [*Fundamentals of German grammar*]. Berlin: Akademie-Verlag.
Isačenko, Alexander V.
1965 "Das syntaktische Verhältnis der Bezeichnungen von Körperteilen im Deutschen [The syntactic relation of body part terms in German]", in: *Studia Grammatica V: Syntaktische Studien*. Berlin: Akademie-Verlag, 7-27.
Kiefer, Ferenc – Nicholas Ruwet (eds.)
1973 *Generative Grammar in Europe*. Foundations of Language (Supplementary Series 3) Dordrecht-Holland: D. Reidel.
Manoliu-Manea, Maria
1995 "Inalienability and topicality in Romanian: pragma-semantics of syntax", [this volume.]
Neumann, Dorothea
1987 *Objects and spaces: A study in the syntax and semantics of the German case system*. (Studien zur deutschen Grammatik 32) Tübingen: Gunter Narr.
Nichols, Johanna
1988 "On alienable and inalienable possession", in: William Shipley (ed.), 557-609.
von Polenz, Peter
1969 "Der Pertinenzdativ und seine Satzbaupläne [The dative of pertinence and its sentence patterns]", in: Ulrich Engel – Paul Grebe – Heinz Rupp (eds.), 146-171.
Shipley, William (ed.)
1988 *In honor of Mary Haas. From the Haas Festival conference. Native American linguistics*. Berlin: Mouton.
von Weiss, A.
1962 "Dativ und Genitiv [Dative and genitive]", *Wirkendes Wort*, Sammelband 1: Sprachwissenschaft. Düsseldorf: Schwann, 259-263.
Wierzbicka, Anna
1972 *Semantic primitives*. (Linguistische Forschungen 22) Frankfurt/M.: Athenäum.
1988 *The semantics of grammar*. (Studies in language companion series 18) Amsterdam/Philadelphia: John Benjamins.

Part VII

Languages of Africa

Body parts in Ewe grammar

Felix Ameka

> The semantics corresponding to this grammatical distinction [between "alienable" and "inalienable" possession] are peculiar. (...) Perhaps the most noteworthy observation is that body parts have the morphology of "alienable" possession [in Ewe].
>
> Claudi – Heine (1986: 316).

1. Introduction

1.1. Aims and scope

The purpose of this paper is to describe the treatment of body parts (and other meronyms, that is, part terms in part-whole relationships, (cf. Cruse 1986: 157ff; Winston et al. 1987; Iris et al. 1988)) in the grammar of Ewe, a Kwa language of West Africa.[1] The most significant thing about body parts in this language is that they are consistently presented in the grammar as distinct entities, separated, as it were, from their owners. When body parts are the loci of effect of events they may have the core grammatical roles of subject or object in the clause. In that case, the possessor may occur as a peripheral argument in a prepositional phrase. Furthermore, unlike other inherent relational terms such as kin and spatial orientation terms, body parts (and other part terms) occur in an "alienable" nominal possessive construction. It is in relation to this last grammatical property that Claudi – Heine (1986) make the comment in the quote above.

It seems that this puzzle about body parts concerns the fact that the real world knowledge about the relationship between a body and its parts is not matched by its linguistic representation. It is needless to say that language is not about reality; it is about how people perceive or conceptualise reality. In this paper an attempt will be made to discover the way in which the Ewes conceptualise the relationship between the parts of a body and the 'owner' of the body through a syntactic prism. A semantic explanation will also be offered for the prima facie peculiarity of the "alienable" grammar of body parts in the language.

It is assumed that language and, in this particular instance, grammar, is an embodiment and a reflection of conceptual structures of its speakers (cf. Bally 1926 [1995], Boas 1911, Sapir 1920, Whorf 1956, Wierzbicka 1988, Benjamin 1994, Hale 1986, Lakoff 1987, Langacker 1987 among others). Hence an analysis of the Ewe syntactic structures in which body parts participate should reveal the way in which these items are conceptualised in that language. To this end, the semantics of three adnominal constructions are investigated in section 3: an "alienable" structure signalled by the possessive connective *ɸé*, body parts occur as possessa in this construction; an "inalienable" construction which has no overt marking, body parts do not normally occur in this construction except in some cases with a first or second person singular pronominal possessor; and the syntactic compound marked by a high tone suffix. These compounds may be possessive or classificatory in function, and body parts tend to occur in the latter. The semantic ramifications of the property of body parts to assume grammatical roles distinct from the roles of their 'owners' are also explored.

1.2. Assumptions and methodological framework

The basic assumption of the paper is that grammatical constructions have functions and meanings which can be systematically described. The structures described in 3, for example, serve the function of expressing possession in Ewe. Each of the constructions, however, has an associated meaning (or set of related meanings). The structures code, so to speak, different parts of the semantic domain of possession. Following Seiler (1983: 4) it is assumed that "[S]emantically, the domain of possession can be defined as bio-cultural. It is the relationship between a human being and his kinsmen, his body parts, his material belongings, his cultural and intellectual products. In a more extended sense, it is the relationship between parts and the whole of an organism." (cf. Bally 1926 [1995]). This definition is anthropocentric and characterises the prototypes covered by the semantic space of possession.

In describing the meanings of linguistic structures, I employ the reductive paraphrase and semantic primitives approach espoused by Anna Wierzbicka (see Wierzbicka 1988 in particular for an application of the method to grammar). Essentially this method of analysis involves paraphrasing the grammatical construction being defined in a non arbitrary metalanguage based on a natural language, English. The metalanguage comprises a set of hypothetical indefinables – semantic primitives, namely: *I, you, someone, something, place, when, after, under, this, the same, other, all, no, if, be like, because, can, do, happen, part of, kind of, think of, want, say, know, very, good, bad,* and a few less well established ones such as *big* and *small.* These elements have their own language-independent syntax, i.e. combinatorial properties. It is assumed that the meanings of complex lexical as well as grammatical items can be defined in terms of these elements. Furthermore, the primes are universal, that is an isomorphic set of these indefinables can be drawn for every language. It is required that the paraphrases, that is, the explications or definitions, should be substitutable for the item in all possible contexts *salva significatione,* that is, without changing the sense. This is the test of the adequacy of the definitions. This test could be applied in one go or in stages. If the definition were substituted for some linguistic structures, however, the resulting form would be incomprehensible. Hence, the substitution has to be done in stages. (For a discussion of the principles of the methodology and the primitives, see Bogusławski 1970, 1989, Goddard 1989a, and 1989b; Wierzbicka 1989, 1990, 1991 and 1992 among others).

There are several reasons why this method is favoured in this study. First, for a general semantic description, one needs explications couched in a metalanguage based on a natural language rather than in a formal language (cf. Lyons 1977: 12). Second, the method is a comprehensive and general one which is applicable to all parts of a natural language – lexicon, grammar and even phonology (see the works by Wierzbicka, and Deakin 1981 among others). It is not an ad hoc approach for any specific part of language. Third, the similarities and differences between related constructions which could be thought of as performing similar syntactic and communicative functions are revealed through a comparison of their semantic representations, both within and across languages. Above all the definitions have predictive power, that is the various uses of the constructions can be predicted from their semantic formulae.

Using this method, one could define tentatively the relationship between two entities, X and Y, coded as an "inalienable" one in Ewe as follows:

One can think of X and Y like this:
they are like parts of the same thing
When one thinks of Y
  one cannot not think of X
X cannot do with Y anything X wants to

Thus there are three elements in a prototypical "inalienable" relationship: the items are so inextricably connected that one can think that they are parts of the same thing; the thought of one of the entities evokes the thought of the other; and the use of the possessum by the possessor is constrained. The justification for these features and the formula will become clear in the course of the discussions.

1.3. Structure of the paper

The paper is organised as follows: The major nominal possessive constructions in Ewe are described in section 3. This is accompanied by an investigation of how different categories of nominals are treated with respect to "alienability". The behaviour of body parts in the clausal syntax of Ewe is detailed in section 4. Attention is paid to structures in which the body part and its possessor have different grammatical roles. All this is preceded by an outline of some salient typological features of Ewe as necessary background to appreciating the grammar of body parts in Ewe.

2. Some relevant features of Ewe grammar

2.1. Word order and grammatical relations

Ewe is a grammatical word order language with basic SVO syntax (and subject and object are morphologically unmarked). Alternative orders of OSV, OVS and SOV are systematically linked to this basic one, determined by semantic and pragmatic factors. Core grammatical relations of subject, object1 and object2 are largely defined in positional terms. In an unmarked simple sentence, the subject is the NP that immediately precedes the VP and object1 is the NP that immediately follows the verb. Object2 immediately follows object1 without an intervening preposition. Obliques introduced by various prepositions may follow the verb in intransitive sentences or the objects in transitive ones. Schematically this may be represented as:

NP	V	(NP)	(NP)	(prep NP)	(X)
SUBJ		OBJ1	OBJ2	OBL OBJ	ADJUNCT

Morphologically, Ewe is an isolating language with agglutinative features. It has two tonemes, a high and a non-high. These are realised in context as high, mid, low, rising or falling tones.

2.2. Structure of a nominal phrase

In a nominal phrase, most modifiers come after the nominal head. The internal structure of a nominal phrase is shown below and exemplified in, for example, (1):

$$\text{NP} \rightarrow (\text{INT1}) \begin{Bmatrix} \text{N} \\ \text{PRO} \end{Bmatrix} (\text{Adj})^{*}\ (\text{QT})\ (\text{DET})\ (\text{PL})\ (\text{INT2})^{*}$$

(1) $_{NP}$[*neném ɖeví baɖa eve má- wó kóŋ*]$_{NP}$
$_{NP}$[such child bad two DET PL INT]$_{NP}$
dí- ḿ wó- le.
seek PROG 3PL PRES
'It is *those two very bad boys* they are looking for.'

2.3. Pronouns

Table 1 is a display of Ewe pronouns. The prenominal/independent ones (second row) are the forms used in possessive constructions:

Table 1. Ewe pronouns

	1SG	1PL	2SG	2PL	3SG	3PL
Preverbal	*me-*	*mí(é)*	*(n)e-*	*mi(e)*	wò *é*	*wó*
Prenominal/ Independent	*nye*	*miá(wó)*	*wò*	*mia(wó)*	*é* *eya*	*wó* *woá(wó)*
post verbal/ preposition	*-m*	*mí*	*wò*	*mi*	*-i*	*wó*
logophoric			*ye*	*yewó*	*ye*	*yewó*

The logophoric pronouns are used primarily in reportive contexts to represent the individual (except for the first person) whose speech, thoughts, feelings etc. are reported or are reflected in dependent clauses introduced by *bé(ná)* 'that' (cf. Clements 1979).

2.4. The verbal phrase

The sequence of elements in a verbal phrase can be represented as follows:

(IRR)	(REP)	(MODAL)	(TENSE)	<NP>	VERB	(ASPECT)
(l)á FUT	*ga*	*xa* 'in vain'	*le* PRES			*ḿ* PROG
(n)á SBJV		*nyá* CERT	*nɔ* NPRES			*gé* INGR
		nyá INV				(*n*)*a* HAB

In transitive sentences with imperfective aspect, that is the progressive or the ingressive, the object1 NP comes between the auxiliary verb which indicates tense (or direction) and the main verb. Note the roles of the elements in the following ditransitive verbal sentence with progressive aspect:

	SUBJ	TENSE	OBJ1	VERB	ASPECT	OBJ2		
(2)a.	*é-*	*le*	*akɔ́nta*	*fia*	*ḿ*	*ɖeví*	*á-*	*wó*
	3SG	be:PRES	arithmetic	teach	PROG	child	DEF	PL

'S/he is teaching arithmetic to the children.'

It is useful to think of verbal predicate meanings as expressed by the following structural types in Ewe:

i) simple verb root, e.g. *dzó* 'leave'; *vívi* 'be sweet'
ii) phrasal predicates made up of a verb root and its inherent nominal complement. Some of these can be thought of as semi-lexicalised. The main thing is that the nominal behaves syntactically as an independent nominal, e.g. *ɸo ɖi* (literally: strike dirt) 'be dirty'; *mlɔ́* [*anyí*] 'lie down' (i.e. sleep).
iii) compound verbs are made up of two verbs which together colexicalise a verbal meaning: e.g. *ɖɔ́ kpɔ́* 'wear see = taste'; *xɔ se* 'get hear= believe'
iv) a verb root and its satellite eg: *se ɖá* 'listen'; *kplɔ ɖó* 'follow'

There are two types of clauses involving the concatenation of verbals. These are the serial verb construction and the overlapping clause. In the serial verbal construction, each verb in the series has the same subject, tense, mood and aspect. Subject is only expressed with the first verb. Serialising connectives may be used to link such verbs: *hé* for simultaneous or sequential relations and *ɖa* for purpose relations. For example,

(2)b.

é-	*fɔ́*	*do*	*go*	*le*	*zã*	*me*	*dzáá*
3SG	arise	go	outside	at	night	in	quietly

ɖa-	*ku*	*te*	*ɖa*	*ɖu.*
purp	dig	yam	cook	eat

'He got out quietly at night, dug up yams, cooked them and ate them.'

In the overlapping clause, on the other hand, the subject of the second clause is coreferential with the object of the first clause. Typically, it is used to express simultaneous events. For instance in example (3):

(3)

é-	*da*	*tú-*	*í*	*wò-*	*kú*
3SG	throw	gun	3SG	3SG	die

'S/he shot it dead.'

3. Nominal possessive constructions

3.1. Overview

There are a number of nominal and predicative constructions in Ewe dedicated to the expression of possession (see Ameka 1991, chapter 7). The discussion in this section will focus on three major nominal ones since these are the ones that show the distinction between body parts (and other part terms) and other inherent relational terms most clearly.

Generally speaking, in the nominal constructions the possessor precedes the possessum. The three forms to be described employ different strategies. In the first one, a possessive connective *ϕé* (or its dialect variants) is used to link the possessor to the possessum: NP *ϕé* NP. In the second structure, the possessor and the possessum are juxtaposed to each other: NP NP. In these constructions the NPs can be fully expanded. In the third, the possessor and the possessum form a syntactic compound marked by a high tone suffix. These structures are illustrated below (see examples (4), (5), (6), and (7)):

NP *ɸé* NP:

(4) *agbalẽ- a ɸé akpa*
book DEF poss cover
'the cover of the book'

(5) *ŋútsu má ɸé tú xóxó lá*
man DEM poss gun old DEF
'that man's old gun'

NP NP:

(6) *ame- wó kátã nɔví*
person PL all sibling
'everybody's friend'

(7) *kplɔ̃- a dzí*
table DEF top
'the top of the table'

N - N + high tone suffix:
Note that example (8) here is possessive and (9) is classificatory and involves a body part term (see below for elaboration). The high tone suffix is particularly noticeable when the final syllable has a low tone. The fusion of the low and the high produces a rising tone on the last syllable as seen in example (9):

(8) *viktɔ- 'brá* [possessive]
Victor Abra
'Victor's Abra'

(9) *ame- tǎ* [classificatory]
person head
'human head'

Iconicity principles predict that the linguistic distance between the possessor and the possessum correlates with the degree of "alienability" between them (Haiman 1985: 130 ff.). Thus from a formal point of view, it can be hypothesised that the first construction, the NP *ɸé* NP, is the most "alienable" while the third, the syntactic compound, is the least "alienable". The semantic properties of each of the structures which are described below support this view.

3.2. "Alienable" nominal constructions

3.2.1. NP [POR] *poss NP* [POSS]

The prototypical relation expressed by this construction is that between an animate possessor and a non-relational possessed item. Relational items are nominals which are viewed as having a permanent or inherent association with another nominal. The use of a relational term evokes the thought of an associated nominal. To put it differently, a relational noun can be said to have subcategorised arguments (cf. Seiler 1983; Bally 1926 [1995]; Taylor 1989). Spatial orientation terms (e.g. top, front), kin terms, body parts and other meronymic terms are all relational. To say *bottom* means *bottom of something* and to say *grandfather* means *someone's grandfather.* Similarly, to say *arm* means *the arm of an entity:* someone or something. By contrast, to say *table* or *basket* etc does not imply *someone's table* or *someone's basket*. These are non-relational terms. Such nominals typically occur in the NP *ɸé* NP construction. As mentioned earlier, body parts and other part terms also occur in this construction (see section 3.2.5 below).

Ordinary possessive relations which involve a possessor having ownership, right to use or control over the possessum are expressed by this construction. Consider the noun phrases in examples (10) and (11) involving non-relational nominals :

(10) *kofí ɸé awu vú*
Kofi poss garment tear
'Kofi's garment is torn.'

(11) *ama ɸé tɔme- de- ze gba*
Ama poss river go pot break
'Ama's pot is broken.'

Example (11) could have a number of interpretations. The pot could be the one that Ama made or bought or has been using. In all these cases there is some association between Ama and the pot. The propensities of people and animals may also be coded with this structure. Consider the examples in (12) and (13):

(12) *é- se é- tɔgbui- wó ɸé kalẽwɔwɔ*
3SG hear 3SG grandfather PL poss bravery
tsó ame- wó gbɔ́
from person PL side
'He heard about the bravery of his grandfathers from people.' (Akpatsi 1980: 20)

(13) *alẽ- wó ɸé asokúkú ɖe fu ŋútɔ́*
sheep PL poss foolishness issue trouble much
'The foolishness of sheep is very worrying.'

In all these cases it is reasonable to think of the possessa as attributes of the possessor. The possessa are like parts of the possessors in the sense of being connected or attached to their possessors; not in the sense of their being integral parts of the possessors (cf. Cruse 1986: 157 ff; Winston et al. 1987). A consequence of the association between the possessors and their possessa is that they can do things with their possessions. In the case of abstract attributes like 'bravery' and 'foolishness' (see examples (12) and (13) respectively), it can be argued that, because of these attributes, people have the disposition to do certain things. They use these attributes to do things, hence they can do things with them.

With these considerations in mind, the semantic prototype of this construction may be paraphrased as follows:

X (=POR) *ɸé* Y (=POSS)
One can think of Y like this:
 Y is like a part of X
 X can do things with Y

A stronger form of the last component in the semantic formula could be "X can do what X wants with Y". But this would not be equally applicable to abstract possessa such as *kalẽwɔwɔ* 'bravery' and *asokúkú* 'foolishness' (see (12) and (13)).

3.2.2. PRO poss NP constructions

If the possessor is realised as a pronoun, the independent forms are used (see Table 1). Apart from the first and second person singular forms (discussed below), all other pronouns (including the logophoric *ye*) are linked to the possessed items by the possessive connective *ɸé*. Some of these are illustrated below in examples (14) and (15):

(14) *mía- ɸé agble- xɔ kló*
1PL poss farm house bare
'Our farm house is without a roof.'

(15) *wó- ɸé agbenɔnɔ mé- nyó o*
3PL poss character NEG good NEG
'Their character is not good.'

The possessive link for the first and second person singular forms is indicated by a high tone only. This high tone combines with the low tone of the forms to produce a rising tone on the pronouns:[2]

(16) *nyě (*ɸé) ga bú*
1SG:poss poss money lost
'My money is lost.'

(17) *wǒ (*ɸé) takú dzɔ́*
2SG:poss poss scarf fall
'Your scarf has fallen off.'

3.2.3. NP poss $NP_{\text{[bodypart]}}$

A subtype of the NP poss NP construction is one in which body parts and parts of wholes are the possessa. These terms are relational as explained above. All categories of body part terms – external, internal and personal representation (Chappell – McGregor 1995) – occur in this frame. The only exception is the term *ŋkɔ́* 'name, fame'. This concept, for some reason, is treated as more inherently associated with its owner than body parts as we shall see below. Consider the following examples:

(18) *kofí *(ɸé) ŋkúme ɸo ɖi*
Kofi poss face strike dirt
'Kofi's face is dirty.'

(19) *ɖeví- á *(ɸé) aɖuɖɔ́ biã*
child DEF poss urine orange
'The child's urine is orange in colour.'

(20) *... máwú vu é- ɸé susú me...*
... God open 3SG poss brain in
'... God has opened up his mind ...' (Akpatsi 1980:31)

For the NP *ϕé* NP structure with body part possessa, the following semantic representation is proposed:

> X (=POR) *ϕé* Y (= body part)
> One can think of Y like this:
> Y is a part of X
> When one thinks of Y
>   one cannot not think of X
> X can do things with Y

There are two main differences between this formula and the previous one. First, body parts are not *like* parts but they *are* parts of their possessors. They are organic parts of people and animals. Second, body parts are relational hence the component "When one thinks of Y one cannot not think of X". Notice, however, that the last components in both formulae are identical. The similarity between the two constructions captured in this last component is that the possessor can do things with the possessa. It appears that this is the crucial feature that sets body part terms apart from other inherent relational terms. It is also responsible for the "alienable" grammar they have in adnominal constructions.

Furthermore, other part-whole relations seem to be modelled on the body parts. The only deviation in this case is that inanimate things cannot be said to be able to do things with their parts. Although the relationship between a plant and its parts, (21), an instrument and its parts, (22), or an artefact and its parts, (23), is expressed in the same way as that between a body and its parts, they differ in one aspect. The inanimate wholes lack the right of use of their parts, so to speak.

(21) *ɖuḿ- tí á *(ϕé) alɔ gã ɖeká fé*
'oak' tree DEF poss branch big one split
'One big branch of the 'oak' tree has split off.'

(22) *agblenu gã *(ϕé) atí ŋé*
hoe big poss handle break
'The handle of the big hoe is broken.'

(23) *agbalẽ- á *(ϕé) akpa ɖe*
book DEF poss cover remove
'The cover of the book is torn.'

The explication of these part-whole relations should capture the idea that the possessors cannot necessarily do things with their possessa.

Part of the inherent semantics of the inanimate wholes is that they are [-potent] (cf. Chafe 1970: 109) and this cancels the semantic component of the construction that has to do with the possessor doing things with the possessa. Hence the formula proposed is identical with the one for the body-part relations in all respects except one. Compare this explication with the body part one:

> X (= POR, inanimate whole) *ɸé* Y (= part)
> One can think of Y like this:
> Y is a part of X
> When one thinks of Y
> one cannot not think of X

To sum up, in an NP poss NP construction the possessa are non-relational terms and meronyms – parts in part-whole relations. The semantic protoype of this construction has been characterised. The interaction of the inherent semantics of the different categories of possessa, for example, body parts, with the prototype has led to the recognition of a number of subconstructions. For instance, body parts add the element of their relational nature captured by the component:

> When one thinks of Y, one cannot not think of X.

In specific situations, the semantics of the lexical items involved in the syntactic construction would interact with the prototype defined and yield specific contextual interpretations (cf. the case of inanimate possessors above).

3.3. "Inalienable" nominal constructions

3.3.1. NP $_{[POR]}$ *NP* $_{[POSS]}$

The NP NP construction is used to express spatial, kin, social and cultural relations between a possessor and a possessum. These relationships are in a sense inherent and permanent (Bally 1926 [1995]; Seiler 1983). The nature of the relationship between the possessor and possessum varies from one category of possessa to another. Nevertheless, all the terms that participate in this construction as possessa have one thing in common: they are all relational but not meronymic terms. The various semantic domains covered by this construction are described in turn.

3.3.2. NP NP [Spatial relation terms]

The constructions in which an entity is related to its spatial orientation are generally instantiated as NP N structures because the spatial relational terms are seldom modified (see Duthie in press, Westermann 1930: 52). Even if they are modified, it is usually intensifiers that they take. The spatial relational terms form a subclass of nominals but they are not prototypical ones because they allow a very limited degree of modification. The members of this class vary in their degree of nominality partly because most of them have evolved historically from body part terms (see 3.6 below, and cf. Westermann 1930: 51 ff.; Heine and Reh 1984: 256ff.; Heine 1989).

Spatial relational terms such as *dzí* 'top, above', *gɔme* 'under', *ŋgɔ* 'front' and *gbɔ́* 'side' occur as possessa in this construction as examples (24), (25), and (26) show:

(24) *tsɔ́ ga- a da ɖé kofí gbɔ́*
take money DEF put at Kofi side
'Leave the money with Kofi.'

(25) *atíkpo aɖé le tɔ- a títína*
log INDEF be:at river DEF middle
'There is a log in the middle of the river.'

(26) *kplɔ̃ gã dzí ɸo ɖi*
table big top beat dirt
'The top of the big table is dirty.'

It should be pointed out that one thing that spatial relational, kin and socio-cultural terms have in common is that their possessors cannot do anything with them as they could with body parts and other possessa.

To capture these features of the possessive construction of NP NP constructions in which the possessa are spatial orientation terms, I propose the following explication:

X (=POR) Y (=POSS, spatial orientation)
One can think of X and Y like this:
  One cannot think of all of X and not think of Y
  Y is a part of X
  When one thinks of Y
    one cannot not think of X
  One cannot say
    X can do anything X wants with Y

The second component in this formula indicates that every entity has some spatial relation. Spatial orientations are inseparable parts or elements of any item hence the third component. The last component aims to capture the idea that an entity does not have much control over its spatial relations as it may have with its body parts.

3.3.3. *NP NP*[Kin and social relations]

Kin terms, consanguineal as well as affinal, constitute another category of possessa in the NP NP construction. Consider examples (27) and (28):

(27) *ɖeví- á- wó tɔ́gbé dze dɔ*
child DEF PL grandfather fall sickness
'The grandfather of the children has fallen sick.'

(28) *kofí srɔ̃ eve- á- wó kátã dzi ví*
Kofi spouse two DEF PL all bear child
le ŋkeke ɖeká dzí
at day one top
'The two wives of Kofi bore children on the same day.'

Intuitively, the relationship between an entity and its spatial orientation is different from the one between a human and his/her kin relation. The entity-spatial orientation relation is a kind of whole-part one while the human-kin relation is one where the two terms can be thought of as members of the same socio-cultural group. In other words a kin relative is not a part of the possessor in the same way that a body part, for example, is a part of its possessor. Tentatively, I propose the following formula to capture the kin relationship between a possessor and a possessum:

X (=POR) Y (=POSS, kin term)
One can think of X and Y like this:
X is part of something
Y is part of the same thing as X
When one thinks of Y one thinks of X
One cannot say
X can do anything X wants with Y

In the part-whole and the entity-spatial orientation relations there is an exclusive one-to-one relationship between the possessor and the possessum. This is not the case with respect to the relationship between a human and his/her kin. Normally, several people may have one individual as 'mother' whereas a particular entity has a particular spatial orientation, for instance, 'bottom' (cf. Sapir 1917–1920: 88). This difference is reflected in the formulae in the components that account for the relational nature of the terms; namely, "when one thinks of Y one cannot not think of X" for part-whole relations and "when one thinks of Y one thinks of X" for kin relationships.

Similarly there is no necessary exclusive one-to-one corresponding relationship between a possessor and their social and cultural associates. Thus the attachment between a possessor and their social relations such as *xɔ̃lɔ̃* 'friend', *hati* 'colleague, mate', *lɔ̃lá* 'lover, well-wisher' and *tabiálá* 'suitor' is not a unique one. The same thing can be said of possessions which are thought of as basic in the culture. These are *ŋkɔ́* 'name, reputation', *de* 'native/home land', *aɸé(me)* 'home/dwelling' and *agble* 'farm'. These concepts are merely juxtaposed to their possessors to indicate their connection. Consider examples (29) and (30):

(29) *áma tabiálá- wó wú taɖa*
Ama suitor PL exceed head:hair
'The suitors of Ama are (as) uncountable (as hair).'

(30) *nye- mé- nyá ɖeví má ŋkɔ́ o*
1SG NEG know child DEM name NEG
'I don't know the name of that child.'

Tentatively, the following semantic representation may be proposed for the NP NP structure in which socio-cultural concepts are the possessa:

X (POR)Y Y (= POSS, socio-cultural terms)
One can think of Y like this:
Y is like a part of X
When one thinks of Y one thinks of X
X can do some things with Y
X cannot do with Y anything X wants to

This formula attempts to capture the intuitive idea that socio-cultural possessa are not parts of their possessor the way in which a meronym is a part of its whole. Another feature it tries to capture is that the possessor has limited control over the possessum. This is portrayed in the last two components. In fact, a person's 'use' of these items is constrained by social and cultural norms of behaviour, as the proverb in example (31) teaches about one's behaviour towards one's homeland:

(31) *wó- mé- tsɔ́- á mia- sí*
3PL NEG take HAB left hand
fiá- á ame de o
point HAB person homeland NEG
'One does not point to one's homeland with the left hand.'

For the NP NP construction, when the possessor is first or second person singular, the order of possessor preceding the possessum is reversed (see the discussion on egocentric perspective in section 3.4). This can be schematically represented as follows: NP [POSS] 1/2 PRO [POR] For other pronominals including the logophoric, the normal order is maintained. Compare examples (32) and (33) on the one hand with (34) and (35) on the other:

(32) *ŋkɔ́- nye- é nyé kofí*
name 1SG NFOC be Kofi
'My name is Kofi.'

(33) *srɔ̂- wò fɔ́ a?*
spouse 2SG rise Q
'How is your spouse?'

(34) *miá gbɔ́ fá*
1PL side cool
'There is no bad news around us.' (Literally: 'Our side is cool.')

(35) *ava dzɔ le wó de*
war happen be:at 3PL homeland
'A war broke out in their homeland.'

3.4. The alternation between NP NP and NP poss NP constructions

The three classes of nominals – spatial, kin and socio-cultural relational terms – which occur as possessa in the NP NP structure can also participate in the NP *ɸé* NP construction *pace* Claudi – Heine's (1986: 317-318) assertion that they never do.[3] This happens in marked contexts. Typically, its effect is to de-emphasise the closeness between the entities and to draw attention to the distinct character of either the possessor or the possessum. Some support for this contention comes from the tendency of spatial, kin and socio-cultural terms to occur as possessa in the NP *ɸé* NP structure in focused or emphatic contexts. Consider the following examples, (36) and (37):

(36) *vɔ́nudrɔ̃lá ŋútɔ́ ɸé ví- é nɔ ŋgɔ́*
judge self poss child NFOC NPRES front
ná ví- nye- a háfí wó- de fígbé- a?
to child 1SG DEF before 3PL go thieving Q
'Was it the child of the judge himself who was directing my child when they went to steal?' (Kpodzo 1982: 9)

(37) *du lolo áɖé le eve-nyígbá blibo lá ɸé*
own large INDEF be Ewe:land whole DEF poss
títína tútútú
middle exactly
'There is a large town right in the middle of the whole Ewe territory.' (Nyaku 1984: 7)[4]

Example (36) is particularly significant because it contains two instances of *ví* 'child'. The first one employs the NP *ɸé* NP device while the second occurs in the NP NP structure. Notice that in the former case the two individuals are separately highlighted. The possessor is emphasised by an intensifier and the possessum is focus marked. The message of the structure of the second instance (of *ví*) is the closeness between the possessor and the possessum. These data seem to confirm the view that the NP *φé* NP structure is more "alienable" than the NP NP construction.

While these inherent relational terms can participate in either the NP NP or NP *φé* NP constructions, the other categories of possessa – body parts, other meronyms and material possessions – occur only in the latter. Observe that examples (38) and (39) would be unacceptable if the possessive connective were deleted.

(38)	*kofi*	**(φé)*	*abɔ́*	*ŋé*
	Kofi	poss	arm	break

'Kofi's arm is broken.'

(39)	*kofí*	**(φé)*	*dɔ́*	*gblẽ*
	Kofi	poss	work	spoil

'Kofi's work is ruined.'

There is one caveat, however, with respect to this property. When the possessor is the first or second person singular, it is possible for the relationship between the possessor and these categories of possessa to be expressed using the NP NP strategy. This means that these possessa could occur either in the frame 1/2 PRO (+high tone) [POR] NP [POSS] or in the frame NP [POSS] 1/2 PRO [POR]. The first pattern has already been illustrated in examples (16) and (17). The second pattern is illustrated below in examples (40) and (41):

(40)	*ahuhɔ̃ɛ-*	*nye*	*gba*
	mirror	1SG	break

'My mirror is broken.'

(41)	*afɔkpa-*	*wò*	*nyá*	*kpɔ́-*	*ná*	*ŋútɔ́*
	footwear	2SG	MOD	see	HAB	much

'Your shoes are very nice.'

How can the felicity of these forms be explained? I suggest that this is possible because the use of the first and second person singular pronouns in this construction presents the possessive relationship from an egocentric perspective. That is to say that the configuration of NP PRO is chosen depending on the speaker's empathy or identification with the possessor and the proximity of the possessor to the speaker. Naturally, people tend to be more interested in themselves and in the relations they bear to the participants in the speech situation. It cannot be gainsaid that on the nominal hierarchy the first and second person pronominals are the closest to the speaker in proximity, empathy and topicality (Langacker 1985; Silverstein 1976). In fact, in most cases the speaker and the first person are identical. In these cases the possessor is treated as more important and is of more interest to the speaker than the possessum therefore it is put in the position where the head would normally occur in such a construction. It is not uncommon for languages to treat first and second person possessors differently from other kinds of possessors (see, for example, Seiler 1980; 1983 and Reh et al. 1981 for some examples).

It can be claimed that a syntagm of the form NP (= POSS) - PRO (= 1/2 SG, POR) irrespective of the semantic domain covered by the possessum nominal implies an egocentric orientation to the possessive relation. When the possessum is one which would not normally occur in such a structure the configuration tends to signify the personal dimension as well as the closeness between the possessor and the possessum. It also shows that it is the possessor rather than the possessum who is of focal interest to the speaker.

3.5. Syntactic compounds: N1 - N2 + high tone suffix

3.5.1. Orientation

All relational terms – spatial orientation, kin, socio-cultural and body parts and other meronyms – can be compounded syntactically with other nominals to indicate a permanent relationship between them. Phonologically, the compound is marked by a high tone suffix which is particularly noticeable when the last syllable has a low tone. The coalescence of the low and the high produces a rising tone on the last syllable. For example, the nominals *ŋútsu* 'man' and *afɔkpa* 'footwear' can be compounded and note the tone of the final syllable:

(42) *ŋútsu- 'fɔkpǎ*
man footwear
'men's footwear'

Two functional types of these nominal compounds may be distinguished: a possessive one which is used to express a habitual association between two individuals, and a classificatory one (as in example (42) above) in which the referent of the compound is a member of the class of things denoted by the head nominal, N2. Thus in the above example, the generic nominal 'man' acts as a classifier of the referential nominal 'footwear' to indicate the kind of footwear that the item is. The two functional types are discussed in turn.

3.5.2. Possessive nominal compounds

In this type of syntactic compound, N2 is a kin or social relation of N1. In this case the two people are represented by personal names or by address terms. More specifically, N1 may be the parent or guardian or master or mistress of N2. This is the kind of relationship between the components of the compounds in (43) and (44):

(43) *dzinaku- 'kɔ́sua*
Dzinaku Akɔsua
'Dzinaku's Akɔsua'

(44) *gbede- kɔdzó*
blacksmith Kodzo
'Gbede's Kodzo[5]

The affinal relationship between two individuals may be expressed through the compounding of their names. Typically, the husband is N1 (= possessor) and the wife is N2 (= possessum). For example, (45) and (46):

(45) *klemensi- 'kɔ́suá*
Clemence Akɔsua
'Clemence's Akɔsua'

(46) *viktɔ- 'brã*
Victor Abra
'Victor's Abra'

It should be noted that both nominals are proper names or address titles and are thus both referential and denote specific individuals. On account of this, these compounds are not classificatory (cf. Chappell – McGregor 1989). They encode an "inalienable" relationship between the individuals. Thus they are possessive. This does not imply that there may not be some associated classificatory inference. But these compounds do not say what kind of person the possessum is. Rather they provide a description of the possessum with respect to the possessor such as: child of X, or apprentice of X, or wife of X etc. Several compounds of this type are used extensively in the Ewe address system (see Ameka 1991: chapter 13 on Ewe modes of address). In these compounds, N2 is defined or identified with respect to a specific individual represented by N1. The basis of this identification is a kin or a social relationship that may exist between N1 and N2. N1 is thus determinative of N2 rather than being a classifier of it as is the situation with the classificatory compounds.

Another piece of evidence in support of the possessive rather than classificatory nature of these compounds is that they can be paraphrased as N1 *ɸé* N2. Thus (47) and (48a) are paraphrases of (45) and (46) respectively:

(47) *klémensi ɸé akɔ́suá*
Clemence poss Akɔ́sua
'Clemence's Akɔsua'

(48) a. *gbedé ɸé kɔdzó*
blacksmith poss Kodzo
'Gbede's Kodzo'

Apart from being used to draw attention to the individuals separately, the N1 *ɸé* N2 device tends to imply that N1 has some control over N2. This is particularly evident when it is used to express the association between a husband and a wife. N1 (= wife) - N2 (= husband) syntactic compounds are rare. Such structures are after all grammatical, but their meaning is perceived to be culturally odd. Nevertheless, N1 (= wife) *ɸé* N2 (= husband) expressions have been attested (see example (48b)). The inference carried by such a structure is that the wife dominates the husband which is a poor reflection on the husband.[6]

b. *áma ɸé kɔmla*
Ama poss Komla
'Ama's Komla.'

This piece of evidence would seem to confirm the view that the *ɸé* possessive connective constructions have an element of the possessor having the right to do things with the possessum.

With these considerations in mind, I propose the following semantic representation for the meaning of nominal compounds that express kin and social relations:

X (=N1 [POR]) – Y (= N2 [POSS])
One can think of X and Y like this:
Y is part of something
X is part of the same thing
When one thinks of Y, one cannot not think of X
One can think this: X does things for Y

The first and second components account for the fact that the two individuals could be thought of as belonging to the same socio-cultural group. The third component captures the very close association between the people involved. The last component is meant to represent the idea that the referent of N1 is socially responsible for the referent of N2. Recall that N1 is either a parent, a guardian, a master/mistress or a spouse. All of these people occupy positions of social responsibility with respect to their dependants.

3.5.3. Classificatory compounds

The essential thing about classificatory compounds is that one of the nominals is non-referential and usually generic. This nominal acts as a classifier of the other nominal in the compound. If the generic nominal is made specific the NP *ɸé* NP structure has to be used. In this case both nominals could be modified. Such modification, of course, is not available for the nominals in the compound.

In the N1–N2 compounds which express meronymic relations, N1 represents the whole and N2 the part. Such structures have a generic as opposed to specific interpretations. This can be observed in examples (49) and (50) below, N1 is a generic nominal 'person' and N2 is a term for a part of such nominals. The referent of the compound is a type of head: a human head and N1 tells us what type it is. Compare the following:

(49) *ame- tǎ*
person head
'human head'

(50) *ame- á *(ɸɛ́) ta tsútsɔé má*
person DEF poss head pointed DEM
'the person's pointed head'

Parts of animals, plants and implements can be similarly expressed (see examples (51) and (53) below). The same principles of interpretation apply in these cases as well.

(51) *gbɔ- fɔ̌*
goat leg
'goat leg'

(52) *gbɔ vévẽ má *(φé) afɔ ŋéŋě lá ....*
goat smelly DEM poss leg broken DEF
'that smelly goat's broken leg ...'

(53) *atí- lɔ̌*
tree branch
'tree branch'

(54) *ati gã lá *(φé) a-lɔ ɖeká ...*
tree big DEF poss branch one
'one branch of the big tree ...'

The relationship between N1 and N2 in such compounds may be one of spatial attachment. Thus N2 may denote the spatial orientation of N1. Like the part-whole compounds, these structures also have generic interpretation (see example (55)). It is noteworthy that when the generic terms in the entity-spatial relation compounds are made definite they do not normally take the possessive connective unlike the other types of compounds. This probably suggests that the spatial relation terms are the least "alienable" of the relational nominals. Compare examples (55) and (56):

(55) *kplɔ̃- dzí*
table top
'table-top'

(56) *kplɔ̃- a (*φé) dzí ...*
table DEF poss top
'top of the table ...'

The basic features of the compounds described so far may be captured in the following formula. This explication could be taken as the semantic prototype for the compound structures for meronymic relations:

X (= N1) – Y (= N2)
One can think of X and Y like this:
Y is (always) a part of Xs
One can think this: X and Y are one thing
This thing is a kind of Y

The second component captures the part-whole relation between these terms. It also shows that the first nominal is generic, that is, Xs. The third component represents the idea that the referent of the compound is an entity. The last component tells of the classificatory nature of these compounds.

There are other kinds of classificatory compounds. In some of them N1 represents the user of N2. Thus the item the compound refers to is used by a category of people or entities designated by N1:

(57) *nyɔ́nu – ɖǒ*
woman cloth
'ladies' cloth'

Other compounds are of the specific-generic type. For example,

(58) *du - fiǎ*
town chief
'town chief'

Note that the denotata of these compounds can be persons as in (58) or places.

All these classificatory compounds have two features at least in common: first, they have one referent and second, the compound is a hyponym of the non-referential member of the compound. These features constitute the semantic prototype for the classificatory compounds which can be formulated as follows:

X (= N1) – Y (= N2)
One can think of X and Y like this:
One can think this: X and Y are one thing/place/person
This thing/place/person is a kind of Y

The semantics of the various subtypes are systematically linked to this prototype.

3.6. Summary: the grammar of "alienable" and "inalienable" possession

Three major adnominal constructions (each with various sub-constructions) have been described so far: NP poss NP; NP NP; and N–N. Body part terms occur as possessa in the first construction and as grammatical heads of classificatory compounds. However they do not occur in the second structure. Other relational terms typically occur as possessa in the second construction. These also occur as heads of classificatory compounds – a subtype of the N–N construction.

It has been indicated that the three constructions can be ranked along a scale of decreasing "alienability" as follows (see 3.1):

NP poss NP > NP NP > N - N

This suggestion was based on the structural properties of the forms. This scale can be further supported from the semantics of the constructions. A comparison of the semantic prototypes associated with each construction shows that the conceptual bond between the possessor and the posessum is closest in the N–N compound and least in the NP poss NP construction. In fact in the classificatory compounds, the compound designates only one entity. The distinctiveness of the referents involved in the structures also increases as their "alienability" increases (see Figure 1 below). The "alienability", distinctiveness of referents and conceptual distance between the elements of the nominal possessive constructions with their associated semantic prototypes may be represented as shown in Figure 1.

When one compares the formulae of the various constructions as displayed in the figure above, it becomes evident that the structural "alienability" ordering of the constructions is semantically motivated. For instance, as we move from the bottom of the scale to the top, the wording of the first line of the formulae also changes from 'One can think of X and Y like this' to 'One can think of Y like this'. These components capture in part the increasing distinctness of the referents involved in the constructions.

What is probably harder to substantiate is an "alienability" scale for the classes of possessed nominals. Figure 2 below represents an attempt to capture the linguistic facts as well as the intuitive ideas related to the various categories of possessa.

The Ewe linguistic facts discussed so far suggest that spatial orientation terms are the ones that are most inherently associated with their possessors. Recall that these terms are seldom modified in the NP NP construction. It should also be remembered that their nominal compounds are not paraphrasable as NP *ϕé* NP like other relational terms. In addition the spatial relational terms are constrained in the number of possessive predicative constructions in which they can occur more than the other relational categories (see Ameka (1991) for the details). It will be shown for example in section 4 that the possessors of spatial relation terms cannot be promoted to oblique objects in possessor ascension constructions.

← *distinctiveness of referents* ←

← *increasing conceptual distance* ←

← *increasing alienability* ←

possessive linker		juxtaposition		possessive compound		classificatory compound
NP (=X)*ϕé* NP (=Y)	<	NP (=X) NP (=Y)	<	N1(=X) - N2 (=Y)	<	N1(=X) - N2 (=Y)
One can think of Y like this: Y is like a part of X X can do things with Y		One can think of X and Y like this: One cannot think of all of X and not think of Y Y is a part of X When one thinks of Y one cannot not think of X One cannot say X can do anything X wants with Y		One can think of X and Y like this: Y is part of something X is part of the same thing When one thinks of Y one cannot not think of X One can think this: X does things for Y		One can think of X and Y like this: One can think this: X and Y are one thing/place/person This thing/place/person is a kind of Y

Figure 1

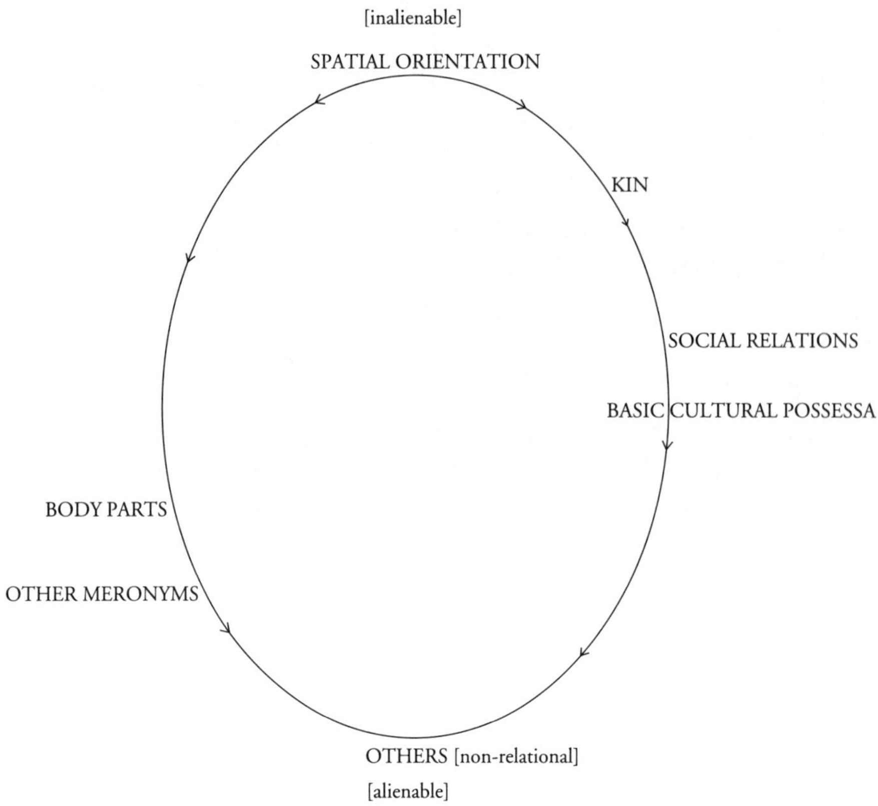

Figure 2. Alienability of semantic classes of possessed items

Conceptually, it is clear that an entity, so to speak, wears its spatial orientations on its sleeves. The validity of this is borne out in Ewe by the fact that spatial relation terms do not normally occur by themselves without an associated nominal.

Thus taking spatial relational terms as the most "inalienable" and as the starting point one can proceed clockwise or anticlockwise around the circle. In either direction, there is increasing "alienability" until one gets to the non-relational terms. One of the advantages of presenting the classes of nominals vis-à-vis "alienability" in this way is that it captures the intuitive, and linguistically valid, connections between spatial relational terms and body part terms. A unidirectional implicational hierarchy such as Figure 3 cannot adequately depict this relationship.

In the circular diagram (Figure 2), body parts and spatial orientation terms are contiguous as they should be. But in the linear diagram they are

spatial relations	> kin	> basic cultural possessa/ social relations	> body parts	> other meronyms	> others

Figure 3. A linear "alienability" hierarchy

separated by other categories as though there were no relation between them.

From the foregoing discussions, it is evident that body part terms have "alienable" grammar in adnominal constructions. This behaviour in a sense is counter-intuitive. It is important to attempt to explain this seemingly odd behaviour.[7] It could be explained in two ways. One can resort to diachrony or one could indulge in some kind of Whorfianism, that is one can explore the possibility that the Ewes conceptualise the relation between a body and its parts differently from what obtains in the real world. These two viewpoints are not necessarily mutually exclusive.

From a historical perspective, the difference between spatial relational terms in particular and body parts could be said to be the result of grammaticalisation and reanalysis. Several spatial relational terms in Ewe have evolved from body part terms. One of the effects of this is that many body parts are homophonous with spatial relational terms. Some examples are given in (59):

(59)		*body part meaning*	>	*spatial relational sense*
	ta	head	>	fore part, top, above
	ŋkúme	face	>	front part
	megbé	back	>	back part, behind
	ŋútí	skin	>	outer surface, near
	nŭ	mouth	>	front, edge
	ɸo	belly	>	surface
	tó	ear	>	edge

It is plausible that body part terms and spatial relational terms receive different grammatical treatment so that they can be more easily distinguished from each other (cf. Reh et al. 1981 for a similar view). Since spatial relational terms would appear to be more inseparable from their possessors than body parts are from their owners, the latter was given "alienable" syntax while the former has "inalienable" grammar. Perhaps this is an instance in which form constrains the grammatical expression of content (cf. Nichols 1986, 1988). The claim being made here is that the grammatical "alienability" of body parts emanates from the homonymy which they share with spatial relational terms for historical reasons. From this viewpoint, it is the phonological form that these body part terms have which determines the grammatical expression and consequently imposes a certain semantics on the grammar of body parts in the language.[8]

From an ethnolinguistic viewpoint, one can claim that the Ewes construe the relationship between body parts and the 'owner' to be one in which the owner can do things with the parts. In this respect the relationship is viewed to be similar to the relationship between, for example, the owner of a basket and the basket. This conceptualisation of the relationship between the body and its owner is closer to the semantic prototype of the NP *ɸé* NP construction than to that of the NP NP structure. Hence body parts take part in the former construction.

Indeed throughout the grammar of Ewe, the treatment of body parts reflects this same view of the relationship between the body and its 'owner'. In the constructions described in section 4, body parts are consistently presented as 'detachable' from their owners, either for the owner's use, or for something to happen to the part.

4. The clausal syntax of body parts

4.1. Body part as locus of effect

4.1.1. Overview

One can present a body part as the area that is affected by a situation independent as it were of the owner. In such constructions the body part has a core grammatical role – subject or object – and the possessor assumes the role of an oblique prepositional object introduced by *ná* 'to, for'. Thus one can represent syntactically, an extralinguistic situation of someone doing something to a part of the body, for instance, twisting an arm, in one of two ways:

(i)	NP	V	[NP	poss	NP (body part)]$_{NP}$
	SUBJ				OBJ

(ii)	NP	V	NP(body part)	[*ná*	NP (POR)]$_{PP}$
	SUBJ		OBJ		OBL OBJ

These structures are illustrated below in examples (60) and (61):

(60) *kwami tró kofí ɸé abɔ́*
Kwami twist Kofi poss arm
'Kwami twisted Kofi's arm.'

(61) *kwami tró abɔ́ ná kofí*
Kwami twist arm to Kofi
'Kwami twisted Kofi's arm.' (Literally: 'Kwami twisted arm for Kofi.')

Similarly, a real world situation in which something happens to a part of someone's body can be expressed using one of the following structures:

(i) [NP poss NP (body part)]$_{NP}$ V
SUBJ

(ii) NP (body part) V [*ná* NP (POR)]$_{PP}$
SUBJ OBL OBJ

For example,

(62) *kofí ɸé abɔ́ tró*
Kofi poss arm twist
'Kofi's arm is twisted.'

(63) *abɔ́ tró ná kofí*
arm twist to Kofi
'Kofi has a twisted arm.'

(I shall revert to using the variables X and Y for possessor and possessum respectively, and I will introduce Z as another variable for other nominals.)

The first structure in each set will be referred to as the possessive construction because it employs the possessive connective. The second may be characterised as a 'possessor ascension construction'. This latter description requires some explanation. Possessor ascension constructions are usually described in terms of structures in which 'the possessor is "promoted" to the status of direct object or dative while the possessed NP is "demoted" to the status of some sort of oblique phrase.' (Fox 1981: 323; cf. Hyman et al. 1970; Hyman 1977; Blake 1984). From this definition, the Ewe construction is not a prototypical possessor ascension structure because the possessed NP is not demoted to any oblique phrase, nor is it promoted to a different role such that the structure might be labelled a possessum promotion construction. Rather the possessum retains the grammatical role of the possessive NP in the alternative structure. The term possessor ascension can however be justified in the sense that the possessor can be thought of as being promoted, as it were, from a genitive modifier in an NP to a dative-oblique in a prepositional phrase. It should be remembered that datives and obliques are higher on the NP accessibility hierarchy à la Keenan – Comrie (1977). Thus the examples in (61) and (63) above may be thought of as 'possessor ascension' constructions. The type of operation involved is the promotion of the possessor of a subject or direct object to an oblique object.[9] It should be noted that the pairs of sentences (60) and (61) versus (62) and (63) respectively describe essentially the same situations. That is, they are referentially the same. The differences between the members of each set lie in how the participants are presented syntactically and viewed semantically. In (60) the possessive NP is the direct object and in (62) it is the subject. In these cases the association between the possessor and the possessum (body part) is explicitly indicated by the appropriate possessive construction. In (61), however, the possessum (body part) retains the object role but its dependent possessor occurs as an oblique object. Similarly, in (63) the body part term has the subject role and the possessor occurs in a prepositional phrase. In sum, in these structures the body part has a core grammatical role while the possessor has a peripheral role.

The preposition *ná* which is used to introduce the promoted possessor in these constructions is homophonous with the verb *ná* 'give, cause, let' from which it has evolved (Ansre 1966; Heine – Reh 1984: 276; Claudi – Heine 1986: 324; Westermann 1930:129 ff). Not surprisingly, this preposition is used to express meanings analogous to those encoded crosslinguistically by the dative case. Thus, in general, the NP governed by *ná* functions as the recipient in sentences. This recipient function may be contextualised in specific instances as benefactive, malefactive or the experiencer of the situation characterised in a clause. Consider examples (64), (65) and (66):

(64) *kofí ɸle avɔ ná áma* [benefactive]
Kofi buy cloth to Ama
'Kofi bought a piece of cloth for Ama.'

(65) *gatagbadze- a vú awu- a ná- m* [malefactive]
nail DEF tear dress DEF to 1SG
'The nail tore the dress for me.'

(66) *akɔɖú lá víví ná ɖeví- á* [experiencer]
banana DEF sweet to child DEF
'The banana was sweet for the child.'

It is needless to say that these contextual senses are not necessarily mutually exclusive of each other but may overlap.

It is instructive that in the possessor ascension construction, the possessor is governed by the preposition *ná*. From the semantics of *ná* it can be said that the possessor is being presented as the 'recipient' in the clause and contextualises perhaps as the goal and/or the experiencer of the situation. Nevertheless, the possessor is coded syntactically as a peripheral argument while the body part term is assigned a core role.

The raison d'être of this strategy of assigning the body part a core role and its possessor a peripheral role appears to be the presentation of the body part as a central participant in the situation. In every situation there is one argument which is critically involved in its realisation. This argument tends to be the object in transitive clauses and the subject in intransitive clauses (cf. Halliday (1985: 146ff) on medium; Starosta (1988: 128) on the patient centrality hypothesis in Lexicase grammar). It can be argued that when body parts are construed as being the most critical arguments involved in the situation, then the possessor ascension constructions are used in such structures, the body parts assume subject or object roles. Thus these constructions are used to specify the body part as the locus of effect of an event and the possessor as the recipient by virtue of the connection it has with the body part. The linguistic separation of the possessor from the body part in these structures through its promotion serves to de-emphasise the partitive nature of the body part and its close association or attachment to the possessor and highlights its distinct nature.

The marking of the promoted possessor by *ná* is consistent in a way with the coding of possessors in general in one of the stative predicative possessive constructions found in Ewe (see Ameka (1991) for a full semantic analysis of this structure). The syntagmatic formula for the relevant construction is: Y [POSS] V [copula] (Z) *ná* X [POR]. This structure is used for the expression of any kind of possession. As Westermann (1930: 93-94) puts it, this expression is used to show that "a thing is possessed naturally, is a part of oneself" and also for 'real' possessions. For instance, examples (67) and (68):

(67) *fofó le- e ná- m*
father be 3SG to 1SG
'I have a father.'

(68) *xɔ le gɛ̃ nɛ*
house be Accra to:3SG
'S/he has a house in Accra.'

It is plausible to claim that the *ná* preposition embodies an element of the possessive relation both in this structure and in the possessor ascension constructions. (There are crosss-linguistic parallels of the dative being used to code possession, see Ultan 1978.)

4.1.2. *Body part as object in possessor ascension constructions*

To return to the possessor ascension constructions of the frame:

Z V Y (body part) *ná* X (= POR)

It should be noted that such forms are quite frequent in the language. In particular, it is a conventionalised pattern for talking about familiar ritual practices such as circumcision (69) and ear piercing (70) which involve something being done to a specific part of one's body.

(69) *atíkewɔlá tso ava (gã lá) ná kofí*
doctor cut penis (big DEF) to Kofi
'The doctor circumcised (the big penis for) Kofi.'

(70) *mamá ŋɔ́ tó (gbaba má- wó) ná vídzĩ lá*
grandma pierce ear (flat DEM PL) to child DE
'Grandmother pierced (those flat) ear(s) for the child.'

Notice that the body part terms can be modified even in these constructions for conventionalised practices. This suggests, I think, that in such constructions the body part is individuated – conceptualised and treated as a separate entity distinct from the possessor.

The construction pattern under discussion here is not used only for the sorts of formulaic practices described above. It can be used in any situation where the specificity of the body part is highly relevant and where the speaker wants to communicate the view that a body part is the critical participant in a situation, be it a state, a process or an event. Consider the occurrences described in the following examples:

(71)a. *avu lá ɖu afɔ ná kofí*
dog DEF eat leg to Kofi
'The dog bit the leg of Kofi.'

b. *kofí le tume (gbadza má) ku- ḿ ná áma*
Kofi PRES back (broad DEM) scratch PROG to Ama
'Kofi is scratching (the broad) back of Ama.'

The possessor NP in the *ná* phrase could be reflexive, that is coreferential with the subject of the clause (as seen in example (72)):

(72) *kofí ŋé afɔ ná é- ɖókui*
Kofi break leg to 3SG REFL
'Kofi broke his own leg.'

Furthermore, the construction is not restricted only to human body parts but applies also to parts of animals as exemplified in (73).

(73) *ka blá afɔ (ŋéŋé lá) ná alẽ á*
cord tie leg (broken DEF) to sheep DEF
'The (broken) leg of the sheep is entangled by a cord.' (Literally: 'A cord has entangled the (broken) leg for the sheep.')

(74) *ka blá alẽ á ɸé afɔ (ŋéŋé lá)*
cord tie sheep DEF poss leg (broken DEF)
'The sheep's (broken) leg is entangled by a cord.' (Literally: 'A cord has entangled the sheep's (broken) leg.')

On the basis of the discussion so far, one can represent the meaning of the construction as follows:

Z V Y[body part] *ná* X [POR]
  I want to say something about Z
  I say: Z did something to Y
  One can think of Y as part of X's body
  Because of that, one can think of
    what happened like this:
    Z did something to X
  X could feel something because of that

This formula reflects two essential things about this construction: first, the undergoer of the event is the body part as an entity in itself; second, it captures the feature that the possessor of the body part is only indirectly affected by the event by virtue of the fact that it is related to the primary undergoer, the body part.

The message of the structure with body part alone as direct object is different from that of a construction in which the object role is filled by an NP *ɸé* NP structure. Thus, although (60) and (61) (repeated below) are pragmatically similar, they have different semantic values. It appears that in a Z V X *ɸé* Y construction, X and Y together are the single undergoers of the event. In this case the body part is presented as a part of the possessor and not necessarily as an entity distinct from it.

(60) *kwami tró kofí ɸé abɔ́*
Kwami twist Kofi poss arm
'Kwami twisted Kofi's arm.'

(61) *kwami tró abɔ ná kofí*
Kwami twist arm to Kofi
'Kwami twisted Kofi's arm.' (Literally: 'Kwami twisted arm for Kofi.')

Perhaps this construction could be paraphrased simply as:

Z V X *ɸé* Y
I want to say something about Z
I say: Z did something to part Y of X's body
[One could think that Z did it to X and Y][10]

This analysis of the constructions may be supported in part by the behaviour of spatial orientation terms in relation to possessor ascension constructions. Spatial relation terms cannot occur by themselves as direct objects in such constructions. Compare the acceptability of these pairs of sentences (75) versus (77) and (76) versus (78).

(75) *áma tútú kofí ɸé ŋkúme*
Ama wipe Kofi poss face
'Ama wiped Kofi's face.'

(76) *áma tútú ŋkúme ná kofí*
Ama wipe face to Kofi
'Ama wiped the face for Kofi.'

(77) *áma tútú kplɔ̃- a* dzí
Ama wipe table DEF top
'Ama wiped the table top.'

(78) **áma tútú dzí ná kplɔ̃- a*
*Ama wipe top to table DEF
'Ama wiped the top for the table.'

From the premise that spatial relational terms are "inalienable" terms, one can draw two inferences from the unacceptability of (78). First, the possessor construed to be in an "inalienable" possessive relation cannot be promoted to a higher grammatical role. Consequently, it could be said that possessor ascension constructions are used when the relation between the possessor and the possessum is viewed in a specific instance as not being a close one. Indeed, the possessors of non-relational terms can also occur in such structures. Compare the following examples, (79) and (80):

(79) *é- vú nyě abaká lá* [possessive]
3SG tear 1SG:poss basket DEF
'S/he broke my basket.'

(80) *é- vú abaká lá ná- m* [possessor ascension]
3SG tear basket DEF to 1SG
'S/he broke the basket on me.'

Second, the inseparability of an entity and its spatial orientation suggests that where the possessive phrase occurs as the direct object, the situation should be interpreted as one in which the possessor and the possessum take part in the event together. This point is consistent with the view that adnominal possessive constructions are a kind of coordination or multiplication of the entities involved. (Cf. Jakobson's (1971: 149) explanation of the syncretism of nominative plural and genitive singular in some Russian paradigms in terms of semantic quantification. In fact, the same kind of explanation could be offered for the homonymy in the northern dialects of Ewe between the dialectal variant of the possessive connective which is *wó*, the plural morpheme *wó* and the third person plural form *wó*.)

The behaviour of kin terms in possessor ascension constructions is somewhat different from that of body parts in the same constructions. Structurally, the kin constructions are analogous with the body part constructions. Schematically, the kin term possessor ascension constructions can be represented as: Z V Y (kin) *ná* X (POR). It has been previously stated that the body part possessor ascension constructions and their corresponding possessive variants are by and large synonymous. However, for some of the structures involving kin terms, the possessor ascension and possessive constructions are not referentially the same. Compare the pairs of sentences in examples (81), (82), (83) and (84):

(81) *papá ɖe srɔ̃ (tsitsǐ má) ná kofí*
'Papa' take spouse (old DEM for Kofi
'Papa took (that old) wife for Kofi.'

(82) *papá ɖe kofí srɔ̃ (tsitsǐ má)*
'Papa' take Kofi spouse (old DEM
'Papa took Kofi's old wife.'

(83) *é- dzu ví- nye*
3SG insult child 1SG
'S/he insulted my child.'

(84) *é- dzu ví ná- m*
3SG insult child to me
'S/he insulted my childlessness.' (Literally: 'S/he insulted child to me.')

Notice that there is no synonymic relation between the two types of constructions even when the kin term is modified (see (81) and (82) above). One could explain the incongruence of these kin term constructions by saying that the possessor ascension constructions have become lexicalised for the expression of specific meanings.

Apart from the absence of synonymy between the kin posseessor ascension and possessive constructions, some instances of the possessive constructions of kin terms do not seem to have possessor ascension counterparts. It is not entirely clear why this is the case. However from the hypothesis put forward earlier that the possessor ascension constructions tend to present the possessor and possessum as being conceptually distant and individuated, it may be that in the instances where the possessor cannot be promoted, the conceptual bond of the entities is at issue. Observe that in examples (85) and (86) the kin term and the predicate can occur in possessor ascension constructions, so the unacceptability of (86) may not be due to the lexical items per se.

(85) *áma ɸo kofi srɔ̃*
Ama beat Kofi spouse
'Ama beat Kofi's wife.'

(86) **áma ɸo srɔ̃ ná kofí*
*Ama beat spouse to Kofi
'Ama beat the wife of Kofi on him'

From these observations one can only conclude that kin and spatial relation and body part terms behave differently with respect to possessor ascension constructions. Spatial relation terms do not participate in them. Kin terms can occur in the possessor ascension constructions with severe semantic and lexical restrictions. Body parts do not have anything constraining their occurrence in these structures. It has already been argued in connection with adnominal constructions that these three categories of possessa are treated differently in terms of "alienability"; the spatial orientation terms being the least "alienable" and the kin terms less "alienable" than body part terms (see Figure 2). If one attempts to correlate the degree of "alienability" of the categories of possessa with their behaviour in possessor ascension constructions, a discernible pattern emerges: the possessors of the least "alienable" possessa, the spatial orientation terms, are the least accessible to the ascension constructions and the possessors of the most "alienable", the body parts, are the most accessible to promotion. This correlation can be more generally stated as follows: the degree of "alienability" assigned to a possessed nominal in a particular situation corresponds to the potential of its possessor to be promoted to an oblique dative. This generalisation is consistent with one of the effects of the possessor ascension construction which has been noted, namely, that of de-emphasising the dependence of the possessor on the possessum and placing emphasis on the distinct character of each of the related entities.

Further support for the generalisation is provided by the fact that the wholes of other meronyms which do not receive "inalienable" grammatical treatment behave like the 'owners' of body parts with respect to possessor ascension. For example, (87) involves an instrument and one of its parts, and it can be paraphrased as (88) to describe the same external reality.

(87) *kofí kpa agblenú- á ɸé atí*
Kofi carve hoe DEF poss stick
'Kofi carved the handle of the hoe.'

(88) *kofí kpa atí ná agblenú- á*
Kofi carve stick to hoe DEF
'Kofi made a handle for the hoe.'

4.1.3. Body part as subject in possessor ascension constructions

Similar observations can be made with respect to the promotion of a possessor of a subject to an oblique object. Consider examples (89) and (90):

(89) *áma ɸé ŋkú (toɖoé má- wó) gba*
Ama poss eye (round DEM PL break
'Ama is blind.' (literally: 'Ama's (round) eyes are broken.')

(90) *ŋkú (toɖoé má- wó) gba ná áma*
eye (round DEM PL break to Ama
'(Those round) eyes are broken to Ama.' (= Ama is blind).

The essential difference between examples (89) and (90) stems from the way the body part-possessor relation is viewed. In (89) the body part in conjunction with the possessor serve as the grammatical subject. In (90) the role of subject is singularly filled by the body part term. It is presented as a distinct participant and its relation to the possessor is an indirect one. Notice that in either construction the body part term could be qualified as demonstrated in examples (89) and (90) above.

With these considerations in mind, one could characterise the semantics of the constructions as follows:

X *ɸé* Y [body part] V (see example (89))
I want to say something about part Y of X ('s body)
I say: something happened to part Y of X ('s body)
[One could think that it happened to X and Y]

Y[body part] V *ná* X (see example (90))
I want to say something about Y
I say: something happened to Y
One can think of Y like this:
Y is a part of X('s body)
Because of that one can think of what happened like this:
it happened to X
X could feel something because of that

It should be mentioned here that when body parts are used metaphorically to express emotions and sensations they tend to occur as grammatical subjects. The experiencer of the emotion or sensation, however, occurs as the direct object. As is evident from examples (91) and (92) below, body parts used metaphorically can be modified.

(91) *dzi (gã aɖé) le kú ye- ḿ*
heart (big INDEF) PRES die 1SG PROG
'I am very angry.'

(92) *ta (sésẽ aɖé) le áma ɖu- ḿ*
head (hard INDEF) PRES Ama eat PROG
'Ama has a severe headache.'

Spatial orientation terms do not occur by themselves as the grammatical subject of process and event verbs.[11] Kin terms like *tɔ́* 'father' *nɔ* 'mother' and *ví* 'child' could occur as subjects of event verbs like *kú* 'die' as exemplified below in example (93):

(93) *ví geɖe- wó kú ná nyɔ́nu má*
child several PL die to woman DEM
'Several children have died on that woman.'

These constructions are rather restricted. It is hard to get any other acceptable utterances with kin terms alone as subjects. It is perhaps worth noting that the kin terms involved in this construction are used only referentially and not in address. This may suggest that they are less endearing and could be used in distance creating situations as this construction is perceived to be.

Be that as it may, this section has presented data to show that by and large body parts are set apart from other relational terms. In particular they can be more readily presented as distinct and individual entities which are connected to the whole in some way in possessor ascension constructions. I maintain that this treatment is consistent with the conceptualisation of body parts as things that the owner could do things with. This claim is further supported in the next section with respect to the constructions in which body parts are presented as instruments.

4.2. Body part as instrument

Just as body parts can be isolated and focused on as locus of effect in possessor ascension constructions, they can also be presented as the instruments used by the possessor to perform activities. Instrumental NPs may be marked in one of two ways in Ewe: (1) they may occur as direct objects of the verb *tsɔ́* 'take' in serial verbal constructions; (2) they may occur as oblique objects of the preposition *kplé* 'with'. Body parts with or without qualifiers may occur by themselves in either of these constructions. Or the instrumental role could be filled by a possessive phrase in which a body part is the possessed item. These instances are illustrated in (94) and (95) below:

(94) *kofí tsɔ́ (é- ɸé) ŋgó (glamɛ lá) ɸú gli*
Kofi take (3SG poss) forehead (pointed DEF) hit wall
'Kofi hit a wall with (his pointed) forehead.'/'Kofi hit (his pointed) forehead against a wall.'

(95) *me- lɔ̃ wò kplé (nyě) dzi (blíbo)*
1SG love 2SG with (1SG:poss) heart whole
'I love you with (all my) heart.'

In this respect body parts behave like any ordinary nominal. Compare the forms in (96) and (97) with the last two examples. The relevant nominals in (96) and (97) are non-relational terms.

(96) *kofi tsɔ́ (é- ɸé) hɛ (ɖáɖɛ́ lá) lã te- á*
Kofi take (3SG poss) knife (sharp DEF) cut yam DEF
'Kofi took (his sharp) knife and cut the yam.'

(97) *áma ɸle aɸé- á kplé (é- ɸé) ga (katã)*
Ama buy house DEF with 3SG poss money all
'Ama bought the house with (all her) money.'

By contrast, the terms used for spatial orientation in example (98) and for kin in examples (99) and (100) cannot be separated from their possessors in these constructions. That is, unlike body parts which can occur by themselves as instruments, only multiple phrases with kin or spatial relation terms as heads are allowed as objects of the instrunmental verb *tsɔ́*. Notice that the presence or absence of a qualifier such as the definite article in (98) does not affect this constraint.

(98) *é- tsɔ́ *(kplɔ̂- á) dzí wɔ dɔ- ɸé*
3SG take *(table DEF) top make sleep place
'S/he has made the top of the table his bed.'

The definite article, used in examples (99) and (100) below, is the possessive article. It is generally used with kin terms. If the possessor is a third person singular pronominal, the pronoun may be elided and the article becomes the only signal for possession, as is the case in the examples below. Note that the article cannot be omitted in these examples without making them ungrammatical:

(99) *é- tsɔ́ vi- *(a) xɔ nzima*
3SG take child (DEF) get juju
'S/he used his/her child to get money.' (That is, 'S/he sacrificed his/her child to a fetish for money.')

(100) *é- xɔ nzima kplé vi- *(á)*
3SG get juju with child (DEF)
'S/he used his/her child to get money.' (That is, 'S/he performed a sacrifice with his/her child for money.')

The main point about the instrumental constructions is that body parts are treated as entities which can be handled like other "alienable" possessions. They also show that body parts behave differently from other relational concepts in this context as well.

In some instances, body parts can be presented as effectors (Foley – Van Valin 1984: 53) of situations. Effectors are non-volitional inanimate entities that can do things. Compare (101) and (102).

(101) *kofí nye abɔ xlá ɖé nyě ŋkú*
Kofi swing arm hit at 1SG:poss eye
'Kofi swung his arm and hit my eye.'

(102) *kofi nye abɔ wò- xlá ɖé nyě ŋkú*
Kofi swing arm 3SG hit at 1SG:poss eye
'Kofi swung his arm and it hit my eye.'

Structurally, (101) is a serial verb construction in which the subject of the first verb is identical with the subject of the second verb. Hence there is no formal marking of subject on the second verb. By contrast, (102) is an overlapping clause (Duthie, in press) in which the subject of the second verb is coreferential with the object of the first verb. Hence there is a pronominal marking for subject on the second verb. The semantic implication of this structural difference is that in (101) 'Kofi' is the agent of both actions, but in (102) 'Kofi' is the agent of the first event (swinging of the arm) and the 'arm' is the effector of the second – hitting the eye. Thus in such constructions body parts can be presented as entities that can bring about some situations.

4.3. 'Possessor deletion'

In this section, the conditions under which a possessor can be deleted will be considered. Different principles seem to apply to different categories of possessa. In a clause where the possessor does something to part of their body, the possessor could be deleted. Consider examples (103) and (104):

(103) *kofí klɔ́ ŋkúme (ɸoɖi má) háfí yi tɔme*
Kofi wash face (dirty DEM) before go river:side
'Kofi washed his (dirty) face before going to the riverside.'

(104) *do asi ɖá*
send hand forward
'Stretch out your hand.'

In example (103), the possessor is deleted under coreference with the subject of the clause. In example (104), it is obvious that the possessor of 'hand' is the addressee of the imperative. Thus in both examples, it can be said that the deleted possessor of a body part term is recoverable from the syntactic environment.

The deleted possessor of a spatial relational term however is only recoverable from the extra-linguistic context of the utterance as can be deduced from examples (105) and (106) below. Observe that in (106), for instance, which is an imperative, the deleted possessor of *gbɔ́* 'side' is not the addressee but something else.

(105) *me- tútú ŋú vɔ*
1SG wipe surface PFV
'I have wiped its surface.'

(106) *nɔ gbɔ́ m- á- vá*
stay side 1SG SBJV come
'Stay at its side till I come.'

For kin terms, one has to distinguish between those that are used referentially, for example, *tɔ́* 'father'; *nɔ* 'mother' and *nɔví* 'brother/sister', and those that may be used both in address and referentially. For instance, *mamá* 'grandmother', *tɔ́gbé* 'grandfather' and *papá* 'father'. If the possessor of a referentially used kin term is omitted, it is interpreted as being identical with the addressee.

(107) *kofí le nɔví (tsitsitɔ) ɸo- ḿ*
Kofi PRES sibling elder beat PROG
'Kofi is beating your elder brother/sister.'

(108) *kpɔ́ nɔ ɖá*
see mother there
'Look at your mother.'

Indeterminacy arises in the interpretation of the deleted possessor of kin terms which can be used referentially and in address. In (109), for instance, the possessor of *tɔ́*gbé 'grandfather' can be understood to be the addressee or the speaker or someone else.

(109) *me- kpɔ́ tɔ́gbé le mɔ́- á dzí*
1SG see grandfather on way DEF top
'I saw grandfather on the way.'

It can be said that the interpretation of possessor deletion operates on different principles for body parts and for other relational terms. For body parts, the deleted possessor is by and large recoverable from the syntactic context. For spatial orientation and kin terms, however, the identification of the deleted possessor is determined by extra-linguistic factors.

5. Conclusion

The foregoing has been a survey of the behaviour of body parts in Ewe grammar. It is evident that body parts tend to be treated differently from other categories of relational terms such as spatial orientation and kin terms. Body parts primarily participate in the "alienable" adnominal possessive constructions unlike kin and spatial relation terms which occur in the "inalienable" nominal structure. Body parts could be assigned core grammatical roles distinct from their possessors in some constructions.

What is the motivation for body parts to be set apart from other relational nouns in their treatment in Ewe grammar? The suggestions that have been made concerning the syntax of body parts in universal grammar do not seem to be applicable to the Ewe situation. Hopper – Thompson (1985: 167), for example, contend that body part nominals tend to be marginal members of the nominal category and they are treated in grammar and discourse as 'dependent, unindividuated entities' because they are 'physically undifferentiated from their "possessors"'. Body part nominals in Ewe are central members of the nominal class; they have all the formal properties of a prototypical noun. Furthermore, the Ewe data suggest that there are instances in the grammar where body parts are treated as individuated entities, independent of their possessors. It appears that for Ewe, Hopper – Thompson's explanation fits spatial orientation terms better than body parts. Spatial relational terms are marginal nominals. Recall that they do not normally take any modifiers in the "inalienable" nominal construction. Besides, they are physically undifferentiated from their possessors.

Similarly, the "alienable" nominal morphology of body parts in Ewe grammar does not seem to be 'a reflection of the real-world fact that body parts are physically contiguous with their "possessors"' (Fox 1981: 323).

It seems that the treatment of body parts in the grammar is the result of two things: (i) the grammaticalisation of body part terms to form spatial orientation terms, and (ii) the way the Ewes conceptualise body parts and their relationship with their possessors. It appears that the language resolved the homonymy between body parts and spatial relation terms that resulted from the evolution of the latter from the former by treating body parts in the grammar as "alienable" entities. Since spatial orientation terms are less separable from their possessors than body parts are from their owners, spatial orientation terms have an "inalienable" grammar. Furthermore, and as a consequence of the reanalysis of the grammar of body parts, they have come to be construed as entities with which their possessors can do things in the synchronic grammar of Ewe. To understand the body part phenomenon in Ewe one should perhaps pay attention to the fact that 'language is not about scenes [that is, real-world facts F.A.]; it is about how people see scenes' (Gárcia 1975: 300).

6. Appendix: Summary of constructions and explications

6.1. "Alienable" nominal constructions

X (= POR) *ɸé* Y (= POSS)

(110) *áma ɸé tɔme- de- ze*
Ama poss river go pot
'Ama's pot'

The semantic prototype (cf. 3.2.1 and 3.2.2)

X (=POR) *ɸé* Y (= POSS)
One can think of Y like this:
Y is like a part of X
X can do things with Y

Body part constructions (cf. 3.2.3)

(111) *kofí ɸé ŋkúme*
Kofi poss face
'Kofi's face'

X (= POR) *ɸé* Y (= body part)
One can think of Y like this:
Y is a part of X
When one thinks of Y
one cannot not think of X
X can do things with Y

Other part-whole constructions (cf. 3.2.3)

(112) *agbalẽ- á ɸé akpa*
book DEF poss cover
'The cover of the book'

X (= POR, inanimate whole) *ɸé* Y (= part)
One can think of Y like this:
Y is a part of X
When one thinks of Y
one cannot not think of X

6.2. "Inalienable" nominal constructions

X (= POR) Y (= POSS)

Spatial relation term constructions (cf. 3.3.2)

(113) *tɔ- á títína*
river DEF middle
'middle of the river'

X (= POR) Y (= POSS, spatial orientation)
One can think of X and Y like this:
When one thinks of X one cannot think of Y
When one thinks of Y
one cannot not think of X
Y is a part of X
One cannot say
X can do anything X wants with Y

Kin term constructions (cf. 3.3.3)

(114) *ɖeví- á- wó tɔ́gbé*
child DEF PL grandfather
'The grandfather of the children'

X (= POR) Y (= POSS, kin term)
One can think of X and Y like this:
X is part of something
Y is part of the same thing as X
When one thinks of Y one thinks of X
One cannot say
X can do anything X wants with Y

Socio-cultural term constructions (cf. 3.3.3)

(115) *miá de*
1PL homeland
'Our homeland'

X (= POR) Y (= POSS, socio-cultural terms)
One can think of Y like this:
Y is like a part of X
When one thinks of Y one thinks of X
X can do some things with Y
X cannot do with Y anything X wants to

6.3. Syntactic compounds

X (= N1) -Y (=N2) + high tone suffix

Possessive compounds (cf. 3.5.2)

(116) *klemensi* - *'kɔ́suá*
Clemence Akɔsua
'Clemence's Akɔsua'

X (= N1 [POR]) – Y (= N2 [POSS])
One can think of X and Y like this:
Y is part of something
X is part of the same thing
When one thinks of Y
one cannot not think of X
One can think this:
X does things for Y

Classificatory compounds (cf. 3.5.3)

(117) *nyɔ́nu* - *ɖǒ*
woman cloth
'ladies' cloth'

The semantic prototype: (cf. 3.5.3)

X (= N1) – Y (= N2)
One can think of X and Y like this:
One can think that X and Y are one thing/place/person
This thing/place/person is a kind of Y

Meronymic compounds (cf. 3.5.3)

(118) *gbɔ- fɔ̆*
goat leg
'goat leg'

X (= N1) – Y (= N2)
One can think of X and Y like this:
Y is (always) a part of X
One can think that X and Y are one thing
This thing is a kind of Y

6.4. Body part as locus of effect constructions

Possessor ascension: body part as object (cf. 4.1.2)

(119) *núfiálá ɸo gbíme ná ɖeví lá*
teacher beat buttocks to child DEF
'The teacher smacked the bottom for the child.'

Z V Y [body part] *ná* X [POR]
I want to say something about Z
I say: Z did something to Y
One can think of Y as part of X's body
Because of that, one can think of
what happened like this:
Z did something to X
X could feel something because of this

Possessive construction: possessive phrase as object (cf. 4.1.2)

(120) *núfiálá ɸo ɖeví lá ɸé gbíme*
teacher beat child DEF poss buttocks
'The teacher smacked the child's bottom.'

Z V X *ɸé* Y
I want to say something about Z
I say: Z did something to part Y of X's body
[One could think that Z did it to X and Y]

Possessive construction: possessive phrase as subject (cf. 4.1.3)

(121) *kofí ɸé abɔ́ tró*
Kofi poss arm twist
'Kofi's arm is twisted.'
X *ɸé* Y V
I want to say something about part Y of X('s body)
I say: something happened to part Y of X('s body)
[One could think that it happened to X and Y]

Possessor ascension: body part as subject (cf. 4.1.3)

(122) *abɔ tró ná kofí*
arm twist to Kofi
'Kofi has a twisted arm.'
Y V *ná* X
I want to say something about Y
I say: something happened to Y
One can think of Y like this:
Y is a part of X('s body)
Because of that one can think of what happened like this:
it happened to X
X could feel something because of this

Notes

1. Ewe, also spelt Eue, is spoken in the south eastern corner of Ghana across southern Togo as far as and just across the Togo-Benin border into Benin. It belongs to the Gbe subgroup of the Kwa branch of Niger-Congo (see 2.0 for a typological sketch). The normal orthography based on the African alphabet is used throughout the paper with the following modifications: i) all high tones are marked with an acute accent [́] in addition to the low tones which are customarily marked by a grave accent [̀]. ii) ɸ is used for *f*. iii) *v* is the form for the phonetic ß. iv) hyphens are introduced to show morpheme boundaries where relevant. Examples used in this study are drawn from both spoken and written standard Ewe. Some of the examples have been culled from prose fiction and drama written by native speakers of Ewe (see references). Others come from radio and television news broadcasts and other programmes, and some also come from casual conversation. The author, being a native speaker, has also constructed some of the examples. Most of the examples have been checked with other native speakers for their acceptability and semantic interpretation. All interlinear and free translations are those of the author unless otherwise stated.

The following abbreviations are used in the paper: Adj = adjective; CERT = certainty; COMP = complementiser; DEF = definite article; DEM = Demonstrative; DET

= determiner; NFOC = nominal focus marker; FUT = future; HAB = habitual; INGR = ingressive; INDEF = indefinite; INT = intensifier; INV = inverse marker; IRR = irrealis; LOG = logophoric pronoun; MOD = modal; N = nominal; NEG = negative; NP = nominal phrase; NPRES = non-present; PFV = perfective; PL = plural marker; POR (=X) = possessor; POSS (=Y) = possessum; poss = possessive linker; PRES = present; PRO = pronominal; PROG = progressive; purp = purposive; Q = question; QT = quantifier; REP = repetitive; SG = singular; SBJV = subjunctive; VP = verbal phrase; (x)* = x is recursive; 1 = first person; 2 = second person; 3 = third person; (*x) = unacceptable if x included; *(x) = unacceptable if x omitted

2. It has been suggested that this suprasegment is a remnant of the high tone on the possessive linker ɸé after the segmental forms have been lost.
3. It must be conceded though that there are semantic constraints on the spatial orientation terms that can occur in this structure. For example, those spatial relation terms that have lost most of their nominality and are fully grammaticalised as postpositions e.g. *me* 'inside', *té* 'under' and *dzi* 'top' cannot be connected to their possessors by the possessive linker *ɸé*. Heine (1989: 115ff) makes a similar observation, but does not take back the 1986 assertion.
4. It is interesting to note that in the next paragraph the same author has a similar sentence in which the possessive marker is not used to link Ewe 'land' and 'middle':

(123) *ési wò- nyé eve- nyigbá blibo lá títína tútútú- é*
when 3SG be Ewe land whole DEF middle exactly NFOC
dunyo le ta lá...
Dunyo be because TP
'Since it is the case that Dunyo is right in the middle of the whole Ewe territory... '

Here the author does not wish to emphasise the location of this large town, rather he is emphasising the reason for the next piece of information that is contained in the main clause of the sentence. Hence he uses the unmarked means of expressing spatial relations.

5. The form *gbedé* means 'blacksmith' and it is used as a proper address term for blacksmiths (see Ameka (1991) on modes of address in Ewe).
6. In Ewe traditional culture, a husband is the one who has authority in the family and provides for his spouse and children. Wives are expected to submit to their husbands and not to dominate them. A husband who is ruled by his wife is said to be effeminate. Given this cultural background it should be clear why these structures are rare and are perceived to be odd.
7. Perhaps it is worth mentioning that Ewe does not seem to be the only language in which body parts have an 'alienable' grammar and kin terms have an 'inalienable' grammar. Haiman (1985: 135) discusses some examples. One of such languages "is Menya, a Papuan language of the Angan family... In Menya, kin terms are possessed by pronominal prefixes: *t- apiqu* "your father". But all other nouns including body parts require the interposition of a genitive /possessive suffix between possessor and possessum: *t- qa angä* "you gen house" *t- qa hanguä* "you-gen shoulder" " Haiman ibid.
8. This view might sound a bit unorthodox in the light of current emphasis on 'grammatical semantics' and the cognitive basis of grammar (cf. e.g. Lakoff 1987, Langacker

1987 and Wierzbicka 1988). The main tenet of this perspective on language is that syntax follows from semantics or conceptual structures. However, it seems that if we want to understand synchronic grammar in its entirety, we need to approach the matter not only from a semantic angle but also from the perspective of the processes and consequences of grammaticalisation as they impinge on the forms in the grammar. The historical perspective helps to provide explanations for synchronic structure, but each grammatical structure has its own associated semantics which may be independent of that construction's history. Thus, although I share the spirit of the suggestion by Nichols (1988: 557) "that form can determine (or shape, or limit) meaning", I think that this occurs at the beginning of the grammaticalisation process. Once the grammaticalisation is complete, the reverse situation of content determining form, at least in usage, is what obtains.

9. This operation is perhaps similar to the type of 'possessor ascension' proposed for languages like Albanian, Chocktaw and Georgian (Blake 1984: 438).
10. If a component is enclosed in square brackets in the semantic formulae, it implies that a definite decision has not been reached as to whether or not that component is a part of the meaning of the construction.
11. Spatial relational terms like all other categories of possessa could function as the subject of the locative copula in the predicative possessive construction Y [POSS] V [copula] (Z) *ná* X [POR]. This structure is used for the expression of any kind of possession as discussed earlier (p. 37). In the following example the spatial orientation term *gbɔ́* 'side' is the subject:

(124) *gbɔ- é le ná wò.*
side NFOC be for 2SG
'You have a side.'

The issue here is that these terms do not function as subjects of clauses expressing occurrences.

References

Allan, Edward J.
1975/6 "Inalienable possession in four Ethiopian languages", *Afrika und Übersee* LIX: 300-307.
Ameka, Felix K.
1991 *Ewe: its grammatical constructions and illocutionary devices.* [PhD thesis, Australian National University.]
Anderson, Lloyd B.
1974 "The part-whole squish: main versus subsidiary predications and why grammatical insertion is like lexical insertion", *Chicago Linguistic Society* 10: 1-16.
Ansre, Gilbert
1966 *The grammatical units of Ewe.* [PhD thesis, University of London.]
Auwera, Johan van der – Louis Goossens (eds.)
1987 *Ins and outs of predication.* Dordrecht: Foris.
Auwera, Johan van der – Michel Kefer (eds.)
1992 *Meaning and grammar.* Berlin: Mouton.

Bally, Charles
1926 "L'expression des idées de sphère personelle et de solidarité dans les langues indo-européenes", in: Franz Fankhauser – Jakob Jud (eds.), 68 -78.
[1995] "The expression of the concept of the personal domain and indivisibility in Indo-European languages", [translated by Christine Béal and Hilary Chappell] [this volume.]

Benjamin, Geoffrey
1994 "Grammar and polity: the cultural and political background to standard Malay", in: William Foley (ed.), 341-392.

Blake, Barry
1984 "Problems of possessor ascension: some Australian examples", *Linguistics* 22: 437-453.

Boas, Franz
1918 Introduction. *Handbook of American Indian Languages*. Bureau of American Ethnology. Bulletin 40, part 1: 1-83.

Bogusôawski, Andrzej
1970 "On semantic primitives and meaningfulness", in: Greimas – Jakobson – Mayenowa (eds.), 143-152.
1989 "Knowledge is the lack of lack of knowledge, but what is that lack of?" *Quaderni di Semantica* 10.1: 15-31.

Byarushengo, Ernest Rugwa – Alessandro Duranti – Larry M. Hyman (eds.)
1977 *Haya grammatical structure*. (Southern California Occasional Papers in Linguistics 6) Los Angeles: University of Southern California.

Chafe, Wallace
1970 *Meaning and the structure of language*. Chicago: The University of Chicago Press.

Chappell, Hilary – William McGregor
1989 "Alienability, inalienability and nominal classification", *Berkeley Linguistic Society* 15: 24-36.
1995 "Prolegomena to a theory of inalienability", [this volume.]

Claudi, Ulrike – Bernd Heine
1986 "On the metaphorical base of grammar", *Studies in language* 10: 297-335.
1989 "On the nominal morphology of 'alienability' in some African languages", in: Paul Newman – Robert Botne (eds.), 3-19.

Claudi, Ulrike – Fritz Serzisko
1985 "Possession in Dizi: inalienable or not?" *Journal of African Languages and Linguistics* 7: 131-154.

Clements, George N.
1979 "The logophoric pronoun in Ewe: its role in discourse", *Journal of West African Languages* 10.2: 141-177.

Cruse, David A.
1986 *Lexical semantics*. Cambridge: Cambridge University Press.

Deakin, Greg
1981 Indirect speech acts and intonation. [Unpublished MA thesis, Australian National University.]

Dixon, Robert M. W. (ed.)
1976 *Grammatical categories in Australian languages.* Canberra: Australian Institute of Aboriginal Studies.
Duthie, Alan S.
in press *Introducing Ewe linguistics patterns.* Accra: Ghana Universities Press.
Evens, Martha Walton (ed.)
1988 *Relational models of the lexicon.* Cambridge: Cambridge University Press.
Fankhauser, Franz – Jakob Jud (eds.)
1926 *Festschrift Louis Gauchat.* Aarau: H.R. Sauerländer.
Foley, William (ed.)
1994 *The role of theory in language description.* Berlin: Mouton.
Foley, William A. – Robert D Van Valin
1984 *Functional syntax and universal grammar.* Cambridge: Cambridge University Press.
Fox, Barbara.
1981 "Body part syntax: towards a universal characterisation", *Studies in language* 5: 323-342.
Gárcia, Erica
1975 *The role of theory in linguistic analysis: the Spanish pronoun system.* Amsterdam: North-Holland.
Goddard, Cliff
1989a "Issues in natural semantic metalanguage", *Quaderni di Semantica* 10.1: 51-64.
1989b "The goals and limits of semantic representation", *Quaderni di Semantica* 10.2: 297-308
Greenberg, Joseph (ed.)
1978 *Universals of human language.* Vol 4: *Syntax.* Stanford: Stanford University Press.
Greimas, A. J. – Roman Jakobson – M. R. Mayenowa (eds.)
1970 *Sign, language, culture.* The Hague: Mouton.
Haiman, John
1985 *Natural syntax.* Cambridge: Cambridge University Press.
Haiman, John (ed.)
1985 *Iconicity in syntax.* Amsterdam: John Benjamins.
Hale, Ken
1986 "Notes on world view and semantic categories: some Warlpiri examples" in: Pieter Muysken – Henk van Riemsdijk (eds.), 233-254.
Halliday, Michael A. K.
1985 *An introduction to functional grammar.* London: Edward Arnold.
Heine, Bernd
1989 "Adpositions in African languages", *Linguistique africaine* 2: 77-127.
Heine, Bernd – Mechthild Reh
1984 *Grammaticalisation and reanalysis in African languages.* Hamburg: Helmut Buske Verlag.
Hopper, Paul J. – Sandra A. Thompson
1985 "The iconicity of the universal categories 'noun' and 'verb' ", in: John Haiman (ed.), 151-183.

Hyman, Larry M.
1977 "The syntax of body parts", in: Ernest Rugwa Byarushengo – Alessandro Duranti – Larry M. Hyman (eds.), 99-117.

Hyman, Larry M. – Danny K. Alford — Elizabeth Akpati
1970 "Inalienable possession in Igbo", *The Journal of West African Languages* 7: 85-101.

Iris, Madelyn Anne – Bonnie E. Litowitz – Martha Evens
1988 "Problems of the part-whole relation", in Martha Walton Evens (ed.), 61-288.

Ivič, Milka.
1970 "On the part-whole relation and its linguistic consequences", in: Roman Jakobson – Shigeo Kawamoto (eds.), 281-286.

Jakobson, Roman
1971 "The relationship between the genitive and plural in the declension of Russian nouns", in: *Selected Writings II* : 148-53. The Hague: Mouton.

Jakobson, Roman – Shigeo Kawamoto (eds.)
1970 *Studies in general and oriental linguistics*. Tokyo: TEC Company.

Keenan, Edward L. – Bernard Comrie
1977 "Noun phrase accessibility and universal grammar", *Linguistic Inquiry* 8: 63-99.

Kliffer, Michael D.
1983 "Beyond syntax: Spanish inalienable possession", *Linguistics* 21: 759-794.
1984 "Interpretation of linguistic levels: French inalienable possession", *Lingua* 62: 187-208.

Lakoff, George
1987 *Women, fire and dangerous things*. Chicago: The University of Chicago Press.

Langacker, Ronald
1985 "Observations and speculations on subjectivity", in: John Haiman (ed.), 109-150.
1987 *Foundations of cognitive grammar*. Vol. 1. Stanford: Stanford University Press.

Winfred P. Lehmann (ed.)
1986 *Language typology*. Amsterdam: John Benjamins.

Lyons, John
1977 *Semantics*. Vol 2. Cambridge: Cambridge University Press.

Muysken, Pieter – Henk van Riemsdyk (eds.)
1986 *Features and projections*. Dordrecht: Foris.

Newman, John – Robert Botne (eds.)
1989 *Current approaches to African linguistics*. Vol. 5. Dordrect: Foris.

Nichols, Johanna
1986 "On form and content in typology", in: Winfred P. Lehmann (ed.), 141-162.
1988 "On alienable and inalienable possession." in: William Shipley (ed.), 557-609.

Reh, Mechthild – Bernd Heine – Marcello Lamberti
1981 "Inhärenz in Possessivkonstruktionen Afrikanischer Sprachen", [Unpublished paper, Unversity of Köln.]

Saah, Kofi K.
1988 "Alienable/Inalienable possession in some African languages", [Unpublished paper, Legon: University of Ghana.]

Sapir, Edward
1917-1920 Review of Uhlenbeck C. C. *Het Passieve Karakter van het verbum transitivum van het verbum actionis in Talen von Noord-Amerika. IJAL* 1: 82-90.
1920 *Language*. New York: Harcourt, Brace.
Seiler, Hansjakob
1980 Two types of Cahuilla kinship expressions: inherent and establishing. *Arbeiten des Kölner Universalien – Projekts* 39.
1983 *Possession as an operational dimension of language*. Tübingen: Gunter Narr Verlag.
Shipley, William (ed.)
1988 *In honor of Mary Haas*. Berlin: Mouton.
Silverstein, Michael
1976 "Hierarchy of features and ergativity", in: Robert M. W. Dixon. (ed.), 112-171.
Starosta, Stanley
1988 *The case for lexicase: an outline of lexicase grammatical theory*. London: Pinter.
Taylor, John R.
1989 "Possessive genitives in English", *Linguistics* 27: 663-686.
Ultan, Russell
1978 "Towards a typology of substantival possession" , in: Joseph Greenberg (ed.), 11-50.
Vandeweghe, Willy
1987 "The possessive dative in Dutch: syntactic reanalysis and predicate formation ", in: Johan van der Auwera and Louis Goossens (eds.), 137-151.
Voeltz, Erhard F. K.
1976 "Inalienable possession in Sotho", *Studies in African Linguistics* Supplement 6: 255-266.
Westermann, Diedrich
1930 *A study of the Ewe language*. [translated by A. L. Bickford- Smith] London: Oxford University Press.
Whorf, Benjamin Lee
1956 *Language, thought and reality*. [edited by J. Carroll] Cambridge, Mass: MIT Press.
Wierzbicka, Anna
1988 *The semantics of grammar*. Amsterdam: John Benjamins.
1989 "Semantic primitives and lexical universals", *Quaderni di Semantica* 10.1: 103-121.
1990 "Semantic primitives: the expanding set", *Quaderni di Semantica* 10.2: 133-157.
1991 "Semantic complexity: conceptual primitives and the principle of substitutability", *Theoretical Linguistics* 17: 75-97
1992 "Lexical universals and universals of grammar", in: Johan van der Auwera – Michel Kefer (eds.), 383-415.
Winston, Morton E. — Roger Chaffin – Douglas Herrmann
1987 "A taxonomy of part-whole relations" *Cognitive Science* 11: 417-444.

Ewe Texts
Akpatsi, Robert S.
1980 *ame aɖeke menya etsɔ me o* [Nobody knows tomorrow]. Accra: Bureau of Ghana Languages.

Kpodzo, L. K. W.
1982 *aleke mahe vinyee*? [How shall I educate my child?]. Accra: Bureau of Ghana Languages.
Nyaku, Frank K.
1984 *hotɔwu kple wɔmenɔ* [Hotɔwu and Wɔmenɔ]. Accra: Asempa Publishers.

Body parts in Acholi: alienable and inalienable distinctions and extended uses

Edith L Bavin

1. Introduction

The Western Nilotic languages include Lango, Acholi and Nuer (spoken mainly in Uganda) and Dholuo (spoken in Kenya). These languages are classified by Greenberg (1963) as members of one of the three branches of Nilo-Saharan, the others being Southern Nilotic and Eastern Nilotic (which includes the Teso-Turkan group of languages and the Maa languages, of which Masai is one). According to Dimmendaal (1983), the division into three branches goes back to Köhlers (1955). The Nilotes originated in the southern Sudan; several migrations have resulted in the present distribution of the Western Nilotic languages. The people migrated from the Southern Sudan through the present area of Uganda to Lake Victoria, and more recently into Western Kenya. Through the migrations south, the people have had some contact with Bantu speaking groups.

The Western Nilotic languages include 3 branches (Greenberg 1966): the Barun group, the Dinka-Nuer group and the Shilluk-Luo group. Acholi belongs to the Luo group, and is generally mutually intelligible with Lango. Alur shares a high percentage of lexical items and grammatical features, as does Padhola. The Luo group have SVO word order, verb classes determined on morphological grounds and tone. Estimates of the number of speakers vary. Okot (1971) suggests there are about half a million people who speak either Acholi or Alur, while other estimates give almost half a million for Acholi alone, slightly more for Lango and 200,000 for Alur (Voegelin – Voegelin 1977). The largest group of Luo is the Dholuo group of Kenya with about 800,000 speakers.

One of the features of Western Nilotic languages is a distinction between alienable and inalienable possession, the inalienable applying to body parts (Tucker – Bryan 1966). The paper focuses on a number of aspects relating to body parts in Acholi: the morphological distinctions for expressing body part-whole and possessive (genitive) relations; the juxtaposing of two nouns is used to express part-whole relations as well as nominal classification; and extensions in the use of body parts to locative and spatial concepts, as well as the association of a number of emotions and qualities with different body parts.

In possession constructions in Acholi, a grammatical morpheme (*pa*) is used to link the possessor and possessed; thus it functions as a genitive marker. However, *pa* is not used between body parts and their "owners"; nor is it used between some other part-whole elements, even if the part is separate. The consequence of this is that inalienable expressions for body parts are comparable (morphologically) with compound nouns in the language. With pronominal possessors, alienable-inalienable distinctions result in morphological alternations. Chappell – McGregor (1989) argue that in many languages, inalienable possession and nominal classification (compounding) are frequently marked in the same way, often by zero marking. This is in contrast to alienability, which is distinct from classification, and which is normally morphologically marked (Chappell – McGregor 1989: 24). Acholi follows the pattern outlined by Chappell – McGregor; the inalienable and compound constructions are morphologically equivalent in contrast to the alienable, which is morphologically marked.

Languages follow different paths in selecting sources for locative concepts, frequently using body parts or landscape lexemes (Svorou 1986). In the Western Nilotic pattern, human body parts are extended to locative concepts (Heine – Claudi – Hünnemeyer 1991). Thus, 'on' develops from 'head'; 'under' from 'buttock', 'anus' or 'foot'; 'in' from 'belly' or 'stomach'; 'front' from 'eye', 'face'; 'forehead' or 'chest' and 'back' from 'back' (Heine et al 1989: 21). Heine et al (1989: 25) interpret the process underlying grammaticalisation as a problem-solving strategy whereby "abstract" concepts are described or understood in terms of less "abstract" concepts. This involves transfer from one cognitive domain to another, as from object to space. As in other languages, the body terms in Acholi have been extended to objects and to locative and spatial domains (less concrete concepts).

In Acholi, various parts of the body are associated with qualities and emotions, and this will be illustrated in the final section of the paper.[1]

2. The pronominal affixes

Before discussing the alienable, inalienable and nominal classification distinctions in Acholi, I will illustrate the basic sentence and agreement patterns of the language.[2] They are relevant for understanding the morphology of the possessive constructions and the alienable-inalienable distinctions.

A transitive sentence consists of a subject, verb and object in that order. However, Acholi is a PRO-drop language in that a pronominal subject is not usually overt in the spoken language. It is used mainly for emphasis and contrastive information, and in the first clause of connected discourse. The verb is inflected to agree with the person and number of the subject noun phrase, as illustrated in (1) and (2).[3]

(1) *(an) a-cito*
I 1:SG-went
'I went.'

(2) *latin awobe o-cito*
child boy 3:SG-went
'The boy went.'

With the exception of third person singular, the subject agreement prefixes are clearly related to the independent pronoun forms. Table 1 lists the forms for the independent pronouns and the subject affixes. The third person singular affix varies between *o* and ø depending on tense, and in limited contexts the form *e* is used (see Bavin 1982 for a discussion of the bound pronominal forms in Lango and Acholi). The third plural affix is *gi*, or *gu* in past contexts. The form *gu* is a contraction of *gi+o*, which represents a double marking. It is not usual to find plural marked nouns in discourse, even though some plural forms exist (one is illustrated in example (8)); usually the singular form of the noun is used. Thus the use of *gi+o* shows double marking with the *o* agreeing with the noun form and the *gi* giving plural reference.

Table 1. Independent pronouns and subject affixes in Acholi

person	singular		plural	
	pro	affix	pro	affix
1	*an*	*a-*	*wan*	*wa-*
2	*yin*	*i-*	*wun*	*wu-*
3	*en*	*o/ø/e*	*gin*	*gi-/gu*

In sentences with a pronominal object, a suffix on the verb indicates the person and number of the object, and no independent pronoun is used. The suffix takes the same form as the subject affix, except third singular is *e*; this is also used as the reflexive form. The pronominal object suffix replaces the verb final vowel *o*, the transitive marker. This vowel is present with or without overt object noun phrases. Examples are given in (3) and (4).

(3) *latin awobe o-nen-o gwok*
child boy 3:SG-saw-T dog
'The boy saw a dog.'

(4) *latin awobe o-nen-a*
child boy 3:SG-saw-1:SG
'The boy saw me.'

The forms used as pronominal objects are also used as the possessive suffixes on nouns when the possessor is pronominal. Possessive constructions are described in section 3.

3. Noun possessors and compounds

3.1. Part-whole relations

Inalienable possession in Acholi is indicated by juxtaposing a noun possessor to the possessed nouns. The examples in (5) illustrate this construction for body parts.

(5) *cing dano*
hand person
'person's hand'

cing lyec
hand elephant
'elephant's trunk'

yewi laco
hair man
'man's hair'

wang *dako*
eye/face woman
'woman's eye/face'

pyen *lagwa*
skin zebra
'zebra's skin'

ping *wang* *laco-ni*
juice eye man-DET
'this man's tears'

Body part terms which appear in this construction include external parts, internal organs, body fluids and hair. Some of these are listed in (6).

(6)	*dok*	'mouth'	*opwuny*	'ankle'
	tyen	'leg/foot'	*bad*	'arm'
	um	'nose'	*kor*	'chest'
	wic	'head'	*nget*	'side'
	ic	'stomach'	*nyim*	'forehead'
	tik	'chin'	*kom*	'body'
	ngut	'neck'	*ngec*	'back'
	adunu	'heart'	*leb*	'tongue'
	obi	'lung'	*it*	'ear'
	cwiny	'liver/heart'	*lak*	'tooth'
	remo	'blood'	*balgwok*	'shoulder'
	em	'thigh'		

There is no distinction made between parts that are still attached to the person or animate being and separated parts, so *yewi laco* 'man's hair' is used whether the hair is still on the man or cut off.

Other nominals associated with the body are treated as body parts, in that the noun is juxtaposed with the possessor with no overt marking of possession. These include 'name' and 'character', which are illustrated in (7), (8) and (9).

(7) *nying* *latin*
name child
'child's name'

(8) *nying dyel-li nga?*
name goat-PL who
'What are your goats' names?'

(9) *cal latin*
character child
'character/likeness of child'

The same construction (juxtaposition) is used when reference is made to parts of inanimate objects, such as branches of trees, or seeds of fruit. Whether the part is separated or not is not relevant. Consider the examples listed in (10):

(10) *jang yat*
branch tree
'branch of a tree'

pok lemun
rind orange
'rind of an orange'

leb tong
tongue spear
'blade of a spear'

tyen kom
leg chair
'leg of a chair'

weng agwata
handle calabash
'handle of a calabash scoop'

dogola motoka
door car
'door of a car'

lak kweri
tooth hoe
'blade of a hoe'

oboke cai
leaf tea
'tea leaf'

3.2 The genitive construction

Acholi treats places and property associated with animates quite distinctly in terms of the construction used. Compare the examples in (5) with those in (11), in which the possessed nominal is linked to the possessor with the morpheme *pa*.

(11) *lok pa laco-ni*
word POSS man-det
'this man's words'

cam pa dano
food POSS people
'people's food'

dyang pa nga?
cow POSS who
'whose cow?'

dyang pa rwot
cow POSS chief
'chief's cow'

gang pa rwot
village POSS chief
'chief's village'

jami tuku pa latin
things play POSS child
'child's play things'

bongo pa latin
clothes POSS child
'child's clothes'

lac pa gado
wide POSS road
'width of the road'

The last example above is included to show that while most N *pa* N constructions express an owner – object relation, *pa* can be extended in usage to link a quality with an inanimate object.

Example (9) illustrated that *cal* 'character/likeness' may be treated as an inalienable possession, but (12) shows *cal* linked with *pa*. Speakers are not consistent in putting in the linking morpheme *pa*. The combination *cal pa* may also be realised as *cal-la*. The assimilation of a morpheme initial consonant to a preceding consonant is a common feature of the language. *Tibu* 'shadow' is considered alienable and is linked with *pa*, but not for inanimate "possessors" (as in (14)).

(12) *cal pa latin ma ki-maka*
likeness POSS child REL PFV-caught
'photograph of the child'

(13) *tibu pa laco-ni bor*
shadow POSS man-DET long
'This man's shadow is long'

(14) *tibu yat*
shadow tree
'tree's shadow'

Note that a person's relatives and children require the possessive marker *pa*, as illustrated below. It is not possible to say 'the woman's child' without *pa*, even in reference to an unborn child. The child is a separate entity.

(15) *jo gang pa laco-ni*
people village POSS man-DET
'this man's relatives'

(16) *latin pa dako*
child POSS woman
'woman's child'

(17) *min pa latin*
mother POSS child
'child's mother'

The examples illustrate that the primary function of *pa* is to distinguish between parts and possessions for animates. In that *pa* is used to distinguish possessive relations from part-whole ones, the *pa* construction can be viewed as a marked genitive construction.

3.3 Other compounds

Many two-noun constructions without *pa* which do not represent a part-whole relationship may be viewed as compounds. For example 'head covering', 'eating utensil', 'ribs', 'mask', 'oil pot', 'cooking pot', and 'water pot' are a few of the numerous compounds found in Acholi. Many of these express an 'X for Y' relationship, as with 'a covering for the head', or 'a pot for oil'. These also express a generic-specific relation. The type of thing is given first and the specific follows. However, not all the combinations are generic-specific; some are more idiomatic, as with 'ribs'.

(18) *gin wic*
thing head
'head covering' (covering for the head)

gin cam
thing eat
'eating utensil' (utensil for eating)

abino moo
pot oil
'oil pot' (pot for oil)

agulu deko
pot food
'cooking pot' (pot for cooking)

agulu pi
pot water
'water pot' (pot for water)

koro gweno
shelter chicken
'chicken house' (shelter for chickens)

won poto
father field
'elder of the field'

lak nget
teeth side
'ribs'

cal *wang*
likeness face
'mask'

The names of seeds are expressed with *nying* 'seed' followed by the plant name to which the seed is associated; these are also generic-specific relations. Similarly 'cow's milk' and 'hen's egg' may be viewed as generic-specific (or part-whole) relations. Millet flour is a product, since flour is not in existence before processing, so the relation is not one of part-whole; rather it is generic-specific. The distinction is semantic.

(19) *nying kal*
seed millet
'millet seed'

cak dyang
milk cow
'cow's milk'

moko kal
flour millet
'millet flour'

latin dyang
child cow
'calf'

These double noun constructions are treated as single entities in possessive constructions. So 'this small cooking pot' is expressed by modifying *agulu deko* 'cooking pot' as in (20). The modifier is post-nominal and consists of *ma*, the relative marker, together with the lexical stem; an optional determiner follows. In (21), a determiner is added to a double noun unit; it is not possible to add the determiner after the first part of the compound.

(20) *agulu deko ma-tidi-ni*
pot food REL-small-DET
'this small cooking pot'

(21) *gin cam-mi*
thing eat-det
'this thing for eating'

3.4 The *me* construction

A number of examples were given above to show that two nouns often expressed an X for Y relation, as with 'water pot' or 'oil pot'. For other combinations expressing a similar, functional association, a linking morpheme, *me* 'for', is required. This is illustrated below in (22). There seems to be no semantic reason why 'cooking pot' and 'cooking fire' require the morpheme *me*, while other two-noun constructions appear with no linking morpheme. I can only hypothesise that there is a general understanding of the function of fires and skins (and other objects), and the use of *me* is to introduce an alternative function, while pots have several functions and, when necessary, the type of pot can be specified using the juxtaposed construction. Although the semantic relations may be similar, speakers of Acholi distinguish two constructions morphologically.

(22) *pyen me buto*
skin for sleep
'sleeping skin'

caa me tic
hour for work
'work time'

mac me tedo
fire for cook
'cooking fire'

4. Pronominal possessors

When possessors are pronominal, a bound pronominal form is suffixed to the possessed noun, as illustrated in the following examples. As noted in section 2, the form of the object suffix on a verb, and the possessive suffix on a noun are identical (except the singular possessive suffixes each have two allomorphs as discussed below). They are distinguished by distribution. Compare the *a* suffix in (23) with that in (24). In pronominal possessor constructions, an alienable-inalienable distinction is made, depending on whether the possessed is a body part or an object in a state of temporary possession. This is discussed below.

(23) *dako* *o-nen-a*
woman 3:SG-saw-1:SG
'The woman saw me.'

(24) *dako* *o-nen-o* *cing-a*
woman 3:SG-saw-T hand-1:SG:POSS
'The woman saw my hand.'

(25) *latin* *camo* *lwer-e*
child ate nail-3:SG
'This child sucked/bit his nails.'

(26) *kiyu-na* *o-to*
glasses-1:SG 3:SG-broke
(*onong-o* *tye* *i-wang-a*)
(found-T be LOC-face-1:SG:POSS)
'My glasses broke (on my face).'

In (26), the possessed noun *wang* 'face' is consonant final, and the possessor suffix is just a vowel, but *kiyu* 'glasses' is vowel final, and the possessor is *na*. In (25), the possessive suffix also consists of a vowel alone. Each of the singular possessive pronouns has two forms, one consisting of the appropriate vowel, and the other of the vowel plus a "linking" consonant *n* (*a*/*na* '1SG'; *i*/*ni* '2SG'; *e*/*ne* '3SG'). The plural forms only have one representation, C+vowel (*wa* '1PL', *wu* '2PL', *gi* '3PL').

The final sound of the possessed noun is one conditioning factor for the allomorph alternations for the singular possessive pronouns, but this interacts with another factor: the relationship between the two words. Inalienable parts with a consonant-final stem require the vowel allomorph, and those ending in a vowel, the C+V affix, but inalienable possessions always require the C+V suffix, even if the part noun ends in a consonant. For example, when an alienable body part is attributed to a pronominal possessor, as with 'my leg' meaning the leg of a slaughtered animal that I have been given, rather than a part of my body, *na* 'my' is added to *bad* 'arm' giving *bad+na*, but an assimilation rule produces a geminate consonant in place of the *d+n* combination. This results in *bad-da* 'my arm (leg)'. This contrasts with *bad-a* 'my arm' (part of my body). Note that an animal leg is generally referred to as *bad* 'arm', although *tyen* 'leg' is used to contrast the back legs from the front.

'My house' is *ot+na*, giving *od-da*. The final stop of *ot* becomes voiced before the nasal, and then the assimilation rule applies giving a geminate *d.*[4] Similarly, 'my book' is *buk-ka*. In slow speech, however, *ot-na* and *buk-na* are sometimes heard, so the assimilation of the nasal to the final stop of the possessed item is a variant, but a common feature of spoken language. There are a few lexical items that have an alternate final consonant, voiced or voiceless. The final consonant of *wot* 'son' and *dok* 'mouth' are voiced before other sounds, but are voiceless in word final position. Thus 'my mouth' is *dog-a*, and 'our mouth(s)' is *dog-wa*.

Because a voiceless consonant weakens between vowels at a normal speech rate, with the few lexical exceptions discussed in the previous paragraph, it is possible to tell when a geminate consonant is present. A geminate does not weaken between vowels. Thus **bux-a* would not be 'my book' since 'my book' is an alienable relation and the form is *buk-ka*, but *tik-a* 'my chin' becomes *tix-a* in fast speech.

For body parts ending in vowels, the CV suffix is always added, so the alienable/inalienable distinction is neutralised. For example, *remo-na* is 'my blood (alienable and inalienable), and *obo-ni* is 'your lung'. With all plural nouns, the alienable/inalienable distinction is neutralised, as in *cwiny-gi* 'their liver/heart (alienable or inalienable). So, in summary, it is only with singular possessors that the alienable-inalienable distinction is detectable morphologically.

There is a *y/c* alternation in the final consonant for a number of nouns. Word final palatal *c* (which varies from a stop to a fricative across speakers) is lost when the word is the first part of part of a compound, and it weakens to *y* before a vowel. So from the word *ngec* 'back' is the word *ngey-a* 'my back' (inalienable). 'Head' is *wic* in elicited form, but as part of a compound it is lost, as in *wi yat* 'tree top'. The following sentence illustrates a change of *c* to *y* for *wic* 'head'. This is itself part of a compound with *yer* 'hair', the final *r* of *yer* being lost before *w*.

(27)	*dako-ni*	*o-lwok-o*	*yerwiy-e*
	woman-DET	3:SG-washed-T	hair-3:SG:POSS
	'Woman washed her hair.'		

5. Part-whole in syntactic contexts

As mentioned above, parts may be separated (for example, a branch separated from a tree, or a lid from a pot) and they are referred to in the same way as if not separated. However, in talking about the removal of a part, it is quite usual, although not obligatory, to mention the location from which the part originates. For example, leaves come from the head (top) of the tree, or the body of the tree, as indicated in the the examples in (28). Similarly, the location from which a possession is removed can be named, as in (29) and (30).

(28) a. *oboke o-pot-o ki-i-kom yat*
leaves 3:SG-fell-T with-LOC-body tree
'The leaves fell from the tree.'

b. *oboke o-pot-o ki-i-wi yat*
leaves 3:SG-fell-T with-LOC-head tree
'The leaves fell from the tree.'

(29) *a-yec-o bongo pa latin (i-kom-e)*
1:SG-tore-T clothes POSS child (LOC-body-3:SG)
'I tore the child's clothes (from his body).'

(30) *a-yec-o bongo pa latin a-tyek-o ko*
1:SG-tore-T clothes POSS child 1:SG-finish-T COM
ki-i-kom-e
with-LOC-body-3:SG:POSS
'I tore the child's clothes off into pieces.'

6. Actions involving body parts

In using body parts to carry out actions it is not necessary to state both the whole and part, only the part. So the part is the agent, as in the following examples. Note that there is no distinction made between deliberate action and accidental when body parts are agents, so sentence (31) can imply deliberate hitting or accidental touching.

(31) *cing-e o-yot-o yat*
hand-3:SG:POSS 3:SG-hit-T tree
'His hand hit the tree.'

If a body part is affected by some action, the part bears the syntactic role of object; it never appears as a locative argument as in 'The woman hit the child on the hand'.

Many verbs imply the body part involved. It is assumed that the hand is used for hitting and the foot for kicking. However, in addition, a number of specific verbs imply the instrument of contact. In using specific verbs, the body part or instrument involved is not mentioned. For example, compare (32) in which the verb implies contact with the hand, (33) in which the verb implies contact with a stick, and (34) in which the verb implies contact with a small object.

(32) *o-dong-o* *bad-a*
3:SG-hit:with:hand-T arm-1:SG:POSS
'Someone used his hand to hit me on my arm.'

(33) *o-goy-o* *gwok*
3:SG-hit:with:stick-T dog
'Someone hit the dog.'

(34) *o-yok-o* *gwok*
3:SG-hit:with:small:thing:like:a:stone-T dog
'Someone hit the dog.'

7. Locative uses of body parts

Many of the body terms are extended to objects, and also to spatial concepts. For example, *nget* 'side' can also be used for 'edge' as in the edge of the table; *wic* 'head' can be extended to mean 'top', and with the locative prefix *i*, (*i-wi*) it means 'on top'; *nyim* 'forehead' with the locative prefix *i* (*i-nyim*) is 'in front, ahead', while with the prefix *a* (*a-nyim*), it is 'the front'; *ngec* is 'back' and with the prefix *i*, (*i-ngec*) it is 'at the back, behind', *a-ngec* is 'backwards', and *i-nge* is 'after'. For 'under' there is an extension of *te* 'buttocks', and *ic* 'stomach' is used for 'inside'.

Note the use of body parts in the following. Examples (36)-(43) illustrate the extension of body parts to inanimates, the last two based on perceptual similarities between teeth and a blade, and between a nose and a peaked roof.

(35) *wi* *ature*
head flower
'flower head'

(36) *kor ature*
chest flower
'stalk'

(37) *kor yat*
chest tree
'tree trunk'

(38) *kom ot*
body house
'structure of the house'

(39) *kor ot*
chest house
'walls'

(40) *wang ot*
eye/face house
'windows'

(41) *wi ot*
head house
'roof'

(42) *lak cumeno*
tooth saw
'blade of the saw'

(43) *um dero*
nose granary
'roof of the granary' (peaked)

While (44) refers to the front side of the house, with *i-nyim*, the sentence would indicate a locational relation 'in front of the house'. The example in (45) shows a compound derived from a body part to indicate a location, and (46) an idiomatic interpretation of 'head' to represent a location. Examples (47)-(51) illustrate more idiomatic uses of body parts.

(44) *nyim ot*
forehead house
'front of the house'

(45) *ka nget*
place side
'beside'

(46) *wi poto*
head field
'the other end of the field'

(47) *ka wang*
place eye
'in place of'

(48) *nget lok*
side words
'talking, not coming to the point'

(49) *wang lobo*
eye/face earth
'country, district'

(50) *wi lok*
head word
'the real issue'

(51) *apar wiy-e acel*
ten head-3sg one
'eleven'

8. Emotions, qualities and body parts

Body parts are used to show the location of emotions and qualities. Many of these utilise *cwiny* which can be interpreted as 'heart' as well as 'liver'. The quality/emotion label consists of an attribute followed by the body part (A+P). When the characteristic is used in reference to a particular individual, the body part is suffixed with a pronominal possessor and the attribute follows (P-poss+A). Thus the part is the experiencer, using a literal interpretation. The liver is associated with most emotions and some qualities, and the eye/face, mouth, body, and hand are associated with other qualities.

Liver

(52) a. *yom cwiny*
soft liver
'happiness/generosity'

b. *cwiny-e yom*
liver-3:SG:POSS soft
'He is happy/generous.'

(53) a. *cwero cwiny*
leak liver
'sadness'

b. *cwiny-e cwer*
liver-3:SG:POSS leak
'He is sad.'

(54) a. *wang cwiny*
burn liver
'sadness/worries'

b. *cwiny-e wang*
liver-3:SG:POSS burn
'He is sad/worried.'

(55) a. *tek cwiny*
hard liver
'courage'

b. *cwiny-e tek*
liver-3:SG:POSS hard
'He is courageous/ strong headed.'

(56) a. *kec cwiny*
bitter liver
'unhappy'

b. *cwiny-e kec*
liver-3:SG:POSS bitter
'He is unhappy/angry/unfriendly.'

(57) a. *myel cwiny*
dance liver
'alarm/fear'

b. *cwiny-e myel*
liver-3:SG:POSS dance
'He is alarmed.'

(58) a. *leng cwiny*
clean liver
'honesty/generosity'

b. *cwiny-e leng*
liver-3:SG:POSS clean
'He is honest/ generous.'

Note also the following, for which I was able to elicit either the emotion (59), or an attribute for a particular individual (60) and (61), but not both.

(59) *cwiny peke*
liver nothing
'despair'

(60) *cwiny-e acel*
liver-3:SG:POSS one
'He is reliable.'

(61) *cwiny-a lem-a*
liver-1:SG:POSS hurt-1sg
'I feel nauseous.' (my liver hurts me.)

Eye/face

(62) a. *nwang wang*
durable eye
'durable'

b. *wang-e nwang*
eye-3:SG:POSS durable
'He is hard (can't be deceived).'

(63) a. *ber wang*
good eye
'beauty'

b. *wang-e* *ber*
eye-3:SG:POSS good
'She is beautiful.'

(64) a. *leng wang*
clean eye
'beauty'

b. *wang-e* *leng*
eye-3:SG:POSS clean
'She is beautiful.'

(65) a. *bit wang*
sharp eye
'sharp sighted'

b. *wang-e* *bit*
eye-3:SG:POSS sharp
'He is perceptive.'

Mouth

(66) a. *yot dog*
light mouth
'approachable'

b. *dog-e* *yot*
mouth-3:SG:POSS light
'He is sociable.'

(67) *en o-wac-o dog-e*
3:SG 3:SG-spoke-T mouth-3:SG:POSS
'He gave his opinion.' (he spoke his mouth)

Body

(68) a. *yot kom*
light body
'health'

b. *kom-e* *yot*
body-3:SG:POSS light
'He is healthy.'

(69) a. *lit kom*
pain body
'sickness'

b. *kom-e lit*
body-3:SG:POSS pain
'He is sick.'

(70) a. *wac kom*
lazy body
'laziness'

b. *kom-e wac*
body-3:SG:POSS lazy
'He is lazy.'

(71) a. *gum kom*
luck body
'luck'

b. *kom-e gum*
body-3:SG:POSS luck
'He is lucky.'

(72) a. *kec kom*
bitter body
'bad luck'

b. *kom-e kec*
body-3:SG:POSS bitter
'He is unlucky'

(73) *dako cano kom-e pek*
woman that:one body-3:SG:POSS heavy/slow
'That woman is pregnant.'

Hand

(74) a. *yot cing*
soft hand
'skill'

b. *cing-e* *yot*
hand-3:SG:POSS soft
'He is skillful in crafts.'

(75) a. *mit* *cing*
quick hand
'skill'

b. *cing-e* *mite*
hand-3:SG:POSS quick
'He is skillful at a specific craft.'

(76) a. *bit* *cing*
sharp hand
'accuracy/skill'

b. *cing-e* *bit*
hand-3s:SG:POSS sharp
'He is accurate in spear throwing.'

Ear

(77) *it-e* *pe*
ear-3:SG:POSS NEG
'He is deaf.' (Literally: 'his ear is not')

9. Conclusion

A morphological distinction is made in Acholi for alienable versus inalienable possession. Compounds, like inalienables, have zero linking morphology. However, compounds express relations that are best described as classification rather than part-whole, in that they express a generic-specific relation: types are specified. They do not employ pronominal morphology; rather they are restricted to two-noun constructions. Alienable relations are marked with a genitive construction, and they apply to animate possessors.

An inalienable/alienable distinction is maintained with pronominal possessors. However, the morphological distinction is not apparent in two phonetic environments (with pronominal possessors): when the possessed noun is vowel-final, and with plural pronominals. Recall that the plural possessive suffixes, unlike the singular, are consonant initial. Under either of these conditions the marked contrast is lost. Some speakers avoid alienable possessive construction (for example, 'my tail' for a piece of an animal); instead they name the animal or use a verbal construction (for example, 'I cut off the tail').

Acholi, unlike other languages described in Chappell – McGregor (1995), allows the body part to be an agent. Body parts are also experiencers, and states are attributed to them (as in 'my liver is sick'). Thus the body part and the whole are one.

Like many other languages, Acholi has extended the names of body parts to objects and to spatial relations, often using the general locative prefix *i*. So from a construction <at – body part – object> a label for the space removed from the object has developed. For example, *i-nying ot* 'at forehead house' is interpreted as the space in front, not part of the house.

Notes

1. Examples for the paper have been taken from a corpus of Acholi data collected over a period of 8 years, with the assistance of five native speakers. I am particularly grateful to Simon Olwach, Charles Okello, and Bosco Nyeko. Abbreviations used in the paper are as follows: COM = completive; DET = determiner; NEG = negative; PFV = perfective; POSS = possessive; PL = plural; PRO = pronoun; REL = relative marker; SG = singular; T = transitive suffix.
2. A description of Acholi is given by Crazzolara (1955).
3. The convential orthography for Acholi has been followed in the paper, except *ng* has been used in place of the velar nasal symbol. Tone is not relevant to any of the distinctions discussed in this paper and so it has not been marked.
4. Speakers also allow *otta* in free variation with *odda*. Note that Okoth-Okombo (1982) in describing the genitive construction in Dholuo, spoken in Kenya, lists a regular alternation in word final consonants for specific nouns in the genitive case, However, the same alternations are not found to apply in forming possessives in Acholi.

References

Bavin, Edith L.
1982 "Aspects of morphological and syntactic divergence in Lango and Acholi", *Studies in African Linguistics* 13 (3): 231-248.
Chappell, Hilary – William McGregor
1989 "Alienability, inalienability and nominal classification", *Proceedings of the Berkeley Linguistic Society* 15: 24-36.

1995 "Prologomena to a theory of inalienability", [this volume.]
Crazzolara, J. Pasquale
1955 *A study of the Acholi language*. London: Oxford University Press.
Dimmendaal, Gerrit J.
1983 "The two morphological verb classes in Nilotic", in: Rainer Voßen – Marianne Bechhaus-Gerst (eds.), 269-309.
Greenberg, Joseph H.
1963 *The languages of Africa*. The Hague: Mouton & Co.
Heine, Bernd – Ulrike Claudi – Friederike Hünnemeyer
1991 "From cognition to grammar", in: Elisabeth Traugott – Bernd Heine (eds.) Vol 1, 149-187.
Okot, p'Bitek
1971 *Religion of the central Luo*. Nairobi: East Africa Literature Bureau.
Okoth-Okombo, Duncan
1982 *Dholuo morphophonemics in a generative framework*. Berlin: Dietrich Reimer Verlag.
Svorou, S. Roula
1986 "On the evolutionary paths of locative expressions", *Proceedings of the Berkeley Linguistics Society* 12: 515-527.
Traugott, Elisabeth – Bernd Heine (eds.)
1991 *Approaches to grammaticalization.* 2 vols. Amsterdam/Philadelphia: John Benjamins.
Tucker, Archibald N. – Margaret Bryan
1966 *Linguistic analysis: the non-Bantu languages of North-Eastern Africa*. London: Oxford University Press.
Voegelin , Charles Frederick – Florence Maria Voegelin
1977 *Classification and index of the world's languages*. New York: Elsevier.
Voßen, Rainer – Marianne Bechhaus-Gerst (eds.)
1983 *Nilotic Studies (2): Proceedings of the International Symposium on Languages and History of the Nilotic peoples*. Berlin: Dietrich Reimer Verlag [Kölner Beiträge zur Afrikanistik 10].

The syntax of body parts in Haya

Larry M. Hyman

1. Introduction[1]

In Hyman – Alford – Akpati (1970), an attempt was made to study in Igbo, a Kwa language spoken in Eastern Nigeria, the kinds of syntactic constructions that seemed to be made possible by the presence of a body part. The well-known French examples in (1) and (2) were cited as illustrating two prevalent tendencies in languages from all parts of the world.

(1) *Il a levé le bras*
he has raised the arm
'He raised his arm' (literally: he raised the arm)

(2) *il m' a cassé le bras*
he me has broken the arm
'He broke my arm.' (literally: he broke to me the arm)

In (1) we see that a body part, arm, occurs without an overt possessive marker, a situation we might compare with the English sentence in (3), taken from Fillmore (1968).

(3) *Mary pinched John on the nose.* (that is, on his nose)

In (2) we see that instead of an overt possessive marker, we find an indirect or dative clitic pronoun *me* (*m'*). The archaic constructions in (4),

(4) a. *Look me in the eye!*

b. *Don't look a gift horse in the mouth!*

where 'eye' and 'gift horse' are datives, were said to be remnants of what was once a more widespread construction in English.

In the 1970 study it was demonstrated that Igbo, is characterised by the same phenomena as in (1) and (2), which I shall henceforth refer to as *possessor deletion* and *possessor promotion*, respectively, as seen in (5) and (6).[2]

(5) *ó sàrà ákà*
he washed hand
'He washed his hand(s).'

(6) *ó sàrà m ákà*
he washed me hand
'He washed my hands.' [cf. *ákà ḿ* 'my hand(s)']

Since that time I have surveyed these constructions in a number of other West African, and more recently, Bantu languages and have concluded that there is in fact widespread agreement on the way in which body parts are syntacticised. My aim in the present paper is to examine in considerable detail the syntax of body parts in Haya, a Bantu language spoken in Tanzania, especially as concerns the phenomena of possessor deletion and possessor promotion. I shall focus both on the grammatical properties of these constructions as well as the semantic considerations that make speakers likely to use them. Since I have found similar phenomena in Luganda, Kinyarwanda, Sesotho, Sukuma, Ciluba, and Punu, languages which are representative of the vast Bantu region of Sub-Saharan Africa, it is likely that much of what I shall have to say about Haya will be applicable to Bantu as a whole. In discussing the Haya data it will be instructive to include occasional parallels from certain European languages, e.g. French and Italian. I shall conclude this study with a brief reconsideration of the notion of "inalienable possession" and point out some of the areas in which it might be fruitful to conduct further investigations.

2. Possessor deletion

Turning now to the syntactic treatment of body parts in Haya, we see in (7) that possessor deletion occurs upon identity with the subject.

(7) *n-k-óógy' émikôno*
I-P3-wash hands (P3 = distant past)
'I washed my hands.' (Literally: 'I washed hands')

The corresponding sentence in (8), where the possessive pronoun is explicitly expressed, is unacceptable, since it would imply that the 'hands' were not part of my body, but rather were detached items.

(8) *?n-k-óógy' émikóno yange*
I-P3-wash hands my
'I washed my (detached) hands.'

In (9) we see that the sentence corresponding to (8) is perfectly fine, if the object is not a body part.

(9) *n-k-óógy' émótoká yange*
I-P3-wash car my
'I washed my car.'

On the other hand, if we attempt to delete the possessive pronoun of 'car', as in (10),

(10) *n-k-óógy' émótoka*
I-P3-wash car
'I washed the/a car.'

it is not understood that the subject is also the possessor, but rather the possessor is not specified, or perhaps there is no possessor.

In (11), we see that an object can also control possessor deletion, this time occurring within a locative phrase.

(11) *n-ka-téél' ómwáán' áha-mútwe*
I-P3-hit child on-head
'I hit the child on the head.'

In the English gloss we observe that English also can allow possessor deletion in such cases. Let us assume for the moment that prepositonal phrases such as 'stabbed in the heart', 'pain in the neck' and 'to go by foot' involve either a greater recoverability factor, in terms of identifying the possessor, or even, that the possessor is indeterminate or irrelevant. I shall come back to possessor deletion in the lengthy discussion of possessor promotion below, but let us first take note of the fact that deletability is in part conditioned by the nature of the verb. Thus, as seen in (12),

(12) *n-ka-bón' émikôno*
I-P3-see hands
'I saw (the) hands.'

it is not clear whether the hands were mine, someone else's, or no-one's. I shall provide an explanation for the disparity between (12) and (7) after considering possessor promotion.

3. Possessor promotion

3.1. Promotion to object

Possessor promotion in Haya, as well as every other Bantu language I have studied, transforms the possessor into a direct object, if the verb is transitive. An example is seen in (13).

(13) *n-ka-hénd' ómwáán' ómukôno*
I-P3-break child arm
'I broke the child's arm.' (Literally: 'I broke the child the arm')

Failure to promote the possessor invariably implies that the body part has previously been severed from the whole. Thus, in (14),

(14) ?*n-ka-hénd' ómukono gw' ómwáana*
I-P3-break arm of child
'I broke the (detached) arm of the child.'

the normal implication is that a child came in with an arm s/he had found, and I broke it. With an ordinary object, of course, promotion is impossible. Thus, 'breaking a child's stick' must be realised as in (15), not as in (16).

(15) *n-ka-hénd' énkoni y' ómwáana*
I-P3-break stick of child
'I broke the stick of the child.'

(16) **n-ka-hénd' ómwáán' énkoni*
I-P3-break child stick
(Literally: 'I broke the child the stick')

3.1.1. Grammatical characterisation

Before entering into the various semantic and pragmatic factors which make promotion possible, let us consider the syntax of sentences such as (13). The notion has already been advanced that the promoted possessor is a direct object. In Haya, as in Bantu in general (cf. Duranti – Byarushengo 1977), the three criteria in (17) are frequently employed to establish whether any one of a number of noun phrases following the verbal complex is a direct object, with multiple DO's being possible:

(17) Criteria for direct object (DO) status:
a. accessibility to immediate post-verbal position
b. accessibility to subject of passive
c. accessibility to object prefix position (as a pronoun)

The last qualification "as a pronoun" is added, since object markers (OM's) can arise as the result of pronominalisation or agreement of different types (see Morolong – Hyman 1977 for a discussion and critique of these criteria). Let us apply (17a-c) to the noun 'child' in (13).

As already seen in (13), 'child' can occupy the position immediately following the verbal complex. It therefore satisfies the first criterion for direct object status. In (18), it is furthermore seen that it can occur as the subject of a corresponding passive sentence, where the agent is not expressed for reasons of simplicity only.

(18) *omwáán' a-ka-hénd-w' ómukôno*
child he-P3-broke-PASS arm
'The child's arm was broken.' (Literally: 'The child was broken the arm')

The passive marker *-w-* occurs as an extension on the verb just preceding the final vowel *-a* (which is elided and marked by ' in the above example). In (19) we observe that a corresponding pronominal OM is also possible replacing the noun.

(19) *n-ka-mu-hénd' ómukôno*
I-P3-him-break arm
'I broke his arm' (Literally: 'I broke him arm')

Thus, 'child' passes all of the tests, and we establish it as the DO of 'break' in (13).

This leads us to ask what the status is of 'arm' in (13). Applying the same three criteria we find, first in (20),

(20) **n-ka-hénd' ómukón' ómwáana*
I-P3-break arm child
'I broke the child's arm.'

that it cannot occur directly after the verbal complex. In (21),

(21) **omukóno gú-ka-hénd-w' ómwáana*
arm it-P3-break-PASS child
'The child's arm was broken.' (Literally: 'The arm was broken the child')

we see that it cannot occur as the subject of the corresponding passive. Finally, in (22),

(22) **n-ka-gu-hénd-w' ómwáana*
I-P3-it-break-PASS child
'I broke that of the child.' (Literally: 'I broke it the child')

'arm' is seen not to be pronominalisable into the OM slot. We thus conclude that 'arm' in (13) is not a DO, even though it seems to be the accusative complement of the verb 'break'.

At this point it may be worthwhile to consider briefly the nature of verbal extensions in Haya (and in Bantu as a whole). Certain extensions such as the *-il-/-el-* applied suffix, or the *-is-/-es-* or *-i-* causative/instrumental suffix have the effect of introducing or making possible additional verbal arguments. Thus, we have the simple sentence in (23), as compared to the corresponding extended verb in (24).

(23) *n-ka-cumb' ébitooke*
I-P3-cook bananas
'I cooked (the) bananas.'

(24) *n-ka-cumb-il' ómwáán' ébitooke*
I-P3-cook-APP child bananas
'I cooked the child bananas.'

The benefactive noun 'child' is possible in (24) only because of the presence of the applicative extension, as seen by the ungrammaticalilty of (25).

(25) **n-ka-cumb' ómwáán' ébitooke*
I-P3-cook child bananas
'I cooked child bananas.'

One might suggest that 'arm' is not accorded the DO properties in (13) because the verb 'break' does not have such an extension. Note, first, in (26),

(26) ?*n-ka-hénd-el' ómwáán' ómukôno*
I-P3-break-APP child arm
'I broke the/my arm for the child.'

that when the applied extension *-el-* is added to (13), two readings are possible, neither being equivalent to (13). First, the arm can have already been severed from a body, in which case there is no possessor or the possessor is indeterminate or irrelevant. Second, the unmarked noun 'arm' can be construed as belonging to the subject, 'me', with the possessive 'my' having undergone deletion of the type we have already discussed. It is impossible to interpret (26) as meaning 'the child's arm', since the subject has, as it were, prior claim in controlling possessor deletion.

A second reason why the non-object status of 'arm' cannot be attributed to the lack of an extension is that there is a limited set of verbs in Haya, such as *-há-* 'give', which allow two basic DO's without verb marking (cf. Duranti – Byarushengo 1977). In (27), for instance, both word orders are possible, as opposed to the unacceptable word order in (20).

(27) a. *n-ka-h' ómwáán' ébitooke*
I-P3-give child bananas
'I gave the child bananas.'

b. *n-ka-h' ébitook' ómwáana*
I-P3-give bananas child
'I gave the child bananas.'

In (28) we see that either noun can occur as the subject of the corresponding passive.

(28) a. *omwáán' a-ka-háá-bw' ébitooke*
child he-P3-give-PASS bananas
'The child was given the bananas.'

b. *ebitooke bí-ka-háá-bw' ómwáana*
bananas they-P3-give-PASS child
'The bananas were given (to) the child.'

Finally, in (29) it is seen that either noun can be pronominalised in the OM position.

(29) a. *n-ka-mú-h' ébitooke*
I-P3-him-give bananas
'I gave him (the) bananas.'

b. *n-ka-bí-h' ómwáana*
I-P3-them-give child
'I gave them (to) the child.'

In fact, Haya is one of the Bantu languages which allows more than one pronoun in the OM position, as seen in (30).

(30) *n-ka-bi-mú-h-a*
I-P3-them-him-give
'I gave them to him.'

It thus appears, as argued by Duranti – Byarushengo (1977), that more than one DO can occur in Haya. We therefore are faced with the task of explaining why 'I broke child arm' behaves differently, syntactically, from 'I gave child bananas'. Since Haya can in principle support two post-verbal DO's, the non-object status of the body part must be due to the nature of possessor promotion, which it is now instructive to approach more systematically.

3.1.2. Semantic characterisation

Given an input structure something like that in (31),

(31) SUBJECT VERBAL COMPLEX X OF Y

we are concerned with the conditions under which the structure in (32) evolves:

(32) SUBJECT VERBAL COMPLEX Y - X

What I would like to suggest at this point is that the obtaining of structure (32) instead of structure (31) is brought about *in situations in which the possessor is at the same time the experiencer of the action of the verb.* The object which is created by possessor promotion experiences or, in other words, is *affected* by the action of the verb (cf. Diffloth 1974 for French; Garcia 1975 for Spanish). The evidence stems from the observation that a possessor object will be created only if certain conditions are met, as outlined in (33).

(33) Creation of a possessor object depends on:
a. the nature of the possessed noun
b. the nature of the verb
c. the nature of the possessor

We have already seen the relevance of (33a). In order for possessor promotion to take place, a certain relationship must exist between the possessed and the possessor. Body parts are ideal, since when an action is performed on them, the possessor is necessarily also affected. When we are dealing with possessed nouns other than body parts, this connection is not as strong. The important factor is the extent to which by affecting the possessed noun we are affecting the possessor as a whole. In (34) and (35) we observe that it is possible to have both constructions when the possessed noun is an item of clothing.

(34) *n-ka-teemul' éshaati y' ómwáana*
I-P3-tear shirt of child
'I tore the shirt of the child.'

(35) *n-ka-teemul' ómwáán' éshaati*
I-P3-tear child shirt
'I tore the shirt of(f) the child.' (Literally: 'I tore the child the shirt')

In (34) it is understood either that the shirt was not on the child when I tore it, or that I am not particularly focusing on the effect my shirt-tearing has on the child as a whole. In (35), on the other hand, the only situation in which this construction would be acceptable is if the child is wearing the shirt and the child is therefore necessarily bodily affected. Where the possessor's body is not physically affected, Haya (but not all other languages) finds it a little more difficult to accept possessor promotion. Thus, the sentence we saw in (16) is not acceptable, even if we try to focus on the effect of stick-breaking on the child, rather than on the effect of the breaking on the stick. In summary, with other things being equal, the structure in (31) will be found when the speaker is concerned with the effect of the action on X (the possession); the structure in (32) will be found when the speaker is concerned with the effect of the action on Y (the person or animate entity).

Turning to (33b), if the notion of an experiencer or affected object is correct, we should expect action verbs to provide an environment more conducive to possessor promotion than verbs of state, for example, sensory verbs. Thus, with the verb 'like', the sentence in (36) is much more acceptable than the sentence with the verb 'break' in (14).

(36) *ñ-ñend' ámáisho g' ómwáana*
I-like eyes of child
'I like the eyes of the child'

If we attempt to promote the possessor, as in (37),

(37) *ñ-ñend' ómwáán' ámáisho*
I-like child eyes
'I like the child's *eyes*.' (That is, what I like about the child is the eyes)

the nuance which is added is that I *only* like the child's eyes, consistent with the tendency for asserted information to occur late or at the end of the utterance in Haya. If we compare these sentences with sentences involving the verb 'see', as in (38) and (39), a similar nuance is obtained.

(38) *n-ka-bón' ómukono gw' ómwáana*
I-P3-see arm of child
'I saw the arm of the child.'

(39) *n-ka-bón' ómwáán' ómukôno*
I-P3-see child arm
'I saw the child's *arm*.'

Sentence (38) is neutral, and would in any case have to be used if the child were not involved in person (either because I was viewing his arm in a photograph, or because the arm had been detached). Sentence (39) implies that what I saw of the child was only his *arm*, possibly also that I wasn't supposed to see it. What is important in obtaining (39) instead of (38) is that the child be *involved* in the action, as opposed to being incidental or totally unimportant. With verbs such as 'to like', and 'to see' an object body part can be readily perceived without implying the involvement or affectedness of the whole. Actions such as washing and breaking are necessarily different. In Haya, when you break someone's arm, you break that person. In fact, in context, a person can utter 'he has broken me', probably pointing to one's arm. Thus, when a body part changes its state (for example, becomes washed or broken), we can expect this to affect the person more than when that body part is merely perceived (for example, liked or seen).

Turning, finally, to (33c), the nature of the possessor is of considerable importance also to determining whether possessor promotion will obtain. We have found in Haya that first and second person possessors are much more prone to promotion than third person (human) possessors. Thus, (40) is considerably worse than (41).

(40) ??*bá-ka-hénd' ómukóno gwange*
they-P3-break arm my
'They broke my arm.'

(41) ?*bá-ka-hénd' ómukonó gwe*
they-P3-break arm his
'They broke his arm.'

Both, of course, are expected to undergo possessor promotion, as seen in (42) and (43), although (41) seems more tolerable than (40).

(42) *bá-ka-m-pénd' ómukôno*
they-P3-me-break arm
'They broke my arm.' (Literally: 'They broke me arm')

(43) *bá-ka-mu-hénd' ómukôno*
they-P3-him-break arm
'They broke his arm.' (Literally: 'They broke him arm')

This hierarchy can be extended, as seen in (44).

(44) 1st pers. > 2nd pers. > 3rd pers. human > 3rd pers. animal > 3rd pers. inanimate

What is involved here is a personal hierarchy identical to that proposed in another context for another Bantu language, Shona (Hawkinson – Hyman 1974). Highest in the hierarchy are those persons closest to the discourse (the speaker and hearer), followed by humans, then animals, then inanimates. Call it a natural person hierarchy or a hierarchy of egocentricity, efficacy, or empathy, but the same fact remains, namely that persons higher in (44) will be more susceptible to possessor promotion. This does not mean that inanimates cannot be promoted. On the contrary, Voeltz (1976) has shown that the same kind of possessor promotion must be extended in Sesotho, another Bantu language, in order to include all part-whole relationships. A corresponding example to those cited by Voeltz is given from Haya in (45).

(45) *n-ka-hénd' ómuty' éítáagi*
I-P3-break tree branch
'I broke the branch of(f) the tree.' (Literally: 'I broke the tree the branch')

By breaking the branch, I have necessarily affected the tree, which then undergoes possessor promotion. If promotion does not occur, as in (46),

(46) *n-ka-hénd' éítáagi ly' ómúti*
I-P3-break branch of tree
'I broke the branch of the tree/the tree-branch.'

this means that the branch of the tree (or the 'tree-branch') has already been severed, and I later snapped it in two. Since branches more frequently occur severed from trees than do arms occur severed from human or animal bodies, (46) should occur with greater frequency, and make greater sense, than (14), where I have broken the previously severed arm of the child. Stated perhaps in reversed terms, since speaker/hearers are less concerned with general welfare of trees, as opposed to human beings, we should expect (13) to be more mandatory than (45). In Haya, we *do* have the option of making the tree into an involved participant, if in fact it is affected by what happens to its parts. Thus, we can conclude that any possessor can potentially undergo promotion, provided that a part-whole relationship obtains, and that by affecting the part, one is also affecting the whole. The likelihood of affecting the whole depends of course also on the nature of the verb, as we have seen.

Final support for the hypothesis that promotion applies to an affected or experiencer possessor is seen from comparisons such as (47) and (48).

(47) *n-ka-ly' ókugulu kw' émbûzi*
I-P3-eat leg of goat
'I ate the leg of goat.'

(48) *n-ka-ly' émbúzy' ókugulu*
I-P3-eat goat leg
'I ate the leg off the goat.' (Literally: 'I ate the goat the leg')

The normal sentence is (47), where possessor promotion fails to obtain because (a) the goat is normally dead before it is cooked and eaten and, therefore, cannot be an experiencer; and (b) the leg is normally severed off the goat prior to consumption (if not also prior to cooking). Thus it is highly unlikely that speaker/hearers will be concerned about the welfare of the goat at this point. In (48), however, I have treated the goat as an experiencer, and as can be expected, the only appropriate interpretation of this sentence is that I ate the leg *off* of the live goat. Put a lion in subject position, if you prefer!

Now compare the two sentences (49) and (50).

(49) *n-ka-bák' éíbaba ly' éñoñi*
I-P3-catch wing of bird
'I caught the wing of the bird/the bird wing.'

(50) *n-ka-bák' éñoñy' éíbaba*
I-P3-catch bird wing
'I caught the bird (by) the wing.'

In (49) the wing had previously been severed before I could do any catching. A bird could have lost a wing, or someone could have thrown a bird-wing in the air, which I caught. In (50) on the other hand, where the possessor is promoted, the wing must necessarily have had to be attached in order for the bird to be an experiencer. The result is that this sentence means 'I caught the bird by the wing'. In (51) and (52), a similar distinction is observed, accompanied by a possible slight difference in the meaning of the verb itself.

(51) *n-ka-kwáát' ókugulu kw' ómwáana*
I-P3-hold leg of child
'I held the leg of the child.'
'I touched the leg of the child'

(52) *n-ka-kwáát' ómwáán' ókugulu*
I-P3-hold child leg
'I caught/held the child (by) the leg.'

In (51) the verb *okukwáata* 'to hold' can mean, when possessor promotion does not occur, either that I held the child's leg or that I simply *touched* it (i.e. without having an effect on the whole body of the child). In (52), however, the child as a whole was necessarily held or caught. In (51) we are interested in the effect of holding on the leg; in (52) we are interested in the effect of holding on the child. Both are possible, depending on the roles assigned to the two participants on the basis of the reality of tlhe situation as seen through the eyes of the speaker.

3.2. Promotion to subject

In all of the cases of possessor promotion presented thus far, the possessor turns up as a DO. This is made possible by the transitivity of the verb in each sentence. If the verb is intransitive, on the other hand, the possessor is promoted into *subject* position. This is seen in (53) and (54).

(53) ??*omutwe gw' ómwáána ni-gu-sháash-a*
head of child PRES-it-ache
'The head of the child is aching.'

(54) *omwáána n-aa-sháásh' ómútwe*
child PRES-he-ache head
'The child has a headache.' (Literally: 'The child is aching the head')

Example (53) is unacceptable because the head is treated as having already been severed from the whole, which would make it difficult for it to be aching. Example (54), however, is much preferred, since when the head is aching, the child is in fact aching. A similar pair of sentences is observed in (55) and (56).

(55) ??*omukono gw' ómwáána ni-gu-núuk-a*
hand of child PRES-it-smell
'The hand of the child smells.'

(56) *omwáána n-aa-núúk' ómukôno*
child PRES-he-smell hand
'The hand of the child smells.' (Literally: 'The child smells [with respect to] the hand')

Again, (55) gives the impression of the hand not being part of the child's body. It does not seem as bad as (53), however, since it is conceivable for the child to have a severed hand which happens to smell (but not, of course, to ache, unless the severed hand has a soul of its own!). In (56) we see that possessor promotion to subject position has occurred, since it necessarily follows from the smelling [intransitive] of the hand that the child itself smells.

Final confirmation of this interpretation of intransitives with possessor promotion is seen in (57) and (58).

(57) *omukono gw' ómwáána gú-ka-gw-a*
arm of child it-P3-fall
'The arm of the child fell.'

(58) **omwáán' a-ka-gw' ómukôno*
child he-P3-fall arm
(Literally: 'The child fell [with respect to] the arm')

In (57) we imagine a situation where the child was holding up his arm, which then suddenly fell. The child in (57) is not an experiencer, but rather the *agent* of the action, unlike (54) and (56), where he is the experiencer. In (58), where I have attempted to treat the child as the experiencer of the arm-falling, the sentence is not acceptable. This

I interpret to mean that when the child's arm falls, the child does not himself fall. Thus, the action of the part does not affect or predict the same action with respect to the whole.[3]

4. A possible solution

It would appear from the preceding discussion that a possessor is promoted when it represents the experiencer of an action deriving from a part-whole relationship such as the one body parts normally take part in. The promotion is to *object* if the verb is transitive, or to *subject* if the verb is intransitive. We know from studies of grammatical relations in Haya and other Bantu languages that these functions, subject and DO, are exploited for the manifestation of important arguments of a verb, or as Hawkinson (1977) has suggested in her study of five Eastern Bantu languages, important participants in a discourse.[4] In fact, the notion of higher versus lower participants in an event or action may help to understand the sentence in (56), where 'child' is not a true experiencer, though his whole body/self is "affected" by a property of the lower participant (i.e. his hand). Whether a participant will be "high" enough to be syntacticised as a DO of a transitive verb or the subject of an intransitive verb depends crucially on three factors in Bantu, outlined in (59).

(59) Syntacticisation of a participant as a DO will depend on:
  a. case relations: ben > dat > acc > instr (incomplete)
  b. the personal hierarchy (see (44))
  c. reference: given versus new (and/or presupposed versus asserted); definite versus indefinite; specificity

In (59a) a benefactive semantic case will make a participant more likely to be ranked over a dative, accusative, and instrumental in acquiring DO properties (see (17) above). In addition, the participants higher in the personal hierarchy given in (44) will receive more DO properties than participants lower in the hierarchy. Finally, given/presupposed, definite, and specific referents may be accorded higher DO status than new/asserted, indefinite, and nonspecific referents. To the extent that the three factors outlined in (59) line up there will be no problem in determining which one or ones of several participants will qualify as having DO status. However, when there is conflict, Bantu languages move in a variety of non-arbitrary directions in order to contain the conflict. In transitive sentences we have seen the possessor promoted to DO, and the possession demoted to non-object status. Because the possessor is an experiencer, it occurs somewhere in the case hierarchy on a par with benefactives or datives, i.e. higher than accusatives or patients. Because it is normally human, or at least animate, it also qualifies by the animacy factor (that is, our personal hierarchy) for higher grammatical status than the accusative, which in the case of body parts is inanimate. Finally, we had at least one hint (in (37) and (39) above) that when the possessor is given information, i.e. discourse-old, it may also have a greater chance at being promoted to DO.[5] The promotion part of possessor promotion is thus well-motivated.

However, in order to explain the *de*motion part of possessor promotion it appears in Haya that the body part is not high enough in terms of the hierarchies in (59) to maintain its objecthood in the face of the rising possessor. I would like to propose that possessor promotion involves the change from a structure such as that in (60) to a structure such as that in (61) or in a model that does not involve actual derivations, non-promoted versus promoted possessor constructions differ as indicated in these two structures.

(60)

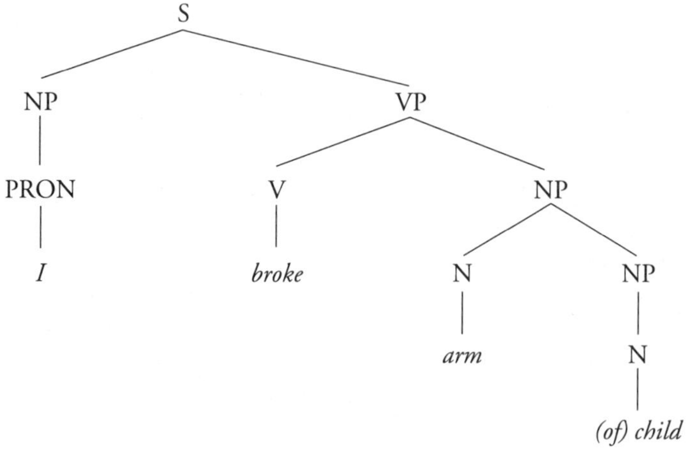

(61)

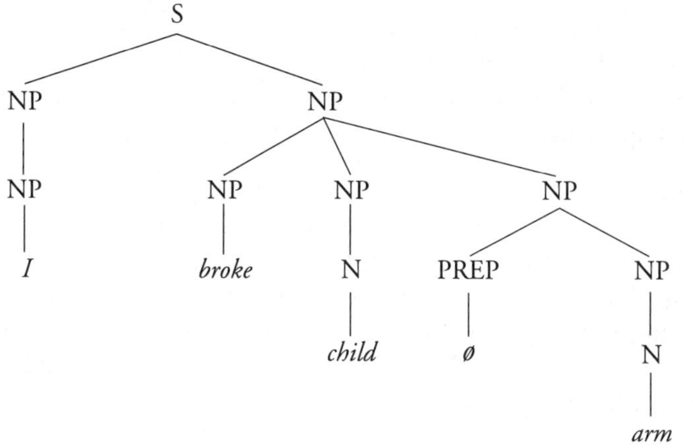

In (60) the possessor is expressed through a genitive NP, while in (61) it is represented as an NP within PP with a zero head, that is, as a "prepositionless oblique". The result is a kind of "switching" between the body part and the affected possessor, with the latter winding up a "chômeur" in relational grammar terms.[6] An argument for the prepositionless oblique hypothesis is the parallel that obtains between the non-object body part and the agent of the passive in Haya. As seen in (62),

(62) *omwáán' a-ka-téél-w' ómusháija*
child he-P3-hit-PASS man
'The child was hit by the man.'

there is no agentive 'by' preposition in the language. The agent noun 'man' in (62) is not an object, however, since it cannot be pronominalised into the OM position, as seen in (63).

(63) **òmwáán' a-ka-mu-téél-w-a*
child he-P3-him-hit-PASS
('The child was hit by him.')

Further support for the notion of a 0 preposition is seen from the fact that the demoted *subject* body part in sentences with intransitive verbs does not become an object. As seen in (64) and (65), the body parts in (54) and (56) cannot occur as corresponding OM pronouns.

(64) **omwáána n-aa-gu-sháash-a*
child PRES-he-it-ache
('The child's [head] is aching.') (Literally: the child is aching it)

(65) **omwáána n-aa-gu-núuk-a*
child PRES-he-it-ache
('The child's [hand] smell.') (Literally: 'The child smells [with respect to] it')

The reason why (64) and (65) are ungrammatical is that they treat the demoted body part as a DO, whereas the intransitive verbs 'ache' and 'smell' do not take an object. That is, possessor promotion involves the *inversion* of the roles of the (body) part and the affected possessor into pre-existing grammatical relations in Haya, rather than the creation of *new* grammatical slots. This not only explains the ungrammaticality of (64) and (65), but also that of (67).

(66) *embw' é-ka-gw' áha-mutwe gw' ómwáana*
dog it-P3-fall on-head of child
'The dog fell on the head of the child.'

(67) ?**embw' é-ka-gw' ómwáán' áha-mútwe*
dog it-P3-fall child on-head
(Literally: 'The dog fell the child on the head')

In (66) we have the intransitive verb 'fall' followed by a locative phrase 'on the head of the child'. In (67), however, where we have attempted to apply possessor promotion to the preceding sentence, the result is ungrammatical (or at least, extremely awkward). The reason is that 'child' in (67) would have to be treated as a DO, but the verb 'fall' is intransitive. Since possessor promotion does not create a new DO slot, but rather switches the participants into pre-existing grammatical slots, (67) is not acceptable.

We conclude, then, that demoted body parts (or tree parts) are in an oblique relation to the verb, are best conceptualised as having a ø preposition. Syntactically they behave no differently from other prepositional phrases which also fail the tests for direct objecthood given in (17) above. Our test sentences are therefore best translated as 'I broke the child with respect to the arm' and 'The child aches with respect to the head'. It remains only to be seen why possessor promotion should not be structure-creating as a process. The answer, as argued by Morolong – Hyman (1977), is that a prepositionless oblique (or "dead object" or "chômeur") will be created in many if not most Bantu languages if the promotion process which accompanies the demotion *does not receive the backing of the grammar*. Possessor promotion is motivated on purely semantic grounds and there is no overt grammatical marking that accompanies it. In the case of the verb 'to give', the grammar (in this case, the lexicon) allows two objects. In the case of 'to cook' with an applied extension allowing a second (benefactive) DO, the grammar *supports* what many see as the promotion of a benefactive to DO. Thus, it may be the case (a) that we must distinguish between semantically versus grammatically and semantically motivated promotions, and (b) that the former may not have the "strength" to create new grammatical relations (for more discussion, see Morolong – Hyman 1977, where a parallel is drawn between body part syntax and cases of promotion based solely on the humanness of a participant).

5. Parallels with other languages

What is striking about the Haya data is the extent to which it conforms with everything that is known about body part syntax in distant West Africa as well as in Europe and even in Mon Khmer (Diffloth 1974). Returning, for example, to the French examples in (1) and (2), we see the same possessor deletion and possessor promotion at work, although promotion is from a possessive to a dative or indirect object in French. Compare (1) with the French sentence in (68).

(68) *il s' est cassé le bras*
he REFL is broken the arm
'He$_i$ broke his$_i$ arm.' (Literally: 'He broke to himself the arm')

Example (68) differs significantly from (1) in that a reflexive pronoun is required. When we remove the reflexive in (68) or insert one in (1), unacceptable sentences result, as seen in (69) and (70).

(69) *?il a cassé le bras*
he has broken the arm
'He broke the (severed) arm.'

(70) **il s' est levé le bras*
he REFL is raised the arm
(Literally: 'He raised the arm to himself')

Langacker (1968) pointed out that a semantic difference is found between verbs like 'break' and verbs like 'raise'. With arm-breaking or hand-washing an action is performed onto a body part. With arm-raising and shoulder-shrugging the body part performs its own action rather than being acted upon. The difference is clearest with washing vs. raising. If I am washing my hands and someone asks me what I am doing, I can reply, "I am washing myself". If I raise my hand, and someone asks me what I have done, I cannot reply, "I have raised myself". The question concerns the extent to which the action performed by or to the body part can be attributed to the body as a whole. In other words, is the body (or its possessor) the agent *and* the experiencer (as in the case of washing and breaking) or solely the agent (as in the case of raising one's arm or shrugging one's shoulders). We saw this same distinction with the verb 'to fall' in (57) as opposed to the verbs 'to ache' and 'to smell' in (54) and (56).

Some languages that promote a possessor/experiencer extend the process to other kinds of possession and/or experiencers. In French, and even more so in Italian, the dative is frequently used to indicate the presence of a possessive experiencer, even when there is no part-whole relationship. Consider the sentences in (71) and (72).[7]

(71) *ho mangiato la tua bistecca*
I ate the your steak
'I ate your steak' [for example, the one that you bought]

(72) *ti ho mangiato la bistecca*
you I ate the steak
'I ate your steak on you.' [That is, what I have done to you is eat your steak]

In (71) we have the neutral possessive construction which would normally be expected. With a dative clitic pronoun in (72), however, the nuance is added that I have deprived you of the steak you were going to eat. Thus, eating the steak destined for you (as opposed to simply a steak that may have been in your possession in (71)) has caused you to become an "affected" object, represented by the dative in Italian. Syntacticising a participant as a dative in Romance languages may mean that it is a benefactive, a dative-recipient, or an experiencer. And the ability to make a possessor a dative may eventually be generalised to any situation in which the possessor is affected directly by the action and still later may be generalised to all possessive situations, as has happened comparing Nupe versus Gwari, for instance (see Hyman – Magaji 1970). Finally, before concluding, let us reconsider in light of the above the differences noted by Martine Mazaudon (see Hyman – Zimmer 1976) with respect to the following French causative sentences in (73) and (74).

(73) *j' ai fait nettoyer la cuisine par Jean*
I made clean the kitchen by John
'I had the kitchen cleaned by John.'

(74) *j' ai fait nettoyer la cuisine à Jean*
I made clean the kitchen to John
'I made John clean the kitchen.'

In (73) my primary intention was to do something to the *kitchen* (get it cleaned) and John plays only a secondary role – it could have been done by anyone, as far as I was concerned. In (74), however, since the dative preposition *à* is used, it is clear that my primary intention was to do something to *John*, and the unpleasant task that I chose for him was cleaning the kitchen. In (73) John is thus an agent, but in (74) he is also an experiencer. He undergoes the cleaning of the kitchen as well as bringing it about.

6. Summary and conclusion

To recapitulate, we have seen that possessor promotion can take place only if the possessor is affected by the action of the verb. We can generalise to possessor deletion and say that if this process is made possible by the recoverability of the deleted possessor, then in instances where the possessor is an experiencer, it will more easily delete under identity with a subject (or object) noun. Thus, 'I washed hands' is better than 'I saw hands', if both mean 'my hands'. When I wash the hands, I wash myself, but when I see the hands, I do not necessarily see myself. Because of such parallels, it is tempting to try to collapse the two phenomena into one statement. Perhaps we can say that possessor deletion is a kind of promotion under identity, where the promoted possessor is "absorbed" into the coreferential higher participant (for example, the subject or DO). Alternatively, perhaps we can say that possessor promotion is a kind of deletion, if we start with an "ethical" dative in Romance, extended presumably as an ethical DO in Haya, and delete the possessor upon identity with this new element. In this case, 'I broke child arm' would "come from" 'I broke child arm of child' with deletion (rather than promotion) of 'of child'. Whether such talk of collapsing the two into one can have any empirical consequences is not clear, and I shall therefore not discuss this issue any further.

In none of the foregoing have I made use of the notion of "inalienable possession". In Hyman – Alford – Akpati (1970) this term was used to cover both syntactic and morphological characteristics of body parts. Morphologically, many languages of the world do not permit a body part to be cited without an overt possessor. This phenomenon has generally been attributed to the fact that a body part does not occur without a possessor, that is, without a body, in nature, and this fact has been encoded into the language.[8] In the morphological case we are dealing with notions that are inherently "relational" as opposed to "free" (cf. Welmers 1973). That is, body parts combine with kinship terms and certain locatives (for example, 'top of') which can only be conceived *in relation to* something. Into this category might be put a few cultural items such as 'canoe', if everyone must have a canoe, or if no canoe exists without a possessor. In the syntactic case, body parts combine with part-whole relationships (cf. Voeltz 1976) frequently involving such alienable concepts as 'the branch of a tree' etc. Kinship terms as frequently as not do not parallel the *syntactic* characteristics of body parts. It is thus odd to say the French sentence in (75),

(75) *?je lui ai embrassé la soeur*
I him have kissed the sister
'I kissed his sister.'

even if in stretching one's imagination, one saw this action as somehow affecting 'him', the brother. In fact, not only is the class membership for morphological versus syntactic cases of body part marking different, but so are the hierarchies and implicational relationships. Thus, there are many languages which must morphologically mark kinship terms, but not body parts. There are no languages, apparently, which must mark kinship terms syntactically (as with possessor promotion), but do not do likewise with body parts. It seems that in a relational continuum, kinship terms would have to come higher than body parts, since it is harder to conceive of a 'mother' or 'father' except in relation to someone, while it is possible to conceive of a stray body part. On the other hand, body parts must come higher in the experiential (syntactic) hierarchy, since doing something to someone's body part will affect that person more directly (that is, more physically, if you will) that doing something to that person's kin. Finally, in this so-called experiential continuum, animates must of course come higher than inanimate possessors in part-whole relationships. Thus, in Italian, a dative experiencer is required for humans, but somewhat stranger for inanimates, as seen in the following sentences:

(76) ?*ho tagliato i capelli del bamb ino*
I cut the hair of:the child
'I cut the hair of the child.'

(77) *ho tagliato i capelli al bambin o*
I cut the hair to:the child
'I cut the hair off the child.'

(78) *ho tagliato il ramo dell' alber o*
I cut the branch of:the tree
'I cut the branch of the tree.'

(79) ?*ho tagliato il ramo all' alber o*
I cut the branch to:the tree
'I cut the branch off the tree.'

Because 'child' is high on the personal (here, experiential) hierarchy, (77) must be preferred over (76). The language must, therefore, treat 'child' as an affected object, put into the dative. But since 'tree' is much lower on this same hierarchy, Italian treats it somewhat differently, preferring the structure in (78) over that in (79). We should expect that different languages will cut the hierarchies in different places, but that there will be implicatonal relationships of the type: if X is treated as an affected object, so will X+1. This would seem to be a fruitful area for further research.

The final point I would like to briefly discuss is the difference between promotion to DO in Haya (and Bantu in general) versus promotion to IO (indirect or dative object) in Romance. We have seen that with the promotion of the possessor to DO status comes a concomitant *demotion* of the basic or accusative object. With the promotion of a possessor to an IO in Romance, the underlying ACC is allowed to remain DO to the verb. Romance thus seems to respect the grammatical relation of the basic object and chooses simply to "upgrade" the possessor to a DAT, which recall from (59a), is higher in the case hierarchy than an ACC. Perhaps because the affected possessor receives higher case status, it is not necessary to demote the basic DO. In Bantu, however, when the possessor is promoted, it assumes the grammatical relation of the basic DO, which then is demoted as a prepositionless oblique.

The reason for this difference, I would like to suggest, is that Romance marks grammatical relations on *nouns*, whereas Bantu marks grammatical relations on *verbs*. We have seen that Bantu allows multiple DO's (cf. Duranti – Byarushengo 1977). Since this is the case, the only way one DO can be seen as having higher status is if the verb is "oriented" toward one of them. This can be accomplished in one of two ways. In a few cases where a verb like 'to give' can take two lexical DO's, the verb is *lexically* oriented towards the semantic DAT (perhaps because it is normally animate). In other cases, the verb is grammatically oriented towards a DO by means of an extension (for example, the applied suffix, which, among other uses, orients a verb towards a benefactive or dative object). We have already seen in (26) that the affected possessor cannot be singled out by means of the applied suffix. This means, as was pointed out, that the promotion of the affected possessor to DO must take place without the overt backing of the grammar (in this case, the morphology). If Bantu were to treat the two post-verbal nouns in (13) 'I broke child arm' as DO's, the two would not be differentiated with respect to participant status, in which case, one might ask why promotion of the possessor should take place at all.[9] The only alternative left is to promote the affected possessor and to demote the possessed part. By so doing, the same basic grammatical relations are kept, but, as seen in (60) and (61), the participants have been reversed. Whether the distinction I have made between semantically-motivated promotions versus semantically and grammatically-motivated promotions holds up will have to be tested against data from other Bantu (and ultimately, non-Bantu) languages.

Notes

1. This paper was first published under the title "The syntax of body parts" in Ernest Rugwa Byarushengo – Alessandro Duranti – Larry M. Hyman (eds.), 99-117, and appears here in the same form, except for minor corrections and editorial changes. An earlier version of this paper was presented to the U.S.C. Historical Bantu Syntax seminar in the Fall of 1976, and at the University of California, Berkeley, on May 19, 1977. I wish to acknowledge the helpful comments I received at both of these meetings, and especially thank Ernest Byarushengo, Alessandro Duranti, Annie Hawkinson, and Karl Zimmer for making specific suggestions which I may or may not have been able to follow up. I am particularly indebted to Ernest Byarushengo, on whose insightful intuitions this paper is based.

2. In this study I do not assume that one sentence is literally derived from another sentence, but rather that both are "obtained" from a combination of syntactic, semantic and pragmatic considerations, a number of which are dealt with further below.
3. Notice as in (56) that while the property (the smell) of the hand is transferred to the whole body (that is, the body is *affected*), one cannot say that the child is a true experiencer (as also pointed out to me by Henry Thompson). I return to this below.
4. The theoretical framework on which Hawkinson's work is based is available in Garcia (1975). I am particularly endebted to Robert Kirsner for several discussions on this approach to syntax, which I hope I have not misinterpreted or misapplied.
5. Sentences such as (37) and (39) can be used either when the child is discourse-old or when enumerating qualities, for example, 'I like the child the eyes', but 'I like the woman the hair'.
6. The structure in (61) seems quite general in Bantu. In Kinyarwanda, however, Kimenyi (1976) reports that the demoted body part remains a second object, and hence a different analysis is required.
7. I would like to thank Francesco Antinucci, Alessandro Duranti, Paola Giunchi and Gian Maria Lojacono for their intuitions on such sentences (and others) in Italian on many occasions.
8. Inalienable and alienable types of possession can also be distinguished morphologically by requiring different possessive markers or different sets of possessive pronouns.
9. It is an open question whether two participants can be of equal status, or whether communication is such that speakers are forced to hierarchise. The situation in Kinyarwanda, where both the affected possessor and the possessed part are DO's (Kimenyi 1976), would seem to be a likely candidate in providing two participants of equal status.

References

Emmon Bach – Robert Harms (eds.)
1968 *Universals in linguistic theory*. New York: Holt, Rinehart and Winston.

Byarushengo, Ernest Rugwa – Alessandro Duranti – Larry M. Hyman (eds.)
1977 *Haya grammatical structure*. Southern California Occasional Papers in Linguistics. No. 6 Los Angeles: University of Southern California.

Diffloth, Gerard
1974 "Body moves in Semai and French", in: *Papers from the Tenth Regional Meeting of the Chicago Linguistic Society*, 128-138.

Duranti, Alessandro – Ernest Rugwa Byarushengo
1977 "On the notion of 'direct object", in: Ernest Byarushengo *et al* (eds.), 45-71.

Fillmore, Charles J.
1968 "The case for case", in: Emmon Bach – Robert Harms (eds.), 1-88.

Garcia, Erica C.
1975 *The role of theory in linguistic analysis: the Spanish pronoun system*. Amsterdam: North-Holland.

Hawkinson, Anne K.
1977 "Verbal agreement in Eastern Bantu", [Unpublished MS., University of California, Berkeley.]

Hawkinson, Anne K. – Larry M. Hyman
1974 "Hierarchies of natural topic in Shona", *Studies in African Linguistics* 5: 147-170.

Hyman, Larry M. – Danny Keith Alford – Elizabeth Akpati
1970 "Inalienable possession in Igbo", *Journal of West African Languages* 7: 85-101.

Hyman, Larry M. – Daniel J. Magaji
1970 *Essentials of Gwari grammar*. Ibadan, Nigeria: University of Ibadan Press.

Hyman, Larry M. – Karl E. Zimmer
1976 "Embedded topic in French", in: Charles N. Li (ed.), 189-211.

Hyman, Larry M. – Leon C. Jacobson – Russell G. Schuh (eds.) .
1976 *Papers in African linguistics in honor of Wm. E. Welmers*. (Studies in African linguistics 6).

Kimenyi, Alexandre
1976 *A relational grammar of Kinyarwanda*. Berkeley and Los Angeles: University of California Press.

Langacker, Ronald
1968 "Observations on French possessives", *Language* 44: 51-75.

Charles N. Li (ed.)
1976 *Subject and topic*. New York: Academic Press.

Morolong, Malillo – Larry M. Hyman
1977 "Animacy, objects, and clitics in Sesotho", *Studies in African Linguistics* 8: 199-218.

Voeltz, Erhard F.K.
1976 "Inalienable possession in Sotho", in: Larry M. Hyman – Leon C. Jacobson–Russell G. Schuh (eds.), 255-266.

Welmers, Wm. E.
1973 *African language structures*. Berkeley and Los Angeles: University of California Press.

Part VIII
Bibliography

Bibliography on Inalienability

Hilary Chappell and William McGregor

This bibliography is largely compiled from our reading and research in the area of inalienable possession with many additions sent in or provided by the contributing authors of the volume. The aim has been to provide an extensive listing of monographs and articles on the topic of inalienability rather than an exhaustive one, which may serve as a starting point for other researchers in this general area. We have included, therefore, references on morphological and syntactic strategies which are used to code (in)alienability such as adnominal apposition, genitive markers, noun classes, nominal classifiers, noun incorporation, possessor ascension, double subject constructions and dative constructions as well as references on lexical analyses, for example, on body part terminology. Descriptive and functional grammars which contain sections on the expression of possession have, in the main, not been listed. The precise page references for many specialist works of this type can be found in the individual contributions to this volume.

Abraham, Werner
1973 "The ethic dative in German", in: Ferenc Kiefer – Nicolas Ruwet (eds.), 1-19.
Aissen, Judith L.
1979 "Possessor ascension in Tzotzil", in: L. Martin (ed.), 89-108.
Alieva, Natalia
1980 "The possessive syntactical type and the Malay language", *Bijdragen tot de Taal-, Land-, en Volkenkunde van Nederlandsch-Indiè* 136: 417-425.
Allan, Edward J.
1975-1976 "Inalienable possession in four Ethiopian languages", *Afrika und Übersee* 59: 300-307.
Allen, Barbara J. – Donald G. Gardiner – Donna B. Frantz
1984 "Noun incorporation in Southern Tiwa", *International Journal of American Linguistics* 50: 292-311.
1990 "Verb agreement, possessor ascension, and multistratal representation in Southern Tiwa", in: Paul M. Postal – Brian D. Joseph (eds.), 321-382.
Allen, W. Sidney
1964 "Transitivity and possession", *Language* 40: 337-343.
1972 "Transitivity and possession", in: Fred W. Householder (ed.), 82-90.
Ameka, Felix
1995 "Body parts in Ewe grammar", [this volume.]
Andersen, Elaine S.
1978 "Lexical universals of body-part terminology", in: Joseph H. Greenberg (ed.), Vol. 3: 337-368.
Anderson, Lloyd B.
1974 "The part-whole squish, main vs. subsidiary predications, and why 'grammatical insertion' is like 'lexical insertion' ", *Papers from the Tenth Regional Meeting of the Chicago Linguistic Society* 10: 1-16.

Andrews, Barry J.
1987 "Article défini ou article possessif?", *Review of Applied Linguistics* 75: 37-102.
Asher, Robert E.
1968 "Existential, possessive, locative and copulative sentences in Malayalam", in: John W. M. Verhaar (ed.), Part 2: 88-111.
Austin, Peter
1982 "Transitivity and cognate objects in Australian languages", in: Paul J. Hopper – Sandra A.Thompson (eds.), 37-47.
Auwera, Johan van der – Louis Goossens (eds.)
1987 *Ins and outs of the predication.* Dordrecht: Foris.
Axelrod, Melissa
1990 "Incorporation in Koyukon Athabaskan", *International Journal of American Linguistics* 56: 179-195.
Bach, Emmon – Robert Harms (eds.)
1968 *Universals in linguistic theory.* New York: Holt, Rinehart & Winston.
Bahr, Donald M.
1986 "Pima-Papago *-ga*, 'alienability' ", *International Journal of American Linguistics* 52: 161-171.
Baker, Mark
1988 *Incorporation: A theory of grammatical function changing.* Chicago: University of Chicago Press.
Bally, Charles
1926 "L'expression des idées de sphère personnelle et de solidarité dans les langues indo-européennes", in: Franz Fankhauser – Jakob Jud (eds.), 68-78.
[1995] [Translated and reprinted in this volume.]
Bauer, Robert
1991 "Sino-Tibetan *tongue and *lick", *Linguistics of the Sino-Tibetan Area* 11: 144-165.
1992 "*Winkin', blinkin', and nod*: a study in historical-comparative semantics of Southeast Asian languages. *Linguistics of the Sino-Tibetan Area* 15: 151-184.
Bavin, Edith L.
1995 "Body parts in Acholi: Alienable and inalienable distinctions, and extended uses", [this volume.]
Bell, Sarah
1983 "Advances and ascensions in Cebuano", in: David Perlmutter (ed.), 143-218.
Benveniste, Emile
1950 "Actif et moyen dans le verbe", *Journal de Psychologie Normale et Pathologique* 43: 121-129.
1962 "Pour l'analyse des fonctions casuelles: Le génitif latin", *Lingua* 11: 10-18. [Also in: Emile Benveniste (ed.), Vol I: 140-150.]
Benveniste, Emile (ed.)
1966 *Problèmes de linguistique générale.* Paris: Gallimard.
1971 *Problems in general linguistics.* [Translated by Mary E. Meek.] Coral Gables: University of Miami Press.
Bird, Charles S.
1972 The syntax and semantics of possession in Bambara. (Mimeograph of the Indiana University Linguistics Club.) Bloomington: Linguistics, Indiana University.

Bibovic, Ljiljana
1976 "On the notion of body part instrument", *Folia Linguistica* 9: 311-324.
Birner, Betty J.
1988 "Possessives vs. indefinites: Pragmatic inference and determiner choice in English", *Papers in Pragmatics* 2: 136-146.
Blake, Barry
1984 "Problems for possessor ascension: some Australian examples", *Linguistics* 22: 437-453.
Blau, Joshua
1979 "Redundant pronominal suffixes denoting intrinsic possession", *The Journal of the Ancient Near Eastern Society of Columbia University* 11: 31-37.
Bodewitz, H. W.
1992 "Belly, pelvis, buttocks or cheeks? Vedic Kuksi (Dual)", *Indo-Iranian Journal* 35: 19-31.
Bonvillain, Nancy
1973 "Noun incorporation in Mohawk", in: Michael K. Foster (ed.).
1989a "Body, mind, and idea: Semantics of noun incorporation in Akwesasne Mohawk", *International Journal of American Linguistics* 55: 341-358.
1989b "Noun incorporation and metaphor: semantic process in Akwesasne Mohawk, *Anthropological Linguistics* 31: 173-194.
Booker, Karen
1981 "Incorporation in Muskogean", *Kansas Working Papers in Linguistics* 6: 55-69.
Bravmann, Meïr Moshe
1977 "The idea of 'possession' in linguistic expression", in Meïr Bravmann (ed.), 357-373.
Bravmann, Meïr Moshe (ed.)
1977 *Studies in Semitic Philology.* Leiden: E.J. Brill.
Brightman, Robert A.
1985 "The indefinite possessor prefix in Woods Cree", *International Journal of American Linguistics* 51: 353-356.
Brown, Cecil H.
n.d. Folk anatomy and language change. [Unpublished MS.]
1976 "General principles of human anatomical partonymy and speculations on the growth of partonymic nomenclature", *American Ethnologist* 3: 400-424.
Brugman, Claudia
1983 "The use of body part terms as locatives in Chalcatongo Mixtec", in: *Report No. 4 of the Survey of California and Other Indian Languages.* Berkeley: University of California, 235-290.
Brugman, Claudia – Monica Macaulay
1986 "Interacting semantic systems: Mixtec expressions of location", *Proceedings of the Twelfth Annual Meeting of the Berkeley Linguistics Society* 12: 315-327.
Burridge, Kate
1995 "Degenerate cases of body parts in Middle Dutch", [this volume.]
Burston, Jack L.
1981 "Body parts in French: The determination of identity", *International Review of Applied Linguistics* 19: 129-136.

Byarushengo, Ernest Rugwa – Alessandro Duranti – Larry M. Hyman (eds.)
1977 *Haya grammatical structure.* (Southern California Occasional Papers in Linguistics 6) Los Angeles: University of Southern California.
Chappell, Hilary
1986 "The passive of bodily effect in standard Chinese", *Studies in Language* 10: 271-296.
1995 "Inalienability and the personal domain in Mandarin Chinese discourse", [this volume.]
Chappell, Hilary – William McGregor
1989 "Alienability, inalienability and nominal classification", *Proceedings of the Fifteenth Annual Meeting of the Berkeley Linguistics Society* 15: 24-36.
1995 "Prolegomena to a theory of inalienability", [this volume.]
Chappell, Hilary – Sandra A. Thompson
1992 "Semantics and pragmatics of associative *de* in Mandarin Chinese", *Cahiers de Linguistique Asie Orientale* 21: 199-229.
Chevalier, Jean-Claude – Maurice Gross (eds.)
1976 *Méthodes en grammaire française.* Paris: Klincksieck.
Chu, Chauncey
1976 "Conceptual wholeness and the retained object", *Journal of Chinese Linguistics* 4.1: 14-23.
Clark, Eve V.
1978 "Locationals: Existential, locative, and possessive constructions", in: Joseph H. Greenberg (ed.), Vol. 4: 85-126.
Clark, Marybeth
1995 "Where do you feel? – Stative verbs and body part terms in mainland Southeast Asia", [this volume.]
Clasen, Bernd
1981 "Inhärenz und Etablierung", (Arbeiten des Kölner Universalien Projekts, 41.) Köln: Institut für Sprachwissenschaft.
Claudi, Ulrike – Bernd Heine
1986 "On the metaphorical base of grammar", *Studies in Language* 10: 297-335.
1989 "On the nominal morphology of 'alienability' in some African languages", in: Paul Newman – Robert D. Botne (eds.), 3-19.
Claudi, Ulrike – Fritz Serzisko
1985 "Possession in Dizi: Inalienable or not?", *Journal of African Languages and Linguistics* 7: 131-154.
Crain, C.
1979 "Advancement and ascension to direct object in Chamorro", in: P. Hubbard – Peter Tiersma (eds.).
Cristea, Theodora
1974 "Remarques sur le datif possessif en roumain et en français", *Bulletin de la Société Roumaine de Linguistique Romane* 10: 5-14.
Croft, William
1985 "Indirect object lowering", *Proceedings of the Eleventh Annual Meeting of the Berkeley Linguistics Society* 11: 39-51.

Crowley, Terry
1991 "Linked nouns in Paamese: grammatical iconicity in part-whole constructions", *Language and linguistics in Melanesia* 22.
1995 "Inalienable possession in Paamese grammar", [this volume.]
Cruse, David A.
1979 "On the transitivity of the part-whole relation", *Journal of Linguistics* 15: 29-38.
Davies, William D.
1981 "Possessor ascension in Choctaw", *Proceedings of the Eleventh Annual Meeting of the North East Linguistic Society* 11: 38-57.
1984 "Inalienable possession and Choctaw referential coding", *International Journal of American Linguistics* 50: 384-402.
Demonte, Violeta
1988 "El 'articulo en lugar del posesivo' y el control de los sintagmas nominales" [The 'article in place of the possessive' and the control of nominal syntagms], *Nueva Revista de Filologia Hispanica* 36: 89-108.
Diem, Werner
1986 "Alienable und inalienable Possession im Semitischen", *Zeitschrift der deutschen Morgenländischen Gesellschaft* 136: 227-291.
1987 "Alienable und inalienable Relation im Semitischen", in: Hermann Jungraithmayr – Walter W. Müller (eds.), 497-507.
Diffloth, Gérard
1974 "Body moves in Semai and in French", in: *Papers from the Tenth Regional Chicago Linguistic Society*, 128-138.
Doerfer, Gerhard
1984 "Die Körperteilbezeichnungen des Kili (Ein Beitrag zur Frage der dialektologischen Klassifikation)", *Nyelvtudomanyi Kozlemenyek* 86: 238-246.
Downing, Pamela – Michael Noonan (eds.)
in press *Word order in discourse*. Amsterdam: John Benjamins.
Egerod, Søren
1985 "To have and have not – the category of possession in East Asian languages and the role it plays in the syntactic core", [Unpublished MS.] Paper presented at the 18th International Conference on Sino-Tibetan Languages and Linguistics, Bangkok.
Elbert Samuel H.
1965/6 "The 127 Renellese possessives", *Acta Linguistica Hafniensa* 9: 16-24.
Engels, Ulrich – Paul Grebe – Heinz Rupp (eds.)
1969 *Festschrift für Hugo Moser*. Düsseldorf: Schwann.
Evans, Nicholas
1995 "The syntax and semantics of body part incorporation in Mayali", [this volume.]
Evens, Martha Walton (ed.)
1988 *Relational models of the lexicon*. Cambridge: Cambridge University Press.
Fankhauser, Franz – Jakob Jud (eds.)
1926 *Festschrift Louis Gauchat*. Aarau: H.R. Sauerländer.
Fillmore, Charles
1968 "The case for case", in: Emmon Bach – Robert Harms (eds.), 1-88.

Fleischmann, Eberhard
1983 "Der semantische Lokativ. Eine Untersuchung an Hand russischer Wortfügungen", *Zeitschrift für Slavistik* 28: 1-8.
Foster, Michael K. (ed.)
1973 *Papers in Linguistics from the 1972 Conference on Iroquoian Research.* (Mercury Series, Ethnology Division, No. 10.) Ottawa: National Museum of Manitoba.
Fox, Barbara
1981 "Body part syntax: towards a universal characterisation", *Studies in Language* 5: 323-342.
Freeze, Ray A.
1976 "Possession in K'ekchi (Maya)", *International Journal of American Linguistics* 42: 113-125.
Frei, Henri
1939 "Sylvie est jolie des yeux", in: Albert Séchehaye et al. (eds.), 185-192.
Gaatone, David
1991 "A syntactic play on words in French", *Journal of French Language Studies* 1: 45-53.
Garr, W. Randall
1992 "The grammar and interpretation of Exod 6:3", *Journal of Biblical Literature* 111: 385-408.
Golovaceva, A.V.
1981 "Semantiko-sintaksiceskij sposb vyrazenija kategorii posessivnosti", in: Vjac. Vs. Ivanov – T. M. Sudnik – T.V. Civ'jan (eds.), 8-13.
Golovaceva, A.V. – Vjac. Vs. Ivanov – T. N. Molosnaja – T. M. Nikolaeva – T.N. Svesnikova – E. A. Xelimskij
1981 "Pritjazatel'nost' (posessivnost') i sposoby ee vryazenija", in: Vjac. Vs. Ivanov – T. M. Sudnik – T.V. Civ'jan (eds.), 1-6.
Gonzalez, Omar
1988 " 'Yo soy arbol', o la identidad cuerpo-naturaleza" ['I am a tree', or the body-nature identity], *Glotta* 3: 8-13.
Greenberg, Joseph H. (ed.)
1978 *Universals of human language.* Vols. 3-4. Stanford: Stanford University Press.
Gregersen, Edgar A.
1976 "A note on the Manam language of Papua New Guinea", *Anthropological Linguistics* 18: 95-111.
Guéron, Jacqueline
1985 "Inalienable possession, pro-inclusion and lexical chains", in: Jacqueline Guéron – Hans-Georg Obenauer – Jean-Yves Pollock (eds.), 43-86.
Guéron, Jacqueline – Hans-Georg Obenauer – Jean-Yves Pollock (eds.)
1985 *Grammatical representation.* Dordrecht: Foris
Haas, Mary
1941 "Noun incorporation in the Muskogean languages", *Language* 17: 311-315.
Hahn, E.
1963 "Partitive apposition", in: Christine Mohrmann – F. Norman – Alf Sommerfelt (eds.), 784-794.
Haiman, John
1983 "Iconic and economic motivation", *Language* 59: 781-819.

1985 *Natural syntax*. Cambridge: Cambridge University Press.

Haiman, John (ed.)

1985 *Iconicity in syntax*. Amsterdam: John Benjamins.

Hainsworth, Joan

1972 "Narak noun possession and morphophonemic rules", *Anthropological Linguistics* 14: 46-61.

Hale, Kenneth

1981 "Preliminary remarks on the grammar of part-whole relations in Warlpiri", in: John Hollyman – Andrew Pawley (eds.), 333-344.

Halim, Amran – Lois Carrington – Stephen A. Wurm (eds.)

1982 *Papers from the Third International Conference on Austronesian Linguistics*. Volume 1: Currents in Oceanic (Pacific Linguistics Series C, No. 74.) Canberra: Australian National University.

Hamp, Eric P.

1984 Toward Indo-European reflexive reference", *Folia Linguistica Historica* 5: 195-196.

Hanon, Suzanne

1988 "Qui à quoi? Réflexions sur la possession inaliénable et le verbe *avoir* en français", *Revue Romane* 23: 161-177.

Hantson, A.

1979 "Some remarks on possessives and the pronoun *one*", *English Studies* 60: 762-769.

Harvey, Mark

1995 "Body parts in Warray", [this volume.]

Harning, K.E.

1980 "The analytic genitive in the Modern Arabic dialects", *Orientalia Gothoburgensia* 5.

Hatcher, Anna Granville

1944a "Il tend les mains vs Il tend ses mains", *Studies in Philology* 41: 457-481.

1944b "Il me prend les bras vs Il prend mon bras", *Romanic Review* 35: 156-164.

Hayes, La Vaughn H.

1992 "On the common morphological origins of word families in Southeast Asia", *Linguistics of the Tibeto-Burman Area* 15: 185-202.

Heine, Bernd

1975 Körperteile in afrikanischen Sprachen. [Unpublished MS.]

1989 "Adpositions in African languages", *Linguistique africaine* 2: 77-127.

1995 *Possession: cognitive sources, forces and grammaticalization*. Cambridge: Cambridge Universtiy Press.

Heine, Bernd – Mechthild Reh

1984 *Grammaticalisation and reanalysis in African languages*. Hamburg: Helmut Buske Verlag.

Herschensohn, Julia

1975 An interpretive approach to inalienable possession in French. (ERIC Microfiche no. ED 123912.) Washington: Centre for Applied Linguistics.

Herslund, Michael

1983 "Le datif de la possession inaliénable en français", in: Michael Herslund – O. Mordrup – F. Sorensen (eds.), 99-115.

1988 *Le datif en français*. Louvain, Paris: Peeters.
Herslund, Michael– O. Mordrup – F. Sorensen (eds.)
1983 *Analyses grammaticales du français. Etudes publiées à l'occasion du 50e anniversaire de Carl Vikner*. Copenhagen: Akademisk Forlag.
Hinnebusch, Thomas J. – Robert S. Kirsner
1980 "On the inference of 'inalienable possession' in Swahili", *Journal of African Languages and Linguistics* 2: 1-16.
Hollyman, John – Andrew Pawley (eds.)
1981 *Studies in Pacific linguistics in honor of Bruce Biggs*. Auckland: Linguistic Society of New Zealand.
Hopper, Paul J. – Sandra A. Thompson.
1980 "Transitivity in grammar and discourse", *Language* 56: 251-299.
1985 "The iconicity of the universal categories 'noun' and 'verb' ", in: John Haiman (ed.), 151-183.
Hopper, Paul J. – Sandra A. Thompson (eds.)
1982 *Studies in transitivity*. (Syntax and Semantics 15.) New York: Academic Press.
Horst, Johannes Martinus van der
1985 "Verkenning van de onpersoonlijke constructies", *Tijdschrift voor Nederlandse Taal-en Letterkunde* 101: 34-63, 81-92.
Hosokawa, Komei
1995 " 'My face *am* burning!': Quasi-passive, body parts and related issues in Yawuru grammar and cultural concepts", [this volume.]
Householder, Fred W. (ed.)
1972 *Syntactic theory 1*: Structuralist: Selected readings. Harmondsworth: Penguin.
Hubbard, P. – Peter Tiersma (eds.)
1979 *Linguistic Notes from La Jolla*. Vol. 6. San Diego: University of California Linguistics Department.
Hyman, Larry M.
1977 "The syntax of body parts", in: Ernest Rugwa Byarushengo – Alessandro Duranti – Larry Hyman (eds.), 99-117.
1995 "The syntax of body parts in Haya", [this volume.]
Hyman, Larry M. – Danny Keith Alford – Elizabeth Akpati
1970 "Inalienable possession in Igbo", *Journal of West African Languages* 7: 85-101.
Hyman, Larry M. – Leon C. Jacobson – Russell G. Schuh (eds.)
1976 *Papers in African linguistics in honor of Wm. E. Welmers*. (Studies in African linguistics: Supplement 6.)
Iris, Madelyn Anne – Bonnie E. Litowitz – Martha Evens
1988 "Problems of the part-whole relation", in: Martha Walton Evens (ed.), 261-268.
Isačenko, Alexander V.
1965 "Das syntaktische Verhältnis der Bezeichnungen von Körperteilen im Deutschen", *Studia Grammatica V: Syntaktische Studien*. Berlin: Akademie Verlag, 7-27.
Ispas, Bujorel
1985 "Modele metonimice iterative in structura semantica a denumirilor partilor corpului omenesc in limbile rusa si romana" [Regular metonymic models in the semantic structure of body-part names in Russian and Romanian], *Analele Universitatii din Timisoara, Serie stiinte filologice* 23: 37-41.

Ivanov, Vjac. Vs. – T. M. Sudnik – T.V. Civ'jan (eds.)
1981 *Struktura teksta -81. Tezisy simpoziuma*. Moscow: Slavic and Balkan Institute, Academy of Sciences.

Ivič, Milka
1970 "On the part-whole relation and its linguistic consequences", in: Roman Jakobson – Shigeo Kawamato (eds.), 281-286.

Jacobsen Jr., William H.
1987 Washo nominal possession: Form and function. [Unpublished MS.]

Jacquinod, Bernard
1989 *Le double accusatif en grec d'Homère à la fin du Ve siècle avant J.-C.* (Bibliothèque des Cahiers de l'Institut de Linguistique de Louvain 54.) Louvain: Peeters.

Jaisser, Annie C.
1990 "Del*iver*ing an introduction to psycho-collocations with *siab* in Hmong", *Linguistics of the Tibeto-Burman Area* 13: 159-178.

Jakobson, Roman
1971 "The relationship between the genitive and plural in the declension of Russian nouns", in: Roman Jakobson (ed.), 148-153.

Jakobson, Roman (ed.)
1971 *Selected writings*: Vol. II. The Hague: Mouton.

Jakobson, Roman – Shigeo Kawamato (eds.)
1970 *Studies in general and oriental linguistics in honor of Shiro Hattori.* Tokyo: TEC Co.

Jarceva, V.N. (ed.)
1977 *Kategorii bytija i obladanija v jazyke*. Moscow: Nauka.

Johnson, Mark
1987 *The body in the mind*. Chicago: University of Chicago Press.

Jungraithmayr, Hermann – Walter W. Müller (eds.)
1987 *Proceedings of the Fourth International Hamito-Semitic Congress*. Amsterdam: John Benjamins.

Kachru, Braj – Henry Kahane – Renée Kahane. (eds.)
1973 *Issues in Linguistics. Papers in Honor of Henry and Renée Kahane*. Urbana: University of Illinois Press.

Kapeliuk, O.
1973 "Traitement spéciale du corps et de l'âme dans la syntaxe éthiopienne", *Journal of Ethiopian Studies* 11: 143-160.

Kay, Paul – Karl Zimmer
1976 "On the semantics of compounds and genitives in English", *Sixth California Linguistics Association Conference Proceedings:* 29-35.

Kiefer, Ferenc – Nicolas Ruwet (eds.)
1973 *Generative Grammar in Europe*. (Foundations of Language – supplementary series.) Dordrecht: Reidel.

Kimenyi, Alexandre
1977 "Possessor objectivization in Kinyarwanda", in: Paul F. A. Kotey – Haig Der-Houssikan (eds.), 303-314.

Kliffer, Michael D.
1979 "Levanto la mano/Me lavo las manos/Me levanto la mano", *The Canadian Modern Language Review/ La Revue Canadienne des langues vivantes* 35: 217-226.
1981 "Inalienable possession in Romance: Grammar or inference?" (ERIC Microfiche no. 210942.) Washington D.C.: Center for Applied Linguistics.
1983 "Beyond syntax: Spanish inalienable possession" *Linguistics* 21: 759-794.
1984 "Interpenetration of linguistic levels: French inalienable possession", *Lingua* 62: 187-208.
Kotey, Paul F.A. – Haig Der-Houssikan (eds.)
1977 *Language and linguistic problems in Africa*. Columbia: Hornbeam Press.
Kovecses, Zoltan
1986 *Metaphors of anger, pride and love: A lexical approach to the structure of concepts*. Amsterdam: John Benjamins.
Kudrnacova, Nadezda
1990 "On the semantics of constructions expressing body part movements", *Brno Studies in English* 18: 75-91.
Labahn, Thomas (ed.)
1983 *Proceedings of the Second International Congress of Somali Studies, University of Hamburg, August 1-6, 1983*. Hamburg: Helmut Buske Verlag.
Landar, Herbert J. – Joseph B. Casagrande
1962 "Navaho anatomical reference", *Ethnology* 1: 370-373.
Langacker, Ronald W.
1968 "Observations on French possessives", *Language* 44: 51-75.
1972 "Possessives in Classical Nahuatl", *International Journal of American Linguistics* 38: 173-186.
1991 Possession and possessive constructions. [Unpublished MS.]
Langdon, Margaret
1988 Possession and kinship in Yuman languages. [Unpublished MS.].
Le Centre d'Analyse Syntaxique de l'Université de Metz
1976 *Etudes de syntaxique du moyen français* [Studies of the syntax of Middle French]. Paris: Klincksieck.
Leclère, Christian
1976 "Datifs syntaxiques et datif éthique", in: Jean-Claude Chevalier – Maurice Gross (eds.), 73-96.
1978 "Sur une classe de verbes datifs", *Langue Française* 39: 66-75.
Leeding, Velma J.
1995 "Body parts and possession in Anindilyakwa", [this volume.]
Lehrer, Adrienne
1974 "Extended meanings of body-part terms", *International Journal of American Linguistics* 40: 135-137.
Levine, James S.
1984 "On the dative of possession in contemporary Russian", *Slavic and East European Journal* 28: 493-501.
Lévy-Bruhl, Lucien
1914 "L'expression de la possession dans les langues mélanésiennes", *Mémoire de la Société de Linguistique de Paris* 19: 96-104.

Lewis, Gilbert
1974 "Gnau anatomy and vocabulary for illnesses", *Oceania* 45: 50-78.
Lichtenberk, Frantisek
1985 "Possessive constructions in Oceanic languages and in Proto-Oceanic", in: Andrew Pawley – Lois Carrington (eds.), 93-140.
Liston, Jerry L.
1972 "The semantic structure of body-part terms in Serbo-Croatian: The part-whole hierarchy", *Anthropological Linguistics* 14: 328-338.
Lunt, Horace G. (ed.)
1964 *Proceedings of the Ninth International Congress of Linguistics.* Cambridge, Mass., August 27-31, 1962. (Janua Linguarum, Series Maior 12.) The Hague: Mouton.
Lupardius, Karen Jacque
1984 "Muskogean possessive affixes", *Ryudai Review of Language and Literature* 29: 99-115.
Lynch, John
1973 "Verbal aspects of possession in Melanesian languages", *Oceanic Linguistics* 12: 69-102.
1982 "Towards a theory of the origin of the Oceanic possessive construction", in: Amram Halim – Lois Carrington – Stephen A. Wurm (eds.), 243-268.
Lyons, John
1967 "A note on possessive, existential, and locative sentences", *Foundations of Language* 3: 390-396.
1968 "Existence, location, possession, and transitivity", in: B. van Rootselaar – J.F. Staal (eds.), 495-509.
McCawley, James
1976 "From OE/ME 'impersonal' to 'personal' constructions: What is a subject-less S", in: Sanford B. Steever – Carol A. Walker – Salikoko S. Mufwene (eds.), 192-204.
McCawley, James (ed.)
1976 *Notes from the linguistic underground.* (Syntax and Semantics 7.) New York: Academic Press.
McClure, Erica F.
1975 "Ethno-anatomy: The structure of the domain", *Anthropological Linguistics* 17: 78-88.
McGregor, William
1985 "Body parts in Kuniyanti grammar", *Australian Journal of Linguistics* 5: 209-232.
1990 *A functional grammar of Gooniyandi.* Amsterdam: John Benjamins.
1991 The morphology and semantics of noun classes in Unggumi. [Unpublished MS.]
1995 "The grammar of nominal prefixing in Nyulnyul", [this volume.]
McGregor, William (ed.)
forthcoming. *Studies in Kimberley languages in honour of Howard Coate.* München: Lincom Europa.
McKay, Graham R.
1995 "Body parts, possession marking and nominal classes in Ndjébbana", [this volume.]

MacLaury, Robert E.
1989 "Zapotec body-part locatives: Prototypes and metaphoric extensions", *International Journal of American Linguistics* 55: 119-154.
Malczewska, Beata – William McGregor
1990 Body parts in Polish grammar. [Unpublished MS.]
1991 Incorporating notions of the personal domain and inseparability into systemic theory. [Unpublished MS.]
Manessy, Gabriel
1964 "La relation génitive dans quelque langues mandé", in: Horace G. Lunt (ed.), 467-475.
Manoliu-Manea, Maria
1975 "Ma seaca la inima: Romanian vs. Romance: The objectivization of the experiencer", *Revue Romaine de Linguistique* 20: 531-533.
1977 "Levou-me o chapéu. Le possessif-objet entre syntaxe, lexique et actes du langages", *Revue Romaine de Linguistique* 22: 187-191.
1982 "Un cas désespéré de la syntaxe romane: Les possessifs. Syntaxe et pragmatique du langage", in: Aina Moll – Jaume Vincens (eds.), 259-268.
1995 "Inalienability and topicality in Romanian. Pragma-semantics of syntax", [this volume.]
Mardirussian, Galust
1975 "Noun incorporation in universal grammar", *Papers from the Eleventh Regional Meeting of the Chicago Linguistic Society* 11: 383-389.
Martin, Jack
1993 " 'Inalienable possession' in Creek (and its possible origin)", *International Journal of American Linguistics* 59: 442-452.
Martin, L. (ed.)
1979 *Papers in Mayan linguistics.* Columbia: Lucas.
Matisoff, James A.
1978 *Variational semantics in Tibeto-Burman. The "organic" approach to linguistic comparison.* (Occasional Papers of the Wolfenden Society on Tibeto-Burman Linguistics, VI.) Philadelphia: Institute for the Study of Human Issues.
1985 "Out on a limb: ARM, HAND, and WING in Sino-Tibetan", in: Graham Thurgood – James A. Matisoff – David Bradley (eds.), 421-450.
1986 "Hearts and minds in Southeast Asian languages and English: an essay in the comparative lexical semantics of psycho-collocations." *Cahiers de Linguistique Asie Orientale* 15: 5-57.
1992 "Following the marrow: two parallel Sino-Tibetan etymologies", *Linguistics of the Tibeto-Burman Area* 15: 159-177.
Matisoff, James A. (ed.)
in prep. *Sino-Tibetan etymological dictionary and thesaurus. Fascicule 1: Body part terminology.* University of California at Berkeley.
Merlan, Francesca
1976 "Noun incorporation and discourse reference in modern Nahuatl", *International Journal of American Linguistics* 42: 177-191.
Mey, Jacob
1970 "Possessive and transitive in Eskimo", *Journal of Linguistics* 6: 47-56.

Mikami, Akira
1960 *Zoo wa hana ga nagai* [The elephant has a long nose]. Tookyoo: Kuroshio.
Miner, Kenneth
1982 "A note on noun incorporation", *Algonquian Linguistics* 7: 36-37.
1983 "Noun stripping and loose incorporation in Zuni", *Kansas Working Papers in Linguistics* 8: 83-93.
Mithun, Marianne
1984 "The evolution of noun incorporation", *Language* 60: 847-894.
1986 "On the nature of noun incorporation", *Language* 62: 32-37.
1995 "Multiple reflections of inalienability in Mohawk", [this volume.]
Modini, Paul
1981 "Inalienable possession and the 'double subject' constructions in East Asian", *Cahiers de Linguistique Asie Orientale* 9: 5-15.
Mohrmann, Christine – Alf Sommerfelt – Joshua Whatmough (eds.)
1963 *Trends in European and American Linguistics 1930-1960*: Edited on the Occasion of the 9th International Congress of Linguists, Cambridge, Mass., 27 August – 1 September, 1962. Utrecht, Netherlands: Spectrum.
Moll, Aina – Jaume Vincens (eds.)
1982 *Actes, XVI Congrés Internacional de Lingüística i Filología Romàniques*, Palma de Mallorca, 7-12 d'Abril de 1980: 1. Sessions Plenariès i Taules Rodones. Palma de Mallorca: Moll.
Monnerie-Goarin, Annie
1985 "Avoir les yeux bleus ou des yeux bleus?", *Le Français dans le Monde* 25: 48-51.
Morgan, Lawrence R.
1986 Alienable and inalienable possession in Kootenay. [Unpublished MS.]
Morolong, Malillo – Larry M. Hyman
1977 "Animacy, objects and clitics in Sesotho", *Studies in African Linguistics* 8: 199-218.
Mosel, Ulrike
1980 A few remarks on possessive phrases in Austronesian languages. [Unpublished MS.]
1981 Possessive phrases in Tolai. [Unpublished MS.]
1982 *Possessive constructions in Tolai.* (Arbeiten des Kölner Universalien – Projekts 44) Köln: Institut für Sprachwissenschaft.
1983 *Adnominal and predicative possessive constructions in Melanesian languages.* (Arbeiten des Kölner Universalien – Projekts 50.) Köln: Institut für Sprachwissenschaft.
Mullie, Joseph M.
1933 "Le double nominatif en chinois", *T'oung Pao* 30: 231-236.
Munro, Pamela
1984 "The syntactic status of object possessor raising in Western Muskogean", *Proceedings of the 10th annual meeting of the Berkeley Linguistics Society* 10: 634-649.
1993 "The Muskogean II prefixes and their significance for classification", *International Journal of American Linguistics* 59: 374-404.
Myhill, John
1988 "Nominal agent incorporation in Indonesian", *Journal of Linguistics* 24: 111-136.

Neumann, Dorothea
1987 *Objects and spaces: A study in the syntax and semantics of the German case system.* (Studien zur Deutschen Grammatik, 32.) Tübingen: Gunter Narr.
1995 "The dative and the grammar of body parts in German", [this volume.]
Newman, Paul – Robert D. Botne (eds.)
1989 *Current approaches to African linguistics.* Vol. 5. (Publications in African languages and linguistics, 8.) Dordrecht: Foris.
Nichols, Johanna
1986 "Head-marking and dependent-marking grammar", *Language* 62: 56-119.
1988 "On alienable and inalienable possession", in: William Shipley (ed.), 557-609.
Oey, Eric M.
1990 " 'Psycho-collocations' in Malay", *Linguistics of the Tibeto-Burman Area* 13: 141-158.
Ostrowski, Manfred
1981 Zu Fragen der Possession, der Lokalisation und der Verb-Valenz in drei kaukasischen Sprachen (Tscherkessisch, Georgisch und Tabbassaranisch). [Unpublished MS.]
Osumi, Midori
1995 "Body parts in Tinrin", [this volume.]
Palmer, Gary B. – Lawrence Nicodemus
1985 "Coeur d'Alene exceptions to proposed universals of anatomical nomenclature", *American Ethnologist* 12: 341-359.
Pawley, Andrew – Lois Carrington (eds.)
1985 *Austronesian Linguistics at the 15th Pacific Science Congress.* (Pacific Linguistics Series C, No. 88) Canberra: Australian National University.
Payne, Doris L.
1982 "Chickasaw agreement morphology: a functional explanation", in: Paul J. Hopper – Sandra A. Thompson (eds.), 351-378.
Payne, Doris L. (ed.)
1990 *Amazonian linguistics: Studies in lowland South American languages.* Austin: University of Texas Press.
Peile, Anthony R.
nd Body and Soul – an Australian Aboriginal view. [Unpublished Ms.]
forthcoming "Kukatja botanical terms and concepts", to appear in William McGregor (ed.).
Perlmutter, David
1978 "Impersonal passives and the unaccusative hypothesis", *Proceedings of the Fourth Annual Meeting of the Berkeley Linguistics Society* 4: 157-189.
Perlmutter, David (ed.)
1983 *Studies in relational grammar 1.* Chicago: University of Chicago Press.
Petruck, Miriam Rachel
1987 *Body part terminology in Hebrew: A study in lexical semantics.* [Doctoral dissertation, University of California at Berkeley.]
Polenz, Peter von
1969 "Der Pertinenzdativ und seine Satzbaupläne", in: Ulrich Engels – Paul Grebe – Heinz Rupp (eds.), 146-171.

Popescu-Ramirez, Liliana – Liliane Tasmowski de Ryck
1988 Thématicité et possessivité en roumain. [Unpublished MS.]
Postal, Paul M. – Brian D. Joseph
1990 *Studies in relational grammar 3*. Chicago: Chicago University Press.
Quaireau, Andre
1988 "Structures des énoncés agni incluant des termes référant au corps", *Verbum* 11: 143-155.
Reh, Mechthild – Bernd Heine – Marcello Lamberti
1981 Inhärenz in Possessivkonstruktionen afrikanischer Sprachen. [Unpublished MS.]
Riegel, M.
1984 Pour une redéfinition linguistique des relations dites de 'possession' et d'"appartenance' ", *L'Information Grammaticale* 23: 3-7.
Roldan, Mercedes
1972 The Romance possessives revisited. [Unpublished MS.]
Rootselaar, B. van – J.F. Staal (eds.)
1968 *Logic, methodology, and philosophy of science, III*. Amsterdam: North-Holland.
Rosén, Hajim B.
1959 "Die Ausdrucksform für 'veräußerlichen' und 'unveräußerlichen Besitz' im Frühgriechischen. Das Funktionsfeld von homer phílos", *Lingua* 8: 264-293.
Rosen, Sarah
1989 "Two types of noun incorporation", *Language* 65: 294-317.
Saah, Kofi K.
1988 Alienable/inalienable possession in some African languages. [Unpublished MS.]
Sadock, Jerold
1980 "Noun incorporation in Greenlandic", *Language* 56: 300-319.
Sapir, Edward
1911 "The problem of noun incorporation in American languages", *American Anthropologist* 13: 250-282.
1917-1920 Review of "Het identificeerend Karakter der Possessive Flexie in Talen van Noord-Amerika" by C.C.Uhlenbeck 1917. *International Journal of American Linguistics* 1: 82-90.
Saunders, Ross
1975 "Bella Coola referential suffixes", *International Journal of American Linguistics* 41: 355-368.
Séchehaye, Albert et al. (eds.)
1939 *Mélanges de linguistiques offerts à Charles Bally*. Geneva: Georg.
Seiler, Hansjakob
1973a "On the semanto-syntactic configuration 'possessor of an act' ", in: Braj Kachru – Henry Kahane – Renée Kahane (eds.), 836-853.
1973b "Zum Problem der sprachlichen Possessivität", *Folia Linguistica* VI: 231-250.
1977 Two systems of Cahuilla kinship expressions: Labelling and descriptive. (Arbeiten des Kölner Universalien – Projekts 27.) Köln: Institut für Sprachwissenschaft.
1982a "Two systems of Cahuilla kinship expressions: Inherent and establishing", *International Journal of American Linguistics* 48: 185-196.
1982b *Possessor, subject and object*. (Arbeiten des Kölner Universalien–Projekts 43.) Köln: Institut für Sprachwissenschaft.

1983a *Possession as an operational dimension of language.* (Language Universals Series Vol. 2) Tübingen: Narr. [Also (Arbeiten des Kölner Universalien–Projekts 42.) Köln: Institut für Sprachwissenschaft.]
1983b "Possessivity, subject and object", *Studies in Language* 7: 89-117.

Seliverstova, O. N.
1977 "Semanticeskij analiz existencial'nych i posessivnych konstrukcij v anglijskom jazyke", in: V.N. Jarceva (ed.), 5-6.

Serzisko, Fritz
1983a "Possession in Somali", in: Thomas Labahn (ed.), 101-114.
1983b *Der Ausdruck der Possession in Somali.* Tübingen: Gunter Narr Verlag.

Shen, Jiaxuan
1987 "Subject function and double subject construction in Mandarin Chinese", *Cahiers de Linguistique Asie Orientale* 16: 195-211.

Shipley, William (ed.)
1988 *In honor of Mary Haas: From the Haas Festival Conference on Native American Linguistics.* Berlin: Mouton de Gruyter.

Stark, Louisa
1969 "The lexical structure of Quechua body parts", *Anthropological Linguistics* 11: 1-15.

Steele, Ross – Terry Threadgold (eds.)
1987 *Language topics. Essays in honour of Michael Halliday.* Vols 1-2. Amsterdam: John Benjamins.

Steele, Susan
1977 "On being possessed", *Proceedings of the Third Annual Meeting of the Berkeley Linguistics Society* 3: 114-131.

Steever, Sanford B. – Carol A. Walker – Salikọko S. Mufwene (eds.)
1976 *Papers from the parasession on diachronic syntax.* Chicago: Chicago Linguistic Society.

Sugita, Hiroshi
1973 "Semitransitive verbs and object incorporation in Micronesian languages", *Oceanic Linguistics* 12: 393-416.

Sullivan, Paul
1984 "Noun incorporation in Yucatec Maya", *Anthropological Linguistics* 26: 138-160.

Takahashi, Taroo
1975 "Bunchuu ni arawareru shozokukankei no shushusoo" [Aspects of possession relationship expressed in sentences], *Kokugogaku* 103: 1-17.

Taylor, John R.
1989 "Possessive genitives in English", *Linguistics* 27: 663-686.

Teng, Shou-hsin
1974 "Double nominatives in Chinese", *Language* 50: 455-473.

Thompson, Chad
1991 "The low topicality pronoun *k'i* in Koyukon", *Studies in Language* 15: 59-84.
1995 "On the grammar of body parts in Koyukon Athbaskan", [this volume.]

Thompson, Sandra A.
1973 "Transitivity and some problems with the *ba* construction in Mandarin Chinese", *Journal of Chinese Linguistics* 1: 208-221.

Thurgood, Graham – James A. Matisoff – David Bradley (eds.)
1985 *Linguistics of the Sino-Tibetan area: The state of the art.* (Pacific Linguistics Series C, No. 87.) Canberra: Australian National University.
Tsao, Feng-fu
1982 "The double nominative construction in Mandarin Chinese", *Tsing Hua Journal of Chinese Studies* 14: 275-297.
Tsunoda, Tasaku
1995 "The possession cline in Japanese and other languages", [this volume.]
Tuggy, David
1980 "Ethical dative and possessor omission *si*, possessor ascension No!", *Working Papers of the Summer Institute of Linguistics* 24: 97-141.
Uhlenbeck, C.C.
1917 "Het identificeerend Karakter der Possessive Flexie in Talen van Noord-Amerika"[Reprinted from *Verslagen en Mededeelingen der Koninklijke Akademie van Wetenschapen.* Afdeeling Letterkunde 5e Reeks 2e Deel, 3e Stuk. Amsterdam (1916): 187-216.]
Ultan, Russell
1975 *Descriptivity grading of Finnish body-part terms.* (Arbeiten des Kölner Universalien – Projekts 16.) Köln: Institut für Sprachwissenschaft.
1976 *Descriptivity in the domain of body-part terms.* (Arbeiten des Kölner Universalien – Projekts 21.) Köln: Institut für Sprachwissenschaft.
1978 "Towards a typology of substantival possession", in: Greenberg, Joseph H. (ed.), Vol. 4: 11-50.
Vandeweghe, Willy
1987 "The possessive dative in Dutch: Syntactic reanalysis and predicate formation", in: Johan van der Auwera – Louis Goossens (eds.), 137-151.
Velázquez-Castillo, Maura
1989 "Guarani possessive constructions", *The Fourth Annual Meeting of the Pacific Linguistic Conference.* Eugene, Oregon, 489-506.
1991 "The semantics of Guarani agreement markers", *Proceedings of the Seventeenth Annual Meeting of the Berkeley Linguistics Society* 17: 324-335.
1993 *The grammar of inalienability: Possession and noun incorporation in Paraguayan Guarani.* [Doctoral dissertation, University of California at San Diego.]
in press "Noun incorporation and object placement in discourse", in: Pamela Downing – Michael Noonan (eds.).
Verhaar, John W. M.
1978 "Syntactic (in)alienability in Indonesian", in: Stephen A. Wurm. – Lois Carrington (eds.), 317-325.
Verhaar, John W. M. (ed.)
1968 *The verb* be *and its synonyms, part 2.* (Foundations of language, supplementary series 6.) Dordrecht: Reidel.
Voeltz, Erhard F.K.
1976 "Inalienable possession in Sotho", in: Larry M. Hyman – Leon C. Jacobson – Russell G. Schuh (eds.), 255-266.
Vol'f, E.M.
1977 "Nekotorye osobennosti mestoimennych posessivnych konstrukcij (ivero-romanskie jazyki)", in : V.N. Jarceva (ed.), 144-193.

Wackernagel, Jakob
1908 "Genitiv und Adjektiv", in: *Mélanges de linguistique offerts à Ferdinand de Saussure*, Société Linguistique de Paris, Paris: H. Champion, 125-152.
Walsh, Michael
1987 "The impersonal verb construction in Australian Languages", in: Ross Steele – Terry Threadgold (eds.), Vol. 1: 425-438.
1995 "Body parts in Murrinh-Patha: Incorporation, grammar and metaphor", [this volume.]
Watkins, Calvert
1966 "Remarks on the genitive", in: Calvin Watkins (ed.), Vol. III: 2191-2198.
Watkins, Calvin (ed.)
1966 *To honor Roman Jakobson*. Vol. 3. The Hague: Mouton.
Weggelaar, C.
1986 "Noun incorporation in Dutch", *International Journal of American Linguistics* 52: 301-305.
Weir, Helen
1990 "Incorporation in Nadeb", in: Doris L. Payne (ed.), 321-363.
Weiss, A. von
1962 "Dativ und Genitiv", in: A. von Weiss (ed.), 159-163.
Weiss, A. von (ed.)
1962 *Wirkendes Wort, Sammelband I: Sprachwissenschaft*. Düsseldorf: Schwann.
Werner, Oswald – Kenneth Y. Begishe
1970 "A lexemic typology of Navajo anatomical terms, I: The foot", *International Journal of American Linguistics* 36: 247-265.
Wierzbicka, Anna
1976 "Mind and body from a semantic point of view", in: James McCawley (ed.), 129-158.
1979 "Ethnosyntax and the philosophy of grammar", *Studies in Language* 3: 313-383.
Wilson, William H.
1982 *Proto-Polynesian possessive marking*. (Pacific Linguistics Series B, No. 85.) Canberra: Australian National University.
Winston, Morton E.
1987 "A taxonomy of part-whole relations", *Cognitive Science* 1: 417-444.
Woodbury, Hanni
1975 "Onondaga noun incorporation", *International Journal of American Linguistics* 41: 10-21.
Wotjak, Barbara
1985 "Zu Inhalts- und Ausdrucksstruktur ausgewählter somatischer Phraseolexeme", *Deutsch als Fremdsprache* 22: 216-223.
Wotjak, Gerd
1986 "Zur Bedeutung ausgewählter verbaler Phraeologismen des Deutschen", *Zeitschrift für Germanistik* 7: 183-200.
Wunderli, Peter
1976 "Les structures du possessif en moyen français", in: Le Centre d'Analyse Syntaxique de l'Université de Metz, 111-150.
Yamamoto, Akira Y.
1978 "Syntax and semantics of possession", *Gengo Kenkyu* 73: 62-72.

Young, Katherine
1989 "Disembodiment: The phenomenology of the body in medical examinations", *Semiotica* 73: 43-66.
Z'graggen, John A.
1965 "Possessor-possessed relationship in the Saker language, northeastern New Guinea", *Oceanic Linguistics* 4: 119-126.
Zubiri, Jesus Olza
1985 "Los nombres relativos en un idioma Aruaca o Maipure, el Guajiro" [Relative nouns in Guajiro, an Aruaca (Maipure) language], *Montalban* 16: 227-276.
Zurinskaja, M. A.
1977 "Immenye posessivnye konstrukcii i problema neottorzimoj prinadleznosti", in V. N. Jarceva (ed.), 194-258.

Note We would like to thank Aveline Peréz, Linguistics Department, La Trobe University for her invaluable assistance in compiling this bibliography.

Authors' Addresses

1. Felix Ameka

Vakgroep Afrikaanse Taalkunde,
Rijksuniversiteit te Leiden,
Postbus 9515,
2300 RA Leiden
The Netherlands

2. Edith Bavin

School of Linguistics,
La Trobe University,
Bundoora, Melbourne VIC 3083
Australia

3. Christine Béal

School of European Studies,
La Trobe University
Bundoora, Melbourne VIC 3083
Australia

4. Kate Burridge

School of Linguistics,
La Trobe University,
Bundoora, Melbourne VIC 3083
Australia

5. Hilary Chappell

School of Linguistics,
La Trobe University,
Bundoora, Melbourne VIC 3083,
Australia

6. Marybeth Clark

c/- Arnold
2425 Ellentown Rd
La Jolla, San Diego
CA 92037 USA

7. Terry Crowley

Department of Linguistics
University of Waikato
Private Bag 3105
Hamilton
New Zealand

8. Nicholas Evans

Department of Linguistics,
University of Melbourne
Parkville, Melbourne VIC 3052
Australia

9. Mark Harvey

Linguistics Department
University of Newcastle,
Callaghan, NSW 2308
Australia

10. Komei Hosokawa

Department of Anthropology,
College of Arts,
University of Saga,
Saga City, 840
Japan

11. Larry Hyman

Linguistics Department
University of California at Berkeley
Berkeley CA 94720
USA

12. Velma Leeding

c/- Umbakumba School,
Umbakumba,
Via Darwin,
N.T. 0822,
Australia

13. William McGregor

Department of Linguistics,
University of Melbourne,
Parkville, Melbourne VIC 3052,
Australia

14. Graham McKay

Department of Language Studies
0984,
Edith Cowan University
2 Bradford St,
Mount Lawley WA 6050
Australia

15. Maria Manoliu-Manea

Department of French and Italian,
University of California at Davis,
Davis CA 95616,
USA

16. Marianne Mithun

Linguistics Department
University of California at
Santa Barbara
Santa Barbara CA 93106
USA

17. Dorothea Neumann

20 Talbot St
Forrest ACT 2603
Australia

18. Midori Osumi

Department of International
Cultural Studies,
Meiō University,
1220-1 Bimata Nago,
Okinawa 905,
Japan

19. Chad Thompson

Department of English and
Linguistics
Indiana University /
Purdue University at Fort Wayne
Fort Wayne IN 46805
USA

20. Tasaku Tsunoda

Faculty of Letters,
University of Tokyo,
Bunkyo-ku,
Tokyo 113,
Japan

21. Michael Walsh

Linguistics Department,
University of Sydney,
Sydney NSW 2006
Australia

Subject Index

Language Index

Author Index